Stability of Metal Structures

A World View

Second Edition

Lynn S. Beedle

Editor-in-Chief

Structural Stability Research Council

Sponsored in part by

NATIONAL SCIENCE FOUNDATION

and

UNITED STATES TRADE

AND DEVELOPMENT PROGRAM

U.S. International Development Cooperation Agency

Bruce G. Johnston
1905 - 1989

Dedication

Bruce G. Johnston was an inspiration not only to the research workers who had the privilege of learning from him, but he was an equal stimulus to the design professionals with whom he came in touch. As a founding member of the Column Research Council (CRC), predecessor to the Structural Stability Research Council (SSRC), his dedication, devotion, and persistence led to a viable organization, and one that eventually embraced the full scope of structural stability problems in metal structures. This dedication is doubly appropriate in view of the emphasis he always placed upon application of research results to design. And he consistently took an active role in international organizations – a prelude to an eventual focus of SSRC.

He was an active member of the AISC Specification Committee for over 30 years, and was consultant to the Association of Iron and Steel Engineers in the development of specifications for steel mill ladles, overhead traveling cranes, and mill building structures. So it was natural that he was one of the first individuals appointed to the team of editors for this book.

Born in Detroit, Michigan, on October 13, 1905, he married Ruth Barker in August of 1939 and they had three children: Sterling, Carol, and David. He received the degree in Civil Engineering from the University of Illinois in 1930, being awarded the Ira Baker First Place Award in Civil Engineering. In 1932, he was appointed Research Fellow at Lehigh University in a program supported by Bethlehem Steel Corporation, "Structural Beams in Torsion". After receiving the MS degree in 1934, he became an instructor at Columbia University, where he was awarded his PhD in 1938.

As Assistant Director, then Associate Director, and finally as Director, Bruce Johnston was in charge of the structural research programs at Lehigh's Fritz Engineering Laboratory from 1938 to 1950, except for interruption during World War II. In 1950 he became Professor of Structural Engineering at the University of Michigan. In 1968 he was named Professor Emeritus, following which he accepted the call of the University of Arizona to be Special Lecturer in Civil Engineering.

In 1969 he received ASCE's highest award, that of Honorary Member. In 1979 he was elected to the National Academy of Engineering. In 1981 he received the "Alumni Honor Award for Distinguished Service in Engineering" from the University of Illinois.

As part of the graduate program which he developed at Fritz Engineering Laboratory following World War II, Bruce Johnston initiated research on the effect of residual stress on column strength and demonstrated the importance of the strain-hardening modulus upon the behavior of steel columns. He also initiated Lehigh's pioneer research program on plastic design, a major part of which involved the solution of stability problems and the development of design rules essential to practical application.

It was in 1944 that he played a leading role in the organization of the Column Research Council, serving as its chairman from 1956 to 1962. He was editor of the first three editions of the Council's "Guide to Stability Design Criteria for Metal Structures". He remained an active member of the Council until his death in 1989 at age 84. It is in his memory that this Second Edition of the "World View" is dedicated.

STABILITY OF METAL STRUCTURES

A World View

2nd Edition

COORDINATING EDITORS

Lynn S. Beedle	Lehigh University
Gerard F. Fox	Howard Needles Tammen & Bergendoff
Yixian Gu	Lehigh University
Duiliu Sfintesco (Hon.)	Centre Technique Industriel de la Construction Metallique

REGIONAL COORDINATORS

Shao-Fan Chen	Xian Institute of Metallurgy & Construction Engineering
John E. Harding	University of Surrey
Jerome S. B. Iffland	Iffland Kavanagh Waterbury, P.C.
Ben Kato	Toyo University
Jindrich J. Melcher	Technical University of Brno
John Springfield	Carruthers & Wallace Limited
Nicholas S. Trahair	The University of Sydney

TOPICAL EDITORS

Russell Q. Bridge	The University of Sydney
Franklin Y. Cheng	University of Missouri-Rolla
Patrick J. Dowling	Imperial College of Science & Technology
Amr S. Elnashai	Imperial College of Science & Technology
Leo Finzi	Politecnico di Milano
Yuhshi Fukumoto	Osaka University
John E. Harding	University of Surrey
Miklos Ivanyi	Technical University of Budapest
Ben Kato	Toyo University
Shigeru Kuranishi	Tohoku University
Le-Wu Lu	Lehigh University
Pavel J. Marek	San Jose State University
David A. Nethercot	The University of Nottingham
Alexis Ostapenko	Lehigh University
Gerald W. Schulz	University of Innsbruck
Donald R. Sherman	University of Wisconsin-Milwaukee
John Springfield	Carruthers & Wallace Limited
Ton Toma	TNO Institute for Building Materials and Structures
Carlo Urbano	Technical University of Milan

COGNIZANT ORGANIZATIONS

AUSTRALIAN INSTITUTE OF STEEL CONSTRUCTION
CHINA TECHNICAL COMMITTEE FOR STANDARDS OF STEEL STRUCTURES
EAST EUROPEAN COOPERATING GROUPS
EUROPEAN CONVENTION FOR CONSTRUCTIONAL STEELWORK
JAPAN SOCIETY OF STEEL CONSTRUCTION, COLUMN RESEARCH COMMITTEE OF JAPAN
STRUCTURAL STABILITY RESEARCH COUNCIL

This document is not intended as a design standard. It should be used only as a reference. For actual design the user should consult the original specifications.

ISBN 1-879749-50-5

Manuscript assembled by
Structural Stability Research Council
Fritz Engineering Laboratory
Lehigh University 13
Bethlehem, Pennsylvania 18015, U.S.A.

Printed in the United States of America

Foreword

The World View focuses on stability specification design provisions in an attempt to identify the differences in codes and specifications and explain why such differences exist. It is a second effort, expanding on what was begun in the First Edition. The reader will see that although considerable progress has been made, there is still much that could be done. It turned out to be a considerable task just to identify the differences, let alone explain why they exist. The first edition of the World View, published in 1982, was based upon an examination of various approaches and design philosophies that was the feature of the first two International Colloquia in Stability led by Duiliu Sfintesco. Their purpose was to encourage further coordination and progress in the field of stability of metal structures.

The Coordinating Editors pay tribute to the group of over 90 editors and contributors who have completed a remarkable job in documenting the essence of the provisions and identifying the differences. That has not been easy. In many cases, translations were required. Almost without exception, each reference specification was arranged in a different way. Differences were encountered in units, symbols, abbreviations and glossary.

The table of contents reflects the careful decisions that were made regarding the "characteristic topics" into which specifications are arranged. These "characteristic topics" were selected to cover the major content of a subject while eliminating all of the minor variations and special cases.

The headquarters staff of the Structural Stability Research Council was responsible for the overall coordination of the effort, for the production editing, and for the publication of this document. The preliminary work was started when Mr. Graham S. Stewart was SSRC Technical Secretary. Special acknowledgment is due Mr. Yixian Gu, the present Technical Secretary, who carried the major responsibility of producing this volume, including the computer programming of text and illustrations. Particular appreciation goes to Mrs. Lesleigh Federinic, Administrative Secretary of SSRC, who provided important leadership in various phases of the work. Also, grateful appreciation goes to the editorial and production staff: Miss Tina Federinic, Mr. Yongmin Li, and Mrs. Furong Su. Others who provided assistance at various stages: Mr. Ahmed Asif, Mr. Craig Constantine, Mr. Yuan-Peir Gau, Mrs. Jean Polzer, Mr. Nawtej Singh Sehra, Mr. Mark Tamaro, Miss Alaana Verdi, Mrs. Diana Walsh, Mr. Ming Xie, and Mr. Shih Yuan.

The Coordinating Editors extend a special note of thanks to Mr. Jerome S. B. Iffland. As Vice Chairman of SSRC Task Group 11, he contributed many of the concepts that led to the successful completion of this work.

We are grateful to the National Science Foundation and Program Director John B. Scalzi who sponsored the 1987 workshop in St. Remy, France, and the 1989 North America Session of the 4th International Colloquium, which allowed experts from all over the world to come together and compare code differences. Appreciation is due the Sponsors, Participating Organizations, and Participating Firms of SSRC for their financial contributions.

The final publication of this document was made possible by a grant from the UNITED STATES TRADE AND DEVELOPMENT PROGRAM, U.S. International Development Cooperation Agency, who made possible two additional workshops in 1989. Its significant support is gratefully acknowledged.

Lynn S. Beedle
Gerard F. Fox
Yixian Gu
Duiliu Sfintesco

Preface

The objectives of this Second Edition

1. To present relevant design provisions for the stability of metal structures on a world-wide basis. This includes tentative specifications that are considered likely to be adopted.

2. To provide essential background information.

3. To compare and contrast the design provisions of each characteristic topic by region, and where necessary, within the regions on a national or organizational basis.

4. To compare and contrast the design provisions of each characteristic topic internationally. The effort here is to identify the differences and to explain why those differences exist.

5. To arrange the document so that it can be easily updated and expanded to include new regions, topics, and developments.

Why a New Edition

The First Edition of this book was published in 1982. While it was a valuable initial effort, there are several reasons why a Second Edition is needed. These include the following:

1. The First Edition did not complete one of the original objectives -- to compare and contrast design approaches in different regions of the world. Although comparisons were made and contrasts developed, the study of these differences needed to be extended.

2. The new edition also gives opportunity to expand the areas of coverage: both topically and geographically. Topically, the subjects of arches and tubular members have been added. Geographically, the expanded coverage includes Australia and the People's Republic of China.

3. The exercise of preparing a new edition can only result in reducing the differences and identifying regional problems. The last 20 years of similar efforts has proved that this approach has resulted in closer regional agreement. The result has been improved stability requirements for everyone.

4. There have been substantial developments in many of the topical areas over the past few years. The revised provisions are presented and the background of these provisions explained.

Audience of the New Edition

This new edition is written in part for researchers and the academic establishment. It will also be a resource document for practicing engineers and contractors, and enable them to assess practices in other regions of the world as well as to extend their businesses to new regions. Perhaps most important, it will provide concise provisions and explanations for specifications writers, alerting them to potential improvements.

This document is not intended as a design standard. It should be used only as a reference. For actual design the user should consult the original specifications.

Topics

The topics and their scope are described in the following tabulation:

1. Compression Members
   Columns subjected to centrally applied loads (excluding built-up, tubular, cold-formed, and composite columns)

2. Built-Up Members
   Members assembled from structural shapes, plates, and bars (excluding planar and space trusses)

3. Beams
   Flexural members (excluding tubular, cold-formed, and composite beams)

4. Plate and Box Girders
   Fabricated plate and box girders (excluding composite girders)

5. Beam-Columns
   Columns subjected to axial loading as well as uniaxial or biaxial bending (excluding tubular, cold-formed, and composite members)

6. Frames
   Structural frameworks and columns as frame members

7. Arches
   Vertically curved members

8. Triangulated Structures
   Planar trusses and space frames (excluding shell-like reticulated frameworks)

9. Tubular Structures
   Manufactured and fabricated tubular members (excluding composite members)

10. Shells
    Conical shells, doubly curved shells and shell-like structures (including reticulated frameworks)

11. Cold-Formed Members
    Members that are cold-formed to shape from hot- or cold-rolled plate, sheet or strip

12. Composite Members
    Composite columns, beams, beam-columns, and girders

Appendix 1: Stability Under Seismic Loading
    Stability requirements under earthquake loading

Appendix 2: General Provision and Design Requirements
    Limits of applicability, materials, loads, load combinations, and design basis

Arrangement

The following single character abbreviations for countries or regions have been used for the tables, charts, and matrices:

A – Australia
C – People's Republic of China
E – Eastern Europe
J – Japan
N – North America
W – Western Europe

Non-dimensional units have been used wherever possible. Where dimensions are necessary, SI units are provided. A conversion table for units is included for reader convenience.

A word or two is in order about the page numbering. Since the focus of the book is on the chapter, the major heading is the "Characteristic Topic". Using the running head designation, 3a(J1) as an example, the "3" designates the chapter number, "a" the characteristic topic "Lateral-Torsional Buckling", "J" the Japanese contribution to "Specification Provisions and Explanation", and "1", page 1 of the Japanese part. 3a(J1) is printed at the inner corner ("gutter") at the top of the page as a part of the running head.

The applicable reference standards are identified in the agreed-upon format: a condensed alphabetical code, any necessary identification code, and the year of publication (two digits only). For example, AISC LRFD 86 represents the AISC Manual of Steel Construction, "Load & Resistance Factor Design", 1986, by the American Institute of Steel Construction. Complete citations for each specification will be found in the appendix – Codes, Specifications, and Standards.

Each cited specification presents a brief statement of its provision, followed in some cases by commentary. The explanation includes certain within-region comparisons.

After all of the specification provisions have been given for the first characteristic topic (3a, for example), some regional comparisons and then international comparisons are presented. First comes the section "What are the differences?". Then, "Why are there differences?". With the completion of Topic 3a one goes on to Topic 3b, "Flange Local Buckling", and so on.

Not all of the specifications are treated in every chapter. In some cases there are a limited number of specifications applicable to a particular subject, such as Chapter 7 – Arches. In some cases there are so many specifications applicable to a subject that only a few were selected for inclusion.

The appendix, General Provisions and Design Requirements, provides general information on the background and philosophy for each specification.

Based on the agreed-upon format, symbols used in specifications are kept, as much as possible, the same as in the original specification. This will help when one wants to look at the original specification. However, in the regional comparison or international comparison sections, symbols are unified to make comparison among specifications possible. There is a symbol list following each chapter.

The editorial group is identified in two lists of contributors. The Chapter Editors provided the basic text for their region within each chapter, including commentary and comparison within the region. The Topical Editors were responsible for the individual chapters, provided for overall coordination between the different regions of the world within his topical area, and prepared Section III – International Comparisons & Comment. The Regional Coordinators had the responsibility for the work in his region, and contributed the material for "General Provisions and Design Requirements". The Coordinating Editors were responsible for bringing all of the work together.

The reader is reminded that there are major changes underway in the European codes and specifications. Eurocode 3 is a unified approach that is undergoing extensive development and this accounts for the reference in different chapters to the various editions: 1988, 89, 90.

Table of Contents

Contributors

REGION		AUSTRALIA	CHINA	EAST EUROPE	JAPAN	NORTH AMERICA	WEST EUROPE	TOPICAL EDITORS
GPDR & Regional Coordinator		N. S. Trahair	S. F. Chen	J. J. Melcher	B. Kato K. Takanashi	J. Iffland J. Springfield	J. Harding	J. Harding P. Marek J. Springfield
CHAPTER		CHAPTER EDITORS						
No	Title							
1	Compression Members	R. Q. Bridge	M. Z. Wei W. W. Zhao B. K. He	J. J. Melcher G. E. Belskij	M.Wakabayashi	L. Tall	G.W. Schulz	R. Q. Bridge G.W. Schulz
2	Built-up Members	R. Q. Bridge	J. Chen	J. J. Melcher G. E. Belskij	T. Suzuki T. Ogawa	Z. Razzaq B. G. Johnston	L. Finzi	L. Finzi
3	Beams	M.A.Bradford	Z. B. Xia	H. D. Glas H. E. Goeben	Y. Fukumoto J. Sakamoto	T. V. Galambos	J. Lindner	Y. Fukumoto
4	Plate & Box Girders	N. W. Murray	X. H. Ren X. L. Liu	M. Skaloud J. Djubek	F. Nishino S. Komatsu	A. Ostapenko	P. Dubas	A. Ostapenko
5	Beam-Columns	M.A.Bradford	S. F. Chen Z. Y. Shen B. K. He	D. Mateescu I. Caraba M. Ivanyi	K. Takanashi T. Nakamura	W. F. Chen	D. A. Nethercot	D. A. Nethercot
6	Frames	R. Q. Bridge	H. B. Wu	M. Ivanyi	S. Morino	L. W. Lu J. Springfield	J. Springfield	L. W. Lu J. Springfield
7	Arches	(None)	(None)	L. Kollar	S. Kuranishi	S. Vinnakota	S. Kuranishi	S. Kuranishi
8	Triangulated Structures	L. Schmidt	Z. T. Guo	M. Ivanyi	H. Akiyama	N. F. Morris	P. Dubas C. Urbano	C. Urbano
9	Tubular Structures	G. J. Hancock	(None)	D. Mateescu	B. Kato	D. R. Sherman	J. Rondal	B. Kato D. R. Sherman
10	Shells	J. M. Rotter	(None)	M. Ivanyi Z. Mendera	(None)	N.F. Morris	D. Vandepitte	M. Ivanyi
11	Cold-Formed Members	G. J. Hancock	S. J. Wang Z. Q. Zhang	J. J. Melcher V. Gioncu	T. Suzuki T. Ono	T. Pekoz	T. Toma J. Lindner	T. Toma
12	Composite Members	P. Ansourian R. Q. Bridge	S. T. Zhong	J. Lapos N. Streleckij	M. Wakabayashi	R. W. Furlong S. Daresalem-wala	P. J. Dowling A. S. Elnashai R. Bergmann K. Roik	P. J. Dowling A. S. Elnashai
AP	Stability under Seismic Loading	N. S. Trahair J. Butterworth	S. F. Chen	E. Juhásová	K. Takanashi	S. T. Mau F. Y. Cheng	F. Massolani	F. Y. Cheng

Coordinating Editors:

Lynn S. Beedle	Structural Stability Research Council, Lehigh University
Gerard F. Fox	Howard Needles Tammen & Bergendoff
Yixian Gu	Structural Stability Research Council, Lehigh University
Duiliu Sfintesco (Hon.)	Centre Technique Industriel de la Construction Metallique

Contributor Name List

Hiroshi Akiyama	University of Tokyo
David Anderson	University of Warwick
Peter Ansourian	The University of Sydney
Lynn S. Beedle	Lehigh University
Georgij E. Belskij	CNIISK, Moscow
Mark A. Bradford	University of New South Wales
Russell Q. Bridge	The University of Sydney
J. W. Butterworth	The University of Auckland
Ioan Caraba	Polytechnic Institute of Timisoara
Alexander Chajes	University of Massachusetts
Ji Chen	Xian Institute of Metallurgy & Construction Engineering
Shao-Fan Chen	Xian Institute of Metallurgy & Construction Engineering
Wai-Fah Chen	Purdue University
Franklin Y. Cheng	University of Missouri-Rolla
Saif Daresalemwala	Howard Needles Tammen & Bergendoff
Alan G. Davenport	The University of Western Ontario
Jozef Djubek	Slovak Academy of Science
Patrick J. Dowling	Imperial College of Science & Technology
Pierre Dubas	Swiss Federal Institute of Technology
Ahmed Elghazouli	Imperial College of Science & Technology
Amr S. Elnashai	Imperial College of Science & Technology
Leo Finzi	Politecnico di Milano
Gerard F. Fox	Howard Needles Tammen & Bergendoff
Yuhshi Fukumoto	Osaka University
Richard W. Furlong	University of Texas
Theodore V. Galambos	University of Minnesota
Victor Gioncu	Polytechnic Institute of Timisoara
Hans-Dieter Glas	Technical University of Leipzig
Hans E. Goeben	Technical University of Leipzig
Yixian Gu	Lehigh University
Zaitian Guo	Xian Institute of Metallurgy & Construction Engineering
Gregory J. Hancock	The University of Sydney
John E. Harding	University of Surrey
Akio Hasegawa	University of Tokyo
Bao Kang He	Xian Institute of Metallurgy & Construction Engineering
Jerome S. B. Iffland	Iffland Kavanagh Waterbury, P.C.
Miklos Ivanyi	Technical University of Budapest
Bruce G. Johnston*	Consultant
Emilia Juhăsovä	Slovak Academy of Sciences

* Deceased

Ben Kato	Toyo University
Sritiwat Kitipornchai	University of Queensland
Shigeo Kobayashi	Tokyo Institute of Science & Technology
Lajos Kollar	Technical University of Budapest
Sadao Komatsu	Osaka Sangyo University
Shigeru Kuranishi	Tohoku University
Jozef Lapos	Slovak Technical University
Joachim Lindner	Technische Universitat Berlin
Xiliang Liu	Tianjin University
Le-Wu Lu	Lehigh University
Pavel Marek	San Jose State University
Dan Mateescu	Polytechnic Institute of Timisoara
Seng-Taur Mau	University of Houston
Federico M. Massolani	Universita Di Napoli
Jindrich J. Melcher	Technical University of Brno
Zbigniew Mendera	Politechnika Krakowska
Clarence D. Miller	CBI Technical Services Company
Shosuke Morino	Mie University
Nicholas F. Morris	Manhattan College
Noel W. Murray	Monash University
Takeshi Nakamura	Kyoto Institute of Tech.
David A. Nethercot	The University of Nottingham
Fumio Nishino	University of Tokyo
Toshiyuki Ogawa	Tokyo Institute of Technology
Tetsuro Ono	Nagoya Institute of Technology
Alexis Ostapenko	Lehigh University
Teoman Pekoz	Cornell University
Zia Razzaq	Old Dominion University
Xinhua Ren	Tianjin University
Jacques Rondal	Universite de Liege
J. Michael Rotter	The University of Edinburgh
Jun Sakamoto	Nagoya University
Lewis Schmidt	University of Wollongong
Gerald W. Schulz	University of Innsbruck
Duiliu Sfintesco	Centre Technique Industriel de la Construction Metallique
Zu-Yan Shen	Tong-Ji University
Donald R. Sherman	University of Wisconsin-Milwaukee
Miroslav Skaloud	Czechoslovak Academy of Sciences
John Springfield	Carruthers & Wallace Limited
Graham S. Steward	Dolan Contractors
Nikolaj Streleckij	CNIPSK, Moscow
Toshiro Suzuki	Tokyo Institute of Technology
Koichi Takanashi	The University of Tokyo
Lambert Tall	Florida International University
Ton Toma	TNO Institute for Building Materials and Structures
Nicholas S. Trahair	The University of Sydney

Carlo Urbano	Technical University of Milan
Daniel Vandepitte	Ghent University
Sriramulu Vinnakota	Marquette University
Minoru Wakabayashi	General Building Research Corporation
Shiji Wang	Hunan University
Mingzhong Wei	Chongqing Architectural Institute
Huibi Wu	Chongqing Architectural Institute
Zhi Bin Xia	Zhejiang University
Zhong Quan Zhang	Zhongnan Building Design Institute
Wen Wei Zhao	Tsinghua University
Shan-tong Zhong	Harbin Architectural & Civil Engineering Institute

Chapter 1

Compression Members

TOPICAL EDITORS

Russell Q. Bridge, The University of Sydney
Gerald W. Schulz, University of Innsbruck

CHAPTER EDITORS

Georgij E. Belskij, CNIISK
Russell Q. Bridge, The University of Sydney
John E. Harding, University of Surrey
Bao Kang He, Xian Institute of Metallurgy & Construction Engineering
Jindrich J. Melcher, Technical University of Brno
Gerald W. Schulz, University of Innsbruck
Lambert Tall, Florida International University
Minoru Wakabayashi, General Building Research Corporation
Mingzhong Wei, Chongqing Architectural Institute
Wen Wei Zhao, Quinghuo University

1a. Column Strength

SPECIFICATION PROVISIONS & EXPLANATION

AUSTRALIA

CHAPTER EDITOR

Russell Q. Bridge, The University of Sydney

AS DR87164 87

The nominal capacity N_c for the strength limit state of an axially loaded compression member is given by

$$N_c = \alpha_c N_s$$

where

N_s = the section capacity of a short column (see Eq. 1c.A1)

$$\alpha_c = \xi \left\{1 - \sqrt{1 - 90^2/(\xi^2\lambda^2)}\right\} \tag{1a.A1}$$

$$\eta = 0.00326(\lambda - 13.5) \geq 0 \tag{1a.A2}$$

$$\lambda = \lambda_n + \alpha_b \alpha_a$$

$$\alpha_a = \frac{2100(\lambda_n - 13.5)}{\lambda_n^2 - 15.3\lambda_n + 2050} \tag{1a.A3}$$

α_b = the appropriate section constant given in Table 1a.A1

$$\lambda_n = \sqrt{Q(L_e/r)}\ \sqrt{(\sigma_y/250)} \tag{1a.A4}$$

$$\xi = \frac{(\lambda/90)^2+1+\eta}{2(\lambda/90)^2} \quad (1a.A5)$$

σ_y = yield strength (MPa)

Typical column curves are shown in Fig. 1a.A1 for categories of columns described in Table 1a.A1.

COMMENTARY

Early in the development of rules for column strength, the value of the multiple column curve concept was recognized (Rotter, 1982). Following reviews of the American SSRC recommendations (Johnston, 1976), the ECCS rules derived from the work of Beer and Schulz (1970), and simplifications of these by Dwight (1975) and Rondal and Macquoi (1979), it was decided to seek a multiple column curve formulation with the following features:

(a) a wide separation of the curves so that the designer would feel that the additional complexity of multiple curves was worthwhile,
(b) a simple algebraic description for easy coding into microcomputers and calculators,
(c) an easy means of providing transitions between curves, and
(d) a way of setting out the curves in tabular form (including transitions between the curves) without several tables being required.

Table 1a.A1(a) Values of Section Constant α_b for Q=1.0

α_b	Section description
−1.0	– Hot-formed RHS and CHS – Cold-formed (stress relieved) RHS and CHS
−0.5	– Cold-formed (non-stress relieved) RHS and CHS
0	– Hot-rolled UB and UC sections (flange thickness up to 40 mm) – Welded H and I sections fabricated from flame-cut plates – Welded box sections
0.5	– Angles and tees flame-cut from universal sections – Hot-rolled channels – Welded H and I sections fabricated from as-rolled plates (flange thickness up to 40 mm) – Other sections not listed in this table
1.0	– Hot-rolled UB and UC sections (flange thickness over 40 mm) – Welded H and I sections fabricated from as-rolled plates (flange thickness over 40 mm)

The solution adopted was to use a single central reference curve, which would apply to many column sections, and to make use of a single transforming function that could shift the reference curve up or down by amounts under the control of the user. This gave rise to the "modifying factor" method for multiple column curves (Rotter, 1982).

The reference curve (for which α_b=0) is given by a Perry equation fit to the SSRC curve.

The value of the shift parameter α_b is defined according to a section assignment table that is similar to those found in the ECCS and SSRC recommendations, but that tends toward a slightly more conservative choice wherever uncertainty arises. The resulting curves are very similar in form to the SSRC curves, but are simpler and more flexible.

Table 1a.A1(b) Values of Section Constant α_b for Q < 1.0

α_b	Section description
−0.5	– Hot formed RHS and CHS – Cold-formed RHS and CHS (stress relieved) – Cold-formed RHS and CHS (non-stressed relieved)
0	– Hot-rolled UB and UC sections (flange thickness up to 40 mm) – Welded box sections
0.5	– Welded H and I sections (flange thickness up to 40 mm)
1.0	– Other sections not listed in this table

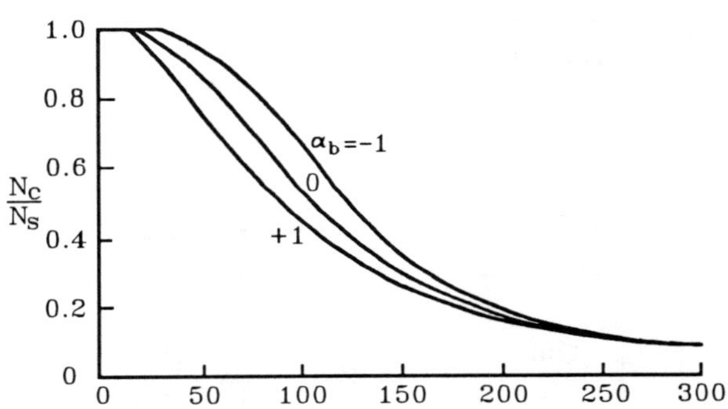

Fig. 1a.A1 Column Strength Curves for Various Section Constants

CHINA

CHAPTER EDITORS
Bao Kang He, Xian Institute of Metallurgy & Construction Engineering
Mingzhong Wei, Chongqing Architectural Institute
Wen Wei Zhao, Quinghuo University

GBJ 17 88

The stability of centrally compressed members should be checked by

$$\frac{N}{\phi A} \leq f \qquad (1a.C1)$$

where

- A = cross-sectional area
- f = design normal stress of steel
- N = design value of axial force
- ϕ = strength reduction factor of centrally compressed members; that is, the ratio of ultimate strength of compression member to the specified minimum yield stress

The ϕ curves are shown in Fig. 1a.C1 and their expressions are

for $\overline{\lambda} \leq 0.215$

$$\phi = 1-\alpha_1\overline{\lambda}^2 \qquad (1a.C2)$$

for $\overline{\lambda} > 0.215$

$$\phi = \frac{1}{2\overline{\lambda}^2}\left\{(\alpha_2+\alpha_3\overline{\lambda}+\overline{\lambda}^2)-\sqrt{(\alpha_2+\alpha_3\overline{\lambda}+\overline{\lambda}^2)^2-4\overline{\lambda}^2}\right\} \qquad (1a.C3)$$

where

- f_y = specified yield stress, f_y=235 N/mm^2 for A3 steel, 345 N/mm^2 for 16Mn steel, 390 N/mm^2 for 15MnV steel
- α_1, α_2, α_3 = coefficients shown in Table 1a.C1
- λ = slenderness ratio
- $\overline{\lambda}$ = $\frac{\lambda}{\pi}\sqrt{\frac{f_y}{E}}$, dimensionless slenderness ratio

Table 1a.C1 Coefficients α_1, α_2, α_3

curve		α_1	α_2	α_3
a		0.41	0.986	0.152
b		0.65	0.965	0.300
c	$\overline{\lambda} \leq 1.05$	0.73	0.906	0.595
	$\overline{\lambda} > 1.05$		1.216	0.302

COMMENTARY

Equation 1a.C1 is based on the limit states design method. The load factors for loads N are 1.2 for dead loads and 1.4 for live loads. $f = f_y/\gamma_R$, where γ_R, the resistance factor is 1.087 for A3 and 16Mn, and 1.111 for 15MnV.

The $\phi-\overline{\lambda}$ curves in Fig. 1a.C1 are based on the ultimate strength theory. In these curves, the members are considered as pinned ended, with initial deflection at midspan of $\ell/1000$ (ℓ is the length of members), with different types and sizes of sections and different patterns of residual stresses.

The 96 calculated $\phi-\overline{\lambda}$ curves are divided into three groups according to the types of sections and load-carrying capacities. In each group the values corresponding to 50% probability are taken as the representative curves; that is, curves a, b, c in Fig. 1a.C1. Equations 1a.C2 and 1a.C3 are the equivalent expressions for the curves. Taking the Perry formula as the equivalent model of curves a, b, c (that is, the model for the extreme fiber yield criterion), Eq. 1a.C3 is obtained from fitting to curves a, b, c. The equivalent formula is

$$\phi = \frac{1}{2\overline{\lambda}^2}\left\{(1+\xi_o+\overline{\lambda}^2) - \sqrt{(1+\xi_o+\overline{\lambda}^2)^2 - 4\overline{\lambda}^2}\right\} \tag{1a.C4}$$

where

V_o = equivalent initial deflection

W = elastic section modulus

$\xi_o = V_o\frac{A}{W}$, equivalent eccentricity ratio

$= 0.152\overline{\lambda}-0.014$ for curve a

$= 0.300\overline{\lambda}-0.035$ for curve b

$= 0.595\overline{\lambda}-0.094$ for curve c when $\overline{\lambda} \leq 1.05$

$= 0.302\overline{\lambda}+0.216$ for curve c when $\overline{\lambda} > 1.05$

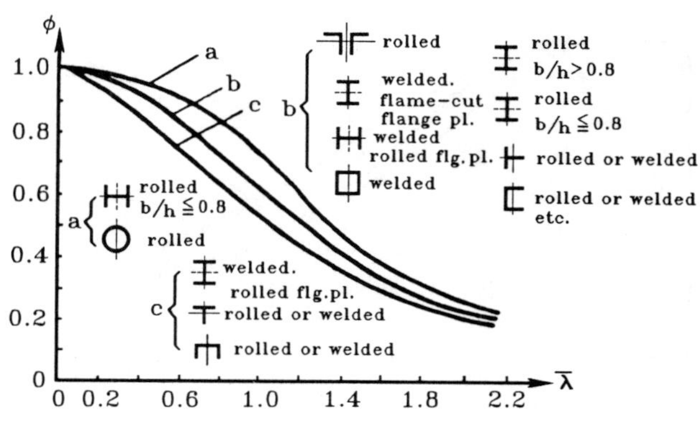

Fig. 1a.C1 Column Strength Curves a, b, and c

TBJ 2 85

The stability of centrally compressed members should be checked by the formula

$$\frac{N}{A} \leq \phi\sigma \qquad (1a.C5)$$

where

f_y = specified minimum yield stress
= 230 N/mm^2 for A3 and 16q steel
= 340 N/mm^2 for 16 Mn and 16Mnq steel
N = nominal axial force
σ = allowable stress of steel
σ_u = the ultimate strength of compression members
ϕ = σ_u/f_y

The $\phi-\lambda$ curves are shown in Fig. 1a.C2.

COMMENTARY

Equation 1a.C5 is based on the allowable stress method and here $\sigma = f_y/K$. K, a factor for safety, is taken as 1.7 for the main load combination.

The $\phi-\lambda$ curves in Fig. 1a.C2 are based on ultimate strength theory. These curves take into account different shapes, various residual stress patterns, and initial member crookedness. The initial member crookedness is 0.001 ℓ for $\lambda \leq 45$, and 0.0008 ℓ for $\lambda > 45$, where ℓ is the member length.

When $\lambda \leq 30$ (for curves a′, b, b′) or 40 (for curve a) the value of ϕ is reduced to 0.9. The reason is that the design is controlled by strength of the net area, and the strength of net area need not be checked according to this specification.

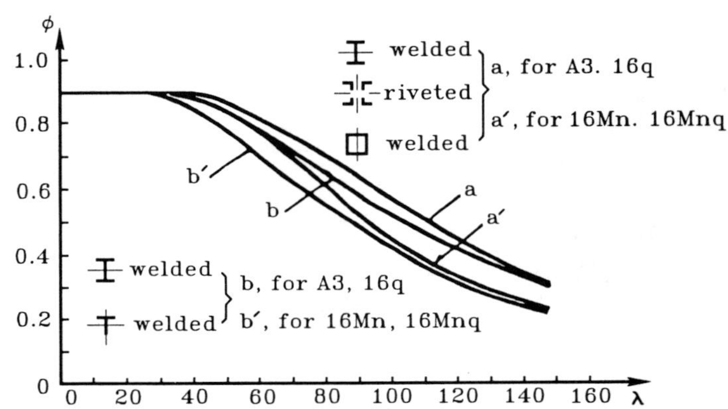

Fig. 1a.C2 Column Strength Curves a, a′, b, and b′

EAST EUROPE

CHAPTER EDITORS
Georgij E. Belskij, CNIISK, Moscow
Jindrich J. Melcher, Technical University of Brno

ČSN 73 1401 84

The limit states design approach and design criterion conformable to Eq. 1a.E11 are used. Also the general formula for the buckling curves can be expressed by Eq. 1a.E13, but the corresponding imperfection parameter m_o in this case is

$$m_o = a_o \, \overline{\lambda}^2 \tag{1a.E1}$$

Two basic column curve parameters for flexural buckling are introduced:

for round or rectangular thin-walled tubes

$$a_o = 0.17$$

for all other structural shapes and thin-walled or solid sections

$$a_o = 0.26$$

The material reliability factor is $\gamma_m = 1.15$ for steel grades of $R_y \leq 300$ MPa and $\gamma_m = 1.25$ for $R_y > 300$ MPa, respectively.

COMMENTARY

The design load of centrally compressed members is derived by the conventional procedure using the onset of yielding in the column with a specified initial curvature (expressing the influence of all imperfections) as the limit state. In general, one column curve (b) (Fig. 1a.E1) is used except for hollow sections, where column curve (a) is recommended. There is also another curve (c) defined by

$$\varphi = 1 - \overline{\lambda}/2.16 \tag{1a.E2}$$

For $\lambda(R_y/210)^{1/2} \geq 126$ curve (c) is identical with the curve (b). Column curve (c) can be used for approximate analysis of flexural-torsional buckling of angle, T, and U sections when only the flexural bending is considered. Thus this curve (c) does not belong to the system of multiple column curves for flexural buckling.

MSZ 15024/1 85

The limit states design approach is used. For $\overline{\lambda} > 0.2$ the buckling curves are given by the general formula

$$\varphi = \beta - \sqrt{\beta^2 - \frac{1}{\overline{\lambda}^2}} \tag{1a.E3}$$

where β is $\dfrac{1+\alpha(\overline{\lambda}-0.2)+\overline{\lambda}^2}{2\overline{\lambda}^2}$ (1a.E4)

and the parameter of imperfections α is given in Table 1a.E1 for four different curves (Fig. 1a.E1).

Table 1a.E1 Imperfection Parameter

curve	a	b	c	d
α	0.21	0.34	0.49	0.76

For $\alpha \leq 0.2$ the buckling coefficient $\varphi = 1$.

COMMENTARY

The multiple column curve concept and the grouping of the sections are applied in the same form as proposed in the EUROCODE 3.

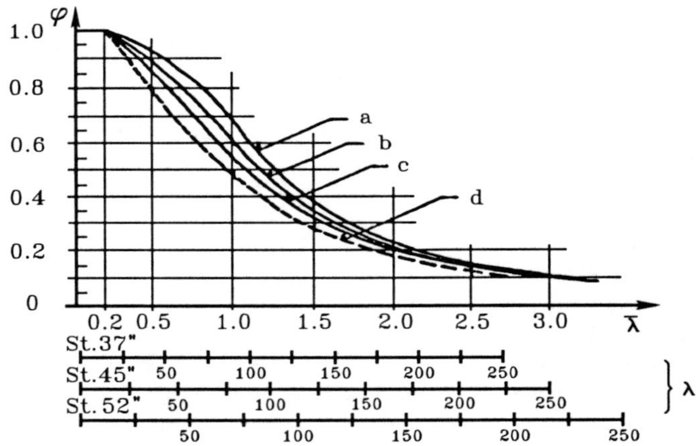

Fig. 1a.E1 Column Strength Curves a, b, c, and d

PN 80/B-03200 84

The limit states design approach with the basic criterion according to Eq. 1a.E11 is used. The corresponding material reliability factor in Eq. 1a.E12 in this case is

for $R_y \leq 355$ MPa

$$\gamma_m = 1.15 \tag{1a.E5}$$

for 355 MPa $< R_y \leq 460$ MPa

$$\gamma_m = 1.20 \tag{1a.E6}$$

for 460 MPa $< R_y \leq 590$ MPa

$$\gamma_m = 1.25 \tag{1a.E7}$$

Only one buckling strength curve is recommended. This curve consists of two branches and can be expressed as follows:

for $\overline{\lambda} \leq 1.22$

$$\varphi = 1 - 0.164\overline{\lambda} - 0.067\overline{\lambda}^2 - 0.110\overline{\lambda}^3 \qquad \text{(1a.E8)}$$

for $\overline{\lambda} > 1.22$

$$\varphi = 0.744/\overline{\lambda}^2 \qquad \text{(1a.E9)}$$

COMMENTARY

Currently, a new edition of the Polish specification is under preparation. The multiple column curve concept has been adopted and the proposed formula of uniform curves is

$$\varphi = \frac{1}{\sqrt[n]{1 + (\overline{\lambda})^{2n}}} \qquad \text{(1a.E10)}$$

with the parameter of imperfection n according to the Table 1a.E2.

A similar approach with three different column curves has been applied also in the Rumanian specification STAS 10108/0 78 since 1978.

Table 1a.E2 Imperfection Parameter

curve	a	b	c
n	1.60	1.20	1.00

SNiP II-23-81 82

The limit states design approach based on the Load and Resistance Factor Design method is used. The general format of the design criterion for compression members with cross-sectional area A is

$$N_d \leq \varphi \, A \, R_d \qquad \text{(1a.E11)}$$

where

N_d = limit (design) axial load
φ = buckling coefficient

The design stress R_d is defined as

$$R_d = R_y \; / \; \gamma_m \qquad \text{(1a.E12)}$$

where

R_y = specified yield stress
γ_m = material reliability factor considered in the range of 1.025 to 1.15

The multiple column curve concept is not applicable for the practical design of compression members and only one buckling strength curve is used.

The corresponding analytical formulation for the single column curve is given by

$$\varphi = \frac{1}{2\overline{\lambda}^2}[1+m_o+\overline{\lambda}^2-\sqrt{(1+m_o+\overline{\lambda}^2)^2-4\overline{\lambda}^2}\,] \tag{1a.E13}$$

where the modified slenderness $\overline{\lambda}$ is given by

$$\overline{\lambda} = \frac{\lambda}{\pi}\sqrt{R_y/E} \tag{1a.E14}$$

and the imperfection parameter m_o is given by

$$m_o = c_1 + c_2\overline{\lambda} \tag{1a.E15}$$

The coefficients c_1 and c_2 in Eq. 1a.E15 are

for $\overline{\lambda} \le 1.20$

$$c_1 = -0.05, \qquad c_2 = 0.314 \tag{1a.E16}$$

for $\overline{\lambda} > 1.20$

$$c_1 = -0.8, \qquad c_2 = 0.94 \tag{1a.E17}$$

About thirty standard types of steel grades with the yield stress range from 185 to 590 MPa are used in the steel building industry. The corresponding single buckling curve is presented in Fig. 1a.E2 as influenced by the various steel grades.

COMMENTARY

The design provisions for the flexural buckling of centrally compressed members are based on the strength analysis of a column with initial out-of-straightness and initial eccentricity of applied load. The magnitude of geometrical imperfections is obtained from specified values of allowed fabrication and erection deviations. The design load carrying capacity is derived using the inelastic column buckling approach.

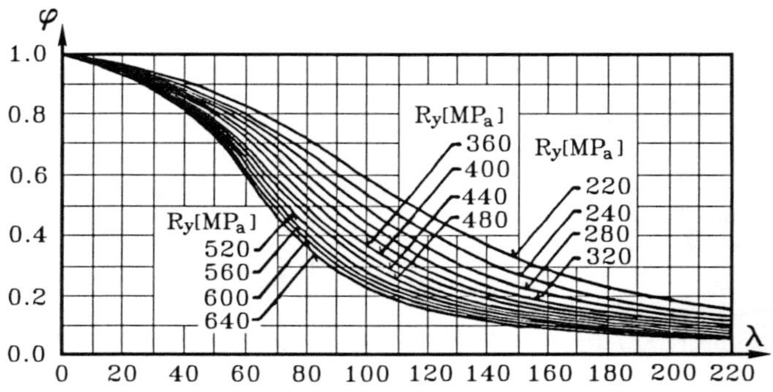

Fig. 1a.E2 Column Strength Curves for Various Yield Stresses

TGL 13503/02 82

The allowable stress design concept with a safety factor is used. It is probable that from 1990 this concept will be changed to that of a limit state design approach using partial load and resistance safety factors.

The general format of the design criterion is

$$N \leq \varphi \cdot A \cdot \sigma_{all} \tag{1a.E18}$$

where N is the actual (specified) axial load

The allowable stress σ_{all} is defined by the relationship

$$\sigma_{all} = R_y/\gamma \tag{1a.E19}$$

where

γ = safety factor
= 1.50 (for principal load)
= 1.33 (for principal and secondary loads)
= 1.20 (for secondary loads)

For the buckling curves, the general Eq. 1a.E13 with Eq. 1a.E14 and Eq. 1a.E15 can be used as well.

The corresponding coefficients c_1 and c_2 are given in Table 1a.E3 for the four column strength curves of Fig. 1a.E1: (a), (b), (c) and (d). In the specification the more detailed commentary to Table 1a.E3 is presented making the choice of the suitable buckling curve possible according to the influence of shape of cross section, manufacturing procedures, and the direction of buckling.

COMMENTARY

The multiple column curve concept is based on the ECCS theoretical studies and experiments resulting in the stability provisions of the European Recommendations for Steel Construction.

Table 1a.E3 Coefficients c_1 and c_2

Residual stresses		Cross section (examples)					
		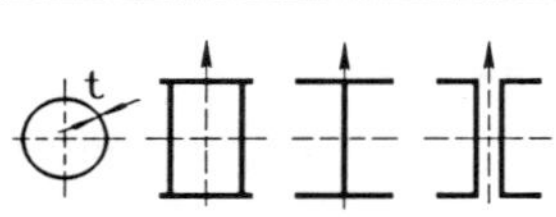			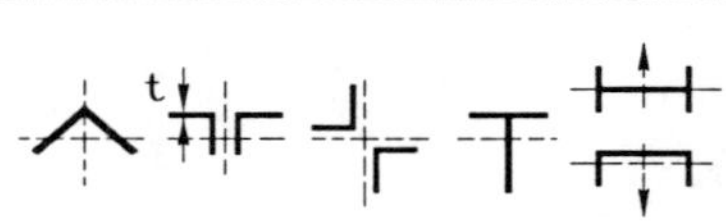		
		Curve	c_1	c_2	Curve	c_1	c_2
slight		a	−0.030	0.186	b	−0.031	0.291
fair	t ≤ 40	b	−0.031	0.291	c	−0.045	0.423
(mm)	t > 40	c	−0.045	0.423	d	−0.063	0.581

REGIONAL COMPARISON

The regional view of the specification provisions presented in Fig. 1a.E3 indicates that there is not yet a unified approach to the design of compression members from the point of view of the initial imperfections and their influence on the behavior of a real strut incorporated into an actual structural system.

In many countries the multiple column curve concept is accepted as a significant result of scientific research, but some important reservations as to its general adoption for practical design are discussed. One of these problems is the influence of loading history upon the residual stress distribution. The multiple column curve concept is based on only one cycle of loading. We may hardly anticipate that the column curve relationship derived for one loading cycle would also apply for the subsequent loading cycles in general when the residual stress distribution under axial force after several loading cycles must be totally different from the one assumed for an initially unloaded member.

Another problem is the application of the spectrum of column curves for one steel grade to many different kinds of steel used in the building industry (compare the examples of multi-column curve system in Fig. 1a.E3 to the single curve in Fig. 1a.E2). In this connection the problem of design economy of structural members made of different steel grades also must be considered.

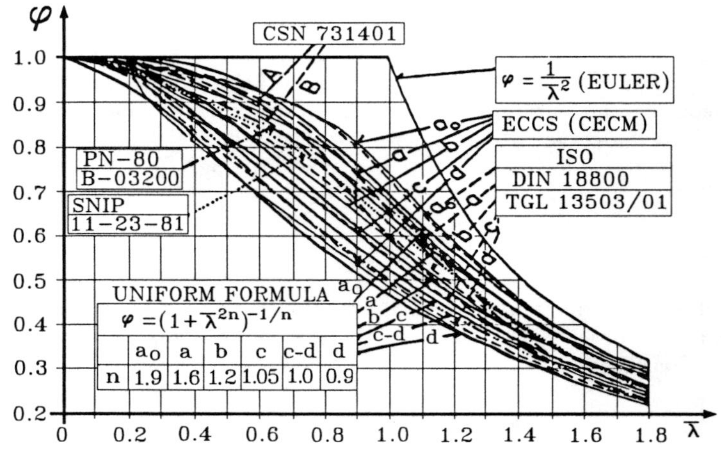

Fig. 1a.E3 Column Curve Comparison for Different Codes

JAPAN

CHAPTER EDITOR
Minoru Wakabayashi, General Building Research Corporation

AIJ 70

The AIJ Standard allows the following equations to stipulate the allowable flexural stress f_c of compression members

for $\lambda \leq \Lambda$

$$f_c = \{1-0.4(\lambda/\Lambda)^2\}F/\{3/2+2/3(\lambda/\Lambda)^2\} \qquad (1a.J1)$$

for $\lambda > \Lambda$

$$f_c = 0.277F(\lambda/\Lambda)^2 \qquad (1a.J2)$$

where

F = specified yield stress

$\Lambda = \sqrt{\pi^2 E/0.6F}$

$\lambda = \ell/r$, where ℓ and r are the effective length and radius of gyration of the member, respectively

AIJ 75

The AIJ Guide for Plastic Design stipulates the ultimate flexural strength N_{cr} of compression members as

for $0 \leq \bar{\lambda} \leq 0.3$

$$N_{cr} = N_y \qquad (1a.J3)$$

for $0.3 < \bar{\lambda} \leq 1.3$

$$N_{cr} = \{1-0.545(\bar{\lambda}-0.3)\}N_y \qquad (1a.J4)$$

for $\bar{\lambda} > 1.3$

$$N_{cr} = 1/(1.3\bar{\lambda}^2)N_y \qquad (1a.J5)$$

where

N_y = specified yield force

$\lambda = \ell/r$ where ℓ and r are the effective length and radius of gyration of the member, respectively

$\bar{\lambda} = \frac{\lambda}{\pi}\sqrt{\sigma_y/E}$

σ_y = specified yield stress

AIJ 89b

Equations given by the AIJ LSD Guide for specifying the flexural strength (N_c) of compression members are for $\lambda_c \leq {}_p\lambda_c$

$$N_c = N_y$$
$$\phi_c = 0.9 \qquad (1a.J6)$$

for ${}_p\lambda_c < \lambda_c \leq {}_e\lambda_c$

$$N_c = \{1-0.50(\lambda_c-{}_p\lambda_c)\}N_y/({}_e\lambda_c-{}_p\lambda_c)$$
$$\phi_c = 0.90-0.05(\lambda_c-{}_p\lambda_c)/({}_e\lambda_c-{}_p\lambda_c) \qquad (1a.J7)$$

for ${}_e\lambda_c < \lambda$

$$N_c = 1/(1.2\lambda^2)N_y$$
$$\phi_c = 0.85 \qquad (1a.J8)$$

where

N_e = Euler's buckling force ($\pi^2 EI/\ell^2$ with I and ℓ as the moment of inertia and effective length of the member)
N_y = specified yield force
r_c = $\sqrt{N_y/N_e}$
${}_e\lambda_c$ = limiting slenderness ratio for elastic buckling given as $1/\sqrt{0.6}$
${}_p\lambda_c$ = limiting slenderness ratio for plastic analysis given as 0.15
ϕ_c = resistance factor

COMMENTARY

The Architectural Institute of Japan (AIJ) provides three general standards or guides for designing steel building structures. They are the AIJ Standard for Steel Structures (the AIJ Standard), the AIJ Standard for Plastic Design of Steel Structures (the AIJ Guide for Plastic Design), and the AIJ Standard for Limit State Design of Steel Structures (the AIJ LSD Standard). The AIJ Standard adopts the allowable stress design concept and is the one used most commonly when designing steel building structures constructed in Japan. The AIJ Guide for Plastic Design is based on ultimate strength. This Guide is often referred to in evaluating the earthquake force resisting capacity of steel buildings, because the Japanese seismic design specification requires that tall buildings need be investigated for their strength and ductility performance against earthquake loading. The AIJ LSD Standard follows a reliability based design format.

Equations 1a.J1 to 1a.J8 are compared in Fig. 1a.J1 in the form of column curves. All the curves consist of an elastic portion representing the linear buckling and an inelastic portion representing the nonlinear buckling. The linear portion is a factored Euler's buckling curve, whereas the nonlinear portion a parabola in the AIJ Standard and a straight line in the AIJ Guide for Plastic Design and the AIJ LSD Standard. Considering the geometrical and material imperfections, in all column curves, the elastic portion is terminated at a stress (force) smaller than $F\,(N_y)$. The maximum force governed by the elastic portion is 0.6F in the AIJ Standard, $0.455N_y$ for the AIJ Guide for Plastic Design, and $0.5N_y$ for the AIJ LSD Standard, respectively. In the AIJ Standard, the factor of safety (the denominator of Eq. 1a.J1) is given as 1.5 for a member with zero length and increases with the increase of the length, considering the observation that the strength of a longer member is more sensitive to geometrical and material imperfections. The maximum factor of safety (governing the elastic portion) is set at 2.17. In the AIJ Standard for Plastic Design, $N_c=N_y$ for $\bar{\lambda}$ less than 0.3, which reflects the observation that, for short members, decrease in the maximum resistance caused by various imperfections is overshadowed by significant strain hardening. The factor 1.3 appearing in Eq. 1a.J5 is approximately equal to 2.17/1.65, where 2.17 and 1.65 are the maximum factor of safety in the AIJ Standard and the load factor adopted in the AIJ Guide for Plastic Design, respectively. The resistance factor in the AIJ LSD Standard varies from 0.9 for short members to 0.85 for long members, again considering the observation that the strength of a long member is more sensitive to various imperfections. As shown in Eq. 1a.J7, in the intermediate range, the resistance factor is taken to decrease linearly with the increase in λ_c.

JRA 80

The following equations are given for the allowable flexural stress (σ_{cag}) (in N/mm^2) of compression members.

SS41, SM41, SMA41:

for $0 \leq \lambda \leq 20$

$$\sigma_{cag} = 137$$

for $20 < \lambda \leq 93$

$$\sigma_{cag} = 137-0.824(\lambda-20) \tag{1a.J9}$$

for $93 < \lambda$

$$\sigma_{cag} = 1180000/(6700+\lambda^2)$$

SS50:

for $0 \leq \lambda \leq 17$

$$\sigma_{cag} = 167$$

for $17 < \lambda \leq 86$

$$\sigma_{cag} = 167-1.11(\lambda-17) \tag{1a.J10}$$

for $86 < \lambda$

$$\sigma_{cag} = 1180000/(5700+\lambda^2)$$

SM50:

for $0 \leq \lambda \leq 15$

$$\sigma_{cag} = 186$$

for $15 < \lambda \leq 80$

$$\sigma_{cag} = 186-1.27(\lambda-15) \tag{1a.J11}$$

for $80 < \lambda$

$$\sigma_{cag} = 1180000/(5000+\lambda^2)$$

SM50Y, SM53, SMA50:

for $0 \leq \lambda \leq 14$

$$\sigma_{cag} = 206$$

for $14 < \lambda \leq 76$

$$\sigma_{cag} = 206-1.47(\lambda-14) \tag{1a.J12}$$

for $76 < \lambda$

$$\sigma_{cag} = 726000/(4500+\lambda^2)$$

SM58, SMA58:

for $0 \leq \lambda \leq 14$

$$\sigma_{cag} = 255$$

for $14 < \lambda \le 67$

$$\sigma_{cag} = 255-2.06(\lambda-14) \qquad (1a.J13)$$

for $67 < \lambda$

$$\sigma_{cag} = 1180000/(3600+\lambda^2)$$

In the above equations, λ is the effective slenderness ratio.

COMMENTARY

Two sets of specifications are made available for designing steel bridges constructed in Japan. They are the Japan Society of Civil Engineering (JSCE) Specification for Steel Railway Bridges (JSCE Specifications) and the Japan Road Association (JRA) Specification for Highway Bridges (JRA Specifications). Both specifications adopt the allowable stress design concept in accordance with the material strength employed.

The allowable stress provisions stipulated in the JSCE Specification have their basis on a column curve shown in Fig. 1a.J2. The column curve consists of three portions: that is, a portion expressed by the Euler buckling curve and applied in the range $\sigma_{cr}/\sigma_y \le 0.5$ (σ_{cr} is compressive stress and σ_y is specified yield stress), a straight line representing inelastic buckling and terminating at $\bar{\lambda}=0.3$ where $\sigma_{cr}/\sigma_y=1.0$, and a line showing the yield stress for $\bar{\lambda} \le 0.3$, in which $\bar{\lambda}$ is the normalized effective slenderness ratio defined as $\frac{\ell}{r}\frac{1}{\pi}\sqrt{\frac{\sigma_y}{E}}$ with ℓ and r as the effective length and the radius of gyration of the member concerned. The column curve thus specified is considered to represent the flexural buckling strength of columns having large residual stresses such as those built-up by welding. The factor of safety adopted in the Specifications varies from 1.9 for $\bar{\lambda} \le 0.3$ to 2.8 for $\bar{\lambda} \ge \sqrt{2}$.

Fig. 1a.J3 shows the allowable stress (that is, Eqs. 1a.J9 to 1a.J16) in nondimensionalized form. The column curve considered in the JRA Specifications was obtained based on numerical analysis. In this analysis, a column with an I-shaped cross section was assumed, and a combination of maximum compressive residual stress of $0.4\sigma_y$ and out-of-straightness of $\ell/1000$ were taken into account. The flexural strength thus computed (σ_{cr}) was simplified into the following two equations, which constitute the column curve for compression members:

for $\bar{\lambda} \le 1$

$$\sigma_{cr}/\sigma_y = 1-0.136\bar{\lambda}-0.300\bar{\lambda}^2 \qquad (1a.J14)$$

for $\bar{\lambda} > 1$

$$\sigma_{cr}/\sigma_y = 1.276-0.888\bar{\lambda}+0.176\bar{\lambda}^2 \qquad (1a.J15)$$

where $\bar{\lambda}$ is the normalized effective slenderness ratio defined as: $\ell/r/\pi\sqrt{\sigma_y/E}$. This column curve is shown in Fig. 1a.J3. To account for the allowable stress, the first equation (1a.J14) is approximated by two straight lines, and the second equation (1a.J15) by an equation

$$\sigma_{cr}/\sigma_y = 1/(0.773+\bar{\lambda}^2) \qquad (1a.J16)$$

Further, a uniform safety factor of 1.7 is used in the JRA Specifications. The allowable stress thus enforced is also illustrated in Fig. 1a.J3.

JSCE 83

The allowable flexural stress σ_w of compression members (in N/mm^2) is expressed by the following equations for the different steels.

Material	Slenderness Ratio	Allowable Stress
SS41, SM41, SMA41	for $0 \leq \lambda \leq 28$	$\sigma_w = 123$
	for $28 < \lambda \leq 130$	$\sigma_w = 123-0.785(\lambda-28)$
	for $130 < \lambda$	$\sigma_w = 726000/\lambda^2$
SM50	for $0 \leq \lambda \leq 24$	$\sigma_w = 167$
	for $24 < \lambda \leq 115$	$\sigma_w = 167-1.23(\lambda-24)$
	for $115 < \lambda$	$\sigma_w = 726000/\lambda^2$
SM50Y, SM53, SMA50	for $0 \leq \lambda \leq 22$	$\sigma_w = 186$
	for $22 < \lambda \leq 105$	$\sigma_w = 186-1.45(\lambda-22)$
	for $105 < \lambda$	$\sigma_w = 726000/\lambda^2$
SM58, SMA58	for $0 \leq \lambda \leq 20$	$\sigma_w = 235$
	for $20 < \lambda \leq 95$	$\sigma_w = 235-2.07(\lambda-20)$
	for $95 < \lambda$	$\sigma_w = 726000/\lambda^2$

In the above equations, λ is the effective slenderness ratio.

REGIONAL COMPARISON

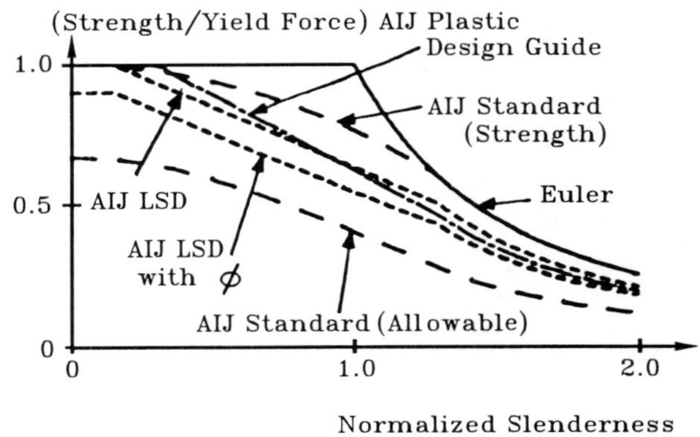

Fig. 1a.J1 Column Curves and Allowable Stresses Considered in AIJ Standard, AIJ Guide for Plastic Design, and AIJ LSD Guide

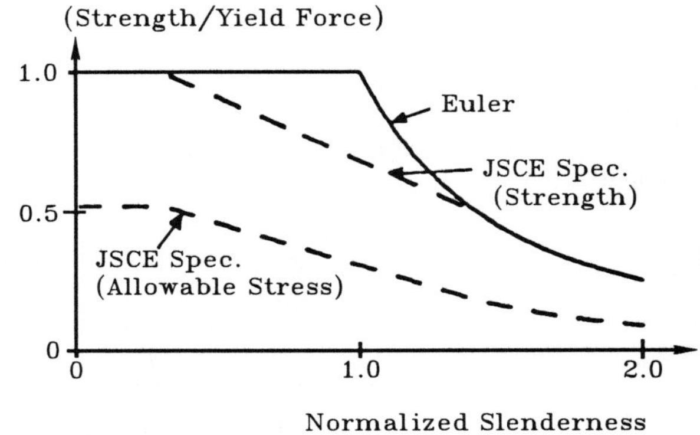

Fig. 1a.J2 Column Curves and Allowable Stresses Considered in JSCE Specifications

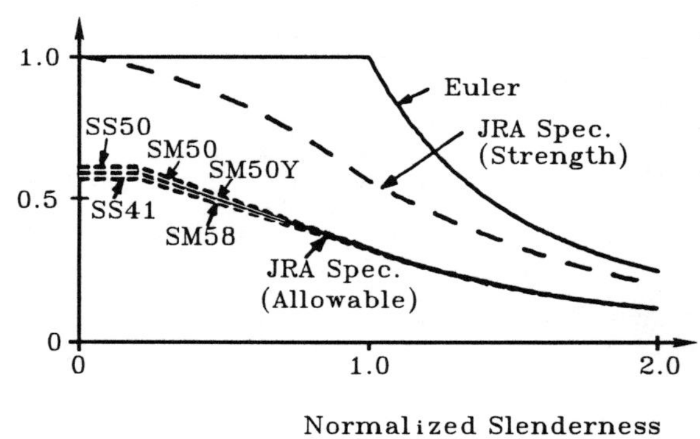

Fig. 1a.J3 Column Curves and Allowable Stresses Considered in JRA Specifications

NORTH AMERICA

CHAPTER EDITOR
Lambert Tall, Florida International University

AA 86

The column curve is a straight line between the yield strength and the Euler curve, and is a simplification of the tangent modulus formula. The constants in Eqs. 1a.N1, N2, and N3 are dependent on the particular aluminum alloy and temper under consideration, and these are enumerated in detail in the specification. A constant Factor of Safety is used, 1.95 for buildings and 2.20 for bridges, or, for short columns, 1.65 and 1.85 respectively.

The allowable compressive stress, F_a, is

for $\frac{L}{r} \le S_1$

$$F_a = \frac{F_{cy}}{k_c\, n_y} \tag{1a.N1}$$

for $S_1 < \frac{L}{r} \le S_2$

$$F_a = \frac{1}{n_u}\left(B_c - D_c \frac{L}{r}\right) \tag{1a.N2}$$

for $S_2 < \frac{L}{r}$

$$F_a = \frac{\pi^2 E}{n_u \left(\frac{L}{r}\right)^2} \tag{1a.N3}$$

where

B_c = intercept (a function of yield; tabulated)
D_c = slope (function of yield and E; tabulated)
F_{cy} = compressive yield stress
k_c = coefficient for compression members (tabulated according to alloy and temper)
n_u = factor of safety on ultimate strength (1.95 or 2.20)
n_y = factor of safety on yield strength (1.65 or 1.85)

$$S_1 = \frac{B_c - \frac{n_u F_{cy}}{k_c\, n_y}}{D_c}$$

$$S_2 = C_c = 0.41 \frac{B_c}{D_c}$$

COMMENTARY

The specification notes that the use of n_u and the neglect of end restraint compensate for the possible weakening effects of accidental eccentricity and crookedness in columns.

Equations 1a.N1, N2, and N3 are tabulated with their actual values for each alloy and temper for both buildings and bridges, so that their use is simple. For example, for 2014–T6 and T651 alloy or Alclad, these equations become for $\frac{L}{r} \leq 12$ $\quad F_a = 28$ ksi

for $12 < \frac{L}{r} \leq 55$ $\quad F_a = 30.7 - 0.23 \frac{L}{r}$ ksi

for $55 < \frac{L}{r}$ $\quad F_a = \frac{5400}{(\frac{L}{r})^2}$ ksi

AASHTO 87

The allowable stress F_a in concentrically loaded columns is

$$\left.\begin{array}{ll} \text{for } \frac{kL}{r} \leq C_c = \sqrt{\frac{2\pi^2 E}{F_y}} & F_a = \frac{F_y}{F.S.}\left(1 - \frac{(\frac{kL}{r})^2 F_y}{4\pi^2 E}\right) \\ \text{for } \frac{kL}{r} > C_c & F_a = \frac{\pi^2 E}{F.S.(\frac{kL}{r})^2} = \frac{135{,}000{,}740}{(\frac{kL}{r})^2} \end{array}\right\} \qquad (1a.N4)$$

where F.S. is 2.12

Equation 1a.N4 is identical with the AISC in that the CRC Column Strength Curve is used as the basis. However, AASHTO uses a constant Factor of Safety. Yield strengths of 36 ksi and above are covered.

The maximum strength of a concentrically-loaded column is

$$P_u = 0.85\ A_s F_{cr} \qquad (1a.N5)$$

where

F_{cr} = allowable stress (see Eq. 1a.N4)
F_{cr} = F_a (F.S.)

AISC ASD 89

The allowable stress F_a on the gross section of axially loaded compression members is

for $\frac{kL}{r} \leq C_c = \sqrt{\frac{2\pi^2 E}{F_y}}$

$$F_a = \frac{\left(1 - \frac{(\frac{kL}{r})^2}{2C_c^2}\right) F_y}{\frac{5}{3} + \frac{3}{8} \cdot \frac{\frac{kL}{r}}{C_c} - \frac{(\frac{kL}{r})^3}{8C_c^3}} \qquad (1a.N6)$$

for $\frac{kL}{r} > C_c$

$$F_a = \frac{12\pi^2 E}{23(\frac{kL}{r})^2} \qquad (1a.N7)$$

The plastic design provisions are

$$P_{cr} = 1.7\ A\ F_a \tag{1a.N8}$$

where F_a are values defined by Eq. 1a.N6

COMMENTARY

Equation 1a.N6 is the CRC Column Strength Curve (Galambos, 1988; Johnston, 1966; Tall, 1974), a parabolic curve introduced in 1960, and is an expression based on the tangent modulus theory taking into consideration the effect of residual stresses and is a compromise curve for both strong and weak axis buckling for small and medium size rolled shapes. The denominator of the equation is the Factor of Safety, which is variable. The limiting slenderness ratio C_c marks the position where the parabola meets the Euler curve, and corresponds to an assumed compressive residual stress equal to $F_y/2$. The yield strength is 36 ksi and above.

AISC LRFD 86

The design strength of compression members is $\phi_c P_n$

$$\phi_c = 0.85 \tag{1a.N9}$$

$$P_n = A_g F_{cr}$$

for $\lambda_c \leq 1.5$

$$F_{cr} = (0.658^{\lambda_c^2})\ F_y \tag{1a.N10}$$

for $\lambda_c < 1.5$

$$F_{cr} = \left(\frac{0.877}{\lambda_c^2}\right) F_y \tag{1a.N11}$$

where $\lambda_c = \frac{kL}{\pi r}\sqrt{\frac{F_y}{E}}$

COMMENTARY

The curve given by Eqs. 1a.N10 and 1a.N11 is a simplification of SSRC Curve 2P, (Bjorhovde, 1972, Bjorhovde and Tall, 1971, 1975a, 1975b) and is based on a probabilistic analysis of column strength. Plastic analysis is permitted for $\lambda_c \leq 1.5K$.

AREA 89

The allowable stress in axial compression for ASTM A36 steel ($F_y = 36$ ksi, or 250 MPa) is

$$\left.\begin{array}{ll} 20{,}000 \text{ psi} & \text{when } \frac{kL}{r} \leq 15 \\ 21{,}500 - 100\,\frac{kL}{r} \text{ psi} & \text{when } 15 \leq \frac{kL}{r} \leq 143 \\ \dfrac{147{,}000{,}000}{(\frac{kL}{r})^2} \text{ psi} & \text{when } \frac{kL}{r} > 143 \end{array}\right\} \tag{1a.N12}$$

For high-strength structural steel, the allowable stress is

$$\left.\begin{array}{ll} 0.55\ F_y & \text{when } \frac{kL}{r} \le \frac{3388}{\sqrt{F_y}} \\ 0.6\ F_y - \left(\frac{F_y}{1662}\right)^{3/2} \cdot \frac{kL}{r} & \text{when } \frac{3388}{\sqrt{F_y}} \le \frac{kL}{r} \le \frac{27111}{\sqrt{F_y}} \\ \frac{147{,}000{,}000}{\left(\frac{kL}{r}\right)^2} & \text{when } \frac{kL}{r} > \frac{27111}{\sqrt{F_y}} \end{array}\right\} \quad (1a.N13)$$

where all stresses are in psi.

Equations 1a.N12 and 1a.N13 are straight line variations between the yield strength and the Euler curve and correspond approximately to the AISC Eq. 1a.N6, although neither residual stresses nor the tangent modulus were included in their development.

CSA S6 M89 84a

The allowable stress F_a in axial compression is

$$\left.\begin{array}{ll} \text{for } \frac{kL}{r} \le C_o & F_a = 0.6F_y \\ \text{for } C_o < \frac{kL}{r} \le C_p & F_a = 0.60F_y - m\left(\frac{kL}{r} - C_o\right) \\ \text{for } C_p < \frac{kL}{r} & F_a = \frac{1{,}030{,}000}{\left(\frac{kL}{r}\right)^2} \end{array}\right\} \quad (1a.N14)$$

where

$$C_o = 30 - \frac{F_y}{35} \qquad \text{but not more than 20}$$

$$C_p = 1000\sqrt{\frac{1.97}{F_y - 90}} \qquad \text{but not less than 78}$$

$$m = \frac{0.60F_y - 1{,}030{,}000/(C_p)^2}{C_p - C_o}$$

and where all stresses are in MPa.

These are straight-line formulas between the yield strength and the Euler curve, and neither residual stress nor tangent modulus were considered.

CSA S16.1 M89 84b

The standard resistance factor, $\phi = 0.90$. Multiple column curves are used.

For stress-relieved shapes, the factored axial compressive resistance is

$$
\left.
\begin{array}{ll}
\text{for } 0 \le \lambda \le 0.15 & C_r = \phi A F_y \\
\text{for } 0.15 < \lambda \le 1.2 & C_r = \phi A F_y (0.990 + 0.122\lambda - 0.367\lambda^2) \\
\text{for } 1.2 < \lambda \le 1.8 & C_r = \phi A F_y (0.051 + 0.801\lambda^{-2}) \\
\text{for } 1.8 < \lambda \le 2.8 & C_r = \phi A F_y (0.008 + 0.942\lambda^{-2}) \\
\text{for } 2.8 < \lambda & C_r = \phi A F_y \lambda^{-2} = \phi A \left(\frac{1{,}970{,}000}{(\frac{kL}{r})^2}\right)
\end{array}
\right\} \qquad (1a.N15)
$$

where $\lambda = \frac{kL}{r}\sqrt{\frac{F_y}{\pi^2 E}}$

For hollow and wide flange shapes other than stress-relieved, the factored axial compressive resistance is

$$
\left.
\begin{array}{ll}
\text{for } 0 \le \lambda \le 0.15 & C_r = \phi A F_y \\
\text{for } 0.15 < \lambda \le 1.0 & C_r = \phi A F_y (1.035 + 0.202\lambda - 0.222\lambda^2) \\
\text{for } 1.0 < \lambda \le 2.0 & C_r = \phi A F_y (-0.111 + 0.636\lambda^{-1} + 0.087\lambda^{-2}) \\
\text{for } 2.0 < \lambda \le 3.6 & C_r = \phi A F_y (-0.009 + 0.877\lambda^{-2}) \\
\text{for } 3.6 < \lambda & C_r = \phi A F_y \lambda^{-2} = \phi A \left(\frac{1{,}970{,}000}{(\frac{kL}{r})^2}\right)
\end{array}
\right\} \qquad (1a.N16)
$$

For heavy shapes (greater than 4.5 cm thick), the compressive resistance is calculated from either of the respective compressions, Eqs. 1a.N15 or 1a.N16, with the area A taken as the effective area defined in accordance with CSA S16.1 M89 84b.

Eqs. 1a.N15 and 1a.N16 are SSRC curves 1 and 2, the deterministic column curves. (Galambos, 1988; Bjorhovde, 1972, Bjorhovde and Tall, 1971, 1975a, 1975b)

Plastic analysis is permitted, with some restrictions as to the load-strain curve, width-thickness ratios, and lateral bracing.

OHBDC 83

The factored compressive resistance, C_r, of a member subjected to axial compressive force is

$$
\left.
\begin{aligned}
&\text{for } 0 \le \lambda \le 1.0 && C_r = \phi A F_y (1.035 + 0.202\lambda - 0.222\lambda^2) \\
&\text{for } 1.0 < \lambda \le 2.0 && C_r = \phi A F_y (-0.111 + 0.636\lambda^{-1} + 0.087\lambda^{-2}) \\
&\text{for } 2.0 < \lambda \le 3.0 && C_r = \phi A F_y (-0.009 + 0.877\lambda^{-2})
\end{aligned}
\right\} \quad (1a.N17)
$$

where ϕ is 0.90.

Eq. 1a.N17 is a slightly simplified version of SSRC Column 2, and thus corresponds to Eq. 1a.N16 above.

COMMENTARY

The specifications are compared for the design of steel and aluminum compression members of the main U.S. and Canadian code-writing bodies. It will be seen that there are few differences, and what differences there are, are not major. The specifications consider both Allowable Stress Design (ASD) and Limit States Design (LSD).

REGIONAL COMPARISON

The column strength of the allowable stress specifications (AISC ASD 89, AREA 89, AASHTO 87, CSA S16.1 M89 84b, and AA 86) are approximations to the tangent modulus strength, with a parabola (CRC column strength curve) for AISC ASD 89 and AASHTO 87, and a straight line for AREA 89, CSA S16.1 M89 84b, and AA 86. The column formulas use the yield strength and the Euler curve as limits, with the Euler curve indicating elastic buckling.

The CRC column strength curve (1960) was developed for small rolled shapes – the curve is a compromise as it is used for bending about both axes, based on the tangent modulus buckling concept, and considering residual stresses. (Johnston, 1976; Tall, 1974) It was felt that the loads carried by such columns do not exceed the tangent modulus load enough to warrant the use of a post-buckling strength analysis. (The CRC curve is basically the Johnston parabola used by many specifications in the past century – its current use is based on a much more complete understanding of column behavior.) (Tall, 1974)

In the 1960s, attention was focused on the fact that a maximum strength analysis would be a better representation of column strength for practical columns (Tall, 1964), and in 1979 the SSRC formalized this philosophy with its Technical Memorandum 5, General Principles for Stability Design of Flexural Structures. (Galambos, 1988) The concept of "multiple column curves" was discussed in the early 1960s (Tall, 1964) as a better representation to cover the broad scatter of all column curves. SSRC curves 1, 2, and 3, and 1P, 2P, and 3P are maximum strength curves. (Galambos, 1988; Bjorhovde, 1972, Bjorhovde and Tall, 1975a and 1975b)

The column strength for the allowable stress specifications are plotted in Fig. 1a.N1. Limit states design is a "method of proportioning structures such that no applicable limit state is exceeded when the structure is subjected to all appropriate load combinations" (AISC LRFD 86). The AISC uses the term "Load and Resistance Factor Design", or LRFD. Both elastic and plastic analysis is permitted in Limit States Design.

The bases of the column formulas of the limit states design specifications (AISC LRFD 86, AASHTO 87, CSA S16.1 M89 84b, and OHBDC 83) show a somewhat greater difference than for the allowable stress design formulas. AASHTO 87 uses the same buckling load parabola (CRC column strength curve) as in its allowable stress design provisions, and the other specifications are based on maximum strength provisions. The formulas of AISC LRFD 86 and OHBDC 83, and the basic formula of CSA S16.1 M89 84b, use slight modifications of SSRC Curves 2 and 2P, both of which are essentially identical to each other although based on probabilistic and deterministic considerations of maximum strength, respectively. (Galambos, 1988; Bjorhovde, 1972, Bjorhovde and Tall, 1975a and 1975b)

CSA S16.1 M89 84b uses a multiple column curve approach, with two curves, SSRC Curves 1 and 2, the deterministic curves based on an initial out-of-straightness of L/1000. OHBDC 83 uses SSRC Curve 2, while AISC LRFD 86 uses an approximation to SSRC Curve 2P with an initial out-of-straightness of L/1500. The column formulas for the limit states design specifications are plotted in Fig. 1a.N2.

CSA S16.1 M89 84b notes that welded H–shapes should have flange edges flame-cut. This is to ensure a more favorable residual stress distribution with tension at the flange tips, (Tall, 1974) so that the column strength achieved corresponds to curve 2.

The SSRC curves (1, 2, 3, and 1P, 2P, 3P) are defined by three equations in addition to the yield and Euler values. This is used directly in CSA S16.1 M89 84b and AISC LRFD 86, but modified into two equations in AISC LRFD 86 for inelastic and elastic buckling respectively.

The resistance factor for compression, ϕ, is 0.85 for both U.S. codes, AISC LRFD 86 and AASHTO 87, and is 0.90 for the Canadian codes CSA S16.1 M89 84b and OHBDC 83.

The Commentary of AISC LRFD 86 refers to the reliability index, β, with the target value of 3.0, and actual values varying from 2.6 to 3.3. CSA S16.1 M89 84b also has the target value of 3.0 with actual values varying from 2.7 to 4.2.

All the column formulas, both ASD and LSD, take into account the availability of different yield strengths, and in the case of aluminum, of different alloys.

In Canada, Limit States Design was introduced in 1974, and for 10 years was used concurrently and alternatively with the existing Allowable Stress Design until the 1984 edition was introduced. Allowable Stress Design is no longer covered. Multiple column curves were also introduced in the 1984 edition, with the original curve (SSRC curve 2) retained, and the additional curve (SSRC curve 1) added for stress-relieved shapes, respectively. (Kennedy, 1984)

In the U.S.A., AISC LRFD 86 was introduced in draft form in 1983 as the LRFD Specification for use on a trial basis, and in 1986 in its final form. Currently, the ASD (AISC ASD 89) and LSD (AISC LRFD 86) specifications are both in use.

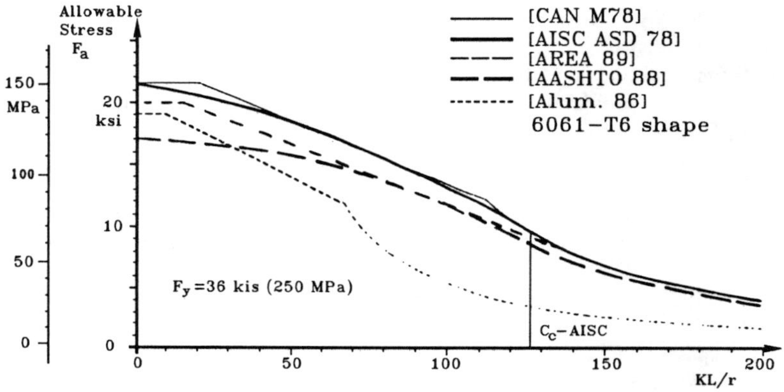

Fig. 1a.N1 Comparison of Column Strength Curves (Allowable Stress Design)

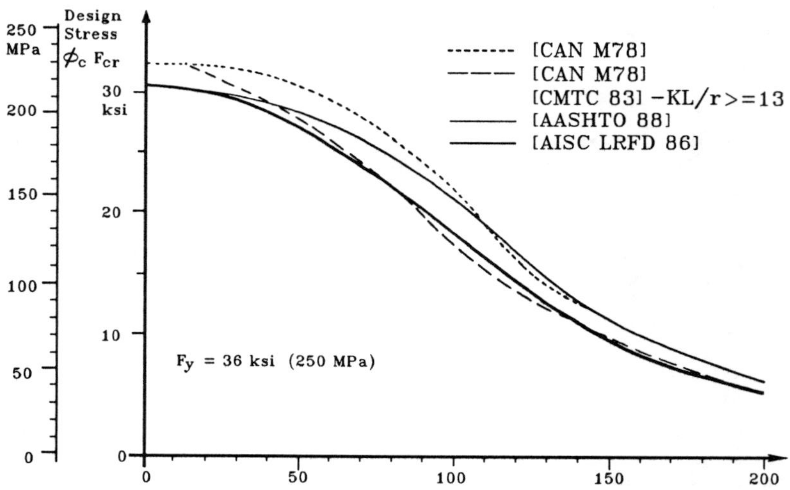

Fig. 1a.N2 Comparison of Column Strength Curves (Limit States Design)

WEST EUROPE

CHAPTER EDITOR
Gerald W. Schulz, University of Innsbruck

BS 5400 P.3 82

Axial load $\leq \frac{A_e \sigma_c}{\gamma_m \gamma_{f3}}$

where

A_e = effective area
σ_c = buckling strength based on multiple column curves

$$\frac{\sigma_c}{\sigma_y} = 0.5\left(\left\{1+(1+\eta)\,\frac{5700}{\lambda^2}\right\} - \sqrt{\left\{1+(1+\eta)\,\frac{5700}{\lambda^2}\right\}^2 - \frac{22800}{\lambda^2}}\right) \quad \text{(1a.W1)}$$

where

α = 0.0025 for curve A
0.0045 for curve B
0.0062 for curve C
0.0083 for curve D

η = 0 when $\lambda \leq 15$
= $\alpha(\lambda-15)$ $\lambda > 15$

$\lambda = \frac{\ell_e}{r}\sqrt{\frac{\sigma_y}{355}}$

σ_y = yield strength (N/mm^2)

Curve		
A	rolled or stress relieved	$r/y \geq 0.70$
B	rolled or stress relieved	$r/y = 0.60$
	welded	$r/y \geq 0.70$
	rolled or stress relieved	$r/y = 0.50$
C	welded	$r/y < 0.70$
	rolled or stress relieved	$r/y \leq 0.45$
D	rolled with flange thickness over 40 mm	
A	not finished hollow sections	

BS 5950 P.5 85

Section Resistance

(A) plastic, compact or semi-compact
$P_c = A_g \rho_c$

(B) slender
$P_c = A_g \rho_c \times$ stress reduction factor (see width/thickness ratios section)

Column Strength

ρ_y should be reduced by 20 MPa for welded sections

$$\rho_c = \frac{\rho_e \rho_y}{\phi + (\phi^2 - \rho_e \rho_y)^{1/2}} \tag{1a.W2}$$

where

η = Perry factor
η = $0.001\ a\ (\lambda - \lambda_o) \geq 0$ (a is the Robertson constant)
λ = L_e/r
λ_o = $0.2\left(\frac{\pi^2 E}{\rho_y}\right)^{1/2}$
ρ_e = Euler strength = $\pi^2 E/\lambda^2$
ρ_y = design strength
ϕ = $\frac{\rho_y + (\eta+1)\rho_e}{2}$

Values of a			Axis of buckling
a = 2.0	for	Hot-rolled hollow sections	(major and minor)
		Rolled I-sections	(major)
a = 3.5	for	Rolled I-sections	(minor)
		Rolled H-sections up to 40 mm	(major)
		Welded I or H sections	(major)
		Welded box sections up to 40 mm	(major and minor)
a = 5.5	for	Rolled H-sections up to 40 mm	(minor)
		Rolled H-sections over 40 mm	(major)
		Welded I or H sections up to 40 mm	(major)
		Welded box sections over 40 mm	(major and minor)
a = 8.0	for	Rolled H-section over 40 mm	(minor)
		Welded I or H sections over 40 mm	(minor)

EC 3 88

The ultimate limit state verification for flexural buckling is carried out with the formula

$$\frac{N}{\chi N_{pl}} \leq 1 \tag{1a.W3}$$

where

f_y = specified yield stress, applicable for $t \leq 40$ mm;
for 40 mm $< t \leq 80$ mm, f_y shall be reduced by 10%
N = axial force under design load
N_{pl} = Af_y, ultimate axial capacity

λ_{pl} = $\pi\sqrt{\frac{E}{f_y}}$
$\bar{\lambda}$ = $\frac{\lambda}{\lambda_{pl}} = \sqrt{\frac{N_{pl}}{N_{ki}}}$ = relative slenderness

χ = reduction coefficient, depending on the reference slenderness ratio $\bar{\lambda}$ and the buckling curve (Fig. 1a.W1), given by Eq. 1a.W4 and Table 1a.W1, or by Table 1a.W2.

$$\chi = \frac{1+\alpha(\bar{\lambda}-0.2)+\bar{\lambda}^2}{2\bar{\lambda}^2} - \frac{1}{2\bar{\lambda}^2}\sqrt{(1+\alpha(\bar{\lambda}-0.2)+\bar{\lambda}^2)^2-4\bar{\lambda}^2} \tag{1a.W4}$$

The appropriate buckling curve (a to d) for a section shall be taken from Table 1a.W3.

The design criteria according to Eq. 1a.W3 is valid for class 1, class 2, and class 3 sections, dependent on their width-thickness ratio.

For class 4 sections the interaction between column buckling and plate buckling may be considered by replacing f_y with the limiting buckling stresses σ_u for calculating the values of N_{pl} and $\overline{\lambda}$ or by using f_y with the effective width. In the case of torsional buckling, the slenderness $\overline{\lambda}$ has to be calculated using the relevant value of N_{ki}.

COMMENTARY

EUROCODE 3, which will be, from 1992 on, the main Steel Specification of the European Communities and of its associated countries in West Europe, has adopted the ECCS multiple column curve concept.

In addition most Western European countries have already replaced in their national codes the "one column curve" concept by introducing the ECCS column curves in the original or in a modified version.

The column curves a^o, a, b, c and d as shown in Fig. 1a.W1 are based on the ultimate strength concept of ECCS. The theoretical model is a hinged strut with an initial out-of-straightness at midspan of $\ell/1000$ and, depending on shape of cross section and manufacturing procedures, with residual stresses.

Table 1a.W1 Coefficient α in Eq. 1a.W4

Buckling curve	a^o	a	b	c	d
α	0.13	0.21	0.34	0.49	0.76

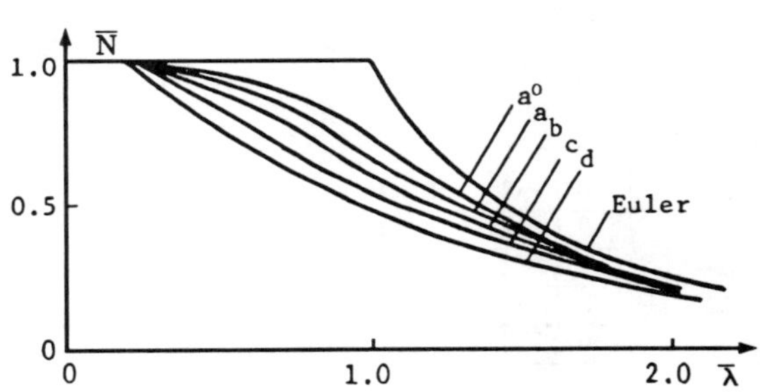

Fig. 1a.W1 European column curves for steel

The characteristic strength, as determined experimentally, was defined by ECCS as mean value minus twice the standard deviation of a series of test results. The parameters for the theoretical analysis were chosen in such a way that the same safety level was obtained as defined for the experimental investigation.

Recent comparisons showed, that the column curves comply well with the safety provisions as outlined in EUROCODE 1.

The column curves are used as design curves for single, hinged members (simple members, built-up members, composite members) and for members in trusses. They are used as references curves for the design of beam-columns.

Table 1a.W2 Reduction Coefficients χ

1	2	3	4	5	6
	χ for buckling curve				
$\bar{\lambda}$	a^o	a	b	c	d
0.2	1.0000	1.0000	1.0000	1.0000	1.0000
0.3	0.9859	0.9775	0.9641	0.9491	0.9235
0.4	0.9701	0.9528	0.9261	0.8973	0.8504
0.5	0.9513	0.9243	0.8842	0.8430	0.7793
0.6	0.9276	0.8900	0.8371	0.7854	0.7100
0.7	0.8961	0.8477	0.7837	0.7247	0.6431
0.8	0.8533	0.7957	0.7245	0.6622	0.5797
0.9	0.7961	0.7339	0.6612	0.5998	0.5208
1.0	0.7253	0.6656	0.5970	0.5399	0.4671
1.1	0.6482	0.5960	0.5352	0.4842	0.4189
1.2	0.5732	0.5300	0.4781	0.4338	0.3762
1.3	0.5053	0.4703	0.4269	0.3888	0.3385
1.4	0.4461	0.4179	0.3817	0.3492	0.3055
1.5	0.3953	0.3724	0.3422	0.3145	0.2766
1.6	0.3520	0.3332	0.3079	0.2842	0.2512
1.7	0.3150	0.2994	0.2781	0.2577	0.2289
1.8	0.2833	0.2702	0.2521	0.2345	0.2093
1.9	0.2559	0.2449	0.2294	0.2141	0.1920
2.0	0.2323	0.2229	0.2095	0.1962	0.1766
2.1	0.2117	0.2036	0.1920	0.1803	0.1630
2.2	0.1937	0.1867	0.1765	0.1662	0.1508
2.3	0.1779	0.1717	0.1628	0.1537	0.1399
2.4	0.1639	0.1585	0.1506	0.1425	0.1302
2.5	0.1515	0.1467	0.1397	0.1325	0.1214
2.6	0.1404	0.1362	0.1299	0.1234	0.1134
2.7	0.1305	0.1267	0.1211	0.1153	0.1062
2.8	0.1216	0.1182	0.1132	0.1079	0.0997
2.9	0.1136	0.1105	0.1060	0.1012	0.0937
3.0	0.1063	0.1036	0.0994	0.0951	0.0882

Table 1a.W3 Selection of buckling curve for a cross-section

Cross section		buckling perpendicular to axis	buckling curve
hollow section — Any increase in the yield strength which might be caused by a cold forming process shall be ignored 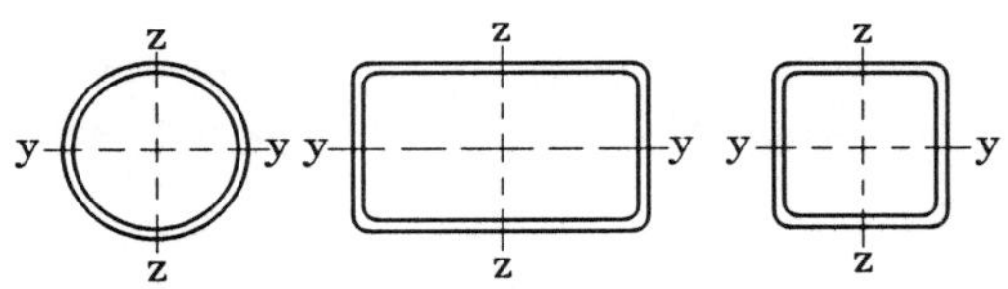		y – y z – z	a
welded box sections 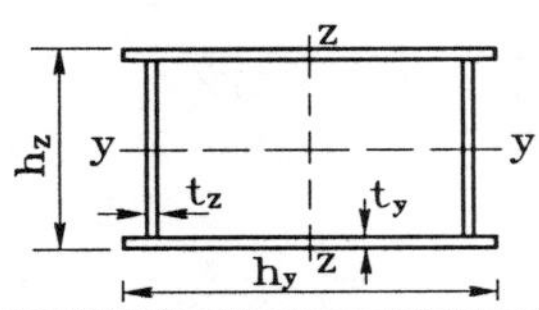		y – y z – z	b
	thick welds and $h_y/t_y < 30$, $h_z/t_z < 30$	y – y z – z	c
rolled I-sections 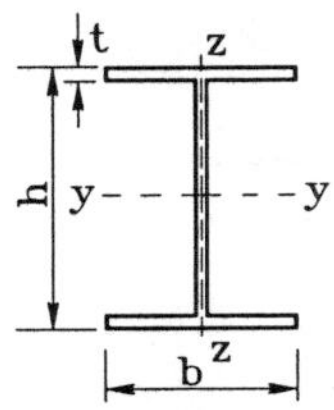	$\frac{h}{b} > 1.2$; $t \leq 40$ mm	y – y z – z	a (a_o) b (a)
	$\frac{h}{b} > 1.2$; $40 < t \leq 80$mm and $\frac{h}{b} \leq 1.2$; $t \leq 80$ mm	y – y z – z	b (a) c (b)
	$t > 80$ mm	y – y z – z	d d
welded I-sections 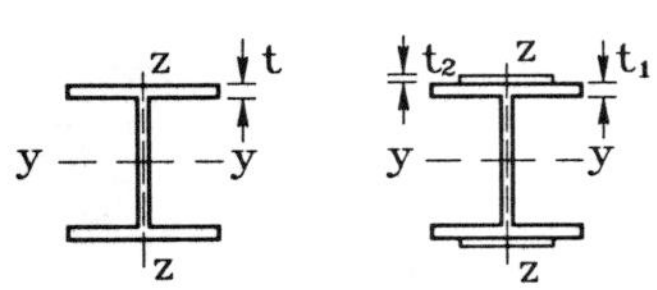	$t_1 \leq 40$ mm	y – y z – z	b c
	$t_1 > 40$ mm	y – y z – z	c d
U-, L-, T- and solid sections 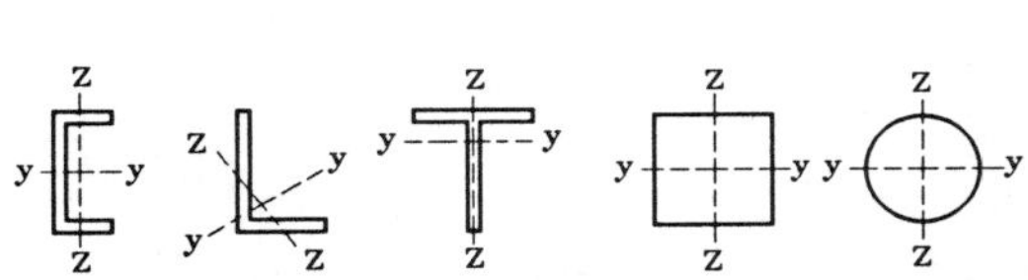		y – y z – z	c

Sections not contained in this table shall be classified analogously.

The buckling curves put in brackets may be assumed for high-strength steels with $f_y \geq 430$ N/mm^2 and $t \leq 40$ mm.

For transmission towers and similar structures where sufficient experimental data exist, alternative procedures can be used.

INTERNATIONAL COMPARISONS & COMMENT

WHAT ARE THE DIFFERENCES?

The basic case of the centrally compressed column reflects well the general changes in the theoretical treatment of stability problems as well as the status of the subsequent implementation of the achieved theoretical progress in the design practice throughout the world.

For a more realistic assessment of column strength, the tangent modulus method based on bifurcation theory has been replaced by a theoretical model, where

Table 1a.1 Concept Matrix Comparison

REGION		A	C		E		
CODES		AS DR 87164 87	GBJ 17 88	TBJ 2 85	SNiP II-23-81 82	ČSN 73 1401 84	TGL 13503/ 02 82
Design Concept	Allowable stress			X			X
	Limit States	X	X		X	X	
Strength Concept	Tangent Modulus						
	Ultimate Strength	X	X	X	X	X	X
Local Buckling	Section with b/t ratio		X	X	X	X	X
	Effective widths	X					
	Slenderness adjustment	X					
Column Curve Concept	Single column curve				X		
	Multiple column curve(No.)	5	3	4		2	4
Analytical Expressions	For entire slender-range	X				X	X
	For interval slender-range		2	2	2,λ=1.2		
Parameters: Residual Stresses	Measured as manufactured	X	X	X			X
	Modified by cold-straightening						
	Modified by loading cycles						
Geometrical Imperfections	Specification tolerances				X	X	
	Average measured value						
	f=ℓ/1000	X	X	X	X		X
	f<ℓ/1000						
	With eccentricity			X	X		
	Represent of geo. imperf.				X	X	
	Variation over slenderness			X	X		
	Modifications						
Safety Level	Safety factor ν (ASD)			1.7			1.2-1.5
	R factor γ_m(LRFD)	0.9	1.1		1.02-1.15	1.15	
	Safety index β	3.0					
	Mean value m		X				
	m - 2s (test)				X	X	X
	Comparison with past practice						
	Concepts						

stability is considered as a problem of divergence of equilibrium and is treated as inelastic second order analysis of columns with initial geometrical and material imperfections. This ultimate strength analysis of imperfect structural members has become the commonly accepted basis of all recent research and of the corresponding design clauses in limit states design concepts.

These principles are also the basis of the design recommendations of the Structural Stability Research Council (SSRC), the European Convention of Constructional Steelwork (ECCS), and the Institutul de Cercetari in Constructii si Economia

Table 1a.1 (Cont'd) Concept Matrix Comparison

REGION		E		J			
CODES		PN 80/ B-03200 84	MSZ 15024/ 1 85	AIJ 70	AIJ 75	AIJ 89b	JSCE 83
Design Concept	Allowable stress			X			X
	Limit States	X	X		X	X	
Strength Concept	Tangent Modulus			X			X
	Ultimate Strength	X	X		X	X	
Local Buckling	Section with b/t ratio	X	X	X	X	X	X
	Effective widths						
	Slenderness adjustment						
Column Curve Concept	Single column curve			X	X	X	X
	Multiple column curve(No.)	3	4				
Analytical Expressions	For entire slender-range	X	X				
	For interval slender-range			2	3	3	3
Parameters: Residual Stresses	Measured as manufactured		X	X	X	X	X
	Modified by cold-straightening						
	Modified by loading cycles						
Geometrical Imperfections	Specification tolerances						
	Average measured value						
	$f=\ell/1000$	X	X				
	$f<\ell/1000$						
	With eccentricity	X					
	Represent of geo. imperfe.	X					
	Variation over slenderness						
	Modifications						
Safety Level	Safety factor ν (ASD)			1.5-2.17			1.9-2.8
	R factor γ_m(LRFD)	1.15-1.25				0.85-0.9	
	Safety index β						
	Mean value m						
	m - 2s (test)		X				
	Comparison with past practice						
	Concepts						

Constructiitor (INCERC). Although the design clauses in the three recommendations differ in details, they all provide multiple column curves for the design of centrally compressed columns.

In Australia, The People's Republic of China, and in West East Europe this idea of multiple ultimate strength curves has been widely accepted and has been adopted in the course of introduction of limit states design in most of the national codes. The North American and Japanese design practice, still partly allowable stress design oriented, relates the column strength throughout the various building and bridge codes in the majority to one column curve, based either on tangent modulus theory or ultimate strength theory.

Table 1a.1 (Cont'd) Concept Matrix Comparison

REGION		J	N				
CODES		JRA 80	AISC LRFD 86	AREA 89	AAS-HTO 87	CSA S16.1 M89 84b	AAS-HTO 87
Design Concept	Allowable stress	X		X	X	X	
	Limit States		X				X
Strength Concept	Tangent Modulus				X		X
	Ultimate Strength	X	X				
Local Buckling	Section with b/t ratio	X		X	X	X	X
	Effective widths		X				
	Slenderness adjustment						
Column Curve Concept	Single column curve	X	X	X	X	X	X
	Multiple column curve(No.)					2	
Analytical Expressions	For entire slender-range						
	For interval slender-range	2	2	3	2	5	2
Parameters: Residual Stresses	Measured as manufactured	X			X		X
	Modified by cold-straightening						
	Modified by loading cycles						
Geometrical Imperfections	Specification tolerances						
	Average measured value						
	$f=\ell/1000$	X				X	
	$f<\ell/1000$		1/1500				
	With eccentricity						
	Represent of geo. imperfe.						
	Variation over slenderness						
	Modifications						
Safety Level	Safety factor ν (ASD)	1.7			2.12		
	R factor γ_m(LRFD)		.85			0.9	
	Safety index β						
	Mean value m						
	m - 2s (test)						
	Comparison with past practice						
	Concepts						

This continued use of one column design curve with the tangent modulus concept has to be understood from the role that the column has in an allowable stress design concept. Since stability is treated as a bifurcation problem, the design of structural members and of systems can be related by means of the effective length method to the column curve of the hinged column. The critical attitude in such a design concept toward the use of multiple column curves has to be seen with regard to their application for beam column and frame design. For such applications, the significance of multiple column curves for the design results can eventually diminish, in particular when the structural members underconsiderations are stocky or of low slenderness ratio. However, the dogmatic adherence to one column curve is not justified.

Table 1a.1 (Cont'd) Concept Matrix Comparison

REGION		N		W		
CODES		OHBDC 83	AISC ASD 89	EC 3 88	BS 5950 P.5 85	BS 5400 P.3 82
Design Concept	Allowable stress		X			
	Limit States	X		X	X	X
Strength Concept	Tangent Modulus		X			
	Ultimate Strength	X		X	X	X
Local Buckling	Section with b/t ratio	X				
	Effective widths		X	X	X	
	Slenderness adjustment			X		
Column Curve Concept	Single column curve	X	X			
	Multiple column curve(No.)			4	4	4
Analytical Expressions	For entire slender-range			X	X	X
	For interval slender-range	3	2			
Parameters: Residual Stresses	Measured as manufactured	X	X	X	X	X
	Modified by cold-straightening					
	Modified by loading cycles					
Geometrical Imperfections	Specification tolerances			X	X	X
	Average measured value					
	$f=\ell/1000$	X		X	X	X
	$f<\ell/1000$					
	With eccentricity					
	Represent of geo. imperfe.					
	Variation over slenderness					
	Modifications					
Safety Level	Safety factor ν (ASD)		varies			varies
	R factor γ_m(LRFD)	0.9		1.1		
	Safety index β			3.8		
	Mean value m					
	m - 2s (test)			X		
	Comparison with past practice					
	Concepts					

Quite different is the role of multiple column curves in a limit states design concept based on ultimate strength principles. The application as immediate design curve is mainly restricted to the design of axially simple and built-up members and of triangulated structures, where the different assessment of column strength through multiple column curves significantly influences the design. For the design of frames in such a limit states design concept, the column curves act only as reference curves for the determination of initial geometrical imperfections for a subsequent analysis according to a second order plastic hinge method.

The introduction of a multiple column curve design scheme based on ultimate strength principles depends therefore not only on the state of the art of column research, but of the readiness to introduce similar advanced methods for the design of other structural members of systems.

Design Concept

Of the specifications examined, about half are based on allowable stress design and about half on limit states design.

Strength Concept

The more recent codes have shifted from the tangent modulus concept (or the basic imperfection concept) to an ultimate strength concept that includes both initial geometric imperfections and material non-linearity with residual stresses.

Local Buckling

Some codes limit the width to thickness ratio for plates that can be used to form the column. Australia, Japan, Western Europe, and North America use an effective width concept to determine section strength as a means of accounting for plate buckling and column buckling interaction. Australia uses a further modification to the shape of the column curve to account for this interaction.

Column Curve Concept

Of the specifications examined about half use a single column curve, but the trend internationally is toward multiple column curves ranging in number from 2 to 5. For each classification of column, the specifications vary widely in the choice of curve and shape of curve.

Analytical Expressions

For any one given column curve, most specifications use more than one analytical expression to cover the entire slenderness range.

Residual Stresses

If considered at all, the column curves are based on measured stresses as manufactured. Some specifications do not consider residual stresses at all, either directly or implied.

Geometrical Imperfections

There is a wide variation in what is taken into account, either directly or implied, including specified or measured tolerances, additional eccentricities, variation in imperfection with slenderness and the shape of the imperfection.

Safety Level

There are wide differences in approach with the use of safety factors in ASD, load factors and resistance factors in LSD, variations in target safety index β, the use of tests to determine reliability and comparisons with existing practice.

WHY ARE THERE DIFFERENCES?

Design Concept

In some cases, the transition is under way and it takes time to make the change from ASD to LSD. The advantages of LSD are not immediately apparent to the designer. The traditional conservatism of the government bureaucracy can retard development. The investment, both educationally and financially in existing design procedures, is often influential in maintaining the status quo while mitigating against improved methods. Some specifications overcome this, others are less successful.

Strength Concept

The "local influence of ancient masters" is undoubtedly a factor when considering column strength. There is difficulty in getting current research in an "easy-to-understand" form before the committees preparing specifications.

It is only recently that there has been the capability of developing analytical procedures to describe the complex behavior of columns accounting for residual stress patterns, geometrical imperfections of any form, and non-linear material behavior. Prior to this, simpler approximate theories were used to account for non-linear behavior.

Where fatigue is a consideration, there is a resistance to considering yield as a parameter and plastic behavior as an implied requirement.

Local Buckling

The "local influence of young tyros" in promoting their latest theory and research before a receptive specification committee can result in new provisions being adopted.

Where seismic considerations are important, as in Japan, there is not the interest in slender plate elements although provision is made for reduction in strength where plate slenderness exceed limiting values.

Column Curve Concept

The USA uses one curve instead of three or more because the documentation of variability of column strength does not yet satisfy some specification committees. The presentation of multiple column curves has sometimes appeared complex that has received resistance.

As columns in practice are most likely to be beam-columns with bending moment, the influence of a small change in column axial strength is even less significant and any economy is reduced.

The designer may not be aware of the manufacturing process and therefore cannot make a decision about the appropriate column curve.

Some products would be disadvantaged by being placed on a lower curve, therefore the single curve was more attractive.

In other cases, testing has shown that a higher curve was more appropriate and the multiple curve concept was more attractive.

Analytical Expressions

The use of more than one expression reflects a difference in interest in curve fitting. Some specifications use expressions derived on a semi-rational basis in an attempt to be transparent. The expressions are often approximations to a different data base.

The thought that the elastic buckling Euler curve is sacrosanct has influenced the development of the expressions. Expressions have been adopted to calibrate with past practice.

Residual Stresses

None of the specifications allow for modification to the residual stress patterns as a result of cold-straightening or loading cycles.

Geometrical Imperfections

The differences reflect the relative importance that specifications place on geometrical imperfections over the other parameters that influence strength.

In those cases where the geometrical imperfection for the column curve is taken as less than the specification fabrication tolerance, the justification is that the use of peak values for the column strength parameters in combination is thought to be too conservative.

Safety Level

Although only two basic design concepts are used (ASD and LSD), each specification has its own method of incorporating safety factors or load and resistant factors in their design provisions. Some use combinations of practical load and material factors. Different statistical methods have been used for determining β safety indices. Safety levels are proving hard to define.

1b. Effective Length

SPECIFICATION PROVISIONS & EXPLANATION

AUSTRALIA

CHAPTER EDITOR
Russell Q. Bridge, The University of Sydney

AS DR87164 87

Effective lengths are considered as part of frame analysis (see Chapter 6 – Frames). Both alignment charts for restrained members and rational buckling analysis are permitted. A distinction is made between the effective lengths obtained from the two methods.

CHINA

CHAPTER EDITORS

Bao Kang He, Xian Institute of Metallurgy & Construction Engineering
Mingzhong Wei, Chongqing Architectural Institute
Wen Wei Zhao, Quinghuo University

GBJ 17 88

Effective lengths for sway prevented and sway uninhibited frames are obtained from the geometric length L modified by an effective length parameter μ from alignment tables. (See Chapter 8 for truss members)

Table 1b.C1 Column effective length factor μ for frames without joint translation

	0	0.05	0.1	0.3	0.5	1.0	3.0	5.0	10.0	20.0	∞
0	1.000	0.990	0.981	0.949	0.922	0.875	0.791	0.760	0.732	0.716	0.699
0.05	0.990	0.981	0.971	0.940	0.914	0.867	0.784	0.754	0.726	0.711	0.694
0.1	0.981	0.971	0.962	0.931	0.906	0.860	0.778	0.748	0.721	0.705	0.689
0.2	0.964	0.955	0.946	0.916	0.891	0.846	0.767	0.737	0.711	0.696	0.679
0.3	0.949	0.940	0.931	0.902	0.878	0.834	0.756	0.728	0.701	0.687	0.671
0.4	0.935	0.926	0.918	0.889	0.866	0.823	0.747	0.719	0.693	0.678	0.663
0.5	0.922	0.914	0.906	0.878	0.855	0.813	0.738	0.710	0.685	0.671	0.656
1.0	0.875	0.867	0.860	0.834	0.813	0.774	0.704	0.677	0.654	0.640	0.626
2.0	0.820	0.814	0.807	0.784	0.765	0.729	0.663	0.638	0.615	0.603	0.590
3.0	0.791	0.784	0.778	0.756	0.738	0.704	0.640	0.616	0.593	0.581	0.568
4.0	0.773	0.766	0.760	0.739	0.721	0.688	0.625	0.601	0.580	0.568	0.555
5.0	0.760	0.754	0.748	0.728	0.710	0.677	0.616	0.592	0.570	0.558	0.546
10.0	0.732	0.726	0.721	0.701	0.685	0.654	0.593	0.570	0.549	0.537	0.524
20.0	0.716	0.711	0.705	0.687	0.671	0.640	0.581	0.558	0.537	0.525	0.512
∞	0.699	0.694	0.689	0.671	0.656	0.626	0.568	0.546	0.524	0.512	0.500

Notes

1. Values of factor μ are calculated from the equation

$$\left(\left(\frac{\pi}{\mu}\right)^2+2(K_1+K_2)-4\,K_1K_2\right)\frac{\pi}{\mu}\cdot\sin\frac{\pi}{\mu} - 2\left((K_1+K_2)\left(\frac{\pi}{\mu}\right)^2+4\,K_1\,K_2\right)\cos\frac{\pi}{\mu}+8\,K_1K_2=0 \quad (1b.C1)$$

where

K_1, K_2 = ratio of the sum of girders' linear stiffness to that of columns' linear stiffness at the top and bottom of the relevant column, respectively.

2. The linear stiffness of a girder is taken as zero when the girder is simply supported on the column.

3. For columns of the bottom story: take K_2=0 for columns hinged to the foundation, and $K_2=\infty$ for columns rigidly fixed to the foundation.

Table 1b.C2 Column effective length factor μ for frames with joint translation

	0	0.05	0.1	0.3	0.5	1.0	3.0	5.0	10.0	20.0	∞
0	∞	6.02	4.46	3.01	2.64	2.33	2.11	2.07	2.03	2.02	2.00
0.05	6.02	4.16	3.47	2.58	2.31	2.07	1.90	1.86	1.83	1.82	1.80
0.1	4.46	3.47	3.01	2.33	2.11	1.90	1.75	1.72	1.70	1.63	1.67
0.2	3.42	2.86	2.56	2.05	1.87	1.70	1.57	1.54	1.52	1.51	1.50
0.3	3.01	2.58	2.33	1.90	1.74	1.58	1.46	1.44	1.42	1.41	1.40
0.4	2.78	2.42	2.20	1.80	1.65	1.50	1.39	1.37	1.35	1.34	1.33
0.5	2.64	2.31	2.11	1.74	1.59	1.45	1.34	1.32	1.30	1.29	1.28
1.0	2.33	2.07	1.90	1.58	1.45	1.32	1.21	1.19	1.17	1.17	1.16
2.0	2.17	1.94	1.79	1.49	1.37	1.24	1.14	1.12	1.10	1.09	1.08
3.0	2.11	1.90	1.75	1.46	1.34	1.21	1.11	1.09	1.07	1.06	1.06
4.0	2.08	1.87	1.73	1.45	1.32	1.20	1.10	1.08	1.06	1.05	1.04
5.0	2.07	1.86	1.72	1.44	1.32	1.19	1.09	1.07	1.05	1.04	1.03
10.0	2.03	1.83	1.70	1.42	1.30	1.17	1.07	1.05	1.03	1.03	1.02
20.0	2.02	1.82	1.68	1.41	1.29	1.17	1.06	1.04	1.03	1.02	1.01
∞	2.00	1.80	1.67	1.40	1.28	1.16	1.06	1.03	1.02	1.01	1.00

Notes

1. Values of factor μ are calculated from the Eq.

$$\left(36\, K_1\, K_2 - \left(\frac{\pi}{\mu}\right)^2\right) \sin\frac{\pi}{\mu} + 6\,(K_1 + K_2)\,\frac{\pi}{\mu} \cdot \cos\frac{\pi}{\mu} = 0 \qquad \text{(1b.C2)}$$

2. The linear stiffness of a girder is taken as zero when the girder is simply supported on the column.

3. For columns of the bottom story: take K_2=0 for columns hinged to the foundation, and $K_2=\infty$ for columns rigidly fixed to the foundation.

TBJ 2 85

The effective lengths are the same as in specification GBJ 17 88 for braced (sway-prevented) members. There are no provisions for unbraced (sway-uninhibited) member as this specification is for bridges.

JAPAN

CHAPTER EDITOR
Minoru Wakabayashi, General Building Research Corporation

AIJ 70, AIJ 75, AIJ 89b, JRA 80, and JSCE 83

The effective length factor, k, has been specified based upon the elastic behavior of individual members and frames. This k factor may be determined from exact analysis, but use of an approximate value (for example, using alignment charts) is also accepted. All the five standard/specifications/guide introduced in (1a. Column Strength) adopt the concept of effective length when considering the strength of compression members. (See Chapter 6)

NORTH AMERICA

CHAPTER EDITOR
Lambert Tall, Florida International University

AA 86

K is not specifically mentioned. However, the definition for L implies the use of a rational procedure.

L is the length of compression member between points of lateral support, or twice the length of a cantilever column (except where analysis shows that a shorter length can be used).

AASHTO 87

K = 0.75 for riveted, bolted, or welded end connections
K = 0.875 for pinned ends.

For members having ends not fully supported laterally, the K-factor shall be determined by a rational procedure (Galambos, 1988, Johnston, 1976).

AISC ASD 89

K is equal to 1.0 for compression members in braced frames unless analysis shows that a smaller value may be used.

KL in unbraced frames is to be determined by a rational method, and shall not be less than the unbraced length.

Rational methods to determine K are discussed in the Commentary to the Specification, and are the same as those of the SSRC (Galambos, 1988).

Plastic Design generally considers the strength of the complete structure taking into account its failure mode. Where K-values are used, they are the same as in ASD.

AISC LRFD 86

Braced frames: K = 1.0 unless structural analysis shows that a smaller value may be used.
Unbraced frame: K is to be determined by structural analysis and shall not be less than unity.

This is identical with allowable stress design, (AISC ASD 89).

AREA 89

$K = \frac{7}{8}$ for members with pin-end connections

$K = \frac{3}{4}$ for members with riveted, bolted, or welded end connections.

CSA S6 M89 84a

K cannot be less than the value indicated by test or rational analysis. If restraint by supporting elements cannot be predicted, K shall not be less than unity.

CSA S16.1 M89 84b

K is equal to 1.0 for non-sway frames and for trusses, unless analysis shows that a lower value is applicable. For sway frames, use a computed K, but not less than 1.0. Reference is made to the Appendices, which give the SSRC-recommended methods, either the simple chart, or the use of alignment charts.

OHBDC 83

The statements are identical with those in CSA S6.1 M89 84b, above.

COMMENTARY

Effective length factors are used to take into account the end conditions of the compression members.

REGIONAL COMPARISON

Effective length factors (K-factors). In general, all specifications allow the use of a rational procedure to determine K, and most refer to the SSRC.

AREA specifies K = 7/8 or 3/4 for pin-end or fixed conditions, but adds the phrase "under normal conditions" without any further explanation. AASHTO 87 uses the same values, but also specifies the use of a rational procedure for ends not fully supported laterally, and makes reference to the SSRC. AA 86 does not use K, but it is implied in the definition of L, as is the use of a rational procedure.

WEST EUROPE

CHAPTER EDITOR
Gerald W. Schulz, University of Innsbruck

BS 5400 P.3 82

	L_e
Held in position and restraint in directions at both ends	0.7L
Held in position at both ends and restrained in direction at one end	0.85L
Held in position at both ends only	L
Held in position and restrained in direction at one end; partially restrained in direction at other end only	1.5L
Held in position and restrained in direction at one end only	2.0L

An elastic critical buckling analysis may be used. For a strut held in position and fully or partially restrained in direction an effective length factor can be calculated from a graph as a function of the smaller value, at either end, of rotation stiffness relative to bending stiffness.

BS 5950 P.5 85

Standard Condition of Restraint		L_e
Held in position at both ends.	Restrained in direction at both ends	0.7L
	Partially restrained in direction at both ends	0.85L
	Restrained in direction at one end	0.85L
	Not restrained in direction at either end	1.00L
Effectively held in position and restrained in direction at one end.	Other end restrained in direction only	1.2L
	Other end partially restrained in direction only	1.5L
	Other end not restrained in direction	2.0L

Stanchions for Single Story Buildings

Guidance is given pictorically of restraint conditions leading to effective lengths.

Side Stanchions	Side Stanchions with restraints	Side Stanchions with crane gantry
Minor axis 0.85L Major axis 1.5L	related to restraint separation 1.5L	0.85 x crane girder height 1.5L

Effective lengths are also given for compound side stanchions with crane gantry.

Struts in Rigid Frames

For non-sway frames and frames with unrestricted side sway graphs are presented for effective column length as a function of the relative total column and beam stiffness at the top and bottom of the column in question.

For frames with intermediate values of partial sway bracing the code evaluates the effective bracing stiffness and provides similar graphs for two intermediate values. Interpolation between graphs can be used.

For beams not supporting a concrete slab the beam stiffness should take due account of the degree of flexibility at its far end in the relevant buckling mode of the frame. For non-sway frames the beam stiffness is $\frac{0.5\,I}{L}$ and for sway frames $\frac{1.5\,I}{L}$. For beams with concrete slabs the beam stiffness is $\frac{I}{L}$.

Elastic critical load analysis for the frame can be used if preferred to obtain the effective column lengths.

EC 3 88

EUROCODE 3 allows the use of the effective length concept for triangulated structures as well as for braced and unbraced frames. The buckling length ℓ may be determined from Appendix E in EC 3 88.

For frames deemed to be braced according to a stiffness criterion, effective lengths may be determined using an alignment graph in which the stiffness of the column to the total stiffness at the joint at each level is used as the parameter for selecting values.

For sway frames, a similar alignment graph is provided. The influence of axial force on restraint stiffness is considered.

INTERNATIONAL COMPARISONS & COMMENT

WHAT ARE THE DIFFERENCES?

All specifications use the effective length concept with an effective length factor (K). Most base the K-factors on an elastic buckling analysis and provide alignment charts or graphs for the sway permitted and sway prevented cases. In some, the results of an elastic analysis of the whole frame can be used directly to obtain the K-factors.

The process has been evolutionary, starting with K-factors for standard cases with simplified end boundary conditions through to incorporating stability effects directly in approximate or "exact" frame analyses. This has seen some specifications using K=1 for sway frames when second-order elastic analysis is used. More recent developments eliminate the use of K altogether when all possible parameters affecting the strength are included in the analysis. Table 1b.1 lists K-factor comparisons among specifications with the topics explained in more detail as follows:

Not required. Where a rigorous analysis including all possible parameters affecting the strength is performed, AS DR87164 87 does not require a member check, therefore no requirement for effective lengths.

Exact Buckling Analysis - Inelastic. Presently, no specifications include directly the effects of plasticity on the effective length. Some research has shown the influence of plasticity but more work is needed to derive provisions that can be made acceptable for specifications.

Table 1b.1 Concept Matrix Comparison

TOPICS \ SPECIFICATIONS	AS DR87164 87	GBJ 17 88	TBJ 2 85	AIJ 70	AIJ 75	AIJ 89b	JSCE	JRA 80	AISC ASD 89	AREA 89
Not required part of frame analysis	X									
Exact buckling analysis – inelastic										
Exact buckling analysis – elastic	X			X	X	X	X	X	X	
K-factor translational restraint										
K-factor alignment chart									X	
K-factor alignment graph	X			X	X	X	X	X	X	
K-factor tables		X	X							
K-factor standard cases	X	X	X	X	X					X
K-factor K=1 braced frame				X	X	X	X	X	X	
K-factor end condition restraints	X									
K-factor axial force in restraints	X									
K-factor by test										

Exact Buckling Analysis - Elastic. Most specifications permit the use of a rational elastic buckling analysis to derive the effective lengths.

K-factor Translational Restraint. Very few specifications account for the stiffness of the bracing system in the determination of effective length. Most use the two limiting cases of sway prevented and sway uninhibited.

K-factor Alignment Charts, Graphs, Tables. Most all specifications use K-factors derived from the stability of a restrained member in a sub-assemblage for sway prevented and sway uninhibited cases. The derived characteristic stability equations have been presented in a variety of forms: alignment charts; alignment graphs; or as tables of values.

K-factor Standard Cases. Standard cases for limiting restraint conditions (fixed or free) are provided in most specifications.

K-factor of Unity. For braced frames, some specifications use a K value of unity only for braced frames.

K-factor Restraint Influences. In a few specifications, the influence of far end conditions and axial force in restrained members is accounted for directly in the provisions rather than implied.

K-factor by Test. One specification permits testing to establish effective length.

Table 1b.1 (Cont'd) Concept Matrix Comparison

SPECIFICATIONS / TOPICS	AASHTO 87	CSA S6 M89 84a	AA 86	AISC LRFD 86	CSA S6.1 M89 84b	OHBDC 83	EC 3 88	BS 5950 P.5 85	BS 5400 P.3 82
Not required part of frame analysis									
Exact buckling analysis – inelastic									
Exact buckling analysis – elastic	X	X		X	X	X		X	X
K-factor translational restraint							X	X	
K-factor alignment chart		X		X	X	X			
K-factor alignment graph							X	X	X
K-factor tables									
K-factor standard cases	X	X	X		X	X		X	X
K-factor K=1 braced frame		X		X	X	X		X	X
K-factor end condition restraints		X			X	X	X	X	
K-factor axial force in restraints							X		
K-factor by test		X							

WHY ARE THERE DIFFERENCES?

One of the reasons for differences has been the transition from calculating effective lengths for standard cases through determining values from rational elastic buckling analysis to eventual elimination altogether, incorporating the effect in frame analysis. Specifications are at different stages in this process.

The use of member effective lengths to account for the effects of frame buckling on the strength of frame has proven to be paradoxical, particularly as current design provisions are essentially checks on member strength rather than the strength of the frame.

The inclusion of member effective length in design is a topic which has concerned, perhaps frustrated, researchers in many countries and this has led to some confusion for designers who use the specifications. This may well remain until multi-tiered specifications become the norm where rational frame analyses can be used directly to determine the strength of the frame.

1c. Width/Thickness Ratio

SPECIFICATION PROVISIONS & EXPLANATION

AUSTRALIA

CHAPTER EDITOR
Russell Q. Bridge, The University of Sydney

AS DR87164 87

The section capacity N_s is taken as

$$N_s = Q\, A_n\, \sigma_y \tag{1c.A1}$$

where

Q = form factor for local buckling of slender plate elements,

$$Q = A_e/A \tag{1c.A2}$$

The effective area A_e is calculated by summing the effective areas of individual elements of thickness t whose effective widths b_e are given by

$$b_e = (\lambda_{cy}/\lambda_c)\, b \ \leq\ b \tag{1c.A3}$$

The section slenderness λ_c is given by

$$\lambda_c = (b/t)\sqrt{\sigma_y/250} \tag{1c.A4}$$

where

b = clear width of the plate element
λ_{cy} = limiting plate slenderness given in Table 1c.A1
σ_y = yield strength (MPa)

COMMENTARY

For compression members fabricated from slender plates such that the form factor Q is less than unity, appropriate column curves that account for the interaction of local buckling with overall buckling have been established for I-sections (Davids, 1985) and rectangular hollow sections (Key, 1985). These curves are represented by an appropriate value of the shift parameter α_b in Table 1a.A1(b). Although the curves are not the correct shape for interaction buckling, particularly for the shorter columns, the method is conservative (Hancock, 1987).

Table 1c.A1 Values of Section Slenderness Limits λ_{cy}

Plate type	Edges supported	Residual stresses	slenderness limit λ_{cy}
Flat	One	SR	16
		HR	16
		LW, CF	15
		HW	14
	Both	SR	45
		HR	45
		LW, CF	40
		HW	35
Circular hollow sections		HR, CF	82
		LW	82
		HW	82

Notes to Table 1c.A1:

SR: Stress relieved
CF: Cold formed
LW: Lightly welded
HW: Heavily welded
HR: Hot rolled or hot finished

Note: Members whose compression residual stresses are less than 40 MPa may be considered lightly welded.

CHINA

CHAPTER EDITORS

Bao Kang He, Xian Institute of Metallurgy & Construction Engineering
Mingzhong Wei, Chongqing Architectural Institute
Wen Wei Zhao, Quinghuo University

GBJ 17 88

To prevent local buckling of plate elements in the axially loaded members, the limitations of ratio of b/t or h_o/t_w (see Fig. 1c.C1) are provided in the specification, which are:

for I-section members

$$b/t \leq (10 + 0.1\ \lambda)\sqrt{\frac{235}{f_y}} \tag{1c.C1}$$

$$h_o/t_w \leq (25 + 0.5\ \lambda)\sqrt{\frac{235}{f_y}} \tag{1c.C2}$$

where

f_y = specified minimum yield stress of materials

$\lambda = \frac{KL}{r}$, slenderness ratio of members, taking λ=30 for $\lambda < 30$ and λ=100 for $\lambda > 100$.

for box-section members

$$b_o/t \leq 40\sqrt{\frac{235}{f_y}} \tag{1c.C3}$$

$$h_o/t_w \leq 40\sqrt{\frac{235}{f_y}} \tag{1c.C4}$$

where b_o is the width between webs.

When the b_o/t ratio exceeds the values above, only a width of $20t_w\sqrt{\frac{235}{f_y}}$ at each edge of the web is considered effective.

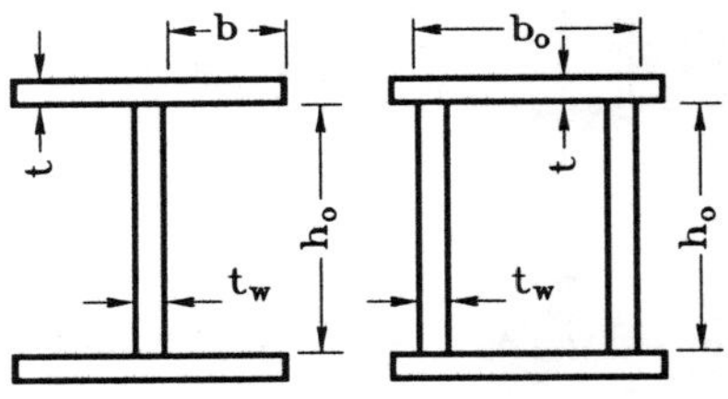

Fig. 1c.C1 Box and I-section Members

COMMENTARY

To prevent the interaction between local plate and overall member buckling, the ratio of b/t and h_o/t_w of plates is limited to a value that keeps the buckling stress of members (Eqs. 1c.C1 and 1c.C2), to the yield point of materials (Eqs. 1c.C3 and 1c.C4).

In developing Eqs. 1c.C1 and 1c.C2, the residual stress and restraining effect between adjoining plate elements have been considered. In the inelastic range, the tangent modulus factor $\eta = E_t/E$ is taken from the test data of axially loaded members.

In deriving Eqs. 1c.C3 and 1c.C4, the local buckling stress of plates is taken as $0.95f_y$ and tangent modulus factor η taken as 0.4. No restraining effects have been taken into consideration in evaluating the buckling stress of plate elements.

TBJ 2 85

The width-to-thickness ratio b/t for plate elements in the centrally compressed members is limited and it can be taken from Table 1c.C1. The definition of ratio of b_1/t_1, b_2/t_2, b_3/t_3, and b_4/t_4 is shown in Fig. 1c.C2.

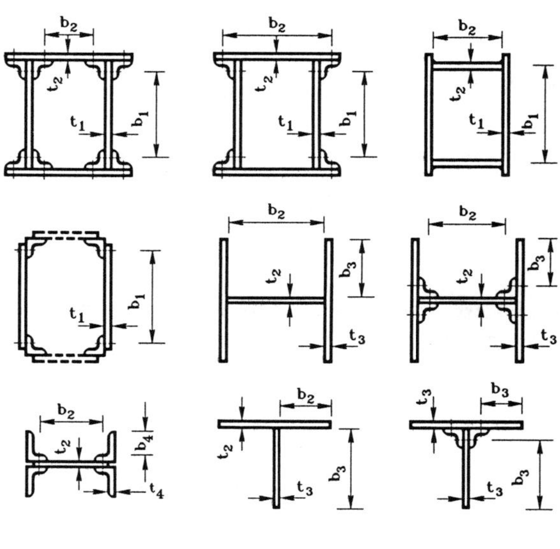

Fig. 1c.C2 Arrangements of Plate Elements

COMMENTARY

The width-to-thickness ratio of b_1/t_1, b_2/t_2, and b_3/t_3 listed in Table 1c.C1 is established by taking the local buckling stress σ_{cr} of the component plate equal to or less than the overall buckling stress σ of the member. The restrained effects between plates are taken into account in evaluating the critical stress coefficient for the web plate.

Table 1c.C1 The Maximum Permitted Width-to-Thickness Ratios for Plate Elements Subject to Uniform Compression in the Centrally Compressed Members

Type of Section	Slenderness ratio $\lambda = \frac{KL}{r}$	b/t	
		16q	16Mnq
Box & H-Sections	≤ 60	$b_1/t_1 \leq 35$	$b_1/t_1 \leq 30$
	> 60	$b_1/t_1 \leq 0.6\lambda$ but less than 50	$b_1/t_1 \leq 0.5\lambda$ but less than 45
	≤ 50	$b_2/t_2 \leq 35$	$b_2/t_2 \leq 30$
	> 50	$b_2/t_2 \leq 0.6\lambda + 5$ but less than 50	$b_2/t_2 \leq 0.5\lambda + 5$ but less than 45
Riveted H & Tee Sections		$b_3/t_3 \leq 12$	$b_3/t_3 \leq 10$
Welded H & Tee Sections	≤ 60	$b_3/t_3 \leq 14$	$b_3/t_3 \leq 12$
	> 60	$b_3/t_3 \leq 0.15\lambda + 5$	$b_3/t_3 \leq 0.2\lambda$
		but less than 18 for primary members and 20 for secondary members	
Outstanding Leg of Riveted Angles		$b_4/t_4 \leq 12$ for primary members	
		$b_4/t_4 \leq 16$ for bracing and secondary members	

Note: Except for b_4/t_4, the limiting value b/t given in Table 1c.C1, can be increased if the computed compressive stress σ is less than the allowable stress $\phi\sigma$ (see Eq. 1a.C5), provided by overall buckling of the members.

JAPAN

CHAPTER EDITOR
Minoru Wakabayashi, General Building Research Corporation

AIJ 70

Following provisions are given with respect to the width-to-thickness ratio of basic structural elements

for outstanding legs of angles

$$b/t \leq 20/\sqrt{F} \tag{1c.J1}$$

for flanges of I-shaped members

$$b/t \leq 24/\sqrt{F} \tag{1c.J2}$$

for webs in axial or combined axial and flexural compression

$$d/t \leq 74/\sqrt{F} \tag{1c.J3}$$

for webs in flexural compression

$$d/t \leq 110/\sqrt{F} \tag{1c.J4}$$

where b, d, t, and F are the flange width, web width, thickness, and specified yield stress in ton/cm^2.

AIJ 75

The AIJ Guide for Plastic Design provides following equations to limit the width-to-thickness ratio of elements forming plastic hinges.

for flanges of I-shaped members

$$\begin{aligned} &\text{SS41: } b/t_f \leq 10 \\ &\text{SS50: } b/t_f \leq 8.5 \end{aligned} \tag{1c.J5}$$

for webs in compression

If $N/N_y \leq A_w/A$:

$$\begin{aligned} &\text{SS41: } d/t_w \leq 71 - 26A/A_w \cdot N/N_y \\ &\text{SS50: } d/t_w \leq 61 - 22A/A_w \cdot N/N_y \end{aligned} \tag{1c.J6}$$

If $N/N_y > A_w/A$:

$$\begin{aligned} &\text{SS41: } d/t_w \leq 45 \\ &\text{SS50: } d/t_w \leq 39 \end{aligned} \tag{1c.J7}$$

where

A = cross sectional area of the section
A_w = cross sectional area of the web
t_f = flange thickness
t_w = web thickness

AIJ 89b

The AIJ LSD Guide provides a set of limit width-to-thickness ratios, in which interaction between the flange and web elements is taken into account as shown in Table 1c.J1.

Table 1c.J1 Plate b/t ratio limits

		P–I	P–II
A. H Section		$\frac{(b/t_f)^2}{(20\sqrt{F_{yf}})^2}+\frac{(b/t_w)^2}{(127/\sqrt{F_{yw}})^2}\le 1$	$\frac{(b/t_w)^2}{(21\sqrt{F_{yf}})^2}+\frac{(b/t_w)^2}{(136/\sqrt{F_{yw}})^2}\le 1$
		but $d/t_w \le 100\sqrt{F_{yw}}$	
B. H Column		$\frac{(b/t_f)^2}{(20\sqrt{F_{yf}})^2}+\frac{(b/t_w)^2}{(93/\sqrt{F_{yw}})^2}\le 1$	$\frac{(b/t_f)^2}{(21\sqrt{F_{yf}})^2}+\frac{(b/t_w)^2}{(99/\sqrt{F_{yw}})^2}\le 1$
		but $d/t_w \le 71\sqrt{F_{yw}}$	
C. Angle Section	Rolled	$\frac{B}{t}\le\frac{63}{\sqrt{F_y}}$	$\frac{B}{t}\le\frac{75}{\sqrt{F_y}}$
	Cold-formed	$\frac{B}{t}\le 30$	$\frac{B}{t}\le 35$
D. Angle Columns	Rolled	$\frac{B}{t}\le\frac{50}{\sqrt{F_y}}$	$\frac{B}{t}\le\frac{61}{\sqrt{F_y}}$
	Cold-formed	$\frac{B}{t}\le 24$	$\frac{B}{t}\le 28$
E. Cold-formed Channel		$D/t \le 38$	$D/t \le 56$
F. Cold-formed Columns		$D/t \le 36$	$D/t \le 54$

JRA 80

The limit width-to-thickness ratios stipulated in the JRA Specifications are:

For outstanding legs

SS41, SM41, SMA41: $b/t \le 13$ (1c.J8)

SS50, SM50: $b/t \le 12$ (1c.J9)

SM50Y, SM53, SMA50: $b/t \le 11$ (1c.J10)

SM58, SMA58: $b/t \le 10$ (1c.J11)

For elements constrained at both ends

SS41, SM41,SMA41: $b/t \le 40$ in axial compression (1c.J12)

$b/t \le 26\phi^2 + 4\phi + 40$

in combined axial and flexural compression

SS50:	$b/t \le 36$ in axial compression	(1c.J13)
	$b/t \le 24\phi^2 + 4\phi + 36$	
	in combined axial and flexural compression	
SM50:	$b/t \le 34$ in axial compression	(1c.J14)
	$b/t \le 22\phi^2 + 4\phi + 34$	
	in combined axial and flexural compression	
SM50Y, SM53, SMA50:	$b/t \le 32$ in axial compression	(1c.J15)
	$b/t \le 21\phi^2 + 3\phi + 32$	
	in combined axial and flexural compression	
SM58, SMA58:	$b/t \le 28$ in axial compression	(1c.J16)
	$b/t \le 19\phi^2 + 3\phi + 28$	
	in combined axial and flexural compression	

where

σ_1 = stress at the edge where compressive stress is applied by bending moment

σ_2 = stress at the edge where tensile stress is applied by bending moment.

$\phi = (\sigma_1 - \sigma_2)/\sigma_1$

Table 1c.J1 (Cont'd) Plate b/t ratio limits

	P–III		P–IV
A. H Section	$\frac{(b/t_f)^2}{(23\sqrt{F_{yf}})^2} + \frac{(b/t_w)^2}{(148/\sqrt{F_{yw}})^2} \le 1$		$17/\sqrt{F_{yf}} < b/t_f \le 37/\sqrt{F_{yf}}$
	but $d/t_w \le 100\sqrt{F_{yw}}$		$100/\sqrt{F_{yw}} < b/t_w \le 276/\sqrt{F_{yw}}$
B. H Column	$\frac{(b/t_f)^2}{(23\sqrt{F_{yf}})^2} + \frac{(b/t_w)^2}{(106/\sqrt{F_{yw}})^2} \le 1$		$17/\sqrt{F_{yf}} < b/t_f \le 37/\sqrt{F_{yf}}$
	but $d/t_w \le 71\sqrt{F_{yw}}$		$71/\sqrt{F_{yw}} < b/t_w \le 159/\sqrt{F_{yw}}$
C. Angle Section	Rolled	$\frac{B}{t} \le \frac{92}{\sqrt{F_y}}$	$92/\sqrt{F_y} < B/t_f \le 105/\sqrt{F_y}$
	Cold-formed	$\frac{B}{t} \le 43$	$\frac{B}{t} \le 43$
D. Angle Columns	Rolled	$\frac{B}{t} \le \frac{74}{\sqrt{F_y}}$	$74/\sqrt{F_y} < B/t_f \le 84/\sqrt{F_y}$
	Cold-formed	$\frac{B}{t} \le 36$	$\frac{B}{t} \le 36$
E. Cold-formed Channel		$D/t \le 92$	$D/t \le 92$
F. Cold-formed Columns		$D/t \le 90$	$D/t \le 90$

JSCE 83

The JSCE Specifications provide a set of limit width-to-thickness ratios shown in Table 1c.J2.

Table 1c.J2 Plate b/t ratio limits

Boundary Conditions	Maximum b/t ratio	examples
Supported at one edge	$450/\sqrt{\sigma}$ but $\leq$ 16	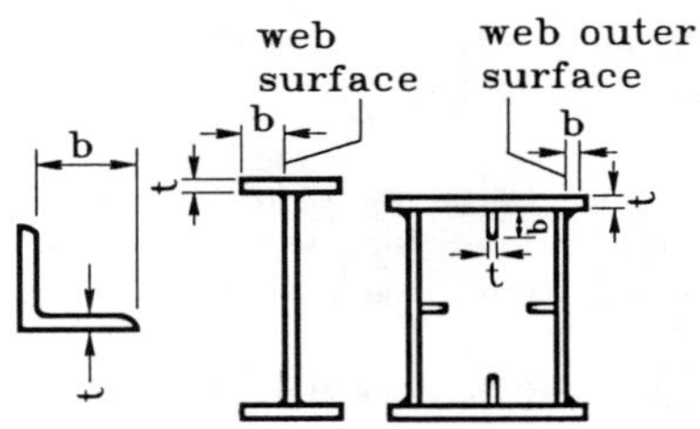
Supported at both edges	$1400/\sqrt{\sigma}$ but $\leq$ 50	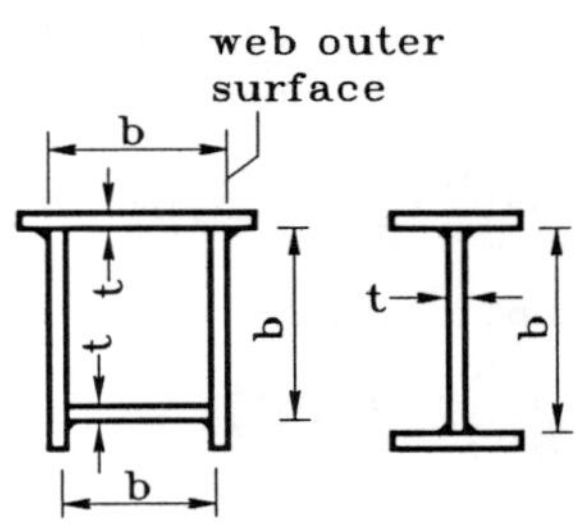
Supported at both edges, and with a stiffener at each division point for n segments	$1400n/\sqrt{\sigma}$ but $\leq$ 50n	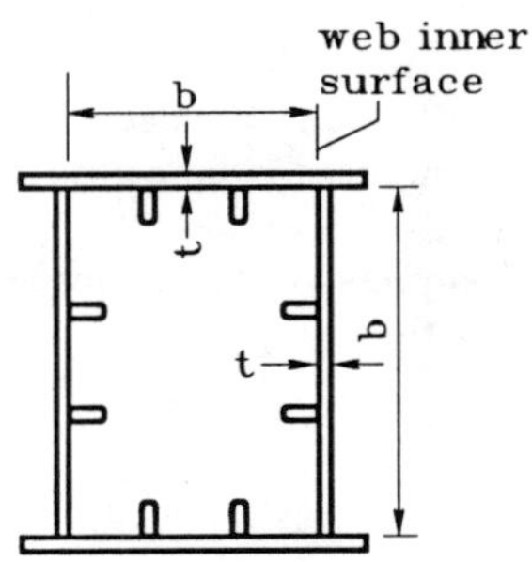

where

- b = plate width
- n = number of segments divided by stiffeners
- t = plate thickness
- σ = maximum working stress in compression

NORTH AMERICA

CHAPTER EDITOR
Lambert Tall, Florida International University

AA 86

Local buckling of aluminum shapes is given in terms of allowable stress formulas. Thus, there are formulas similar to those for compression members, Eqs. 1a.N3, 1a.N9, and 1a.N10, where the limits for each formula are given by the b/t ratio rather than the L/r ratio.

Thus, the design of compression members requires the determination of the allowable stress F for the column as a whole, for the flange, and for the web. The smallest F from these is the critical one, and the allowable load = F · A.

It is assumed that the strength of columns with outstanding flanges will be limited by the local buckling strength of the flanges. No allowance is made for post-buckling strength in such members. It is also noted that unsymmetrical open shapes may be subject to failure by combined torsion and flexure, and should not be used unless a special analysis is made.

AASHTO 87

The limitations are given under TRUSSES; Compression Members:

$$\frac{b}{t} \leq \frac{1625}{\sqrt{f_a}} \qquad (1c.N1)$$

but no greater than 12 for main members
16 for secondary members

where

b = distance from edge of plate to point of support
f_a = computed axial compressive stress (psi)

When the compressive stress equals the limiting factor of $0.44F_y$, the $\frac{b}{t}$ ratio shall not be greater than

F_y = 36,000 psi $\frac{b}{t}$ = 12

F_y = 50,000 psi $\frac{b}{t}$ = 11

F_y = 90,000 psi $\frac{b}{t}$ = 8

F_y = 100,000 psi $\frac{b}{t}$ = 7.5

for webs

$$\frac{b}{t} \leq \frac{4000}{\sqrt{f_a}} \quad \text{with a maximum} = 45 \qquad (1c.N2)$$

the limitations for the web are

$$F_y = 36{,}000 \text{ psi} \qquad \frac{b}{t} = 32$$

$$F_y = 50{,}000 \text{ psi} \qquad \frac{b}{t} = 27$$

$$F_y = 90{,}000 \text{ psi} \qquad \frac{b}{t} = 20$$

$$F_y = 100{,}000 \text{ psi} \qquad \frac{b}{t} = 19$$

No specific requirements are given for columns. For beams and compact sections,

$$\frac{b}{t} \leq \frac{1600}{\sqrt{F_y}} \tag{1c.N3}$$

where b is projecting compression flange element. (For $F_y = 36$ ksi, the limitation is 8.4.)

AISC ASD 89

$$\left.\begin{aligned} &\text{for flange} \quad \frac{b}{2t_f} \leq \frac{95}{\sqrt{F_y}} \\ &\text{for web} \quad \frac{d}{t_w} \leq \frac{253}{\sqrt{F_y}} \end{aligned}\right\} \tag{1c.N4}$$

where F_y is in ksi.

These values may be exceeded, provided the allowable compressive stress is reduced, as given in AISC ASD 89 Appendix C.

For A36 steel ($F_y = 36\text{ksi} = 250$ MPa), the above values become

$$\frac{b}{2t_f} \leq 15.8 \qquad \frac{d}{t_w} \leq 42.2$$

These requirements have been determined from a theoretical consideration of buckling, such that the plate buckling stress is less than the yield stress: that is, buckling is not allowed to occur until the yield stress is attained. (Johnston, 1976) (These requirements correspond to the traditional values for structural carbon steel, 16 and 42, used earlier and based on actual usage experience of close to a century.)

Plastic Design: The allowable $b/2t_f$ values for flanges are tabulated, and vary from 8.5 for $F_y = 36$ ksi to 6.0 for $F_y = 65$ ksi. The d/t_w value is limited to $190/\sqrt{F_y}$

These values were based on studies (ASCE, 1969) such that sufficient plastic hinge rotation would occur without reduction in M_p.

AISC LRFD 86

The b/t requirements are identical with those for Allowable Stress Design, Eq. 1c.N4.

When the requirements are exceeded, the members shall be proportioned according to AISC Appendix B, which gives reduction factors.

AREA 89

For outstanding elements in compression (for main members carrying axial stress)

$$\frac{b}{t} \leq 12 \tag{1c.N5}$$

For web plates in compression members consisting of segments connected by lacing or by solid cover plates,

$$\frac{b}{t} \leq 32\sqrt{\frac{p_c}{f}} \tag{1c.N6}$$

where

- b = unsupported distance between the nearest lines of fasteners or welds, or between the roots of rolled flanges
- f = actual stress in compression
- p_c = allowable stress for the member in axial compression

$$\sqrt{\frac{p_c}{f}} \leq 2$$

for $p_c = f$

$$b/t \leq 32$$

For high-strength structural steel, Eqs. 1c.N5 and 1c.N6 become

$$\frac{b}{t} \leq \frac{2300}{\sqrt{F_y}} \tag{1c.N7}$$

and

$$\frac{b}{t} \leq \frac{6000}{\sqrt{F_y}}\sqrt{\frac{p_c}{f}} \tag{1c.N8}$$

where $\sqrt{\frac{p_c}{f}}$ is less than or equal to 2.

All stresses are in psi.

CSA S6 M89 84a

The width-thickness ratio of projecting elements of compression members is $260/\sqrt{F_y}$ for plates projecting from columns. (For F_y = 250 MPa, b/t = 16.4) The width-thickness ratio is limited to 16 when the slenderness ratio of the member exceeds 100.

For webs, the b/t limit is $670/\sqrt{F_y}$. (This is 42.4 for $F_y = 250$ MPa.) However, if the unsupported width exceeds $670/\sqrt{F_y}$ times its thickness, but a portion of its width no greater than $670/\sqrt{F_y}$ times the thickness would satisfy the stress requirements, the member will be considered acceptable.

CSA S16.1 M89 84b

Structural sections are designated into classes, depending on the maximum width-thickness ratios of their elements in compression. The limitations given below show the values for 250 MPa steel in parentheses.

	flange		web		
for Class 1 (plastic design)	$\frac{b}{t} \le \frac{145}{\sqrt{F_y}}$	(9.2)	$\frac{h}{w} \le \frac{670}{\sqrt{F_y}}$	(42.4)	(1c.N9)
for Class 2 (compact)	$\frac{b}{t} \le \frac{170}{\sqrt{F_y}}$	(10.8)	$\frac{h}{w} \le \frac{670}{\sqrt{F_y}}$	(42.4)	(1c.N10)
for Class 3 (non-compact)	$\frac{b}{t} \le \frac{200}{\sqrt{F_y}}$	(12.6)	$\frac{h}{w} \le \frac{670}{\sqrt{F_y}}$	(42.4)	(1c.N11)

where b is half the flange width and F_y is in MPa.

OHBDC 83

There are two classifications for sections: compact and non-compact. The limitations below are for flange and web, with values in parentheses for 250 MPa steel.

compact	$\frac{b}{t} \le \frac{170}{\sqrt{F_y}}$	(10.8)	$\frac{h}{w} \le \frac{670}{\sqrt{F_y}}$	(42.4)	(1c.N12)
non-compact	$\frac{b}{t} \le \frac{260}{\sqrt{F_y}}$	(12.6)	$\frac{h}{w} \le \frac{670}{\sqrt{F_y}}$	(42.4)	(1c.N13)

COMMENTARY

b/t limitations are used to prevent local buckling and the premature failure of the member. The local buckling limitations for structural shapes are given for various elements of the shapes used. For a wide-flange column shape, the limitations of interest are those for flange and web, which are discussed here.

REGIONAL COMPARISON

Width-thickness ratios are specified to prevent local buckling and thus premature failure. The North American building specifications, both ASD and LSD, are similar to each other and reflect the SSRC recommendations, as do the Canadian highway codes, while the U.S. highway and railway codes, AREA and AASHTO, are somewhat conservative and limited in their scope. Plastic design limitations are more strict to ensure the achievement of the required rotation capacity. The AA 86 recommendations are different, so that no direct comparison can be made with steel.

The steel codes are compared below for a steel of F_y = 36 ksi (250MPa), where b is half the flange width.

Allowable Stress Design	flange	web
AISC ASD 89	$\frac{b}{t} \leq \frac{95}{\sqrt{F_y}} = 15.8$	$\frac{d}{w} \leq \frac{253}{\sqrt{F_y}} = 42.2$
plastic design	$\frac{b}{t} = 8.5$	$\frac{d}{w} \leq \frac{190}{\sqrt{F_y}} = 31.6$
AREA 89	$\frac{b}{t} \leq \frac{2300}{\sqrt{F_y}} = 12$	$\frac{d}{w} \leq \frac{190}{\sqrt{F_y}}\sqrt{\frac{P_c}{f}} = 32$ when $p_c = f$
AASHTO 87	$\frac{b}{t} \leq \frac{1625}{\sqrt{f_a}} = 12$ 12 is limit for F_y=36	$\frac{d}{w} \leq \frac{4000}{\sqrt{f_a}} = 32$ 32 is limit for $F_y = 36$
CSA S6 M89 84a	$\frac{b}{t} \leq \frac{260}{\sqrt{F_y}} = 16.4$	$\frac{d}{w} \leq \frac{670}{\sqrt{F_y}} = 42.4$
Limit States Design		
AISC LRFD 86	same as AISC ASD 89	
AASHTO 87 for beams	$\frac{b}{t} \leq \frac{1600}{\sqrt{f_y}} = 8.4$	
CSA S16.1 M89 84b		
Class 1 (Plastic Design)	$\frac{b}{t} \leq \frac{145}{\sqrt{F_y}} = 9.2$	$\frac{d}{w} \leq \frac{670}{\sqrt{F_y}} = 42.4$
Class 2 (Compact)	$\frac{170}{\sqrt{F_y}} = 10.8$	$\frac{670}{\sqrt{F_y}} = 42.4$
Class 3 (Non-Compact)	$\frac{200}{\sqrt{F_y}} = 12.6$	$\frac{670}{\sqrt{F_y}} = 42.4$
OHBDC 83		
Compact	$\frac{b}{t} \leq \frac{170}{\sqrt{F_y}} = 10.8$	$\frac{d}{w} \leq \frac{670}{\sqrt{F_y}} = 42.4$
Non-Compact	$\frac{260}{\sqrt{F_y}} = 16.4$	$\frac{670}{\sqrt{F_y}} = 42.4$

WEST EUROPE

CHAPTER EDITORS
John E. Harding, University of Surrey
Gerald W. Schulz, University of Innsbruck

BS 5400 P.3 82

Section Restrictions

Unstiffened outstand $\frac{b}{t} \leq 12\sqrt{\frac{355}{\sigma_y}}$

Circular hollow section $\frac{D}{t} \leq 100\sqrt{\frac{355}{\sigma_y}}$

where D is the outside diameter and σ_y is in MPa.

For other component plates an effective width is specified as a function of the local slenderness: that is, for plates with both edges supported.

For circular hollow sections the effective area should be reduced for

$$\frac{D}{t}\sqrt{\frac{\sigma_y}{355}} > 50 \leq 100 \text{ to } A_c\left(1.15 - 0.003\,\frac{D}{t}\sqrt{\frac{\sigma_y}{355}}\right)$$

where A_c is the gross area with allowance for holes.

BS 5950 P.5 85

Classification of Section Shapes

Class 1 – Plastic

A plastic hinge can be developed with sufficient rotation to allow redistribution of moments.

Class 2 – Compact

Sections can be used without restriction but not for plastic design.

Class 3 – Semi-compact

Extreme fiber stress can reach design strength but not full plastic moment.

Class 4 – Slender

Stress reduction factor applied to design strength.

Limiting width to thickness ratios			(1)	(2)	(3)
flange outstand	welded	$\frac{b}{t}$	$\leq 7.5\epsilon$	$\leq 8.5\epsilon$	$\leq 13\epsilon$
	rolled		$\leq 8.5\epsilon$	$\leq 9.5\epsilon$	$\leq 15\epsilon$
internal flange element	welded		$\leq 23\epsilon$	$\leq 25\epsilon$	$\leq 28\epsilon$
	rolled		$\leq 26\epsilon$	$\leq 32\epsilon$	$\leq 39\epsilon$

Limiting width to thickness ratios (cont'd)		(1)	(2)	(3)
web in compression	welded			$\frac{d}{t} \leq 28\epsilon$
	rolled			$\leq 39\epsilon$
circular tube	welded or rolled	$\frac{D}{t} \leq 40\epsilon^2$	$\leq 57\epsilon^2$	$\leq 80\epsilon^2$

where

$$\epsilon = \sqrt{\frac{275}{p_y}}$$

p_y = yield stress in MPa

Stress reduction factor for Class 4 sections

flange outstand	welded	$\frac{10}{\frac{b}{t\epsilon} - 3}$
	rolled	$\frac{11}{\frac{b}{t\epsilon} - 4}$
internal flange element	welded	$\frac{21}{\frac{b}{t\epsilon} - 7}$
	rolled	$\frac{31}{\frac{b}{t\epsilon} - 8}$

EC 3 88

EUROCODE 3 distinguishes between 4 different types of cross-sections and related modes of structural behavior:

(1) Class 1 - plastic cross-sections are those that can form a plastic hinge with the rotation capacity required for plastic analysis.
(2) Class 2 - compact cross-sections are those that can develop their plastic resistance moment, but have limited rotation capacity.
(3) Class 3 - semi-compact cross-sections are those, in which the calculated stress in the extreme compression fiber of the steel member can reach its yield strength, but local buckling is liable to prevent development of the plastic resistance.
(4) Class 4 - slender cross-sections are those, in which local buckling is liable to prevent the calculated stress in the extreme compressive fiber of the steel member reaching its yield strength.

The maximum b/t ratios for elements of cross-sections in uniform compression have been extracted from a more comprehensive table.

For webs of rolled I-sections and webs and flanges of welded box-sections:

Class 1	$b/t < 33\sqrt{235/f_y}$
Class 2	$b/t < 38\sqrt{235/f_y}$
Class 3	$b/t < 42\sqrt{235/f_y}$

For outstand flanges:

	Welded I-sections	Rolled I-sections
Class 1	$b/t < 9\sqrt{235/f_y}$	$b/t < 10\sqrt{235/f_y}$
Class 2	$b/t < 10\sqrt{235/f_y}$	$b/t < 11\sqrt{235/f_y}$
Class 3	$b/t < 14\sqrt{235/f_y}$	$b/t < 15\sqrt{235/f_y}$

Local buckling is taken into account in the design of Class 4 cross-sections by means of the effective width method. An interactive method based on the calculated maximum stress in place of the yield strength f_y is permitted. All stresses are in MPa.

INTERNATIONAL COMPARISONS & COMMENT

WHAT ARE THE DIFFERENCES?

Although the concept of local buckling is well understood, its influence on plate strength is open to interpretation. A number of plate tests have been carried out around the world but the results are somewhat variable as boundary conditions have not always been well-defined or achieved in the tests. Hence, different plate strength curves, or different interpretations of the same data, have been used to define the width/thickness limits.

The interaction of plate assemblages (e.g., web flange) can have significant effects on the individual plate strengths. This is not normally taken into account in specifications.

The specifications that allow slender plate elements use an effective width approach to account for the reduction in capacity due to local buckling. A number of formulations are possible and this has lead to variations in effective widths.

The differences in width/thickness limits are not that great and it should be possible, with international cooperation, to harmonize the limits between performance classifications. Table 1c.1 lists the comparison among specifications with the topics explained in more detail as follows:

Classification by member or section

All specifications classify by section (for example, web, flange) although some have large tables of cross-section types to define the plate elements. Some specifications include yield stress non-dimensionally, usually expressed in terms of the square root of the ratio of the yield stress to a standard value although some tabulate values for various stresses.

Table 1c.1 Concept Matrix Comparison

TOPICS \ SPECIFICATIONS	AS DR87164 87	GBJ 17 88	TBJ 2 85	AIJ 70	AIJ 75	AIJ 89b	JSCE	JRA 80	AISC ASD 89	AREA 89
Classification by member or section	X	X	X	X	X	X	X	X	X	X
No classification										
Plastic					X	X			X	
Compact	X					X				
Non-compact	X					X				
Slender	X					X				
Web-flange interaction	X					X				
Slenderness reduction factor	X	X	X	X		X	X		X	X
Variation with column slenderness		X	X							

Classification by performance
A number of specifications have no specified classification although the trend is toward specific definitions for example, plastic, compact, non-compact and slender.

There is a considerable variation in the values of the width/thickness ratios that define the boundaries between the various classifications.

Web/Flange Interaction
Only two specifications allow for the interaction local buckling between plate elements forming an assemblage. The rest consider each plate element independently.

Slenderness Reduction Factor
All specifications allow for reduction in section strength for members comprising slender plate elements usually in the form of an effective width or an effective area formulation although some use other forms of empirical formulation.

Variation with Column Slenderness
Only two specifications considers the variation of width/thickness ratio with overall column slenderness.

Table 1c.1 (Cont'd) Concept Matrix Comparison

SPECIFICATIONS / TOPICS	AASHTO 87	CSA S6 M89 84a	AA 86	AISC LRFD 86	CSA S6.1 M89 84b	OHBDC 83	EC 3 88	BS 5950 P.5 85	BS 5400 P.3 82
Classification by member or section	X	X		X	X	X	X	X	X
No classification									
Plastic					X		X	X	
Compact				X	X	X	X	X	
Non-compact				X	X	X	X	X	
Slender				X	X	X	X	X	
Web-flange interaction									
Slenderness reduction factor	X	X	X	X	X	X	X	X	X
Variation with column slenderness									

WHY ARE THERE DIFFERENCES?

The width/thickness limits for the specifications have in many cases been established from different research bases or use different interpretations of the same data base.

There are different interpretations as to what is required to meet the expected performance criteria. This is particularly the case for plastic sections where sufficient rotation capacity for plastic hinge action has been defined in a number of different ways. There is also the question of an upper limit to prevent local buckling under service loading conditions.

While most are satisfied to consider the summation of individual plate elements as a sufficient basis for design, some are willing to use the knowledge of plate buckling of assemblages as an improved basis for strength. This reflects the knowledge base that exists in a region.

1d. Maximum Slenderness Ratio

SPECIFICATION PROVISIONS & EXPLANATION

AUSTRALIA

CHAPTER EDITOR
Russell Q. Bridge, The University of Sydney

AS DR87164 87

AS DR87164 87 sets no limitations for the maximum slenderness ratio.

CHINA

CHAPTER EDITORS
Bao Kang He, Xian Institute of Metallurgy & Construction Engineering
Mingzhong Wei, Chongqing Architectural Institute
Wen Wei Zhao, Quinghuo University

GBJ 17 88

The maximum permitted values of slenderness λ are 150 for columns or main compression members and 200 for secondary compression members.

TBJ 2 85

The maximum permitted values of slenderness λ are 100 for main compression members and $130 \sim 150$ for secondary ones.

JAPAN

CHAPTER EDITOR
Minoru Wakabayashi, General Building Research Corporation

AIJ 70

The maximum slenderness ratio is 250 for compression members in general but limited to 200 when the members are used as structural columns.

AIJ 75

AIJ 89b

The maximum slenderness ratio is 200 for column members.

JRA 80

The maximum slenderness ratio is 150 for compression members but limited to 120 when used as truss members.

JSCE 83

The maximum slenderness ratio is 100 for major compression members and 120 for secondary members.

NORTH AMERICA

CHAPTER EDITOR
Lambert Tall, Florida International University

AA 86

There is no limitation specified on column length – however, this is accomplished indirectly to some extent by the reduction of the allowable stress in compression when local buckling lowers column strength.

AASHTO 87

$$\frac{KL}{r} \leq 120 \tag{1d.N1}$$

AISC ASD 89

$$\frac{KL}{r} \leq 200 \tag{1d.N2}$$

Plastic Design: In the plane of bending of columns that would develop a plastic hinge at ultimate load, L/r shall not exceed C_c. (See Eq. 1a.N7 for C_c.)

AISC LRFD 86

$$\frac{KL}{r} \leq 200 \tag{1d.N3}$$

Plastic Analysis:

$$\lambda_c \leq 1.5K \tag{1d.N4}$$

AREA 89

$$\frac{L}{r} \leq 100 \tag{1d.N5}$$

CSA S6 M89 84a

$$\frac{KL}{r} \leq 120 \tag{1d.N6}$$

CSA S16.1 M89 84b

$$\frac{KL}{r} \leq 200 \tag{1d.N7}$$

OHBDC 83

$$\frac{KL}{r} \leq 120 \tag{1d.N8}$$

COMMENTARY

The slenderness ratio is limited for ease of handling and for load-carrying capacity. Main members only are given here, secondary members are allowed larger values.

REGIONAL COMPARISON

The slenderness ratio limitations for all North American specifications for main members is 120 for highway structures, 200 for building structures, and 100 for railway structures. For plastic design, KL/r is limited to C_c, and for aluminum columns a limitation on column length is achieved indirectly by a formula that reduces the allowable stress in compression when local buckling affects column strength.

The maximum slenderness ratio is limited for ease of handling and for load-carrying capacity. Main members only are given here, secondary members are allowed larger values.

WEST EUROPE

CHAPTER EDITOR
Gerald W. Schulz, University of Innsbruck

BS 5400 P.3 82

No limit on ℓ/r

Graphs apply to $\frac{\ell_e}{r}\sqrt{\frac{\sigma_y}{355}} = 300$ but formula unlimited.

BS 5950 P.5 85

λ should not exceed the following

For members resisting loads other than wind loads	180
For members resisting self weight and wind loads only	250
For tie members subject to stress due to wind	350

EC 3 88

EUROCODE 3 sets no limitation for the slenderness ratio.

INTERNATIONAL COMPARISONS & COMMENT

WHAT ARE THE DIFFERENCES?

Only a few specifications have no limit at all. For main members, most are 200 except bridges, in which they vary from 100 to 150.

For secondary members, only a few are concerned with values. For buildings, the value varies from 200 to 350, while for bridges, 120 to 150.

One specification uses loading as a classification for the ratio.

Table 1d.1 Concept Matrix Comparison

TOPICS \ SPECIFICATIONS			AS DR87164 87	GBJ 17 88	TBJ 2 85	AIJ 70	AIJ 75	AIJ 89b	JSCE	JRA 80	AISC ASD 89	AREA 89
No limits			X									
Main Members		200				X	X	X			X	
		180										
		150		X						X		
		120										
		100			X				X			X
Secondary members		350										
		250				X						
		200		X								
		130–150			X							
Trusses		120							X			
Plastic design										X		
Loading type	Total	180									X	
	Wind	250										
	Wind reversal	350										

WHY ARE THERE DIFFERENCES?

There are probably misconceptions as to the purpose of this provision and this is reflected in the wide variation of values.

Where limitations are based on professional judgment, there are likely to be as many values as there are opinions. They are often used as a convenient starting point in the selection of member size.

The traditional conservatism (perhaps warranted) of bridge specifications is reflected in generally lower values.

The ratios are historic in origin and are gradually being raised and simplified. For example, the distinction between main and secondary member is no longer clear.

It would appear that the trend is toward eliminating this requirement.

Table 1d.1 (Cont'd) Concept Matrix Comparison

TOPICS \ SPECIFICATIONS			AASHTO 87	CSA S6 M89 84a	AA 86	AISC LRFD 86	CSA S6.1 M89 84b	OHBDC 83	EC 3 88	BS 5950 P.5 85	BS 5400 P.3 82
No limits									X		X
Main Members		200				X	X				
		180									
		150									
		120	X	X	X			X			
		100									
Secondary members		350									
		250									
		200									
		130–150									
Trusses		120									
Plastic design						X					
Loading type	Total	180								X	
	Wind	250								X	
	Wind reversal	350								X	

Symbol List for Chapter 1 -- Compression Members

A = cross section area (1a.C) & (1c.J)
A_e = effective area (1a.W)
A_g = cross section area (1a.N) & (1a.W)
A_n = net section area (1c.A)
A_s = cross section area (1a.N)
A_w = cross section area of the web (1c.J)
a_o = imperfection coefficent (1a.E)
B = leg width of angle (1c.J)
B_c = intercept (a function of yield; tabulated) (1a.N)
b = clear width of the plate element (1c.A)
b = flange width (1c.J)
b = distance from edge of plate to point of support (1c.N)
b = unsupported distance between the nearest lines of fasteners or welds, or between the roots of rolled flanges (1c.N)
b_e = effective width (1c.A)
b_o = width between webs (1c.C)
b_1, b_2, b_3, b_4 = see Fig. 1c.C2 (1c.C)
C_c = limiting slenderness ratio (1a.N)
C_o = limiting slenderness ratio (1a.N)
C_p = limiting slenderness ratio (1a.N)
C_r = factored axial compressive resistance (1a.N)
c_1, c_2 = coefficients (1a.E)
D = web height of channel (1c.J)
D = outside diameter (1c.W)
D_c = slope (function of yield and E; tabulated) (1a.N)
d = web width (1c.J)
E = elastic modulus (1a.E)
F = specified yield stress in ton/cm^2 (1a.J) & (1c.J)
F_a = allowable compressive stress (1a.N)
F_{cr} = allowable stress (see Eq. 1a.N4) (1a.N)
F_{cy} = compressive yield stress (1a.N)
F_y = yield stress (1a.N)
F_{yf} = yield stress of flange (1c.J)
F_{yw} = yield stress of web (1c.J)
f = design normal stress of steel (1a.C)
f = actual stress in compression (1c.N)
f_a = computed axial compressive stress (1c.N)
f_c = allowable flexural stress (1a.J)
f_y = specified yield stress, applicable for $t \leq 40$ mm; for 40 mm $< t \leq 80$ mm f_y shall be reduced by 10% (1a.W)
f_y = specified minimum yield stress (1a.C) & (1c.A)
h_o = clear height of web (1c.C)
K = effective length factor (1b.N)
K = factor for safety (1a.C)
k = effective length factor (1a.N)
k_c = coefficient for compression members (tabulated according to alloy and temper) (1a.N)
L = length of the member (1a.J)

L_e = effective length (1a.A) & (1a.W)
ℓ = length of member (1a.C)
ℓ = effective length (1a.J)
ℓ_e = effective length of column (1a.W)
m = coefficient (1a.N)
m_o = imperfection factor (1a.E)
N = design value of axial force (1a.C) & (1a.W)
N_c = nominal capacity of compression member (1a.A)
N_c = flexural strength of compression member (1a.J)
N_{cr} = ultimate flexural strength (1a.J)
N_d = limit (design) axial load (1a.E)
N_e = Euler's buckling force (1a.J)
N_{ki} = Euler buckling load (1a.W)
N_{pl} = ultimate axial capacity (1a.W)
N_s = section capacity of a short column (1a.A)
N_y = specified yield force (1a.J)
n = parameter of imperfection (1a.E)
n = number of segments divided by stiffners (1c.J)
n_u = factor of safety on ultimate strength (1a.N)
n_y = factor of safety on yield strength (1a.N)
P_c = section resistance (1a.W)
P_{cr} = plastic design strength (1a.N)
P_n = nominal strength of compression member (1a.N)
P_u = maximum strength of concentrically loaded column (1a.N)
p_c = allowable stress for the member in axial compression (1c.N)
Q = form factor for local buckling of slender plate elements (1c.A)
R = effective length coefficient (1a.N)
R_d = design stress (1a.E)
R_y = specified yield stress (1a.E)
r = radius of gyration (1a.J) & (1a.N)
r = radius of gyration of gross section (1a.W)
S_1 = limit slenderness ratio (1a.N)
S_2 = limit slenderness ratio (1a.N)
t = thickness of plate element (1a.E)
t_f = flange thickness (1c.J)
t_w = web thickness (1c.J)
t_1, t_2, t_3, t_4 = see Fig. 1c.C2 (1c.C)
V_o = equivalent initial deflection (1a.C)
W = elastic section modulus (1a.C)
w = thickness of web (1c.N)
y = distance from neutral axis to extreme fibre (1a.W)
α = parameter of imperfection (1a.E)
α = column curve coefficient (1a.W)
α_1, α_2, α_3 = coefficients shown in Table 1a.C1 (1a.C)
γ = safety factor (1a.E)
γ_{f3} = partial load factor = 1.1 (1a.W)
γ_m = material reliability factor considered in the range of 1.025 to 1.15 (1a.E)
γ_m = partial material factor = 1.2 (1a.W)
γ_R = resistance factor (1a.C)
η = Perry factor (1a.W)

η = coefficient (1a.W)
η = tangent modulus factor (1c.C)
λ = slenderness ratio (1a.C) & (1a.E)
λ = effective slenderness ratio (1a.J) & (1a.W)
λ = modified slenderness ratio (1a.N) & (1a.W)
λ = $\frac{\mathrm{KL}}{\mathrm{r}}$, slenderness ratio of members (1c.A)
$\overline{\lambda}$ = modified slenderness ratio (1a.C), (1a.E), & (1a.J)
$\overline{\lambda}$ = relative slenderness (1a.W)
λ_c = modified slenderness ratio (1a.J) & (1a.N)
λ_c = section slenderness (1c.A)
λ_{cy} = limiting plate slenderness (1c.A)
λ_o = limiting slenderness ratio (1a.W)
λ_{pl} = limiting slenderness ratio (1a.W)
${}_e\lambda_c$ = limiting slenderness ratio for elastic buckling given as $1/\sqrt{0.6}$ (1a.J)
${}_p\lambda_c$ = limiting slenderness ratio for plastic analysis given as 0.15 (1a.J)
μ = effective length factor (1b.C)
ξ_o = equivalent eccentricity ratio (1a.C)
ρ_c = design stress of column in compression (1a.W)
ρ_e = Euler buckling stress (1a.W)
ρ_y = yield stress (1a.W)
σ = allowable stress of steel (1a.C)
σ = maximum working stress in compression (1c.J)
σ_{all} = allowable stress (1a.E)
σ_c = buckling strength based on multiple column curves (1a.W)
σ_{cag} = allowable flexural stress for compression members (1a.J)
σ_{cr} = flexural strength for compression members (1a.J)
σ_{cr} = local buckling stress (1c.C)
σ_w = allowable flexural stress for compression members (1a.J)
σ_y = specified yield stress (1a.J)
σ_1 = stress at the edge where compressive stress is applied by bending moment (1c.J)
σ_2 = stress at the edge where tensile stress is applied by bending moment (1c.J)
ϕ = strength reduction factor of centrally compressed members (1a.C)
ϕ = resistance factor (1a.N)
ϕ_c = resistance factor (1a.J)
φ = buckling coefficient (1a.E)
χ = reduction coefficient (1a.W)

Chapter 2

Built-Up Members

TOPICAL EDITOR

Leo Finzi, Politecnico di Milano

CHAPTER EDITORS

Georgij E. Belskij, CNIISK, Moscow
Russell Q. Bridge, The University of Sydney
Ji Chen, Xian Institute of Metallurgy & Construction Engineering
Leo Finzi, Politecnico di Milano
Bruce G. Johnston, Consultant
Jindrich J. Melcher, Technical University of Brno
Toshiyuki Ogawa, Tokyo Institute of Technology
Zia Razzaq, Old Dominion University
Toshiro Suzuki, Tokyo Institute of Technology

2a. Column Strength

SPECIFICATION PROVISIONS & EXPLANATION

AUSTRALIA

CHAPTER EDITOR

Russell Q. Bridge, The University of Sydney

AS 1250 81

Laced Members and Battened Members

Where adequate batten elements, lacing elements or connections are provided, a column composed of two or more main components which are parallel to one another are designed to act as a single integral member with effective slenderness λ_{ef}

$$\lambda_{ef} = (\lambda^2+\lambda_f^2)^{1/2} \tag{2a.A1}$$

where

λ = slenderness of the whole member

λ_f = slenderness of a main component between adjacent connections.

CHINA

CHAPTER EDITOR
Ji Chen, Xian Institute of Metallurgy & Construction Engineering

GBJ 17 88

The slenderness ratio about the solid-web axis of built-up members shall be computed on the same basis as prescribed for solid compression members.

The slenderness ratio about the open-web axis of built-up members shall be computed by using an effective slenderness ratio as follows:

Members Comprised of Two Longitudinal Components as Shown in Fig. 2a.C1-a

for members with tie plates

$$\lambda_{ox} = \sqrt{\lambda_x^2 + \lambda_1^2} \quad (2a.C1)$$

for members with lacing bars

$$\lambda_{ox} = \sqrt{\lambda_x^2 + 27\frac{A}{A_{1x}}} \quad (2a.C2)$$

Members Comprised of Four Components as Shown in Fig. 2a.C1-b

for members with tie plates

$$\lambda_{ox} = \sqrt{\lambda_x^2 + \lambda_1^2} \quad (2a.C3)$$

$$\lambda_{oy} = \sqrt{\lambda_y^2 + \lambda_1^2} \quad (2a.C4)$$

for members with lacing bars

$$\lambda_{ox} = \sqrt{\lambda_x^2 + 40\frac{A}{A_{1x}}} \quad (2a.C5)$$

$$\lambda_{oy} = \sqrt{\lambda_y^2 + 40\frac{A}{A_{1y}}} \quad (2a.C6)$$

Members Comprised of Three Components as Shown in Fig. 2a.C1-c

$$\lambda_{ox} = \sqrt{\lambda_x^2 + \frac{42A}{A_1(1.5 - \cos^2\theta)}} \quad (2a.C7)$$

$$\lambda_{oy} = \sqrt{\lambda_y^2 + \frac{42A}{(A_1\cos^2\theta)}} \quad (2a.C8)$$

where

λ_{ox} and λ_{oy} = effective slenderness ratios of the whole member

λ_x and λ_y = slenderness ratios of built-up members assumed to react as solid members about x and y axes, respectively

λ_1 = slenderness ratio of component

$$\lambda_1 = l_1/i_1$$

where

A	=	cross-sectional area of the built-up member
A_1	=	sum of the gross sectional areas of all the lacings in a section of the strut
A_{1x}, A_{1y}	=	sectional areas of lacing elements perpendicular to x and y axes, respectively
i_1	=	the smallest radius of gyration of component
l_1	=	spacing of tie plates
θ	=	the half-angle between adjacent faces of the members as shown in Fig. 2a.C1-c

COMMENTARY

The formulas for effective slenderness ratios are used to evaluate the reduction in column capacity due to shear deformation in built-up members.

The slenderness ratio of components shall not exceed 0.7 λ_{max} and 0.5 λ_{max}, respectively, for built-up members with lacing bars and those with tie plates, where λ_{max} is the maximum slenderness ratio of the built-up member.

The stiffness of tie plates shall be $\geq$ 6 times that of the larger component of built-up members. The slenderness ratio of lacing shall not exceed 150.

The slenderness ratio limitations of components account for the effects of the geometrical imperfections.

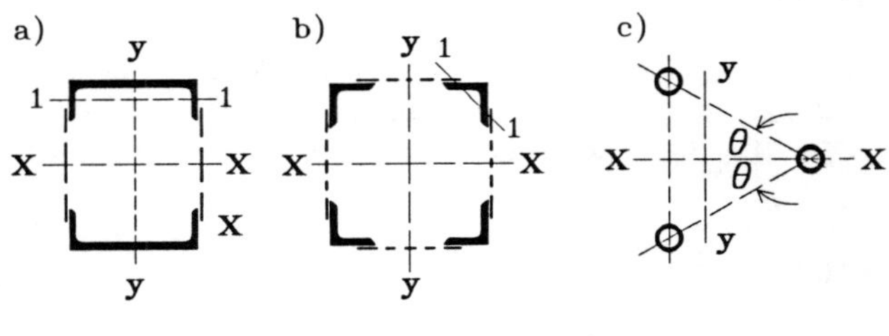

Fig 2a.C1 Components of Built-Up Members

EAST EUROPE

CHAPTER EDITORS

Georgij E. Belskij, CNIISK, Moscow
Jindrich J. Melcher, Technical University of Brno

ČSN 73 1401 84

The design provision conformable to Eq. 2a.E2 for the effective slenderness is used.

Generally the design criterion of limit states format with the buckling coefficient

$$\varphi = \varphi_1 \varphi_r \tag{2a.E1}$$

is used (Eq. 2a.E8). The factor φ_r in Eq. 2a.E1 is derived from the modified slenderness $\lambda_{ef}\sqrt{\varphi_1}$.

COMMENTARY

For built-up members composed of single-angle components, the slenderness λ_1 used for the determination of the buckling coefficient φ_1 in Eq. 2a.E1 shall be derived from the smallest radius of gyration of one single angle. In Eq. 2a.E4 the slenderness ratio λ_1 shall be derived using the radius of gyration of a single angle for its central axis perpendicular to the open web plane.

PN 80/B-03200 87

The formulas for the calculation of effective slenderness can be adapted to the general form of Eq. 2a.E2. The corresponding slenderness λ_1 shall be determined from Eq. 2a.E5 or Eq. 2a.E7 when the limitation $L_1/i_1 \leq 60$ is used.

COMMENTARY

For built-up members composed of single angle components more detailed design provision are available in the complete specification.

SNiP II-23-81 82

The following rules are given for built-up members in compression made up of two, three, or four parallel chords interconnected either by lacing or by batten plates placed into two, three (triangle section) or four (box section) planes.

Slenderness ratios about the solid-web axis of the built-up members shall be determined on the same basis as prescribed for solid compression members.

The general format of the design criterion for the buckling of built-up compression members in their open-web planes is of the same form as for solid members (see Eq. 1a.E2 in Chapter 1), except for the determination of buckling coefficient the effective slenderness ratio λ_{ef} shall be used.

For the basic example of a built-up section shown in Fig. 2a.E1 the slenderness ratio λ_{ef} is in general

$$\lambda_{ef} = \sqrt{\lambda_y^2+\lambda_1^2} \tag{2a.E2}$$

where

λ_1 = defined by Eq. 2a.E5

λ_y = slenderness ratio of a built-up member assumed to buckle as an integrated member

For columns with batten plates the slenderness ratio is

for $n = I_1 b/(I_p L_1) > 0.2$

$$\lambda_1 = (L_1/i_1)\sqrt{0.82(1+n)} \tag{2a.E3}$$

for $n \leq L_1/i_1$

$$\lambda_1 = L_1/i_1 \tag{2a.E4}$$

where

I_p = moment of inertia of batten plate section about its strong central axis

I_1 = moment of inertia of a single chord

i_1 = radius of gyration of a single chord

L_1 = spacing of batten plates

For laced columns

$$\lambda_1 = \pi\sqrt{\frac{A}{ZA_d}\,\frac{d^3}{L_1 b^2}} \tag{2a.E5}$$

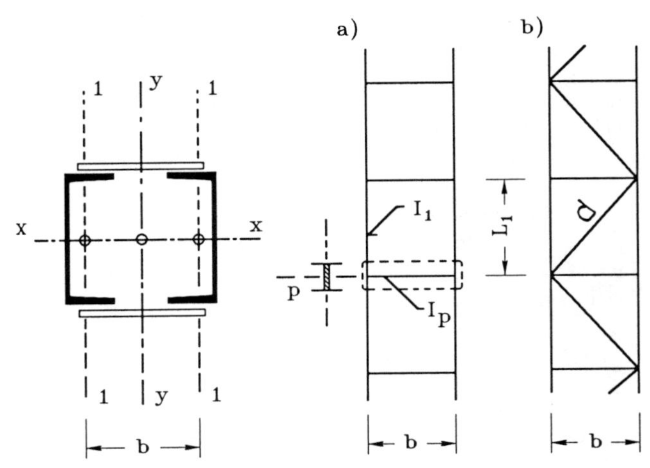

Fig. 2a.E1 Built-Up Members

where

A = cross-sectional area of the built-up member
A_d = cross-sectional area of the diagonal
Z = number of structural planes formed by lacing

The slenderness limits of $\lambda_1 < 40$ and $\lambda_1 < 80$ are required for columns with batten plates and laced columns, respectively, or generally $\lambda_1 < \lambda_{ef}$.

COMMENTARY

The slenderness ratio limitations of individual sub-members are to account for the influence of geometrical imperfections upon the value of effective slenderness of a built-up member which are not otherwise considered. Generally the limit states design approach is used with factored loads and resistance.

TGL 13503/01 82

The effective slenderness of built-up members regarding buckling in the open web planes may be approximated by the formula

$$\lambda_{ef} = \sqrt{\lambda_y^2 + \frac{m}{2}\lambda_1^2} \qquad (2a.E6)$$

where

m = number of components or groups of components connected by lacing or batten plates (m = 2 for the arrangement in Fig. 2a.E2)
λ_1 = shall be calculated according to the type of the fasteners

For batten plates or separators

$$\lambda_1 = L_1/i_1 \leq 0.5\lambda_x \quad \text{or} \quad \lambda_1 \leq 50 \qquad (2a.E7)$$

where

i_1 = smallest radius of gyration of components
L_1 = spacing of connection elements
λ_x = slenderness ratio of the built-up member about the solid-web axis.

For laced or latticed members Eq. 2a.E5 shall be used.

If the limitations of Eq. 2a.E7 for the individual sub-members are not satisfied, the buckling coefficient

$$\varphi = \varphi_{ef}\varphi_1 \qquad (2a.E8)$$

shall be used.

In Eq. 2a.E8 the coefficient φ_{ef} shall be derived from the effective slenderness λ_{ef} using column curve (b)(Table 1a.E3). In case of non-welded shapes applied for single components the column curve (a) (Table 1a.E3) should be used. The coefficient φ_1 depends on slenderness λ_1, and the column curve appropriate to the section of the individual sub-member under consideration.

JAPAN

CHAPTER EDITORS
Toshiyuki Ogawa, Tokyo Institute of Technology
Toshiro Suzuki, Tokyo Institute of Technology

AIJ 70

Laced Members and Battened Members

Slenderness ratios of built-up compression members

(A) Slenderness ratios about the solid-web axis of built-up members shall be calculated on the same basis as prescribed for compression member composed of single element.

(B) Calculations regarding buckling of built-up compression members in their open-web planes shall be computed by using an effective slenderness ratio as follows

$$\lambda_{ye} = \sqrt{\lambda_y^2 + \frac{m}{2}\lambda_1^2} \qquad (2a.J1)$$

If, in the above formula, $\lambda_1 \leq 20$, Eq. 2a.J2 may be used.

$$\lambda_{ye} = \lambda_y \qquad (2a.J2)$$

where

m = number of components or groups of components connected by lacing or tie plates

λ_y = the slenderness ratio of built-up members assumed to buckle as an integrated member

λ_{ye} = effective slenderness ratio

The value of λ_1 shall be calculated using one of the following formulas, depending on the type of built-up member under consideration.

Members with Tie Plate Separators

$$\lambda_1 = \frac{l_1}{i_1} \qquad (2a.J3)$$

where

i_1 = the smallest radius of gyration of components

l_1 = spacing of separator batten plates

Laced or Latticed Members

$$\lambda_1 = \pi\sqrt{\frac{A}{nA_d}\frac{l_d^3}{l_2 e^2}} \qquad (2a.J4)$$

where

A = sum of sectional areas of column components

A_d = sectional area of lacing lattice, or total sectional area of a pair of lacing elements in the case of double lacing

e = distance between axes of gravity of column components

l_d = length of lacing or lattice

l_2 = component length of lacing or lattice as projected on a plane parallel to the longitudinal axis of a built-up member

n = number of structural planes formed by lacing or batten plate

Limit values for the local slenderness are as follows:

(1) The slenderness ratio λ_1 of components of battened members shall not exceed 50.

(2) The slenderness ratio of each component of laced members shall not exceed the slenderness ratio of a built-up member about whichever of the two principle axes that results in the greater value.

(3) The slenderness ratio of lattices shall not exceed 160.

Other Members

Members with cover plates having holes (perforated plates)

$$\lambda_1 = 1.7\sqrt{\frac{l_1}{p}\frac{l_1}{i_1}} \tag{2a.J5}$$

where

i_1 = the smallest radius of gyration of components

l_1 = length of hole

p = spacing of holes

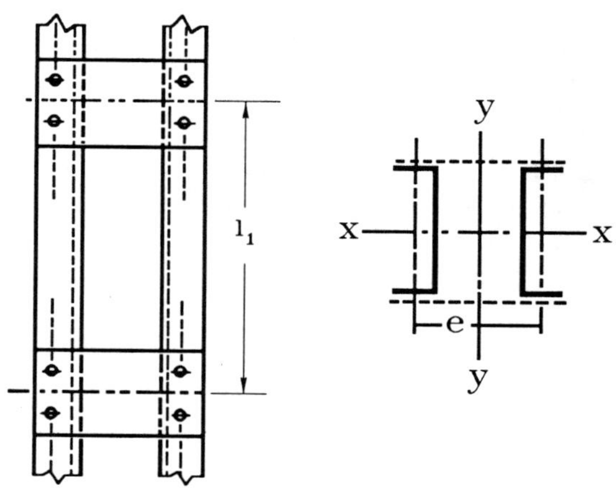

Fig. 2a.J1 Battened Members

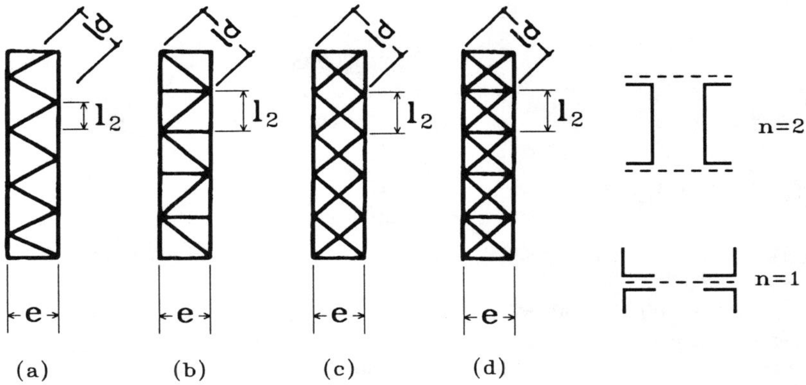

Fig. 2a.J2 Laced Members

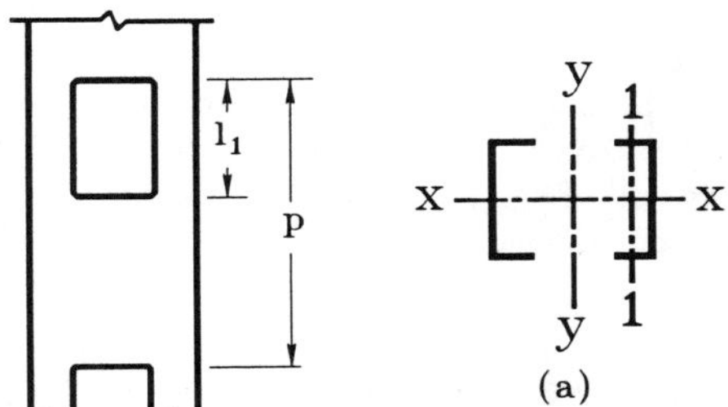

Fig. 2a.J3 Perforated Plates

NORTH AMERICA

CHAPTER EDITORS
Bruce G. Johnston, Consultant
Zia Razzaq, Old Dominion University

AASHTO 87

Permitted types of steel have yield points ranging from 248 to 690 MPa.

No reduction in allowable axial stress is specified to allow for reduced column strength caused by shear deformation in members with lacing or perforated cover plates. The allowable stress formulas for columns given in Chapter 1 are also used for built-up members.

The provisions for perforated cover plates, stay plates, and lacing bars are as follows.

Perforated Cover Plates
The ratio of length, in direction of stress, to width of perforation, shall not exceed two. The clear distance between perforations in the direction of stress shall not be less than the distance between points of support. The clear distance between the end perforation and the end of the cover plate shall not be less than 1.25 times the distance between points of support.

Stay Plates
Where the open sides of compression members are not connected by perforated plates, such members shall be provided with lacing bars and shall have stay plates as near each end as practicable. Stay plates shall be provided at intermediate points where the lacing is interrupted. In main members, the length of the end stay plates between end fasteners shall be not less than 1 1/4 times the distance between points of support and the length of intermediate stay plates not less than 3/4 of that distance. In lateral struts and other secondary members, the over-all length of end and intermediate stay plates shall be not less than 3/4 of the distance between points of support.

Lacing Bars
Lacing bars of compression members shall be so spaced that the slenderness ratio of the portion of the flange included between the lacing bar connections will not be more than 40 or more than 2/3 of the slenderness ratio of the member. The section of the lacing bars shall be determined by the formula for axial compression in which ℓ is taken as the distance along the bar between its connections to the main segments for single lacing and as 70 percent of that distance for double lacing. If the distance across the member between fastener lines in the flanges is more than 15 inches and a bar with a single fastener in the connection is used, the lacing shall be double and fastened at the intersections.

AISC ASD 89

Requirements for built-up compression members are similar (Table 2a.N1) to those for AASHTO and AREA bridge specifications and the same range of steel yield points are permitted (248 to 690 MPa).

The allowable stress formulas are given in Chapter 1. Other key provisions are as follows:

Compression members composed of two or more rolled shapes separated by intermittent fillers shall be connected at these fillers at intervals such that the slenderness ratio Kl/r of either shape, between the fasteners, does not exceed 3/4 times the governing slenderness ratio of the built-up member. The least radius of gyration r shall be used in computing the slenderness ratio of each component part. At least two intermediate connectors shall be used along the length of the built-up member.

Open sides of compression members built up from plates or shapes shall be provided with lacing having tie plates at each end and at intermediate points if the lacing is interrupted. Lacing, including flat bars, angles, channels or other shapes employed as lacing, shall be so spaced that the ratio l/r of the flange included between their connections shall not exceed 3/4 times the governing ratio for the member as a whole.

AISC LRFD 86

The design strength formulas are given in Chapter 1. Other provisions are as follows, and are summarized briefly in Table 2a.N1.

The design strength of built-up members composed of two or more shapes is determined in accordance with the AISC LRFD 86 formulas for columns, including the flexural torsional effects if present, and subject to the following modification. If the buckling mode involves a relative deformation that produces shear deformation forces in the connectors between individual shapes, the usual slenderness ratio, Kl/r, for compression members must be replaced by a modified slenderness ratio $(Kl/r)_m$ in determining the compressive strength of built-up members as follows:

for snug-tight bolted connections

$$\left(\frac{Kl}{r}\right)_m = \sqrt{\left(\frac{Kl}{r}\right)_o^2 + \left(\frac{a}{r_i}\right)^2} \qquad \text{(2a.N1)}$$

for welded connectors and for fully tightened bolted connectors as required for slip-critical joints

$$\text{with } \frac{a}{r_i} > 50 \qquad \left(\frac{Kl}{r}\right)_m = \sqrt{\left(\frac{Kl}{r}\right)_o^2 + \left(\frac{a}{r_i} - 50\right)^2} \qquad \text{(2a.N2)}$$

$$\text{with } \frac{a}{r_i} \leq 50 \qquad \left(\frac{Kl}{r}\right)_m = \left(\frac{Kl}{r}\right)_o \qquad \text{(2a.N3)}$$

where

a = distance between connectors
a/r_i = largest column slenderness of individual components
$(\frac{Kl}{r})_o$ = column slenderness of built-up member acting as a unit
$(\frac{Kl}{r})_m$ = modified column slenderness of built-up member
r_i = minimum radius of gyration of individual component

COMMENTARY

Eq. 2a.N1 is the same as that used in the Italian code as well as other European specifications. AISC LRFD 86 uses the modified Eq. 2a.N2 for $a/r_i > 50$.

AREA 87

For allowable compressive stresses, refer to Chapter 1.

The provisions for perforated plates, stay plates, and lacing bars are as follows:

Perforated Cover Plates
Perforations shall be ovaloid or elliptical. The length of perforation shall not be more than twice its width. For compression members the ratio of the length of perforation to the radius of gyration of the half member at the center of perforation about its own axis shall not be more than 20 nor more than one-third of the slenderness ratio of the member about its axis perpendicular to the perforation.

Stay Plates
On the open sides of compression members, the segments shall be connected by lacing bars and there shall be stay plates as near end as practicable. There shall be stay plates at intermediate points where the lacing is interrupted. In main members, the length of the end stay plates shall not be less than 1 1/4 times the distance between the lines of connections to the outer flanges.

Lacing
Lacing bars of compression members shall be so spaced that the slenderness ratio of the portion of the flange included between lacing-bar connections will not be more than 40 nor more than 2/3 of the slenderness ratio of the member. The main provisions are summarized in Table 2a.N1.

CSA S6 M89 84a

Members with lacing bars, perforated cover plates, or battens are permitted. Battened members are allowed only when the member does not carry calculated bending moments in the plane of the battens. Detailed provisions for proportioning members with lacing or perforated cover plates are similar to those of the AASHTO specification (see Table 2a.N1).

CSA S16.1 M89 84b

Requirements for built-up compression members are similar (see Table 2a.N1) to those of the AISC specification for buildings, and the CSA specifications for bridges. Batten plates are required to have a thickness of not less than 1/60 of the distance between lines of bolts or welds connecting them to main components of the member.

CSA S16.1 M89 84c

For compression members composed of two or more sections connected together at discrete points, such as double angles and batten channels, the factored compressive resistance for buckling about the built-up member axis is determined by using

$$F_p = 0.833\ F_e \tag{2a.N4}$$

where

a = fastener spacing

a/r_1 = slenderness ratio of an individual section between points of connection about an axis parallel to the built-up member axis

E = Young's modulus (203,000 MPa)

$$F_e = \frac{\pi^2 E}{(Kl/r)^2 + (a/r_1)^2} \tag{2a.N5}$$

F_p = buckling stress

Kl/r = overall slenderness ratio of the complete section about the built-up member axis

r_1 = radius of gyration of the fully effective cross sectional area of an individual section in a built-up member

REGIONAL COMPARISON

Two types of built-up columns are used for comparison, one using lacing and tie plates, and the other using perforated plates, shown in Fig. 2a.N1. Built-up columns using battens could not be compared since AISC, AASHTO, and AREA do not cover them. As Table 2a.N1 shows, many design requirements of the different specifications are the same, wherein both Canadian specifications are presented in their metric form. An area of difference between specifications is in the calculation of the design compressive stress. Each specification gives formulas by which to calculate this stress (see Chapter 1). The AISC 89 and AASHTO 87 calculate this stress in the same manner except that the latter uses a fixed factor of safety of 2.12, while the former uses a factor of safety as a function of the slenderness ratio resulting in an upper limit of 1.92.

Also, while AISC LRFD 86 and CSA 3 S16.1 use the same slenderness parameter, λ, the former has only two ranges of formulas as compared to five formulas for the latter. The AISC LRFD 86 uses load factors of 1.2 and 1.6 for dead and live loads, respectively, and a load factor of 1.4 when only the dead load is present. The CSA 3 S16.1 uses 1.25 and 1.50 as dead and live load factors, respectively.

The AREA and CSA 3 S6 have three ranges of formulas dependent upon the value of the slenderness ratio.

Another area of difference between the specifications is the slenderness ratio limits. Referring to Table 2a.N1, AREA has the most conservative limit (100), with AASHTO and CSA 3 S6 becoming less conservative, respectively. A consensus exists amongst the U.S. specifications as to the slenderness ratio limits for lacing. The Canadian specifications are most conservative for double lacing, with CSA 3 S6 being slightly less conservative.

Two chapters in the SSRC Guide (SSRC, 1988) deal with built-up columns:
Chapter 11. Columns with lacing, battens, or perforated cover plates
Chapter 12. Mill-building columns

Chapter 11 provides formulas, with varying degrees of empiricism, for evaluating the required reduction in allowable column stress to account for shear deformation in built-up laced or battened columns. Guidelines for the design of columns with perforated plates, and lattice towers and masts are also given. The three types of built-up columns treated in this chapter are illustrated in Fig. 2a.N2. Chapter 11 also considers the limiting case of the battened column, where the battens act only as spacers and are ineffective in resisting column shear. Without end batten plates, the critical load for such a member is no greater than the sum of the pinned-end strength of the individual component elements. But with adequate end tie plates, the longitudinal components are forced to buckle in a modified second mode with theoretical critical loads that may approach four times that of the pinned-end mode.

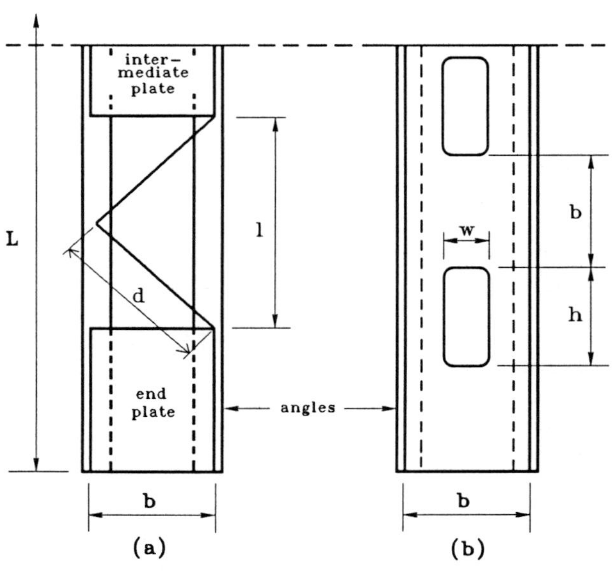

Fig. 2a.N1 Key Dimensional Symbols in Table 2a.N1

Mill-building columns are treated in Chapter 12 of the Guide. Some common types of mill-building columns are illustrated in Fig. 2a.N3. The column segment supporting the crane runway girders in a stepped mill-building column may be attached to the lower part of the building column by lacing or by ties which may be the equivalent of battens, or may simply act as spacers that force the two segments to deflect in the same shape. The importance of adequate diaphragms or tie plates at the top and bottom of such columns has been demonstrated by the spaced column studies. Chapter 12 also provides a detailed procedure for the overall design of a stepped mill-building column by means of an adaptation of the

Table 2a.N1 Detailed Provisions of North American Specifications for Built-Up Steel Members

Subject	AISC 89	AISC LRFD 86	AREA 87	AASHTO 87	CSA S16.1 M89	CSA S6 M89
Column						
KL/R		modified				
Main	200	200	100	120	200	120
Secondary	200	200	120	140	200	160
Flange ≤ KL/R	KL/R	KL/R	40 or	2/3 KL/R	KL/R	60 or 3/4 KL/R
Lacing						
kl/r	where ℓ=d, see Fig. 2a.N1					
Single	140	140	140	140	140	160
Double	200	200	200	200	140	160
K-value						
Single	1	1	1	1	1	1
Double	0.7	0.7	0.7	0.7	0.7	0.7
Angle						
Single ≥	60	60	60	60	45	45
Double ≥						
ℓ/d >						
Single Bars			1/40	1/40		1/40
Double Bars			1/60	1/60		1/60
b/t < $s/\sqrt{F_y}$						
Single Angle	76	76			200	
Double Angle	95	95			200	
Tie Plates						
End-length	b	b	1.25 b	1.25 b	b	1.25 b
Intermediate	1/2 b	1/2 b	3/4 b	3/4 b	1/2 b	3/4 b
t/b >	1/50	1/50	1/50	1/50	1/60	1/50
Shear load	2%	2%	1.6.4.1	16.36	2.5%	2.5%
Perfor. Plate						
h/w hole ≤	2	2	2	2	2	2
Clear dist ≥	b	b	b	b	b	b
Min. radius	1.5′	1.5′	1.5′	1.5′	40 mm	40 mm
t/b >			1/50			
b/t < $s/\sqrt{F_y}$	317	317		288	840	840
Allowable Comp. Stress		SEE CHAPTER 1				
Battens	N/A	N/A	N/A	N/A	YES	YES

AISC specification for buildings, and the use of the AISC guide for the design and construction of mill-buildings.

Major differences in comparison of AISC ASD 89 with bridge design (AASHTO 87 and AREA) include (1) a lesser requirement for added shear force for design of lacing at 2% of the actual axial load and (2) a less conservative limit for L/r of main member components between lacing bar connections, which may be as great as the governing slenderness ratio of the member instead of 2/3 of that amount. Members with battens are not permitted.

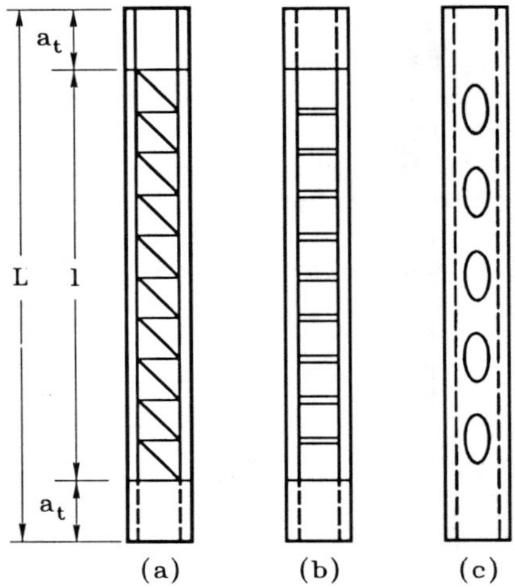

Fig. 2a.N2 Built-Up Columns

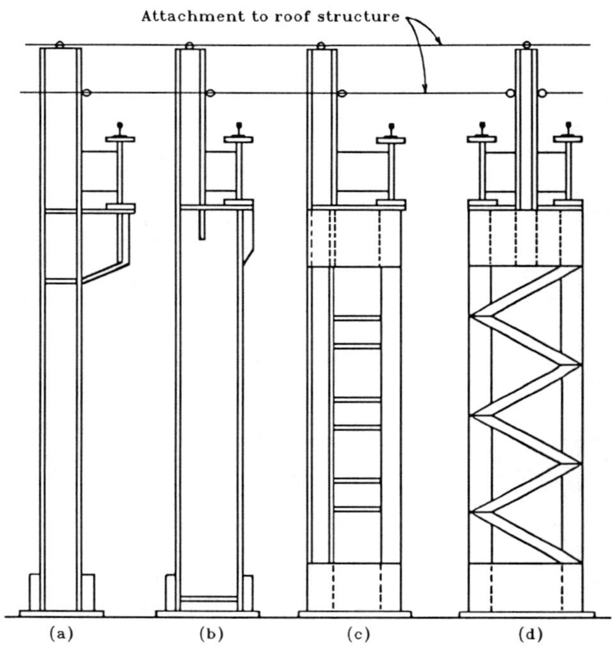

Fig. 2a.N3 Types of Mill-Building Columns

WEST EUROPE

CHAPTER EDITOR
Leo Finzi, Politecnico di Milano

EC 3 88

Built-up compression members consisting of two or more main components connected together at intervals to form a single integral member are designed incorporating an equivalent geometric imperfection comprising an initial bow (w_o) of not less than $\ell/500$, where ℓ is the length of the member.

The deformation of the built-up member is taken into account in determining the internal forces and moments in the main components, internal connections, and any subsidiary components such as lacings or battens.

The design of the main and subsidiary components shall be checked using the methods given for struts.

In addition to the axial forces, allowance must also be made for any other forces or moments applied to the member, such as the effects of its self-weight or wind loading acting on the member.

Laced Members

The design procedure is for a compressive force N applied to a built-up member consisting of two similar parallel chords of uniform cross section, with a fully triangular system of lacing which is uniform throughout the length of the member.

The chords may be solid members or may themselves be laced or battened in the perpendicular plane.

Tie panels must be provided at the ends of lacing system at points where the lacing is interrupted and at connections with other members.

The lacing shall be positively connected to the main components, either by fasteners or by welding.

(A) Moment of inertia

The effective moment of inertia I of a laced compression member with two main components must be taken as

$$I = 0.5\, h_o^2\, A_f \qquad (2a.W1)$$

where

A_f = the cross-sectional area of one chord
h_o = the distance between centroids of chords

(B) Chord forces at mid-length
The chord force N_f at mid-length is to be determined from

$$N_f = 0.5\ N + M_s/h_o \tag{2a.W2}$$

where

E	=	modulus of elasticity
ℓ	=	buckling length
M_s	=	$N\,w_o/(1 - N/N_{cr} - N/S_v)$
N_{cr}	=	$\pi^2\ EI/\ell^2$
S_v	=	the shear stiffness of the lacing (the shear force required to produce unit shear deformation)
w_o	=	$\ell/500$

Values of S_v for various lacing systems are given in Fig. 2a.W1.

(C) Buckling resistance of chords
The buckling length of a chord in the plane of a lacing system should be taken as the system length a between lacing connections.

In a member with four chords made of angles with lacing in both directions, the buckling length for buckling about the weak axis depends on the arrangement of the lacing, see Fig. 2a.W2.

System	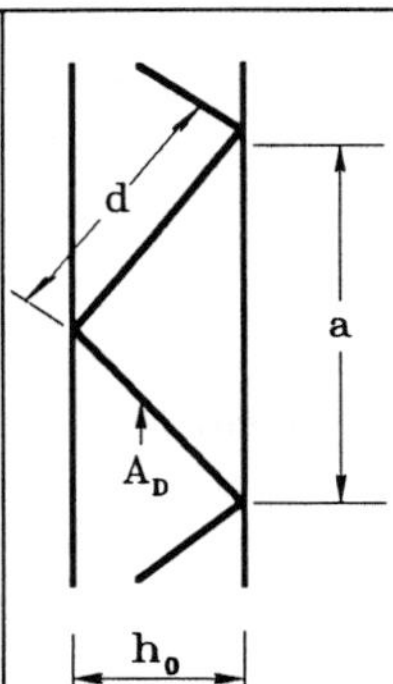	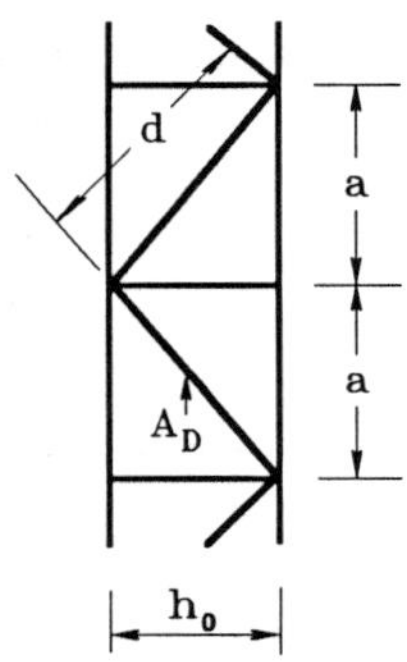	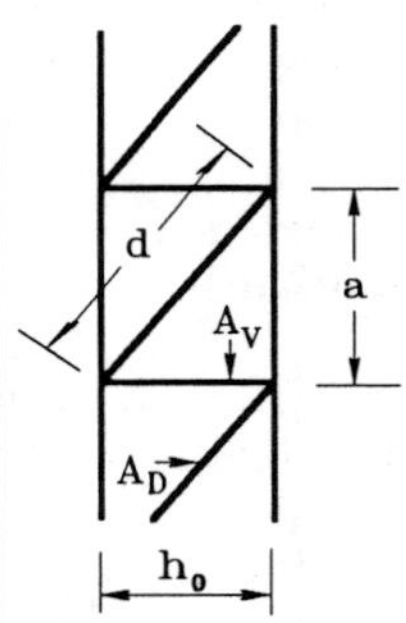
S_V	$\dfrac{m\,E\,A_D\,a\,h_o^2}{2\,d^3}$	$\dfrac{m\,E\,A_D\,a\,h_o^2}{d^3}$	$\dfrac{m\,E\,A_D\,a\,h_o^2}{d^3\left(1+\dfrac{A_o\,h_o^3}{A_V\,d^3}\right)}$

m is the number of planes of lacings
A_D and A_V refer to a single plane

Fig. 2a.W1 Laced Compression Members

Battened Members

The design procedure is for a compressive force N applied to a built-up member consisting of two similar parallel chords of uniform cross-section, spaced apart and inter-connected by means of battens, which are rigidly connected to the chords and uniformly spaced throughout the length of the member.

The chords may be solid members or may themselves be laced or battened in the perpendicular plane.

Battens shall be provided at each end of the member. Battens shall also be provided at intermediate points where load or lateral restraint is applied.

Intermediate battens should divide the length of the member into at least 3 panels and there should be at least 3 panels between points which are laterally restrained in the plane of the battens. In general, intermediate battens should be spaced and proportioned uniformly throughout the length of the member.

Where parallel planes of battens are supplied, the battens in each plane should be arranged opposite one another.

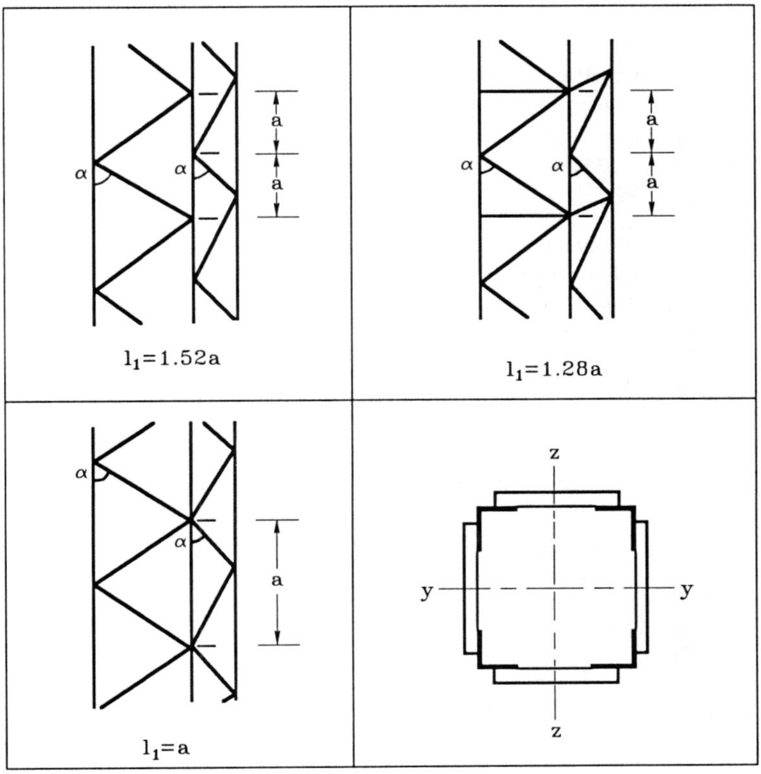

Fig. 2a.W2 Buckling Length of Angle Chords in Laced Members

When S_v is evaluated by disregarding the flexibility of the batten plates, the width of an end batten along the length of the member should not be less than h_o, the distance between the centroids of the chords. The width of an intermediate batten should not be less than 0.5 h_o.

Unless the flexibility of the batten plates is explicitly taken into account in the evaluation of S_v, the battens should also satisfy

$$\frac{m I_b}{h_o} \geq \frac{10 I_f}{a} \tag{2a.W3}$$

where

- a = system length between center lines of battens
- h_o = distance between centroids of chords
- I_b = moment of inertia of one batten
- I_f = moment of inertia of one chord
- m = number of planes of battens

(A) Moment of inertia
The effective moment of inertia I of a battened compression member with two main components should be taken as

$$I = 0.5 h_o^2 A_f + 2 \mu I_f \tag{2a.W4}$$

with μ obtained from the following

$$\lambda \leq 75 : \quad \mu = 1$$
$$75 < \lambda < 150 : \quad \mu = 2 - \lambda/75$$
$$\lambda \geq 150 : \quad \mu = 0$$

where

- A_f = cross-sectional area of one chord
- h_o = distance between centroids of chords
- I_b = moment of inertia of the batten plate
- I_f = moment of inertia of a chord
- I_1 = value of I with $\mu = 1$
- $i_o = \sqrt{0.5 I_1 / A_f}$ (2a.W5)
- $\lambda = \ell / i_o$

(B) Chord forces at mid-length should be determined from

$$N_f = 0.5(N + M_s h_o A_f / I) \tag{2a.W6}$$

where

- $M_s = N w_o / (1 - N/N_{cr} - N/S_v)$
- $N_{cr} = \pi^2 EI/\ell^2$
- $w_o = \ell/500$

The design shear stiffness is given by

$$S_v = 2\pi^2 E I_f / a^2 \tag{2a.W7}$$

provided that

$$\frac{m I_b}{h_o} \geq \frac{10 I_f}{a}$$

When this criterion is not satisfied, the flexibility of the batten plates should be taken into account by obtaining S_v from

$$S_v = \frac{24\,E I_f}{a^2\left[1 + \dfrac{2 I_f}{m I_b} \cdot \dfrac{h_o}{a}\right]} \quad \text{but} \geq \frac{2\pi^2 E I_f}{a^2} \qquad (2a.W8)$$

(C) Buckling resistance of chords

The buckling length of a chord in the plane of battens should be taken as the system length a between center lines of battens.

Other Members

Star-battened angle members

Built-up compression members consisting of two similar angle members connected by pairs of battens in two perpendicular planes as shown in Fig. 2a.W3 may be checked for buckling about the x-x axis as a single integral member, provided that the buckling lengths in the two perpendicular planes x-x and y-y are equal and provided that the spacing of pairs of battens is not more than 70 i_{min}.

In the case of unequal-leg angles it may be assumed that

$$i_x = i_o/1.15 \qquad (2a.W9)$$

where

i_{min} = the minimum radius of gyration of one angle
i_o = the minimum radius of gyration of the built-up member

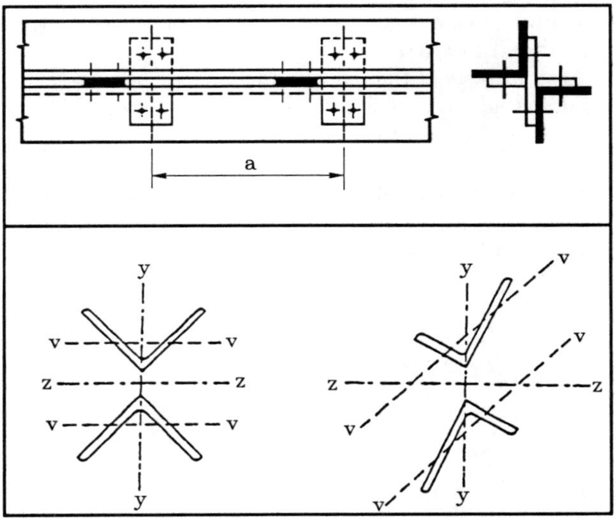

Fig. 2a.W3 Star-Battened Angle Members

INTERNATIONAL COMPARISONS & COMMENT

WHAT ARE THE DIFFERENCES?

Column Strength of Laced Members (see Fig. 2a.1)

Laced Members can be designed by using usual buckling curves with the λ_d defined in the following table:

A	$\lambda_d = (\lambda^2+\lambda_f^2)^{1/2}$	$\lambda_f = a/i_f$
C	$\lambda_d = (\lambda^2+nA/A_d)^{1/2}$	(n = 27 for 2 chords, n = 40 for 4 chords)
E	$\lambda_d = (\lambda^2+\frac{m}{2}\lambda_f^2)^{1/2}$	with $\lambda_f = \pi(A\,d^3/n\,A_d\,a\,h_o^2)^{1/2}$
J	$\lambda_d = (\lambda^2+\frac{m}{2}\lambda_f^2)^{1/2}$	with $\lambda_f = \pi(A\,d^3/n\,A_d\,a\,h_o^2)^{1/2}$
	$\lambda_d = \lambda$	when $\lambda_f \leq 20$
N	$\lambda_d = (\lambda^2+(K_m\lambda_f)^2)^{1/2}$	$K_m = 1$ for snug-tight bolted connections $K_m = 0.65$ for welded or slip-critical connections
W	The idea of a design slenderness λ_d is not used. The buckling length of the chord, is taken as equal to a with $N_f = 0.5N+M/h$ where $M = Nw_{max}$ and $w_{max} = w_o/(1-N/N_{cr})$	

A = total cross section area of the member
A_d = cross section area of diagonal bracings
A_f = cross section area of one chord
m = number of chords connected by lacing or batten plates
N_{cr} = $1/(\ell^2/\pi^2EI+1/S^*)$
S^* = shear stiffness of the built-up member
λ = total slenderness of the element
λ_f = local slenderness of the element

Column Strength of Battened Members (see Fig. 2a.1)

Battened members can be designed with λ_d defined in the following table.

A	$\lambda_d = (\lambda^2+\lambda_f^2)^{1/2}$	$\lambda_f = a/i_f$
C	$\lambda_d = (\lambda^2+\lambda_f^2)^{1/2}$	$\lambda_f = a/i_f$
E	$\lambda_d = (\lambda^2+\frac{m}{2}\lambda_f^2)^{1/2}$	$\lambda_f = a/i_f(0.82(1+I_f h_o/I_{bp}\,a))^{1/2}$ if $I_f h_o/I_{bp}\,a > 0.2$
J	$\lambda_d = (\lambda^2+\frac{m}{2}\lambda_f^2)^{1/2}$	$\lambda_f = a/i_f$
	$\lambda_d = \lambda$	when $\lambda_f \leq 20$
N	$\lambda_d = (\lambda^2+(k_m\lambda_f)^2)^{1/2}$	
W	Same as for laced members	

Limit Values for Local Slenderness

A	$\lambda_f \leq 0.6\ \lambda_d$ or $\lambda_f \leq 50$
C	$\lambda_f \leq 0.7\ \lambda_d$ for laced members
	$\lambda_f \leq 0.5\ \lambda_d$ or $\lambda_f < 40$ for battened members
E	$\lambda_f \leq \lambda_d$
	$\lambda_f < 80$ for laced members
	$\lambda_f < 40$ (USSR)(50 other E. Europe countries) or $\lambda_f \leq 0.5\ \lambda_x$ for battened members
J	$\lambda_f \leq 50$ for battened members
	$\lambda_i \leq \max(\lambda_d, \lambda_x)$ for laced members
	$\lambda_l \leq 160$
N	$\lambda_x \leq 0.75\ \lambda_d$
W	Not applicable

λ_i = the slenderness ratio of each component (a/i_f)
λ_l = the slenderness ratio of lattices (d/i_d)
λ_x = the slenderness ratio of a built-up member about the solid web axis

Perforated Plates

	$\lambda_d = (8\lambda^2 + \frac{m}{2}\lambda_f^2)^{1/2}$ $\quad$ $\lambda_d = \lambda$ when $\lambda_f \leq 20$
J	Japanese codes take care of members with cover plates having holes.
	In this case $\quad$ $\lambda_f = 1.7(l_l/a)^{1/2}\ l_l/i_f$

Closely Spaced Built-Up Members

A	Interconnecting fasteners designed to transmit a longitudinal shear force per connector of $0.25\ V_d \lambda_f$
C	Designed as solid compression members
E	Designed as solid compression members
J	Designed as solid compression members
N	Interconnecting fasteners designed to transmit a longitudinal shear force of $0.05\ N_f$ per connection
W	Treated as a single integral member if connected at more than $15\ i_l$ or, if not, designed as normally spaced built-up members

Star-Battened Angle Members

A	Designed as battened built-up members
C	Designed as solid compression members
E	Designed as solid compression members
J	Designed as solid compression members
N	Designed as solid compression members
W	Designed for λ_x if the spacing of connectors is not greater than $70\ i_f$.

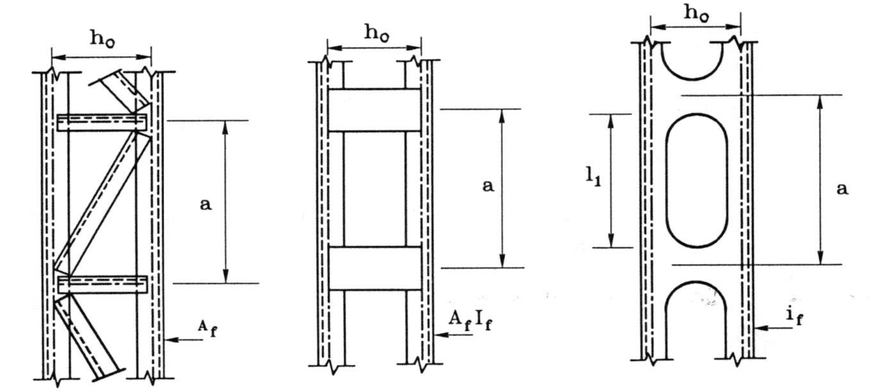

Laced Member Battened Member Perforated Plates

	Laced members			Battened members	
	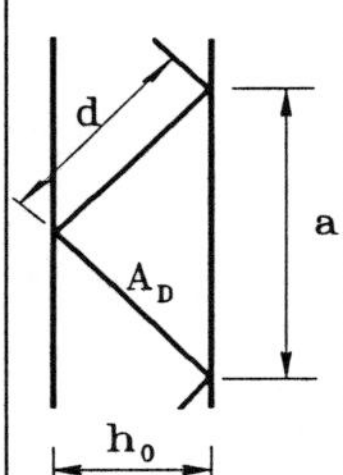	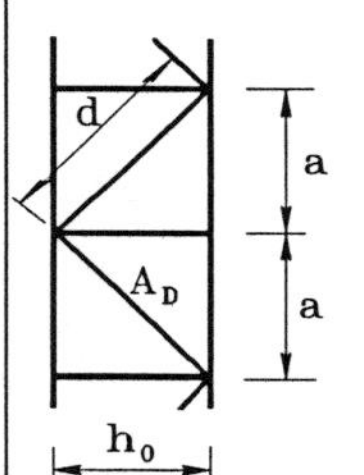	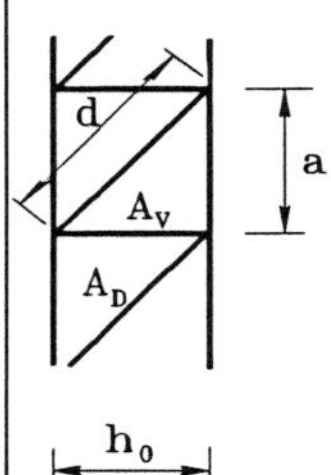	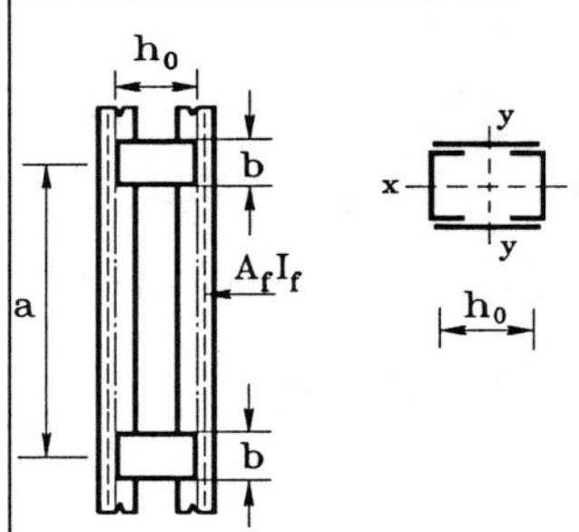	
S*	$\frac{mEA_D a h_o^2}{2d^3}$	$\frac{mEA_D a h_o^2}{d^3}$	$\frac{mEA_D a h_o^2}{d^3\left(1+\frac{A_D h_o^3}{A_V d^3}\right)}$	$\frac{2\pi^2 EI_f}{a^2}$	Here, the contribution to the deformations due to any bending of shear stiffness of the batten plates is neglected
	m is the number of active lacings				

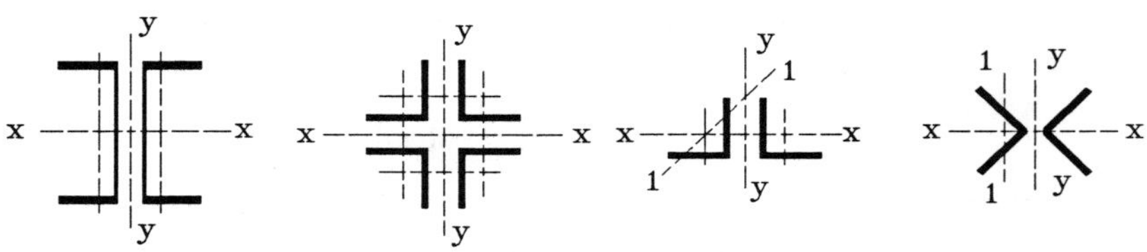

Fig. 2a.1 Built-Up Members

WHY ARE THERE DIFFERENCES?

Different assessment schemes are applied in different specifications all over the world. Namely, the bifurcation theory is also now the basis for some codes whereas the second order theory considering initial geometrical and mechanical imperfections has been or is being adopted all over Europe. This holds independently from the deterministic or probabilistic approach considered.

Further differences arise from taking advantage or ignoring the scatter between the first yielding and the collapse of the structure.

Another source of differences is the evaluation of initial imperfections and their application in the design formulas.

Only one West European code (namely EC 3 88) is mentioned because, from 1992 on, codes of the twelve member countries will be, in practice, replaced by EC 3 88.

2b. Shear Strength

SPECIFICATION PROVISIONS & EXPLANATION

AUSTRALIA

CHAPTER EDITOR
Russell Q. Bridge, The University of Sydney

AS 1250 81

The main components, bracing elements, and their connections are proportioned to resist a design transverse shear force V_d applied at any point along the length of the member in the most unfavorable direction, where

$$V_d = \pi(N_s/N_u - 1)N_d/\lambda_{ef} \tag{2b.A1}$$

but not less than $0.02N_d$.

where

N_d = design axial force
N_s = section capacity of the column
N_u = member capacity of the column accounting for slenderness effects as obtained from the design column curve

The batten element and its connections, are designed to transmit simultaneously a longitudinal shear force V_1 given by

$$V_1 = \frac{V_d\ a}{n_{bp}\ h_{con}} \tag{2b.A2}$$

and a moment M_d to the main component given by

$$M_d = \frac{V_d\ a}{2\ n_{bp}} \tag{2b.A3}$$

where

a = longitudinal center-to-center distance between the battens
n_{bp} = number of parallel planes of battens
h_{con} = lateral distance between the centroids of the welds or fastenings.

For columns composed of two angle, channel, or tee-section members discontinuously separated back to back by a distance not exceeding that required for end gusset connection, the interconnecting fasteners are designed to transmit a longitudinal shear per connection of $0.25V_d\lambda_f$.

CHINA

CHAPTER EDITOR
Ji Chen, Xian Institute of Metallurgy & Construction Engineering

GBJ 17 88 and TBJ 2 86

The shear force to be taken into account is either that due to an external force or a shear force calculated by the expression

$$V = \frac{Af}{85}\sqrt{\frac{f_y}{235}} \qquad (2b.C1)$$

where

f = design stress of the material
f_y = yield stress of the material (MPa)

COMMENTARY

The force V is formulated by taking into account the second order effect of an initially crooked member.

Shear force of built-up compression members with tie plates is calculated by the formula

$$V = \alpha A[\sigma]\frac{\varphi_{min}}{\varphi} \qquad (2b.C2)$$

where

A = sectional areas of components
α = a coefficient
= 0.015 for steel grade 16 q
= 0.017 for steel grade 16 M_nq.
(see General Provisions GPb(C1))
$[\sigma]$ = allowable stress
φ = strength reduction factor of the centrally compressed member about the open web axis
φ_{min} = the least value of strength reduction factors of the same members

EAST EUROPE

CHAPTER EDITORS
Georgij E. Belskij, CNIISK, Moscow
Jindrich J. Melcher, Technical University of Brno

SNiP II-23-81 82

A design shear force must be considered for checking the fasteners defined by the formula

$$V = 7.15 \times 10^{-6} \frac{N}{\varphi}(2330 - \frac{E}{R_y}) \qquad (2b.E1)$$

where

N = design axial load of the built-up member
R_y = specified yield stress
φ = buckling coefficient derived from the effective slenderness ratio λ_{ef}.

TGL 13503/01 82

A shear force for the design of fasteners is given by

$$V = \frac{N}{100} \frac{1}{1 - \nu\sigma_c/\sigma_{ef}} \qquad (2b.E2)$$

or approximately

$$V = A\ \sigma_{all}/65 \qquad (2b.E3)$$

where

$$\sigma_c = N/A \qquad (2b.E4)$$
$$\sigma_{ef} = \pi^2 E/\lambda_{ef}^2 \qquad (2b.E5)$$

for the allowable stress design concept

$$\sigma_{all} = R_y/\nu \qquad (2b.E6)$$

with the safety factor $\nu = 1.5$ (principal load), $\nu = 1.33$ (principal and additional loads) and $\nu = 1.2$ (exceptional loads).

ČSN 73 1401 84

A conventional shear force for the design of fasteners is given by

$$V = \frac{\pi A R_d}{\lambda_y}(1 - \varphi) \qquad (2b.E7)$$

where

R_d = design stress (Eq. 1a.E12)
φ = derived from the effective slenderness ratio λ_{ef}.

PN 80/B-03200 87

A conventional shear force is given by

$$V = 0.012N/\varphi \tag{2b.E8}$$

where φ is the buckling coefficient.

REGIONAL COMPARISON

Common unified rules for assessment of stability of compression built-up members are not yet available in East European specifications; however, the individual approaches to the solution, currently applied, are very similar. The second order theory (considering initial geometrical and physical imperfections) in the basis for all mentioned codes.

JAPAN

CHAPTER EDITORS
Toshiyuki Ogawa, Tokyo Institute of Technology
Toshiro Suzuki, Tokyo Institute of Technology

AIJ 70

The transverse shear force to be taken into account is $V_d = 0.02N$.

COMMENTARY

By means of experiments of steel truss members and frames, the deformation capacity and lateral-torsional stability of steel truss frames are currently being investigated by Suzuki (1983, 1986) and Matsui (1986).

NORTH AMERICA

CHAPTER EDITORS
Bruce G. Johnston, Consultant
Zia Razzaq, Old Dominion University

AASHTO 87

Lacing members are to be designed for the compression induced by the shearing force normal to the axis of the member.

In addition to the shear due to weight of member or external forces, a shear force is to be added to account for accidental load eccentricity and deflection as given by

$$V = \frac{F}{100}\left(\frac{100}{L/r+10} + \frac{(L/r)\sigma_y}{22{,}754}\right) \tag{2b.N1}$$

where

- F = allowable compressive axial load, kN
- L = length of member
- r = radius of gyration about the axis perpendicular to the planes of lacing
- V = normal shear force (kN)
- σ_y = specified yield strength (MPa)

The bridge specifications provide explicit and detailed provisions governing the design and proportions of perforated cover plates and lacing bars in compression members. Table 2a.N1 gives a summary.

The shear force to be added to that due to external forces and weight of member in the design of lacing is taken as 2.5% of the axial force in the member at working load.

Battens are required to resist simultaneously a longitudinal shear force q given by

$$q = Vd/na \tag{2b.N2}$$

and a moment M given by

$$M = Vd/2n \tag{2b.N3}$$

where

- a = minimum distance between centroids of groups of fasteners or welds
- d = longitudinal distance center-to-center of battens
- n = number of parallel planes of battens
- V = 2.5% of the total axial force in the member at working load

Batten plates are required to have a thickness of not less than 1/50 of the minimum distance between the innermost lines of connecting bolts or weld. Batten plates with stiffened edges, or rolled shapes with flanges perpendicular to the main components are not subject to this requirement.

AISC ASD 89

See Table 2a.N1. Lacing shall be proportioned to resist a shearing stress normal to the axis of the member equal to 2% of the total compressive stress in the member. The ratio l/r for lacing bars arranged in single systems shall not exceed 140. For double lacing this ratio shall not exceed 200.

AISC LRFD 86

See Table 2a.N1. Open sides of compression members built up from plates or shapes shall be provided with continuous cover plates perforated with a succession of access holes. The unsupported width of such plates at access holes is assumed to contribute to the design strength provided that:

a. The width-thickness ratio limitations are satisfied.
b. The ratio of length (in direction of stress) to width of hole shall not exceed two.
c. The clear distance between holes in the direction of stress shall be not less than the transverse distance between nearest lines of connecting fasteners or welds.
d. The periphery of the holes at all points shall have a minimum radius of 1 1/2 in.

The function of perforated cover plates may be performed by lacing with tie plates at each end and at intermediate points if the lacing is interrupted. Tie plates shall be as near the ends as practicable. In main members providing design strength, the end tie plates shall have a length of not less than the distance between the lines of fasteners or welds connecting them to the components of the member. Intermediate tie plates shall have a length not less than 1/2 of this distance. The thickness of tie plates shall be not less than 1/50 of the distance between lines of welds or fasteners connecting them to the segments of the members. In welded construction, the welding on each line connecting a tie plate shall aggregate not less than 1/3 the length of the plate. In bolted and riveted construction, the spacing in the direction of stress in tie plates shall be not more than six diameters and the tie plates shall be connected to each segment by at least three fasteners.

Lacing, including flat bars, angles, channels or other shapes employed as lacing, shall be so spaced that the L/r ratio of the flange included between their connections shall not exceed the governing slenderness ratio for the member as a whole. Lacing shall be proportioned to provide a shearing strength normal to the axis of the member equal to 2% of the compressive design strength of the member. The L/r ratio for lacing bars arranged in single systems shall not exceed 140. For double lacing this ratio shall not exceed 200. Double lacing bars shall be joined at their intersections. For lacing bars in compression, L may be taken as the unsupported length of the lacing bar between welds or fasteners connecting it to the components of the built-up member for single lacing, and 70% of that distance for double lacing. The inclination of lacing bars to the axis of the member shall preferably be not less than 60° for single lacing and 45° for double lacing. When the distance between the lines of welds or fasteners in the flanges is more than 15 in., the lacing shall preferably be double or be made of angles.

AREA 87

The shear force provisions are as follows:

The shear force normal to the member in the planes of lacing or continuous plates with or without perforations shall be assumed divided equally between all such parallel planes. The total shear force shall include any force due to the weight of the member and to other forces and, for compression members, also that obtained by the following formula

$$V = \frac{P}{100}\left(\frac{100}{l/r + 10} + \frac{(l/r)\,\sigma_y}{24{,}822}\right) \tag{2b.N4}$$

where

l = length of member, in.
P = allowable compressive axial force in member
r = radius of gyration of section about the axis perpendicular to the plane of lacing or plates, in.
V = shear force
σ_y = specified yield stress (MPa)

CSA S6 M89 84a

Same as AASHTO 87 - see Table 2a.N1.

CSA S136 M89 84c

Same as AASHTO 87 - see Table 2a.N1.

Each discrete connection is expected to be capable of transmitting a longitudinal shear force of 5% of the force in one section of a built-up member.

COMMENTARY

The various specifications, account for shear load by including member weight, external forces and added shear due to the greatest axial load. Both AISC's and CSA's specifications determine the added shear load to be 2% and 2.5%, respectively, of the axial load. AREA and AASHTO use similar formulas to calculate the shear load based on the slenderness ratio. AASHTO further refines the formula by accounting for different yield stresses. A comparison of the various other design aspects can be observed by studying Table 2a.N1.

WEST EUROPE

CHAPTER EDITOR
Leo Finzi, Politecnico di Milano

EC 3 88

Laced members
The lacing forces adjacent to the ends of the member should be derived from the internal shear force V_s taken as

$$V_s = \pi M_s/\ell \qquad (2b.W1)$$

where

ℓ = the buckling length
M_s = see Eq. 2a.W1

The force N_D in a diagonal lacing is given by

$$N_D = \frac{V_s d}{m h_o} \qquad (2b.W2)$$

with d, m, and h_o from Fig. 2a.W1

Battened members
The battens and their connections to the chords themselves should be checked for the moments and forces in the end panel indicated in Fig. 2b.W1 in which the internal shear force V_s is taken as

$$V_s = \pi M_s/\ell \qquad (2b.W3)$$

where

ℓ = the buckling length
M_s = see Eq. 2a.W1

For the purpose of this check, the axial force in each chord may be taken as 0.5N even when there are only three panels in the length of the member.

Other members
Built-up compression members such as those shown in Fig. 2b.W2, with main components in contact or closely spaced and connected through packing plates, need not be treated as battened members provided that they are connected together by bolts or welds at a spacing of not more than $15i_{min}$, where i_{min} is the minimum radius of gyration of a main component.

The interconnecting bolts or welds should be designed to transmit the longitudinal shear between the main components derived from the internal shear force V_s which should be taken as 2.5% of the axial force in the member.

Alternatively, V_s, may be determined as for battened members. The longitudinal shear per interconnection may be taken as $0.25 V_s a/i_{min}$, where a is the system length of the main component center-to-center of interconnections.

REGIONAL COMPARISON

Presently most of the national codes ask for an elastic evaluation of the buckling load, by reference to an effective slenderness ratio

$$\lambda_{eff} = \sqrt{\lambda^2 + \lambda_1^2} \tag{2b.W4}$$

Some codes ask that the slenderness ratio of the individual sub-members do not go beyond a fixed value. A conventional shear force V_d of about 1% of the design axial force N_d in the strut must be considered for checking the fasteners.

A new code is nevertheless being edited in Germany (DIN 18800 P.2 86) and is fully in compliance with EC 3 88.

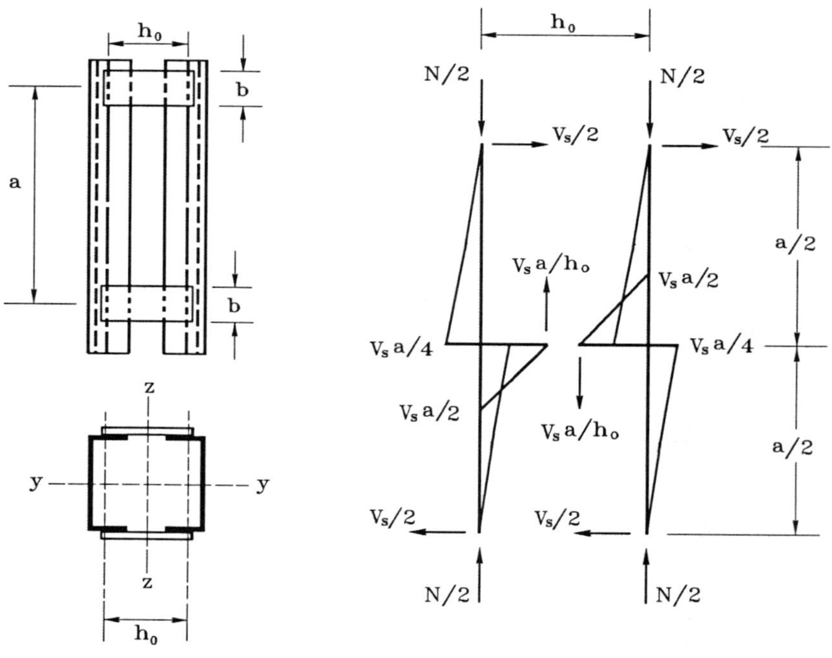

Fig. 2b.W1 Battened compression member

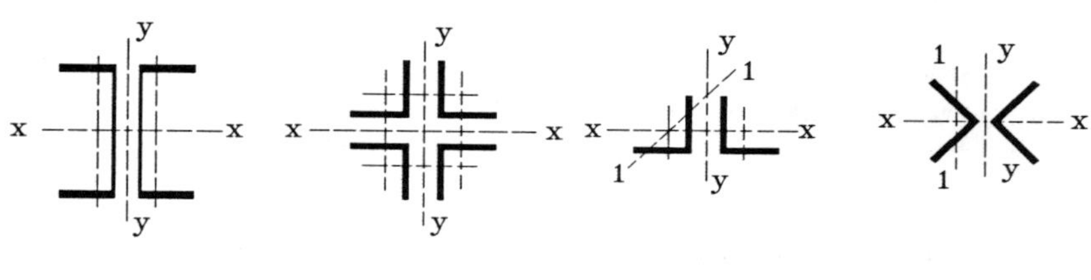

Fig. 2b.W2 Closely spaced built-up members

INTERNATIONAL COMPARISONS & COMMENT

WHAT ARE THE DIFFERENCES?

Design Shear V_d to be Considered[1]

A	$V_d = \pi(N_u/N_{ki} - 1)\, N_d/\lambda_d$ but not less then 0.02 N
C	$V_d = A f_d/85\ (f_y/235)^{0.5}$ (f_d = design stress) (f_y in MPa) or $V_d = 0.015\, A f_d\, \varphi_{min}/\varphi_d$ for Steel 16q (0.017 for 16 M_{ng}) (φ_d is the strength reduction factor of the member for slenderness λ_d and φ_{min} the least value of the strength reduction factors of the same member)
E	$V_d = 7.15 \times 10^{-6} N/\varphi_d\ (2330 - E/f_y)$ or $V_d = N/100 \dfrac{1}{1 - \gamma \frac{N}{A} / \frac{\pi^2 E}{\lambda_d^2}}$ or $V_d = A f_d/65$ or $V_d = 0.012\, N/\varphi_d$
J	$V_d = 0.02N$
N	$V_d = N_d/100\, [100/(\lambda+10) + \lambda\, f_y/22754\ (MPa)]$ (f_y in MPa) $V_d = 0.02$ N (0.025 N for Canada)
W	$V_d = \pi\, M/l$

Limit Values for Batten Sizes[1]

A	. . . .
C	$I_b/h_o \geq 6\, I_f/a$; $b = 0.75\, h_o$ ($1.25\, h_o$ at end)
E	. . . .
J	. . . .
N	thickness not less than 1/50 of the minimum distance between the innermost lines of connecting bolts or welds (1/60 for Canada)
W	$I_b/h_o \geq 10\, I_f/ma$

[1] Note: Unified symbols are applied in the table.

WHY ARE THERE DIFFERENCES?

As one compares the rules given in the different codes one sees that rules tend to be related in many countries to the Eulerian bifurcation concept with minor adaptations. In Europe more attention has been given recently to a more realistic elastic plastic collapse mechanism of compressed members, emphasizing the role played by both geometrical and mechanical imperfections.

The limit values given for local slenderness and batten sizes appear to be more related to the wish of simplifying the design formulas (for instance disregarding the batten plate deformations) than to a real need.

It should be possible for someone, starting from existing experimental results, to put together an international updated set of simple rules which are not based on useless a priori limitations.

Some codes require stiff connection elements to obtain simplified models and simplified corresponding formulas. Some codes require the exclusive use of welded or friction bolted connections. Fabrication and erection requirements need more liberal tolerances for connecting methods.

In any case, it seems that the differences are also related to the major or minor need of simplification and unification of structural detailing standards.

Symbol List for Chapter 2 -- Built-Up Members

A = cross-sectional area of the built-up member (2a.E)
A_d = sectional area of lacing lattice (2a.E) & (2a.J)
A_f = cross-sectional area of one chord (2a.W)
A_1 = sum of the gross sectional areas of all the lacings in a section of the strut (2a.C)
A_{1x} = sectional area of lacing elements perpendicular to x-axis (2a.C)
A_{1y} = sectional area of lacing elements perpendicular to y-axis (2a.C)
a = longitudinal center-to-center distance between the battens or fastener spacing (2a.N) & (2b.A)
b = distance as defined in Fig. 2a.E1 (2a.E)
d = longitudinal distance center-to-center of battens (2b.N)
d = length of diagonal lacing (2b.W)
E = modulus of elasticity (2a.W)
e = distance between axes of gravity of column components (2a.J)
F = allowable compressive axial load (2b.N)
F_e = critical compressive axial load (2a.N)
F_p = buckling stress (2a.N)
f = design stress of the material (2b.C)
f_d = design stress (2b.1)
f_y = yield stress of the material (2b.C)
h_{con} = lateral distance between the centroids of the welds or fastenings (2b.A)
h_o = distance between centroids of chords (2a.W)
I = effective moment of inertia (2a.W)
I_b = moment of inertia of one batten (2a.W)
I_f = moment of inertia of one chord (2a.W)
I_p = moment of inertia of batten plate section about its strong central axis (2a.E)
I_1 = moment of inertia of a single chord (2a.E)
i_{min} = the minimum radius of gyration of one angle (2a.W)
i_o = the minimum radius of gyration of the built-up member (2a.W)
i_1 = radius of gyration of a single chord (2a.C), (2a.E), & (2a.J)
K = effective length factor (2a.N)
K_m = local effective length factor (2a.N)
L = total length of built-up member (2a.N)
L_1 = spacing of connection elements (2a.E)
l = length of member (2b.N)
l_d = length of lacing or lattice (2a.J)
l_1 = spacing of separator batten plates (2a.J)
l_2 = component length of lacing or lattice as projected on a plane parallel to the longitudinal axis of a built-up member (2a.J)
ℓ = buckling length of a built-up member (2a.W)
M = bending moment (2b.N)
M_s = mid-span bending moment (2a.W)
m = number of planes of battens, lacings, or tie plates (2a.E), (2a.J), & (2a.W)
N = design axial load of the built-up member (2b.E)
N_{cr} = axial buckling load (2a.W) & (2a.1)
N_D = axial force in diagonal lacing (2b.W)
N_d = design axial force (2b.A)

N_f = chord force at mid-length (2a.W)
N_{ki} = critical load of the column (2b.1)
N_s = section capacity of the column (2b.A)
N_u = squash load of the column (2b.A) & (2b.1)
n = number of structural planes formed by lacing or batten plate(2a.J) & (2b.N)
n = stiffness ratio (2a.E)
n_{bp} = number of parallel planes of battens (2b.A)
p = spacing of holes (2a.J)
q = longitudinal shear force (2b.N)
R_y = yield stress (2b.E)
r = radius of gyration of overall built-up member about the axis perpendicular to the planes of lacing (2a.N) & (2b.N)
r_i = minimum radius of gyration of individual component (2a.N)
r_1 = radius of gyration of the fully effective cross sectional area of an individual section in a built-up member (2a.N)
S^* = shear stiffness of the built-up member (2a.1)
S_v = shear stiffness of the lacing (2a.W)
V = design shear force (2b.C), (2b.E), & (2b.N)
V_d = design shear force (2a.2)
V_s = internal shear force (2b.W)
V_1 = longitudinal shear force (2a.2)
w_{max} = effective out of straightness (2a.1)
w_o = initial out of straightness (2a.W)
Z = number of structural planes formed by lacing (2a.E)
α = a coefficient depending on steel grade (2b.C)
θ = the half-angle between adjacent faces of the members as shown in Fig. 2a.C1-c (2a.C)
λ = slenderness of the whole member (2a.A) & (2a.W)
λ_d = design slenderness (2a.1)
λ_{ef} = effective slenderness ratio (2a.J)
λ_{eff} = effective slenderness ratio (2b.W)
λ_f = slenderness of a main component between adjacent connections (2a.A)
λ_{max} = maximum slenderness ratio of a built-up member (2a.C)
λ_{ox} = effective slenderness ratio of the whole member about x-axis (2a.C)
λ_{oy} = effective slenderness ratio of the whole member about y-axis (2a.C)
λ_x = slenderness ratio of built-up members assumed to react as solid members about x-axis (2a.C) & (2a.E)
λ_y = slenderness ratio of built-up members assumed to react as solid members about y-axis (2a.C) & (2a.E)
λ_{ye} = effective slenderness ratio (2a.J)
λ_1 = slenderness ratio of component (2a.C)
μ = coefficient (2a.W)
ν = safety factor (2b.E)
σ_{all} = allowable design stress (2b.E)
σ_c = mean axial stress (2b.E)
σ_{ef} = critical compressive stress (2b.E)
σ_y = specified yield strength, MPa (2b.N)
$[\sigma]$ = allowable stress (2b.C)
ϕ = buckling coefficient derived from the effective slenderness ratio (2b.E)

φ = strength reduction factor of the centrally compression member about the open web axis (2b.C)
φ_d = strength reduction factor (2b.1)
φ_{ef} = coefficient derived using columns curve (b) (2a.E)
φ_r = factor derived from the modified slenderness (2a.E)
φ_{min} = least value of strength reduction factors of the same members (2b.C)
φ_1 = buckling coefficient derived from the smallest radius of gyration of a single angle (2a.E)

Chapter 3

Beams

TOPICAL EDITOR

Yuhshi Fukumoto, Osaka University

CHAPTER EDITORS

Mark A. Bradford, The University of New South Wales
Yuhshi Fukumoto, Osaka University
Theodore V. Galambos, University of Minnesota
Hans-Dieter Glas, Technical University of Leipzig
Hans E. Goeben, Technical University of Leipzig
Joachim Lindner, Technische Universität Berlin
Jun Sakamoto, Nagoya University
Zhi Bin Xia, Zhejiang University

This chapter presents the ultimate moment capacity of beams as it is affected by lateral-torsional buckling and local flange or web buckling. All specification formulas are stripped of factors of safety or resistance factors, so that direct comparisons can be made between all of them.

3a. Lateral-Torsional Buckling

SPECIFICATION PROVISIONS & EXPLANATION

AUSTRALIA

CHAPTER EDITOR

Mark A. Bradford, The University of New South Wales

AS DR87164 87

The nominal beam moment capacity M_b is given by

$$M_b = \alpha_m \alpha_s M_s \leq M_s \qquad (3a.A1)$$

where

F_y = yield stress
L_e = effective length
$M_s = F_y Z_e$
Z_e = effective section modulus (See Flange Local Buckling)
α_m = moment distribution factor given in Table 3a.A1
α_s = strength reduction factor given as follows

$$\alpha_s = 0.6\Big([(M_s/M_0)^2 + 3]^{1/2} - M_s/M_0\Big) \qquad (3a.A2)$$

$$M_0 = \frac{\pi}{L_e}\sqrt{EI_y GJ}\sqrt{1 + \frac{\pi^2 EI_w}{L_e{}^2 GJ}} \qquad (3a.A3)$$

Note: Commonly used terms (E, I_y, G, J, I_w) are defined in the list of symbols

For monosymmetric beams, the elastic critical moment M_o is given by

$$M_o = \frac{\pi}{L}\sqrt{EI_yGJ}\left\{\sqrt{1+\frac{\pi^2 EI_w}{GJL^2}+\left(\frac{\pi\gamma_m}{2}\right)^2}+\frac{\pi\gamma_m}{2}\right\} \tag{3a.A4}$$

where

β_x = the monosymmetry parameter (SSRC, 1988)

$\gamma_m = \frac{\beta_x}{L}\sqrt{\frac{EI_y}{GJ}}$

For tapered beams, the elastic critical moment M_o is obtained by multiplying the elastic critical moment based on the most critical cross-section by α_{st}

$$\alpha_{st} = 1.0 - 1.2\left(\frac{L_m}{L}\right)\left\{1-\left(0.6+0.4\,\frac{D_m}{D_c}\right)\frac{A_m}{A_c}\right\} \tag{3a.A5}$$

where

A_m, A_c = the flange areas at the minimum and critical cross-sections, respectively

D_m, D_c = the section depths at the minimum and critical cross-section, respectively

L_m = 0.5L

Table 3a.A1 Moment Modification Factors for Supported Beams

Beam segment	Moment distribution	α_m	Range
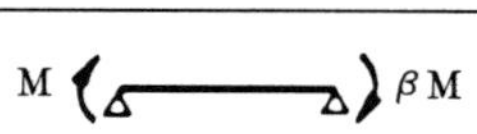	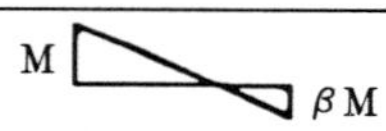	$1.75+1.05\beta+0.3\beta^2$ 2.5	$-1 \le \beta \le 0.6$ $0.6 < \beta \le 1$
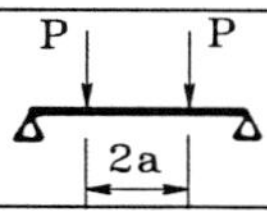	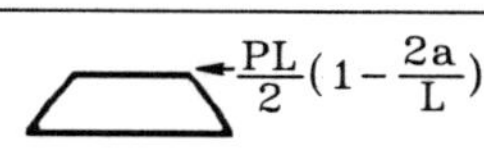	$1.0+0.35(1-2a/L)^2$	$0 \le \frac{2a}{L} \le 1$
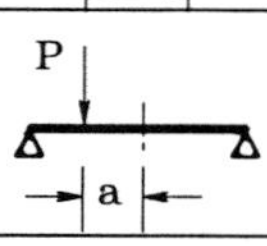	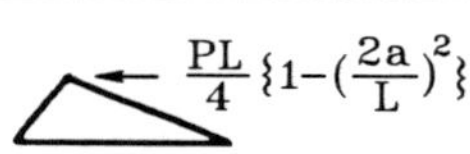	$1.35+0.4(2a/L)^2$	$0 \le \frac{2a}{L} \le 1$
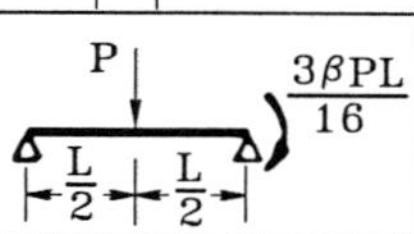	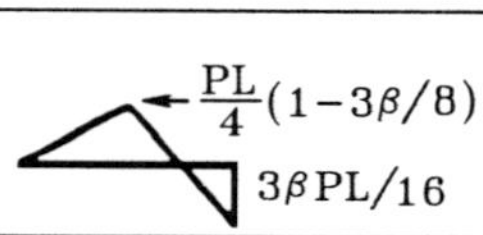	$1.35 + 0.15\beta$ $-1.2 + 3.0\beta$	$0 \le \beta < 0.9$ $0.9 \le \beta \le 1$
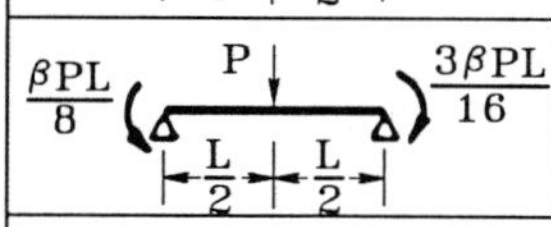	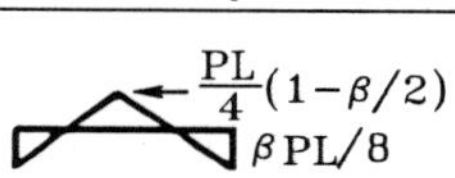	$1.35 + 0.36\beta$	$0 \le \beta \le 1$
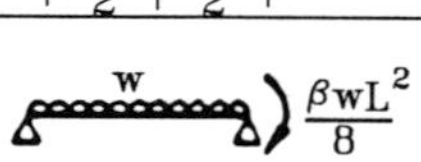	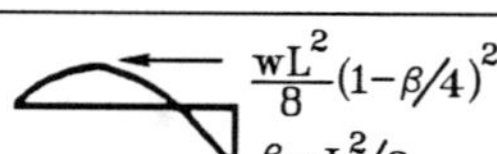	$1.13 + 0.10\beta$ $-1.25 + 3.5\beta$	$0 \le \beta \le 0.7$ $0.7 \le \beta \le 1$
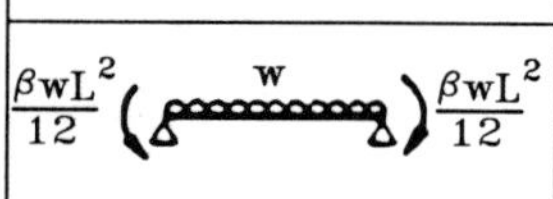	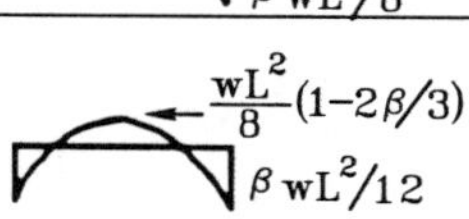	$1.13 + 0.12\beta$ $-2.38 + 4.8\beta$	$0 \le \beta \le 0.75$ $0.75 \le \beta \le 1$

CHINA

CHAPTER EDITOR
Zhi Bin Xia, Zhejiang University

GBJ 17 88

The maximum ℓ_1/b_1 value of I-section simply supported beams for preventing lateral-torsional buckling is given in Table 3a.C1, where ℓ_1 is the distance between lateral supports of beams and b_1 is the width of top flange.

A member bent in its maximum principal plane shall be checked for overall stability as follows

$$\frac{M_x}{\phi_b W_x} \leq f \qquad (3a.C1)$$

where

- f = nominal strength of steel
- M_x = factored maximum bending moment about the x-axis
- ϕ_b = overall stability factor

Welded I-section Simple Beam

The overall stability factor ϕ_b of simple beams with welded I-section shall be determined by the following formula

$$\phi_b = \beta_b \frac{4320}{\lambda_y{}^2} \frac{Ah}{W_x} \left(\sqrt{1 + \left(\frac{\lambda_y t_1}{4.4h}\right)^2} + \eta_b \right) \frac{235}{f_y} \qquad (3a.C2)$$

where

- f_y = yield strength (MPa)
- t_1 = thickness of compression flange
- α_b = $I_1/(I_1+I_2)$
- β_b = factor of equivalent bending moment given in Table 3a.C2
- λ_y = slenderness ratio about the weak axis y-y
- η_b = factor of unsymmetry of the section
  - $= 0$ for doubly symmetric I-section
  - $= 0.8(2\alpha_b-1)$ for beams with enlarged compression flange
  - $= 2\alpha_b-1$ for beams with enlarged tension flange

ϕ_b shall be replaced by ϕ_b' calculated by the following formula in case its value given by Eq. 3a.C2 is larger than 0.6.

$$\phi_b' = 1.1 - \frac{0.4646}{\phi_b} + \frac{0.1269}{\phi_b^{1.5}} \qquad (3a.C3)$$

Rolled I-section Simple Beam

The ϕ_b factor for overall stability of rolled I-section simple beams shall be taken from Table 3a.C3. Values of ϕ_b larger than 0.6 shall be replaced by the corresponding value of ϕ_b' .

Table 3a.C1 Maximum ℓ_1/b_1 values of I-section simply supported beams for preventing lateral-torsional buckling

Steel Grade	Beams without intermediate lateral support, load acting at the upper flange	Beams without intermediate lateral support, load acting at the lower flange	Beam with intermediate lateral support, load anywhere
Grade 3	13	20	16
16 Mn. 16 Mnq	11	17	13
15MnV. 15MnVq	10	16	12

Notes:

(1) The maximum ℓ_1/b_1 value of beams made of steel grade other than those shown in Table 3a.C1 shall be that of grade 3 steel multiplied by $\sqrt{235/f_y}$.

(2) Detailing measures shall be taken to prevent torsion of the end section at the beam supports.

Table 3a.C2 Factor β_b for I-section simple beams

Item	Lateral support	Loading $\xi = \ell_1 t_1/b_1 h$		$\xi \leq 2.0$	$\xi > 2.0$	Applicability
1	No intermediate lateral support	Uniform loading acting at	upper flange	$0.69+0.13\xi$	0.95	Only for sections a & b in Fig. 3a.C1
2			lower flange	$1.73-0.20\xi$	1.33	
3		Concentrated loading acting at	upper flange	$0.73+0.18\xi$	1.09	
4			lower flange	$2.23-0.28\xi$	1.67	
5	One lateral support at mid-span	Uniform loading acting at	upper flange	1.15		For all sections in Fig. 3a.C1
6			lower flange	1.40		
7		Concentrated loading acting at any level		1.75		
8	Not less than two interme-diate supports equally spaced	Uniform or concentrated loading between lateral supports acting at	upper flange	1.20		
9			lower flange	1.40		
10	No transverse loading between lateral supports		$1.75-1.05(M_1/M_2)+0.3(M_1/M_2)^2$ but not larger than 2.3			

Notes:

(1) Concentrated loading in Items 3, 4 and 7 denotes one single load or a few concentrated loads near the mid-span.

(2) In Items 8 and 9, β_b shall be taken equal to 1.20 when concentrated loads are acting on the laterally supported points.

(3) For I-sections with enlarged compression flanges so as $\alpha_b > 0.8$, the value of β_b shall be multiplied by a coefficient as follows

For Item 1, 0.95 in case $\xi \leq 1.0$

For Item 3, 0.90 in case $\xi \leq 0.5$

0.95 in case $0.5 \leq \xi \leq 1.0$

Plastic design is allowed for fixed ended beams, continuous beams, and low-rise rigid frames with solid web members, provided that they do not carry dynamic loads directly.

For plastic design, the load factors are the same as those used in elastic design, but an additional resistance factor, γ_p=1/1.111, should be used; that is, f_p=0.9f should be used to replace f in investigation formulas for strength.

Table 3a.C3 ϕ_b values of rolled I-section simple beams

Item	Lateral support and loading			I-section Number	ℓ_1 (m)				
					2	4	6	8	10
1	No intermediate lateral support	Concentrated loading acting at	upper flange	10—20	2.0	0.99	0.68	0.53	0.43
				22—32	2.4	1.09	0.72	0.54	0.45
				36—63	2.8	1.07	0.68	0.50	0.40
2			lower flange	10—20	3.1	1.34	0.82	0.63	0.52
				22—40	5.5	1.84	1.07	0.73	0.56
				45—63	7.3	2.30	1.20	0.80	0.60
3		Uniform loading acting at	upper flange	10—20	1.7	0.84	0.57	0.45	0.37
				22—40	2.1	0.93	0.60	0.45	0.36
				45—63	2.6	0.97	0.59	0.44	0.35
4			lower flange	10—20	2.5	1.08	0.68	0.52	0.42
				22—40	4.0	1.45	0.85	0.60	0.46
				45—63	5.6	1.80	0.95	0.65	0.49
5	With intermediate lateral support(no matter what loading is acting)			10—20	2.2	1.01	0.66	0.52	0.42
				22—40	3.0	1.24	0.76	0.56	0.43
				45—63	4.0	1.38	0.80	0.56	0.43

Notes:

(1) Concentrated loading in Items 3, 4 and 7 denotes one single load or a few concentrated loads near the mid-span.

(2) Values of φ_b in this table apply to steel grade 3 and shall be multiplied by $\frac{235}{f_y}$ for other steel grades.

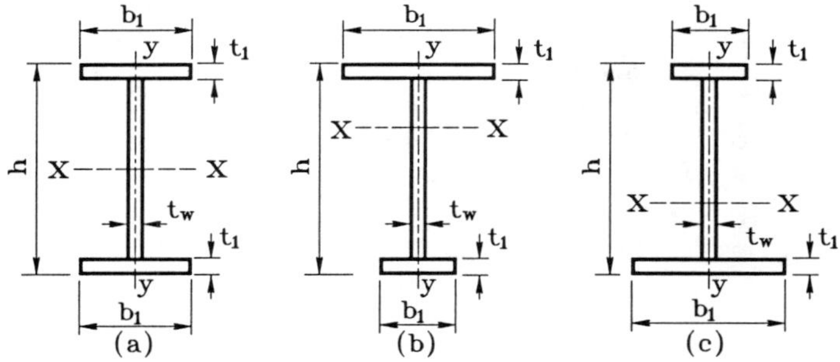

Fig. 3a.C1 Welded I-section (a) doubly symmetric; (b) with enlarged compression flange; (c) with enlarged tension flange

If plastic design is employed, the cross section of the member should be "compact." At the locations where plastic hinges yield, lateral bracing should be provided and at a distance no greater than ℓ_{cr} to either side of the hinge.

$$\text{for} \quad -1 \le \frac{M_1}{W_{px}f_p} \le 0.5 \qquad \left(\frac{\ell}{i_y}\right)_{cr} \le \left(60 - 40\frac{M_1}{W_{px}f_p}\right)\sqrt{\frac{235}{f_y}} \tag{3a.C4}$$

$$\text{for} \quad 0.5 \le \frac{M_1}{W_{px}f_p} \le 1.0 \qquad \left(\frac{\ell}{i_y}\right)_{cr} \le \left(45 - 10\frac{M_1}{W_{px}f_p}\right)\sqrt{\frac{235}{f_y}} \tag{3a.C5}$$

where

- f_p = plastic design strength
- ℓ/i_y = λ_y = slenderness ratio in the plane perpendicular to that in which the moment is acting
- M_1 = moment at the section braced at a distance ℓ_{cr} from the hinge. M_1 is taken as negative when the curvature of the segment within the length ℓ_{cr} changes direction, and vice versa
- W_{px} = plastic moment about bending axis

The distance between the points of lateral bracing, where no plastic hinge develops, may be determined by investigation of lateral torsional buckling as in elastic design.

TBJ 2 85

Allowable stress design is used, in which the allowable extreme fiber compressive stress for I-beams subjected to major axis bending while investigating lateral-torsional buckling is expressed by $\varphi_b f$, and is taken equal to the allowable stress of an axially compressed column with an equivalent slenderness ratio

$$\lambda_e = \alpha\frac{\ell_b}{h}\cdot\frac{i_x}{i_y} \tag{3a.C6}$$

where

- f = f_y/F.S., the basic allowable normal stress of steel in axial compression where f_y is the specified minimum yield stress equal to 340 N/mm^2 for steel 16Mnq, a low alloy steel available for bridges, and F.S. is the factor of safety taken as 1.7
- ℓ_b = lateral unsupported length of the compression flange of the beam
- α = coefficient, 1.8 for welded beams and 2.0 for riveted beams
- φ_b = σ_{cr}/f_y, the overall stability factor taken from the column curve as a function of λ_e (Fig. 3a.C2)

COMMENTARY

Equation 3a.C6 is derived from the elastic lateral buckling strength of simply supported beams with doubly symmetric I-section under uniform moment.

The coefficient α is given by

$$\alpha = 2\left(1 + \frac{4\,\ell_b^2\,\mathrm{GJ}}{\pi^2\,\mathrm{h}^2\,\mathrm{EI}_y}\right)^{-0.25} \tag{3a.C7}$$

where GJ and EI_y are the torsional constant and lateral flexural rigidity of the cross section, respectively. TBJ 2 recommends that the α may be taken as a constant as shown in Eq. 3a.C6 for welded and for riveted I-beams.

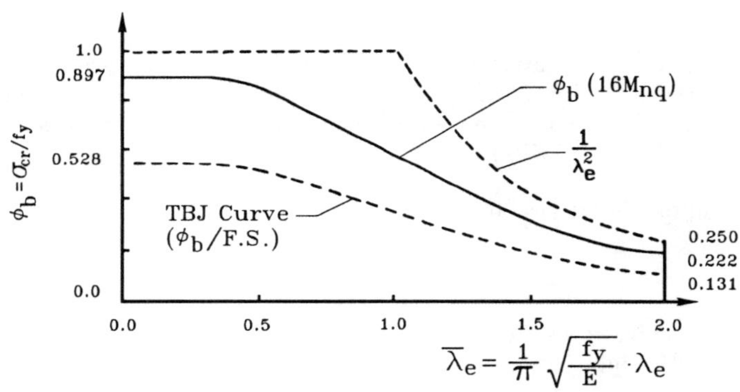

Fig. 3a.C2 TBJ allowable design stress curve

EAST EUROPE

CHAPTER EDITORS
Hans-Dieter Glas, Technical University of Leipzig
Hans E. Goeben, Technical University of Leipzig

CMEA RS131 74

The proof of stability against buckling is not required if

- the compression flange of the beam is fixed to a rigid plate or
- the spacing l_1 of the lateral supports of the compression flange fulfills the following condition:

For I-sections $$l_1 \leq 40\ i_y \sqrt{\frac{240}{R}} \tag{3a.E1}$$

where

i_y = radius of gyration of the compressed parts of the section relative to the web axis (y-axis); the compression flange and one third of the depth of web are considered the compressed parts of the section

R = yield point (MPa)

Proof of stability against lateral buckling

$$\frac{M_x}{\phi_b\ W_x} \leq R \tag{3a.E2}$$

where ϕ_b is coefficient of lateral buckling, which is determined according to the following:

For Beams with Lateral Support of the Compression Flange Whose Spacing Exceeds the Maximum Values of Eq. 3a.E1 and for Beams with End Moments

The beams are subdivided into one of 9 groups, depending on their shape of section and steel grade. For each group, one table for determining ϕ_b as a function of λ_{pr} is available in the specification.

The value of λ_{pr} for built-up I-sections whose y-axis is the axis of symmetry and for beams with end moments is given by

$$\lambda_{pr} = \frac{l_1}{i_y \sqrt{C_b}} \tag{3a.E3}$$

where

C_b = factor for taking into account the moment shape; it is calculated using the amounts of bending moments of two adjacent beam parts between supports

l_1 = spacing of the lateral supports of the compression flange

The value of λ_{pr} for rolled I-sections is given by

$$\lambda_{pr} = 3.9 \sqrt{\frac{l_1 \text{ h}}{\text{b t C}_b}} \tag{3a.E4}$$

where l_1, C_b are the same as defined in Eq. 3a.E3.

Determination of ϕ_b for Beams without Lateral Support of the Compression Flange within the Span (General Method)

$$\phi_b = \frac{\sigma_{kr,M}}{\sigma_t} \tag{3a.E5}$$

$$\sigma_{kr,\,M} = \frac{M_{kr}}{W_x} \tag{3a.E6}$$

where

$$C = \sqrt{\frac{(\mu_y/\mu_w)\ I_w + 0.039\ (\mu_y\ l)^2\ I_k}{I_y}} \tag{3a.E7}$$

e = distance of the x-axis from the center line of the compression flange

$$r_x = \int \frac{y\ (x^2+y^2)\ dA}{I_x} \tag{3a.E8}$$

v = distance of the point of application of the transverse loads from the gravity axis of the section

y_c = distance of the shear center from the center of gravity

μ_w = value of fixation for preventing warping

μ_y = value of fixation for preventing bending about the y-axis

μ_y, μ_w = factors taking into account the support conditions at the ends of the beam

μ_y, μ_w = 1 for simply supported beams

μ_y, μ_w = 0.5 for full fixation at the two ends

μ_y, μ_w = 0.7 for full fixation of one end and fork support at the other

ξ = factor for taking into account the moment shape

$$\sigma_{kr,M} = \frac{\xi \pi^2 E I_y e}{(\mu_y l)^2\ I_x} \left[\sqrt{\left(\frac{5\ \mu_y^{\,2} v}{\pi^2} + \frac{r_x}{3} - y_c\right) + C^2} - \left(\frac{5 \mu_y^{\,2} v}{\pi^2} + \frac{r_x}{3} - y_c\right)\right] \tag{3a.E9}$$

σ_t = yield stress

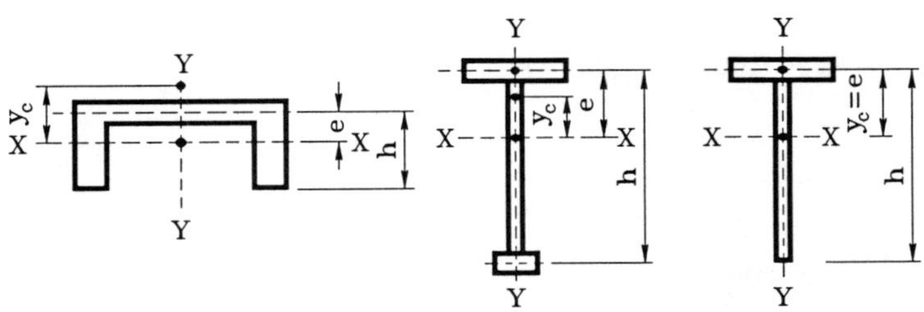

Fig. 3a.E1 Some singly symmetric sections and coordinate axes

ČSN 73 1401 84

The concept of limit states design is used.
The proof of stability against lateral buckling is not required if

a) the section of the beam is torsionally stiff (for example, closed)
b) the beam is subjected to bending about the y-axis (see Fig. 3a.E2)
c) the spacing of the supports of the compression flange of the beam is smaller than the radius of gyration of the flange multiplied by 40

$$c < 40\, i_{y1} \tag{3a.E10}$$

$$i_{y1} = \frac{i_y}{i_x} \cdot \frac{h}{2} \tag{3a.E11}$$

where

- c = spacing between supports of compression flange
- h = distance between the gravity centers of the flanges in the y-direction (Fig. 3a.E2)
- $i_x > i_y$ = radii of gyration to the principal axes of the beam section
- i_{y1} = radius of gyration of the compression flange

Proof of stability against lateral buckling

$$\frac{M}{\phi_o W_1} \leq R \tag{3a.E12}$$

For a flange subject to tension, the following condition is additionally required

$$\frac{M}{W_{o2}} \leq R \tag{3a.E13}$$

where

- R = nominal strength of material (MPa)
- W_1 = section modulus, with the openings subtracted, relative to the compression flange
- W_{o2} = section modulus, with the openings subtracted, relative to the tension flange
- ϕ_o = factor of lateral buckling for the critical slenderness ratio of the compression flange according to Eq. 3a.E14 or Table 3a.E1

for $\lambda \sqrt{\frac{R}{210}} \leq \frac{93}{\sqrt{0.8}} = 104$

$$\phi_o = 0.8 - 0.02\left(\frac{\lambda}{93}\sqrt{\frac{R}{210}}\right)^2 + \sqrt{\left[0.8 - 0.02\left(\frac{\lambda}{93}\sqrt{\frac{R}{210}}\right)^2\right]^2 - 0.6} \tag{3a.E14}$$

Table 3a.E1 Factor of lateral buckling ϕ_o

$\lambda\sqrt{R/210}$	≤ 40	50	60	70	80	90	100	104	> 104
ϕ_o	1.0	0.99	0.96	0.94	0.91	0.88	0.84	0.80	$[(93/\lambda) \cdot \sqrt{210/R}\,]^2$

λ = critical slenderness ratio of the compression flange

$$\lambda = \gamma \cdot \frac{\beta \cdot L_{y1}}{i_{y1}} \tag{3a.E15}$$

Referring to Table 3a.E1, for deep welded or riveted beams

$$\lambda \geq \lambda_1 = \frac{\beta \cdot a}{i_{y1}} \tag{3a.E16}$$

a = distance between the cross ribs in the zone of the maximum moment
i_{y1} = radius of gyration of the compression flange according to Eq. 3a.E11
L_{y1} = spacing of the lateral supports of the compression flange preventing buckling out of the bending plane. For cantilever beams, the double value of the cantilever length is used
β = factor of lateral buckling length in relation to the moment distribution; given in a table in the specification
γ = slenderness ratio

$$\gamma = \frac{1}{\sqrt{k \cdot \frac{a_o + e}{a_2} + \sqrt{\left(k \cdot \frac{a_o + e}{a_2}\right)^2 + \left(\frac{c}{a_2}\right)^2}}} \tag{3a.E17}$$

$$c = \frac{h}{2} \cdot \sqrt{\frac{I_w}{I_y} \cdot \left(\frac{2}{h}\right)^2 \cdot \left(\frac{L_{y1}}{L_w}\right)^2 + \left(\frac{2a_t}{\pi}\right)^2} \tag{3a.E18}$$

where

a_o = distance of the gravity center from the shear center
a_t = torsional parameter

$$a_t = 0.62\,\frac{L_{y1}}{h}\sqrt{\frac{I_t}{I_y}} \tag{3a.E19}$$

a_2 = distance of the tension flange from the shear center
e = distance of the point of application of load from the shear center
I_t = torsional resistance
I_w = warping constant
k = 1.0 for beams subject to end moments
k = 0.5 for beams subject to transverse loads
L_w = buckling length in case of torsion of the beam

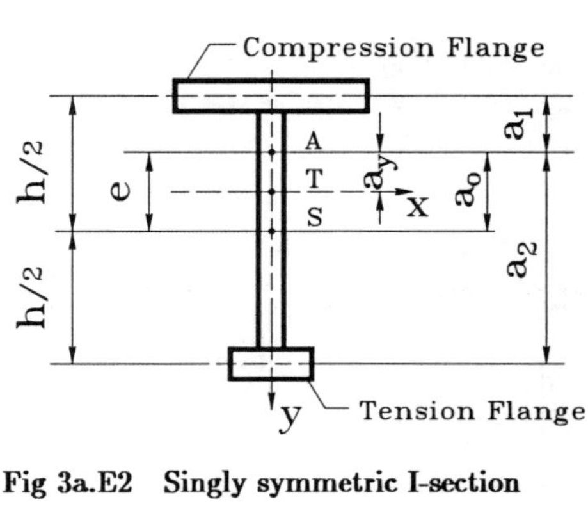

Fig 3a.E2 Singly symmetric I-section

MSZ 15024/1 85

The concept of allowable stress design is used.

The proof of stability against lateral buckling is not required if
- a) the flange is supported along its length by a rigid plate to prevent lateral displacement
- b) the spacing of the lateral supports does not exceed the values given in Table 3a.E2

Rough proof of stability against lateral buckling

$$\sigma_{max,\ ny} \leq \sigma_{kif,\,\mathrm{H}} \qquad \text{(3a.E20)}$$

where

ℓ_o = length of buckling. For a beam, laterally supported at its two ends only, it can be determined like that of compression members owing to the compressive stress varying over the beam length

$$\ell_o = \nu \cdot \ell \qquad \text{(3a.E21)}$$

$\lambda_{y,\ddot{o}v}$ = ℓ_o / i_y (3a.E22)
= slenderness ratio of the compression flange relative to the y-axis is given in a figure in the standard

$\sigma_{kif,\ \mathrm{H}}$ = limit stress in lateral buckling is given in a table in the specification

$\sigma_{max,\ ny}$ = maximum edge compressive stress relative to the gross area of cross section

The factor of buckling length, ν, is determined based on the proof of stability against buckling.

Table 3a.E2 Maximum spacing of lateral supports

class of strength (kN/cm^2)	37	45	52
maximum spacing of lateral supports	$46\mathrm{i}_y$	$42\mathrm{i}_y$	$37\mathrm{i}_y$

where i_y is radius of gyration of the compression flange about the web, the parts of the web which, in addition to the compression flange, are taken into account for determining i_y are shown in Fig. 3a.E3.

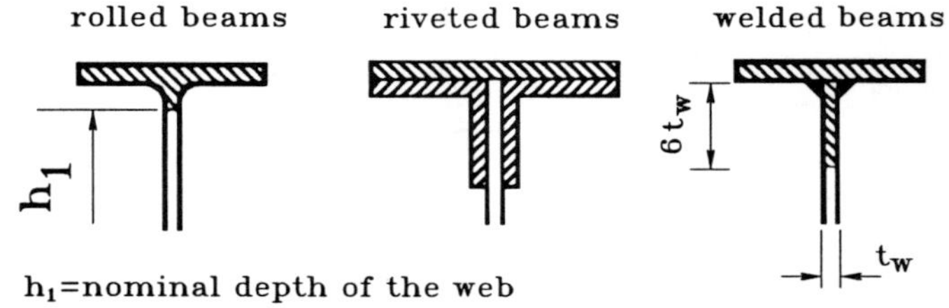

Fig 3a.E3 Compressive portion of beams

Strict proof of stability against lateral buckling

$$\sigma_{max,ny} \leq \sigma_{kif,\mathrm{H}} \tag{3a.E23}$$

where

$\sigma_{kif,\mathrm{H}}$ = limit stress in lateral buckling according to a table in this specification as a function of λ_{ki} and the class of strength

$\sigma_{max,ny}$ = see Eq. 3a.E20

For all the cases not included, the ideal slenderness ratio for lateral buckling, λ_{ki}, can be calculated using Eq. 3a.E24 and assuming an ideally elastic behavior of material.

$$\lambda_{ki} = \pi \sqrt{\frac{\mathrm{E}}{\sigma_{ki}}} \tag{3a.E24}$$

NiP 565/86 87

The Bulgarian Standard is based on the concept of limit states design using resistance factors.

The proof of stability against lateral buckling is not required, if

a) the uniformly distributed static load is transmitted to the beam by a rigid plate that is continuously fixed to the compression flange of the beam (reinforced concrete slabs or light-weight concrete slabs, plain or grooved metal plates, asbestos cement plates, etc.)

b) the ratio of the beam length l to the width of the compression flange b_n does not exceed the values of Table 3a.E3

For other steel grades Table 3a.E3 may be used, if the maximum values of l/b_n are multiplied by the factor $\sqrt{210/\mathrm{R}}$ (R in MPa).

Proof of stability against lateral buckling

$$\frac{\mathrm{M}}{\phi_r \cdot \mathrm{W}} \leq \mathrm{m} \cdot \mathrm{R} \tag{3a.E25}$$

where

m = safety factor depending on the structural element, $0.75 \leq \mathrm{m} \leq 0.9$

R = computational strength of material

ϕ_r = factor of lateral buckling, explained in the following

ϕ_r for doubly symmetric I-sections

$$\phi_r = \psi \frac{\mathrm{I}_y}{\mathrm{I}_x} \left(\frac{\mathrm{h}}{\ell}\right)^2 \cdot 10^3 \tag{3a.E26}$$

ψ is given in tables in the standard as a function of α

Table 3a.E3 Maximum values of l/b_n for beams made of steel St3 or St4 for which stability check against lateral buckling is not required.

Beam type	$\frac{h}{b_n}$	Maximum values of l/b_n for beams with					
		$h/\delta_n = 100$			$h/\delta_n = 50$		
		Load applied to the		Between fixed supports*	Load applied to the		Between fixed supports
		upper flange	lower flange		upper flange	lower flange	
Welded	2	18	28	22	19	30	23
	4	16	26	19	18	27	21
	6	15	24	18	16	25	19
Riveted	2	21	30	24	30	42	33
	4	18	28	21	25	35	27
	6	16	25	19	21	32	24

*These beam parts between fixed supports are independent of load location

where

b_n, δ_n = width, thickness of the compression flange, respectively

l = spacing of lateral supports of the upper flange (e.g., nodal points of the bracing or fixing points of the floor slabs) or distance between the cross connections preventing the torsion of the beam section. If there are no lateral supports, l is the span length of the beam

PN 80/B-03200 87

The proof of stability against lateral buckling is not required if,

a) the compression flange of the beam is connected with a rigid plate

b) the ratio l/b does not exceed the values given in Table 3a.E4

For all the steel grades not given in the table, the maximum value of l/b is obtained by multiplying the appropriate values given in the column "steel St3" of Table 3a.E4 by $\sqrt{215/R}$. (R in MPa.)

Table 3a.E4 Maximum value of l/b for proof of stability against lateral buckling to be omitted

h/b	Maximum values of l/b			
	steel St3		steel 18G2	
	Load applied to the			
	upper flange	lower flange	upper flange	lower flange
2	12	19	10	16
4	11	17	9	15
6	10	16	8	13

where h is the depth of the beam according to Fig. 3a.E4

Proof of stability against lateral buckling in case of bending in one plane

$$\sigma = 1.15 \frac{M \cdot m_z}{W_{cbr}} \leq R \tag{3a.E27}$$

Proof of stability against lateral buckling in case of bending in two planes

$$\sigma = 0.87 \frac{M \cdot m_z}{W_{xbr}} + \frac{M_y}{W_{ybr}} \leq R \tag{3a.E28}$$

where

R = computational strength of material; given in a table in the standard
m_z = factor of lateral buckling
W_{cbr} = gross section modulus relative to the compression flange
W_{xbr}, W_{ybr} = gross moments of resistance about the axes x, y

The factor of lateral buckling m_z is to be taken from Table 3a.E5 using the ratio λ/λ_p. The following equations are used

$$\lambda = \frac{l}{h}\sqrt{\frac{I_x}{I_y}} \tag{3a.E29}$$

$$\lambda_p = \frac{K_z}{\sqrt{R}} \tag{3a.E30}$$

where

I_s = torsional resistance
K_z = coefficient depending on the type of loading and on the section of beam

$$\lambda_s = \frac{l}{h}\sqrt{\frac{I_s}{I_y}} \tag{3a.E31}$$

1) K_z for doubly symmetric I-sections.
K_z is given in table of this code for simply supported and for cantilever beams depending on the kind of loading and on λ_s.

Table 3a.E5 Factor m_z of lateral buckling

λ/λ_p	m_z	λ/λ_p	m_z	λ/λ_p	m_z
0.0	1.00	1.0	2.00	2.0	8.00
0.1	1.02	1.1	2.42	2.1	8.82
0.2	1.05	1.2	2.88	2.2	9.68
0.3	1.08	1.3	3.38	2.3	10.58
0.4	1.12	1.4	3.92	2.4	11.52
0.5	1.18	1.5	4.50	2.5	12.50
0.6	1.25	1.6	5.12	2.6	13.52
0.7	1.35	1.7	5.78	2.7	14.58
0.8	1.48	1.8	6.48	2.8	15.68
0.9	1.68	1.9	7.22	2.9	16.82

2) K_z for monosymmetric I- and T-sections

$$K_z = k\sqrt{a + \sqrt{a^2 + c}} \qquad (3a.E32)$$

The values a,c,k are given in table of this code.

$$n = \frac{I_1}{I_1 + I_2} \qquad (3a.E33)$$

where I_1, I_2 are moments of inertia of the larger, smaller flange, respectively, relative to the axis of symmetry of the section ($I_1 + I_2 \approx I_y$)

For the remaining notations see Fig. 3a.E4.

If $n > 0.7$ and $l/b_1 \geq 5$ where $l/b_1 \leq 25$, K_z is multiplied by the factor ϕ_1

$$\phi_1 = 1 - 0.025 \cdot (n-0.7) \frac{l}{b_1} \qquad (3a.E34)$$

If $n > 0.9$ and $\lambda_s < 5$, K_z is multiplied by the factor ϕ_2

$$\phi_2 = 1 -(n - 0.9) \cdot (l-0.03\lambda_s{}^2) \qquad (3a.E35)$$

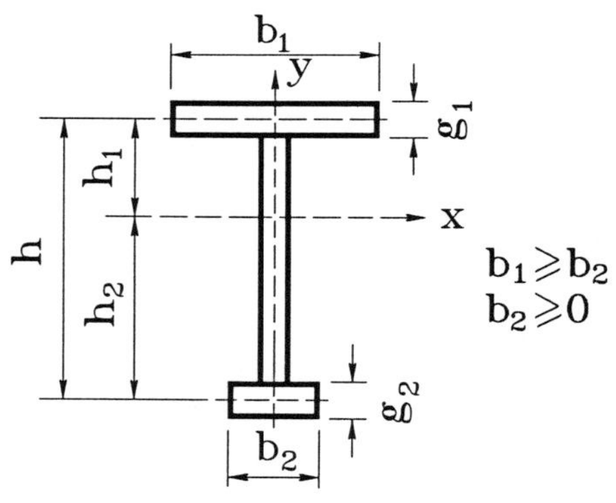

Fig. 3a.E4 Singly symmetric I-sections

SNiP II-23-81 82

Proof of stability against lateral buckling is not required if

a) the load is applied to the beam by a massive, rigid plate that is permanently supported by and continuously fixed to the compressive flange of the beam (reinforced concrete slabs, plain and grooved metal plates, corrugated iron, etc.).

b) the ratio of the computational length of the beam l_{ef} to the width of the compression flange b does not exceed the values of Table 3a.E6 (the condition is valid for both doubly symmetric and mono-symmetric I-sections; width of tension flange $\geq$ 75% of the compression flange width).

For beams with flanges connected by high-strength bolts, the maximum value of l_{ef}/b calculated from Table 3a.E6 is multiplied by the factor 1.2.

Proof of stability against lateral buckling

$$\frac{M}{\phi_b W_c} \leq R_y \gamma_c \tag{3a.E36}$$

where

l_{ef} = distance between the nodal points of the longitudinal and transverse bracing or distance between the fixing points of the rigid plate

- $l_{ef} = l$ (span length of the beam) if there are no supports of the upper flange
- $l_{ef} = l$ (length of the cantilever beam) if there is no horizontal support of the compression flange at the end of the cantilever beam. Distance of the fixing points of the compression flange with horizontal support of the compression flange at the end of the cantilever beam

$R_y = R_{yn}/\gamma_m$ (3a.E37)

R_{yn} = standard nominal strength of the material

W_c = section modulus relative to the compression flange

γ_c = resistance factor depending on the structural element

For plate elements of plane trusses or space frameworks, $0.7 \leq \gamma_c \leq 0.9$.

For elements of tension members, $\gamma_c = 0.9$ or $\gamma_c = 0.95$

In the standard, the resistance factor γ_c is given for 10 kinds of structural elements, where not in any case all values of γ_c must be taken into account at once. For all cases not arranged, γ_c must be equal to 1.0.

γ_m = safety factor for the material, $1.025 \leq \gamma_m \leq 1.15$

Table 3a.E6 Maximum value of l_{ef}/b for proof of stability against lateral buckling

point of application for force	maximum value of l_{ef}/b, allowing the proof of stability against lateral buckling to be omitted if for these rolled and welded beams $1 \leq h/b < 6$ and $15 \leq b/t \leq 35$
at the upper flange	$l_{ef}/b=(0.35+0.0032b/t+(0.76-0.02b/t)b/h)\sqrt{E/R_y}$
at the lower flange	$l_{ef}/b=(0.57+0.0032b/t+(0.92-0.02b/t)b/h)\sqrt{E/R_y}$
independent of the point of application of force if parts of beam between trusses are calculated and for pure bending	$l_{ef}/b=(0.41+0.0032b/t+(0.73-0.016b/t)b/h)\sqrt{E/R_y}$

where

b,t = width, thickness of the compression flange, respectively

h = distance between the axes of the flange plates

R_y, l_{ef} = see proof of stability against lateral buckling

Lateral Buckling Coefficients for Doubly Symmetric I-Sections

$$\phi_1 = \psi \frac{I_y}{I_x}\left(\frac{h}{l_{ef}}\right)^2 \frac{E}{R_y} \tag{3a.E38}$$

for $\phi_1 \leq 0.85$

$$\phi_b = \phi_1 \tag{3a.E39}$$

for $\phi_1 > 0.85$

$$\phi_b = 0.68 + 0.21\,\phi_1 \leq 1.0 \tag{3a.E40}$$

where ψ is the factor dependent on the kind of loading and on parameter α, given in tables of this code

a) α for rolled I-sections

$$\alpha = 1.54 \frac{I_t}{I_y}\left(\frac{l_{ef}}{h}\right)^2 \tag{3a.E41}$$

where I_t is torsional resistance

b) α for welded I-sections composed of 3 plates and for I-beams with flanges connected by high strength bolts

$$\alpha = 8\left(\frac{l_{ef}\, t_1}{h\, b_f}\right)^2 \left(1 + \frac{a\, t^3}{b_f\, t_1^3}\right) \tag{3a.E42}$$

where for welded I-sections
- t_1 = thickness of flange plate
- t = web thickness
- a = 0.5h

for I-beams with flanges connected by high-strength bolts:
- t_1 = sum of the thicknesses of the flange plates and of the horizontal leg of angle
- a = sum of the thickness of the flange plates and the length of the vertical leg of angle
- t = sum of the web thickness and thickness of the vertical legs of angles

Factor of Lateral Buckling for Monosymmetric I-Sections

$$\phi_1 = \psi \frac{I_y}{I_x} \frac{2\,h\,h_1}{l_{ef}^2} \frac{E}{R_y} \tag{3a.E43}$$

$$\phi_2 = \psi \frac{I_y}{I_x} \frac{2\,h\,h_2}{l_{ef}^2} \frac{E}{R_y} \tag{3a.E44}$$

$$\psi = D(B + \sqrt{B^2 + C}) \tag{3a.E45}$$

where
- B,D,C = coefficients given in Tables of this code
- h_1, h_2 = distances of the center of gravity of the section from the axis of the larger or smaller flange

Factor of Lateral Buckling for Channel-Sections

ϕ_b is calculated analogously to a doubly symmetric I-section.

Note: α is calculated using Eq. 3a.E42

ϕ_1 is multiplied by 0.7.

TGL 13503 82

The standards are based on the concept of allowable stress design utilizing a safety factor. For any beam that is not obviously safe from lateral buckling

$$M \le \frac{M_{kr}}{\nu_{kr}} = \frac{M_T \, \phi_M}{\nu_{kr}} \tag{3a.E46}$$

or

$$\sigma_b = M/W_T \le \sigma_{all} \; \phi_M \tag{3a.E47}$$

where

M_F = yield bending moment

$$= W_{el} \, \sigma_F \tag{3a.E48}$$

M_T = modified load-bearing moment

$$= (M_F + M_{pl})/2 \le 1.2 \; M_F \tag{3a.E49}$$

M_{pl} = moment at full plasticity

$$= W_{pl} \, \sigma_F \tag{3a.E50}$$

W_T = modified section modulus

$$= (W_{el} + W_{pl})/2 \le 1.2 \; W_{el} \tag{3a.E51}$$

W_{el} = elastic section modulus

W_{pl} = plastic section modulus

μ_M = imperfection factor

$$= \frac{92.93 \; \overline{\lambda}_M - c_1}{c_2} \tag{3a.E52}$$

ν_{kr} = safety factor to prevent reaching the critical load in case of partial plasticization cases of critical load
= 1.50 for principal loads
= 1.33 for principal and additional loads
= 1.20 for special loads

σ_F = yield stress

ϕ_M = factor of lateral buckling

$$= \frac{M_{kr}}{M_T} = \frac{1}{2}\,(\frac{1+\mu_M}{\overline{\lambda}_M^2} + 1) - \sqrt{[\,\frac{1}{2}(\frac{1+\mu_M}{\overline{\lambda}_M^2} + 1)\,]^2 - \frac{1}{\overline{\lambda}_M^2}} \tag{3a.E53}$$

The factors of lateral buckling are fixed for four groups of beams depending on the magnitude of the imperfections (Table 3a.E7 and Fig. 3a.E5).

$\overline{\lambda}_M$ is relative slenderness ratio

$$\overline{\lambda}_M = \sqrt{\frac{M_T}{M_{ki}}} \tag{3a.E54}$$

There are two different methods of determining $\overline{\lambda}_M$

Rough Calculation of $\overline{\lambda}_M$ for Simply Supported Beams

$$\overline{\lambda}_M = \frac{\Theta}{\lambda_s} \frac{l_k}{i_G} \tag{3a.E55}$$

where

Θ = factor for taking into account the point of application of load
= 1.25 if the load is applied to the upper flange of welded section
= 1.0 if the load is applied to the upper flange of rolled section
= 0.8 if the load is applied to the lower flange

If the load is applied to the beam axis, interpolation is possible.

λ_s= slenderness ratio for $\sigma_{ki} = \sigma_F$

$$\lambda_s = \pi \sqrt{\frac{E}{\sigma_F}} \tag{3a.E56}$$

l_k= spacing of the supports of the upper flange preventing lateral displacement (it can be reduced if the bending moment is varying over the beam length)

Table 3a.E7 Classifying beams into the curves of buckling moment

	curve of overturning moment	C_1	C_2
not welded, with or without flange plate welded on	a	40	900
welded, $h_s/s \leq 150$ or with cross ribs adjacent to the flange, at 1/3	b	25	600
$150 \leq h_s/s \leq 250$ without cross ribs or welded under the condition of restricted shrinkage, $h_s/s \leq 150$	c	15	400
$h_s/s > 250$ without cross ribs or welded under the condition of restricted shrinkage, $h_s/s > 150$	d	10	220
h, s are depth, thickness of the web, respectively. If the material has been stress-poor welded or annealed, the beam is classified into the next higher curve.			

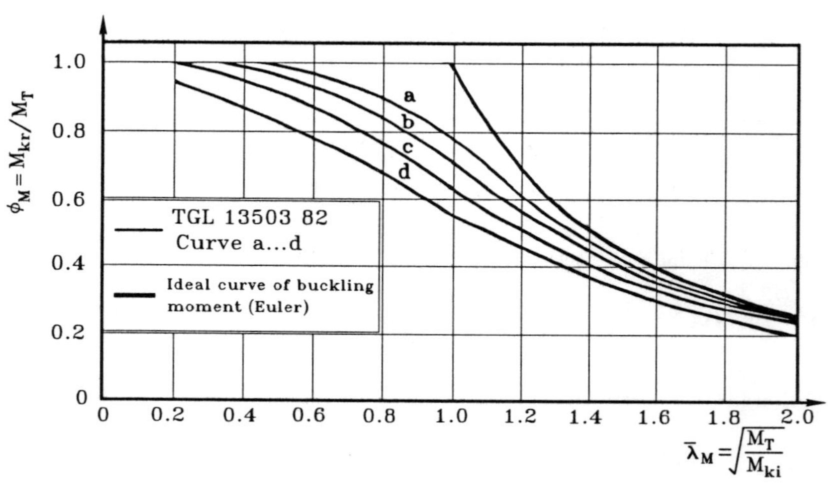

Fig. 3a.E5 Lateral buckling strength curves a, b, c, and d

i_G= principal radius of gyration of the compression flange relative to the web plane is calculated based on the area of the compression flange plus 2/5 of the web area in compression

Referring to Fig. 3a.E5, the proof of stability against lateral buckling is not required if for a beam

according to the curve a of buckling moment $\bar{\lambda}_M \leq 0.6$
according to the curve b of buckling moment $\bar{\lambda}_M \leq 0.4$
according to the curve c of buckling moment $\bar{\lambda}_M \leq 0.28$

Exact Calculation of $\bar{\lambda}_M$

Eq. 3a.E54 holds.

M_{ki}, the buckling moment under ideal conditions, is given by

$$M_{ki}= \frac{k}{\beta\, l}\sqrt{E\, I_y\, G\, J_D} \tag{3a.E57}$$

where

GJ_D = torsional rigidity
k = factor depending on the support conditions of the beam
β = factor of lateral buckling length taking into account the degree of the elastic end-restraint

COMMENTARY

A new limit states specification, TGL 13503 82 is presently under consideration. The new standards are based on the concept of limit states design using incremental load factors. The internal forces and stresses are calculated by taking into account the load factors and combination of load cases.

Eqs. 3a.E46 and 3a.E47 are changed to the following

$$M \leq \frac{M_{kr}}{\gamma_m} = \frac{M_T}{\gamma_m}\, \phi_M \tag{3a.E58}$$

$$\sigma_b = \frac{M}{W_T} \leq \frac{R_n}{\gamma_m}\, \phi_M \tag{3a.E59}$$

where

M, σ = design moment, design stress (with load factor, combination factor and factor γ_n)
R_e = yield stress
R_n = standard strength of material = $R_e\, \gamma_d$ (3a.E60)
γ_d = resistance factor $\gamma_d \leq 1$
γ_m = safety factor for the material, $\gamma_m = 1.15$
γ_n = safety factor depending on the structural element and on consequences in case of destruction, $0.9 \leq \gamma_n \leq 1.1$

REGIONAL COMPARISON

Fig. 3a.E6 shows the lateral-torsional buckling strength of a laterally unsupported simple beam I-300x10-200x15 (TGL 26088) under a concentrated load applied at the mid-span lower flange. Fig. 3a.E7 shows the strength ratios of different specifications based on TGL 13503 82 as 100%. From these figures, for short span range, MSZ 15024/1 85 gives the highest moment, and for median and long span range, SNiP II-23-81 82 specifies the highest moment while PN 80/B-03200 87 gives the most conservative moment. In the long span range, a large variation of 75%-130% can be observed from the figures.

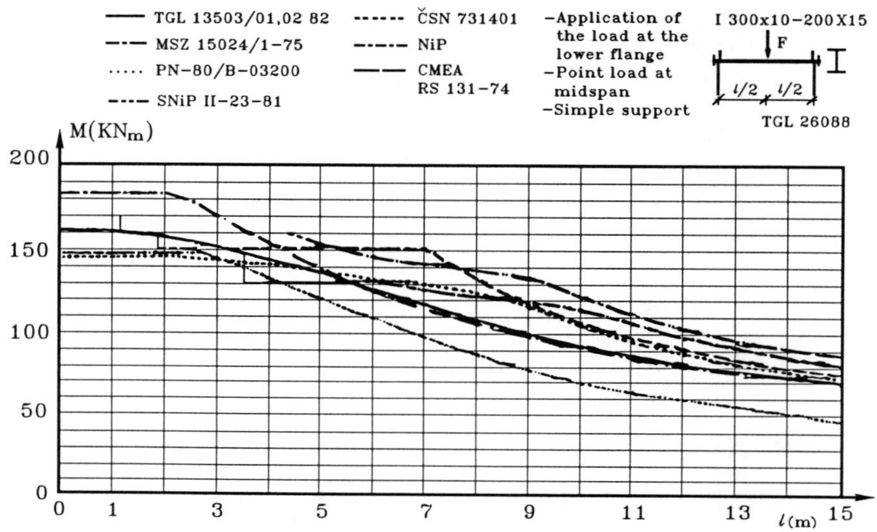

Fig. 3a.E6 Comparison of different specifications of East Europe

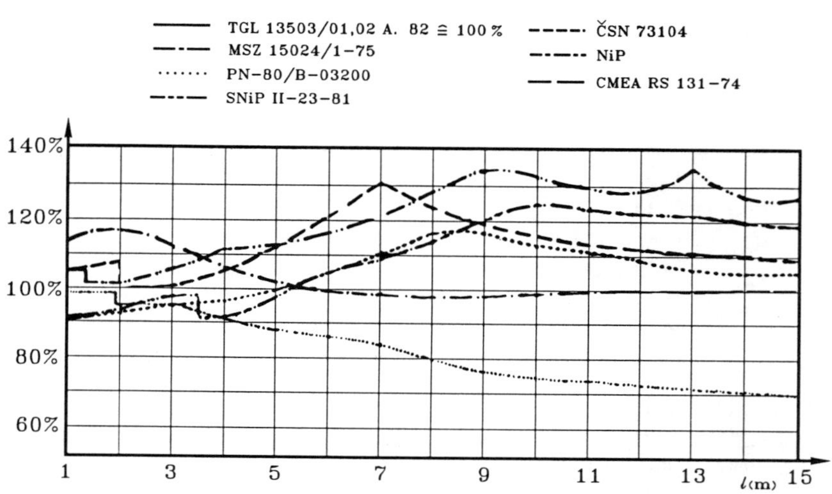

Fig. 3a.E7 Strength ratio comparison of different specifications of East Europe

JAPAN

CHAPTER EDITORS
Yuhshi Fukumoto, Osaka University
Jun Sakamoto, Nagoya University

AIJ 70

The allowable stress for rolled shapes and plate girders with H-section subjected to strong-axis bending is the larger one of the results obtained from the following two equations and shall not exceed the allowable tensile stress f_t.

$$f_b = \left[1 - 0.4\frac{\lambda_b^2}{C\,\Lambda^2}\right] f_t \qquad (3a.J1)$$

$$f_b = \frac{0.429\ E}{(l_b h)/A_f} \qquad (3a.J2)$$

where

A_f = sectional area of compression flange
C = $1.75 - 1.05(M_2/M_1) + 0.3(M_2/M_1)^2 \leq 2.3$
f_t = $\sigma_y/1.5$ (for long-term loading)
= σ_y (for short-term loading)
i = radius of gyration of T-section composed of compression flange and one-sixth of web about web axis
l_b = unsupported length of a compression flange
$\Lambda = \sqrt{\dfrac{\pi^2 E}{0.6\sigma_y}}$
$\lambda_b = l_b/i$

M_1 and M_2 are end moments, the larger of which is M_1. (M_2/M_1) is taken positive under single curvature bending and negative under double curvature bending. When the intermediate moment is greater than M_1, the C-value should be unity.

When the beam lacks possibility of lateral buckling as in an H-section beam bent about the weak axis of the section, f_b shall be f_t.

COMMENTARY

Equation 3a.J1 represents the flexural resistance of the compression flange to lateral instability and takes almost the same expression as the column curve applicable to short columns stressed beyond the elastic limit.

Equation 3a.J2 indicates torsional resistance of beams and practically applies to long beams. Both equations are originally applicable to uniformly bent beams and their use is expanded to general cases by introducing modification factor C.

AIJ 90b

Design strength is represented as ϕM, where ϕ is the resistant factor and M is the nominal strength.

For Beams Subjected to End Moment

Moment capacity is given by

for $\lambda_b \leq {}_p\lambda_b$

$$\phi_b = 0.90$$
$$M_{ux} = M_{px} \tag{3a.J3}$$

for ${}_p\lambda_b \leq \lambda_b \leq {}_e\lambda_b$

$$\phi_b = 0.90 - 0.05 \frac{\lambda_b - {}_p\lambda_b}{{}_e\lambda_b - {}_p\lambda_b} \tag{3a.J4}$$

$$M_{ux} = \{1 - 0.4 \frac{\lambda_b - {}_p\lambda_b}{{}_e\lambda_b - {}_p\lambda_b}\} M_{px} \tag{3a.J5}$$

for ${}_e\lambda_b \leq \lambda_b$

$$\phi_b = 0.85 \tag{3a.J6}$$

$$M_{ux} = \frac{M_{px}}{\lambda_b^2} \tag{3a.J7}$$

where

M_{ex} = elastic lateral buckling moment, is equal to α M_{exo}

M_{exo} = elastic lateral buckling moment under uniform moment

$= \sqrt{\pi^4 EI_y\, EI_w/\ell_b^4 + \pi^2 EI_y GJ/\ell_b^2}$

M_{px} = plastic moment about the strong axis

M_{ux} = moment capacity about the strong axis

α = modification factor, is equal to $1.75 - 1.05\beta_x + 0.3\beta_x^2 \leq 2.3$

β_x = M_2/M_1 (See Fig. 3a.J1)

λ_b = $\sqrt{M_{px}/M_{ex}}$

${}_p\lambda_b$ = $0.6 - 0.3\beta_x$

${}_e\lambda_b$ = $1/\sqrt{0.6}$

For Rectangular Hollow Sections and Rectangular Sections

Elastic lateral buckling moment is evaluated by

$$M_{exo} = \sqrt{\pi^2 EI_y GJ/\ell_b^2} \tag{3a.J8}$$

For Beams with Rectangular Sections and H-beams Subjected to Bending about the Weak Axis That Do not Buckle Laterally

Moment capacity is obtained by

$$M_p = \sigma_y Z_p \tag{3a.J9}$$

Strength and Stiffness Requirements for the Lateral Supports

(A) lateral supports along compression members

stiffness $K \geq 4/\ell_1 \cdot N_c$

strength $F \geq 0.02 \cdot N_c$

(B) lateral supports along the compression flange of beams

stiffness	$K \geq (4/\ell_1 \sim 5/\ell_1) \cdot M_p/h$
strength	$F \geq (0.02 \sim 0.03) \cdot M_p/h$

(C) lateral supports along beam-columns

stiffness	$K \geq b/\ell_1 \cdot N_y$
strength	$F \geq 0.03 \cdot N_y$

where

F = required strength of lateral supports
K = required stiffness of lateral supports
ℓ_1 = unsupported length
N_c = buckling strength of compression members
N_y = yield axial force

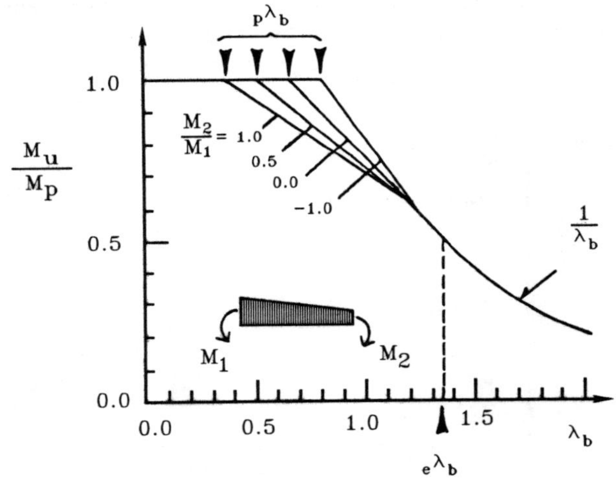

Fig. 3a.J1 Limit states of Lateral-Torsional Buckling

JRA 80

The allowable stress design approach is used, in which the equation giving the allowable stress in the compression flange is

for $\alpha > 0.2$

$$\sigma_b = \frac{\sigma_y}{1.7}[1-0.412(\alpha-0.2)] \tag{3a.J10}$$

for $\alpha \leq 0.2$

$$\sigma_b = \frac{\sigma_y}{1.7}$$

where

A_c = gross area of compression flange
A_w = gross area of web
b = compression flange width (cm)
L = unsupported flange length (cm)

$$\alpha = \frac{2}{\pi} K \frac{L}{b} \sqrt{3 + \frac{A_w}{2A_c}} \qquad \text{if } \frac{A_w}{A_c} > 2$$

$$\alpha = 2 \qquad \text{if } \frac{A_w}{A_c} \leq 2 \tag{3a.J11}$$

but L/b is not greater than 30.

COMMENTARY

Equation 3a.J10 is a modification of the lateral buckling strength of a simply supported I-beam under uniform moment. For unequal end moments, the bending stress σ_b is increased by M/M_{eq}, with not greater than maximum allowable compression stress in bending, in which M is the bending moment in the section being designed and $M_{eq} = 0.6\ M_1 + 0.4M_2$ but not less than $0.4\ M_1$, where $M_1 \geq M_2$.

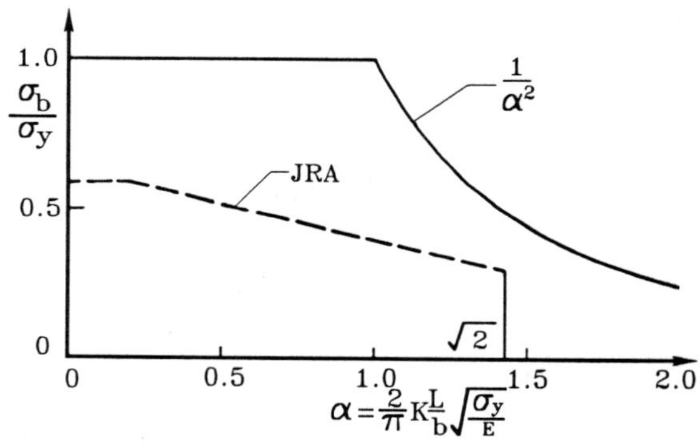

Fig. 3a.J2 JRA allowable stress design curve

JSCE 83

The allowable stress design approach is used, in which the equation giving the allowable stress in the compression flange is

for $0 \leq \bar{\lambda}_e \leq 0.3$

$$\sigma_b = \sigma_y / 1.9$$

for $0.3 \leq \bar{\lambda}_e \leq \sqrt{2}$

$$\sigma_b = (\sigma_y / \text{F.S.})[1 - 0.449(\bar{\lambda}_e - 0.3)] \tag{3a.J12}$$

for $\overline{\lambda}_e \geq \sqrt{2}$

$$\sigma_b = \frac{1}{2.8}\frac{\pi^2 E}{(L/r)_e^2}$$

where

b	= flange width(cm)	
F	$= \sqrt{12+2\beta/\alpha}$	(for I-beam)
F	$= 1.3\sqrt{3\alpha+\beta}\,\sqrt{b/L}$	if $\alpha \leq 2$ (for box)
F	$= 1.3\sqrt{6+\beta}\,\sqrt{b/L}$	if α=2 (for box)
L	= unsupported length of compression flange(cm)	
$(L/r)_e$	= F(L/b)	
α	=(flange thickness)/(web thickness)	
β	= (beam height)/b	
$\overline{\lambda}_e$	$= (1/\pi)\sqrt{\sigma_y/E}\,(L/r)_e$	(3a.J13)

COMMENTARY

This formula is a modification of the lateral buckling strength of simply supported beams under uniform moment. The factor of safety varies from 1.9 to 2.8 between $\overline{\lambda}_e$ =0.3 and $\sqrt{2}$.

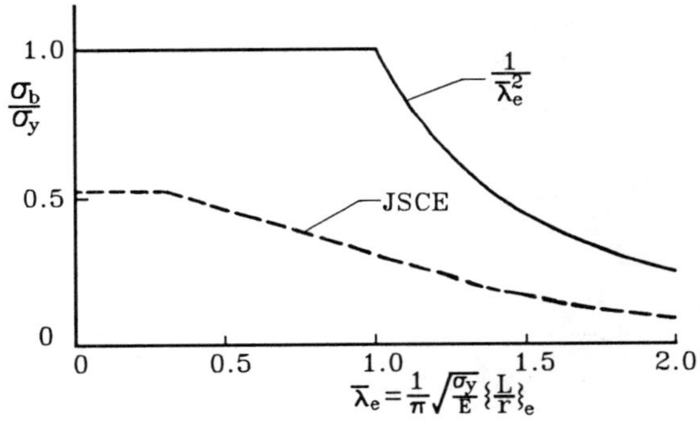

Fig. 3a.J3 JSCE allowable stress design curve

NORTH AMERICA

CHAPTER EDITOR
Theodore V. Galambos, University of Minnesota

The following symbols apply to the North American criteria (additional commonly used symbols are included in the chapter symbol list):

A_f = compression flange area
A_w = web area
b_f = flange width
C_b = $1.75 - 1.05(M_1/M_2) + 0.3(M_1/M_2)^2 \leq 2.3$
C_w = warping constant
F_{cy} = yield stress in compression for aluminum
F_r = maximum compressive residual stress in flange
F_{rR} = 10 ksi (69MPa), F_r for rolled shapes
F_{rW} = 16.5 ksi (114 MPa), F_r for welded shapes
F_y = yield stress
L_b = unbraced length
M_p = plastic moment
M_y = yield moment
M_1/M_2 = ratio of smaller to larger end moments in the unbraced length of the beam, positive when bending is reversed
r_t = radius of gyration of the compression flange plus 1/6 of the web
S_x = elastic section modulus
t_f = flange thickness
t_w = web thickness
Z_x = plastic section modulus

AA 87

Table 3a.N1 Definitions

Temper designation	-0, -H, $-T_1$, $-T_2$, $-T_3$, $-T_4$	$-T_5$, $-T_6$, $-T_7$, $-T_8$, $-T_9$
B_c	$\left(1 + \sqrt{10\ F_{cy}/E}\right)F_{cy}$	$\left(1 + \sqrt{4.44\ F_{cy}/E}\right)F_{cy}$
D_c	$\frac{B_c}{20}\sqrt{6B_c/E}$	$\frac{B_c}{10}\sqrt{B_c/E}$
C_c	$\frac{2B_c}{3D_c}$	$\frac{0.41\ B_c}{D_c}$

$$S_1 = \frac{L_{b1}}{r_y} = \frac{1.2(B_c - F_{cy})}{D_c} \qquad S_2 = \frac{L_{b2}}{r_y} = 1.2C_c \qquad K_b{}^2 \simeq C_b$$

$$r_{yeff} = \frac{K_b}{1.7}\sqrt{\frac{I_y d}{S_x}\sqrt{1 + 0.152\left(\frac{J}{I_y}\right)\left(\frac{L_b}{d}\right)^2}} \tag{3a.N1}$$

Ultimate Moment

for $\frac{L_b}{r_y} \le S_1$ $\quad M_u = F_{cy}S_x$

for $S_1 < \frac{L_b}{r_y} \le S_2$ $\quad M_u = \left(B_c - \frac{D_c}{1.2}\frac{L_b}{r_y}\right)S_x$ (3a.N2)

for $\frac{L_b}{r_y} > S_2$ $\quad M_u = \frac{\pi^2 E S_x}{\left(\frac{L_b}{1.2r_y}\right)^2}$

For a more exact solution replace r_y by r_{yeff}.

AASHTO 89

Definitions

for $C_b \le 1.16$ $\quad L_b^* = 1.300 r_y\sqrt{E/F_y}$

for $C_b > 1.16$ $\quad L_b^* = 2.228 r_y\sqrt{E/F_y}$ (3a.N3)

Ultimate Moment

for $L_b \le L_b^*$ $\quad M_u = F_y Z_x$

for $L_b^* < L_b \le \frac{0.690\ A_f E}{dF_y}$ $\quad M_u = F_y S_x$

for $L_b > \frac{0.690\ A_f E}{dF_y}$ $\quad M_u = K_c\left(1 - \frac{3}{\pi^2}\left(\frac{L_b}{b_f}\right)^2\left(\frac{F_y}{E}\right)\right)F_y S_x$ (3a.N4)

where

for $C_b \le 1.16$ $\quad K_c = 1.00$

for $C_b > 1.16$ $\quad K_c = 1.20$

AISC ASD 89

Definitions

$$L_{bc} = \text{MIN}\left[0.446 b_f\sqrt{E/F_y}\ ,\ \frac{0.690A_fE}{dF_y}\right] \tag{3a.N5}$$

$$L_{b1} = 1.875 r_t\sqrt{C_bE/F_y} \tag{3a.N6}$$

$$L_{b2} = 4.194 r_t\sqrt{C_bE/F_y} \tag{3a.N7}$$

$$L_{be} = \frac{0.690A_fC_bE}{dF_y} \tag{3a.N8}$$

for $L_b \le L_{b1}$ $\quad M_u = F_y S_x$

for $L_{b1} < L_b \le L_{b2}$ $\quad M_u = \left(1.111 - 0.0317(\frac{F_y}{E})(\frac{1}{C_b})(\frac{L_b}{r_t})^2\right) F_y\, S_x$

for $L_b > L_{b2}$

$$M_u = \frac{9.770 C_b E S_x}{(L_b/r_t)^2} \qquad (3a.N9)$$

for $L_b \le L_{be}$ $\quad M_{ue} = F_y S_x$

for $L_b > L_{be}$

$$M_{ue} = \frac{0.690 A_f C_b E S_x}{d L_b} \qquad (3a.N10)$$

Ultimate Moment

for $L_b \le L_{bc}$

$$M_u = 1.100 S_x F_y \qquad (3a.N11)$$

for $L_b > L_{bc}$ $\quad M_u = MAX(M_u, M_{ue})$

AISC LRFD 86

Definitions

$$\lambda = L_b/r_y, \quad M_p = F_y Z_x, \quad M_r = (F_y - F_r)\, S_x, \quad \lambda_p = 1.762\sqrt{E/F_y} \qquad (3a.N12)$$

$$X_1 = \frac{\pi}{S_x}\sqrt{\frac{EGJA}{2}}, \quad X_2 = \frac{4C_w}{I_y}\left(\frac{S_x}{GJ}\right)^2, \quad \lambda_r = \frac{X_1}{(F_y - F_r)}\sqrt{1+\sqrt{1+X_2(F_y - F_r)^2}} \qquad (3a.N13)$$

$$F_{cr} = \frac{\sqrt{2} C_b X_1}{\lambda}\sqrt{1 + \frac{X_1^2 X_2}{2\lambda^2}} \qquad (3a.N14)$$

Ultimate Moment

for $\lambda \le \lambda_p$ $\quad M_u = M_p$

for $\lambda_p \le \lambda \le \lambda_r$ $\quad M_u = C_b\{M_p - (M_p - M_r)\left(\frac{\lambda - \lambda_p}{\lambda_r - \lambda_p}\right)\} \le M_p$

for $\lambda > \lambda_r$

$$M_u = S_x F_{cr} \qquad (3a.N15)$$

AREA 87

Definitions

$M_{u1} = F_y\, S_x$ $\qquad M_{u2} = \left(1 - \frac{F_y}{62.07E}\left(\frac{L_b}{r_y}\right)^2\right) F_y\, S_x$

$M_{u3} = \frac{0.652\, A_f E S_x}{d\, L_b}$ $\qquad M_{u4} = MAX\,(M_{u2}, M_{u3})$

Ultimate Moment

$$M_u = MIN\,(M_{u1}, M_{u4}) \qquad (3a.N16)$$

CSA S16.1 M89 84b

Definitions

$$M_e = \frac{C_b\pi}{L_b}\sqrt{EI_yGJ + \frac{\pi^2E^2I_yC_w}{L_b^2}} \qquad M_p = F_y\,Z_x \qquad M_y = F_yS_x \qquad (3a.N17)$$

Ultimate Moment

for $\frac{b_f}{2t_f} \le 0.376\sqrt{E/F_y}$ and $\frac{d-2t_f}{t_w} \le 3.054\sqrt{E/F_y}$

for $M_e > 2M_p/3$

$$M_u = 1.15M_p\left(1 - \frac{0.28M_p}{M_e}\right) \le M_p$$

for $M_e \le 2M_p/3$

$$M_u = M_e$$

for $0.376\sqrt{E/F_y} \le \frac{b_f}{2t_f} \le 0.587\sqrt{E/F_y}$

and $3.054\sqrt{E/F_y} \le \frac{d-2t_f}{t_w} \le 4.052\sqrt{E/F_y}$

for $M_e > 2M_y/3$

$$M_u = 1.15M_y\left(1 - \frac{0.28M_y}{M_e}\right) \le M_y \qquad (3a.N18)$$

for $M_e \le 2M_y/3$

$$M_u = M_e$$

CSA S6 M89 84a

Definitions

$$M_y = S_x\,F_y \qquad M_1 = (1.1 - 0.004L/r_t)M_y \qquad M_2 = \frac{0.692A_fES_x}{dL_b}$$

$$M_3 = \frac{\pi^2ES_x}{(L_b/r_t)^2} \qquad M_4 = MIN(M_1,M_3) \qquad M_5 = MAX(M_2,M_4)$$

Ultimate Moment

for $b_f/2t_f \le 0.447\sqrt{E/F_y}$ and $(d-2t_f)/t_w \le 1.498\sqrt{E/F_y}$ (3a.N19)

$$M_u = MIN(M_y,\ M_5)$$

OHBDC 83

Definitions

for $M_1/M_2 > 0.5$

$$L_p = 1.230r_y\sqrt{E/F_y}$$

for $M_1/M_2 \le 0.5$

$$L_p = 2.192 r_y \sqrt{E/F_y} \tag{3a.N20}$$

$$M_p = F_y Z_x \qquad M_y = F_y S_x \qquad M_e = \frac{\pi}{L_b}\sqrt{E I_y G J + \frac{\pi^2 E^2 I_y C_w}{L_b^2}} \tag{3a.N21}$$

$$L_{r1} = \frac{0.700 A_f E}{d\, F_y} \qquad L_{r2} = 2.915 r_t \sqrt{E/F_y} \qquad L_r = \mathrm{MAX}(L_{r1}, L_{r2})$$

(A) Compact shape

$$\frac{b_f}{2t_f} \le 0.376\sqrt{E/F_y} \quad \text{and} \quad \frac{d - 2t_f}{t_w} \le 3.054\sqrt{E/F_y}$$

(B) Non-compact shape

$$0.376\sqrt{E/F_y} \le b_f/2t_f \le 0.587\sqrt{E/F_y} \quad \text{and}$$

$$3.054\sqrt{E/F_y} \le \frac{d - 2t_f}{t_w} \le 4.052\sqrt{E/F_y}$$

Ultimate Moment

(A) Compact shapes

for $L_b \le L_p$

$$M_u = M_p$$

for $L_b > L_r$

for $M_e > 2M_p/3$

$$M_u = 1.15 M_p\left(1 - \frac{0.28 M_p}{M_e}\right) \le M_p \tag{3a.N22}$$

for $M_e \le 2M_p/3$

$$M_u = M_e$$

(B) Non-compact shapes

for $L_b \le L_r$

$$M_u = M_y$$

for $L_b > L_r$

for $M_e > 2M_p/3$

$$M_u = 1.15 M_y\left(1 - \frac{0.28 M_y}{M_e}\right) \le M_y \tag{3a.N23}$$

for $M_e \le 2M_p/3$

$$M_u = M_e$$

REGIONAL COMPARISON

The lateral-torsional buckling strength is characterized by a reduced moment capacity as the slenderness ratio L_b/r_y increases (see Fig. 3a.N1). The maximum strength is terminated either by the plastic moment M_p (AISC ASD 89, AISC LRFD 86, AASHTO 89, CSA S16.1 M89 84b and OHBDC 83) or by the yield moment M_y (AREA 87, AA 87 and CSA S6 M89 84b). At the other end of the length scale, elastic buckling governs, typified by the characteristic Euler hyperbola. The exact theoretical elastic buckling formulas are used by AISC LRFD 86, CSA S16.1 M89 84b and OHBDC 83, while the other specifications use conservative approximations which are sometimes very conservative (see Fig. 3a.N1). None of the specifications use an effective length approach to account for end restraint. The greatest spread between the different specifications is in the transition range between elastic and plastic buckling. Some use step changes

(AISC ASD 89, AASHTO 89) while others have linear transitions (AA 87, see Fig. 3a.N2 and AISC LRFD 86) or non-linear transitions (CSA S16.1 M89 84b, OHBDC 83). The remaining two specifications (AREA 87, CSA S6 M89 84b) do not permit plastification.

The liberalizing effect of moment gradient is accomplished by the C_b factor in different ways by AISC ASD 89, AISC LRFD 86, AA 86, AASHTO 89 and CSA S16.1 M89 84b, (see Fig. 3a.N3), whereas AREA 87, CSA S6 M89 84a and

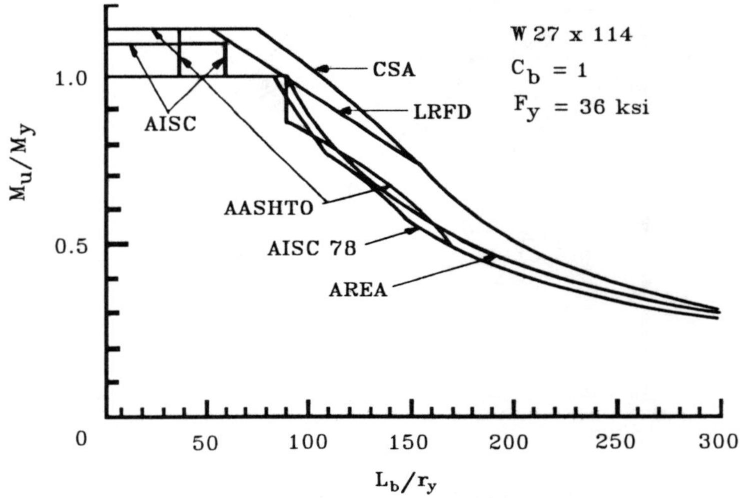

Fig. 3a.N1 Limit states of lateral torsional buckling

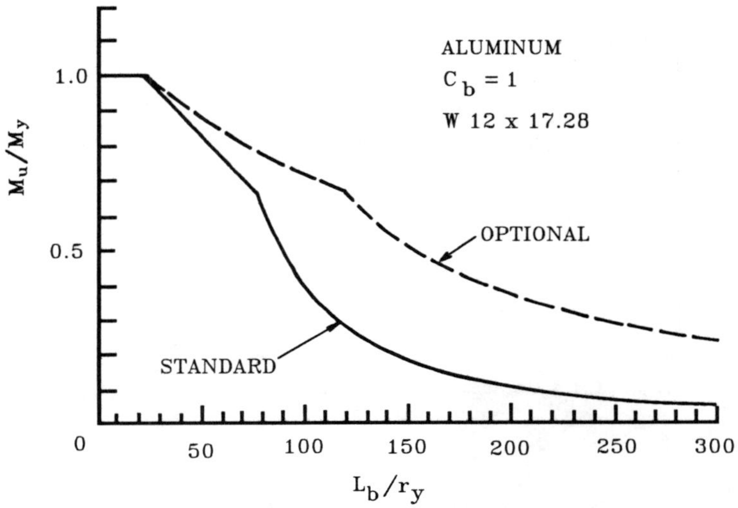

Fig. 3a.N2 Limit states of lateral torsional buckling

OHBDC 83 do not provide a permission to utilize this effect. The aluminum specification (AA 87) provides a conservative lower bound capacity and an optional method based on the exact elastic buckling curve (Fig. 3a.N4).

For each of these specifications the moments are determined by elastic analysis if lateral-torsional buckling controls. For plastic analysis further criteria apply.

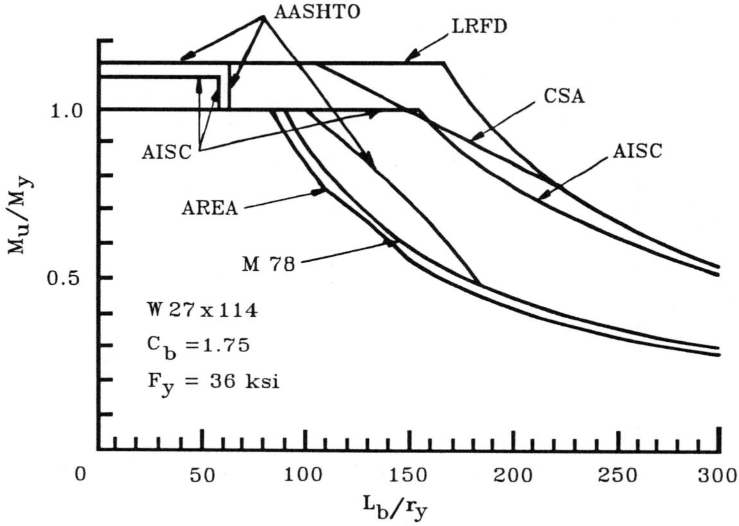

Fig. 3a.N3 Limit states of lateral torsional buckling

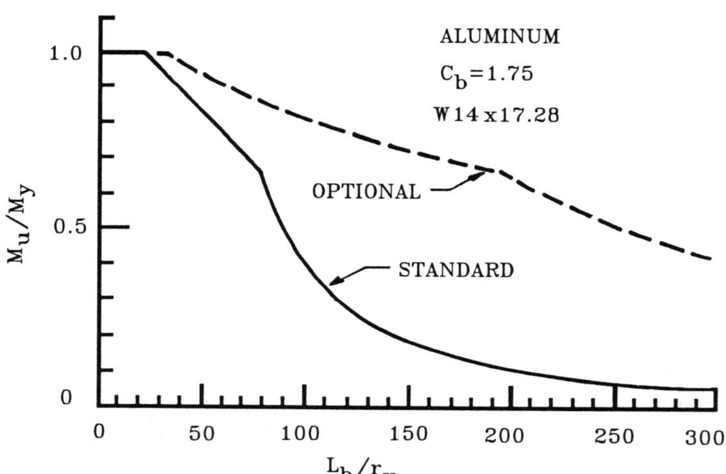

Fig. 3a.N4 Limit states of lateral torsional buckling

WEST EUROPE

CHAPTER EDITOR

Joachim Lindner, Technische Universität Berlin

BS 5400 P.3 82

For compact sections the limiting compressive stress should be taken from Fig. 3a.W1, depending on the slenderness parameter λ_{LT}.

The value of λ_{LT} should be determined appropriate to the type of beam. For uniform I-sections Eq. 3a.W1 should be used

$$\lambda_{LT} = \frac{l}{r_y} k_4 \eta v \tag{3a.W1}$$

where

$k_4 =$ 0.9 for rolled I or channel sections
$=$ 1.0 for all other beams
$v =$ slenderness factor, dependent on the shape of the beam, for beams with equal flanges ranging between 0.468 and 1.0, normally around 0.8
$\eta =$ 1.0 for beams restrained by U-frames or continuously by a deck

Eq. 3a.W1 corresponds to Eq. 3a.W3 where $u = k_4$ and $n = \eta$.
For other sections the limiting compression stress should be taken from Fig. 3a.W2.

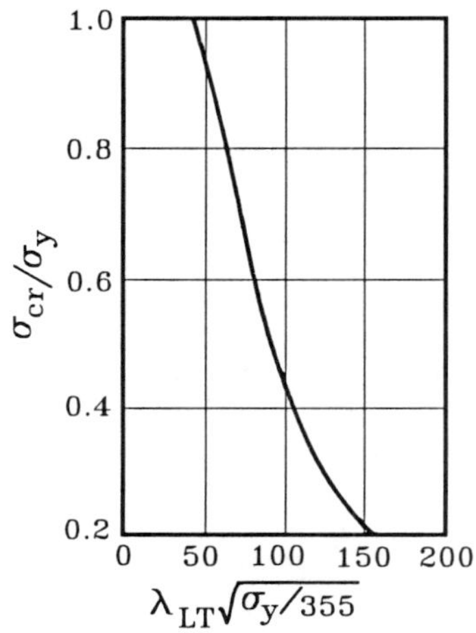

Fig. 3a.W1 Critical compressive stress for compact sections in BS 5400

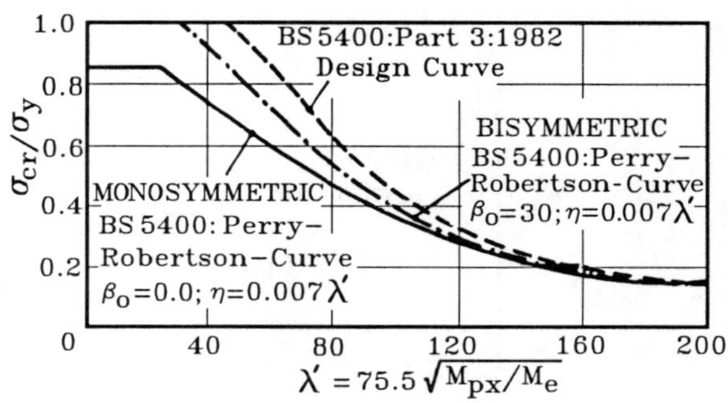

Fig. 3a.W2 Critical compressive stresses in BS 5400

BS 5950 P.5 85

The following conditions should be satisfied

$$m\,M_x \leq M_{cr} \qquad (3a.W2)$$

where

M_{cr} = lateral torsional buckling resistance moment
= $Z_x\,\sigma_{cr}$ which σ_{cr} is obtained from different tables

m = an equivalent uniform moment factor, normally m=1 for variable end moments applied to beams of uniform sections with equal flanges, not loaded in the length between adjacent lateral restraint:
= $0.57 + 0.33\beta + 0.10\beta^2 \geq 0.43$

The buckling moment M_{cr} can also be obtained from Fig. 3a.W3. The Perry coefficient η_{LT} is given for rolled sections and for welded sections, respectively (see Fig. 3a.W3).

$$\lambda_{L0} = 0.4\sqrt{\frac{\pi^2 E}{\sigma_y}}$$

$$\lambda_{LT} = \text{Equivalent slenderness} = n\,u\,v\,\lambda \qquad (3a.W3)$$

where

n = slenderness correction factor, normally 1.0, for members with applied loading within the unrestrained length between 0.65 and 1.0.

u = buckling parameter, from tables, formulae or conservatively
= 0.9 rolled sections
= 1.0 other sections

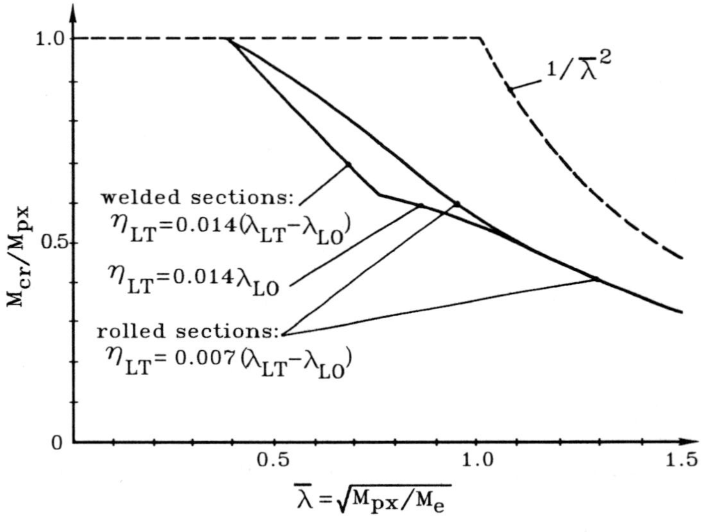

Fig. 3a.W3 Buckling moment M_{cr} in BS 5950

v = slenderness factor depending especially on the torsional index and the actual slenderness, for beams with equal flanges normally between 0.8 and 1.0.

λ = l/r_y

COMMENTARY

1. Application of the rules of BS 5950 85 is facilitated by the provision of tables. Direct member selection is possible using tables prepared by Constrado (1985).
2. The approach of BS 5950 allows effects such as variations in the condition of loading and/or lateral support, non-uniform sections, and unequal flanged sections to be treated by modifications to the basic format of Eq. 3a.W2. Provision is made within the methods of both BS 5950 and BS 5400 for using research data within the framework of the code rules.
3. The basic design curves for the procedures of both BS 5950 and BS 5400 correspond to a Perry-type fit to available test data. The curves are based on an approximate lower bound.
4. Application of the rules of BS 5400 for non-compact sections requires care, particularly when unequal flanged girders are used. New tests by O'hEachteirn-Nethercot (1985) on welded monosymmetric girders have led to the revised proposal in Fig. 3a.W2.

BSK 87

Bracing Requirement

Effect of lateral-torsional buckling on the bending moment capacity is ignored if Eq. 3a.W4 is satisfied.

$$\frac{\ell_1}{b_{tot}} \leq [0.6-0.2\ M_2/M_1-0.1(M_2/M_1)^2]\ \sqrt{\frac{E_k}{f_{yk}}} \tag{3a.W4}$$

where

b_{tot} = the total width of the compression flange

ℓ_1 = the distance between the brace at the plastic hinge and the nearest brace

M_1 = the numerically largest moment within the beam portion

M_2 = the moment at the opposite end of the beam portion. The ratio M_2/M_1 is positive if the moments are in the same direction, and negative otherwise

If forces and moments are calculated by elastic theory, the distance between bracing points can be 20% greater than the limiting value of Eq. 3a.W4.

Bending Moment Capacity M_{Rcd}

M_{Rcd} due to lateral-torsional buckling is calculated by Eq. 3a.W5.

$$M_{Rcd} = \omega_b \cdot \eta_c \cdot W_c \cdot f_{yd} \tag{3a.W5}$$

where

M_{cr} = elastic lateral-torsional buckling moment (See Handbook Bygg)

W_c = section modulus of gross section for extreme compression fiber

η_c = shape factor for extreme compression fiber

$$\lambda_b = \sqrt{\frac{\eta_c W_c f_{yk}}{M_{cr}}} \tag{3a.W6}$$

ω_b = the reduction factor for lateral-torsional buckling. ω_b is calculated from the following formulas or from Fig. 3a.W4 for channel, unisymmetric or bisymmetric I-section.

1) Hot rolled beam

$$\omega_b = \frac{1.02}{\sqrt{1+\lambda_b^4}} \leq 1.0$$

2) Welded beam in Cross Section Class 1 (plastic hinge)

$$\omega_b = \frac{1.04}{(1+\lambda_b^3)^{2/3}} \leq 1.0$$

3) Welded beam in Cross Section Class 2 (yield strength)

$$\omega_b = \frac{1.16}{1+\lambda_b^2} \leq 1.0$$

For bisymmetric I-beams, λ_b in Eq. 3a.W7 is used.

$$\lambda_b = \sqrt{\frac{\eta_c W_c f_{yk}}{\kappa_m M_{1cr}}} \tag{3a.W7}$$

where

$$M_{1cr} = \frac{\pi}{\kappa_{cr} \ell_1} \sqrt{B_y\left(C + \frac{\pi^2 C_w}{\ell_1^2}\right)}$$

κ_{cr} = $0.6+0.3\ M_2/M_1+0.1(M_2/M_1)^2$

κ_m = $0.8+0.2(M_2/M_1)$, but at least 0.8

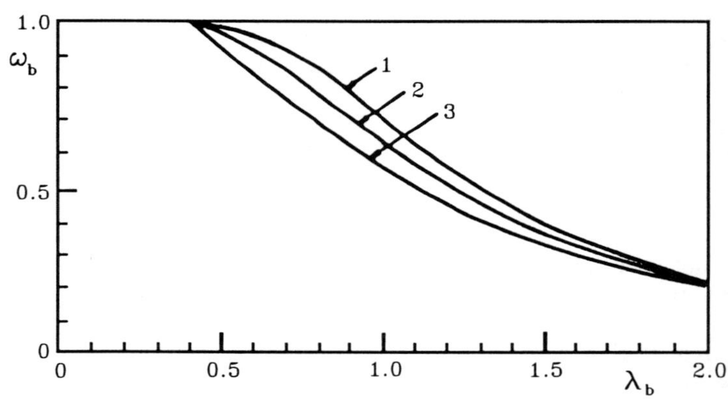

Fig. 3a.W4 The Reduction Factor ω_b for
1. Hot Rolled Beam
2. Welded Beam (compact section)
3. Welded Beam (non-compact section)

The design condition is that $\kappa_m M_1$ should be less than M_{Rcd} and that M_1 should be less than M_{Rcd} with ω_b=1.

CNR 10011 85

General Verification

The verification of sufficient lateral-torsional buckling safety shall be carried out by Eq. 3a.W8.

$$\sigma_b \leq \frac{\sigma_{cr}}{\nu} \qquad (3a.W8)$$

where

ν = safety factor

$\sigma_b = \frac{M_x}{Z_x}$ = actual bending stress

$\sigma_{cr} = \frac{M_{cr}}{Z_x}$ = critical lateral-torsional buckling stress

Especially for doubly symmetrical I-sections, Eq. 3a.W9 is given

$$\sigma_{cr} = \alpha\, \sigma_y \frac{\sigma_e}{\mathrm{n}\sqrt{\sigma_e^n + (\alpha\, \sigma_y)^n}} \qquad (3a.W9)$$

where

n = system factor
= 1.5 for rolled sections
= 1.0 for welded sections

α = Z_x/S_x shape factor

σ_e = elastic lateral-torsional buckling stress

Compression Flange

The lateral-torsional buckling safety may be verified by a check of the compression flange, Eq. 3a.W10

$$\frac{\omega_1\, M_{eq}}{Z_x} \leq \sigma_{ca} \qquad (3a.W10)$$

where

M_{eq}= equivalent moment

$$\omega_1 = \frac{\sigma_y}{0.585\ \mathrm{E}} \frac{\mathrm{d\ L}}{\mathrm{b\ t}_r} \qquad (3a.W11)$$

Values for ω_1 are given in a table for Fe 360, Fe 430, Fe 510.

COMMENTARY

1. The basic equation Eq. 3a.W9 corresponds to the relevant one of EC 3, Eq. 3a.W20. The system factor n is chosen as a very small value due to the lower limit statical evaluations. This does not seem to be necessary.

2. The simplified approach for the compression flange corresponds also in principle to Eq. 3a.W22 of Eurocode 3. A more conservative value of 300 instead of 250 was chosen for the limitation of no reduction.

CTICM 80

General Verification

The verification of sufficient lateral-torsional buckling shall be carried out by Eq. 3a.W15. The reduction factor κ_m is calculated by Eq. 3a.W16.

where

$$n = \text{system factor} = 2.0 \text{ for rolled sections} = 1.5 \text{ for welded sections} \quad (3a.W12)$$

Spacing of Supports for Lateral Buckling

No special check with regard to general verification is necessary if the distance L_o between two supports for the lateral bracing of members in bending satisfies Eq. 3a.W13 or Eq. 3a.W14

for $-1 \leq \beta \leq 0.625$

$$L_o \leq (60-40)\, i_y \sqrt{\frac{235}{\sigma_y}} \quad (3a.W13)$$

for $0.625 \leq \beta \leq 1.0$

$$L_o \leq 35\, i_y \sqrt{\frac{235}{\sigma_y}} \quad (3a.W14)$$

σ_y = yield stress (MPa)

DIN 18800 P.2 86

General Verification

For beams with I-sections and channel sections without torsion, Eq. 3a.W15 should be fulfilled

$$\frac{M_x}{\kappa_M\, M_{px}} \leq 1 \quad (3a.W15)$$

where

n = system factor
= 2.5 for rolled sections, 2.0 for welded and coped sections, 1.5 for castellated beams,
= $0.7+1.8\, h_{min}/h_{max}$ for haunched beams
= For uniform moment n is multiplied by 0.8

κ_M = reduction factor

$$\kappa_M = 1 \quad \text{for} \quad \overline{\lambda} \leq 0.4$$
$$\kappa_M = \left(\frac{1}{1+\overline{\lambda}^{2n}}\right)^{1/n} \quad \text{for} \quad \overline{\lambda} > 0.4 \quad (3a.W16)$$

$\overline{\lambda}$ = non-dimensional slenderness $= \sqrt{\frac{M_{px}}{M_e}}$

Additional formulas are given in the commentary for EC 3 84 to calculate the elastic critical torsional buckling moment M_e.

Compression Flange

If the compression flange of I-beams with regard to the web is supported at spacing c, Eq. 3a.W17 should be fulfilled

$$\frac{c\ k_c}{\bar{r}_y\ \lambda_a} \leq 0.5 \qquad \text{(3a.W17)}$$

where

k_c = for compression force pattern
= 1.0 constant distribution
= 0.94 parabolic distribution
= $1/(1.33-0.3\ \beta)$ linear distribution

$\bar{r}_y$ = radius of gyration for the compression flange, including 1/5 of the web area

β_s = elastic limit stress

λ_a = $\pi\ \sqrt{E/\beta_s}$

If Eq. 3a.W17 is not fulfilled, the simplified check of Eq. 3a.W18 may be used

$$\frac{0.843\ M_x}{\kappa\ M_{px}} \leq 1 \qquad \text{(3a.W18)}$$

where κ = reduction factor for out-of-plane buckling, using the European buckling curve c and the slenderness given by Eq. 3a.W17, left hand side

Torsional Restraint Coefficient for Hot Rolled Beams

Refer to Eq. 3a.W26.

COMMENTARY

1. Equation 3a.W16 is based on the interpretation of theoretical values taking into account inelastic material behavior, residual stresses, and initial imperfections from test data. Special attention in the tests, carried out in Germany, was given to rolled sections used in building structures. This leads to the system factor n=2.5 as Lindner (1980) has reported. The smaller system factors for the other beam types were also established by test results. See Gietzelt (1983) and Lindner (1985).

2. Equation 3a.W17 was established by comparing this equation with the limitation of Eq. 3a.W16. The factor k_c was found by comparing the lateral-torsional buckling load by taking into account top flange loading with the out-of-plane buckling load for compressed members.

   Equation 3a.W18 is similar to that of a member under axial loading but taking into account the before- mentioned limitation of Eq. 3a.W17 that leads (for $\overline{\lambda} = 0.5$) to a buckling reduction factor of 0.843.

3. Especially in building structures, beams are very often restrained torsionally by adjacent members. The torsional restraint coefficient of the designed construction shall be determined by taking into account the flexible connection between the supported beam and the element providing the torsional restraint.

EC 3 84

General Verification

The verification of sufficient lateral-torsional buckling safety shall be carried out by Eq. 3a.W19, which is seen also in Fig. 3a.W5

$$M_x/\kappa_m M_{px} \leq 1 \qquad (3a.W19)$$

where

$$\kappa_m = \left(\frac{1}{1+\overline{\lambda}^5}\right)^{0.4}, \text{ reduction coefficient} \qquad (3a.W20)$$

$$\overline{\lambda} = \sqrt{M_{px}/M_e}, \text{ non-dimensional slenderness} \qquad (3a.W21)$$

For single beams with IPE- or HE-sections, the simplified approach for M_e (Eq. 3a.W22) may be used by

$$M_e = Z_x\ 0.66\ \frac{E\ t_r\ b}{L\ d} \qquad (3a.W22)$$

Furthermore instead of Eqs. 3a.W19 the following Eq. 3a.W23 may be used

$$\frac{L\ d}{b\ t_r} \leq 250\sqrt{\frac{235}{\sigma_y}} \quad \text{: no check necessary} \qquad (3a.W23)$$

Spacing of Supports for Lateral Buckling

No special check with regard to Eq. 3a.W19 is necessary if the distance L_0 between two supports for the lateral bracing of members in bending satisfy Eqs. 3a.W24 or 3a.W25.

for $-1 \leq \beta < 0.5$

$$L_0 \leq (60-40)\ i_y\sqrt{\frac{235}{\sigma_y}} \qquad (3a.W24)$$

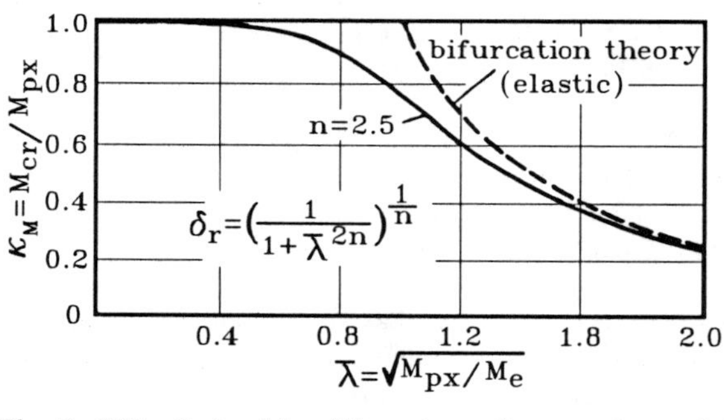

Fig. 3a.W5 Lateral buckling strength curve for n=2.5

for $\beta > 0.5$

$$L_0 \leq 40\, i_y \sqrt{\frac{235}{\sigma_y}} \tag{3a.W25}$$

where β is the ratio of smaller to larger end moment.

Torsional Restraint Coefficient for Hot Rolled Beams

Providing the torsional restraint supplied by floor systems and/or bracing to hot rolled beams of IPE or HE sections provide torsional restraint of a stiffness at least equal to that defined by Eq. 3a.W26 they may be designed based on simple plastic theory.

$$c_\phi \geq \frac{M_{px}^2}{E\, I_y} k_\phi \tag{3a.W26}$$

where

c_ϕ = torsional restraint coefficient
k_ϕ = coefficient, depending on the moment distribution, varying from 1.0 to 4.0

Compression Flange

The lateral-torsional buckling safety may be verified by the check of the compression flange, Eq. 3a.W27

$$\sigma_b \leq \frac{\kappa\, \sigma_y}{0.85} \tag{3a.W27}$$

where

κ = buckling coefficient of the European buckling curve c, depending on $\overline{\lambda}$
$\overline{\lambda}$ = non-dimensional buckling slenderness
σ_b = bending stress in the center of compression flange

COMMENTARY

1. Eq. 3a.W19 is based on the interpretation of theoretical studies in which allowance has been made for inelastic material behavior, residual stresses, initial imperfection, as well as test data. In establishing this curve the following are assumed:
   - The cross section is of hot-rolled I-shape or similar. Welded cross sections with heavy welds such as used in Japan, are not taken into account in evaluating test data.
   - The cross section is constant along the beam length.
   - All eccentricities are accidental.
   - Distortion and local buckling are prevented.
   - Plate buckling effects are not taken into account. Therefore the relevant b/t-ratios must be fulfilled.

   The most important parameter is the system factor n that leads to the numbers 0.4 and 5 in Eq. 3a.W20. Here the choice of n=2.5 is intended to ensure that the design curve corresponds to a mean value for all available test results (including welded section), rather than a lower bound. The design curve corresponds to a lower bound evaluating rolled sections only with nominal yield strength and dimensions.

2. The simplified approach for M_e neglects the effect of warping constant I_w and the direct consideration of the level of application of the load. Fig. 3a.W6 shows comparisons between the accurate solution of M_e and the approximate value of Eq. 3a.W22.

$$F_1 = \frac{M_e\ (\text{approx.})}{M_e\ (\text{accurate})}$$

3. Equations 3a.W24 and 3a.W25 are similar to those used in North America, avoiding a jump. Comparison to ultimate load calculations, see Fig. 3a.W7, shows that for $\beta < -0.7$ Eq. 3a.W24 is slightly on the unconservative side.

4. Values for torsional restraint coefficients (Eq. 3a.W26) take into account especially for initial imperfection of beams whereas lower values published earlier neglected this effect. The actual coefficient c of the designed construction shall be determined by taking into account the flexible connection between the supported beam and the element providing the torsional restraint. (Lindner/Gregull, 1986).

5. Equation 3a.W27 was established by comparing values of this equation to the limitation of $\overline{\lambda} = 0.4$. Therefore no further check by Eq. 3a.W19 is necessary.

6. The final form of these criteria has not yet been completed as of 1989. It has been decided to use one of the European column curves instead of Eq. 3a.W19. The parameter $\overline{\lambda}$ from Eq. 3a.W21 is used.

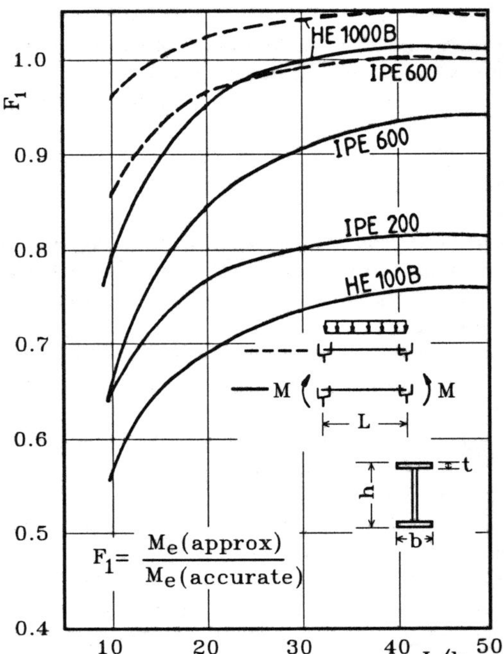

Fig. 3a.W6 Comparison of elastic critical moment M_e

NBN B51002 88

General Verification

An equivalent slenderness should be calculated from Eqs. 3a.W28 or 3a.W29.

$$\lambda_{ed} = \frac{L_o}{r_y\, C_d} \qquad (3a.W28)$$

$$\lambda_{ed} = 3.90 \sqrt{\frac{L\, d}{b\, t_r} \frac{1}{C_d}} \qquad (3a.W29)$$

where

$C = 0.6 + 0.4\, M_2/M_1 \geq 0.4$
C_d = correction factor for non-uniform moment $= 1/C$
β, r_y = the same as in Eurocode 3

From the equivalent slenderness a non-dimensional slenderness is calculated $\overline{\lambda} = \lambda_{ed}/\lambda_s$ and this leads to a reduction factor κ using the European buckling curve that would be applied to flexural buckling of the beam about its weak axis (see Chapter 1: Compression Members)

Spacing of Supports for Lateral Bracing

The regulation is the same as in Eurocode 3.

COMMENTARY

Equation 3a.W29 goes back to the solution of Eq. 3a.W22. In an appendix of the code it is shown that the more accurate formula for $M_{e,cr}$ can be reduced to that of Eq. 3a.W29. The use of buckling curves seems conservative because the influence of the torsional rigidities J, I_w are underestimated.

NEN 8351 72

General Verification

For elastically designed statically determinate beams with support, the largest compressive stress due to the design load should not exceed σ_{cr} calculated by Eq. 3a.W30

for $\frac{L_o\, d}{b\, t} \leq 250$

$$\sigma_{cr} = \sigma_y$$

for $250 < \frac{L_o\, d}{b\, t} < \frac{0.66\, E}{0.7\, \sigma_y}$

$$\sigma_{cr} = \sigma_y \left(1 - \frac{\frac{L_o\, d}{b\, t} - 250}{\frac{0.66\, E}{0.7\, \sigma_y} - 250}\, 0.3\right)$$

for $\frac{L_o\, d}{b\, t} \geq \frac{0.66\, E}{0.7\, \sigma_y}$

$$\sigma_{cr} = \frac{0.66\, E}{\frac{L_o\, d}{b\, t}} \qquad (3a.W30)$$

where

L_o = unbraced length
t = thickness of flange

The rules are valid for doubly-symmetrical I-sections only when the point of application of the load must not be higher than 0.1d above the flange. It is also assumed that the cross section is non-deformable, fulfilling the following conditions.

$$\frac{d}{t_w} \leq 75 \qquad (3a.W31)$$

$$\frac{L_o}{d} \geq 1.25\,\frac{b}{t} \qquad (3a.W32)$$

Similar equations are given for statically indeterminate beams where the support consists of connection through the bottom flange only and the beams with deformable cross-sections.

Spacing of Supports for Lateral Bracing of Plastically Designed Beams

For doubly-symmetrical I-sections and non-deformable cross-sections no additional check is necessary if Eq. 3a.W33 and 3a.W34 are fulfilled

$$\frac{L_o}{r_y} \leq (65-50\beta) \qquad (3a.W33)$$

$\geq$ 40, $\leq$ 65 for steel grades Fe 310, 360, 430

$$\frac{L_o}{r_y} \leq (50-40\,\beta) \qquad (3a.W34)$$

$\geq$ 30, $\leq$ 50 for steel grade F 510

This applies to those parts of the member where the moment exceeds $M_e = W_e$. In that case the smallest moment should be adopted for M. For the rest of the member, where M is smaller than M_e, the maximum permissible unsupported length l_{kr} can be calculated by determining the length for which the rules for elastic lateral buckling, as given in 3.7.1 of the specification.

COMMENTARY

1. The simplified approach of Eq. 3a.W30 neglects the effect of the warping constant I_w. The point of application of the load is indirectly included. Both lead to an elastic critical lateral buckling stress of $\sigma_e = 0.66\ E/(L_o\ d/b\ t)$. On the assumption that the maximum residual stress present in the beam is 0.3 σ_y, this formula is used up to the limit σ_e=0.7 σ_y. From a comparison to weak axis buckling of the compression flange it is assumed that there is no risk of lateral-torsional buckling if L d/(b t) $\leq$ 250. There is chosen a linear interpolation between these two limitations. The influence of initial imperfections is not taken into account.

2. Eqs. 3a.W33 and 3a.W34 are similar to those used in other countries. The upper limit seems conservative.

3. A new code is under preparation. There the general verification is proposed to be used with regard to the ECCS recommendations, especially Eurocode 3. This is here mentioned Eurocode 3. Two different system factors (n=2.5, 1.5) are suggested.

NS 3472 84

General Verification
The control of lateral-torsional buckling strength is given by Eq. 3a.W35

$$M_x \leq M_{cr} \tag{3a.W35}$$

where

M_{cr} = lateral-torsional buckling moment capacity

$= \sigma_{cr} S_x$ (3a.W36)

n = 2.0 for rolled sections

σ_{cr} = factored strength

$= \sigma_y \left(\frac{1}{1+\bar{\lambda}^{2n}}\right)^{1/n}$ (3a.W37)

σ_y = factored material resistance.

Compression Flange
The compression flange of the beam is calculated as a single member in compression. The member slenderness is calculated based on the areas of the compression flange plus 1/3 of the web area in compression.

COMMENTARY

1. Compared to other codes, Eqs. 3a.W36 and 3a.W37 seem very conservative. They do not take into account the full plastic moment M_p even for very small slenderness. Furthermore the system factor n is chosen smaller than in other codes. On the other hand this code is based on the factored strength and factored material resistance, the factors are not given in this code.
2. The regulation for the compression flange is similar to that in other codes.

ÖNORM B4600 78

No detailed check of lateral-torsional buckling safety is necessary if Eq. 3a.W38 is fulfilled for the compression flange of beams with I-sections.

$$r_y \geq c/40 \tag{3a.W38}$$

where c is the spacing of supports for lateral bracing.

Otherwise $$\sigma_{cr} = 1.13\,\sigma_k \tag{3a.W39}$$

where σ_k is the critical buckling strength for axial loaded cross sections, depending on c/r_y.

COMMENTARY

The regulation Eq. 3a.W38 is very simple assuming the compression flange as an isolated component. In contrast to other codes (Eurocode, F.R. Germany), the value r_y is calculated without taking into account the destabilizing effect of compression stresses in the web.

SIA 161 79

General Verification

The lateral buckling stress σ_{cr} is calculated by Eq. 3a.W40

$$\sigma_{cr} = \sigma_y (\frac{1}{1+\overline{\lambda}^{4.5}})^{0.45} \qquad (3a.W40)$$

where

$\overline{\lambda}$ = as defined by Eq. 3a.W21 for compact sections.
= $\sqrt{\sigma_y/\sigma_e}$ for non-compact sections.

Additional expressions are given to calculate the elastic critical lateral-torsional buckling moment M_e.

Critical Buckling Strength

If the actual length is smaller than the critical buckling length of Eq. 3a.W41 the resistance bending moment may be used as the full plastic moment M_{px}.

$$\frac{L_0}{r_y} \leq 2.7 \sqrt{\frac{E}{\sigma_y}} (1-0.5\beta) \qquad (3a.W41)$$

COMMENTARY

1. Eq. 3a.W40 is very similar to Eq. 3a.W19. The system factor n is chosen to n=2.25 instead of n=2.5. It should be taken into account that the code is also used for steel bridges, where plate buckling effects will influence the resisting moment.
2. Eq. 3a.W41 is similar to other regulations, such as Eurocode 3. The limiting values for β=1 are 40.5 (St 37) and 33 (St 52). For β=−1 the corresponding values of 121.5 (St 37) and 99 (St 52) seem high, compared to the results of Fig. 3a.W7.

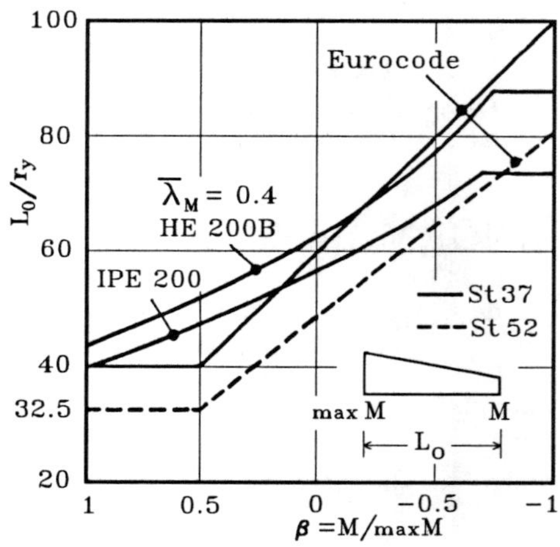

Fig. 3a.W7 Unbraced length vs. moment ratio β

INTERNATIONAL COMPARISONS & COMMENT

WHAT ARE THE DIFFERENCES?

The basic beam strength formulas specified in the referenced specifications are compared graphically with the experimental mean curve (M) and the mean minus two standard deviations curve (M−2S), which are determined through statistical manipulation of test data(NDSS) (Fukumoto and Itoh, 1984) in Figs. 3a.1 through 3a.8.

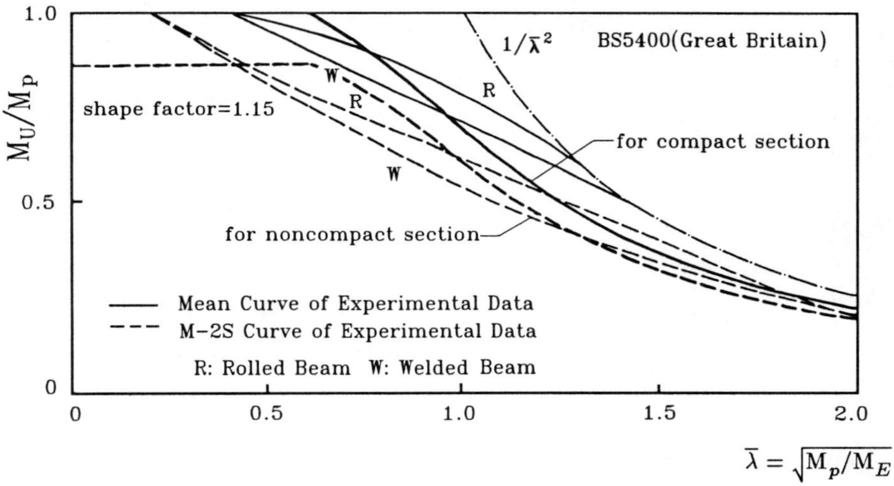

Fig. 3a.1 Comparison between experimental data and BS 5400 82

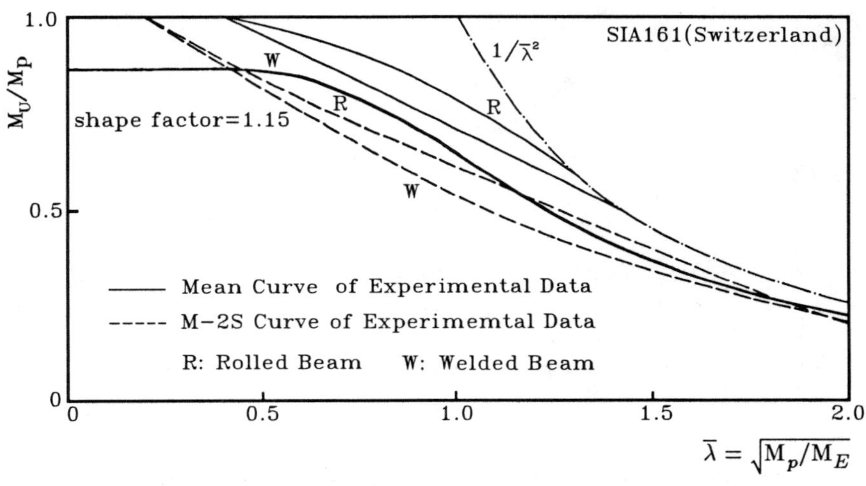

Fig. 3a.2 Comparison between experimental data and SIA 161 79

The local buckling of plate components is not considered.

In all figures, the beam strength M_u is normalized by the full plastic moment M_p, and the beam slenderness is given by $\bar{\lambda} = \sqrt{M_p/M_e}$, where M_e denotes the elastic lateral-torsional buckling moment. In the provisions where the beam strength formulas are normalized by either yield stress σ_y or yield moment M_y, the shape factor of the cross section is assumed as 1.15.

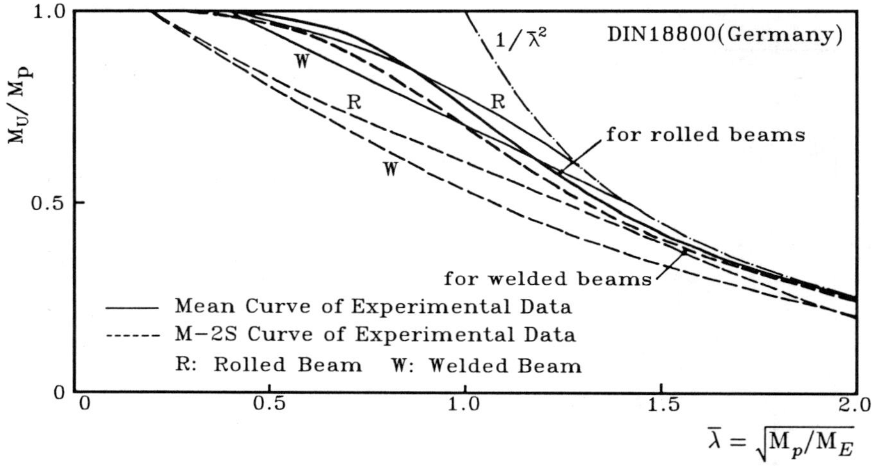

Fig. 3a.3 Comparison between experimental data and DIN 18800 P.2 86

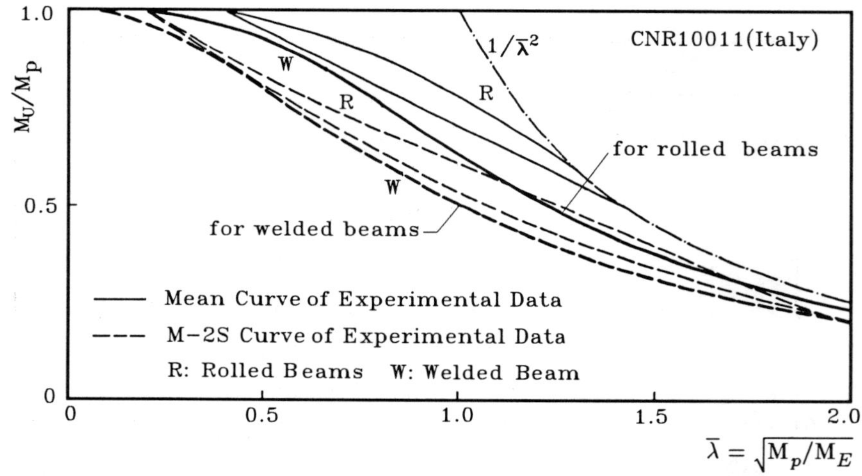

Fig. 3a.4 Comparison between experimental data and CNR 10011 85

All available test data are included in Figs. 3a.1 through 3a.8. These data also include tests where cross-sectional properties, out-of-straightness and yield stress were not reliably reported. The comparison between the experimental statistical representation in these figures should not be considered to be absolutely accurate but only as a reflection of tendencies and trends.

Among the specifications compared in the figures:

(1) AISC LRFD and DIN 18800 provide the highest strength prediction in the transition range of λ and they represent the mean curve of the test data.

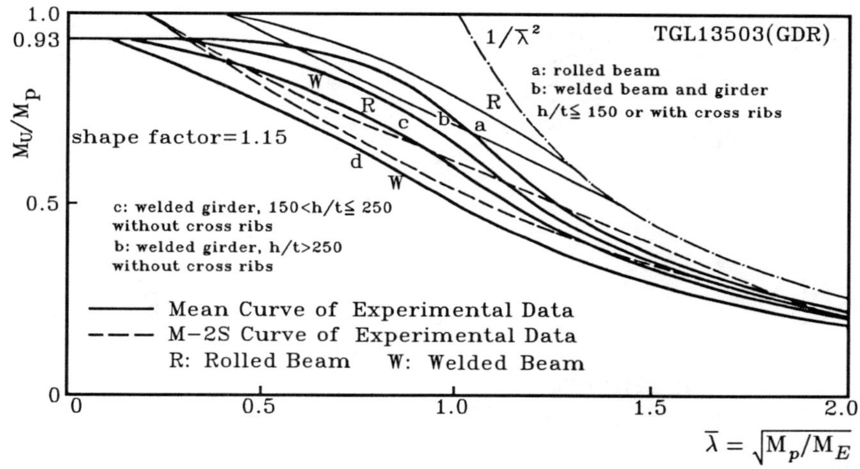

Fig. 3a.5 Comparison between experimental data and TGL 13503 82

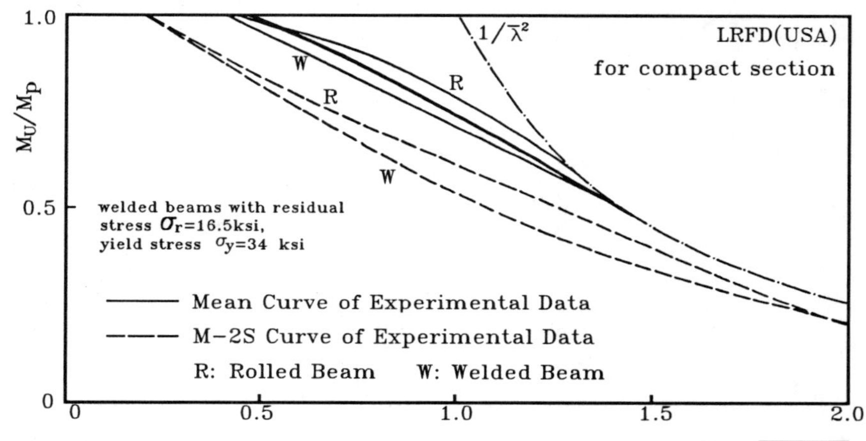

Fig. 3a.6 Comparison between experimental data and AISC LRFD 86

(2) SIA 161 and GBJ 88 give a similar single curve which ranks in the lower bound of the test data.
(3) CNR 10011 for welded beams predicts the lower bound of the welded beam test data.
(4) JRA, which is for welded sections, provides a simple linear strength prediction between the maximum moment capacity M_y and a elastic buckling strength point of $\lambda=\sqrt{2}$.
(5) TGL 13503 provides multi-buckling curves (a $\sim$ d) for rolled or welded beams and for different web slenderness ratios. The lower bound of the welded beam test data falls between curves c and d.

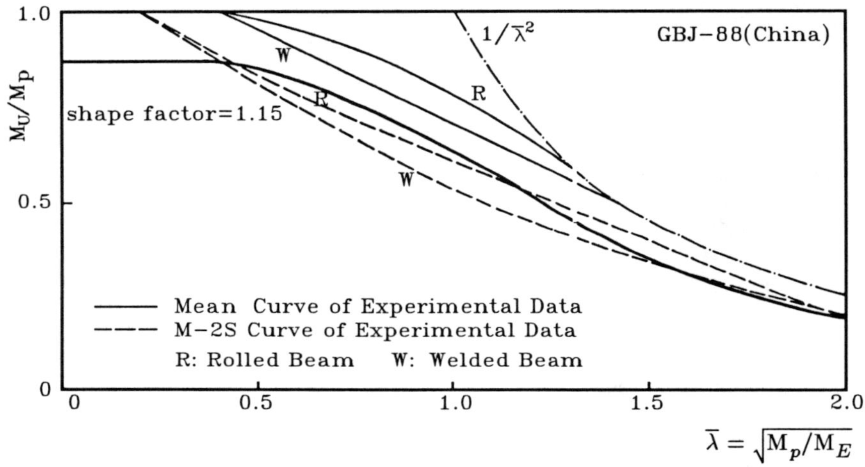

Fig. 3a.7 Comparison between experimental data and GBJ 17 88

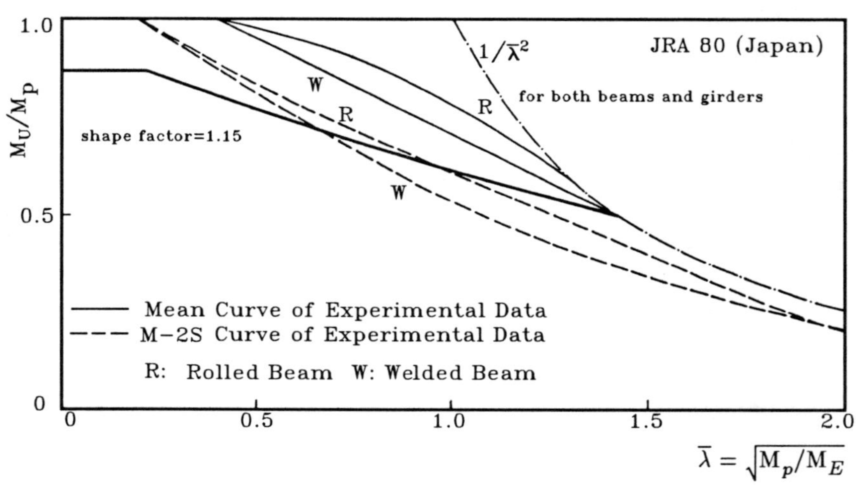

Fig. 3a.8 Comparison between experimental data and JRA 84

The following provisions for flexural members are included in the recommendations, specifications or codes:

Table 3a.1 Comparison of lateral-torsional buckling

Country	Specification	Item								
		(1)	(2)	(3)	(4)	(5)	(6)	(7)	(8)	(9)
Eurocode	EC 3 88	o	o			o				
Great Britain	BS 5400 P.3 82		o		o		o		o	o
	BS 5950 P.5 85	o	o	o	o		o		o	o
Switzerland	SIA 161 79	o	o	o		o	o			
Austria	ÖNORM B4600 78	o	o							
Norway	NS 3472 84		o		o					
FRG	DIN 18800 P.2 86	o	o	o	o	o	o	o		
Netherlands	NEN 8351 72	o	o			o				
Belgium	NBN B51002 88	o	o			o	o			
Sweden	BSK 87	o	o	o	o	o	o			o
Italy	CNR 10011 85	o	o		o					
France	CTICM 80	o	o		o					
CMEA	RS131 74	o	o				o	o	o	o
USSR	SNIP II-23-81 82	o	o		o		o	o		o
GDR	TGL 13503 82		o		o	o	o	o		
Poland	PN 80/B-03200 87	o	o				o	o		o
Czechoslovakia	ČSN 73 1401 84	o	o				o	o	o	o
Hungary	MSZ 15024/1 85		o				o	o	o	o
Bulgaria	NiP 565/86 87	o	o		o		o	o		o
USA	AISC ASD 89	o	o				o			
	AASHTO 89	o	o				o			
	AISC LRFD 86	o	o	o	o		o			
	AA 87		o				o			
Canada	CSA S16.1 M89 84b	o	o				o			
	CSA S6 M89 84a		o				o			
China	GBJ 17 88	o	o	o			o	o		o
Japan	JRA 80		o			o	o			
	AIJ 70		o			o	o			

Explanation for Item in Table 3a.1

(1) slenderness limitation for plastic strength (unsupported length for plastic design).
(2) reduction factor for lateral-torsional buckling (beam strength curve).
(3) distinction between compact section members and noncompact ones.
(4) classification of strength curve due to residual stress level and fabrication process(rolled or welded).
(5) approximate formula to evaluate the slenderness parameter.
(6) effect of moment distribution on lateral-torsional buckling.
(7) effect of loading position on lateral-torsional buckling.
(8) effect of end support conditions on lateral-torsional buckling, explicitly considered by formula.
(9) special provisions for mono-symmetric section members.

WHY ARE THERE DIFFERENCES?

(1) All specifications are based on the well-established theory of elastic lateral-torsional buckling, which is the upper bound for all design criteria.
(2) All specifications have their maximum ultimate capacity at short unbraced lengths either the fully plastic moment, the yield moment, or an intermediate value between the two.
(3) All specifications provide some form of an empirical transition between the maximum capacity at two lengths to the elastic buckling range, with or without some empirical consideration for initial out-of-straightness.
(4) The differences between the various specifications are 1] because of simplifying approximations or 2] because of deliberate introduction of conservative criteria, or 3] because some specifications aim to represent mean predictions while others aim for a lower bound of the tests and the predictions.
(5) Some standards distinguish between rolled and welded sections, while most ignore this difference.
(6) Many standards consider elastic buckling as the limit for long beams, while some incorporate initial out-of-straightness.
(7) Some standards do not distinguish between various loading conditions, while many do account for point- load and moment gradient loads.

Further comparisons are given by Lindner and Galambos in the Proceedings of the SSRC 4th International Colloquium, North American Session of 1989 (New York).

3b. Flange Local Buckling

SPECIFICATION PROVISIONS & EXPLANATION

The limiting slenderness ratios are defined by using different expressions in each specification. In this discussion, two generalized expressions are employed for the unified comparison of each specification.

Limiting Slenderness Ratio $(b/t)_0$

for flanges: $(b/t_f)_0 = C_1\sqrt{(E/\sigma_y)}$

for webs : $(h/t_w)_0 = C_2\sqrt{(E/\sigma_y)}$

Limiting Slenderness Parameter $\overline{\lambda}_0$

$$\overline{\lambda}_0 = \frac{b}{t}\sqrt{\frac{12(1-\nu^2)}{k\pi^2}\;\frac{\sigma_y}{E}}$$

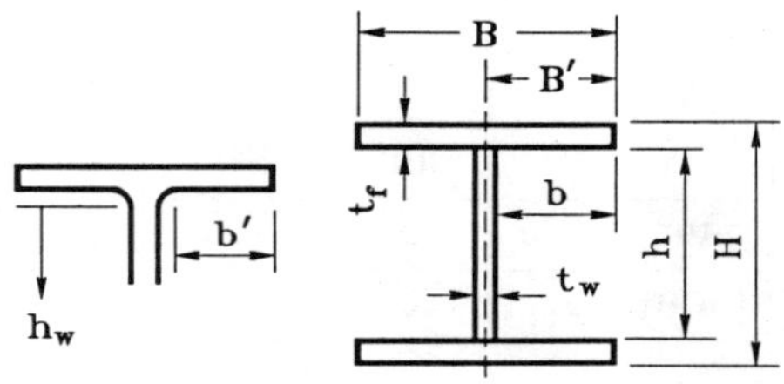

Definition of Cross-Sectional Dimensions

where buckling coefficient k=0.425 for flanges and k=23.9 for webs.

AUSTRALIA

CHAPTER EDITOR

Mark A. Bradford, The University of New South Wales

AS DR87164 87

Limit states design concept, in which the limiting slenderness of compression flanges for full plastic moment and yield moment at a section, and the reduction factor of cross-sectional strength are given as follows:

Limiting Slenderness

The slenderness of the outstanding flange plates is defined as

$$\lambda_s = b/t \cdot \sqrt{(\sigma_y/250)} \qquad (3b.A1)$$

where σ_y is yield strength (MPa)

Reduction Factor (effective section modulus)

for $\overline{\lambda}_{0p} \geq \overline{\lambda}$ $\qquad Z_e = Z_p$

for $\overline{\lambda}_{oy} \geq \overline{\lambda} > \overline{\lambda}_{0p}$ $\qquad Z_e = Z_y + \dfrac{(\overline{\lambda}_{0y}-\overline{\lambda})}{(\overline{\lambda}_{0y}-\overline{\lambda}_{0p})}(Z_o-Z_y)$

for $\bar{\lambda} > \bar{\lambda}_{0y}$

$$Z_e = (\bar{\lambda}_{0y}/\bar{\lambda}) \cdot Z_y$$

where

Z_p = plastic section modulus
Z_y = elastic section modulus

Table 3b.A1 Limiting Slenderness of Flange Plates

Plate Type	λ_{sp}	λ_{sy}	$(b/t_f)_{0p}$	$(b/t_f)_{0y}$	$\bar{\lambda}_{0p}$	$\bar{\lambda}_{0y}$
			$\sqrt{E/\sigma_y}$	$\sqrt{E/\sigma_y}$		
SR	10	17	0.345	0.587	0.553	0.941
HR	9	16	0.311	0.552	0.498	0.886
LW,HW,CF	8	15	0.276	0.518	0.443	0.830

SR: stress relieved, HR: hot-rolled or hot-finished, LW: light welded, HW: heavily welded, CF: cold formed

CHINA

CHAPTER EDITOR

Zhi Bin Xia, Zhejiang University

GBJ 17 88

Flange local buckling of I-sections is prevented by limiting the width-thickness ratio of the compression flange due to bending.

The following are recommended

for beams under direct dynamic loads

$$\frac{b}{t} \leq 30\sqrt{\frac{235}{f_y}} \text{ (MPa)}$$

for beams under indirect dynamic loads or under static loads

$$\frac{b}{t} \leq 26\sqrt{\frac{235}{f_y}}$$

for plastic design

$$\frac{b}{t} \leq 18\sqrt{\frac{235}{f_y}} \qquad (3b.C1)$$

TBJ 2 85

The Specification recommends

$$\frac{b}{t} \leq 20 \text{ (for } 16M_{nq} \text{ steel)} \qquad (3b.C2)$$

where

b = width of the compression flange
t = thickness of the compression flange

JAPAN

CHAPTER EDITORS
Yuhshi Fukumoto, Osaka University
Jun Sakamoto, Nagoya University

AIJ 70

The allowable stress design approach is used, in which the limiting slenderness of compression flanges for yield moment for the compression flange is

Limiting Slenderness

$$\text{LSY:} \quad (b/t_f)_{0y} = 0.524\sqrt{E/\sigma_y} \qquad \bar{\lambda}_{0y} = 0.841 \tag{3b.J1}$$

Effective Width

When the slenderness of the flange plate exceeds the limiting slenderness, the areas of flange plate outside the limiting width are not used.

AIJ 90b

According to the limit states design concept, the limiting slendernesses of plate elements for plastic analysis, full plastic moment and yield moment at a section, and the reduction factor of cross-sectional strength are given as follows:

Classification of Section and Limiting Slenderness

P-I: Plastic analysis can be used in the load combination including seismic load, for $h/t_w \le 100/\sqrt{\sigma_{yw}}$

$$\left(\frac{b/t_f}{20/\sqrt{\sigma_{yf}}}\right)^2 + \left(\frac{h/t_w}{127/\sqrt{\sigma_{yw}}}\right)^2 \le 1 \tag{3b.J2}$$

P-II: Plastic analysis can be used in the load combination without seismic load, for $h/t_w \le 100/\sqrt{\sigma_{yw}}$

$$\left(\frac{b/t_f}{21/\sqrt{\sigma_{yf}}}\right)^2 + \left(\frac{h/t_w}{136/\sqrt{\sigma_{yw}}}\right)^2 \le 1 \tag{3b.J3}$$

P-III: Elastic analysis is used, and the yield moment can be taken as the resistance of section. for $h/t_w \le 100/\sqrt{\sigma_{yw}}$

$$\left(\frac{b/t_f}{23/\sqrt{\sigma_{yf}}}\right)^2 + \left(\frac{h/t_w}{148/\sqrt{\sigma_{yw}}}\right)^2 \le 1 \tag{3b.J4}$$

P-IV: Elastic analysis is used, and the resistance of section is given by multiplying the reduction factor by the yield moment.

$$17/\sqrt{\sigma_{yf}} \le b/t_f \le 37/\sqrt{\sigma_{yf}} \quad \text{or}$$

$$100/\sqrt{\sigma_{yw}} \le h/t_w \le 276/\sqrt{\sigma_{yw}}$$

Reduction Factor

$$M_u = S_x \cdot R_{pg} \cdot R_e \cdot \sigma_{cr} \tag{3b.J5}$$

where

a_r = ratio of web area to compression flange area

h_c = twice the distance from the neutral axis to the inside face of the compression flange

R_e = hybrid girder factor (for non-hybrid girder, $R_e = 1.0$)

$R_{pg} = 1-0.0005 \cdot a_r \cdot \left(\frac{h_c}{t_w} - \frac{276}{\sqrt{\sigma_{yw}}}\right) \leq 1$

σ_{cr} = critical compression flange stress

for $b/t_f \leq 17/\sqrt{\sigma_{yf}}$ $\quad \sigma_{cr} = \sigma_{yf}$

for $17/\sqrt{\sigma_{yf}} \leq b/t_f \leq 37/\sqrt{\sigma_{yf}}$ $\quad \sigma_{cr} = (67 - b/t_f \cdot \sqrt{\sigma_{yf}}) \cdot \sigma_{yf}/50$

for $37/\sqrt{\sigma_{yf}} \leq b/t_f$ $\quad \sigma_{cr} = 800/(b/t_f)^2$

JRA 80

According to the allowable stress design approach, the limiting slenderness of compression flanges for yield moment and the reduction factor of the allowable stress for the compression flange with the slenderness beyond the limiting slenderness are given as follows:

Limiting Slenderness

$$\text{LSY: } (b/t_f)_{0y} = 0.436\sqrt{E/\sigma_y} \qquad \bar{\lambda}_{0y} = 0.7 \tag{3b.J6}$$

Reduction Factor

$$\sigma_u/\sigma_y = (0.7/\bar{\lambda})^2 \leq 1.0 \tag{3b.J7}$$

Maximum slenderness ratio for flange $(b/t_f)_{max}$=16

The allowable stress is given by

$$\sigma_{allow} = \sigma_u/1.7. \tag{3b.J8}$$

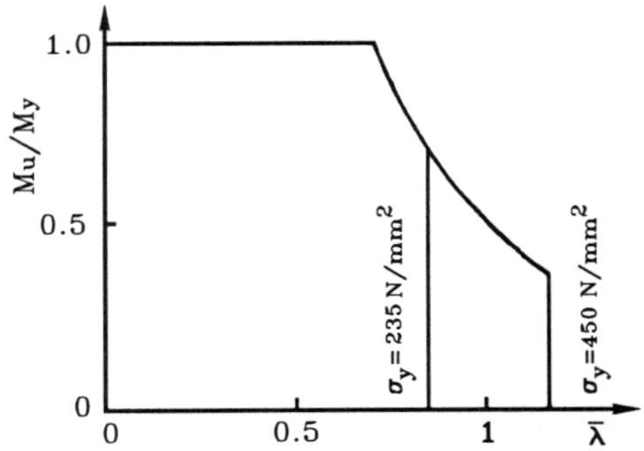

Fig. 3b.J1 Reduction curve for flange local buckling (JRA 80)

JSCE 83

According to the allowable stress design approach, the limiting slenderness of compression flanges for yield moment and the reduction factor of the allowable stress for the compression flange with the slenderness beyond the limiting slenderness are given as follows:

Limiting Slenderness

$$\text{LSY:} \quad (b/t_f)_{0y} = 0.430\sqrt{E/\sigma_y} \qquad \bar{\lambda}_{0y} = 0.7 \tag{3b.J9}$$

Reduction Factor

$$\sigma_u/\sigma_y = (0.7/\bar{\lambda})^2 \leq 1.0 \tag{3b.J10}$$

Maximum slenderness ratio for flange $(b/t_f)_{max} = 16$

The allowable stress is given by

$$\sigma_{allow} = \sigma_u/1.9 \tag{3b.J11}$$

NORTH AMERICA

CHAPTER EDITOR
Theodore V. Galambos, University of Minnesota

AA 87

Table 3b.N1 Definitions

Temper Designation	-0,-H, $-T_1$, $-T_2$, $-T_3$, $-T_4$	$-T_5$, $-T_6$, $-T_7$, $-T_8$, $-T_9$
B_p	$[1+2.385(\frac{\sigma_{cy}}{E})^{\frac{1}{3}}]\sigma_{cy}$	$[1+1.890(\frac{\sigma_{cy}}{E})^{\frac{1}{3}}]\sigma_{cy}$
D_p	$\frac{B_p}{20}\sqrt{\frac{6B_p}{E}}$	$\frac{B_p}{10}\sqrt{\frac{B_p}{E}}$
K_1	0.50	0.35
K_2	2.04	2.27

$$S_1 = \left(\frac{b_f}{2t_f}\right)_1 = \frac{B_p - \sigma_{cy}}{5.1D_p} \qquad S_2 = \left(\frac{b_f}{2t_f}\right) = \frac{K_1 B_p}{5.1D_p} \qquad S = \frac{b_f}{2t_f}$$

Ultimate Moment

for $S \le S_1$ $\quad M_u = \sigma_{cy} S_x$

for $S_1 < S \le S_2$ $\quad M_u = [B_p - 5.1D_p S]\, S_x$

for $S > S_2$

$$M_u = \frac{K_2\, S_x \sqrt{B_p\, E}}{5.1S} \qquad (3b.N1)$$

AASHTO 89

The limit states design concept is used, in which the limiting slendernesses of compression flanges for full plastic moment and yield moment at a section are given as follows:

Limiting Slenderness

$$\text{LSP:} \quad (B/2t_f)_{0p} = 0.297\sqrt{E/\sigma_y} \qquad \overline{\lambda}_{0p} = 0.476 \qquad (3b.N2)$$

$$\text{LSY:} \quad (B/2t_f)_{0y} = 0.371\sqrt{E/\sigma_y} \qquad \overline{\lambda}_{0y} = 0.595 \qquad (3b.N3)$$

Reduction Factor

There is no provision of the strength reduction of outstanding plates.

AISC ASD 89

The allowable stress design concept is used, in which the limiting slendernesses of compression flanges for full plastic moment and yield moment at a section, and the reduction factor of cross-sectional strength are given as follows:

<u>Limiting Slenderness</u>

LSP: $(B/2t_f)_{0p} = 0.382\sqrt{E/\sigma_y}$ $\qquad \bar{\lambda}_{0p} = 0.613$ (3b.N4)

LSY: $(B/2t_f)_{0y} = 0.558\sqrt{E/\sigma_y}$ $\qquad \bar{\lambda}_{oy} = 0.895$ (3b.N5)

<u>Reduction Factor</u>

for $(B/2t_f) \leq 0.382\sqrt{E/\sigma_y}$

$$M_u = 1.100\ S_x\sigma_y$$

for $0.382\sqrt{E/\sigma_y} < (B/2t_f) \leq 0.558\sqrt{E/\sigma_y}$

$$M_u=[1.317-0.568(B/2t_f)\sqrt{E/\sigma_y}]S_x\sigma_y$$

for $0.558\sqrt{E/\sigma_y} < (B/2t_f) \leq 1.034\sqrt{E/\sigma_y}$

$$M_u=[1.415-0.744(B/2t_f)\sqrt{E/\sigma_y}]S_x\sigma_y$$

for $1.034\sqrt{E/\sigma_y} < (B/2t_f)$

$$M_u=[0.690(E/\sigma_y)/(B/2t_f)^2]S_x\sigma_y \qquad (3b.N6)$$

where S_x is the section modulus referred to compression flange.

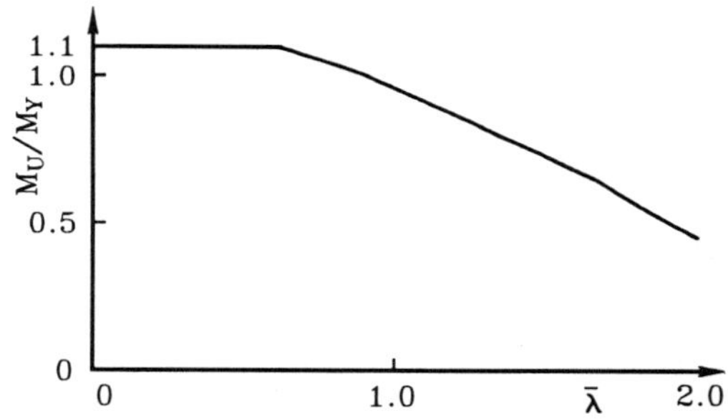

Fig. 3b.N1 Reduction curve for flange local buckling (AISC ASD 89)

AISC LRFD 86

According to the limit states design concept, the limiting slenderness of compression flanges for full plastic moment and yield moment at a section, and the reduction factor of cross-sectional strength are given as follows:

Limiting Slenderness

LSP: $(\mathrm{B}/2\mathrm{t}_f)_{0p} = 0.382\sqrt{\mathrm{E}/\sigma_y}$ $\qquad \overline{\lambda}_{0p} = 0.613$ (3b.N7)

LSY: for rolled shapes:

$$\overline{\lambda}_{0r} = (\mathrm{B}/2\mathrm{t}_f)_{0y} = \frac{0.828\sqrt{\mathrm{E}/\sigma_y}}{\sqrt{1-\sigma_{r\mathrm{R}}/\sigma_y}} \qquad \overline{\lambda}_{0y} = 1.328/\sqrt{1-\sigma_{r\mathrm{R}}/\sigma_y}$$

for welded shapes:

$$\overline{\lambda}_{0r} = (\mathrm{B}/2\mathrm{t}_f)_{0y} = \frac{0.622\sqrt{\mathrm{E}/\sigma_y}}{\sqrt{1-\sigma_{w\mathrm{R}}/\sigma_y}} \qquad \overline{\lambda}_{0y} = 1/\sqrt{1-\sigma_{w\mathrm{R}}/\sigma_y} \qquad (3b.N8)$$

where

$\sigma_{r\mathrm{R}}$ = residual compression stress in flanges of rolled shape
$\sigma_{w\mathrm{R}}$ = residual compression stress in flanges of welded shape

Reduction Factor

$\overline{\lambda} = \mathrm{B}/2\mathrm{t}_f$ $\qquad \mathrm{M}_r = \mathrm{S}_x(\sigma_y - \sigma_r)$

$\overline{\lambda} \leq \overline{\lambda}_{0p}$ $\qquad \mathrm{M}_u = \mathrm{M}_p$

$\overline{\lambda}_{0p} < \overline{\lambda} \leq \overline{\lambda}_{0r}$ $\qquad \mathrm{M}_u = \mathrm{M}_p - (\mathrm{M}_p - \mathrm{M}_r)(\overline{\lambda} - \overline{\lambda}_{0p})/(\overline{\lambda}_{0r} - \overline{\lambda}_{0p})$

$\overline{\lambda}_{0r} < \overline{\lambda}$ $\qquad \mathrm{M}_u = \mathrm{S}_x\sigma_{cr}$ (3b.N9)

where

$\sigma_{cr} = 0.690(\mathrm{E}/\sigma_y)/\overline{\lambda}^2$ for rolled shapes
$\sigma_{cr} = 0.386(\mathrm{E}/\sigma_y)/\overline{\lambda}^2$ for welded shapes

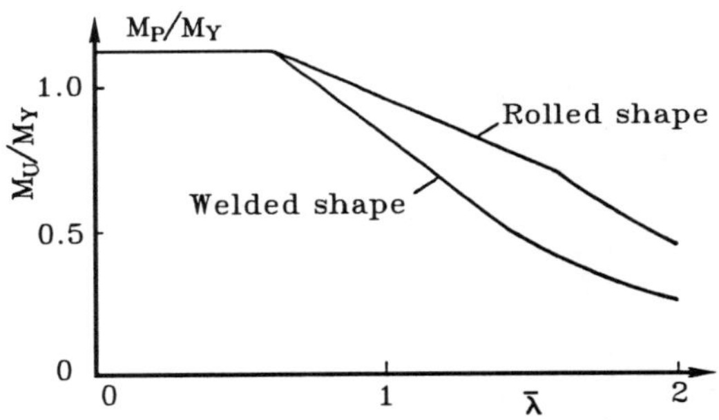

Fig. 3b.N2 Reduction curve for FLB (AISC LRFD 86)

CSA S16.1 M89 84b

The limit states design concept is used, in which the limiting slendernesses of compression flanges for full plastic moment and yield moment at a section, and the reduction factor of cross-sectional strength are given as follows:

Limiting Slenderness

LSP: $(B/2t_f)_{0p} = 0.376\sqrt{E/\sigma_y}$ $\qquad \bar{\lambda}_{0p} = 0.603$ (3b.N10)

LSY: $(B/2t_f)_{0y} = 0.587\sqrt{E/\sigma_y}$ $\qquad \bar{\lambda}_{0y} = 0.942$ (3b.N11)

Reduction Factor

$$0.376\sqrt{E/\sigma_y} < (B/2t_f) \le 0.587\sqrt{E/\sigma_y}$$

$$3.054\sqrt{E/\sigma_y} < (h/t_w) \le 4.052\sqrt{E/\sigma_y}$$

$$M_u = M_y$$

$(B/2t_f) > 0.587\sqrt{E/\sigma_y}$ $\qquad (h/t_w) > 4.052\sqrt{E/\sigma_y}$
design according to CSA-CF

$(B/2t_f) \le 0.587\sqrt{E/\sigma_y}$ $\qquad (h/t_w) > 4.052\sqrt{E/\sigma_y}$
design according to plate girder rules

$(B/2t_f) > 0.587\sqrt{E/\sigma_y}$ $\qquad (h/t_w) \le 4.052\sqrt{E/\sigma_y}$

$$M_u = S_x\sigma_{cr} \qquad (3b.N12)$$

where $B/2t_f \le 0.973\sqrt{E/\sigma_y}$

$$\sigma_{cr} = \{1.46 - 0.814\sqrt{(\sigma_y/E)}\cdot(B/2t_f)\}\sigma_y$$

$B/2t_f > 0.973\sqrt{E/\sigma_y}$

$$\sigma_{cr} = 0.632E/(B/2t_f)^2$$

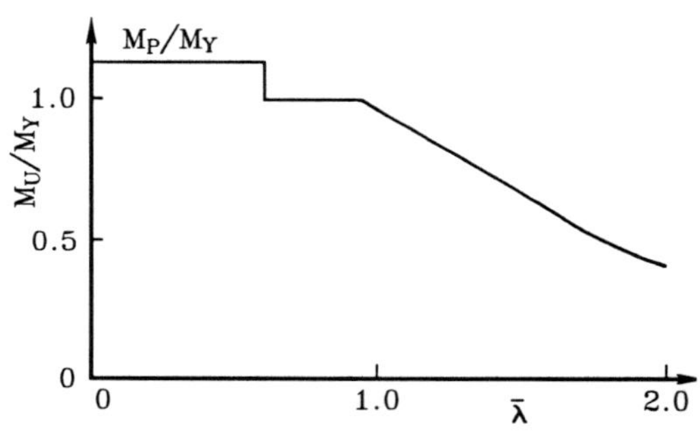

Fig. 3b.N3 Reduction curve for FLB (CSA S16.1 M89 84b)

REGIONAL COMPARISON

Flange local buckling and web local buckling are handled either by a stepwise classification into compactness categories (AREA 87, AASHTO 89, CSA S16.1

M89 84b, CSA S6 M89 84a and OHBDC 83), or a continuous transition is provided between the plastic moment and the elastic buckling curve. Only the AISC LRFD 86 provides for the distinction between rolled and welded shapes due to the different residual stress patterns (see Figs. 3b.N1 and 3b.N2).

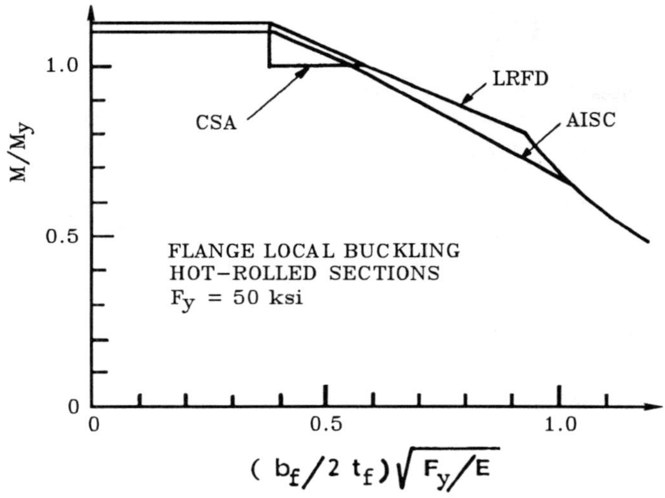

Fig. 3b.N4 Limit states flange local buckling

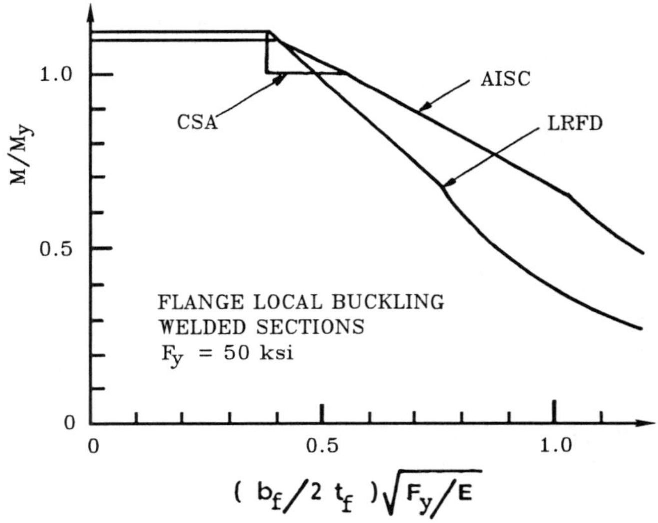

Fig. 3b.N5 Limit states flange local buckling

WEST EUROPE

CHAPTER EDITOR

Joachim Lindner, Technische Universität Berlin

BS 5400 P.3 82

The limit states design concept is used, in which the limiting slendernesses of compression flanges for full plastic moment and yield moment at a section, and the reduction factor of cross-sectional strength are given as follows:

Limiting Slenderness

LSP: (compact section) $(b/t_f)_{0p} = 7\sqrt{355/\sigma_y} = 0.291\sqrt{E/\sigma_y}$

$$\bar{\lambda}_{0p} = 0.466 \tag{3b.W1}$$

LSY: (non compact section) $(b/t_f)_{0y} = 12\sqrt{355/\sigma_y} = 0.498\sqrt{E/\sigma_y}$

$$\bar{\lambda}_{0y} = 0.8 \tag{3b.W2}$$

Maximum slenderness ratio for flange $(b/t_f)_{max} = 16$

Reduction Factor

There is no provision on the strength reduction of outstanding plates.

CTICM 80

The limit states design concept is used, in which the b/t-ratio for plastic analysis is given

For I-sections

LSPA: $$(b'/t_f)_{opa} = 0.335\sqrt{E/f_y} \tag{3b.W3}$$

$$\bar{\lambda}_{opa} = 0.541$$

Further information is given for outstanding flanges and the inner part of box-sections.

DIN 18800 P.2 86

The limit states design concept is used, in which the limiting slendernesses of compression flanges for plastic analysis, full plastic moment and yield moment at a section, and the reduction factor of cross-sectional strength are given as follows:

Limiting Slenderness

LSPA: $$(b'/t_f)_{0pa} = 0.304\sqrt{E/\sigma_y} \qquad \bar{\lambda}_{0pa} = 0.490 \tag{3b.W4}$$

LSP: $(b'/t_f)_{0p} = 0.372\sqrt{E/\sigma_y}$ $\quad \overline{\lambda}_{0p} = 0.600$ (3b.W5)

LSY: $(b'/t_f)_{0y} = 0.436\sqrt{E/\sigma_y}$ $\quad \overline{\lambda}_{0y} = 0.703$ (3b.W6)

where

b' = distance from flange tip to the end of the cross section radius

t_f = web width between tips of the cross section radius

Reduction Factor

for mono-symmetric sections like channels

$b_e/b=1/(\overline{\lambda}^2+0.51)$

for doubly-symmetric sections like I sections

$b_e/b=0.7/\overline{\lambda}$ (3b.W7)

This reduction factor is valid for the case of constant stress occurring normally in mono-symmetric sections like channel section. Doubly-symmetric sections like I-sections are subjected normally to constant strain that leads to a reduction factor

$$b_e/b = 0.7/\overline{\lambda} \tag{3b.W8}$$

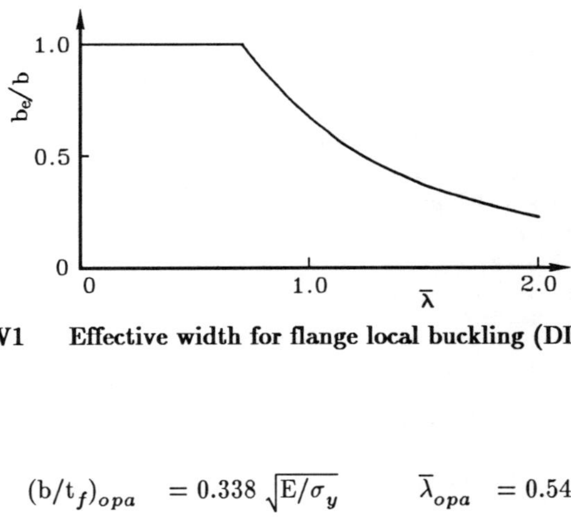

Fig. 3b.W1 Effective width for flange local buckling (DIN 18800)

EC 3 88

LSPA: $(b/t_f)_{opa} = 0.338\sqrt{E/\sigma_y}$ $\quad \overline{\lambda}_{opa} = 0.545$ (3b.W9)

LSP: $(b/t_f)_{op} = 0.372\sqrt{E/\sigma_y}$ $\quad \overline{\lambda}_{op} = 0.600$ (3b.W10)

LSY: $(b/t_f)_{y} = 0.507\sqrt{E/\sigma_y}$ $\quad \overline{\lambda}_{y} = 0.818$ (3b.W11)

NBN B51002 88

The limit states design concept is used, in which the limiting slenderness of compression flanges for plastic analysis, full plastic moment and yield moment at a section are given as follows:

Limiting Slenderness

LSPA: $(b/t_f)_{opa} = 0.335\sqrt{E/\sigma_y}$ $\quad \overline{\lambda}_{opa} = 0.541$ (3b.W12)

LSP: $(b/t_f)_{op} = 0.369\sqrt{E/\sigma_y}$ $\quad \overline{\lambda}_{op} = 0.595$ (3b.W13)

LSY: $(b/t_f)_y = 0.503\sqrt{E/\sigma_y}$ $\quad \overline{\lambda}_y = 0.812$ (3b.W14)

SIA 161 79

The limit states design concept is used, in which the limiting slendernesses of compression flanges for plastic analysis, full plastic moment and yield moment at a section, and the reduction factor of cross-sectional strength are given as follows:

Limiting Slenderness

LSPA: $(b'/t_f)_{0pa} = 0.38\sqrt{E/\sigma_y}$ $\quad \overline{\lambda}_{0pa} = 0.61$ (3b.W15)

LSP: $(b'/t_f)_{0p} = 0.45\sqrt{E/\sigma_y}$ $\quad \overline{\lambda}_{0p} = 0.722$ (3b.W16)

LSY: $(b'/t_f)_{0y} = 0.56\sqrt{E/\sigma_y}$ $\quad \overline{\lambda}_{0y} = 0.9$ (3b.W17)

Reduction Factor

for $\overline{\lambda} > 0.9$

$$b_e = 0.9\, b_c/\overline{\lambda}$$

for flange plates, $b_c = b'$ and k=0.426

$$b_e = 0.558\sqrt{(E/\sigma_y)} \cdot t_f \qquad (3b.W18)$$

Mitigation for the Reduction of Effective Width

If the resistance of a flexural member is determined by a compression stress $\sigma_p < \sigma_y$ (lateral-torsional buckling), the effective width of plate elements may be obtained by replacing the yield stress σ_y by the compression stress $\sqrt{\sigma_y \sigma_p}$. This enables the limiting slenderness ratios and the effective widths of the plate elements of the cross-section to be increased.

$$b_e/b_c = \frac{0.9}{\overline{\lambda}}\sqrt{\frac{\sigma_y}{\sigma_p}} \qquad (3b.W19)$$

σ_p, the lateral torsional buckling strength, is given by

$$\sigma_p = \left(\frac{1}{1+\overline{\lambda}_d^{4.5}}\right)^{0.45} \qquad (3b.W20)$$

where $\overline{\lambda}_d$ is the slenderness parameter for lateral torsional buckling.

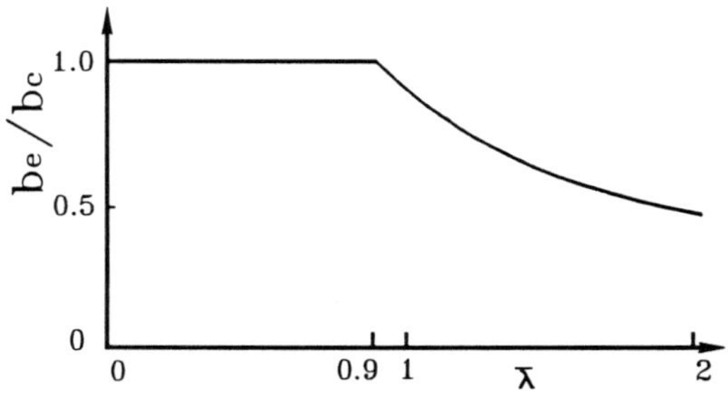

Fig. 3b.W2 Effective width for flange local buckling (SIA161)

INTERNATIONAL COMPARISONS & COMMENT

WHAT ARE THE DIFFERENCES?

Provisions

The provisions governing flange local buckling for I-section beams in each specification is summarized in Table 3b.1. A limiting slenderness ratio of the compression flange is given in all specifications, but in a few specifications the provision of a strength reduction factor is not defined.

Table 3b.1 Existence of provisions for flange local buckling

Specification or code	(1) Limiting Slenderness Ratio	(2) Reduction Factor	(3) Provision on Coupled Buckling
AS DR87164 87	o	o	--
JRA 84	o	o	--
AIJ 70	o	o	--
AIJ 90	o	o	--
AISC ASD 89	o	o	--
AISC LRFD 86	o	o	--
AASHTO 89	o	--	--
CSA S16.1 M89 84b	o	o	--
BS 5400 82	o	--	--
DIN 18800 86	o	o	--
SIA 161 79	o	o	o
NBN B51002 88	o	o	--
CTICM 80	o	o	--

Limiting Slenderness

In limit states design, the limiting slenderness is generally defined for the following three limitations:

1) LSPA: limiting slenderness for plastic analysis, $(b/t)_{0pa}$, $\overline{\lambda}_{0pa}$
2) LSP: limiting slenderness for full plastic moment at a section, $(b/t)_{0p}$, $\overline{\lambda}_{0p}$
3) LSY: limiting slenderness for yield moment, $(b/t)_{0y}$, $\overline{\lambda}_{0y}$

The limiting slenderness parameters in each specification are summarized in Table 3b.2.

Strength Reduction of Section

Fig. 3b.1 shows the strength reduction curve for flange local buckling. And Fig. 3b.2 shows the effective width for flange local buckling.

AISC ASD 89 and AISC LRFD 86 give the linear reduction curve between $\overline{\lambda}_{op}$ to $\overline{\lambda}_{oy}$ for flange local buckling. BS 5400 82, SIA 161 79, CSA S16.1 M89 84b and DIN 18800 86 give the stepped curve from M_p to M_y at $\overline{\lambda}_{op}$.

In the range of $\overline{\lambda} > \overline{\lambda}_{oy}$ for rolled shapes, AISC LRFD 86, AISC ASD 89 and CSA give almost the same strength.

WHY ARE THERE DIFFERENCES?

Limiting Slenderness

Fig. 3b.3 shows the relation between the ultimate strength σ_u/σ_y obtained by the strength tests of the welded cruciform section and the slenderness parameter of the outstanding plates (Takanashi, 1970, Fukumoto, Itoh, 1984). The limiting slenderness parameters of the flange local buckling for yield moment are also shown in the figure. It is observed that AIJ and AASHTO provisions are somewhat conservative, JRA and DIN18800 provisions introduce the limiting slenderness corresponding to the lower bound value, while AS, AIJ, BS5400, AISC ASD, CSA and SIA161 provisions correspond to the mean value of the test data.

As summarized in Table 3b.2, the limiting slenderness ratios $\bar{\lambda}_0$ of the compression flange fall widely in the range of 0.490 to 0.61 for plastic analysis, 0.443 to 0.722 for plastic moment and 0.595 to 0.942 for yield moment. The large variation of the $\bar{\lambda}_0$ values among the codes and standards for the same limit state category indicates that the difficulty exists for specifying the $\bar{\lambda}_0$ value from the theoretical and experimental evaluation. The inelastic ductility behavior is closely related to the width-to-thickness requirements for the compression flange of I-beams and, therefore, further investigation should be needed to establish more reliable and consistent $\bar{\lambda}_{0pa}$, $\bar{\lambda}_{0p}$, and $\bar{\lambda}_{0y}$ values.

Table 3b.2 Comparison of limiting slenderness parameters

Specification		Limiting Slenderness Ratio		
		$\bar{\lambda}_{0pa}$	$\bar{\lambda}_{0p}$	$\bar{\lambda}_{0y}$
AS 87	SR	--	0.553	0.941
	HR	--	0.498	0.886
	LW,HW,CF	--	0.443	0.830
JRA 80		--	--	0.7
AIJ 70		--	--	0.841
AIJ 90		*1)	*1)	0.595
AISC ASD 89		--	0.613	0.895
AISC LRFD 86		--	0.613	*2)
AASHTO 89		--	0.476	0.595
CSA 16.1 M89 84b		--	0.603	0.942
BS 5400 82		--	0.466	0.8
DIN 18800 86		0.490	0.600	0.703
SIA 161 79		0.61	0.722	0.9
NBN B51002 88		0.541	0.595	0.812
EC 3 88		0.545	0.600	0.818

*1) interaction formula of FLB and WLB

*2) for rolled section: $1.328/\sqrt{(1-\sigma_{rR}/\sigma_y)}$,

for welded section: $1/\sqrt{(1-\sigma_{wR}/\sigma_y)}$

Strength Reduction of Section

Fig. 3b.4 shows the strength reduction curves of AISC LRFD 86 and the test results of cruciform sections including both as-welded and annealed. The reduction curve for rolled shapes of AISC LRFD 86 corresponds to the mean strength of the annealed sections. And the reduction curve for welded shapes corresponds to the lower bound of test data for as-welded sections.

Referring to Fig. 3b.1, AISC ASD 89 and CSA 84 provisions, as well as AISC LRFD 86 rolled shape curve closely follow the mean strength of rolled sections. The strength reduction curves of AS DR87164 87 and JRA 80 follow the elastic buckling curve $c/\overline{\lambda}^2$ and are conservative.

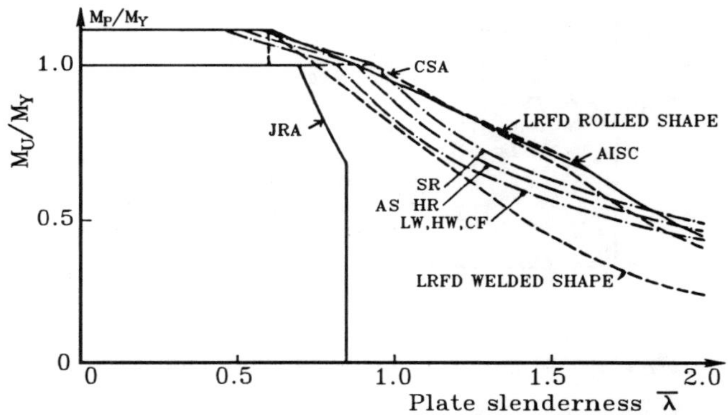

Fig. 3b.1 Comparison of the flexural strength of section (FLB)

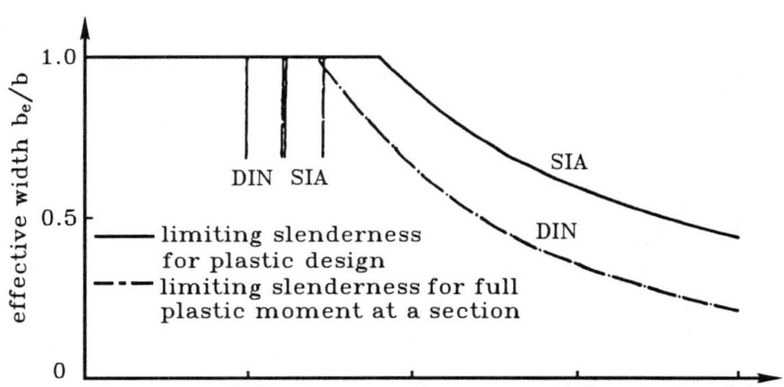

Fig. 3b.2 Comparison of the effective width (FLB)

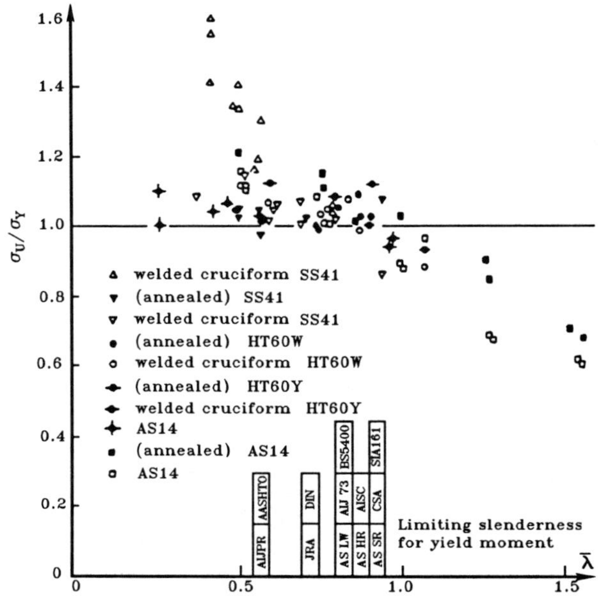

Fig. 3b.3 Test data and the limiting slenderness ratio (FLB)

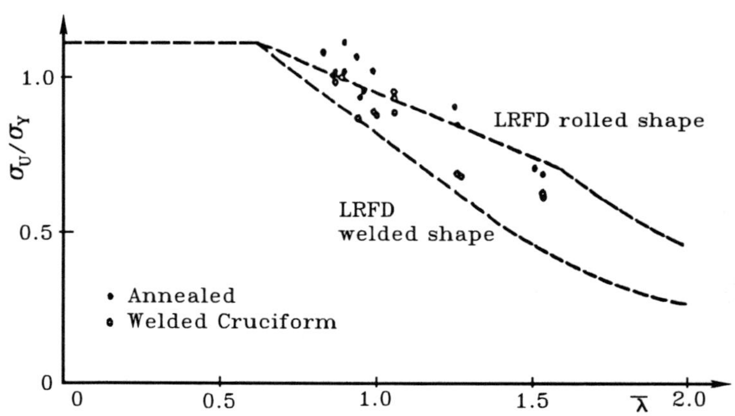

Fig. 3b.4 AISC LRFD 86 reduction curves for flange local buckling compared with test data

3c. Web Local Buckling

SPECIFICATION PROVISIONS & EXPLANATION

AUSTRALIA

CHAPTER EDITOR
Mark A. Bradford, The University of New South Wales

AS DR87164 87

Limiting slenderness ratios for web plates in bending are not defined in this code. However, web slenderness is defined in the provision for plate girder webs (Chapter 4).

CHINA

CHAPTER EDITOR
Zhi Bin Xia, Zhejiang University

GBJ 17 88

The depth-thickness ratio of the web h_o/t_w should be smaller than $80\sqrt{235/f_y}$ (no transverse stiffeners) in allowable stress design. In plastic design they should be smaller than $72\sqrt{235/f_y}$, where f_y (MPa) is the yield stress.

TBJ 2 85

If no transverse stiffeners are used, $h_o/t_w \leq 50$.

JAPAN

CHAPTER EDITORS
Yuhshi Fukumoto, Osaka University
Jun Sakamoto, Nagoya University

AIJ 70

The allowable stress design approach is used, in which the limiting slenderness of web plate in bending for yield moment is

$$(h/t_w)_0 = 2.400\sqrt{E/\sigma_y} \qquad \overline{\lambda}_{0y} = 0.516 \qquad (3c.J1)$$

AIJ 90b

See the provision for flange local buckling.

JRA 80

The allowable stress design approach is used, in which the limiting slenderness of web plate in bending for yield moment and the reduction factor of the allowable stress for web plate with the slenderness beyond the limiting slenderness are given as follows:

Limiting Slenderness

$$(h/t_w)_0 = 5.112\sqrt{E/\sigma_y} \qquad \overline{\lambda}_{0y} = 1.1 \qquad (3c.J2)$$

Maximum slenderness ratio for web $(h/t_w)_{max} = 1.2(h/t_w)_0$

Reduction Factor

$$\sigma_u/\sigma_y = (1.1/\overline{\lambda})^2 \leq 1.0 \qquad (3c.J3)$$

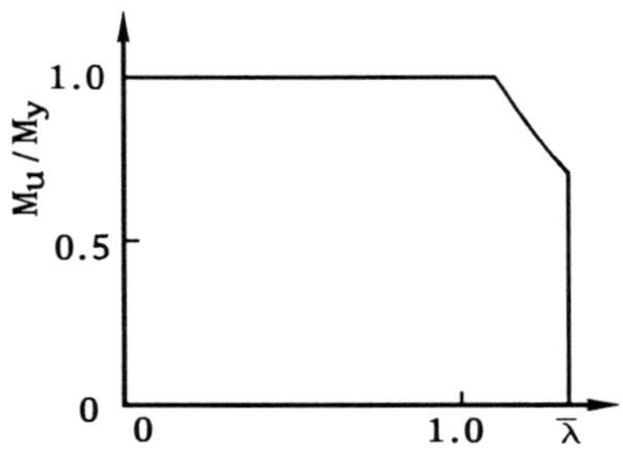

Fig. 3c.J1 Reduction curve for WLB (JRA 80)

JSCE 83

The allowable stress design approach is used, in which the limiting slenderness of web plate in bending for yield moment and the reduction factor of the allowable stress for web plate with the slenderness beyond the limiting slenderness are given as follows:

Limiting Slenderness

$$(h/t_w)_0 = 5.112\sqrt{E/\sigma_y} \qquad \bar{\lambda}_{0y} = 1.1 \tag{3c.J4}$$

Maximum slenderness ratio for web $(h/t_w)_{max} = 200$

Reduction Factor

$$\sigma_u/\sigma_y = (1.1/\bar{\lambda})^2 \leq 1.0 \tag{3c.J5}$$

NORTH AMERICA

CHAPTER EDITOR
Theodore V. Galambos, University of Minnesota

AA 87

Table 3c.N1 Definitions

Temper Designation	-0, -H, $-T_1$, $-T_2$, $-T_3$, $-T_4$	$-T_5$, $-T_6$, $-T_7$, $-T_8$, $-T_9$
B_b	$1.3[1+3.078\left(\frac{\sigma_{cy}}{E}\right)^{1/3}]\sigma_{cy}$	$1.3[1+3.078\left(\frac{\sigma_{cy}}{E}\right)^{1/3}]\sigma_{cy}$
D_b	$\frac{B_b}{20}\sqrt{\frac{6B_b}{E}}$	$\frac{B_b}{20}\sqrt{\frac{6B_b}{E}}$
K_1	0.50	0.50
K_2	2.04	2.04

$$S_1 = \left(\frac{d-2t_f}{t_w}\right)_1 = \frac{B_b-1.3\sigma_{cy}}{0.67D_b} \qquad S_2 = \left(\frac{d-2t_f}{t_w}\right)_2 = \frac{K_1B_b}{0.67D_b} \qquad S = \frac{d-2t_f}{t_w}$$

Ultimate moment

for $S \leq S_1$

$$M_u = 1.3S_x\sigma_{cy}$$

for $S_1 < S \leq S_2$

$$M_u = [B_b - 0.67D_bS]\, S_x$$

for $S > S_2$

$$M_u = \frac{K_2S_x\sqrt{B_bE}}{0.67S} \qquad (3c.N1)$$

AASHTO 89

The limit states design concept is used, in which the limiting slendernesses of web plate in bending for full plastic moment and yield moment at a section are given as follows:

Limiting Slenderness

LSP: $(h/t_w)_{0p} = 2.470\sqrt{E/\sigma_y}$ $\qquad \bar{\lambda}_{0p} = 0.531$ (3c.N2)
LSY: $(h/t_w)_{0y} = 150$ (3c.N3)

Reduction Factor
There is no provision for strength reduction of web plates.

AISC ASD 89

The limit states design concept is used, in which the limiting slendernesses of web plate in bending for full plastic moment and yield moment at a section, and the reduction factor of web plate strength are given as follows:

Limiting Slenderness

LSP: $(h/t_w)_{0p} = 3.758\sqrt{E/\sigma_y}$ $\quad \overline{\lambda}_{0p} = 0.809$ (3c.N4)

LSY: $(h/t_w)_{0y} = 5.762\sqrt{E/\sigma_y}$ $\quad \overline{\lambda}_{0y} = 1.240$ (3c.N5)

Reduction Factor

$(h/t_w) \le 3.758\sqrt{E/\sigma_y}$ $\quad Mu = 1.100\ S_x\sigma_y$

$3.758\sqrt{E/\sigma_y} < (h/t_w) \le 5.762\sqrt{E/\sigma_y}$ $\quad Mu = 1.000\ S_x\sigma_y$

$5.762\sqrt{E/\sigma_y} < (h/t_w)$ $\quad$ Plate girder criteria apply (3c.N6)

AISC LRFD 86

The limit states design concept is used, in which the limiting slendernesses of web plate in bending for full plastic moment and yield moment at a section, and the reduction factor of web plate strength are given as follows:

Limiting Slenderness

LSP: $(h/t_w)_{0p} = 3.758\sqrt{E/\sigma_y}$ $\quad \overline{\lambda}_{0p} = 0.809$ (3c.N7)

LSY: $(h/t_w)_{0y} = 5.696\sqrt{E/\sigma_y}$ $\quad \overline{\lambda}_{0y} = 1.226$ (3c.N8)

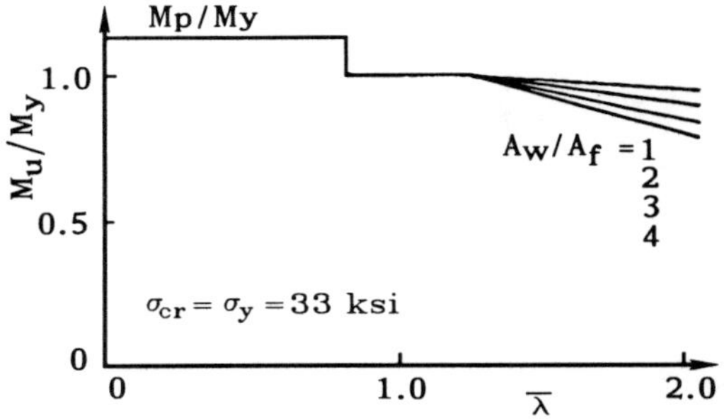

Fig. 3c.N1 Reduction curve for web local buckling (AISC LRFD 86)

Reduction Factor

$$\bar{\lambda} = (H-2t_f)/t_w = h/t_w$$

$$5.696\sqrt{E/\sigma_y} < \bar{\lambda}$$

$$Mu = S_x \cdot R_{pg} \cdot R_e \cdot \sigma_{cr} \qquad (3c.N9)$$

where

a_r = ratio of web area to compression flange area

h_c = twice the distance from the neutral axis to the inside face of the compression flange

R_e = hybrid girder factor (for non-hybrid girder, $R_e = 1.0$)

$R_{pg} = 1-0.0005 \cdot a_r \cdot (\frac{h_c}{t_w} - \frac{970}{\sqrt{\sigma_{cr}}}) \leq 1$

σ_{cr} = critical compression flange stress

CSA S16.1 M89 84b

The limit states design concept is used, in which the limiting slendernesses of web plate in bending for full plastic moment and yield moment at a section, and the reduction factor of web plate strength are given as follows:

Limiting Slenderness

LSP: $(h/t_w)_{0p} = 3.054\sqrt{E/\sigma_y}$ $\quad \bar{\lambda}_{0p} = 0.657$ (3c.N10)

LSY: $(h/t_w)_{0y} = 4.052\sqrt{E/\sigma_y}$ $\quad \bar{\lambda}_{0y} = 0.872$ (3c.N11)

Reduction Factor

See section 3b(N).

WEST EUROPE

CHAPTER EDITOR
Joachim Lindner, Technische Universität Berlin

BS 5400 P.3 82

The limit states design concept is used, in which the limiting slendernesses of web plate in bending for full plastic moment and yield moment at a section, and the reduction factor of web plate strength are given as follows:

Limiting Slenderness

LSP: compact section

$$(h/t_w)_{0p} = 56\sqrt{355/\sigma_y} = 2.324\sqrt{E/\sigma_y}$$
$$\bar{\lambda}_{0p} = 0.5 \qquad (3c.W1)$$

LSY: non-compact section

$$(h/t_w)_{0y} = 136\sqrt{355/\sigma_y} = 5.644\sqrt{E/\sigma_y}$$
$$\bar{\lambda}_{0y} = 1.214 \qquad (3c.W2)$$

Reduction Factor(effective thickness)

$$\left.\begin{array}{ll} \text{for } 68 \geq \frac{y_c}{t}\sqrt{\frac{\sigma_{yw}}{355}} & t_{we} = t_w \\ \text{for } 288 > \frac{y_c}{t}\sqrt{\frac{\sigma_{yw}}{355}} > 68 & t_{we} = t_w\left(1.425 - 0.00625\,\frac{y_c}{t}\sqrt{\frac{\sigma_{yw}}{355}}\right) \\ \text{for } \frac{y_c}{t}\sqrt{\frac{\sigma_{yw}}{355}} \geq 288 & t_{we} = 0 \end{array}\right\} \qquad (3c.W3)$$

where y_c is the depth of web measured in its plane from the elastic neutral axis of the gross section of the beam to the compressive edge of the web.

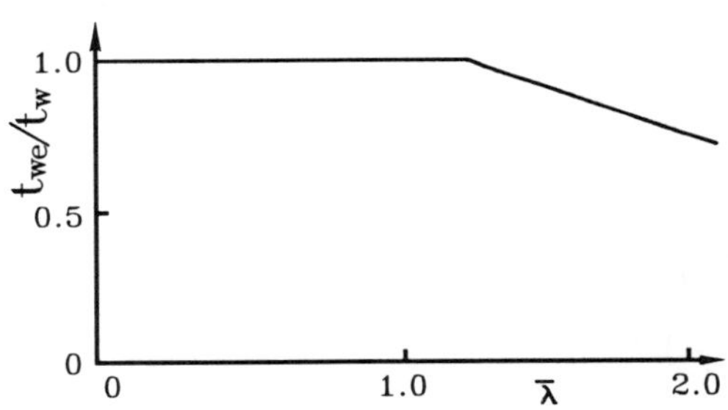

Fig. 3c.W1 Effective thickness of web plate for web local buckling (BS 5400 82)

CTICM 80

No information for bending stress distribution.

DIN 18800 P.2 86

The limit states design concept is used, in which the limiting slenderness of web plate in bending for plastic analysis, full plastic moment and yield moment at a section, and the reduction factor of web plate strength are given as follows:

Limiting Slenderness

LSPA: $(h/t_w)_{0pa} = 2.16\sqrt{E/\sigma_y}$ $\bar{\lambda}_{0pa} = 0.464$ (3c.W4)

LSP: $(h/t_w)_{0p} = 2.50\sqrt{E/\sigma_y}$ $\bar{\lambda}_{0p} = 0.538$ (3c.W5)

LSY: $(h/t_w)_{0y} = 3.25\sqrt{E/\sigma_y}$ $\bar{\lambda}_{0y} = 0.700$ (3c.W6)

Reduction Factor(effective width)

$$h_e/h = 1.25(1/\bar{\lambda} - 0.22/\bar{\lambda}^2) \quad \text{(3c.W7)}$$

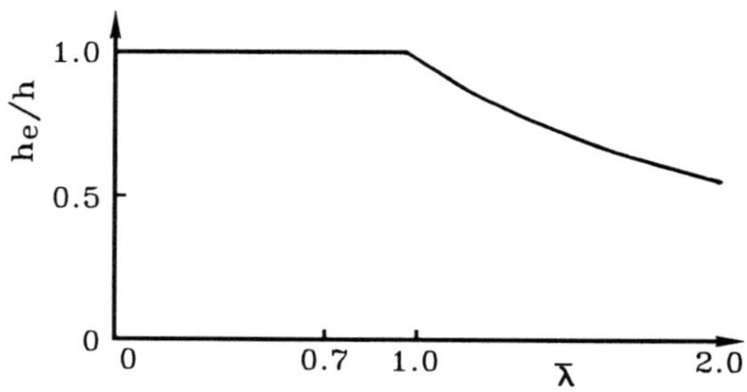

Fig. 3c.W2 Effective width for web local buckling (DIN 18800)

EC 3 88

LSPA: $(b/t_f)_{0pa} = 2.43\sqrt{E/\sigma_y}$ $\bar{\lambda}_{0pa} = 0.523$ (3c.W8)

LSP: $(b/t_f)_{0p} = 3.65\sqrt{E/\sigma_y}$ $\bar{\lambda}_{0p} = 0.785$ (3c.W9)

LSY: $(b/t_f)_{0y} = 4.19\sqrt{E/\sigma_y}$ $\bar{\lambda}_{0y} = 0.902$ (3c.W10)

NBN B51002 88

The limit states design concept is used, in which the limiting slenderness of web plates for plastic analysis, full plastic moment and yield moment at a section are given as follows:

LSPA: $(b/t_f)_{0pa} = 2.21\sqrt{E/\sigma_y}$ $\quad \bar{\lambda}_{0pa} = 0.475$ (3c.W11)

LSP: $(b/t_f)_{0p} = 2.61\sqrt{E/\sigma_y}$ $\quad \bar{\lambda}_{0p} = 0.562$ (3c.W12)

LSY: $(b/t_f)_{0y} = 4.19\sqrt{E/\sigma_y}$ $\quad \bar{\lambda}_{0y} = 0.902$ (3c.W13)

SIA 161 79

The limit states design concept is used, in which the limiting slendernesses of web plate in bending for plastic analysis, full plastic moment and yield moment at a section, and the reduction factor of web plate strength are given as follows

Limiting Slenderness

LSPA: $(h/t_w)_{0pa} = 2.4\sqrt{E/\sigma_y}$ $\quad \bar{\lambda}_{0pa} = 0.516$ (3c.W14)

LSP: $(h/t_w)_{0p} = 3.0\sqrt{E/\sigma_y}$ $\quad \bar{\lambda}_{0p} = 0.645$ (3c.W15)

LSY: $(h/t_w)_{0y} = 4.2\sqrt{E/\sigma_y}$ $\quad \bar{\lambda}_{0y} = 0.9$ (3c.W16)

Reduction Factor

$b_e = 0.9\, b_c/\bar{\lambda}$ $\quad$ if $\bar{\lambda} > 0.9$

$b_c = h/2 \quad k = 23.9 \quad b_e = 2.1\sqrt{E/\sigma_y} \cdot t_w$ (3c.W17)

Fig. 3c.W3 and 3c.W4 identify the effective web section in bending.

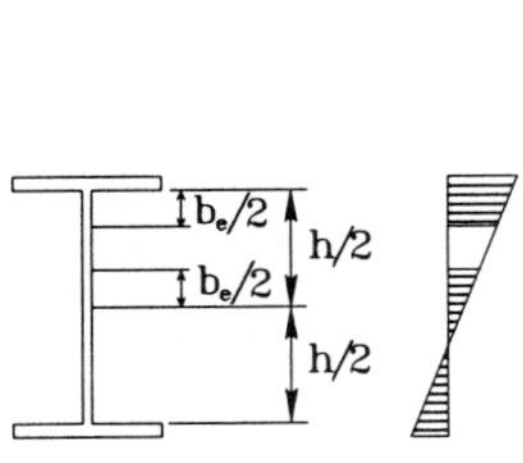

Fig 3c.W3 Effective web area in bending (SIA 161 79)

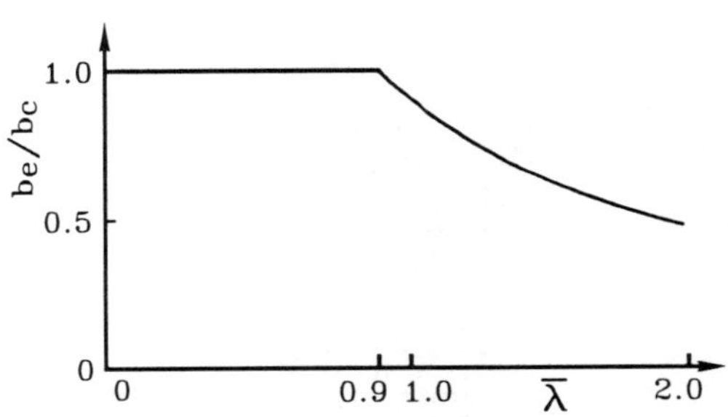

Fig. 3c.W4 Effective width for web local buckling (SIA 161 79)

INTERNATIONAL COMPARISONS & COMMENT

WHAT ARE THE DIFFERENCES?

Provisions
The provisions for web local buckling of I-section beams in each specification are summarized in Table 3c.1. Most provisions on web local buckling are specified in the plate girder rules. Limiting slendernesses of the web plates are given in all specifications. A few specifications don't allow the use of slender web plates exceeding the slenderness limitation for yield moment capacity.

Limiting Slenderness
The limiting slenderness parameters in each code are compared in Table 3c.2.

- The limiting slenderness parameters of plate elements for plastic analysis are defined in the provisions of SIA 161 79 and DIN 18800 86.
- The other codes based on the limit states concept give the limitations for plastic moment of section in the provisions for beam slenderness.
- AISC LRFD 86 gives the limiting slenderness ratios of plate elements considering compression residual stress.

Reduction of Strength
In the range of the slenderness of plate elements beyond the limiting slenderness for yield moment, three types of expression are used for the strength reduction of the cross-section.

They are I) reduction factor: JRA 80, AISC LRFD 86 and CSA 84,
II) effective width: SIA 161 79 and DIN 18800 86,
III) effective thickness: BS 5400 82 (for web plate).

Table 3c.1 Provisions for Web Local Buckling

Specification or code	(1) Limiting Slenderness Ratio	(2) Reduction Factor	(3) Provision on Coupled Buckling
AS DR87164 87	o	--	--
JRA 80	o	o	--
AIJ 70	o	--	--
AIJ 90	o	o	--
AISC ASD 89	o	o	--
AISC LRFD 86	o	o	--
AASHTO 89	o	--	--
CSA S16.1 M89 84b	o	o	--
BS 5400 82	o	o	--
DIN 18800 86	o	o	--
SIA 161 79	o	o	o

In the cases I and II, the flexural capacity of the cross-section must be calculated for the corresponding section.

Fig. 3c.2 shows the strength reduction curves of the cross-section for web local buckling. Herein, the curves except JRA are adopted to the special case of cross-section as shown in Fig. 3c.1.

The strength reduction in the range of $\overline{\lambda} > \overline{\lambda}_{oy}$ is much affected by the ratio of web area to compression flange area.

Table 3c.2 Comparison of limiting slenderness parameters

Specification	Limiting Slenderness Ratio		
	$\overline{\lambda}_{0pa}$	$\overline{\lambda}_{0p}$	$\overline{\lambda}_{0y}$
JRA 80	--	--	1.1
AIJ 70	--	--	0.516
AIJ 90	*1)	*1)	1.296
AISC ASD 89	--	0.809	1.240
AISC LRFD 86	--	0.809	1.226
AASHTO 89	--	0.531	*2)
CSA S16.1 M89 84b	--	0.657	0.872
BS 5400 82	--	0.5	1.214
DIN 18800 86	0.464	0.538	0.700
SIA 161 79	0.516	0.645	0.9
EC 3 88	0.523	0.785	0.902

*1) interaction formula of FLB and web local buckling
*2) $b/t_w = 150$

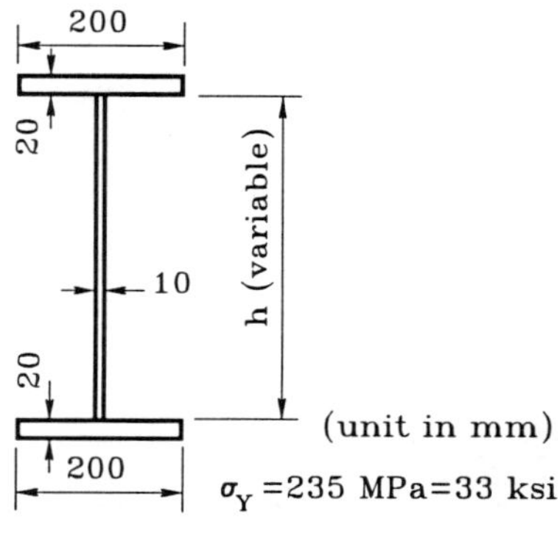

Fig. 3c.1 Dimensions of calculation model

WHY ARE THERE DIFFERENCES?

Limiting Slenderness

Fig. 3c.3 shows AISC LRFD 86 strength curves for web local buckling, the limiting slenderness parameters for yield moment in each specification and the experimental data of plate girders without horizontal stiffeners (Moriwaki, 1977, JSCE). The experimental data show that the limiting slenderness of AISC LRFD 86, BS 5400 82 and AIJ 90 are appropriate. On the contrary, limiting slenderness of AIJ 70 corresponds to the limitation for plate girders without transverse stiffeners. DIN 18800 86 and SIA 161 79 provisions for web local buckling on the strength of sections is comparatively smaller than the reduction effect of flange local buckling. Therefore the limiting slenderness parameter for yield moment is larger than the elastic buckling limit $\overline{\lambda} = 1$.

As summarized in Table 3c.2, the $\overline{\lambda}_0$ values of the web plate vary from 0.464 to 0.516 for plastic analysis, 0.50 to 0.809 for plastic moment and 5.16 to 1.296 for yield moment. By comparing the $\overline{\lambda}_0$ values in order for flange local buckling in each column of $\overline{\lambda}_{0pa}$, $\overline{\lambda}_{0p}$, and $\overline{\lambda}_{0y}$ in Table 3b.2, similar order can be obtained for the $\overline{\lambda}_0$ values for web local buckling. That is, large width-to-thickness (b/t) ratio for the flange plate may also require large b/t ratio for the web plate. Interactive approach to the b/t ratios between flange and web I-beams, like AIJ 90, would be appropriate for the specified inelastic ductility.

Strength Reduction

The dimensions of the test models and the normalized strength of which are plotted in Fig. 3c.3 are listed in Table 3c.3. Table 3c.4 shows the comparison between the strength of test models and the corresponding capacity of those models stipulated by the specifications.

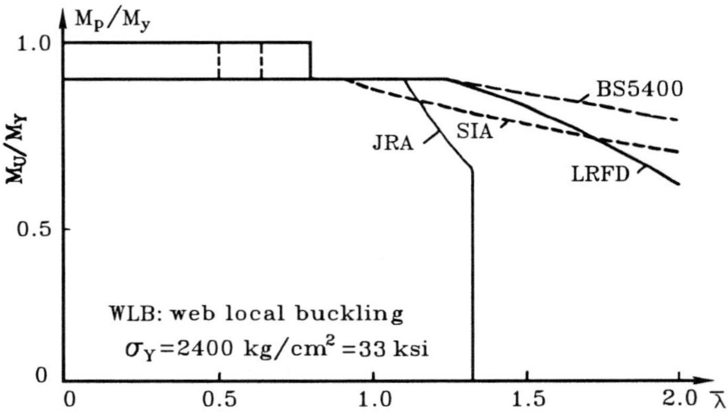

Fig. 3c.2 Comparison of the flexural strength of section (web local buckling)

AISC LRFD 86, BS 5400 82 and SIA 161 79 provisions give almost same values and the stipulated values in these provisions correspond with the upper bound of the test data. Although all test models have stocky flange plates $\overline{\lambda}_f < \overline{\lambda}_{oy}$ and slender web plates $\overline{\lambda}_w > \overline{\lambda}_{oy}$, the collapse mode falls within the torsional buckling type of compression flange (FLB).

Table 3c.3 Dimensions of test models

Model	b_f cm	t_f cm	h cm	t_w cm	σ_{yf} kg/cm^2	σ_{yw} kg/cm^2	$\frac{h}{t_f}$	$\overline{\lambda}_w$	$\frac{b_f}{t_f}$	$\overline{\lambda}_f$	$\frac{A_w}{A_f}$	$\frac{Mu^{ex}}{My}$
1	22.6	0.9	79.6	0.31	3350	2474	256.8	1.896	25.11	0.688	1.21	0.705
2	19.46	1.41	80.4	0.6	5930	5570	134.0	1.484	13.80	0.581	1.76	0.936
3	19.54	1.41	80.45	0.6	5930	5370	134.1	1.459	13.86	0.583	1.75	0.937
4	19.61	1.41	78.9	0.61	5930	5010	129.3	1.359	13.91	0.585	1.74	0.942
5	19.52	1.40	78.9	0.64	5720	6220	123.3	1.444	13.93	0.575	1.85	0.930

Table 3c.4 Comparison between stipulated strength and test data

Ultimate Strength Mu^{ex}/My				
Model	Experiment	LRFD	SIA161	BS5400
1	0.705	0.944	0.978	0.960
2	0.936	0.969	0.978	0.979
3	0.937	0.972	0.979	0.980
4	0.942	0.984	0.982	0.989
5	0.930	0.972	0.978	0.981

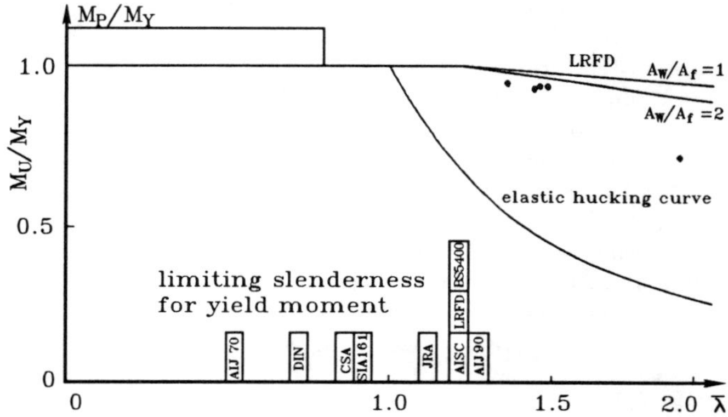

Fig. 3c.3 Comparison of limiting slenderness parameters (web local buckling)

Symbol List for Chapter 3 -- Beams

A = cross-sectional area (3a.N)
A_c = gross area of compression flange (3a.J)
A_c = area of compression flange of tapered beams at the critical cross-sections (3a.A)
A_f = sectional area of compression flange (3a.J) & (3a.N)
A_m = area of compression flange of tapered beams at the minimum critical cross-sections (3a.A)
A_w = web area (3a.N)
a = distance between the cross ribs in the zone of maximum moment (3a.E)
a = sum of the thickness of the flange plates and the length of the vertical leg of angle (3a.E)
a_o = distance of the gravity center from the shear center (3a.E)
a_r = ratio of web area to compression flange area (3c.N) & (3b.J)
a_t = torsional parameter (3a.E)
a_2 = distance of the tension flange from the shear center (3a.E)
B = width of flange (3b.N)
B_b = buckling formula constant of beam webs in bending (3c.N)
B_c = buckling formula constant of columns in compression (3a.N)
B_p = buckling formula constant of beam flanges in compression (3b.N)
b = width of the compression flange (3a.J) & (3b.C)
b' = distance from flange tip to the end of the cross section radius (3b.W)
b_c = width of compression plate (3b.W)
b_e = effective width of compression flange (3b.W)
b_f = flange width (3a.N)
b_n = width of compression flange (3a.E)
b_{tot} = the total width of the compression flange (3a.W)
b_1 = width of compression flange (3a.C)
C = modification factor for non-uniform moment (3a.J) & (3a.W)
C_b = modification factor for non-uniform moment (3a.E) & (3a.N)
C_c = buckling formula constant of columns in compression (3a.N)
C_d = slenderness correction factor for non-uniform moment (3a.W)
C_w = warping constant (3a.N)
c = spacing of supports for lateral bracing (3a.W)
c = modification factor for nonuniform moment (3a.J)
c = supported distance of compression flange with reagrd to the web (3a.W)
c_ϕ = torsional restraint coefficient (3a.W)
D_b = buckling formula constant of beam webs in bending (3c.N)
D_c = buckling formula constant of columns in compression (3a.N)
D_c = section depth at the critical cross-section (3a.A)
D_m = section depth at the minimum cross-section (3a.A)
D_p = buckling formula constant of beam flanges in compression (3b.N)
d = depth of section (3a.N)
E = modulus of elasticity (3a.J), (3a.N), & (3a.E)
E_k = elastic modulus (3a.W)
e = distance of the x-axis from the center line of the compression flange (3a.E)
e = distance of the point of application of load from the shear center (3a.E)
F = required strength of lateral supports (3a.J)
F_{cr} = elastic lateral-torsional buckling stress (3a.N)

F_{cy} = yield stress in compression for aluminum (3a.N)
F_r = maximum compressive residual stress in flange (3a.N)
F_{rR} = maximum compressive residual stress in flange for rolled shapes (3a.N)
F_{rW} = maximum compressive residual stress in flange for welded shapes (3a.N)
F_y = yield stress (3a.N)
f = design normal stress of steel (3a.C)
f_b = allowable bending stress (3a.J)
f_p = plastic design strength (3a.C)
f_t = allowable tensile stress (3a.J)
f_y = yield stress (3a.C)
f_{yd} = yield stress (3a.W)
f_{yk} = yield stress (characteristic value) (3a.W)
G = shear modulus (3a.N)
h = height of section (3a.J)
h_c = twice the distance from the neutral axis to the inside face of the compression flange (3b.J)
h_o = depth of web (3c.C)
h_s = depth of web (3a.E)
h_1 = nominal depth of the web (3a.E)
I_k = torsional resistance (3a.E)
I_s = torsional resistance (3a.E)
I_t = torsional resistance (3a.E)
I_w = warping constant (3a.E)
I_x = moment of inertia about the x-axis (3a.E)
I_y = moment of inertia about the y-axis (3a.E)
I_1 = moment of inertia about the y-axis of compression flange (3a.C) & (3a.E)
I_2 = moment of inertia about the y-axis of tension flange (3a.C) & (3a.E)
i = radius of gyration for the compression flange (3c.W)
i = radius of gyration of T-section composed of compression flange and one-sixth of web about web axis (3a.J)
i_G = principle radius of gyration of compression flange relative to the web plane (3a.E)
i_x = radius of gyration to the principal x-axis of the beam section(3a.C) & (3a.E)
i_y = radius of gyration of the compressed parts of the section relative to the web axis (y-axis); the compression flange and one third of the depth of web are considered the compressed parts of the section (3a.E)
i_y = radius of gyration of the compression flange about the web (3a.E)
i_{y1} = radius of gyration of the compression flange (3a.E)
J = torsion constant (3a.N)
J_D = torsional constant (3a.E)
K = required stiffness of lateral supports (3a.J)
K_z = coefficient depending on the kind of loading and on the section of beam (3a.E)
K_1 = coefficient for crippling of flat plates in compression or bending (3b.N)
K_2 = coefficient for crippling of flat plates in compression or bending (3b.N)
k = factor depending on the support conditions of the beam (3a.E)
k_ϕ = coefficient, depending on the moment distribution, varying from 1.0 to 4.0 (3a.W)
L = distance between lateral supports (3c.W)

L_b = unbraced length (3a.N)

L_b^* = limiting of unbraced length of compression flange (see Eq. 3a.N3) (3a.N)
L_{bc} = limiting of unbraced length of compression flange (see Eq. 3a.N5) (3a.N)
L_{be} = limiting of unbraced length of compression flange (see Eq. 3a.N8) (3a.N)
L_{b1} = limiting of unbraced length of compression flange (see Eq. 3a.N6) (3a.N)
L_{b2} = limiting of unbraced length of compression flange (see Eq. 3a.N7) (3a.N)
L_e = effective length (3a.A)
L_o = unbraced length (3a.W)
L_p = limiting of unbraced length of compression flange (see Eq. 3a.N20) (3a.N)
L_r = limiting of unbraced length of compression flange (see Eq. 3a.N21) (3a.N)
L_{r1} = limiting of unbraced length of compression flange (see Eq. 3a.N21) (3a.N)
L_{r2} = limiting of unbraced length of compression flange (see Eq. 3a.N21) (3a.N)
L_{y1} = spacing of lateral supports (3a.E)
L_w = buckling length in case of torsion of the beam (3a.E)
l_b = unsupported length of a compression flange (3a.J)
l_1 = spacing of the lateral supports of the compression flange (3a.A), (3a.C), & (3a.E)
l = span length of the beam (3a.E)
l_{ef} = distance between the nodal points of the longitudinal and transverse bracing or distance between the fixing points of the rigid plate (3a.E)
ℓ_b = lateral unsupported length of the compression flange of the beam (3a.C)
ℓ_{cr} = critical unbraced length (3a.C)
ℓ_o = length of buckling (3a.E)
ℓ_1 = distance between the brace at the plastic hinge and the nearest brace (3a.W)
ℓ_1 = unsupported length (3a.J)
M = bending moment (3a.N)
M = design moment (3a.E)
M_b = nominal beam moment capacity (3a.A)
M_{cr} = elastic lateral-torsional buckling moment (3a.W)
M_e = elastic lateral-torsional buckling moment (3a.N) & (3a.W)
M_{eq} = equivalent moment (3a.W)
M_{ex} = elastic lateral buckling moment (3a.J)
M_{exo} = elastic lateral buckling moment under uniform moment (3a.J)
M_F = yield bending moment (3a.E)
M_{ki} = buckling moment under ideal conditions (3a.E)
M_{kr} = elastic critical moment for constrained beam (3a.E)
M_o = elastic critical moment (3a.A)
M_p = plastic moment (3a.J) & (3a.N)
M_{pl} = moment at full plasticity (3a.E)
M_{px} = plastic moment about the strong axis (3a.J)
M_{Rcd} = bending moment capacity (3a.W)
M_r = yield bending moment considering the influence of residual stress (3a.N)
M_s = yield bending moment of effective section (3a.A)
M_T = modified load-bearing moment (3a.E)
M_u = ultimate moment (3a.N)
M_{ux} = moment capacity about the strong axis (3a.J)
M_{u1} = ultimate moment, see Eq. 3a.N16 (3a.N)

M_{u2} = ultimate moment, see Eq. 3a.N16 (3a.N)
M_{u3} = ultimate moment, see Eq. 3a.N16 (3a.N)
M_{u4} = ultimate moment, see Eq. 3a.N16 (3a.N)
M_{ue} = elastic ultimate moment (3a.N)
M_x = factored maximum bending moment about the x-axis (3a.C)
M_y = yield bending moment (3a.N)
M_1 = the numerically largest moment within the beam portion (3a.W)
M_1 = moment at the section braced at a distance ℓ_{cr} from the hinge (3a.C)
M_2 = moment at the opposite end of the beam portion (3a.W)
m = equivalent uniform moment factor (3a.W)
m = safety factor depending on the structural element $0.75 \leq m \leq 0.9$ (3a.E)
m_z = factor of lateral buckling (3a.E)
N_c = buckling strength of compression members (3a.J)
N_y = yield axial force (3a.J)
n = slenderness correction factor (3a.W)
n = ratio of moment of inertia of the larger flange to the sum of the moments of inertia of larger and the smaller flanges relative to the axis of symmetry of the section (3a.E)
n = system factor (3a.W)
R = yield strength, in MPa (3a.E)
R_e = hybrid girder factor (3b.J) & (3c.N)
R_e = yield strength (3a.E)
R_n = standard strength of material (3a.E)
R_{pg} = reduction factor (3b.J) & (3c.N)
R_y = computational strength of material (3a.E)
R_{yn} = standard strength of the material (3a.E)
r = slenderness ratio (3a.E)
r_t = radius of gyration of the flange plus 1/6 of the web (3a.N)
r_y = radius of gyration of the entire cross section (3a.N) & (3a.W)
$\bar{r}_y$ = radius of gyration for the compression flange, including 1/5 of the web area (3a.W)
r_{yeff} = effective radius of gyration of the entire cross section (3a.N)
S_x = elastic section modulus (3a.N)
S_1 = limiting slenderness ratio, see Eq. 3a.N1 (3a.N)
S_2 = limiting slenderness ratio, see Eq. 3a.N1 (3a.N)
s = ratio of the width to the thickness of webs (3c.N)
s = ratio of half of the width of the compression flanges to the thickness (3b.N)
s_1 = limiting ratio of width to thickness of compression flanges and webs (3b.N)
s_2 = limiting ratio of width to thickness of compression flanges and webs (3b.N)
t = thickness of flange (3a.W) & (3b.C)
t = web thickness (3a.E)
t_f = web width between tips of the cross section radius (3b.W)
t_f = flange thickness (3a.N)
t_r = thickness of compression flange (3a.W)
t_w = web thickness (3a.N)
t_{we} = effective thickness of web (3c.W)
t_1 = thickness of compression flange (3a.C) & (3a.E)
u = buckling parameter (3a.W)
v = slenderness factor, dependent on the shape of the beam (3a.W)

v = distance of the point of application of the transverse loads from the gravity axis of the section (3a.E)
W = moment of resistance in the plane of greater stiffness relative to the compression flange (3a.E)
W_1 = moment of resistance, with the openings subtracted, relative to the compression flange (3a.E)
W_c = section modulus of gross section for extreme compression fiber (3a.W)
W_{cbr} = gross section modulus relative to the compression flange (3a.E)
W_{el} = elastic section modulus (3a.E)
W_{o2} = moment of resistance, with the openings subtracted, relative to the tension flange (3a.E)
W_{pl} = plastic section modulus (3a.E)
W_{px} = plastic moment about bending axis (3a.C)
W_T = modified section modulus (3a.E)
W_x = elastic section modulus (3a.C) & (3a.E)
W_{xbr} = gross moments of resistance about the x-axis (3a.E)
W_{ybr} = gross moments of resistance about the y-axis (3a.E)
y_c = depth of web measured in its plane from the elastic neutral axis of the gross section of the beam to the compressive edge of the web (3c.W)
y_c = distance of the shear center from the center of gravity (3a.E)
Z_e = effective section modulus (3a.A)
Z_p = plastic section modulus (3b.A & 3a.J)
Z_x = plastic section modulus (3a.N)
Z_y = elastic section modulus (3b.A)
α = modification coefficient, 1.8 for welded beams and 2.0 for riveted beams (3a.C)
α = shape factor (3a.W)
α = non-dimensional slenderness (3c.W)
α = modification factor (3a.J)
α = ratio of flange thickness to web thickness (3a.J)
α_m = moment distribution factor given in Table 3a.A1 (3a.A)
α_s = strength reduction factor (3a.A)
α_{st} = elastic critical moment modified coefficient (3a.A)
β = factor of lateral buckling length taking into account the degree of the elastic end restraint (3a.E)
β = ratio of smaller to larger end moment (3a.W)
β = ratio of beam height to flange width (3a.J)
β_b = factor of equivalent bending moment given in Table 3a.C2 (3a.C)
β_s = ealstic limit stress of steel (3a.N)
β_x = the monosymmetry parameter (3a.A)
β_x = ratio of smaller to larger end moment (3a.J)
γ_c = safety factor depending on the structural element (3a.E)
γ_d = factor of adoption $\gamma_d \leq 1$ (3a.E)
γ_m = safety factor for the material, $1.025 \leq \gamma_m \leq 1.15$ (3a.E)
γ_n = safety factor depending on the structural element and on consequences in case of destruction, $0.9 \leq \gamma_n \leq 1.1$ (3a.E)
γ_p = partial coefficient for resistance (3a.C)
δ_n = thickness of compression flange (3a.E)
η = slenderness factor (3a.W)
η_b = factor of unsymmetry of the section (3a.C)

η_c = shape factor for extreme compression fiber (3a.W)
η_{LT} = Perry coefficient (3a.W)
Θ = factor for taking into account the point of application of load (3a.E)
κ = buckling coefficient of the European buckling curve c (3a.W)
κ = reduction factor for out-of-plane buckling (3a.W)
κ_{cr} = modification factor (3a.W)
κ_m = reduction coefficient (3a.W)
κ_{M} = reduction factor (3a.W)
Λ = limiting of slenderness ratio (3a.J)
λ = slenderness ratio (3a.W)
λ = critical slenderness ratio of compression flange (3a.E)
$\overline{\lambda}$ = non-dimensional buckling slenderness (3a.W)
$\overline{\lambda}$ = ratio of half of the width to thickness of compression flange (3b.N)
λ_b = slenderness ratio of compressed part of beam (3a.J)
$\overline{\lambda}_d$ = slenderness parameter for lateral torsional buckling (3b.W)
$\overline{\lambda}_e$ = equivalent slenderness ratio (3a.C) & (3a.J)
λ_{ed} = equivalent slenderness (3a.W)
λ_{ki} = ideal slenderness ratio for lateral buckling (3a.E)
λ_{LO} = limiting slenderness, see Eq. 3a.W3 (3a.W)
λ_{LT} = equivalent slenderness (3a.W)
$\overline{\lambda}_M$ = relative slenderness ratio (3a.E)
$\overline{\lambda}_o$ = limiting slenderness parameter (3b.A)
$\overline{\lambda}_{op}$ = limiting slenderness of plastic design (3b.A)
$\overline{\lambda}_{opa}$ = limiting slenderness for plastic analysis (3b.W)
$\overline{\lambda}_{or}$ = limiting slenderness of elastic design (3b.N)
$\overline{\lambda}_{oy}$ = limiting slenderness of elastic design (3b.A)
λ_p = limiting slenderness, see Eq. 3a.E28 (3a.E)
λ_p = limiting slenderness of plastic design (3a.N)
λ_{pr} = equivalent slenderness ratio (3a.E)
λ_r = limiting slenderness of elastic design (3a.E)
λ_{rR} = reduction factor of cross-sectional strength (3b.N)
λ_s = limiting slenderness ratio (3a.E)
λ_s = slenderness of the outstanding flange (3b.A)
λ_{sp} = limiting slenderness of compression flange for plastic moment (3b.A)
λ_{sy} = limiting slenderness of compression flange for yield moment (3b.A)
λ_y = slenderness ratio about the weak axis y-y (3a.C)
$\lambda_{y,\ddot{o}v}$ = slenderness ratio of the compression flange relative to the y-axis is given in a figure of the standard (3a.E)
${}_e\lambda_b$ = limiting slenderness of beam for yield moment (3a.J)
${}_p\lambda_b$ = limiting slenderness of beam for plastic moment (3a.J)
μ_{M} = imperfection factor (3a.E)
μ_w = value of fixation for preventing warping (3a.E)
μ_y = value of fixation for preventing bending about the y-axis (3a.E)
ν = safety coefficient (3a.W)
ν = factor of buckling length (3a.E)

ν_{kr} = safety factor to prevent reaching the critical load in case of partial plasticization cases of critical load (3a.E)
ξ = factor for taking into account the moment shape (3a.E)
σ = design stress (3a.E)
σ_{all} = allowable stress (3a.E)
σ_{allow} = allowable stress (3b.J)
σ_b = bending stress in the center of compression flange (3a.W)
σ_b = allowable stress in the compression (3a.J)
σ_{ca} = allowable stress of compression flange (3a.W)
σ_{cr} = critical lateral-torsional buckling stress (3a.W)
σ_{cr} = critical compression flange stress (3b.J), (3b.N), & (3c.N)
σ_{cy} = yield stress of compression flange (3b.N)
σ_d = lateral-torsional buckling strength (3b.W)
σ_e = elastic lateral-torsional buckling stress (3a.W)
σ_F = yield stress (3a.E)
σ_k = critical buckling strength for axial loaded cross sections (3a.W)
σ_{ki} = ideal lateral buckling stress (3a.E)
$\sigma_{kif,H}$ = limit stress in lateral buckling (3a.E)
$\sigma_{kr,M}$ = elastic critical stress for constrained beam (3a.E)
$\sigma_{max,\,ny}$ = maximum edge compressive stress relative to the gross area of cross section (3a.E)
σ_p = lateral-torsional buckling strength (3b.W)
σ_{rR} = residual compression stress in flanges of rolled shape (3b.N)
σ_t = yield stress (3a.E)
σ_u = ultimate stress (3b.J)
σ_{wR} = residual compression stress in flanges of welded shape (3b.N)
σ_y = minimum specified yield stress (3a.J)
σ_y = factored material resistance (3a.W)
σ_{yf} = yield stress of flange (3b.J)
σ_{yw} = yield stress of web (3b.J)
ϕ = resistant factor for beam (3a.J)
ϕ_b = overall stability factor (3a.C)
ϕ_b = coefficient of lateral buckling (3a.E)
ϕ_b = resistant factor for beam (3a.J)
ϕ_b' = overall stability factor (3a.A)
ϕ_o = factor of lateral buckling for the critical slenderness ratio of the compression flange
ϕ_r = factor of lateral buckling (3a.E)
ϕ_M = factor of lateral buckling (3a.E)
ϕ_1 = modified coefficient (3a.E)
ϕ_2 = modified coefficient (3a.E)
φ_b = overall stability factor (3a.C)
ψ = factor dependent on the kind of loading and on parameter α (3a.E)
ω_b = reduction factor for lateral-torsional buckling (3a.W)

Chapter 4

Plate & Box Girders

TOPICAL EDITOR
Alexis Ostapenko, Lehigh University

CHAPTER EDITORS
Jozef Djubek, Slovak Academy of Science
Pierre Dubas, Swiss Federal Institute of Technology
Sadao Komatsu, Osaka Sangyo University
Xiliang Liu, Tianjin University
Noel W. Murray, Monash University
Fumio Nishino, University of Tokyo
Alexis Ostapenko, Lehigh University
Xinhua H. Ren, Tianjin University
Miroslav Škaloud, Czechoslovak Academy of Sciences

Plate and box girders are flexural components which differ from beams (Chapter 3) by being composed of thin plates which may be subject to local buckling. Stiffeners, transverse and longitudinal, may be added to increase the buckling and postbuckling strength. Although many design rules for plate and box girders are the same as for beams, there are many additional specific rules, and these are discussed in this chapter. Of particular interest are the rules for the web plate in plate and box girders and for the compression flange of box girders. The material is sufficiently broad to separate into numerous topics. However, an examination of specifications indicated that the many treatments are so interwoven that only the following three, more or less separable, groups need be designated

Topic 4a. Web Slenderness Limitations
Topic 4b. Web Strength
Topic 4c. Compression Flange of Box Girders

Topic 4b covers the strength of plate and box girder webs subjected to shear, bending or a combination of shear and bending. In many specifications, these loading cases were not treated as separate effects due to a unified presentation, which led to putting all web strength considerations into one comprehensive topic. Topics 4a and 4c are relatively short, self-contained presentations.

To make comparisons easier, SI units are used throughout, and many equations and tables were adjusted by converting relevant constants. For the same purpose, a consistent notation and sign convention were used where possible.

Notation for dimensions of a typical girder segment is shown in Fig. 4.1. Wherever necessary, additional sketches are presented to clarify some characteristic of a particular specification or a case of loading.

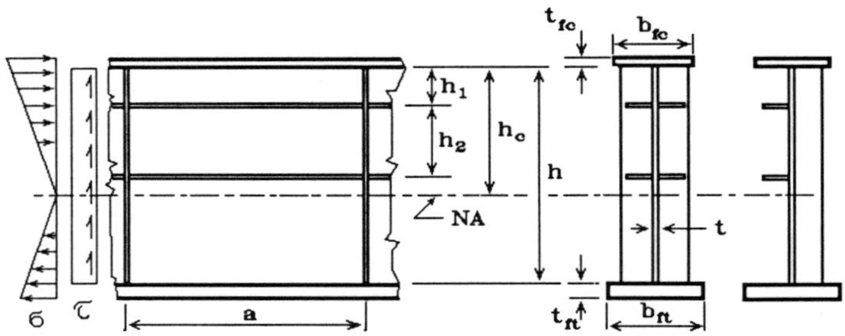

Fig. 4.1 Notation for Dimensions of Girder Segment

where

E = modulus of elasticity (in MPa)

ν = Poisson's ratio

σ_y = yield stress (in MPa)

WHAT ARE THE DIFFERENCES?

The differences among the specification provisions and the apparent reasons for the differences are discussed in subsequent sections topic by topic across the regions. However, it is appropriate to summarize in this introduction the general reasons for them.

WHY ARE THERE DIFFERENCES?

The reasons for the differences can be put into two general groups.

Different Fundamental Theory

For example, buckling vs. ultimate strength. The reasons for accepting one or the other fundamental theory may be ascribed to the following

- Tradition.
- Inertia in incorporating the latest developments and research into design rules.
- Limited use of a particular type of structure (for example, few steel structures) does not warrant a change to new specifications. (At least, this is the impression one gets.)

When the Same Fundamental Theory is used, the differences in the format, procedures and criteria seem to result from the following

- Confidence in current rules from tradition and past experience.
- Emphasis on simplicity rather than accuracy.

4a. Web Slenderness Limitations

SPECIFICATION PROVISIONS & EXPLANATION

Since the slenderness for many typical plate components in plate and box girders, such as, outstanding flange width, width of stiffener plates, etc., are covered in most specifications as applicable to ordinary beams as well and are discussed in this volume in Chapters on Beams and Columns, the emphasis in this chapter is put on the comparison of the requirements for web slenderness.

AUSTRALIA

CHAPTER EDITOR
Noel W. Murray, Monash University

AS DR87164 87

Web slenderness limitations extracted from the specification are given below for specific types of plate girder webs.

Unstiffened webs
The slenderness of an unstiffened web bounded on both longitudinal edges by flanges shall satisfy

$$\frac{h}{t} \leq 180\sqrt{\frac{250}{\sigma_y}} = \frac{2846}{\sqrt{\sigma_y}} \tag{4a.A1}$$

Transversely stiffened webs
The slenderness of a transversely stiffened web panel shall not be greater than

$$\frac{d}{t} \leq 200\sqrt{\frac{250}{\sigma_y}} = \frac{3162}{\sqrt{\sigma_y}} \tag{4a.A2}$$

where d is the smaller of the web depth h or the stiffener spacing a.

Web panels with a > 3h shall be considered to be unstiffened.

Webs with longitudinal and transverse stiffeners the slenderness of which is greater than given by Eq. 4a.A2 but less than

$$\frac{h_c}{t} \leq 250\sqrt{\frac{250}{\sigma_y}} = \frac{3953}{\sqrt{\sigma_y}} \tag{4a.A3}$$

shall have a longitudinal stiffener placed on one or both sides of the web at a distance $0.2h_c$ from the compression flange, where h_c is twice the clear distance from the neutral axis to the compression flange.

If the slenderness is greater than the value from Eq. 4a.A3 but less than

$$\frac{h_c}{t} \leq 400\sqrt{\frac{250}{\sigma_y}} = \frac{6325}{\sqrt{\sigma_y}} \tag{4a.A4}$$

then an additional longitudinal stiffener shall be placed on one or both sides of the web at the neutral axis.

<u>Maximum web panel dimension</u> of a web with transverse or transverse and longitudinal stiffeners shall not exceed 270 times the thickness.

CHINA

CHAPTER EDITORS
Xiliang Liu, Tianjin University
Xinhua H. Ren, Tianjin University

GBJ 18 87

Limit States Design Approach

Transverse stiffeners are required for the web of plate girders with height to thickness ratio falling in the range of

$$80\sqrt{235/\sigma_y} \leq \frac{h}{t} \leq 170\sqrt{235/\sigma_y} \tag{4a.C1a}$$

that is,

$$\frac{1226}{\sqrt{\sigma_y}} \leq \frac{h}{t} \leq \frac{2606}{\sqrt{\sigma_y}} \tag{4a.C1b}$$

Both vertical and horizontal stiffeners are required for webs of plate girders when the height to thickness ratio exceeds $170\sqrt{235/\sigma_y}$.

TBJ 2 85

Allowable Stress Design

Plate girders without intermediate web stiffeners will be used only for $\frac{h}{t} \leq 50$. Plate girders with transverse stiffeners will be used for $140 \geq h/t > 50$.

Plate girders with transverse and longitudinal stiffeners will be used for $250 \geq h/t > 140$.

EAST EUROPE

CHAPTER EDITORS
Jozef Djubek, Slovak Academy of Science
Miroslav Škaloud, Czechoslovak Academy of Sciences

ČSN 73 1401 84 and ČSN 73 6205 84

Specifications ČSN 73 1401 84 and ČSN 73 6205 84 have no specific web slenderness limitations. In practice, web slenderness is governed by
(i) ultimate strength considerations (both in ČSN 73 1401 84 and ČSN 73 6205 84)
(ii) serviceability limit state considerations (ČSN 73 6205 84)

MI 15024/4 85

No web slenderness limitation is directly imposed by this specification. Apparently, it is controlled by the buckling stress requirements.

SNiP II-23-81 82

No transverse stiffeners are required when

$$\frac{h}{t} \leq \frac{K}{\sqrt{\sigma_y/E}} \quad (4a.E1a)$$

where

$K = 3.2$ for static loads
$K = 2.2$ for moving loads (4a.E1b)

When buckling is used as the limit state, transverse stiffeners are to be used with the web slenderness limited to Eq. 4a.E1a with

$$K \leq 6.0 \quad (4a.E2)$$

For static loads, postbuckling strength of the web may be utilized, and the limiting web slenderness increased.

$$\frac{6.0}{\sqrt{\sigma_y/E}} \leq \frac{h}{t} \leq \frac{13.0}{\sqrt{\sigma_y/E}} \quad (4a.E3a)$$

that is,

$$\frac{2683}{\sqrt{\sigma_y}} \leq \frac{h}{t} \leq \frac{5814}{\sqrt{\sigma_y}} \quad (4a.E3b)$$

JAPAN

CHAPTER EDITORS
Sadao Komatsu, Osaka Sangyo University
Fumio Nishino, University of Tokyo

AIJ 86

There are no specified depth/thickness limitations for plate girder webs in building structures. The minimum thickness is controlled by elastic buckling.

JRA 80

Depth-thickness ratio limitation (h/t) for web plates are as shown in Table 4a.J1.

Table 4a.J1 Depth-Thickness Ratio Limitation

No. of longitudinal stiffeners	σ_y(MPa)			
	240	320	360	460
none	152	130	123	110
1	256	220	209	180
2	310	310	294	262

COMMENTARY

The width-thickness limitations for webs without longitudinal stiffeners are specified so that the stability limit is reached when the stress at the compression flange reaches the yield stress. The limitations for webs with horizontal stiffeners are specified under the additional condition that all component plates between the stiffener and the flange and between stiffeners, reach the stability limit simultaneously.

NORTH AMERICA

CHAPTER EDITOR

Alexis Ostapenko, Lehigh University

AASHTO 89

Depth slenderness limitation for symmetrical and unsymmetrical girders with transverse stiffeners is controlled by

$$\frac{2\,h_c}{t} \leq \frac{3030}{\sqrt{\sigma_y}} \qquad (4a.N1)$$

where h_c is twice the portion of the web on the compression side of the centroidal axis or the clear web depth h, whichever is greater. The web slenderness ratio may be doubled by introducing a longitudinal stiffener at $0.2\,h_c$ from the compression flange. However, the increase in the shear strength due to the presence of the longitudinal stiffener is not taken into account.

AISC ASD 89

For girders without transverse stiffeners

$$\frac{h}{t} \leq \frac{5252}{\sqrt{\sigma_y}} \qquad (4a.N2)$$

When transverse stiffeners are used

$$\frac{h}{t} \leq \frac{96{,}527}{\sqrt{\sigma_{yf}(\sigma_{yf}+\sigma_{rc})}} \qquad (4a.N3)$$

where

σ_{rc} = residual compressive stress in flange (MPa)
σ_{yf} = yield stress of flange (MPa)

AISC LRFD 86

The web slenderness limitations for doubly and singly symmetrical plate girders depend on the spacing of transverse stiffeners.

for $\frac{a}{h} \leq 1.5$

$$\frac{h}{t} \leq \frac{5252}{\sqrt{\sigma_{yf}}} \qquad (4a.N4)$$

for $\frac{a}{h} > 1.5$

$$\frac{h}{t} \leq \frac{96{,}527}{\sqrt{\sigma_{yf}(\sigma_{yf}+114)}} \qquad (4a.N5)$$

for unstiffened plate girders

$$\frac{h}{t} \leq 260 \qquad (4a.N6)$$

AREA 89

The AREA 89 Specification applies to railroad bridge girders and the postbuckling web strength is neglected.

$$\frac{h}{t} \leq \frac{2700\sqrt{\sigma_{ba}/\sigma}}{\sqrt{\sigma_y}} \qquad (4a.N7)$$

where

σ = actual bending stress

σ_{ba} = allowable bending stress

CSA S6 M89 84a

The web slenderness limitation of this specification is the same as of the AISC ASD 89.

CSA S16.1 M89 84b

The web slenderness limitation of this specification is the same as of the AASHTO 89.

FHWA TS 80

This specification was developed to provide engineers with a tool for designing box girder bridges of longer spans in which multiple longitudinal stiffeners would be needed.

Depth of the girder web is limited by the following slenderness requirement

$$\frac{h_c}{t} \leq \frac{6.8}{\sqrt{\sigma_y/E}} \qquad (4a.N8)$$

where h_c is twice the clear distance from the neutral axis to the compression flange, but not smaller than h.

WEST EUROPE

CHAPTER EDITOR

Pierre Dubas, Swiss Federal Institute of Technology

BS 5400 P.3 82

Apparently, there is no stated limit on the web slenderness. An indirect practical limit is imposed by specifying the effective thickness of the transversely stiffened web plate to be taken equal to zero in computing its bending capacity when

$$\frac{h_c}{t} = 456\sqrt{\frac{355}{\sigma_y}} = \frac{8600}{\sqrt{\sigma_y}} \qquad (4a.W1)$$

where h_c is twice the distance from the neutral axis to the compression flange. (This means, for example, that a symmetrical section with $\sigma_y = 250$ MPa would have

$$\frac{h_c}{t} = \frac{h}{t} = 543 \qquad (4a.W2)$$

BS 5950 P.5 85

Two web slenderness limitations are set, apparently, assuming a symmetrical section.

1) Serviceability

$\frac{h}{t} \leq 250$ for webs without intermediate transverse stiffeners and for $a/h > 1.0$ (4a.W3a)

$\frac{h}{t} \leq \frac{250}{\sqrt{a/h}}$ for $a/h \leq 1.0$ (4a.W3b)

2) Prevention of compression flange buckling into web

$\frac{h}{t} \leq 250\left(\frac{345}{\sigma_{yf}}\right)$ for webs without intermediate transverse stiffeners and for $a/h > 1.5$ (4a.W4a)

$\frac{h}{t} \leq 250\sqrt{\frac{345}{\sigma_{yf}}}$ for $a/h \leq 1.5$ (4a.W4b)

DASt 015 80

The following web slenderness limitations are imposed for girders under static loads depending on the grade of steel

$h/t \leq 350$ for steel St 37 $(\sigma_y = 240$ MPa) (4a.W5)
$h/t \leq 230$ for steel St 52 $(\sigma_y = 360$ MPa)

(If $\sigma_y = 250$ MPa, the slenderness can be adjusted to $h/t = 343$.)

EC 3 88

Web Slenderness Limit for transversely stiffened girders as set by

$$h/t \leq k(E/\sigma_{yf}) \sqrt{(A_w/A_f)} \tag{4a.W6}$$

where

A_f = area of the compression flange
A_w = h t = area of the web
σ_{yf} = yield stress of the compression flange

The value of factor k, for I-type sections, should be taken as

$$\left.\begin{array}{lll} \text{for} & b/t_f \leq 20\sqrt{235/\sigma_{yf}} & k = 0.3 \\ \text{for} & b/t_f \leq 22\sqrt{235/\sigma_{yf}} & k = 0.4 \\ \text{for} & b/t_f \leq 30\sqrt{235/\sigma_{yf}} & k = 0.55 \end{array}\right\} \tag{4a.W7}$$

No rules have yet been formulated for girders with longitudinal stiffeners.

NBN B51002 88

Web slenderness is limited by elastic buckling. Specific recommendations are given only for unstiffened webs. With the average shear stress τ,

$$\tau \leq \tau_{cr}$$

where τ_{cr} is either the elastic critical shear buckling stress

$$\tau_e \text{, when } \tau_e \leq 0.6\ \sigma_y/\sqrt{3} \tag{4a.W8}$$

or a reduced critical shear buckling stress

$$\tau_{cr,red} = \sqrt{0.6(\sigma_y/\sqrt{3}) \cdot \tau_e} \text{, when } \tau_e > 0.6\ \sigma_y/\sqrt{3} \tag{4a.W9}$$

whichever is the smaller.

Furthermore, the web slenderness of the unstiffened girder is limited to

$$\left.\begin{array}{ll} \text{for } \sigma_y = 235 \text{ (MPa)} & h/t \leq 67\sqrt{136/\tau} \\ \text{for } \sigma_y = 355 \text{ (MPa)} & h/t \leq 54\sqrt{205/\tau} \end{array}\right\} \tag{4a.W10}$$

where the average shear stress τ in the web cannot exceed $\tau_y = \sigma_y/\sqrt{3}$.

SIA 161 79

For girders under predominantly static loads

$$h/t \le 0.4\, E/\sigma_y \tag{4a.W11}$$

When the loads are not predominantly static (the problem of "breathing" or "oil canning"), the smallest dimension of the compression zone of a web panel in any direction shall not exceed 100 times its thickness.

INTERNATIONAL COMPARISONS & COMMENT

WHAT ARE THE DIFFERENCES?

To simplify comparison, only the slenderness limits for a transversely stiffened web plate without longitudinal stiffeners are discussed.

The basis for the limiting slenderness often varies from specification to specification. The principal considerations, as far as can be determined, are the following: (The letters at the end of each item correspond to those used in Column 3 of Table 4a.1)

1) No buckling should occur before yielding under pure bending. (y)

2) The web should not be too slender to prevent development of fatigue cracking caused by "flapping" of the plate near the junction to the flanges or stiffeners. (f)

3) No buckling is permitted due to VERTICAL compressive stresses applied to the web by the flanges when the whole section develops curvature under bending moment. (c)

4) A number of specifications have no particular limits on the web slenderness given. However, the implied limitation is to prevent elastic buckling of the web plate. (In Table 4a.1, these cases are marked with "b" in Columns 3 and 4.)

When the limit is stated, it is usually given either as a specific value or as a function of the yield stress σ_y in the following form

$$\frac{h}{t} \leq \frac{C}{\sqrt{\sigma_y}} \tag{4a.1}$$

where C is the constant stipulated by the specification.

AUSTRALIA

The upper limit for webs without longitudinal stiffeners is

$$\frac{h}{t} \leq \frac{3162}{\sqrt{\sigma_y}} \tag{4a.2}$$

The background of this rule is not clear, but the resultant value is very close to those based on fatigue limitation (Case 2 above), as can be seen in the Table 4a.1 in Column 5.

CHINA

Both specifications are apparently based on Case 1 (yielding) consideration and lead to relatively stocky webs. The specification for buildings and highway bridges is somewhat less conservative with the equation as shown

$$\frac{h}{t} \leq \frac{2606}{\sqrt{\sigma_y}} \tag{4a.3}$$

that is, 165 for σ_y=250 MPa in comparison with the maximum value of 140 for railroad bridges.

Table 4a.1 Comparison of Maximum Web Slenderness Values (h/t)

1 2	3	4	5	6
Region Specification	Criterion	Bldg.	Bridges HW	Bridges RR
Australia				
AS DR87164 87	f?	200	200	
China				
GBJ 18 87	y	165	165	
TBJ 2 85	y			140
East Europe				
ČSN 73 1401 84	?	*		
ČSN 73 6205 84	?		*	
MI 15024/4 85	?	*		
SNiP 82	b	368	170	
Japan				
AIJ 70	b	b		
JRA 80	y	148	148	
North America				
AASHTO 89	f		192	
AISC 89	c	332		
AISC 86	c	332		
AREA 89	y			170
CSA S6 M89 84a	f		192	
CSA S16.1 M89 84b	c	332		
FHWA 80	f		190	
West Europe				
BS 5400 82	b?	543?		
BS 5950 85	c,f	294	250	
DASt 015 80	b	343		
EC 3	f?	~ 250	~ 250	
NBN B51002 88	b	b		
SIA 161 79	b,f	322	(200)	

Comments

Column 3
Web slenderness is controlled by the criterion of
(y) Yielding before buckling;
(f) Fatigue; (c) Curvature buckling; (b) Elastic buckling

When the allowable web slenderness depends on the yield stress, an assumption of σ_y = 250 MPa results in the values shown in Columns 4, 5 and 6 for different types of structures.

Column 4 - Buildings and other structures subjected to static loads

Column 5 - Highway bridges

Column 6 - Railroad bridges

(*) indicates the type of structure covered by the specification if there is no set slenderness limitation.

EAST EUROPE

As shown with (?) in Column 3 of the Comparison Table 4a.1, no numerical limits are given for three specifications, and the slenderness requirements depend on ultimate strength and serviceability considerations. Only SNiP II-23-81 82 has particular limits with C = 5814 for static loads (buildings) and C = 2683 for movable loads (highway bridges).

JAPAN

For highway bridges, the Japanese specification provides a table of the allowable web thickness for various grades of steel and the combinations of none, one or two longitudinal stiffeners. The value of C computed from this table for the web without longitudinal stiffeners falls within the range of 2326 to 2359 with an average of 2344. Thus,

$$\frac{\mathrm{h}}{\mathrm{t}} \leq \frac{2344}{\sqrt{\sigma_y}} \tag{4a.4}$$

This corresponds to Case 1, that is, "no buckling before yielding". The specification for buildings states directly that web slenderness should be controlled by elastic buckling under combined shearing and normal stresses.

NORTH AMERICA

As can be observed in the Comparison Table, the North American specifications neatly fall into three distinct groups; those for buildings, highway bridges, and railroad bridges, with little variation within the groups. Curvature buckling (c - Case 3), fatigue (f - Case 2) and yielding (y - Case 1) are the respective backgrounds for the three types of structures. The formula for buildings is

$$\frac{\mathrm{h}}{\mathrm{t}} \leq \frac{96530}{\sqrt{\sigma_y(\sigma_y+\sigma_{rc})}} \tag{4a.5}$$

where σ_{rc} is the residual compressive stress in the flange. This formula is approximately equivalent to having C = 5265 in Eq. 4a.1. In summary, the approximate values of C for different types of structure are

Buildings	C = 5265	
Highway Bridges	C = 3030	(4a.6)
Railroad Bridges	C = 2690	

As can be seen in Columns 4 to 6 of Table 4a.1, these C's tend to give higher slenderness values than are required by some other specifications. This is especially true for buildings, with a value of 333, which is somewhat below 368 of the SNiP II-23-81 82 and 343 of the DASt 015 80. (Granted, a girder designed only for elastic buckling may end up with higher slenderness.) In the case of railroad bridges, although the basis is the same (no yielding before buckling due to bending), 170 in the Table 4a.1 is to be compared to 140 for China.

WEST EUROPE

As indicated in Table 4a.1 in Column 3, two of the six specifications reviewed for West Europe use curvature buckling (c) for statically loaded girders (BS 5950 85 and DASt 015 80), and two (NBN B51002 88 and SIA 161 79) use elastic buckling (b). Consideration of fatigue ("flapping") is the criterion of web slenderness for highway bridges in three specifications (BS 5950 85, EC 3, and SIA 161 79). BS 5400 82 has no web slenderness limitations stated, but the formulas (Eqs. 4b.W1) for the effective web thickness of plate girders in bending imply a rather high h/t of 543.

WHY ARE THERE DIFFERENCES?

A different fundamental theory behind the slenderness limitations stands out as the most prominent cause of the differences among the regions and specifications. Many non-technical aspects given in the General Introduction also play an inhibiting role in the incorporation of the results of the latest research and experiences into national specifications. An illustration of this is, for example, the convenient form in which the effect of variable σ_y was introduced into the old familiar form of the slenderness limitation as a constant number, $170\sqrt{235/\sigma_y}$ (GBJ 18 87) or $170\sqrt{\sigma_{ba}/\sigma}$ (AREA 89).

4b. Web Strength

SPECIFICATION PROVISIONS & EXPLANATION

AUSTRALIA

CHAPTER EDITOR
Noel W. Murray, Monash University

AS DR87164 87

Bending Strength
Bending strength of plate and box girders is defined by the same procedures as for beams except that, when the web slenderness exceeds the value of

$$\frac{\mathrm{h}}{\mathrm{t}} \geq 115\sqrt{\frac{250}{\sigma_y}} = \frac{1818}{\sqrt{\sigma_y}} \tag{4b.A1}$$

the effective section modulus is to be reduced to

$$\mathrm{S}_{eff} = \mathrm{S}\left(\frac{115}{(\mathrm{h/t})}\right) \tag{4b.A2}$$

where S is the elastic section modulus of the cross section.

Then, the bending strength is

$$\mathrm{M}_u = \sigma_y\,\mathrm{S}_{eff} \tag{4b.A3}$$

The reduction of the bending strength for the effect of lateral-torsional buckling is the same as discussed for beams in Chapter 3. There, the effects of lateral and torsional restraints, of the shape of the moment diagram, and of the variation of the girder depth are taken into account.

Shear Strength
(A) Unstiffened web
The nominal shear capacity V_u for an unstiffened web or a web considered to be unstiffened shall be calculated as follows

$$\mathrm{V}_u = \alpha_v \mathrm{V}_y \leq \mathrm{V}_y \tag{4b.A4}$$

where

$$\mathrm{V}_y = 0.6\,\sigma_y\,\mathrm{h\,t} \tag{4b.A6}$$

$$\alpha_v = \left[\frac{82}{\left(\frac{\mathrm{h}}{\mathrm{t}}\right)\sqrt{\frac{\sigma_y}{250}}}\right]^2 \tag{4b.A5}$$

(B) Transversely stiffened web
The nominal shear buckling capacity V_u for a transversely stiffened web shall be calculated as follows

$$\mathrm{V}_u = \alpha_v \alpha_d \alpha_f \mathrm{V}_y \leq \mathrm{V}_y \tag{4b.A7}$$

where

$$\alpha_v = \left[\frac{82}{\left(\frac{h}{t}\right)\sqrt{\frac{\sigma_y}{250}}}\right]^2 \left[0.75\left(\frac{h}{a}\right)^2 + 1.0\right] \quad \text{when } a/h \geq 1.0 \qquad (4b.A8)$$

or

$$\alpha_v = \left[\frac{82}{\left(\frac{h}{t}\right)\sqrt{\frac{\sigma_y}{250}}}\right]^2 \left[\left(\frac{h}{a}\right)^2 + 0.75\right] \quad \text{when } a/h < 1.0 \qquad (4b.A9)$$

$$\alpha_d = 1 + \frac{1-\alpha_v}{1.15\alpha_v\sqrt{1+\left(\frac{a}{h}\right)^2}} \qquad (4b.A10)$$

α_f = the flange restraint factor to be taken as the smallest of (a), (b), or (c) as follows

(a) $\alpha_f = 1.0$ (4b.A11a)

(b) $\alpha_f = 1.6 - \dfrac{0.6}{\sqrt{1+\left(\dfrac{40b_e t_f^2}{h^2 t}\right)}}$ (4b.A11b)

in which b_e, the effective flange plate width, is the least of the following

i. $\dfrac{12t_f}{\sqrt{\sigma_y/250}}$

ii. the distance from the mid-plane of the web to the nearer edge of the flange (taken as zero if there is no flange outstand)

iii. half the clear distance between the webs if there are two or more webs

or (c) α_f to be determined from a rational analysis (4b.A11c)

where t_f is flange plate thickness.

Interaction of Shear and Bending

Either of the two methods given below is permitted for computing the strength of plate and box girders subjected to the concurrent action of shear and bending.

a) Full shear strength V_u according to Eqs. 4b.A4 to 4b.A10, provided that the moment capacity, M_u^*, is calculated for the flanges alone according to

$$M_u^* = A_{f\ell}\,\sigma_y\,(h + t_f)\,\frac{S_{eff}}{Z} \qquad (4b.A12)$$

where

$A_{f\ell}$ = the smaller gross flange area
S_{eff} = effective section modulus as per Eq. 4b.A2
Z = plastic modulus of the cross section

or b) When the bending moment is assumed to be resisted by the whole cross section, the member shall satisfy

$$V_u^* = V_u \quad \text{for } M_u^* \leq 0.75\,M_u \qquad (4b.A13a)$$

or

$$V_u^* = V_u\left[2.2 - \left(\frac{1.6\,M_u^*}{M_u}\right)\right] \quad \text{for } 0.75\,M_u \leq M_u^* \leq M_u \qquad (4b.A13b)$$

where M_u is the moment strength according to Eq. 4b.A3.

When, in addition to shear, V_u^*, and bending, M_u^*, the girder web is also subjected to a transverse patch load applied to the flange, R^*, and an axial force, N_w^*, the web must satisfy the following yielding and buckling interaction relationships

(A) Yielding Check

$$\left(\frac{R^*}{b_{bf}\ t}\right)^2 - \sigma_w^* \left(\frac{R^*}{b_{bf}\ t}\right) + (\sigma_w^*)^2 + \left(\frac{V_u^*}{h\ t}\right) \le \sigma_y^2 \qquad (4b.A14)$$

where

$$\sigma_w^* = \frac{N_w^*}{h\ t} + 0.77\ \frac{M_w^*}{S_w} \qquad (4b.A15)$$

b_{bf} = width of the patch load on the edge of the web dispersed through the flange thickness at the slope of 2.5:1 on each side

M_w^* = design moment in the web, calculated by elastic theory for sections with non-compact or slender flanges or by plastic theory for sections with compact flanges

S_w = elastic section modulus of the web plate alone = $th^2/6$

(B) Buckling Check

$$\left(\frac{R^*}{R_b}\right) + \left(\frac{N_w^*}{N_{wo}}\right) + \left(\frac{V_w^*}{V_u}\right)^2 + \left(\frac{M_w^*}{M_w}\right)^2 \le 1 \qquad (4b.A16)$$

where

$$d_e = \frac{1.9\ \sqrt{b_{bf}\ h}}{\alpha_w} \qquad (4b.A18c)$$

M_w = nominal bending moment capacity of the web if the moment were resisted by the web alone

N_{wo} = nominal axial load capacity of the web panel if the web panel resisted axial load alone

$$= \frac{51\,h\,t\,\sigma_y}{\left(\frac{h}{t}\right)\sqrt{\frac{\sigma_y}{250}}} \le h\,t\,\sigma_y \qquad (4b.A17)$$

R_b = nominal buckling capacity of a transversely stiffened web in bearing alone

$= \beta_w\, b_{bf}\, t\, \sigma_y$ (4b.A18a)

V_u = nominal shear capacity of the web panel if the web panel resisted shear alone

$$\alpha_w = \left[3.4 + \left(\frac{2.2\ h}{a}\right)\right]\left[0.4 + \left(\frac{0.5\ b_{bf}}{a}\right)\right]$$

$$\beta_w = 0.10 + \frac{20}{\left(\frac{d_e}{t}\right)\sqrt{\frac{\sigma_y}{250}}} \qquad (4b.A18b)$$

Intermediate Tranverse Web Stiffeners

Intermediate transverse stiffeners must satisfy the following requirements:

Minimum area of transverse web stiffeners not subject to external loads or moments shall have an area

$$A_s \geq 0.5\,\gamma h t\,(1-\alpha_v)\left[\left(\frac{a}{h}\right) - \frac{\left(\frac{a}{h}\right)^2}{\left[1+\left(\frac{a}{h}\right)^2\right]}\right] \qquad (4b.A19)$$

where

α_v = the value determined in accordance with Eq. 4b.A8 or 4b.A9
γ = 1.0 for pairs of stiffeners
= 1.8 for single angle stiffeners
= 2.4 for single plate stiffeners

Buckling capacity of an intermediate stiffener shall satisfy

$$R_{sb} \geq V_u \qquad (4b.A20)$$

where

R_{sb} = nominal buckling capacity of the intermediate stiffener treated as a compression member with the effective length (L_e) taken to be

$$L_e = h \qquad (4b.A21)$$

V_u = nominal shear capacity specified by Eq. 4b.A7 for a stiffened web using $\alpha_d = 1.0$ and $\alpha_f = 1.0$

The effective section of the compression member shall be taken as the area of the stiffener together with a portion of web on each side of the centerline not greater than

$$\frac{17.5t}{\sqrt{\frac{\sigma_y}{250}}} \quad \text{or} \quad \left(\frac{a}{2}\right) \qquad (4b.A22)$$

Minimum stiffness of transverse web stiffeners not subject to external loads or moments shall be defined by the minimum moment of inertia (I_s) about the centerline of the web such that

$$I_s \geq 0.75ht^3 \quad \text{for } \frac{a}{h} \leq \sqrt{2} \qquad (4b.A23)$$

or

$$I_s \geq \frac{1.75h^3t}{a^2} \quad \text{for } \frac{a}{h} > \sqrt{2}$$

Longitudinal Web Stiffeners

Longitudinal web stiffeners shall be continuous or shall extend between and be attached to transverse web stiffeners.

When a longitudinal stiffener is required at a distance $0.2h_c$ from the compression flange, it shall have a moment of inertia (I_s) about the face of the web such that

$$I_s \geq 4h_c t^3\left[1+\frac{4A_s}{h_c t}\left(1+\frac{A_s}{h_c t}\right)\right] \qquad (4b.A24)$$

where A_s is the area of the stiffener and h_c is equal to twice the distance between the compression flange and the centroid or to the web depth h, whichever is greater.

When a second longitudinal stiffener is required at the neutral axis of the section, it shall have a moment of inertia (I_s) about the face of the web such that

$$I_s \geq h_c t^3 \qquad (4b.A25)$$

CHINA

CHAPTER EDITORS
Xiliang Liu, Tianjin University
Xinhua H. Ren, Tianjin University

GBJ 18 87

The spacing of transverse stiffeners for girder webs subjected to bending and shear without edge loading, is limited by the following formulas

for $\frac{h}{t}\sqrt{\eta\tau} \leq 1200$

$$a \leq 2h$$

for $1200 < \frac{h}{t}\sqrt{\eta\tau} \leq 1500$

$$a \leq \frac{500\ h}{\frac{h}{t}\sqrt{\eta\tau} - 1000} \tag{4b.C1}$$

for $\frac{h}{t}\sqrt{\eta\tau} > 1500$

$$a \leq \frac{1000\ h}{\frac{h}{t}\sqrt{\eta\tau} - 500} \tag{4b.C2}$$

where

- h = clear distance between the top and bottom flanges
- t = thickness of the web
- σ = bending stress of the compression edge of web at the same cross section as for calculating τ
- τ = average shearing stress due to the maximum shear force within the portion of girder considered

$$\tau = \frac{V}{ht} \tag{4b.C3}$$

$$\eta = \frac{1}{\sqrt{1 - \left(\frac{\sigma}{715}\left(\frac{h}{100\ t}\right)^2\right)^2}} \tag{4b.C4}$$

For girder webs subjected to bending and shear, the distance between the longitudinal stiffener and the compression edge of the web must be limited to

$$h_1 \leq \frac{1120\ t}{\sqrt{\sigma}} \tag{4b.C5}$$

where

$$\sigma = \frac{M_{max}}{I}\frac{h_c}{2} \tag{4b.C6}$$

- I = moment of inertia
- M_{max} = maximum moment in the panel

At the same time, the spacing of transverse stiffeners shall be determined by Eq. 4b.C1 or Eq. 4b.C2, by using $h_2 = h - h_1$ instead of h and taking $\eta = 1$.

If the calculated value of the right hand side of Eqs. 4b.C1 and 4b.C2 is greater than 2h (or $2h_2$) or if the denominator is negative, then $a \leq 2h$ (or $a \leq 2h_2$) should be taken.

The width of intermediate transverse stiffeners for webs without longitudinal stiffener shall be not less than

$$\frac{h}{30} + 40 \text{ mm} \tag{4b.C7}$$

and the thickness not less than

$$\frac{d}{15} \tag{4b.C8}$$

where d is the width of the intermediate transverse stiffener.

For webs with longitudinal stiffeners, the transverse stiffeners shall not only satisfy Eqs. 4b.C7 and 4b.C8 but also have the moment of inertia not less than

$$I_{st} = 3\ h\ t^3 \tag{4b.C9}$$

The moment of inertia of the longitudinal stiffeners must satisfy the following equations

$$I_{s\ell} = 1.5\ ht^3 \qquad \text{for} \qquad \frac{a}{h} \leq 0.85 \tag{4b.C10}$$

$$I_{s\ell} \geq (2.5 - 0.4\ \frac{a}{h})\ (\frac{a}{h})^2\ ht^3 \tag{4b.C11}$$

TBJ 2 85

The spacing of transverse stiffeners should be determined as follows

$$a \leq \frac{950\ t}{\sqrt{\tau}} \leq 2 \text{ m} \tag{4b.C12}$$

The longitudinal stiffener should be placed within the range (1/5 to 1/4) h on the side of the compression flange.

$\tau = \frac{V}{h\ t}$ is the average shearing stress of the web panel

V = the shear on the middle cross section of the web panel

For transverse stiffeners set up in pairs on both sides of the web, the outstanding width of each stiffener on one side of the web should not be smaller than h/30 + 40 mm.

For plate girders with transverse and longitudinal stiffeners, the moment of inertia of the longitudinal stiffener should not be smaller than both $ht^3\ (2.4(a/h)^2 - 0.13)$ and $1.5\ ht^3$. The width-thickness ratio of the outstanding leg of the stiffeners should not be greater than 15.

COMMENTARY

To prevent buckling of the web, Eqs. 4b.C1, 4b.C2, 4b.C4 and 4b.C12 are to be used to determine directly the spacing of transverse stiffeners. These equations are derived through some simplifications of the formulas for elastic buckling analysis.

EAST EUROPE

CHAPTER EDITORS

Jozef Djubek, Slovak Academy of Science
Miroslav Škaloud, Czechoslovak Academy of Sciences

ČSN 73 1401 84

The ultimate strength design of plate girders takes into account both initial imperfections and postbuckling behavior of web plates.

The ultimate values of axial force (N_w), bending moment (M_w) and shear force (V_w) carried by a web plate panel are given by

$$N_w = \psi_n A_w R_{dw} \tag{4b.E1}$$

$$M_w = \psi_m S_w R_{dw} \tag{4b.E2}$$

$$V_w = \psi_v 0.6 A_w R_{dw} \tag{4b.E3}$$

where

A_w = area of the web
R_{dw} = σ_{yw}/γ_m is the design strength (resistance) of the web material (MPa)
S_w = section modulus of the web
ψ_n, ψ_m, ψ_v = the reduction coefficients for the decrease of the limit force due to the postbuckling behavior of the web

Denoting $\eta = \sqrt{210/R_{dw}} = \sqrt{210\ \gamma_m/\sigma_{yw}}$, the reduction coefficients are (4b.E4)

$$\psi_n = 40\ (t/d)\ \eta \le 1 \tag{4b.E5}$$

$$\psi_m = 145\ (t/d)\ \eta \le 1 \tag{4b.E6}$$

$$\psi_v = (70 + 30/\mu^2)\ (t/d)\ \eta \le 1 \tag{4b.E7}$$

where

d/t = the slenderness of the longer side of the web panel to the thickness
μ = the ratio of the longer side of the web panel to the shorter side

The ultimate forces N_u, M_u, V_u, F_u (edge load - Fig 4b.E1) of a plate girder web panel are

Compression

$$N_u = A_e R_d \tag{4b.E8}$$

where

$A = A_{f1} + A_{f2} + ht$ is the area of the cross section
A_{f1}, A_{f2} are the areas of the respective flanges

$$A_e = A\left(1 - \frac{ht}{ht + 2A_f}(1-\psi_n)\right) \tag{4b.E9}$$

Bending - Fig. 4b.E2

$$M_u = S_e\, R_d \tag{4b.E10}$$

where S_e is equal to the unreduced section modulus S_2 with respect to the tension flange or to the effective section modulus S_{1e} with respect to the compression flange.

$$S_{1e} = S_1 \left(1 - \frac{I_{w1}}{I}\,(1-\psi_m)\right) \tag{4b.E11}$$

where

$$I = 2\, A_{f1} \left(\frac{z_1+z_{1w}}{2}\right)^2 + \frac{1}{12}\,(2\, z_{1w})^3\, t$$

$$I_{w1} = \frac{1}{12}\,(2z_{1w})^3\, t$$

Reduction coefficient ψ_m is calculated from Eq. 4b.E6 for the web of the height $h=2\,z_{1w}$ and the design strength R_{dw}.

If the design strength R_{df} of flanges is greater than the design strength R_{dw} of the web, the ultimate force in both cases, a) and b), may be calculated for $R_d = R_{df}$, but, instead of ψ_m, the value $\psi_m R_{dw}/R_{df}$ has to be inserted, while ψ_m is calculated for R_{dw}.

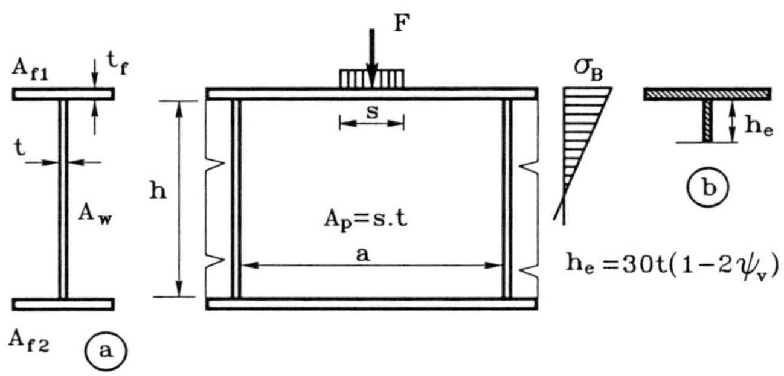

Fig. 4b.E1 Web Panel under Shear, Bending and Edge Load

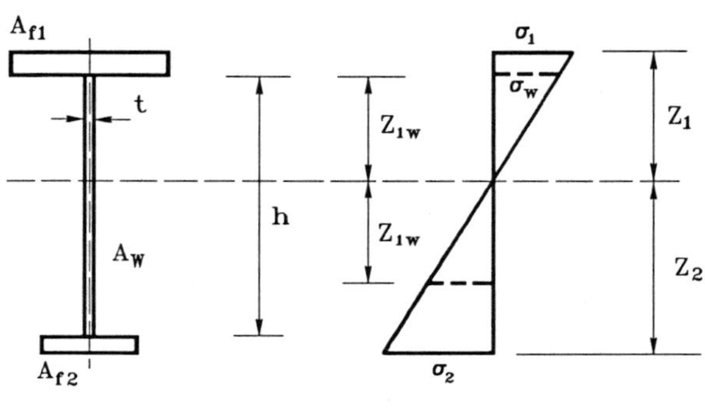

Fig. 4b.E2 Unsymmetrical Section

Shear

$$V_u = (\psi_v + \psi_{vf})\, A_w\, 0.6\, R_{dw} \qquad \text{(4b.E12)}$$

The beneficial effect of flanges upon V_u expressed by ψ_{vf} can be exploited only if the connection of the web to the flanges and stiffeners is fully effective up to the ultimate tension force and the edge load F is not present.

A) Moving loads

$$\psi_{vf} = \frac{1-\psi_v}{1+ C\mu A_w/A_f} \qquad \text{(4b.E13)}$$

where

$A_f \;= (A_{f1}+A_{f2})/2 \quad \leq 1.5\, A_w \sqrt{\mu}$
C $\;= 10$ for end panels
$\;= 2 \sim 3$ for inner panels

B) Stationary loads

$$\psi_{vf} = (1-\psi_n)\sqrt{\frac{16\, S_{f,pl}\, R_{df}}{a\, A_w\, 0.6\, R_{dw}} \cdot \frac{\sqrt{3}}{\mu + 1/\mu}} \qquad \text{(4b.E14)}$$

where $S_{f,pl} R_{df}$ is the ultimate bending moment of the "effective flange" section according to Fig. 4b.E1b.

Edge Load (stationary) - Fig. 4b.E1

$$F_u = \psi_F\, A_p\, R_{dw} \qquad \text{(4b.E15)}$$

where

A_p, s, t_f, σ_B - see Fig. 4b.E1

$$\psi_F = \left(14.6 + 24.4\, \frac{s}{h}\right) \frac{\sqrt{t_f\, t}}{s} \sqrt{1-(\sigma_B/R_{dw})^2} \cdot \eta \qquad \text{(4b.E16)}$$

Combined Loads

The limit state of the plate girder panel subjected to the combined action of loads N, M, V, F is checked by the interaction formula

$$\left(\frac{N}{N_u} + \frac{M}{M_u} + \frac{F}{F_u}\right)^2 - 3\left(\frac{N}{N_y} + \frac{M}{M_y}\right)\frac{F}{F_y} + \left(\frac{V}{V_u}\right)^2 \leq 1 \qquad \text{(4b.E17)}$$

where N_y, M_y, F_y denote the ultimate forces given by Eqs. 4b.E8, 4b.E10 and 4b.E15, respectively, but taking $\psi_n = \psi_m = \psi_F = 1.0$.

Equation 4b.E17 is to be checked with the ultimate forces N_u, M_u and N_y, M_y calculated with respect to the compression as well as to the tension flange.

Stiffeners

The required moment of inertia of stiffeners is given by

$$I_s \geq 0.1\, ht^3\, \chi\, \gamma^* \qquad \text{(4b.E18)}$$

where γ^* is the linear-buckling-theory optimum rigidity

$$\chi = \chi_s\left(\frac{h}{t\,\bar{\eta}} - 1\right) \leq \chi_s \qquad \text{(4b.E19)}$$

where

$\bar{\eta} =$ 40η (compression)
$\bar{\eta} =$ 145η (bending)
$\bar{\eta} =$ 70η (shear)
$\chi_s =$ varies between 3 and 7 depending on the location of the longitudinal stiffener (value h_1/h).

In combined loading

$$\chi\,\gamma^* = \sqrt{\left(\chi_n\,\gamma_n^*\,\frac{N}{N_u} + \chi_m\,\gamma_m^*\,\frac{M}{M_u}\right)^2 + \left(\chi_v\,\gamma_v^*\,\frac{V}{V_u}\right)^2} \tag{4b.E20}$$

where indices refer to the individual kinds of loading.

ČSN 73 6205 84

The ultimate limit state of each web panel subjected to compression σ_a, transverse patch load σ_{loc}, bending σ_m, and shear τ is checked by the following condition

$$\sqrt{\left(\frac{\sigma_a}{\psi_n\,R_{dw}}\right)^2 + \left(\frac{\sigma_{loc}}{\psi_{loc}\,R_{dw}}\right)^2 + \left(\frac{\sigma_m}{\psi_m\,R_{dw}}\right)^2 + \left(\frac{\tau}{\psi_v\,0.6\,R_{dw}}\right)^2} \le 1 \tag{4b.E21}$$

where

R_{dw} = the design strength of the web material
ψ_n, ψ_{loc}, ψ_m, ψ_v = are the reduction buckling coefficients, determined so as to allow for the effect of postbuckling behavior and initial imperfections.

Two kinds of web panels are considered in the analysis, viz.,
I) panels of type I (boundary panels, i.e., panels adjacent to a flange or the end post).
II) panels of type II (the inner panels of webs stiffened by at least two longitudinal stiffeners, i.e., panels surrounded on all sides by other web panels.

The reduction buckling coefficients are then defined as follows

Compression

$$\psi_{n,\mathrm{I}} = 40\,\lambda_n \le 1.0 \tag{4b.E22}$$
$$\psi_{n,\mathrm{II}} = \psi_{n,\mathrm{I}}\,(1.3 - 12\,\lambda_n),\ \text{but} \ge \psi_{n,\mathrm{I}} \text{ and} \le 1.0 \tag{4b.E23}$$

where λ_n is equal to
$$\frac{\sqrt{\dfrac{210}{R_{dw}}}}{\dfrac{h}{t} + 10} \tag{4b.E24}$$

where

h = clear distance between the top and bottom flanges
t = thickness of the web panel

Transverse Patch Loading

The coefficient ψ_{loc} is calculated from the above formula for $\psi_{n,\mathrm{I}}$ but on condition that h be replaced by an equivalent depth

$$h_e = 1.9 \sqrt{\frac{s\,h}{\left(3.4 + 2.2\,\frac{h}{a}\right)\left(0.4 + \frac{s}{2a}\right)}} \tag{4b.E25}$$

s being the patch load width (dispersed through the thickness of the loaded flange at the slope of 1:1) and a the spacing of the transverse stiffeners.

Bending

$$\psi_{m,\mathrm{I}} = 1.3 - \frac{\lambda_m}{2} \leq 1.0 \tag{4b.E26}$$

$$\psi_{m,\mathrm{II}} = 1.2 - \frac{\lambda_m}{3} \leq 1.0$$

where λ_m is equal to $\frac{h}{200t}\sqrt{\frac{R_{dw}}{210}}$ (4b.E27)

Shear

$$\psi_v = \frac{\beta_v}{\lambda_v^{\eta}} \leq 1.0 \tag{4b.E28}$$

where λ_v is equal to $\frac{h}{60t}\sqrt{\frac{R_{dw}}{210}}$ (4b.E29)

Factors β_v and η are given in Table 4b.E1. For intermediate values, linear interpolation is used.

Some provisions are also given for a serviceability check of web panels, but they affect the design only in the case of very slender webs ($\lambda > 180$).

Longitudinal stiffeners must satisfy the following requirement for the moment of inertia $I_{s\ell}$

$$I_{s\ell} \geq 0.1\, ht^3 \kappa\, \gamma^*$$

where

$$\gamma^* = \sqrt{\left(\gamma_n^* \frac{\sigma_a}{\psi_n R_{dw}} + \gamma_m^* \frac{\sigma_m}{\psi_m R_{dw}}\right)^2 + \left(\gamma_v^* \frac{\tau}{\psi_v\, 0.6\, R_{dw}}\right)^2} \tag{4b.E30}$$

γ_m^* = the optimum rigidities according to the linear buckling theory for bending

γ_n^* = the optimum rigidities according to the linear buckling theory for compression

γ_v^* = the optimum rigidities according to the linear buckling theory for shear

Table 4b.E1 Values of β_v and η in Eq. 4b.E28

	Panels of type I				Panels of type II		
$\alpha = a/h$	≤ 0.5	1	2	≥ 3	≤ 0.5	1	≥ 2
β_v	1.344	1.032	0.964	0.953	1.080	1.002	0.978
η	0.514	0.515	0.666	0.988	0.145	0.109	0.148

These are given in the Design Code for a number of types of longitudinal stiffening.

κ is a factor which allows for the effect of postbuckling behavior and initial imperfections on stiffener strength. It is defined in Table 4b.E2

For the values of h/t between h/t=120 and h/t=240, linear interpolation is used.

Transverse stiffeners are designed for rigidity and for strength. The moment of inertia I_{st} of the stiffener must satisfy the condition

$$I_{st} \geq 0.1\, ht^3 \xi\, \gamma^* \tag{4b.E31}$$

where

γ^* = the optimum rigidity according to the linear buckling theory
ξ = factor which allows for the effect of postbuckling behavior and initial imperfections. This factor is given as follows

for $\frac{h}{t} \leq 60$ $\qquad \xi = 0$

for $\frac{h}{t} = 75$ $\qquad \xi = 1$

for $\frac{h}{t} \geq 150$ $\qquad \xi = 3$ (4b.E32)

For intermediate values, linear interpolation is used.

When the web also has longitudinal stiffeners, t in the above formula for I_{st} is replaced by thickness t_e of an equivalent web, the load-carrying capacity in shear of which equals that of the given web with longitudinal stiffeners.

From the point of view of strength, the transverse stiffener must sustain an axial load N_{st} assumed to act in the web mid-plane so that in the case of single-sided stiffeners the corresponding eccentricity of load is considered in the calculation.

$$N_{st} = V - V_o \tag{4b.E33}$$

where

V = the larger of the two shear forces acting on either side of the respective stiffener
V_o = portion of the shear force which the stiffener carries prior to the buckling of the adjacent web panels and which is taken as $V_o = 0.8\, V_{cr}$, with V_{cr} denoting the critical shear load of the web plate.

Table 4b.E2 Values of κ

		$h/t \leq 120$	$h/t \geq 240$
Open-section stiffeners	one stiffener	1.25	4.0
	more stiffeners	1.0	3.2
Closed-section stiffeners	one stiffener	1.25	2.5
	more stiffeners	1.0	2.0

The effect of longitudinal stiffeners, if there are any, on the behavior of transverse stiffeners is taken into account by way of dummy transverse loads assumed to act perpendicular to the web plane and to be of the magnitude equal to 1% of the axial load acting in the longitudinal stiffener.

MI 1502/4 85

The limit states design approach is used, and generally, girder webs are designed for the criterion of buckling under combined normal, shearing and bending stresses. Postbuckling strength under predominantly shearing stresses is also permitted. Both of these approaches are discussed here for a girder panel without longitudinal stiffeners.

Buckling Web Strength Under Combined Loads

Buckling stress for the case of pure axial, bending or shear stress is given by

$$\text{Buckling stress} = k \frac{\pi^2 E}{12(1-\nu^2)} \frac{1}{\left(\frac{h}{t}\right)^2} \qquad (4b.E34)$$

where the buckling coefficient k is equal to k_a for the axial buckling stress σ_{acr}, k_b for the bending buckling stress σ_{bcr}, and k_v for the shear buckling stress τ_{cr}.

With the panel aspect ratio $\alpha = a/h$, the buckling coefficients are to be computed from the following formulas

Axial

for $\alpha \geq 1.0$ $\quad k_a = 4.0$

for $\alpha < 1.0$ $\quad k_a = \left(\alpha + \frac{1}{\alpha}\right)^2$ $\qquad$ (4b.E35)

Bending

for $\alpha \geq 2/3$ $\quad k_b = 23.9$

for $\alpha < 2/3$ $\quad k_b = 15.87 + \frac{1.87}{\alpha^2} + 8.6\alpha^2$ $\qquad$ (4b.E36)

Shear

for $\alpha \geq 1.0$ $\quad k_v = 5.34 + \frac{4.0}{\alpha^2}$

for $\alpha < 1.0$ $\quad k_v = 4.0 + \frac{5.34}{\alpha^2}$ $\qquad$ (4b.E37)

In design, the actual stresses, σ_a, σ_b and τ, are related to each other, and the interaction formula for buckling is taken to be

$$\frac{\sigma_a}{\sigma_{acr}} + \left(\frac{\sigma_b}{\sigma_{bcr}}\right)^2 + \left(\frac{\tau}{\tau_{cr}}\right)^2 \leq 1.0 \qquad (4b.E38)$$

The reduced buckling coefficient for the combined state of stress, k_{red}, is defined by

$$k_{red} = \frac{\sqrt{\sigma_1^2 + 3\tau^2}}{\frac{\sigma_a}{2k_a} + \sqrt{\left(\frac{\sigma_a}{2k_a}\right)^2 + \left(\frac{\sigma_b}{k_b}\right)^2 + \left(\frac{\tau}{k_v}\right)^2}} \tag{4b.E39}$$

where σ_1 is the combined stress at the web edge under compression, that is $\sigma_a + \sigma_b$ for a symmetrical girder.

The relative slenderness of the web plate is defined by

$$\overline{\lambda}_o = \frac{3.3}{\pi}\sqrt{\frac{\sigma_y}{k_{red}E}}\left(\frac{h}{t}\right) \tag{4b.E40}$$

and the buckling reduction coefficient φ_b by

for $\overline{\lambda}_o \leq 0.85$

$$\varphi_b = 1 \tag{4b.E41}$$

for $0.85 \leq \overline{\lambda}_o \leq 1.12$

$$\varphi_b = 1 - 0.74\left(\overline{\lambda}_o - 0.85\right) \tag{4b.E42}$$

for $\overline{\lambda}_o > 1.12$

$$\varphi_b = \frac{1}{\overline{\lambda}_o^2} \tag{4b.E43}$$

The effective design stress for the actual combined stresses must satisfy the requirement for buckling

$$\sigma_{red} = \sqrt{\sigma_1^2 + 3\tau^2} \leq \varphi_b \sigma_w \tag{4b.E44}$$

where the design working stress of the material is

$$\sigma_w = \frac{\sigma_y}{\gamma_n} \tag{4b.E44}$$

For pure shear, the design requirement is thus,

$$\tau \leq \varphi_b \tau_w \tag{4b.E45}$$

where τ_w is equal to $\frac{\sigma_w}{\sqrt{3}}$ (4b.E46)

Ultimate (Postbuckling) Web Strength Under Predominant Shear

Utilization of the ultimate strength requires that the buckling limitations be satisfied under the service loads.

The average shear stress, $\tau = V/ht$ must fulfill the following condition

$$\tau \leq \overline{\varphi}_b \tau_w \tag{4b.E47}$$

where τ_w is the allowable shear stress of Eq. 4b.E46 and

$$\overline{\varphi}_b = C + (1 - C)\,\varphi_b \leq 1 \tag{4b.E48}$$

is a reduction coefficient and

$$C= \frac{0.866}{\sqrt{1+ \alpha^2}} \qquad (4b.E49)$$

Here, φ_b is to be computed from Eqs. 4b.E40 to 4b.E43 with $\sigma_a = \sigma_b = 0$ in Eq. 4b.E39.

If the moment acts interactively with the shear force V, its effect can be accounted for (unless a more accurate solution is employed) by entirely neglecting the web in the calculation of the section modulus and, consequently, in the determination of the maximum bending stress σ.

Transverse Stiffeners

In the design of intermediate transverse stiffeners, the postbuckling behavior of the webs is to be taken into account by analyzing the transverse stiffener as an eccentrically loaded column of length h and having an initial bow of amplitude e_o=h/500 and subjected to a compressive force N_{st}.

$$N_{st} = \left(\left(\alpha \sqrt{1+\alpha^2}\right) - \alpha^2\right) V \qquad (4b.E50)$$

as well as, to a fictitious "disturbing" bending moment

$$M_k= N_h\, e_o \qquad (4b.E51)$$

where

$N_h = (0.23\sigma_1 + 0.17\,\sigma_t)$ ht (4b.E52)
V = governing shear force

with $\sigma_1$and σ_t being, respectively, the compressive and tensile bending stresses at the extreme fibers of the web panel disregarding their signs.

SNiP II-23-81 82

The main body of the SNiP II-23-81 82 Design Code is based on the linear buckling theory. However, it also provides rules for considering postbuckling strength in girders subjected to static loads, that is, girders in buildings.

Stresses σ, τ, σ_{loc} (bending, shear and local) are calculated on the assumption of the elastic behavior of the material.

In the case of girders the tension flange of which is subjected to a local concentrated load F inducing local stress σ_{loc}, e.g., the portion of the tension flange of a continuous girder which is over a support, only two components of the stress state, either σ and τ or σ_{loc} and τ, are to be considered in the analysis to act simultaneously.

The following interaction formulas are used for different cases of loading

Bending and Shear

$$\sqrt{\left(\frac{\sigma}{\sigma_{cr}}\right)^2+\left(\frac{\tau}{\tau_{cr}}\right)^2} \le \gamma_c \qquad (4b.E53)$$

Bending, Shear and Local Edge Load

$$\sqrt{\left(\frac{\sigma}{\sigma_{cr}}+\frac{\sigma_{loc}}{\sigma_{loc,cr}}\right)^2+\left(\frac{\tau}{\tau_{cr}}\right)^2} \le \gamma_c \tag{4b.E54}$$

Longitudinally Stiffened Girder Under Bending, Shear and Local Load

(A) for the web sub-panel between the compression flange and the longitudinal stiffener

$$\frac{\sigma}{\sigma_{cr,\ 1}}+\frac{\sigma_{loc}}{\sigma_{loc,cr,1}}+\left(\frac{\tau}{\tau_{cr,1}}\right)^2 \le \gamma_c \tag{4b.E55}$$

(B) for the web sub-panel between the tension flange and the longitudinal stiffener

$$\sqrt{\left(\frac{\sigma(1-2h_1/h)}{\sigma_{cr,2}}+\frac{\sigma_{loc,2}}{\sigma_{loc,cr,2}}\right)^2-\left(\frac{\tau}{\tau_{cr,2}}\right)^2} \le \gamma_c \tag{4b.E56}$$

where

$\sigma_{loc,2} = 0.4\ \sigma_{loc}$ if the local load acts on the compression flange (4b.E57)
$\sigma_{loc,2} = \sigma_{loc}$ if the local load acts on the tension flange.

Coefficient γ_c has the following values

for plate girders in theatres and cinemas

$$\gamma_c=0.9 \tag{4b.E58}$$

for other cases

$$\gamma_c=1.0$$

Buckling stresses, the stresses subscripted with "cr", are computed according to the procedures described below

$$\sigma_{cr} = \frac{k_b\sigma_y}{\lambda_w^2} \tag{4b.E59}$$

where λ_w is equal to $\frac{h}{t}\sqrt{\frac{\sigma_y}{E}}$ (4b.E60)

the buckling coefficient k_b for bending

$$k_b = 30.0 \tag{4b.E61}$$

The value of k_b can be increased up to 35.5, according to the tables in the specification, to take into account a greater web restraint provided in special cases by the flanges, such as, by the rail in crane girders.

$$\tau_{cr} = 10.3\left(1+\frac{0.76}{\mu^2}\right)\frac{\tau_y}{\lambda_{ef}^2} \tag{4b.E62}$$

where

d = the shorter of the panel sides (a or h)

$\lambda_{ef} = \frac{d}{t}\sqrt{\frac{\sigma_y}{E}}$ (4b.E63)

μ = ratio of the larger to the smaller side dimensions of the web panel (or subpanel) (4b.E64)

The buckling stress $\sigma_{loc,cr}$ due to a locally applied concentrated load is found as follows

For the aspect ratio

$$\alpha = \frac{a}{h} \leq 0.8, \qquad \text{(4b.E65a)}$$

$$\sigma_{loc,cr} = \frac{k_{loc}\sigma_y}{\lambda_a^2} \qquad \text{(4b.E65b)}$$

where λ_a is equal to $\frac{a}{t}\sqrt{\frac{\sigma_y}{E}}$ (4b.E66)

and k_{loc} varies from 11.5 to 83.6, depending on the panel aspect ratio and the restraint by the flanges (Tables are provided in the Specification). For example, for $\alpha = 1.0$, the minimum value k_{loc}= 18.0. A similar process with tables is provided for $\alpha > 0.8$.

An analogous procedure with appropriate tables is given for plate girders with one longitudinal stiffener. Design provision is also made for the use of short transverse stiffeners between the compression flange and the longitudinal stiffener as a reinforcement against edge loading. When the cross section is unsymmetrical, that is, when the compression and tension flanges are not equal, the effective girder depth is taken as twice the distance from the edge of the compression portion of the web to the neutral axis, that is, h_c is to be used.

Transverse stiffeners are designed to have a moment of inertia according to

$$I_{st} \geq 3ht^3 \qquad \text{(4b.E67)}$$

Longitudinal stiffeners are required to have their moment of inertia, $I_{s\ell}$, equal to or larger than the values given by the formulas given in Table 4b.E3. The formula to use depends on the relative location of the longitudinal stiffener h_1/h.

Table 4b.E3 Required $I_{s\ell}$

Position of longitudinal stiffener, h_1/h	$I_{s\ell}$
0.2	$1.5ht^3 \leq [(2.5-0.5\alpha)\ at^2\ \alpha] \leq 7ht^3$
0.25	$1.5ht^3 \leq [(1.5-0.4\alpha)\ at^2\ \alpha] \leq 3.5ht^3$
0.3	$1.5ht^3$

Postbuckling Strength

Transversely stiffened symmetrical plate girders subjected to static loading may also be designed taking into account the postbuckling strength.

The ultimate moment capacity is computed from

$$M_u = \sigma_y th^2\left[\frac{A_f}{th} + \frac{0.85}{\lambda_w}\left(1 - \frac{1}{\lambda_w}\right)\right] \qquad \text{(4b.E68)}$$

and the ultimate shear from

$$V_u = \tau_y th \left[\frac{\tau_{cr}}{\tau_y} + 3.3 \left(1 - \frac{\tau_{cr}}{\tau_y} \right) \frac{\beta\mu}{1+\mu^2} \right] \qquad (4b.E69)$$

where the additional definitions are

$$\tau_y = \sigma_y/\sqrt{3} \qquad (4b.E70)$$

A_f = flange area
β = the coefficient computed according to

$$\text{for } \rho \le 0.03 \qquad \beta = 0.05 + 5\rho \ \ge\ 0.15$$
$$\text{for } 0.03 \le \rho \le 0.1 \qquad \beta = 0.11 + 3\rho \ \le\ 0.40 \qquad (4b.E71)$$

where

S_{min} = the minimum section modulus of the Tee section consisting of the compression flange plate and a portion of the web plate with the width equal to $0.5t\sqrt{E/\sigma_y}$

$$\rho = \frac{8\,S_{min}}{t\,a^2}\,(1+\alpha^2) \qquad (4b.E72)$$

In addition to the moment of inertia requirement, transverse stiffeners for ultimate girder strength must be designed as columns with an effective length of

$$L_e = h(1-\beta), \text{ but } \ge 0.7h \qquad (4b.E73)$$

and subjected to an axial force

$$N = 3.3\,\tau_y th \left(1 - \frac{\tau_{cr}}{\tau_y} \right) \frac{\beta\mu}{1+\mu^2} \qquad (4b.E74)$$

where μ is from Eq. 4b.E64.

Combined Moment and Shear

Combined action of moment and shear must satisfy the following interaction formula

$$(M/M_u)^4 + (V/V_u)^4 \le 1.0 \qquad (4b.E75)$$

where the ultimate moment and shear, M_u and V_u, are from Eqs. 4b.E68 and 4b.E69, respectively.

JAPAN

CHAPTER EDITORS

Sadao Komatsu, Osaka Sangyo University
Fumio Nishino, University of Tokyo

AIJ 86

Webs of plate girders in buildings must satisfy the following buckling requirement for each web subpanel between transverse and longitudinal stiffeners

$$\left(\frac{\sigma}{\sigma_{cr}}\right)^2 + \left(\frac{\tau}{\tau_{cr}}\right) \leq 1 \qquad (4b.J1)$$

where under pure bending

for $h/t \geq 56/C_1$

$$\sigma_{cr} = \frac{1900}{(C_1\ h/t)^2} f_t \qquad (4b.J2a)$$

for $h/t < 56/C_1$

$$\sigma_{cr} = (1.78 - 0.021 C_1\ h/t) f_t \leq f_t \qquad (4b.J2b)$$

for $h/t \leq 210/\sqrt{F/10}$

$$\sigma_{cr} = f_t \qquad (4b.J2c)$$

and where under pure shear

for $h/t \geq 74/C_2$

$$\tau_{cr} = \frac{3300}{(C_2\ h/t)} f_s \qquad (4b.J3a)$$

for $h/t < 74/C_2$

$$\tau_{cr} = (1.74 - 0.0154\ C_2\ h/t)\ f_s \qquad (4b.J3b)$$

Other notation is as follows:

$$C_1 = \sqrt{F/(10k_1)} \qquad (4b.J4)$$

$C_2 = \sqrt{F/(10k_2)}$

$F = \sigma_y/98.7$

$f_s = 0.6 \dfrac{\sigma_y}{1.5\sqrt{3}} = 0.231\ \sigma_y$

$f_t = 0.6 \dfrac{\sigma_y}{1.5} = 0.4\ \sigma_y$

$k_1 = (1 + \psi/6)(\psi^3 + 3\psi^2 + 4)$

k_2 = k_v of Eq. 4b.J21 for webs without longitudinal stiffeners

σ_c = stress at the edge with higher compression (negative sign); thus $\psi=0$ for uniform compression and $\psi=+2$ for pure bending

σ_t = stress at the other edge; tension (positive) or lower compression (negative)

$\psi = 1-(\sigma_t/\sigma_c)$

for $\alpha < 1$

$$k_2 = 4.00 + \frac{5.34}{\alpha^2} + \frac{(n+1)^2\eta}{\alpha}\sqrt{\frac{8\mu}{3\alpha}} \tag{4b.J5a}$$

for $\alpha \geq 3$

$$k_2 = 5.34 + \frac{4.00}{\alpha^2} + \frac{(n+1)^2\eta}{\alpha}\sqrt{\frac{8\mu}{3\alpha}} \tag{4b.J5b}$$

$$\eta = d/h$$

where

- d = smaller distance of length or depth of panels partitioned by longitudinal stiffeners
- $I_{s\ell}$ = moment of inertia of longitudinal stiffener
- n = number of longitudinal stiffeners, 1 or 2
- μ = $10.9\,I_{s\ell}/(ht^3)$

The moment of inertia of intermediate transverse stiffeners for webs without longitudinal stiffeners are

for $\alpha < 1$

$$I_{st} \geq 1.1\ ht^3\left(\frac{1}{\alpha^2} - 0.5\right) \tag{4b.J6a}$$

for $\alpha \geq 1$

$$I_{st} \geq 0.55\ ht^3 \tag{4b.J6b}$$

and for webs with longitudinal stiffeners

for $\alpha' < 1$

$$I_{st} \geq 1.1\ ht^3\left[\frac{1}{(\alpha')^2} - 0.5\right] \tag{4b.J7a}$$

for $\alpha' \geq 1$

$$I_{st} \geq 0.5\ ht^3 \tag{4b.J7b}$$

where α' is determined by solving

for $\alpha < 1$

$$\frac{5.34}{(\alpha')^2} = \frac{5.34}{\alpha^2} + \frac{(n+1)^2\,\eta}{\alpha}\sqrt{\frac{8\mu}{3\alpha}} \tag{4b.J8a}$$

for $\alpha \geq 1$

$$\frac{4.00}{(\alpha')^2} = \frac{4.00}{\alpha^2} + \frac{(n+1)^2\,\eta}{\alpha}\sqrt{\frac{8\mu}{3\alpha}} \tag{4b.J8b}$$

Radius of gyration, r, of longitudinal stiffeners is found from

$$r/t \geq C_m\,[135(0.5 - \eta)^3 + 3]\,\alpha^{2/3} \tag{4b.J9}$$

$$C_m = 0.7\,\frac{1}{200(n+1)}\,\frac{r}{t}\,\frac{1}{\delta} \tag{4b.J10}$$

where

$$0.2 \leq \eta < 0.5 \text{ when n=1} \tag{4b.J11a}$$

$$0.15 < \eta < 0.3 \text{ when n=2} \tag{4b.J11b}$$

$$\delta = A_s/h\,t \tag{4b.J11c}$$

Radius of gyration of longitudinal stiffeners of webs subjected to bending and compression is found from for $\alpha < \left(\frac{2+n^2}{1+n}\right)$

$$r/t \geq C_n\,(1-0.1\,\psi^2)(2+2n)\,\alpha^{2/3} \qquad (4b.J12a)$$

for $\alpha \geq \left(\frac{2+n^2}{1+n}\right)$

$$r/t = C_n\,(1-0.1\,\psi^2)(4+2n^2) \qquad (4b.J12b)$$

where C_n is equal to $0.7 + \frac{1}{100(n+1)(1+\psi/2)}\frac{r}{t}\frac{1}{\delta}$ (4b.J13)

COMMENTARY

Equations 4b.J2a and 4b.J3a are approximate expressions of elastic strength applicable to buckling strength equal to or less than 60 percent of the yield stress with a factor of safety of 1.5. Eqs. 4b.J2b and 4b.J3b are straight line approximations of buckling strength to take into account inelastic action.

Moments of inertia for stiffeners are based on linear elastic buckling analysis, so that the stiffeners act as nodes at the inception of buckling.

JRA 80

This specification provides the allowable stress design approach for highway bridges.

The allowable stress for compression in extreme fibers of I and U sections is given by

for $\lambda \leq 0.2$

$$\sigma_b = \frac{\sigma_y}{1.7} \qquad (4b.J14)$$

for $0.2 < \lambda \leq \frac{4}{3}$

$$\sigma_b = \frac{\sigma_y}{1.7}\left[1-0.412(\lambda-0.2)\right] \qquad (4b.J15)$$

where

A_c = area of the compression flange
A_w = h t, area of the web
b = width of the compression flange

$k = 2$ for $\frac{A_w}{A_c} \leq 2$ (4b.J16)

$k = \sqrt{3+\frac{A_w}{2A_c}}$ for $\frac{A_w}{A_c} > 2$

L_b = unbraced length

$$\lambda = \frac{2}{\pi}\,k\,\frac{L_b}{b}\sqrt{\frac{\sigma_y}{E}} \qquad (4b.J17)$$

When moment distribution is not uniform within the unbraced length, L_b, the allowable stress of Eq. 4b.J15 can be increased by multiplying it by M/M_{eq}, where M is the moment at the design section and $M_{eq}=0.6M_1 + 0.4M_2$ but not less than $0.4\,M_1$, M_1 and M_2 being the moments at the points of bracing ($M_1 \geq M_2$).

For ∏ and box sections

$$\sigma_b = \frac{\sigma_y}{1.7} \qquad (4b.J18)$$

Buckling strength of the web plate of girders subjected to shear and bending is controlled by the following interaction equation

$$\left(\frac{\sigma}{\sigma_{cr}}\right)^2 + \left(\frac{\tau}{\tau_{cr}}\right)^2 \leq \left(\frac{1}{1.25}\right)^2 \qquad (4b.J19)$$

where

$$k_b = 23.9 \qquad (4b.J20)$$

$$k_v = 5.34 + \frac{4.00}{\alpha^2} \quad \text{when } \alpha > 1 \qquad (4b.J21a)$$

$$k_v = 4.00 + \frac{5.34}{\alpha^2} \quad \text{when } \alpha \leq 1 \qquad (4b.J21b)$$

$$\alpha = a/h \qquad (4b.J22)$$

$$\sigma_{cr} = k_b\,\sigma_e \qquad (4b.J23)$$

$$\sigma_e = \frac{\pi^2\,E}{12(1-\nu^2)}\left(\frac{t}{h}\right)^2 \qquad (4b.J24)$$

$$\tau_{cr} = k_v\,\sigma_e \qquad (4b.J25)$$

h is the distance between the tensile and compression flanges. For webs with longitudinal stiffeners, h is the distance between the tensile flange and the stiffener farthest from the compression flange. σ is the compressive stress at the location from which h is defined. a is the distance between transverse stiffeners.

Moment of inertia of stiffeners is proportioned to satisfy

$$I_s \geq \frac{ht^3}{11}\,\gamma \qquad (4b.J26)$$

where

$$\gamma = \frac{8}{\alpha^2} \quad \text{for transverse stiffeners} \qquad (4b.J27)$$

and

$$\gamma = 30\alpha \quad \text{for longitudinal stiffeners} \qquad (4b.J28)$$

COMMENTARY

Equation 4b.J15 is a lower bound fitting of the results of experimental studies for lateral-torsional failure of welded built-up I-section beams. The parameter λ is derived from the formula for elastic lateral-torsional buckling of a beam by neglecting St. Venant's torsional resistance.

Stability of web plates under the action of normal and shear stresses is checked by the interaction formula of Eq. 4b.J19, using a factor of safety of 1.25 reduced from the basic value of 1.7 to account for the larger amount of post-buckling strength. The stiffness parameter values γ specified in Eqs. 4b.J27 and 4b.J28 are approximations of the values obtained by linear elastic analysis to ensure zero deflection nodes at the stiffeners.

NORTH AMERICA

CHAPTER EDITOR
Alexis Ostapenko, Lehigh University

AASHTO 89

The AASHTO 89 offers both the allowable stress and the ultimate strength (load factor design) approaches. However, except for a few minor differences, the allowable stress approach can be shown to be simply a conversion from the ultimate strength to allowable stresses by using a constant factor of safety. Thus, only the ultimate strength formulation is discussed here.

Bending Strength
is limited by the yielding of a flange or the lateral-torsional buckling presented in Chapter 3 on Beams.

Shear Strength

$$V_u = V_y\left[C + \frac{0.87(1-C)}{\sqrt{1+\alpha^2}}\right] \qquad (4b.N1)$$

where

A_w = h t, is the cross-sectional area of web plate
a = distance between transverse stiffeners

$$C = \frac{1495}{\left(\frac{h}{t}\right)}\sqrt{\frac{1+1/\alpha^2}{\sigma_{yw}}} - 0.3 \leq 1.0 \qquad (4b.N2)$$

h = clear distance between top and bottom flanges
t = thickness of web plate
$V_y = 0.58\,\sigma_{yw} A_w$ (4b.N3)
α = a/h is the web panel aspect ratio
σ_{yw} = yield stress of web plate

An intermediate transverse stiffener shall have a gross area of at least

$$A_{st} = \left[0.15\ DA_w\,(1-C)\,\frac{V}{V_u} - 18t^2\right]\frac{\sigma_{yw}}{\sigma_{yst}} \qquad (4b.N4)$$

where

D = 1.0 for stiffeners in pairs
= 1.8 for single angle stiffener
= 2.4 for single plate stiffener
σ_{yst} = yield stress of stiffener

and the moment of inertia should be

$$I_{st} \geq at^3\,J \qquad (4b.N5)$$

where

$$J = 2.5/\alpha^2 - 2 \geq 0.5 \qquad (4b.N6)$$

For longitudinally stiffened girders, h, as used in α and C of Eqs. 4b.N2 to 4b.N6 shall be the depth of the larger subpanel and the transverse stiffeners shall have the section modulus of at least

$$S_{st} \geq \frac{S_{s\ell}}{3\alpha} \tag{4b.N7}$$

where $S_{s\ell}$ is the section modulus of the longitudinal stiffener and S_{st} of the transverse stiffener.

The longitudinal stiffener shall have the moment of inertia

$$I_{s\ell} \geq A_w t^2 \,(2.4\alpha^2 - 0.13) \tag{4b.N8}$$

and the slenderness ratio

$$\frac{a}{r} \leq \frac{1909}{\sqrt{\sigma_y}} \tag{4b.N9}$$

where r is the radius of gyration.

The shear in the end panel shall be

$$V \leq 8275x10^8 \,(1+1/\alpha^2)\, \frac{t^3}{h} \leq V_p \tag{4b.N10}$$

with

$$\alpha \leq 0.5 \tag{4b.N11}$$

The interaction between moment and shear is provided by the relationship according to Eq. 4b.N19.

COMMENTARY

The ultimate strength method for shear computations has the same basis as the AISC Specification discussed next. Computation of the shear strength [Eqs. 4b.N1 to 4b.N3] was made simpler than in the AISC Specification [Eqs. 4b.N14 to 4b.N16] by replacing the two buckling coefficient formulas [Eq. 4b.N16] with a single formula

$$k = 5(1+1/\alpha^2) \tag{4b.N12}$$

and incorporating it into other equations. Also, the more stringent web slenderness limitation, aimed at preventing the development of fatigue cracks due to the "breathing" of the web under repeated load application, made it reasonable to neglect the effect of the redistribution of bending stresses [Eq. 4b.N13].

AISC ASD 89

This specification uses only the allowable stress approach. However, the allowable stresses were obtained from the ultimate strength divided by a factor of safety.

Allowable Bending Stress in the Compression Flange

$$\sigma_b' = \sigma_b \left[1.0 - 0.0005\, a_r \left(\beta - \frac{1996}{\sqrt{\sigma_b}}\right)\right] \leq 0.6\, \sigma_y \tag{4b.N13}$$

where

a_r = ratio of web area to compression flange area

σ_b = basic allowable bending stress in tension = $0.6F_y$

Allowable Shear Stress

$$\tau_a = \frac{\sigma_y}{2.89}\left(C + \frac{1-C}{1.15\sqrt{1+\alpha^2}}\right) \leq 0.4\,\sigma_y \qquad (4b.N14)$$

where

$$\left.\begin{aligned} C &= \frac{310{,}225}{\sigma_y \beta^2}\,k && \text{when } C < 0.8 \\ &= \frac{499}{\beta}\sqrt{\frac{k}{\sigma_y}} && \text{when } C > 0.8 \end{aligned}\right\} \qquad (4b.N15)$$

$$\left.\begin{aligned} k &= 4.00 + \frac{5.34}{\alpha^2} && \text{when } \alpha \leq 1.0 \\ &= 5.34 + \frac{4.00}{\alpha^2} && \text{when } \alpha > 1.0 \end{aligned}\right\} \qquad (4b.N16)$$

β = h/t

The end panels are designed for buckling only

$$\tau_a = (\sigma_y/2.98)\,C \qquad (4b.N17)$$

The required gross area of an intermediate transverse stiffener is

$$A_{st} = \frac{1-C}{2}\left[\alpha - \frac{\alpha^2}{\sqrt{1+\alpha^2}}\right]\frac{\sigma_{yw}}{\sigma_{yf}}\,DA_w \qquad (4b.N18)$$

where D, σ_{yw}, and σ_{yst} are defined as for Eq. 4b.N4

Interaction of Shear and Bending

$$0.455(\tau/\tau_a) + 0.727(\sigma/\sigma_b') \leq 1.0 \qquad (4b.N19)$$

where τ_a and σ_b' are the allowable shear and bending stresses as given by Eqs. 4b.N14 and 4b.N13 for pure shear or bending, τ and σ are the allowable shear and bending stresses when the girder is subjected to combined action.

COMMENTARY

The above formulas were derived by Basler. Factor 2.89 in Eq. 4b.N14 transforms σ_y into the shear yield stress yield stress and incorporates a safety factor of 1.67.

Equation 4b.N13 takes into account the redistribution of bending stresses caused by web buckling.

AISC LRFD 86

Bending Strength

When the web slenderness $\frac{h_c}{t} \leq \frac{2547}{\sqrt{\sigma_y}}$, the plate girder is to be designed as a beam, that is, according to Chapter 3.

The design flexural strength of plate girders with $\frac{h_c}{t} > \frac{2547}{\sqrt{\sigma_y}}$ is given by

$$M_d = \phi_b M_n \qquad (4b.N20)$$

where

M_n = the nominal bending strength given by the lower value of the following

for tension flange yielding

$$M_n = S_t R_e \sigma_{yt} \qquad (4b.N21)$$

for compression flange buckling

$$M_n = S_c R_{PG} R_e \sigma_{cr} \qquad (4b.N22)$$

ϕ_b = 0.90

where

a_r = ratio of web area to compression flange area
h_c = twice the clear distance from the neutral axis to the compression flange

$$R_{PG} = 1 - 0.0005\, a_r \left(\frac{h_c}{t} - \frac{2547}{\sqrt{\sigma_{cr}}} \right) \leq 1.0 \qquad (4b.N23)$$

R_e = hybrid girder factor
= $1.0 - 0.1(1.3 + a_r)(0.81 - m) \leq 1.0$ (for non-hybrid girders, R_e = 1.0)
m = ratio of web yield stress to flange yield stress or to σ_{cr}
S_c = section modulus with respect to compression flange
S_t = section modulus with respect to tension flange
σ_{cr} = critical compression flange stress, see below
σ_{yt} = yield stress of tension flange

The critical stress σ_{cr} depends on the slenderness parameters λ, λ_p, λ_r and C_{PG} as follows

for $\lambda \leq \lambda_p$

$$\sigma_{cr} = \sigma_{yc} \qquad (4b.N24)$$

for $\lambda_p < \lambda \leq \lambda_r$

$$\sigma_{cr} = C_b \sigma_{yc} \left[1 - \frac{1}{2} \left(\frac{\lambda - \lambda_p}{\lambda_r - \lambda_p} \right) \right] \leq \sigma_{yc} \qquad (4b.N25)$$

for $\lambda > \lambda_r$

$$\sigma_{cr} = \frac{C_{PG}}{\lambda^2} \qquad (4b.N26)$$

Two limit states shall be checked: lateral-torsional buckling and flange local buckling, each with its set of λ's. The lower value of σ_{cr} governs.

For the limit state of lateral-torsional buckling

$$\lambda = \frac{L_b}{r_T} \tag{4b.N27}$$

$$\lambda_p = \frac{788}{\sqrt{\sigma_{yc}}} \tag{4b.N28}$$

$$\lambda_r = \frac{1985}{\sqrt{\sigma_{yc}}} \tag{4b.N29}$$

$$C_{PG} = 1{,}971{,}000\, C_b \tag{4b.N30}$$

where

C_b = $1.75 + 1.05(M_1/M_2) + 0.3(M_1/M_2)^2 \leq 2.3$ (4b.N31)
L_b = unbraced length
M_1, M_2 = the smaller and the larger moments at the ends of the unbraced length
r_T = radius of gyration of compression flange plus one-third of the web portion under compression

For the limit state of flange local buckling

$$\lambda = \frac{b_f}{2t_f} \tag{4b.N32}$$

$$\lambda_p = \frac{171}{\sqrt{\sigma_{yc}}} \tag{4b.N33}$$

$$\lambda_r = \frac{394}{\sqrt{\sigma_{yc}}} \tag{4b.N34}$$

$$C_{PG} = 77{,}221\, C_b \tag{4b.N35}$$

$$C_b = 1$$

Shear Strength

The design shear strength is

$$V_d = \phi_v V_n \tag{4b.N36}$$

where ϕ_v=0.90 and the normal shear strength V_n depends on the web slenderness as follows

for $h/t \leq 491\sqrt{k/\sigma_{yw}}$

$$V_n = 0.6\, A_w \sigma_{yw} \tag{4b.N37}$$

for $h/t > 491\sqrt{k/\sigma_{yw}}$

$$V_n = 0.6\, A_w \sigma_{yw}\left(C + \frac{1 - C}{1.15\sqrt{1 + (a/h)^2}}\right) \tag{4b.N38}$$

where

A_w = $h\,t$ = web area
σ_{yw} = yield stress of web material

C_v, the ratio of the critical to yield shear stresses, is computed according to the web slenderness, h/t.

for $491\sqrt{\dfrac{k}{\sigma_{yw}}} \le \dfrac{h}{t} \le 614\sqrt{\dfrac{k}{\sigma_{yw}}}$

$$C = \frac{491\sqrt{k/\sigma_{yw}}}{h/t} \qquad (4b.N39)$$

for $\dfrac{h}{t} > 614\sqrt{\dfrac{k}{\sigma_{yw}}}$

$$C = \frac{303{,}369}{(h/t)^2\,\sigma_{yw}} \qquad (4b.N40)$$

where k is $5 + \dfrac{5}{(a/h)^2}$ (4b.N41)

except that k shall be taken as 5.0 if a/h exceeds 3.0 or $[260/(h/t)]^2$.

For end-panels in non-hybrid plate girders, for all panels in hybrid and web-tapered plate girders and when a/h exceeds 3.0 or $[260/(h/t)]^2$, tension field action is not permitted and

$$V_n = 0.6\,A_w\,\sigma_{yw}\,C \qquad (4b.N42)$$

Interaction of Shear and Bending

When stiffeners are required and

$$\frac{0.6\,V_n}{M_n} \le \frac{V}{M} \le \frac{V_n}{0.75\,M_n} \quad \text{and} \quad M \le M_d,\ V \le V_d \qquad (4b.N43)$$

the following interaction equation must be satisfied

$$\frac{M}{M_n} + 0.625\,\frac{V}{V_n} \le 1.375\,\phi \qquad (4b.N44)$$

where ϕ is 0.90

Intermediate Transverse Stiffeners

Transverse stiffeners are not required in plate girders when $h/t \le 1098/\sqrt{\sigma_{yw}}$, or when the required shear V is less than or equal to $0.6\,\phi\,A_w\,F_{yw}\,C_v$ where ϕ=0.90 and C_v is determined for k=5.

The moment of inertia I_{st} of a transverse stiffener about an axis in the web center for stiffener pairs or about the face in contact with the web plate for single stiffeners shall not be less than $a\,t^3\,J$, where

$$J = \frac{2.5}{(a/h)^2} - 2 \ge 0.5 \qquad (4b.N45)$$

and the stiffener area A_{st}, when designing for tension field action, shall not be less than

$$A_{st} = \frac{\sigma_{yw}}{\sigma_{yst}}\left[0.15\,Dht(1-C)\,\frac{V}{\phi_v V_n} - 18\,t^2\right] \ge 0 \qquad (4b.N46)$$

where

D = 1 for stiffeners in pairs
= 1.8 for single angle stiffeners
= 2.4 for single plate stiffeners

V = required shear at the location of the stiffener

σ_{yst} = yield stress of the stiffener material, ksi

AREA 89

The AREA Specification uses the allowable stress design approach (ASD). It applies to railroad bridge girders, and the postbuckling web strength is neglected.

Bending Strength
is controlled by the lateral-torsional buckling requirement

$$\sigma_{ba} = 0.55\sigma_y \left[1 - \frac{\sigma_y (L_b/r_y)^2}{12{,}411{,}000}\right] \tag{4b.N47}$$

where

L_b = unbraced length of the compression flange
r_y = radius of gyration of the compression portion of the cross section about the web
σ_{ba} = allowable bending stress

Allowable Shear Stress
for $h/t \leq 946.6/\sqrt{\sigma_y}$ (4b.N48)

$$\tau_a = 0.35\sigma_y \tag{4b.N49}$$

When (h/t) is greater, the shear strength is controlled by specifying the distance between transverse stiffeners to be

$$a/t \leq 872/\sqrt{\tau} \tag{4b.N50}$$

where τ is the actual shear stress in the panel.

Each of the two-sided intermediate transverse stiffeners must have the width not less than (50 mm + h/30).

Interaction between bending and shearing stresses is accomplished by keeping the combined diagonal tensile stress below the allowable stress for tension.

COMMENTARY

The AREA Specification (the only railroad specification in the region) is the lone holdout in the region in maintaining the buckling criteria for design of plate girder webs.

CSA S6 M89 84a and CSA S16.1 M89 84b

The Canadian specifications for Buildings and for Highway Bridges are fundamentally the same as the AISC ASD 89 and AASHTO 89 Specifications. Redistribution of the bending stresses is not neglected as in AASHTO 89, and a formula corresponding to Eq. 4b.N13 is used for stresses in the compression flange.

In order to reduce space, the provisions for plate girders are not repeated here and only those pertaining to box girders are discussed under Topic C, especially since they are not given in the description of the AASHTO Specification there.

FHWA TS 80

This specification was developed to provide engineers with a tool for designing box girder bridges of longer spans in which multiple longitudinal stiffeners would be needed.

Web Buckling Stresses

Bending stress for an unsymmetrical section is composed of pure bending and uniform axial contributions. The buckling stress for the combined case can be computed from the pure case components. The formulas for both cases are the following

for $0.65 \leq \lambda < 1.5$

$$\sigma_{cr}/\sigma_y = 0.072(\lambda - 5.62)^2 - 0.78 \qquad (4b.N51)$$

for $\lambda \geq 1.5$

$$\sigma_{cr}/\sigma_y = 1/\lambda^2$$

where λ is $\dfrac{h/t}{0.95}\sqrt{\dfrac{\sigma_y}{Ek}}$ (4b.N52)

and the buckling coefficient k for the axial and bending cases is, respectively,

for $\alpha > 1.0$

$$k_a = 4$$

or for $\alpha < 1.0$

$$k_a = (\alpha + 1/\alpha)^2 \qquad (4b.N53a)$$

or

for $\alpha > 2/3$

$$k_b = 24$$

or for $\alpha < 1.0$

$$k_b = 2 + 73(2/3 - \alpha)^2 \qquad (4b.N53b)$$

The shear buckling stress is given by

for $\lambda_v \leq 0.58$

$$\tau_{cr} = 0.58\sigma_y$$

for $0.58 < \lambda_v < 1.41$

$$\tau_{cr} = [0.58 - 0.357(\lambda_v - 0.58)^{1.18}]\ \sigma_y \qquad (4b.N54)$$

where

$$k_v = 5 + \frac{5}{\alpha^2} \qquad (4b.N55)$$

$$\lambda_v = 0.8\,\frac{h}{t}\sqrt{\frac{\sigma_y}{Ek_v}} \qquad (4b.N56)$$

Buckling under a combination of axial (σ_a), bending (σ_b) and shearing (τ) stresses takes place when

$$\left(\frac{\sigma_a}{\sigma_{cr}}\right) + \left(\frac{\sigma_b}{\sigma_{crb}}\right)^2 + \left(\frac{\tau}{\tau_{cr}}\right)^2 = 1.0 \qquad (4b.N57)$$

The tension field stress which is needed for computing the ultimate shear capacity of the web is given by the following equation

$$\sigma_{tf} = \sigma_y - \sqrt{0.25\sigma_{2w}^2 + 3\tau^2} \tag{4b.N58}$$

where

σ_{2w} = the lower compressive or the tensile bending stress in the web panel (or subpanel)

τ = the shearing buckling stress under combined stress condition from Eq. 4b.N57.

Ultimate Shear Strength of Web

The ultimate shear strength of the web with transverse and with or without longitudinal stiffeners is the sum of the buckling (V_B) and postbuckling tension-field action (V_T) strengths.

$$V_u = V_B + V_T \tag{4b.N59}$$

with

$$V_B = A_w \tau_{min} \tag{4b.N60}$$

$$V_T = \frac{A_w \sigma_{tf}}{2\left[\alpha + \sqrt{1+\alpha^2}\right]} \tag{4b.N61}$$

where

A_w = ht is the web area

α = a/h is the aspect ratio of the whole web panel

σ_{tf} = tension field stress from Eq. 4b.N58

τ_{min} = lowest shear buckling stress of the web subpanels for the combined state of stress according to Eq. 4b.N57

Note that the tension field contribution in V_T is more conservative than in other specifications since the vertical rigidity of box girder flanges is generally lower than of typical plate girders.

The ultimate shear may not be greater than the shear yield strength.

$$V_u \leq 0.58A_w \sqrt{\sigma_y^2 - (2/3\sigma_{av})^2} \tag{4b.N62}$$

where σ_{av} is the average of normal stresses at the top and bottom edges, disregarding their signs.

WEST EUROPE

CHAPTER EDITOR

Pierre Dubas, Swiss Federal Institute of Technology

The numerous recommendations and codes in West Europe for the design of plate and box girders can be divided into two groups. On the one hand, codes based on the linear plate buckling theory (for example, the Austrian code ÖNORM B4600 78, the Italian code CNR 10011 85, Dutch code NEN 3851 72, Belgian code NBN B51002 88, etc.). On the other hand, some countries have introduced new codes based on ultimate strength design models (British BS 5400 P.3 82, Swiss SIA 161 79, and Norwegian NS 3472 84). Discussed here is a sampling of the current codes or recommendations with particular emphasis on the different treatment of the ultimate strength but also giving a representative buckling approach.

BS 5400 P.3 82

Bending Strength

Reduction of the web capacity due to buckling and postbuckling redistribution of stresses is handled through the computation of the effective web thickness t_e rather than through removal of a portion of the web plate as done by most other specifications.

$$\left.\begin{array}{ll} \text{for } \frac{h_c}{t}\sqrt{\frac{\sigma_y}{355}} \le 136 & t_e = t \\ \text{for } 136 < \frac{h_c}{t}\sqrt{\frac{\sigma_y}{355}} < 456 & t_e = \left(1.425 - 0.00625\,\frac{h_c}{t}\sqrt{\frac{\sigma_y}{355}}\right)t \\ \text{for } \frac{h_c}{t}\sqrt{\frac{\sigma_y}{355}} \ge 456 & t_e = 0 \end{array}\right\} \quad (4b.W1)$$

where h_c is twice the clear distance from the neutral axis to the compression flange.

For girders with longitudinal stiffeners

$$t_e = t \quad (4b.W2)$$

The stress distribution in the girder is based on elastic principles with an allowance that up to 60% of the web normal stress may be shed to the flanges from any or all web subpanels as long as overall equilibrium is maintained.

Each web subpanel is subject to yielding and buckling limitations. The buckling limitation is

$$m_c + m_b \le 1 \quad (4b.W3)$$

where

$$m_b = \left(\frac{\sigma_b \gamma_m \gamma_{f3}}{\sigma_y K_b(1-\rho)}\right)^2 \qquad (4b.W4a)$$

$$m_c = \frac{\sigma_c \gamma_m \gamma_{f3}}{\sigma_y K_c(1-\rho)} \qquad (4b.W4b)$$

where

K_c, K_b = strength multipliers which depend on panel slenderness and edge restraints
γ_{f3} = resustance coefficient for bending
γ_m = material resistance coefficient
ρ = proportion of direct stress redistributed from web to flanges for unrestrained panels or 0 for restrained panels
σ_b = bending stress in the web subpanel considered (σ_c and σ_b are reduced to allow for redistribution)
σ_c = mean normal stress in the web subpanel considered (compression is positive)

Internal panels are restrained. Edge panels are unrestrained unless additional requirements are satisfied.

With the relative slenderness of the i-th web subpanel of depth h_i

$$\lambda = \frac{h_i}{t}\sqrt{\frac{\sigma_y}{355}} \qquad (4b.W5)$$

the K_c values are as follows

(A) Restrained subpanels

$$\left.\begin{array}{lll} \text{for } \lambda \le 24 & & K_c = 1 \\ \text{for } 24 < \lambda \le 43 & & K_c = \left(\frac{24}{\lambda}\right)^{0.50} \\ \text{for } 43 < \lambda \le 59 & & K_c = \left(\frac{28}{\lambda}\right)^{0.68} \\ \text{for } 59 < \lambda \le 90 & & K_c = \left(\frac{30}{\lambda}\right)^{0.75} \\ \text{for } 90 < \lambda \le 130 & & K_c = \left(\frac{36}{\lambda}\right)^{0.90} \\ \text{for } 130 < \lambda \le 200 & & K_c = 0.38 - \frac{\lambda}{2000} \\ \text{for } 200 < \lambda \le 300 & & K_c = 0.33 - \frac{\lambda}{4000} \end{array}\right\} \qquad (4b.W6)$$

(B) Unrestrained subpanels

$$\left.\begin{array}{lll} \text{for } \lambda \le 24 & & K_c = 1 \\ \text{for } 24 < \lambda \le 47 & & K_c = \left(\frac{24}{\lambda}\right)^{0.75} \\ \text{for } 47 < \lambda \le 130 & & K_c = \left(\frac{26}{\lambda}\right)^{0.85} \\ \text{for } 130 < \lambda \le 300 & & K_c = 0.274 - \frac{\lambda}{7000} \end{array}\right\} \qquad (4b.W7)$$

and K_b values:

(C) Restrained subpanels
for $\lambda \leq 300$

$$K_b = 1.3 - 0.0019\lambda \tag{4b.W8}$$

(D) Unrestrained subpanels
for $\lambda \leq 300$

$$K_b = 1.3 - 0.0027\lambda \tag{4b.W9}$$

Shear Strength

$$V_u = \frac{t\,h}{\gamma_m \gamma_{f3}} \tau_u \tag{4b.W10}$$

where the ultimate shear stress τ_u is read from the curves or computed as the maximum value in the following procedure

with
$$\lambda = \frac{h}{t}\sqrt{\frac{\sigma_y}{355}} \tag{4b.W11a}$$

yield stress in shear
$$\tau_y = \sigma_y/\sqrt{3} \tag{4b.W11b}$$

and web panel aspect ratio
$$\alpha = a/h \tag{4b.W12}$$

(A) Calculate:

$$\beta = \frac{\lambda}{\sqrt{5 + \frac{5}{\alpha^2}}} \tag{4b.W13}$$

(B) Calculate:

for $\beta \leq 24.55$
$$\frac{\tau_c}{\tau_y} = 1$$

for $24.55 < \beta < 33.62$
$$\frac{\tau_c}{\tau_y} = 1.54 - 0.022\beta \qquad \text{(4b.W14)}$$

for $\beta \geq 33.62$
$$\frac{\tau_c}{\tau_y} = \frac{904}{\beta^2}$$

(C) Assume a value of ϕ such that:

$$0.33 \operatorname{arccot}\alpha \leq \phi \leq 1.33 \operatorname{arccot}\alpha \text{ and also } \leq \frac{\pi}{4} \tag{4b.W15}$$

(D) Calculate:

$$\frac{\sigma_t}{\tau_y} = \sqrt{\left[3 + (2.25 \sin^2 2\phi - 3)\left(\frac{\tau_c}{\tau_y}\right)^2\right]} - 1.5\,\frac{\tau_c}{\tau_y} \sin 2\phi \tag{4b.W16}$$

(E) Calculate:

$$\frac{\tau_u}{\tau_y} = f\left[\frac{\tau_c}{\tau_y} + 5.264\sqrt{m_f \frac{\sigma_t}{\tau_y}}\, \sin\phi + \frac{\sigma_t}{\tau_y}(\cot\phi - \phi)\sin^2\phi\right] \tag{4b.W17}$$

where

for $\lambda \le 56$

$$f = 1$$

for $56 < \lambda < 156$

$$f = \frac{1.15}{1.15 + 0.002\,(\lambda - 56)} \tag{4b.W18}$$

for $156 \le \lambda$

$$f = \frac{1.15}{1.35}$$

and the effect of flange plastic moment

$$m_f = \frac{\sigma_{yf}\, b\, t_f^2}{2\sigma_y\, t_e^2\, t} \tag{4b.W19}$$

using the smaller value of the top or the bottom flange.

(F) Repeat (C) to (E) with different values of ϕ until the maximum value $(\tau_u/\tau_y)_{max}$ is obtained.

Webs with Longitudinal Stiffeners

The shear strength of the web is the sum of the strengths of individual web subpanels.

$$V_u = \Sigma\, V_{ui} \tag{4b.W20}$$

For each i-th subpanel,

$$V_{ui} = h_i\, t\, \frac{\sigma_y}{\sqrt{3}\ \gamma_m \gamma_{f3}}\, K_q \tag{4b.W21}$$

where

h_i = the depth of the i-th subpanel

K_q = the shear strength multiplier given by the following formulas

with

$$\lambda_i = \frac{h_i}{t}\sqrt{\frac{\sigma_y}{355}} \tag{4b.W22}$$

Restrained panels

for $\alpha \le 0.5$

$K_q = 1$ when $\lambda_i \le 80$

$K_q = \left(\frac{80}{\lambda_i}\right)^{0.15}$ when $80 < \lambda_i \le 300$

for $\alpha = 1.0$

$K_q = 1$ when $\lambda_i \le 48$

$K_q = \left(\frac{48}{\lambda_i}\right)^{0.11}$ when $48 < \lambda_i \le 300$ (4b.W23)

for $\alpha > 2.0$

$K_q = 1$ when $\lambda_i \le 40$

$K_q = \left(\frac{40}{\lambda_i}\right)^{0.15}$ when $40 < \lambda_i \le 300$

Unrestrained panels

for $\alpha \le 0.5$

$K_q = 1$ when $\lambda_i \le 80$

$$K_q = 1-0.385\left(\frac{\lambda_i-80}{120}\right)^{0.743} \qquad \text{when } 80 < \lambda_i \le 224$$

$$K_q = 1-0.660\left(\frac{\lambda_i-80}{320}\right)^{0.505} \qquad \text{when } 224 < \lambda_i \le 300$$

for $\alpha = 1.0$

$$K_q = 1 \qquad \text{when } \lambda_i \le 48$$

$$K_q = \left(\frac{48}{\lambda_i}\right)^{0.5} \qquad \text{when } 48 < \lambda_i \le 300$$

for $\alpha = 2.0$ (4b.W24)

$$K_q = 1 \qquad \text{when } \lambda_i \le 40$$

$$K_q = 1-0.385\left(\frac{\lambda_i-40}{60}\right)^{0.743} \qquad \text{when } 40 < \lambda_i \le 112$$

$$K_q = 1-0.660\left(\frac{\lambda_i-40}{160}\right)^{0.505} \qquad \text{when } 112 < \lambda_i \le 200$$

$$K_q = 0.34-0.07\left(\frac{\lambda_i-200}{100}\right)^{0.8} \qquad \text{when } 200 < \lambda_i \le 300$$

for $\alpha \ge 3.0$

$$K_q = 1 \qquad \text{when } \lambda_i \le 38$$

$$K_q = 1-0.555\left(\frac{\lambda_i-38}{82}\right)^{0.823} \qquad \text{when } 38 < \lambda_i \le 125$$

$$K_q = 0.445-0.205\left(\frac{\lambda_i-120}{80}\right)^{0.725} \qquad \text{when } 125 < \lambda_i \le 221$$

$$K_q = 0.24-0.075\left(\frac{\lambda_i-200}{100}\right)^{0.44} \qquad \text{when } 221 < \lambda_i \le 300$$

The transverse and longitudinal stiffeners are designed as columns subjected to appropriate forces.

COMMENTARY

The adjustment factors of this specification, such as, the shear strength multiplier K_q, were developed from curve-fitting strengths of unrestrained (next-to-flange) and restrained (inner) plate subpanels computed by using F.E. and other numerical methods.

Interaction of Bending and Shear

The following equations must be satisfied.

(a) $V \le V_u$

(b) $M \le M_u$

(c) if $M > M_R$, then $\frac{M}{M_u} + \left(1-\frac{M_R}{M_u}\right)\left(\frac{2V}{V_w} - 1\right) \le 1$ (4b.W25)

(d) if $V > V_w$, then $\frac{V}{V_u} + \left(1-\frac{V_w}{V_u}\right)\left(\frac{2M}{M_R} - 1\right) \le 1$

where

M_R = bending strengths of the effective flanges alone divided by $\gamma_m \gamma_{f3}$

V_w = shear strength of the web alone

V_u, M_u = ultimate shear and moment strengths, respectively

Webs with Longitudinal Stiffeners

For each web subpanel the following interaction equation should be satisfied

$$\frac{\sigma_c \gamma_m \gamma_{f3}}{\sigma_y K_c(1-\rho)} + \left[\frac{\sigma_b \gamma_m \gamma_{f3}}{\sigma_y K_b(1-\rho)}\right]^2 + \frac{V_i}{V_{ui}} \leq 1 \qquad (4b.W26)$$

where, for each subpanel,

V_i = shear carried by the i-th web subpanel
σ_b = pure bending stress
σ_c = average compressive stress

BS 5950 P.5 85

Bending Strength

When $h/t \geq 63\sqrt{\frac{275}{\sigma_y}}$ (4b.W27)

The bending strength of plate girders is given as follows:
(A) Normal stresses to resist the applied moment may be assumed to be carried by flanges alone, and the webs are designed for shear.
(B) Webs may be also designed to carry some of the bending moment.

$$\left(\frac{M_w}{M_{cr}}\right)^2 + \frac{\sigma_a}{\sigma_{acr}} \leq 1 \qquad (4b.W28)$$

where M_w is the maximum moment in the panel carried by the web

$$M_{cr} = \sigma_{bcr} S_w \qquad (4b.W29)$$

$$\sigma_{bcr} = \left(\frac{1630}{h/t}\right)^2 \qquad (4b.W30)$$

where

S_w = the section modulus of the web alone
σ_a = the coexistent overall compression on the web (for unsymmetrical sections). Note that σ_a is to be taken negative for tension.

$$\sigma_{acr} = \left(\frac{850}{h/t}\right)^2$$

is the buckling stress under uniform compression.

Shear Strength

(A) Unstiffened webs are designed for buckling capacity.

$$V_u = h\,t\,\tau_{cr} \qquad (4b.W31)$$

for $\lambda_w \leq 0.8$

$$\tau_{cr} = 0.6\,\sigma_y$$

for $0.8 < \lambda_w \leq 1.25$

$$\tau_{cr} = 0.6\,\sigma_y\,[1-0.8(\lambda_w-0.8)] \qquad (4b.W32)$$

for $\lambda_w \geq 1.25$

$$\tau_{cr} = \tau_e$$

where the equivalent slenderness of the web is given by

$$\lambda_w = \sqrt{\frac{0.6\sigma_y}{\tau_e}} \qquad (4b.W33)$$

and the elastic shear buckling stress by

for $\alpha \leq 1.0$

$$\tau_e = \left[0.75+\frac{1}{\alpha^2}\right]\left[\frac{1000}{h/t}\right]^2$$

for $\alpha > 1.0$

$$\tau_e = \left[1+\frac{0.75}{\alpha^2}\right]\left[\frac{1000}{h/t}\right]^2 \qquad (4b.W34)$$

(B) Transversely Stiffened Webs (Tension-field approach is utilized)

$$V_u = ht\left(\tau_b+\tau_f\sqrt{K_f}\right) \leq 0.6\ \sigma_y ht \qquad (4b.W35)$$

where the shear capacity τ_b is

$$\tau_b = \tau_{cr}+\frac{y_b}{2\left[\alpha+\sqrt{1+\alpha^2}\right]} \qquad (4b.W36)$$

with y_b, the basic tension field strength

$$y_b = \sqrt{\sigma_y^2-3\tau_{cr}^2+\phi_t^2} - \phi_t$$

and

$$\phi_t = \frac{1.5\ \tau_{cr}}{\sqrt{1+\alpha^2}} \qquad (4b.W37)$$

Contribution by the flanges in Eq. 4b.W35 is

$$\tau_f = 0.6\ \sigma_y\left[4\sqrt{3}\ \sin\left(\frac{\theta}{2}\right)\sqrt{\frac{y_b}{\sigma_y}}\right] \qquad (4b.W38)$$

where

$$K_f = \frac{M_{pf}}{4M_{pw}}\left(1-\frac{\sigma}{\sigma_{yf}}\right) \qquad (4b.W39)$$

θ = the inclination of the panel diagonal = arccot α

where

M_{pf} = the plastic moment of the smaller flange about its own equal-area axis

M_{pw} = the plastic moment of the web about its equal-area axis

σ = the normal stress in the smaller flange

Interaction of Bending and Shear

is controlled by web buckling strength according to

$$\left(\frac{M_w}{M_{cr}}\right)^2 + \frac{\sigma_a}{\sigma_{acr}} + \left(\frac{\tau}{\tau_b}\right)^2 \leq 1 \qquad (4b.W40)$$

where τ is the shear stress in the panel, $\tau = V/ht$, and the other terms are defined above.

DASt 015 80

This supplement specification was introduced to enable designers to consider postbuckling strength in the design of plate (and box) girders since other German specifications are based on elastic buckling criteria. Discussed here are only plate girders with transverse stiffeners which are to be spaced at a distance a so that $0.5 \leq \alpha = a/h \leq 3.0$.

Bending Strength

Bending strength of plate girders with slender webs is controlled either by the yielding of the tension flange or by the buckling of the compression flange. The moment is assumed to be carried by the flanges alone, that is, by the couple of the flange forces, $N_f = M/(h + t_f)$. The compression flange is assumed to act as a column with the cross section consisting of the flange alone without any contribution from the web plate when the web plate is assigned to carry shear. Considered are the lateral and torsional buckling modes and the lower capacity controls.

The critical stress for lateral buckling of the flange is computed from Eq. 4b.W41.

$$\sigma_{cr,f} = \frac{\sigma_{yc}}{k + \sqrt{k^2 + \overline{\lambda}_k^2}} \leq \sigma_{yc} \qquad (4b.W41)$$

where σ_{yc} is the yield stress of the compression flange and

$$k = 0.5\left[1 + 0.49\,(\overline{\lambda}_k - 0.2) + \overline{\lambda}_k^2\right] \qquad (4b.W42)$$

$$\overline{\lambda}_k = \frac{L_b}{\pi r_f}\sqrt{\frac{\sigma_y}{E}} \qquad (4b.W43)$$

with a radius of gyration of the flange about the plane of the web, $r_f = b/\sqrt{12}$, and the unbraced length L_b.

The critical stress for torsional buckling is given by

$$\sigma_{cr,f} = \frac{0.836\sqrt{E\,\sigma_{yc}}}{b/t_f} \leq \sigma_{yc} \qquad (4b.W44)$$

Shear Strength

The ultimate shear strength is given by the same procedure as in **EC 3 88** for transversely stiffened plate girders, that is, Eqs. 4b.W53 to 4b.W57, but the equations for g and s are to be replaced with the following equations in order to accommodate girders with unequal flanges. The angle of inclination of the tension-field band, ϕ, is assumed to be equal to 2/3 of the inclination of the panel diagonal.

$$g = (\cot\phi - \alpha + \frac{s_1 + s_2}{h})\sin^2\phi \qquad (4b.W45)$$

with the distances to the plastic moments in the flanges, s_1 and s_2, from

$$s_1 = \frac{2}{\sin\phi}\sqrt{\frac{M_{pc1}}{t\sigma_{tf}}} \leq a \qquad (4b.W46)$$

$$s_2 = \frac{2}{\sin\phi}\sqrt{\frac{M_{pc2}}{t\sigma_{tf}}} \leq a \tag{4b.W47}$$

where the reduced plastic moments of the rectangular flanges are

$$M_{pc1} = M_{p1}\,[1 - (N_{f1}/N_{fp1})^2] \tag{4b.W48}$$

$$M_{pc2} = M_{p2}\,[1 - (N_{f2}/N_{fp2})^2] \tag{4b.W49}$$

with N_{f1} and N_{fp1} being the actual and fully-plastified axial forces in the top flange, respectively. M_{p1} is the plastic moment of the top flange $= b\ t_f^2\ \sigma_{yf}/4$. Subscript "2" indicates analogous quantities for the bottom flange.

For box girders, distances s_1 and s_2 are to be taken equal to zero.

Interaction of Bending and Shear

Interaction between bending and shear is embedded into the shear strength through the effect of the axial flange forces on the plastic moments of the flanges in Eqs. 4b.W48 and 4b.W49 and the fact that the web is assumed not to contribute to the bending strength.

EC 3 88

Bending Strength

The bending capacity of plate girders, both without and with longitudinal web stiffeners, is calculated by introducing a reduced section with the following formula for the effective width b_e of the compression zone $h_c/2$ of the web

$$\frac{2\,b_e}{h_c} = \frac{1}{\overline{\lambda}_p}\left[1 - \frac{0.05(3 + \sigma_{min}/\sigma_{max})}{\overline{\lambda}_p}\right] \leq 1.0 \tag{4b.W50}$$

where

$$\overline{\lambda}_p = \sqrt{\frac{\sigma_y}{\sigma_{cr}}} \tag{4b.W51}$$

and $\overline{\lambda}_p$=0.1 for pure bending (instead of 0.2 for uniform compression). For girders without longitudinal stiffeners, the distribution of the effective width b_e is given as $b_{e1}=0.4b_e$ for the part adjacent to the compression flange and $b_{e2}=0.6b_e$ for the part adjacent to the neutral axis.

For pure bending of the web, as in symmetrical plate girders, the width b_e is equal to h/2 and the stress ratio is $\sigma_{min}/\sigma_{max}=-1.0$.

For the web to be fully effective, i.e., $2\,b_e/h_c=1$, Eq. 4b.W50 requires that $\overline{\lambda}_p$ be less than 0.89. Then, with the buckling coefficient k=23.9 for hinged conditions, Eq. 4b.W51 results in

$$h/t \leq 4.15\sqrt{E/\sigma_y} \tag{4b.W52}$$

that is, 125 for steel with σ_y=235 MPa or 100 for steel with σ_y=355 MPa.

Shear Strength

(A) Plate girders without transverse stiffeners except for bearing stiffeners at supports

Such girders may be designed by using a simple postcritical method which gives the shear strength according to

$$V_u = h\,t\,0.9\sqrt{\tau_{cr}\,\sigma_{yw}/\sqrt{3}} \quad \text{for slender webs with } \sigma_{yw}/(\sqrt{3}\,\tau_{cr}) \geq 1.44 \qquad (4b.W53)$$

where τ_{cr} is $k_v \dfrac{0.9E}{(h/t)^2}$ (4b.W54)

with the buckling coefficient k_v from Eq. 4b.W56a for $\alpha = a/h \geq 1.0$.

(B) Plate girders with intermediate transverse stiffeners
The tension field method may be used for these girders when adjacent panels or end posts provide adequate anchorage for the tension field. The shear strength is obtained from

$$V_u = h\,t\,[\tau_{cr} + \sigma_{tf}\,g] \qquad (4b.W55)$$

where

τ_{cr} = according to Eq. 4b.W54 with k_v from
for $\alpha \geq 1$

$$k_v = 5.34 + 4.0/\alpha^2 \qquad (4b.W56a)$$

or for $\alpha < 1$

$$k_v = 4.0 + 5.34/\alpha^2 \qquad (4b.W56b)$$

α = a/h is the aspect ratio of the web plate panel
σ_{tf} = strength of the tension field, obtained from

$$\sigma_{tf} = \sqrt{\sigma_y^{\,2} - \tau_{cr}^{\,2}\,[3 - (1.5\sin 2\phi)^2]} - 1.5\,\tau_{cr}\sin 2\phi \qquad (4b.W57)$$

g = a parameter for the width and inclination of the tension-field band

$$g = (\cot\phi - \alpha + 2s/h)\sin^2\phi \qquad (4b.W58)$$

ϕ is the inclination of the tension field. The appropriate value of ϕ is the one which gives the maximum value of the shear strength V_u (Eq. 4b.W55). As an approximation $\phi = 2/3\,\text{arccot}(\alpha)$ may be assumed. For a symmetrical girder, the anchorage length s of the tension field along the compression and tension flanges is obtained from

$$s = \frac{2}{\sin\phi}\sqrt{\frac{M_{pc}}{t\,\sigma_{tf}}} \leq a \qquad (4b.W59)$$

where M_{pc} is the reduced plastic moment of the flange allowing for the longitudinal force N_f in the flange due to the overall girder moment M.

$$M_{pc} = \frac{b_f t_f^{\,2}}{4}\,\sigma_{yf}\left[1 - \left(\frac{N_f}{N_{pf}}\right)^2\right] \qquad (4b.W60)$$

where

b = width of the flange plate
N_{pf} = $b\,t_f\,\sigma_{yf}$
t_f = thickness of the flange plate

NBN B51002 88

Bending, shear, and combined bending and shear strengths are computed from the buckling strength adjusted by the correction factor c. The value of this factor depends on the stress distribution; it may exceed 1.0 when an appreciable postbuckling strength is available while it is smaller than 1.0 when no postbuckling strength is expected or when the imperfections are known to have a significant detrimental effect.

Buckling checks must be carried out on the following types of components:
a) the transversely stiffened web;
b) the longitudinally stiffened web (between consecutive transverse stiffeners);
c) the unstiffened web subpanels.

The edges are usually assumed to be simply supported; fixed edges are assumed only when closed sections are used for stiffeners.

The check consists in warranting that the actual correction factor c - with regard to plate buckling - is lower or equal to the required value c*.

$$c \le c^* \tag{4b.W61}$$

The actual correction factor $c=\sigma_c/\sigma_{c,cr}^{red}$ is computed as the ratio of the equivalent stress σ_c (derived from the von Mises yield criterion)

$$\sigma_c = \sqrt{\sigma^2+3\tau^2} \tag{4b.W62}$$

to the (reduced) equivalent critical stress

$$\begin{aligned} \sigma_{c,cr}^{red} &= \eta\,\sqrt{\sigma_{cr}+\tau_{cr}^2} \\ &= \eta\,\frac{\sigma_c}{\frac{1+\psi}{4}\frac{\sigma}{\sigma_{cr}^\circ}+\sqrt{\left(\frac{3-\psi}{4}\frac{\sigma}{\sigma_{cr}^\circ}\right)^2+\left(\frac{\tau}{\tau_{cr}^\circ}\right)^2}} \end{aligned} \tag{4b.W63}$$

where

η	$= (20+\sqrt{25-15\beta^2})/(25+\beta^2)$ for $\beta \le 1.25$ with $\beta=\sigma_y/(\tau_{cr}\sqrt{3})$
σ, τ	= stresses under factored loads at the point of the web panel where σ_c reaches a peak value
σ_{cr}, τ_{cr}	= elastic critical buckling stresses, when normal and shear stresses are coincident
σ_{cr}°, τ_{cr}°	= elastic critical buckling stresses, when normal and shear stresses act separately
ψ	= normal stress ratio with respect to the edge under maximum compression (tension is negative)

The required values c* depend on the stress distribution, the actual behavior of the stiffeners and the aspect ratio of the web panels/sub-panels; they are based on experimental considerations.

Either of the following two approaches may be used.

First Approach

Use is made of the postcritical strength reserve of the plate components; therefore the stiffeners are forced to remain straight and thus nearly rigid not only up to the critical load but also in the postcritical range up to the ultimate load. The relative flexural rigidity of the stiffeners.

$$\gamma = 10.92\, I_s/h\, t_w^3 \tag{4b.W64}$$

I_s being the moment of inertia of the stiffener with an effective web plate width, must be such that

$$\gamma \geq m\gamma^* \tag{4b.W65}$$

γ^*, the optimum rigidity according to the linear plate buckling theory, can be taken from buckling charts. The amplification factor m is equal to 4 for open section stiffeners and 2.5 for closed section stiffeners. Thus, the computed relative rigidity γ must be divided by m before entering the buckling charts for reading the value of the buckling coefficient γ^*.

The required values of c^* are given in Table 4b.W1.

Table 4b.W1 Required c*

Stress distribution	Value of c^*
pure compression	0.83
pure bending	1.05
linearly distributed σ	1.05–0.11 $(1+\psi)$
pure shear	1.05

Because the stiffeners of the first approach are designed to be rigid up to collapse, only plate buckling of the unstiffened web subpanels need be checked.

Second Approach

The designer refers to the results of the linear theory of plate buckling and does not account for a full postcritical strength reserve. The buckling charts are used with computed γ values, i.e., without reduction by m factor as in the first approach. However, because the stabilizing effect of the stiffeners is expected to be less than given by the linear plate buckling theory, the required c^* values (Table 4b.W2) are lower in the second approach than in the first.

Table 4b.W2 Required c*

Stress distribution	Value of c^*
pure compression	0.83
pure bending	0.95
linearly distributed σ	0.95–0.06 $(1+\psi)$
pure shear	0.95

Remarks

(A) ψ is the extreme stress ratio $\psi=\sigma_{min}/\sigma_{max}$. σ_{max} = the larger compressive stress at one edge and σ_{min} = the stress at the other edge with the tensile stress taken as negative. ψ must be $\geq (-1)$.

(B) A reduction of c^*=0.83 used for pure compression is necessary when the plate buckling looks like a column buckling problem because any postcritical strength reserve vanishes; such behavior occurs when $\sigma_E/\sigma_{cr} \geq 0.5$, where σ_E is the EULER buckling load of the pin-ended column composed of the stiffener and the associated web plate width.

(C) For coincident normal and shear stresses, the required value of the correction factor c^* is

$$c^*=\sqrt{\frac{(\sigma\, c^*_\sigma/\sigma^\circ_{cr})^2+(\tau\, c^*_\tau/\tau^\circ_{cr})^2}{(\sigma/\sigma^\circ_{cr})^2+(\tau/\tau^\circ_{cr})^2}} \qquad (4b.W66)$$

Design of stiffeners

- All the stiffener components must have a width-to-thickness ratio such that they are fully effective.
- When computing the moment of inertia I_s, an effective width of the web plate is taken into account; it amounts to $1.9\, t\sqrt{E/\sigma_y}$.
- The condition $\gamma \geq m\, \gamma^*$ is equivalent to $I_s \geq I^*_s$
  (A) For transverse stiffeners with spacing a:
  $$I^*_s = (2.5\alpha^2-2)\, a\, t^3 \geq 0.5\, a\, t^3 \quad (\alpha=a/h) \qquad (4b.W67)$$
  For longitudinally stiffened webs, the transverse stiffeners shall have a section modulus S_T of at least, $S_T \geq S_L\alpha/3$
  where S_L is the section modulus of the longitudinal stiffener.
  (B) For longitudinal stiffeners, γ^* is taken from buckling charts with due regard to the characteristics of the stiffened plate and to the location of the longitudinal stiffeners.

SIA 161 79

Bending Strength

The bending strength M_u of plate girders is determined on the basis of their reduced section, (the full section shall be used for the calculation of the internal forces of redundant systems). The width $h_c/2$ of the zone in compression shall be reduced to the effective width b_e according to

$$\frac{2\, b_e}{h_c} = \sqrt{\frac{0.73\, k_b}{\sigma_y/E}}\, \frac{1}{(h/t)} \leq 1.0 \qquad (4b.W68)$$

where k_b is 23.9.

The effective width b_e can be distributed in equal parts on either side of the compression zone.

The limit slenderness ratio of webs under pure bending for which the total width remains effective up to the yield stress is given by

$$\left(\frac{h}{t}\right)_{lim} \leq \frac{4.2}{\sqrt{\sigma_y/E}} \qquad (4b.W69)$$

that is, 126 for σ_y=235 MPa or 102 for σ_y=355 MPa.

Shear Strength

Shear strength of transversely stiffened plate girders is given by

$$V_u = h\,t\left[\tau_{cr} + \left(\frac{\sigma_y}{\sqrt{3}} - \tau_{cr}\right)\frac{\sqrt{3}}{2\sqrt{1+\alpha^2}}\right] \qquad (4b.W70)$$

when

$$\tau_{cr} \le \frac{0.8\,\sigma_y}{\sqrt{3}} \qquad (4b.W71)$$

where

$$k_v = 4 + \frac{5.34}{\alpha^2} \quad \text{for } \alpha < 1.0$$

$$k_v = 5.34 + \frac{4}{\alpha^2} \quad \text{for } \alpha \ge 1.0 \qquad (4b.W72)$$

$$\tau_{cr} = \frac{0.9\,E}{(h/t)^2}\,k_v \qquad (4b.W73)$$

When $\alpha = a/h > 3.0$, $\alpha = 3.0$ may be used.

The end stiffener must be designed to resist a horizontal force

$$H = \left[V_u - h\,t\left\{0.9\sqrt{\frac{\tau_{cr}\,\sigma_y}{\sqrt{3}}}\right\}\right]\sqrt{1+\alpha^2} \qquad (4b.W74)$$

applied at a quarter point from the top. If the end stiffener does not satisfy this requirement, the ultimate shear strength of the end panel should be limited by its buckling strength.

$$V_u = h\,t\left[0.9\sqrt{\frac{\tau_{cr}\,\sigma_y}{\sqrt{3}}}\right] \qquad (4b.W75)$$

with τ_{cr} from Eq. 4b.W73.

INTERNATIONAL COMPARISONS & COMMENT

Although some specifications treat the strength of the web in a unified manner, wherever possible, the comparison is discussed for different types of loading: bending, shear or a combination of bending and shear. For clarity, the case of edge loading is not included in the comparison although it is covered in some of the specification provisions. The comparison table (Table 4b.1) below shows in what aspects specifications differ from each other.

Table 4b.1 Comparison of Web Strength

1 2	3	4	5	6	7	8	9	10	11
Region	Bending Strgth		No.	Shear Strength				Shear & Mom	
Specification	Bucklng	P-B	LS	Buckl	Adjst	T-F	V_u/V_y	Curve	Stress
Australia									
AS DR87164 87		r	2			t	0.85	c,r	
China									
GBJ 18 87	y		1		*		0.80		s
TBJ 2 85	y		1	*			0.32		s
East Europe									
ČSN 73 1401 84		w	*			t	0.82	c	
ČSN 73 6205 84		w	*			p	0.67		s
MI 15024/3 85		r	1			t	0.84	r	
MI 15024/4 85		r	1			t	0.84	r	
SNiP 82	b		*,0	*		t	0.82	c	
Japan									
AIJ 86	y		2		*		0.83		s
JRA 80	y		2		*		0.60		s
North America									
AASHTO 89	y		1			t	0.87	c	
AISC 89		r	0			t	0.84	c	
AISC 86		r	0			t	0.85	c	
AREA 89	y		0	*			0.30		s
CSA S6 M89 84a		r	1			t	0.84	c	
CSA S16.1 M89 84b		r	1			t	0.84	c	
FHWA 80		w	*			t	0.76		s
West Europe									
BS 5400 82		w	*			t, p	0.74	c	
BS 5950 85		r	?			t	0.81		r,s
DASt 015 80		w, r	*,0		*	t	0.75	r	
EC 3		w	*	*		t	0.74		s
NBN B51002 88	b		*		*		0.59		s
SIA 161 79		w	?			t	0.84		s ?

Explanation of Columns in Table 4b.1

1) Region.
2) Specifications in the Region.
3) Bending strength is limited by elastic buckling (b), or by yielding before buckling (y).
4) Bending strength is limited by postbuckling via effective width or effective thickness (w), or via stress redistribution (r).
5) Number of longitudinal stiffeners permitted: Only one (1), up to two (2), more than two (*).
6) Web shear strength is limited by elastic buckling.
7) Postbuckling shear strength is considered by empirically adjusting (reducing) the factor of safety against buckling or by increasing the buckling stress.
8) Ultimate shear strength is based on tension-field action (t) or an equivalent postbuckling consideration (p).
9) Ultimate shear strength (V_u/V_y) for a/h=1.0, h/t=140, σ_y=250 MPa and E=200,000 MPa. (When relevant, flanges are assumed to be very flexible.)
10) Moment/shear interaction is given by an empirical curve or formula (c), or by reassigning the moment to the flanges and the shear to the web (r).
11) Moment/shear interaction is limited by the combination of local stresses to cause yielding or buckling.

WHAT ARE THE DIFFERENCES?

Bending Strength

Since the behavior of a slender web may affect the bending strength of a girder, the necessary design criteria are established by the specifications. Columns 3 and 4 of the Comparison Table (Table 4b.1) indicate with a letter the principal concept used by a particular specification. These are as follows:

(y) Yielding of the web, with the buckling usually prevented by a lower slenderness limitation.
(b) Elastic buckling stress.
(w) Modified cross section with the web contribution reduced to the effective width defined by its postbuckling strength.
(r) Redistribution of stresses in the web after buckling (or due to initial imperfections) leads to a higher stress at the compression flange. This effect is accommodated by computing the stresses on the original cross section and increasing the computed stress or reducing the allowable compressive stress.

In the following, only the main features of the principal regional rules are presented and compared. A more detailed description is given in the Specification Provisions earlier or can be found in the source specifications. To simplify comparison, provisions for longitudinal stiffeners are not discussed; Column 5 of the Comparison Table only shows the number of longitudinal stiffeners permitted by each specification – one, two, none, or as many as may be needed(*):

AUSTRALIA

The bending strength of girders with thin webs is reduced in direct proportion to the excess of the web slenderness over the limiting slenderness for reaching the yield stress before buckling. (See Eq. 4b.A2) This is an indirect way to consider stress redistribution after the buckling of the web (r).

CHINA

The low web slenderness allows reaching the yield stress. (y in Column 3 of Table 4b.1)

EAST EUROPE

On the basis of information available, one can conclude that the concept for considering web strength varies among the specifications of the region as follows:

(A) ČSN 73 1401 84 --- Redistribution and yielding (r,y)
The postbuckling stress redistribution is based on the results of numerical large-deflection elastic analysis. The design formula of the effective section modulus for the compression stress in an unsymmetrical section is recommended. Here, to facilitate comparison, the formulation is adjusted for a symmetrical section and is shown as the ratio of the resultant allowable moment to the yield moment.

$$m = M_{all}/M_y = \left[1 - \frac{I_w}{I}\left(1 - \frac{2101}{(h/t)\sqrt{\sigma_y}}\right)\right] \tag{4b.2}$$

where
I = the moments of inertia of the whole section
I_w = the moments of inertia of the web plate alone

(B) ČSN 73 6205 84 --- Redistribution and yielding (r,y)
The criteria are similar to those of the other specification, ČSN 73 1401 84.

(C) MI 15024/4 85 --- Elastic buckling and redistribution (b,r)
Redistribution to the extent of excluding the web area altogether is recommended when the web is expected to buckle under the combination of bending and shearing stresses. (Similarly to FHWA TS 80)

(D) SNiP II-23-81 82 --- Elastic buckling (b) or postbuckling strength (t)
In the postbuckling range, a formula, analogous to Eq. 4b.2, gives a somewhat higher than the buckling strength. (Eq. 4b.E68)

JAPAN

Web strength depends on elastic buckling assuming simply supported edges (b).

NORTH AMERICA

Two specifications in the area (AASHTO 89 and AREA 89) use yielding as the strength criterion for webs (y).

Four (AISC ASD 89, AISC LRFD 86, CSA S6 M89 84a and CSA S16.1 M89 84b) include the effect of stress redistribution by using the same formula (r). The allowable compressive bending stress or the maximum moment capacity based on the compressive stress is reduced by the following factor

$$\left[1 - 0.0005\,\frac{A_w}{A_f}\left(\frac{h}{t} - \frac{B}{\sqrt{\sigma_b}}\right)\right] \leq 1.0 \tag{4b.3}$$

where

A_f = area of one flange (compression flange)
A_w = area of the web plate
B = 1546 for AISC ASD 89, 2543 for AISC LRFD 86, 1810 for CSA S6 M89 84a, and 1970 for CSA S16.1 M89 84b
σ_b = the yield or the lateral-torsional buckling stress; B has a slightly different value for each specification

FHWA TS 80 specification eliminates portions of the web after buckling, so that additional moment would be carried by the thus reduced section -- a simplified way of introducing postbuckling strength (w). This is analogous to the recommendation of MI 15024/4 85 in East Europe and several specifications in West Europe (EC 3 88, SIA 161 79).

WEST EUROPE

Bending strength limitations vary from elastic buckling (yielding, for low h/t), adjusted width or thickness of the web plate to account for the postbuckling capacity, to the use of empirical reduction coefficients. Some countries provide more than one approach in different specifications. Buckling with some adjustments, remains as the only limit in NBN B51002 88 and most older specifications. It is also the limit for the web strength in DASt 015, but additional bending strength may be provided by the flanges.

EC 3 88 and SIA 161 79 give formulas for reducing the width of the web plate after buckling, and a modified cross section is to be used. BS 5400 P.3 82 calls for the reduction of the thickness of the web after buckling and a modified cross section. The buckling limitation for individual web subpanels is based on an elaborate set of postbuckling contributions of the web or, as an alternative, on a modest contribution via a simple formula.

Shear Strength

There is considerable variation in the fundamental concepts for computing shear strength: elastic buckling, "adjusted buckling" (a lower factor of safety or higher than theoretical buckling coefficient), and postbuckling, mainly, tension-field (p,t). The use of one or the other concept by a particular specification is respectively marked in Columns 6, 7 and 8 of the Comparison Table (Table 4b.1). To simplify presentation, only symmetrical girders without longitudinal stiffeners are discussed here. The number of longitudinal stiffeners permitted by each specification is shown in Column 5 of the Table.

A numerical comparison of the specifications is given in Column 9 of the Table for a particular set of geometrical and material values: a/h=1, h/t=140, σ_y=250 MPa, E=200000 MPa. The resultant strengths (V_u/V_y) vary from the low 0.30 (AREA 89) to the high 0.94 (BS 5400 P.3 82).

AUSTRALIA

Yield shear is adjusted by three factors: α_v, α_d, and α_f. α_v reduces for web buckling. α_d increases for the postbuckling diagonal tension-field action. α_f adjusts for additional shear strength due to plastic hinges in the flanges. The tension-field formula is basically the same as in all North American specifications, as well as, in British BS 5950 P.5 85, Hungarian MI 15024/4 85, and Swiss SIA 161 79. It is based on the Basler model. This is reflected in the value of 0.85 in Column 9 of the comparison table which is essentially the same as in other specifications mentioned here.

CHINA

Both specifications of the region (GBJ 18 87 and TBJ 2 85) have the design approach which, in the past, used to be popular among many specifications (one other is AREA 89). The formula gives the spacing of transverse stiffeners for the specified shear stress and other dimensions. Both use buckling as the design concept, but the one for buildings and HW bridges (GBJ 18 87) has a built-in upward adjustment (presumably, by using a more realistic, higher buckling coefficient).

EAST EUROPE

Due to a considerable variation, specifications of the region are discussed separately.

(A) ČSN 73 1401 84 --- There are two formulations to consider for postbuckling web strength. One, which is based on the results of numerical large-deflection analysis is for moving loads. The strength criterion is the yielding of the web under membrane stresses developing due to large web deformations or the failure of the flange. The other formulation, which is based on the tension-field theory according to Rockey-Škaloud, is for stationary loads. This is the one used to compute the value of 0.82 for Table 4b.1. This value is very close to others based on the tension-field analysis. The formulations are quite elaborate as they include the effect of flange dimensions. In this respect they are more sophisticated than the Basler tension-field theory used in the Australian, Hungarian, several North American, and some West European specifications.

(B) SNiP II-23-81 82 --- Although the principal specifications are based on the buckling criteria, provisions are made for the design of transversely stiffened symmetrical girders by considering postbuckling strength. The tension-field type formula for the postbuckling strength in shear is quite analogous to those in other specifications, but the inclusion of the contribution by the flanges depends on the section modulus rather than on the plastic moment capacity as is done in most of the other specifications which include this effect.

(C) MI 15024/3 85 and MI 15024/4 85 --- The supplementary specification for considering postbuckling strength in transversely stiffened girders corresponds to the Basler model and neglects any direct contribution of flanges. It is thus essentially the same as for North American specifications (0.84 in Column 9 of Table 4b.1) although the computational process is quite different.

JAPAN

Elastic buckling analysis is the basis for both specifications. Postbuckling strength is recognized by using a factor of safety reduced from 1.7 to 1.25. The resultant strength (0.60 in Col. 9 of the Table) falls about half-way between the pure buckling values of 0.30-0.32 and tension-field values of 0.82-0.87.

NORTH AMERICA

Of the seven specifications considered, five use the same Basler's tension-field concept as it was originally included in the AISC 1963 specification. The difference is mainly in notation, presentation, formulas for buckling coefficients, and some adjustments of constants. The computed results in Table 4b.1 vary from 0.84 to 0.87. The same approach has been accepted by the Australian (AS DR87164 87), Hungarian (MI 15024/4 85), and some West European specifications.

The Proposed Specification for highway box girder bridges (FHWA TS 80) is based on a more conservative "pure Basler" concept which completely neglects contribution of the flanges, and it gives 0.76 for the example girder in Table 4b.1.

One specification (AREA 89) uses pure buckling with the correspondingly much lower computed value of 0.30. This is essentially the same as 0.32 according to TBJ 2 85 (China).

WEST EUROPE

The six specifications selected for comparison fall into three groups of treating shear strength. Elastic buckling is used as the limit by NBN B51002 88 as can be seen by the lowest value of 0.59 shown in Column 9 of Table 4b.1.

Four specifications (BS 5400 82, BS 5950 85, EC 3 88, and DASt 015 80) consider the postbuckling web strength by accepting a form of the Rockey-Škaloud tension-field model which includes the effect of the plastic strength of the flanges on the width of the tension field. The differences among these specifications are mainly in the format of presentation, notations, and numerical approximations of combined constants. This goes to the extent that the fundamental similarity is often difficult to detect. For example, the shear buckling coefficient for α=1.0, stated specifically or embedded in other quantities, is given in the form of $[5+5/\alpha^2]$, $[(0.75+1/\alpha^2)\cdot(16/3)]$, $[4.0+5.34/\alpha^2]$ by the BS 5400 82, BS 5950 85, and EC 3 88 with DASt 015 80, respectively. Some specifications introduce the inelastic transition, others do not. The result of these and other variances is reflected in the different ultimate strength values varying from 0.74 to 0.81 in Column 9 of Table 4b.1.

SIA 161 79 specification is based on the Basler tension-field model which does not directly consider the effect of flanges but compensates for their contribution by making the tension field band have the inclination of the panel diagonal. This is the same model as used by most of the North American specifications as can be seen from the values in Column 9.

Interaction of Shear and Bending

Interaction of shear and bending is handled for the most part in one of three ways: by an empirical formula or curve for the moment and shear acting on the section (c), by reassignment of the moment to be carried only by the flanges and of the shear only by the web plate (r) instead of both by the whole section, or by checking a combined state of stress against buckling or yielding at some point (s). The specific way is marked for each specification in Columns 10 and 11 of Table 4b.1. Specifications which use curve/formula (c) and reassignment (r) are marked in Column 10, and those which use the state of stress (s) in Column 11.

AUSTRALIA

Two methods are permitted for considering interaction between moment and shear. In one, the shear is assigned to the web and the moment to the flanges (r); in the other, an interaction formula is given which is based on the same limitations as in many North American specifications.(c)

CHINA

GBJ 18 87 includes the bending stress in the shear stress formula for computing the required spacing of transverse stiffeners.(s) The other specification (TBJ 2 85) apparently checks only against yielding.(s)

EAST EUROPE

(A) ČSN 73 1401 84 --- An interaction formula for gross moment and shear is used.(c)

$$\left(\frac{M}{M_u}\right)^2 + \left(\frac{V}{V_u}\right)^2 \le 1 \tag{4b.4}$$

(B) ČSN 73 6205 84 --- The interaction is incorporated into the combined strength formula.(s)

(C) MI 15024/4 85 --- In postbuckling ultimate strength design, interaction is achieved by completely neglecting the web in the computation of bending stresses after the web is subjected to buckling shearing stresses.(r)

(D) SNiP II-23-81 82 --- An interaction formula of a rather optimistic nature is used in comparison with other specifications, such as Eq. 4b.4 or Eq. 4b.8.(c)

$$\left(\frac{M}{M_u}\right)^4 + \left(\frac{V}{V_u}\right)^4 \leq 1 \tag{4b.5}$$

JAPAN

(A) AIJ 86 --- A somewhat different formula is used in this specification.(s)

$$\left(\frac{\sigma}{\sigma_{cr}}\right)^2 + \left(\frac{\tau}{\tau_{cr}}\right) \leq 1 \tag{4b.6}$$

(B) JRA 80 --- An approximate interaction formula is used for stresses to guard against buckling and to reduce the factor of safety.(s)

$$\left(\frac{\sigma}{\sigma_{cr}}\right)^2 + \left(\frac{\tau}{\tau_{cr}}\right)^2 \leq 1 \tag{4b.7}$$

NORTH AMERICA

Five of the seven specifications discussed use the same tri-linear interaction relationship although in some cases written in different form. These are: AISC ASD 89, AASHTO 89, AISC LRFD 86, CSA S6 M89 84a and CSA S16.1 M89 84b.(c) This is the same relationship as in the Australian specification.

$$0.727\left(\frac{M}{M_u}\right) + 0.455\left(\frac{V}{V_u}\right) \leq 1 \tag{4b.8}$$

(A) AREA 89 --- states that the diagonal tensile stress caused by shearing and maximum bending stresses should not cause yielding.(s)

(B) FHWA TS 80 --- requires that no component of the box girder cross section may exceed its ultimate capacity under the combination of forces produced by bending and shearing forces acting on it.(s)

WEST EUROPE

(A) BS 5400 82 --- Interaction between buckling and shear is defined by a quadrilinear curve. The bending strength of flanges alone (M_R) and the shear strength of the web plate alone (V_R) are the controlling parameters.

(B) BS 5950 85 --- The combined strength is defined either by assigning bending to the capacity of the flanges alone and shear to the web alone or by using an interaction buckling formula for the web plate. (r, s)

(C) DASt 015 80 --- Moment and shear are reassigned to the flanges and the web, respectively.(r) However, the axial forces in the flanges caused by bending have a direct effect on the shear strength by modifying their plastic moments.

(D) EC 3 88 --- It seems, the interaction between bending and shear has not yet been formalized, but apparently the direction seems to be toward considering a combined state of stress. (s)

(E) NBN B51002 88 --- Pure elastic buckling under combined stresses is the basic criterion of strength. An adjustment is provided for the inelastic range. (s)
(F) SIA 161 79 --- It appears that no specific recommendation is provided for the combined case of loading.

WHY ARE THERE DIFFERENCES?

Bending Strength

When the same concept is used, the differences seem to be attributable to the growth of confidence with time; the later the revision of the specification (or, more precisely, revision of a particular provision), the less conservative is the provision. For example, the value of parameter B in Eq. 4b.3 gradually increased from 1547 in 1978 (AISC ASD 78) to 2543 in 1986 (AISC LRFD 86). A significant difference is introduced by the purpose of the structure: buildings -- HW bridges -- RR bridges, from less to more conservatism, and this is rational since each type in this sequence is subjected to relatively higher and more frequent loads.

Shear Strength

Acceptance of one or another design concept appears to be the dominant reason for the differences in the results. The three broad groups are:
(b) Buckling
(ab) "Adjusted" buckling in which the buckling stress is increased to compensate for postbuckling strength.
(t or p) Postbuckling is included directly through tension-field or some other direct method.

As in other aspects of design, a trade-off is often made between the accuracy and the ease of use, that is, simplicity. An illustration of this is the acceptance of the simpler Basler model for postbuckling tension-field computations in favor of many other methods which may be more accurate but are more complicated and labor intensive.

Interaction of Shear and Bending

Differences in handling interaction follow, to a large extent, from the differences in computing the component stresses or forces.

4c. Compression Flange of Box Girders

SPECIFICATION PROVISIONS & EXPLANATION

AUSTRALIA

CHAPTER EDITOR
Noel W. Murray, Monash University

AS DR87164 87

A plate component of the compression flange in a box girder is expected to reach yield stress unless the width-thickness ratio exceeds

$$\left(\frac{b}{t_f}\right)_y = C\sqrt{\frac{250}{\sigma_y}} \tag{4c.A1}$$

where

b = width of the flange plate between longitudinal stiffeners or webs
C = constant as shown below for different types of fabrication
= 45 for stress relieved, hot-rolled and hot-finished
= 40 for lightly welded and cold-formed
= 35 for heavily welded
t_f = thickness of flange plate

When $(b/t_f) > (b/t_f)_y$, the plate contribution to the cross section is to be reduced by the factor of

$$\frac{(b/t_f)_y}{(b/t_f)} = \frac{C\sqrt{\frac{250}{\sigma_y}}}{(b/t_f)} \tag{4c.A2}$$

Alternatively, the effective cross section is determined by omitting from each flat compression element the width in excess of the width corresponding to $(b/t_f)_y$.

No direct provision is made in the specification for the design of longitudinal or transverse stiffeners in the compression flange. Apparently, longitudinal stiffeners are to be designed as columns with the plate contribution to the cross section according to Eqs. 4c.A1 and 4c.A2.

CHINA

CHAPTER EDITORS
Xiliang Liu, Tianjin University
Xinhua H. Ren, Tianjin University

GBJ 18 87 and TBJ 2 85

Neither of these specifications provides guidelines for designing box girders. However, plate buckling formulas can apparently be used for this purpose.

EAST EUROPE

CHAPTER EDITORS
Jozef Djubek, Slovak Academy of Science
Miroslav Škaloud, Czechoslovak Academy of Sciences

ČSN 73 1401 84

The design of longitudinally stiffened compression flanges is based on a simple bar-simulation analogy, that is, the column strength of the stiffeners with the effective width of the flange plate.

ČSN 73 6205 84

The ultimate stress of the stiffened compression flange is

$$\sigma_u = \psi_{1n}\,\psi_{2n}\,R_d \tag{4c.E1}$$

Coefficients ψ_{1n} and ψ_{2n} of global flange buckling

$$\psi_{1n} = 1-\frac{\alpha_2}{2}\left[\sqrt{\left(\frac{\beta_1-\beta_2}{\beta_2}\right)^2+\frac{2\alpha_3^2}{\sqrt{1+\alpha_2^2}-1}}-\frac{\beta_1-\beta_2}{\beta_2}\right] \qquad \text{if } \psi_{1n} > 1-\frac{\alpha_2}{2} \tag{4c.E2}$$

or

$$\psi_{1n} = 1-\alpha_2\,\frac{2\alpha_3^2-\frac{\beta_1-\beta_2}{\beta_2}\left(\sqrt{1+\alpha_2^2}-1\right)}{2\alpha_3^2+\sqrt{1+\alpha_2^2}-1} \qquad \text{if } \psi_{1n} \le 1-\frac{\alpha_2}{2} \tag{4c.E3}$$

where

α = the side ratio of the flange panel (α=a/b)

$$\alpha_2 = 0.835 - 0.04\alpha + 0.01\,\frac{\lambda}{100} \tag{4c.E4}$$

λ = width-to-thickness ratio ($\lambda = b/t_f$)

$$\alpha_3 = 0.14\alpha + 0.127 \tag{4c.E5}$$

$$\beta_1 = 1 + 12\left(\frac{e_{st}}{t_f}\right)^2 + \frac{12\,I_{s\ell}}{b_i t_f^3}\,(1-\nu^2) \tag{4c.E6}$$

where

$A_{s\ell}$ = area of one longitudinal stiffener
b_i = spacing of longitudinal stiffeners
e_{st} = distance of the centroid of a cross section having its area $= A_{s\ell} + b_i t_f$ to flange mid-thickness
$I_{s\ell}$ = moment of inertia of one longitudinal stiffener
R_d = design yield stress
t_f = thickness of the flange plate
ν = Poisson's ratio of the flange material

$$\beta_2 = (1-\nu^2)\left(1+\frac{A_{s\ell}}{b_i t_f}\right)\frac{a^2}{720\ t^2}\frac{R_d}{210} \tag{4c.E7}$$

The local plate buckling coefficient is computed from

$$\psi_{2n} = \frac{\psi_n + A_{s\ell}/(b_i t_f)}{1 + A_{s\ell}/(b_i t_f)} \tag{4c.E8}$$

where, for the boundary flange plate panels,

$$\psi_n = \frac{40}{(b_i/t_f) + 10}\sqrt{\frac{210}{\psi_{1n}\ R_d}} \tag{4c.E9}$$

and for the inner plate panels

$$\psi_n = \frac{40\sqrt{210/\psi_{1n}\ R_d}}{(b_i/t_f) + 10}\left(1.3 - \frac{12\sqrt{210/\psi_{1n}\ R_d}}{(b_i/t_f) + 10}\right) \tag{4c.E10}$$

The simple design approach based on bar-simulation analogy is also permitted.

MI 15024/4 85

No provisions stated.

SNiP II-23-81 82

No direct provisions, but plate buckling limitations can be used.

JAPAN

CHAPTER EDITORS

Sadao Komatsu, Osaka Sangyo University
Fumio Nishino, University of Tokyo

JRA 80

Plate buckling of compression flanges of box girders is prevented by limiting the width-thickness ratio as follows

$$R \leq 0.7 \tag{4c.J1}$$

and for plates in compression stiffened with (n−1) equally spaced longitudinal stiffeners

$$R \leq 0.7n \tag{4c.J2}$$

where

k = 4 for both edges simply supported
k = 0.43 for one free edge

$$R = \frac{b}{t_f}\sqrt{\frac{\sigma_y}{E}\cdot\frac{12(1-\nu^2)}{\pi^2 k}} \tag{4c.J3}$$

The moment of inertia of the longitudinal stiffener should be

$$I_{s\ell} \geq \frac{b\,t_f^3}{11}\gamma \tag{4c.J4}$$

$$\text{where } \gamma \text{ is } 4\alpha^2 n\left(\frac{t_o}{t_f}\right)^2(1+n\,\delta) - \frac{(\alpha^2+1)^2}{n} \tag{4c.J5}$$

when transverse stiffener of sufficient rigidity, I_{st} (from Eq. 4c.J7), is provided within a_o. Otherwise,

$$\gamma = \frac{1}{n}\left\{\left[2n^2\left(\frac{t_o}{t_f}\right)^2(1+n\,\delta) - 1\right]^2 - 1\right\} \tag{4c.J6}$$

where

$A_{s\ell}$ = cross-sectional area of a longitudinal stiffener, and

$$I_{st} \geq \frac{b\,t_f^3/11 + n\,I_L}{4\alpha^3} \tag{4c.J7}$$

$$a_o = b\left[1 + \frac{n\,I_L}{(b\,t_f^3/11)}\right]^{0.25} \tag{4c.J8}$$

a = distance between transverse stiffeners
b = flange plate width
t_f = thickness of the flange plate
t_o = twice the thickness corresponding to Eq. 4c.J2
α = a/b
δ = $A_{s\ell}/b\,t_f$

COMMENTARY

Equation 4c.J1 is based on experimental results of uniformly compressed plates with various boundary conditions, grades of steel, and various manufacturing processes, and under the condition that plates will not lose stability until working stress reaches yield stress. Equation 4c.J2 is an extension of Eq. 4c.J1 for plate elements between longitudinal stiffeners with sufficient rigidity and strength, specified in Eqs. 4c.J4 through 4c.J8. The moments of inertia of longitudinal and transverse stiffeners are specified under the condition that elastic overall buckling strength of stiffened plates becomes equal to the elastic local buckling strength of the component plates.

NORTH AMERICA

CHAPTER EDITOR
Alexis Ostapenko, Lehigh University

AASHTO 89

Box Girders (Composite or with an Orthotropic Deck)

The rules apply to single-cell multiple-box bridges of "moderate span length" requiring not more than one longitudinal web stiffener.

The webs are to satisfy the requirements for plate girders as described above under Topic B. The provisions for compression flanges, unstiffened or stiffened, are essentially the same as in the Canadian specifications CSA S6 M89 84a and CSA S16.1 M89 84b and are discussed there.

The orthotropic deck is to satisfy strength provisions according to a "rational method" and the ribs are to have the following slenderness ratio

$$(L/r)_{max} = \frac{3216}{\sqrt{\sigma_y}} \sqrt{1 - 9/5(\sigma/\sigma_y)} \qquad (4c.N1)$$

where

L = the distance between the transverses

σ = the gross bending compression in the orthotropic flange

AISC ASD 89, AISC LRFD 86, and AREA 89

These three specifications have no direct provisions for designing box girders.

CSA S6 M89 84a and CSA S16.1 M89 84b

Box Girders

CSA S6 M89 84a considers only single-cell multi-box composite bridges of "moderate span", similarly to the AASHTO 87 specification.

Design of the webs is to be according to the rules for plate girders. The concrete deck requires no stability considerations and, thus, is not discussed here. The orthotropic deck is to be designed so that the "slenderness ratio of longitudinal ribs shall be adequate to ensure that overall buckling of the deck will not occur as a result of compression induced by bending of the longitudinal girders." No other stability criteria, except for the general local buckling width-thickness limitations, are imposed on the orthotropic deck.

The bottom flange in compression is to be designed according to the following rules:

Unstiffened Compression Flanges:

(A) When $b/t_f \leq 510/\sqrt{\sigma_y}$

$$M_u = \sigma_y S \qquad (4c.N2)$$

where

- b = flange width
- M_u = ultimate moment capacity
- S = section modulus of the whole box section

(B) When $510/\sqrt{\sigma_y} < b/t_f \leq 1100/\sqrt{\sigma_y}$

$$M_u = \sigma_{cr} S \qquad (4c.N3)$$

where

$$c = (1100 - b/t_f \sqrt{\sigma_y})/590 \qquad (4c.N4)$$

$$\sigma_{cr} = 0.59\sigma_y \left(1 + 0.687 \sin \frac{\pi c}{2}\right) \qquad (4c.N5)$$

(C) When $b/t_f > 1100/\sqrt{\sigma_y}$

$$M_u = \sigma_{cr} S \qquad (4c.N6)$$

where σ_{cr} is $724{,}000/(b/t_f)^2$ (4c.N7)

Compression Flanges Stiffened Longitudinally

(A) When $b_i/t_f \leq 255/\sqrt{\sigma_y/k_1}$

$$M_u = \sigma_y S \qquad (4c.N8)$$

where

- b_i = plate width between longitudinal stiffeners
- k_1 = buckling coefficient $\leq$ 4.0

(B) When $255/\sqrt{\sigma_y/k_1} < b_i/t_f \leq 550/\sqrt{\sigma_y/k_1}$

Use Eqs. 4c.N3 and 4c.N5, but with

$$c = \left(550 - (b_i/t_f)\sqrt{\sigma_y/k_c}\right)/295 \qquad (4c.N9)$$

where $k_c = k_1$

(C) When $b_i/t_f > 550\sqrt{\sigma_y/k_1}$

Use Eq. 4c.N3, with

$$\sigma_{cr} = 180{,}000\ k_1/(b_i/t_f)^2 \qquad (4c.N10)$$

Each longitudinal stiffener shall have a moment of inertia of

$$I_{s\ell} \geq z\, t_f^3\, b_i \qquad (4c.N11)$$

where

n = number of longitudinal stiffeners

$z = 0.125\ k_1^3 f$ for n=1 (4c.N12)

$= 0.07\ k_1^3$ for n > 1 (4c.N13)

At least one transverse stiffener, not smaller than the longitudinal stiffeners, shall be placed near the dead load inflection points.

Compression Flanges Stiffened Longitudinally and Transversely:

(A) When $b_i/t_f \leq 225/\sqrt{\sigma_y/k_2}$

$$M_u = \sigma_y\ S \tag{4c.N14}$$

(B) When $225/\sqrt{\sigma_y/k_2} < b_i/t_f \leq 550/\sqrt{\sigma_y/k_2}$

$$M_u = \sigma_{cr}\ S \tag{4c.N15}$$

where

c = from Eq. 4c.N9
$k_c = k_2$
σ_{cr} = from Eq. 4c.N5

(C) When $b_i/t_f > 550/\sqrt{\sigma_y/k_2}$
Use Eq. 4c.N15 with

$$\sigma_{cr} = 180{,}000\ k_2/(b_i/t_f)^2 \tag{4c.N16}$$

where

$a/b \leq 3.0$

$$k_2 = \frac{[1 + (a/b)^2]^2 + 87.3}{(n+1)^2\ (a/b)^2\ [1 + 0.1(n+1)]} \leq 4.0 \tag{4c.N17}$$

The moment of inertia requirement for the longitudinal stiffeners is

$$I_{s\ell} \geq 8t_f^3\ b_i \tag{4c.N18}$$

and for the transverse stiffeners,

$$I_t \geq 0.055\ (n+1)^3\ b_i^3\ \frac{\sigma_{cr}\ A_f}{Ea} \tag{4c.N19}$$

where

A_f = total area of the bottom flange
E = modulus of elasticity

Bottom Tension Flange

The effective width shall not be greater than 0.2 of the span or of the distance between the inflection points.

FHWA TS 80

When the bottom flange of a box girder highway bridge is in compression, it is subject to stability considerations with the principal features summarized below.

Unstiffened Bottom Flange

The ultimate capacity of the flange, P_u, is computed from

$$P_u = \sigma_u A_f \text{ or } \sigma'_u A_f \tag{4c.N20}$$

where

A_f = cross-sectional area of the flange including the flange projections, b', on each side beyond the webs not to exceed

$$b'_{max} = \frac{134}{\sqrt{\sigma_y}} \tag{4c.N21}$$

σ_u = ultimate stress of the flange
σ'_u = ultimate stress of the flange modified for shear

Plate slenderness parameter, λ_{pl}, is used to define σ_u for a/b $\geq$ 1.0.

$$\lambda_{pl} = \sqrt{\frac{\sigma_y}{\sigma_{cr}}} = \frac{b}{t_f}\frac{\sqrt{\sigma_y}}{850} \tag{4c.N22}$$

where

a = distance between transverse stiffeners
b = width of plate (distance between webs)
σ_{cr} = the elastic buckling stress of the plate (with the buckling coefficient k=4.0)

Then

for $\lambda_{pl} \leq 0.65$

$$\sigma_u = \sigma_y \tag{4c.N23}$$

for $0.65 < \lambda_{pl} < 1.5$

$$\sigma_u = \sigma_y\left(0.50 + 0.43(\lambda_{pl}-1.73)^2\right) \tag{4c.N24}$$

for $\lambda_{pl} \geq 1.5$

$$\sigma_u = \sigma_y\left(0.82 - 0.82\,\lambda_{pl}\right) \tag{4c.N25}$$

In the presence of shearing stresses, σ_u is modified to σ'_u as follows

If $\tau \leq 0.175\,\sigma_y$

$$\sigma'_u = \sigma_u \tag{4c.N26}$$

If $\tau \geq 0.175\,\sigma_y$

$$\sigma'_u = 1.05\,\sigma_u\sqrt{1 - 3\left(\frac{\tau}{\sigma_y}\right)^2} \tag{4c.N27}$$

where τ is the nominal shearing stress

$$\tau = \frac{1}{3}\tau_{max} \tag{4c.N28}$$

for shearing stress caused by girder flexure

$$\tau = \tau_{ave} \tag{4c.N29}$$

for shearing stress caused by torsion and warping

Bottom Flange with Longitudinal Stiffeners

The ultimate capacity of the longitudinally stiffened flange, P_u, is computed from

$$P_u = \sigma_u A_f \text{ or } \sigma'_u A_f \tag{4c.N30}$$

where

A_f = cross-sectional area of the flange including all longitudinal stiffeners and the flange projections, b', as for the unstiffened flange

σ_u = ultimate stress of a stiffener column consisting of one stiffener and a portion of the flange plate width (stiffener spacing) as given by the nomograph of Fig. 4c.N1

σ'_u = ultimate stress of the flange modified for shear according to Eq. 4c.N27

Definitions for Fig. 4c.N1 are

Plate slenderness parameter λ_{pl}

$$\lambda_{pl} = \sqrt{\frac{\sigma_y}{\sigma_{cr}}} = \frac{w}{t_f}\frac{\sqrt{\sigma_y}}{850} \tag{4c.N31}$$

where w is the tributary plate width (spacing of longitudinal stiffeners)

Column slenderness parameter λ_{col}

$$\lambda_{col} = \sqrt{\frac{\sigma_y}{\sigma_{cr}}} = \frac{a}{r}\frac{1}{\pi}\sqrt{\frac{\sigma_y}{E}} = \frac{a}{r}\frac{\sqrt{\sigma_y}}{1405} \tag{4c.N32}$$

where

a = spacing of transverse stiffeners supporting longitudinal stiffeners

r = radius of gyration of stiffener column

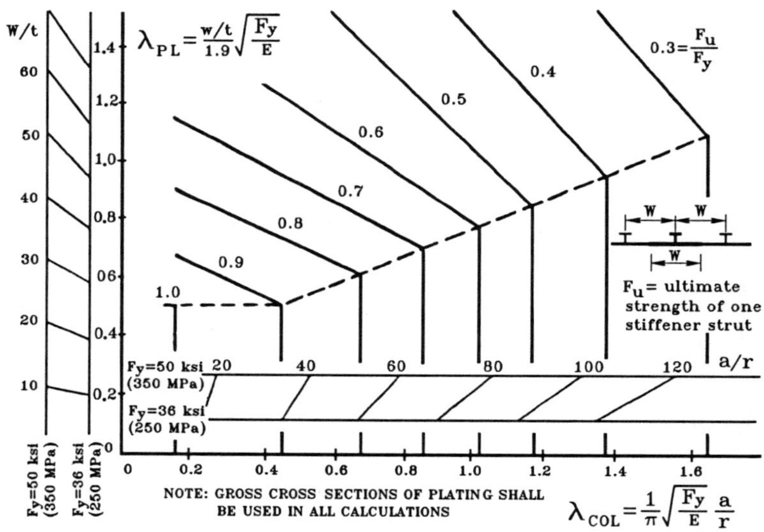

Fig. 4c.N1 Compression Flange Strength (FHWA TS 80)

Longitudinal Stiffeners of Bottom Compression Flange

Longitudinal flange stiffeners must satisfy the following slenderness requirements, C_s to prevent local and torsional buckling

for $\sigma_{max} > 0.5\ \sigma_y$

$$C_s \leq \frac{179}{\sqrt{\sigma_y}} \tag{4c.N33}$$

for $\sigma_{max} < 0.5\ \sigma_y$

$$C_s \leq \frac{291}{\sqrt{\sigma_y}} \tag{4c.N34}$$

where

$$C_s = \frac{d}{1.5\ t_o} + \frac{w}{12\ t_f} \quad \text{for flat-plate stiffeners} \tag{4c.N35}$$

$$C_s = \frac{d}{1.35\ t_o + 0.56\ r_y} + \frac{w}{12\ t_f} \quad \text{for tee or angle stiffeners} \tag{4c.N36}$$

d = stiffener depth

t_o = thickness of flat-plate stiffener or of the web of tee or angle stiffener

σ_{max} = maximum compression stress in stiffened flange panel

For closed stiffeners, the slenderness of plate components should not be less than needed to have the same strength as of the stiffened flange.

COMMENTARY

This specification is based on the latest information for design of steel box girder bridges that was available at the time of its development. Main emphasis was on utilizing the ultimate strength, yet, on being on the safe side. In particular, the nomograph for the stiffened compression flange represents a safe (maybe oversafe) approximation of the test and computed data.

WEST EUROPE

CHAPTER EDITOR
Pierre Dubas, Swiss Federal Institute of Technology

BS 5400 P.3 82

Strength of the compression flange is based on a column approach with a stiffener and associated width of flange plate acting between cross frames. Failure modes toward the flange plate and toward the stiffener are considered.

$$\frac{b_e}{b_i} = K_c \tag{4c.W1}$$

where

b_e, b_i = the effective and actual width of flange plate subpanels, respectively
K_c = computed from Eqs. 4b.W6 or 4b.W7.

The design is to satisfy the following yielding requirements

Stiffener yielding

$$\sigma_a + 2.5\,\tau_1\,k_{s1} < \frac{k_{t1}\,\sigma_{ys}}{\gamma_m\,\gamma_{f3}} \tag{4c.W2}$$

Plate yielding

$$\sigma_a + 2.5\,\tau_1\,k_{s2} < \frac{k_{t2}\,\sigma_{ye}}{\gamma_m\,\gamma_{f3}} \tag{4c.W3}$$

where

γ_m = 1.05
γ_{f3} = 1.1
σ_a = longitudinal stress, positive when compression, at the centroid of the effective section section of the stiffener
τ_1 = in-plane shear stress in the flange plate due to torsion on the girder, always to be taken as positive

$$\sigma_{ye} = \sqrt{\sigma_{yf}^2 - 3\tau^2\,\gamma_m^2\,\gamma_{f3}^2} \tag{4c.W4}$$

$$k_t = 0.5\left[\left\{1 + (1+\eta)\,\frac{5700}{\lambda^2}\right\} - \sqrt{\left\{1 + (1+\eta)\,\frac{5700}{\lambda^2}\right\}^2 - \frac{22800}{\lambda^2}}\,\right] \tag{4c.W5}$$

$$k_t = \frac{1}{1+\eta} \qquad \text{when } \lambda = 0$$

$$k_s = 0.4\,k_t'\,(\eta' + 1.7546 \times 10^{-4}\,\lambda^2) \tag{4c.W6}$$

where

$k'_t = k_t$ with $\eta=\eta'$

$\eta = \frac{y_o \Delta}{r_{se}^2}$ when obtaining k_{t1}

$\eta = \frac{y_z \Delta}{r_{se}^2}$ when obtaining k_{t2}

$\eta' = 0.0083\ (\lambda-15)$ when $\lambda > 15$

$\eta' = 0$ when $\lambda \leq 15$

$\lambda = \frac{a}{r_{se}}\sqrt{\frac{\sigma_{ys}}{355}}$ when obtaining k_{t1} and k_{s1}

$\lambda = \frac{a}{r_{se}}\sqrt{\frac{\sigma_{ye}}{355}}$ when obtaining k_{t2} and k_{s2}

where

a = spacing of cross beams and/or diaphragms

e_o = greatest offset of the flange plate from a straight line of length a, due to specified camber or curvature

r_{se} = radius of gyration of the effective section of a longitudinal stiffener about the centroidal axis parallel to the flange plate

y_{Bs} = distance from the centroid of the effective stiffener section to the neutral axis of the cross section of the beam

y_o = distance from the centroid of the effective stiffener section to the point on the stiffener farthest from the plate

y_z = distance from the centroid of the effective stiffener section to the mid-plane of the flange plate

$$\Delta = \frac{a}{625} + \frac{r_{se}^2}{2\ y_{Bs}} + \frac{e_o}{2}$$

There are also various clauses related to local moments.

BS 5950 P.5 85

The strength of compression flange is based on the column strength of the longitudinal stiffeners with the flanges plate as part of the cross section. The plate capacity is to be reduced if the plate slenderness is greater than a specified value.

$$\text{for } \frac{b_i}{t_f} > 28\sqrt{\frac{275}{\sigma_y}} \qquad (4c.W7)$$

the stress reduction factor is

$$f = \frac{21}{\frac{b_i}{t_f}\sqrt{\frac{\sigma_y}{275}} - 7} \leq 1.0 \qquad (4c.W8)$$

where b_i is the spacing of longitudinal stiffeners.

DASt 015 80

No specific recommendations.

EC 3 88

Recommendations for compression flanges are under development.

NBN B51002 88

Nothing special is said regarding compression flanges of box girders. Their design is conducted as for webs but accounting first for pure compression (ψ=1) and modifying the c* values indicated by Table 4b.W2 and Eqs. 4b.W66 to 4b.W67, as soon as the elastic critical buckling stress σ_E of the column composed of one longitudinal stiffener with the associated flange plate exceeds 0.5 times the elastic critical buckling stress σ°_{cr} of the longitudinally stiffened compression flange. This results in a decrease of c*.

SIA 161 79

The compression flange is designed as a system of columns each consisting of the longitudinal stiffeners and an effective width of the flange plate according to Eq. 4b.W70 modified for uniform compression.

$$\frac{b_e}{b_i} = \sqrt{\frac{0.73\ k_a}{\sigma_y/E}}\ \frac{1}{b_i/t} \leq 1.0 \qquad (4c.W9)$$

where

b_i = distance between longitudinal stiffeners
k_a = 4.0 for uniform compression

The use of linear plate buckling theory for the design of longitudinal stiffeners in compression zones is not allowed.

INTERNATIONAL COMPARISONS & COMMENT

WHAT ARE THE DIFFERENCES?

The bottom flange of box girders in the negative moment area is generally subjected to predominant compression, some shear and, possibly, to some light transverse loading. Usually, the flanges are designed for compression in accordance with specification rules and, if necessary, checked for other effects. The compression flanges can be unstiffened, stiffened longitudinally, or have longitudinal and transverse stiffeners.

Unstiffened compression flanges are treated as long plates under uniform compression and, depending on the width-thickness (b/t_f) ratio, are effective up to the yield or the buckling level of stress. The postbuckling strength is sometimes taken into consideration by reducing the margin of safety or via the effective width concept.

There are two approaches for treating stiffened flanges. In one approach, the stiffened flange is considered to be an orthotropic plate and is analyzed as such with the stiffeners providing additional rigidity in two directions to increase the buckling capacity. In the other, the longitudinal stiffeners with the tributary plate width are treated as axially loaded columns having little or no support in the lateral direction. In general, the column approach leads to heavier stiffeners and a thinner plate, and the strength can be computed to the ultimate capacity without too much difficulty. The orthotropic plate approach is practical up to the buckling strength, but it is "more clean-cut" and familiar.

The Comparison Table (Table 4c.1) shows in Columns 3 and 4 which specifications have accepted the orthotropic plate buckling approach (b) and which the column (ultimate strength) approach (u). The buckling approach with consideration of some postbuckling strength is marked with (p).

Table 4c.1: Compression Flange and Miscellaneous

1 Region	2 Specification	3 Compr Flange Buckl	4 Compr Flange Colmn	5 Edge Ldg El/Ult	6 Edge Ldg Stiff	7 Shear Lag
Australia						
	AS DR87164 87	p		*		
China						
	GBJ 18 87			b	*	
	TBJ 2 85			b	*	
East Europe						
	ČSN 73 1401 84			b	*	
	ČSN 73 6205 84	b				v
	MI 15024/4 85					
	SNiP 82			b	*	
Japan						
	AIJ 70					
	JRA 80	b				
North America						
	AASHTO 89	b		?		*
	AISC 89			b		
	AISC 86			u		
	AREA 89					
	CSA S6 M89 84a	b		b		*
	CSA S16.1 M89 84b			b		
	FHWA 80		u	u		v
West Europe						
	BS 5400 82		c			
	BS 5950 85		c	b		
	DASt 015 80	–				
	EC 3 88	–		u		
	NBN B51002 88	b				
	SIA 161 79		c	u		

Explanation of Columns in Table 4c.1

1) Region.
2) Specification in the Region.

COMPRESSION FLANGE

3) Strength of bottom compression flange is limited by plate buckling (b) or post-buckling effective width (p).
4) Treatment as a column with effective cross section (c) or beam-column at ultimate strength (u).

MISCELLANEOUS

5) Provisions made for edge loading: elastic buckling (b), ultimate (u).
6) Rules for local stiffeners to prevent buckling under edge loading.
7) Consideration of effective flange width for shear lag in box girders; (v) if variable along the girder.

AUSTRALIA -- (p) A brief statement that the postbuckling effective width is used.

EAST EUROPE

ČSN 73 6205 84 --- (b) The buckling stress is given as a product of the overall orthotropic plate buckling and the local buckling of plate subpanels. The formulation considers a transition from the elastic to the inelastic range.

JAPAN

JRA 80 --- (b) The procedure based on elastic analysis and ultimate strength tests assures that the flange with longitudinal and transverse stiffeners does not buckle before yielding. The procedure (Eqs. 4J.5 to 4J.11) is noticeably simpler than the procedure of ČSN 73 6205 84.

NORTH AMERICA

(A) AASHTO 87 and CSA S6 M89 84a --- the two specifications for highway bridges use the same procedures for unstiffened, longitudinally stiffened, and longitudinally and transversely stiffened compression flanges to ensure against buckling. (b)

(B) FHWA TS 80 --- the only specification dedicated to box girders, provides a nomograph for determining the ultimate strength of longitudinal stiffeners treated as columns. The nomograph is based on the results of tests and numerical analysis of a wide range of geometries and material strengths. It incorporates direct interaction between plate and column postbuckling behavior. (u)

WEST EUROPE

Both the orthotropic plate and the column strength approaches are represented among the specifications of the region.

(A) The more modern specifications (BS 5400 82, BS 5950 85, SIA 161 79) treat the compression flange as a series of columns each with a cross section which includes the postbuckling effective width of the plate.

(B) The representative older specification (NBN B51002 88) limits the strength by linear elastic buckling with an inelastic transition.

WHY ARE THERE DIFFERENCES?

The preference for one or the other concept is difficult to explain. As for other topics, seldom is there a discernible rational explanation. Almost as a rule, the more recent the provision, the less conservative it is.

Symbol List for Chapter 4 -- Plate & Box Girders

A_f = area of compression flange or of the average of tension and compression flanges (4b.E), (4b.N) & (4c.N)
$A_{f\ell}$ = area of the smaller flange (4b.A)
A_p = area of patch load on edge of a web (4b.E)
$A_{s\ell}$ = area of longitudinal stiffener (4c.E) & (4c.J)
A_{st} = area of intermediate transverse stiffener (4b.J) & (4b.N)
A_w = area of web plate (4b.E), (4b.J), (4b.N) & (4a.W)
a = distance between transverse stiffeners (in web or in flange) (4b.N), (4c.J)&(4c.N)
a_r = ratio of web area to compression flange area (4b.N)
b = width of flange plate between longitudinal stiffeners or webs (4c.A) & (4c.E)
b = flange width (4b.J), (4b.W), (4c.J) & (4c.N)
b_{bf} = width of patch load on the edge of the web dispersed through flange thickness (4b.A)
b_e = effective width of flange plate or plate subpanels, also of web in compression (4b.A), (4b.W) & (4c.W)
b_{e1} = effective width for web part adjacent to compression flange (4b.W)
b_{e2} = effective width for web part adjacent to neutral axis (4b.W)
b_i = spacing of longitudinal stiffeners in compression flange (4c.E),(4c.N)&(4c.W)
C = modification constant for different position of web panel (4b.E)
C = ratio of critical to yield shear stresses (4b.N)
C_s = slenderness requirement for longitudinal flange stiffeners (4c.N)
c = actual correction factor for plate buckling (4b.W)
c^* = required correction factor for plate buckling (4b.W)
d = smaller distance of width or depth of panels partitioned by transverse and/or longitudinal stiffeners (4b.E) & (4b.J)
d = stiffener depth (4c.N)
e_o = initial out-of-straightness (4b.E) & (4c.W)
e_{st} = distance of the centroid of a combined cross section to flange mid-thickness (4c.E)
F = edge load (4b.E)
F_u = ultimate edge load (4b.E)
g = parameter for the width and inclination of tension-field band (4b.W)
H = horizontal force on end stiffener (4b.W)
h = clear distance between top and bottom flanges (4b.C),(4b.E),(4b.J)&(4b.N)
h_c = twice the clear distance from the neutral axis to the compression flange (4a.A), (4a.N), (4b.N) & (4b.W)
h_e = equivalent depth (4b.E)
h_i = depth of the i-th web subpanel (4b.W)
h_1 = distance between longitudinal stiffener and compression edge of web (4b.C)
I = moment of inertia of full girder section (4b.E) & (4b.1)
I_s = moment of inertia of transverse or longitudinal stiffeners (4b.E), (4b.J) & (4b.W)
I_s^* = required moment inertia of stiffeners, transverse or longitudinal (4b.W)
$I_{s\ell}$ = moment of inertia of longitudinal stiffener (4b.C), (4b.E), (4b.J), (4b.N), (4c.E), (4c.J) & (4c.N)
I_{st} = moment of inertia of transverse stiffener (4b.C), (4b.E), (4b.N)&(4c.J)

I_w = moments of inertia of web plate alone (4b.1)
K = loading type modification coefficient for web slenderness limitations (4a.E)
K_b = strength multiplier for bending (4b.W)
K_c = strength multiplier for compression (4b.W)
K_q = strength multiplier for shear (4b.W)
k = plate buckling coefficient (4b.E)
k = a factor (4a.W)
k_a = buckling coefficient for axial force (4b.E)
k_b = buckling coefficient for bending (4b.E)
k_c = flange plate buckling coefficient (4c.N)
k_{loc} = local edge load buckling coefficient (4b.E)
k_{red} = reduced buckling coefficient for combined state of stress (4b.E)
k_s = parameter for design of flange stiffeners (4c.W)
k_t = parameter for design of flange stiffeners (4c.W)
k_v = buckling coefficient for shear (4b.E) & (4b.J)
k_1 = flange plate buckling coefficient (4c.N)
k_2 = flange plate buckling coefficient (4c.N)
L_b = unbraced length (4b.J) & (4b.N)
L_e = effective length (4b.A) & (4b.E)
M_{all} = allowable moment (4b.1)
M_{cr} = buckling moment (4b.W)
M_d = design flexural strength of plate girder (4b.N)
M_{eq} = equivalent moment (4b.J)
M_k = fictitious "disturbing" bending moment of transverse stiffener (4b.E)
M_n = nominal bending strength (4b.N)
M_{pc} = plastic moment of flange reduced for axial force in flange due to overall girder moment (4b.W)
M_{pc} = reduced plastic moment of flange allowing for longitudinal force N_f (4b.W)
M_{pc1} = reduced plastic moment of top flange (4b.W)
M_{pc2} = reduced plastic moment of bottom flange (4b.W)
M_{pf} = plastic moment of the smaller flange (4b.W)
M_{pw} = plastic moment of web (4b.W)
M_R = bending strengths of effective flanges alone (4b.W)
M_u = ultimate moment capacity (4b.W), (4c.A) & (4c.N)
M_w = maximum moment in panel carried by web (4b.A) & (4b.W)
M_w = ultimate moment (4b.E)
M_y = yield moment (4b.1)
M_u^* = moment capacity under shear and bending (4b.A)
M_w^* = design moment in web (4b.A)
M_1 = smaller moment at ends of unbraced length (4b.N)
M_2 = larger moment at ends of unbraced length (4b.N)
m = ratio of web yield stress to flange yield stress or to buckling stress (4b.N)
m = amplification factor for optimum rigidity (4b.W)
m_f = effect factor of flange plastic moment on ultimate shear strength (4b.W)
N_{f1} = actual axial force in top flange (4b.W)
N_{f2} = actual axial force in bottom flange (4b.W)
N_{fp1} = fully-plastified axial force in top flange (4b.W)
N_{fp2} = fully-plastified axial force in bottom flange (4b.W)
N_{st} = axial load on transverse stiffener (4b.E)

N_u = ultimate axial capacity (4b.E)
N_w = ultimate axial force of web panel (4b.E)
N_{wo} = nominal axial load capacity of web panel (4b.A)
N_y = ultimate force (4b.E)
N_w^* = design axial force of web panel under shear and bending (4b.A)
n = number of longitudinal stiffeners in compression flange (4b.J) & (4c.N)
R = width-to-thickness ratio of compression flange of box girder (4c.J)
R_b = nominal buckling capacity of transversely stiffened web in bearing alone (4b.A)
R_d = design strength of beam material (4b.E)
R_{df} = design strength of flange material (4b.E)
R_{dw} = design strength of web material (4b.E)
R_e = hybrid girder factor (4b.N)
R_{PG} = coefficient to account for postbuckling of web in bending (4b.N)
R_{sb} = nominal buckling capacity of intermediate stiffener (4b.A)
R^* = transverse patch load applied to flange (4b.A)
r = radius of gyration (4b.J) & (4b.N)
r_f = radius of gyration of flange about plane of web (4b.W)
r_{se} = radius of gyration of effective section of longitudinal stiffener about centroidal axis parallel to flange plate (4c.W)
r_T = radius of gyration of compression flange plus one-third of web portion under compression (4b.N)
r_y = radius of gyration of compression portion of cross section about web (4b.N)
S = section modulus of cross section (4a.W) & (4c.N)
S_c = section modulus with respect to compression flange (4b.N)
S_{eff} = effective section modulus (4b.A)
$S_{f,pl}$ = section modulus of "effective flange" section (4b.E)
S_L = section modulus of longitudinal (4b.N)
S_{min} = minimum section modulus of Tee section consisting of compression flange plate and portion of web plate (4b.E)
$S_{s\ell}$ = section modulus of longitudinal stiffener (4b.N)
S_{st} = section modulus of transverse stiffener (4b.N)
S_T = section modulus of transverse stiffener (4b.N)
S_t = section modulus with respect to tension flange (4b.N)
S_w = section modulus of web plate alone (4b.A), (4b.E)&(4b.W)
S_1 = unreduced section modulus with respect to compression flange (4b.E)
S_{1e} = effective section modulus (4b.E)
S_2 = unreduced section modulus with respect to tension flange (4b.E)
s = patch load width (4b.E)
s = anchorage length of tension field along flanges (4b.W)
s_1 = anchorage length of tension field along compression flange (4b.W)
s_2 = anchorage length of tension field along tension flange (4b.W)
t = thickness of web plate (4b.C), (4b.E) & (4b.N)
t_e = effective web thickness (4b.W)
t_f = thickness of flange plate (4b.W), (4b.W), (4c.A), (4c.E) & (4c.J)
t_o = thickness of flat-plate stiffener or of the web of tee or angle stiffener (4c.N)
V = shear force (4b.C), (4b.E) & (4b.N)
V_B = shear strength due to buckling (4b.N)
V_d = design shear strength (4b.N)

V_i = shear carried by the i-th web subpanel (4b.W)
V_n = nominal shear strength (4b.N)
V_o = portion of shear force carried by stiffener prior to buckling of adjacent panels (4b.E)
V_T = shear strength due to postbuckling tension-field action (4b.N)
V_u = ultimate shear strength (4b.A), (4b.E), (4b.N) & (4b.W)
V_w = ultimate shear carried by web (4b.E) & (4b.W)
V_u^* = shear capacity under shear and bending (4b.A)
V_y = yield stress of web (4b.A) & (4b.N)
w = tributary plate width (4c.N)
y_b = basic tension field strength (4b.W)
y_{Bs} = distance from centroid of effective stiffener section to neutral axis of cross section of girder (4c.W)
y_o = distance from centroid of effective stiffener section to point on stiffener furthest from plate (4c.W)
y_z = distance from centroid of effective stiffener section to mid-plane of flange plate (4c.W)
Z = plastic modulus of cross section (4b.A)
z_1 = distance from centroid to extreme fiber of the top flange (4b.E)
z_{1w} = distance from centroid to top edge of web (4b.E)
z_2 = distance from centroid to extreme fiber of the bottom flange (4b.E)
z_{2w} = distance from centroid to bottom edge of web (4b.E)
α = side ratio of plate panel of web or of flange (4b.N), (4b.W) & (4c.E)
α_f = flange restraint factor (4b.A)
γ = relative flexural rigidity of stiffener (4b.W)
γ_c = resistance coefficient for compression (4b.E)
γ_{f3} = resistance coefficient for bending (4b.W)
γ_m = material resistance coefficient (4b.E)
γ^* = optimum stiffener rigidity according to linear buckling theory (4b.E)
γ_m^* = optimum stiffener rigidity according to linear buckling theory for bending (4b.E)
γ_n^* = optimum stiffener rigidity according to linear buckling theory for compression (4b.E)
γ_v^* = optimum rigidity according to linear buckling theory for shear (4b.E)
δ = modification factor for area of transverse stiffener (4b.A)
θ = inclination of panel diagonal (4b.W)
κ = modification factor for effect of postbuckling and initial imperfection on stiffen or strength (4b.E)
λ = width-to-thickness ratio (4c.E)
λ_i = slenderness ratio of i-th sub-panel (4b.W)
$\bar{\lambda}_o$ = relative slenderness of web plate (4b.E)
λ_v = equivalent slenderness for web in shear (4b.N)
λ_w = equivalent slenderness of web plate (4b.W)
μ = ratio of longer side of web panel (or subpanel) to shorter side (4b.E)
ν = Poisson's ratio (4c.E)
ξ = factor which allows for effect of postbuckling behavior and initial imperfections (4b.E)
ρ = porportion of direct stress redistributed from web to flanges (4b.W)

σ = allowable bending stress for girder subjected to shear and bending (4b.N)
σ = normal stress (4a.N), (4b.C), (4b.W) & (4c.N)
σ_a = axial stress (4b.E), (4b.N), (4b.W) & (4c.W)
σ_{acr} = axial buckling stress (4b.E) & (4b.W)
σ_{av} = average of normal stresses at top and bottom edges (4b.N)
σ_B = maximum compressive bending stress (4b.E)
σ_b = bending stress (4b.N) & (4b.W)
σ_b = yield or the lateral-torsional buckling stress (4b.)
σ'_b = allowable stress for pure bending (4b.N)
σ_{ba} = allowable bending stress (4a.N)
σ_{bcr} = bending buckling stress (4b.E) & (4b.W)
σ_c = compressive stress (4b.J) & (4b.W)
σ_{cr} = elastic buckling stress of plate (4b.N) & (4c.N)
σ_{cr} = normal elastic buckling stress when normal and shear stresses are coincident (4b.W)
σ°_{cr} = elastic buckling stress when normal and shear stresses act separately (4b.W)
$\sigma_{cr,f}$ = critical stress for lateral buckling of flange (4b.W)
$\sigma_{cr,1}$ = bending buckling stress of sub-panel between compression flange and longitudinal stiffener (4b.E)
$\sigma_{cr,2}$ = bending buckling stress of sub-panel between tension flange and longitudinal stiffener (4b.E)
σ_E = Euler buckling load of pin-ended column (4b.W)
σ_{loc} = local stress due to a local concetrated load (4b.E)
$\sigma_{loc,cr}$ = local buckling stress for a local concentrated load (4b.E)
$\sigma_{loc,cr,1}$ = local buckling stress of sub-panel between compression flange and longitudinal stiffener for a local concentrated load (4b.E)
$\sigma_{loc,cr,2}$ = local buckling stress of sub-panel between tension flange and longitudinal stiffener for a local concentrated load (4b.E)
$\sigma_{loc,2}$ = ultimate local stress under combined loading (4b.E)
σ_{max} = maximum compressive stress in stiffened flange panel (4c.N)
σ_{min} = minimum compressive stress in stiffened flange panel (4b.W)
σ_{rc} = residual compressive stress in flange (4a.N)
σ_{red} = effective design stress for the actual combined stresses (4b.E)
σ_t = tensile bending stress at extreme fiber of web panel (4b.E) & (4b.J)
σ_{tf} = strength of tension field (4b.N) & (4b.W)
σ_u = ultimate stress of flange (4c.N)
σ'_u = ultimate stress of flange modified for shear (4c.N)
σ_w = design working stress of material (4b.E)
σ_{yc} = yield stress of compression flange (4b.W)
σ_{yf} = yield stress of flange (4a.N) & (4a.W)
σ_{ys} = yield stress of stiffener (4c.W)
σ_{yst} = yield stress of stiffener (4b.N)
σ_{yt} = yield stress of tension flange (4b.N)
σ_{yw} = yield stress of web plate (4b.E) & (4b.N)
σ_1 = compressive stress at extreme fiber of web panel (4b.E)
σ_{2w} = lower compressive or tensile bending stress in web panel (or subpanel) (4b.N)
τ = average shearing stress in web panel (4b.C)

τ = shear buckling stress under combined stress condition (4b.N)
τ_a = allowable shear stress for pure shear (4b.N)
τ_{ave} = average shear stress (4c.N)
τ_b = shear capacity (4b.W)
τ_{cr} = shear buckling stress (4b.E) & (4b.W)
τ_{cr} = elastic shear buckling stress when normal and shear stresses are coincident (4b.W)
τ°_{cr} = elastic shear buckling stress when normal and shear stresses act separately (4b.W)
$\tau_{cr,red}$ = reduced critical shear buckling stress (4a.W)
$\tau_{cr,1}$ = shear buckling stress of sub-panel between compression flange and longitudinal stiffener (4b.E)
$\tau_{cr,2}$ = shear buckling stress of sub-panel between tension flange and longitudinal stiffener (4b.E)
τ_e = elastic shear buckling stress (4a.W) & (4b.W)
τ_{max} = maximum shear stress (4c.N)
τ_{min} = lowest shear buckling stress among web subpanels under combined stresses (4b.N)
τ_w = allowable shear stress (4b.E)
τ_y = shear yield stress of material (4a.W)
τ_1 = in-plane shear stress in the flange plate due to torsion on girder (4c.W)
ϕ = inclination of tension field (4b.W)
ϕ_v = resistance factor for shear (4b.N)
φ_b = buckling reduction coefficient (4b.E)
χ = modification factor of optimum rigidity of stiffeners (4b.E)
χ_m = modification factor of optimum rigidity of stiffeners for bending (4b.E)
χ_n = modification factor of optimum rigidity of stiffeners for axial compression (4b.E)
χ_s = parameter varies between 3 and 7 depending on location of longitudinal stiffener (4b.E)
χ_v = modification factor of optimum rigidity of stiffeners for shear (4b.E)
ψ = normal stress ratio with respect to edge under maximum compression (4b.W)
ψ_{loc} = reduction buckling coefficient of local compressive load (4b.E)
ψ_{vf} = effect of flanges on ultimate shear capacity (4b.E)
ψ_{1n} = global flange buckling coefficient (4c.E)
ψ_{2n} = local plate buckling coefficient (4c.E)

Chapter 5

Beam–Columns

TOPICAL EDITOR
David A. Nethercot, The University of Nottingham

CHAPTER EDITORS
Mark A. Bradford, University of New South Wales
Ioan A. Caraba, Polytechnic Institute of Timisoara
Shao-Fan Chen, Xian Institute of Metallurgy & Construction Engineering
Wai-Fah Chen, Purdue University
Bao Kang He, Xian Institute of Metallurgy & Construction Engineering
Miklos Ivanyi, Technical University of Budapest
Dan Mateescu, Polytechnic Institute of Timisoara
Takeshi Nakamura, Kyoto Institute of Technology
David A. Nethercot, The University of Nottingham
Zu-Yan Shen, Tong-Ji University
Koichi Takanashi, The University of Tokyo

5a. Cross-Sectional Capacity

SPECIFICATION PROVISIONS & EXPLANATION

AUSTRALIA

CHAPTER EDITOR
Mark A. Bradford, University of New South Wales

AS DR87164 87

For a beam-column subjected to an axial force N^* and a moment M^*, the moment M^* must satisfy

$$M^* < \phi M_u \tag{5a.A1}$$

where

M_u = section capacity, the in-plane member capacity or the out-of-plane member capacity, whichever is the least

ϕ = capacity factor is equal to 0.9

Corresponding conditions must be satisfied for a beam-column with axial force N^* and bi-axial moments M_x^* and M_y^*.

The section capacity requirements must be satisfied at each cross-section using the values of N^*, M_x^*, M_y^* for that cross-section. The member capacity requirements must be satisfied for the member as a whole, using the maximum values of N^*, M_x^* and M_y^* in the member length.

The values used for N^*, M_x^* and M_y^* are determined by using an appropriate method of analysis, depending on the relative importance of stability effects. These methods include linear elastic or rigid plastic analysis, amplified linear elastic analysis, and second order elastic analyses incorporating either or both member and frame stability effects.

Section Capacity

For a compact doubly symmetric I-section member, the nominal section moment capacity in uniaxial bending is given by

$$M_{rx} = 1.18 M_{sx}(1 - N^*/\phi N_s) \leq M_{sx} \tag{5a.A2}$$

for bending about the major (x) principal axis, and by

$$M_{ry} = 1.19\, M_{sy}(1 - (N^*/\phi N_s)^2) \leq M_{sy} \tag{5a.A3}$$

for bending about the minor (y) principal axis, where M_s and N_s are the nominal beam and column section capacities.

For any other member, the nominal section moment capacity in uniaxial bending is given by

$$M_r = M_x^* \,(1 - N^*/\phi N_s) \tag{5a.A4}$$

A compact doubly symmetric I-section member in biaxial bending is required to satisfy

$$\left(\frac{M_x^*}{\phi M_{rx}}\right)^{\gamma_s} + \left(\frac{M_y^*}{\phi M_{ry}}\right)^{\gamma_s} \leq 1 \tag{5a.A5}$$

where γ_s is $1.4 + N^*/\phi N_s \leq 2$.

Any other member in biaxial bending is required to satisfy

$$\frac{N^*}{\phi N_s} + \frac{M_x^*}{\phi M_{sx}} + \frac{M_y^*}{\phi M_{sy}} \leq 1 \tag{5a.A6}$$

CHINA

CHAPTER EDITORS

Shao-Fan Chen, Xian Institute of Metallurgy & Construction Engineering
Bao Kang He, Xian Institute of Metallurgy & Construction Engineering
Zu-Yan Shen, Tong-Ji University

GBJ 17 88

Beam-column section capacity is checked by

$$\frac{N}{A_n} \pm \frac{M_x}{\gamma_x W_{nx}} \pm \frac{M_y}{\gamma_y W_{ny}} \le f \tag{5a.C1}$$

where

A_n = net sectional area
f = design normal strength of steel
N, M_x, M_y = design value of axial compression and bending moments (factored)
W_{nx}, W_{ny} = net section modulus about x- and y-axis, respectively
γ_x, γ_y = plastic adaptation coefficient, equal to 1.05 and 1.20 for strong and weak axis bending respectively, but taken as 1.0 for members subject to direct dynamic load.

For plastic design, use

$$\left.\begin{array}{lll} \text{for } \dfrac{N}{A_n f_p} > 0.13 & & M_x \le 1.15\left(1 - \dfrac{N}{A_n f_p}\right) W_{pnx} f_p \\ \text{for } \dfrac{N}{A_n f_p} \le 0.13 & & M_x \le W_{pnx} f_p \end{array}\right\} \tag{5a.C2}$$

where

f_p = design strength of steel for plastic design, taken to be 0.9f
W_{pnx} = net plastic section modulus about the x-axis

The compressive force N shall not exceed $0.6A_n f_p$.

The local buckling of plate elements is prevented by providing the limitation to b/t and h_o/t_w ratios of plates (see Fig. 5a.C1) and the specification provisions are

For flanges of I-sections

$$b/t \le 15\sqrt{235/f_y} \tag{5a.C3}$$

For webs of I-sections

$$\left.\begin{array}{lll} \text{for } 0 \le \alpha_o \le 1.6 & & h_o/t_w \le (16\alpha_o + 0.5\lambda + 25)\sqrt{235/f_y} \\ \text{for } 1.6 < \alpha_o \le 2.0 & & h_o/t_w \le (48\alpha + 0.5\lambda - 26.2)\sqrt{235/f_y} \end{array}\right\} \tag{5a.C4}$$

where

α_o = $(\sigma_{max} - \sigma_{min})/\sigma_{max}$

λ = Kl/r is the slenderness ratio of members in the plane of bending moment, taking $\lambda = 30$ for $\lambda < 30$ and $\lambda = 100$ for $\lambda > 100$ in Eqs. 5a.C4

σ_{max} = maximum compressive edge stress of web

σ_{min} = stress at opposite side of the web (taking compression as positive and tension as negative)

For box-section members

$$b_o/t \leq 40\sqrt{235/f_y} \tag{5a.C5}$$

The ratio h_o/t_w is determined by using 0.8 times the value calculated by Eqs. 5a.C4, but not less than $40\sqrt{235/f_y}$.

The specification also stipulates that if the ratio h_o/t_w of the webs in I- and box-section members does not meet the above requirements, and that no longitudinal stiffener is provided, then an effective width of $20t_w\sqrt{235/f_y}$ at each edge of the web can be used in design.

For plastic design the width-thickness ratio limitation is

Flange of I-section

$$b/t \leq 9\sqrt{235/f_y} \tag{5a.C6}$$

Web of I- and box section

for $\frac{N}{A f_p} < 0.37$

$$h_o/t_w \leq \left(72-100\,\frac{N}{A f_p}\right)\sqrt{235/f_y}$$

for $\frac{N}{A f_p} \geq 0.37$ (5a.C7)

$$h_o/t_w \leq 35\sqrt{235/f_y}$$

Flange of box section

$$b_o/t \leq 30\sqrt{235/f_y} \tag{5a.C8}$$

TBJ 2 85

Section capacity is checked by

$$\frac{N}{A} \pm \left(\frac{M_x}{W_x} + \frac{M_y}{W_y}\right)\cdot\frac{1}{C} \leq [\sigma] \tag{5a.C9}$$

$$C = 1 + 0.3\,\frac{\sigma_{m2}}{\sigma_{m1}} \leq 1.15$$

where

A = gross sectional area

N, M_x, M_y = non-factored axial compression and bending moments

$[\sigma]$ = allowable normal stress of steel

σ_{m1}, σ_{m2} = larger and smaller stresses respectively due to bending moment M_x, M_y

COMMENTARY

Equation 5a.C1 allows partial plastification of beam-column sections except those under direct dynamic load. Eq. 5a.C9 also allows partial plastification but to a lesser extent.

The limiting ratio h_o/t_w of the web (Eq. 5a.C4) in I-section beam-columns is based on the analysis of a simply supported rectangular plate under combined compressive, bending, and shear stresses. In case a beam-column fails by in-plane instability, some part of its web can be yielded. For simplicity, a yielded zone of one quarter of the web depth is assumed. The shear stress is taken to be 0.3 times the maximum stress caused by bending.

Making use of Bijlaard's theory of plastic buckling of simply supported plates, a parametric study of h_o/t_w ratio leads to the following simplified expressions

for $0 \leq \alpha_o \leq 1.6$

$$h_o/t_w = 16\alpha_o+50$$

$$(5a.C10)$$

for $1.6 \leq \alpha_o \leq 2.0$

$$h_o/t_w = 48\alpha_o-1$$

Since the depth of the yielded zone is actually a function of the slenderness ratio λ of the member, these expressions are altered in Eqs. 5a.C4. This revision has the further advantage of coordinating the h_o/t_w ratio with that of axially compressed members.

EAST EUROPE

CHAPTER EDITORS
Ioan A. Caraba, Polytechnic Institute of Timisoara
Miklos Ivanyi, Technical University of Budapest
Dan Mateescu, Polytechnic Institute of Timisoara

CMEA 83

Members subject to compression and bending may be checked by interaction equations corresponding to the ideal bar. Design rules take into account the imperfections of the actual members.

The component walls of bars are assumed to satisfy the limit permissible values for plate slenderness ratios, thus, for the web of double flange sections, the limit plate slenderness is given by

$$\frac{h_o}{t_w} \leq 65\sqrt{\frac{2100}{R_o}} \tag{5a.E1}$$

where

h_o, t_w = dimensions of the web cross-section
R_o = (design strength) minimum specified yield stress

For flanges with unstiffened edges, the plate slenderness is given by

$$\frac{b'}{t_f} \leq 15\sqrt{\frac{2100}{R_o}} \tag{5a.E2}$$

where b' is the outstanding width of the flange (Fig. 5a.E1)

If the wall slendernesses satisfy Eqs. 5b.E1 and 5a.E2, it can be considered that the entire cross section is active; that is, the bearing capacity of the member is not influenced by local buckling.

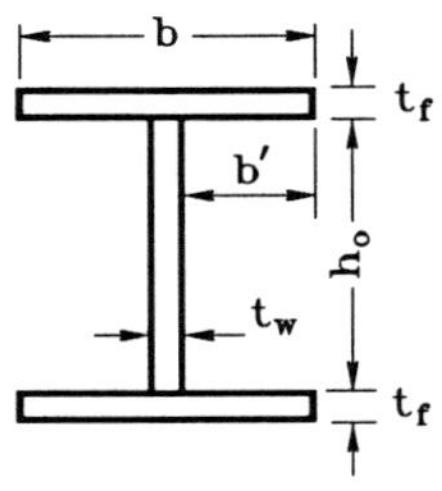

Fig. 5a.E1 Definition of terms

JAPAN

CHAPTER EDITORS
Takeshi Nakamura, Kyoto Institute of Technology
Koichi Takanashi, The University of Tokyo

AIJ 90

Uniaxial Bending

For H-shaped and rectangular hollow members subjected to axial compression force and bending about the strong axis

for $\frac{P}{\phi_p P_y} \leq 0.15$

$$\frac{M}{\phi_p M_p} \leq 1.0 \tag{5a.J1}$$

for $\frac{P}{\phi_p P_y} > 0.15$

$$\frac{P}{\phi_p P_y} + 0.85 \frac{M}{\phi_p M_p} \leq 1.0 \tag{5a.J2}$$

For H-shaped members subjected to axial compression and bending about the weak axis.

for $\frac{P}{\phi_p P_y} \leq 0.4$

$$\frac{M}{\phi_p M_p} \leq 1.0 \tag{5a.J3}$$

for $\frac{P}{\phi_p P_y} > 0.4$

$$\left(\frac{P}{\phi_p P_y}\right)^2 + 0.84 \frac{M}{\phi_p M_p} \leq 1.0 \tag{5a.J4}$$

For circular hollow members

for $\frac{P}{\phi_p P_y} \leq 0.2$

$$\frac{M}{\phi_p M_p} \leq 1.0 \tag{5a.J5}$$

for $\frac{P}{\phi_p P_y} > 0.2$

$$\frac{P}{\phi_p P_y} + 0.8 \frac{M}{\phi_p M_p} \leq 1.0 \tag{5a.J6}$$

where

M = bending moment
M_p = plastic moment
P = axial compression force
P_y = axial compression capacity for yielding
ϕ_p = resistance factor for tension

Uniaxial Bending for Non-compact Sections

$$\frac{P}{\phi_c P_{ux}} + \frac{M}{\phi_b M_c} \leq 1.0 \tag{5a.J7}$$

where

$M_c = F_{cr} Z_e$

P_y = axial compression capacity for yielding

Z_e = elastic section modulus

ϕ_b = resistance factor for bending

ϕ_c = resistance factor for compression

(A) H-shaped sections

$$P_{ux} = \frac{P_y}{0.6 + 1.6 \frac{F_{yf}}{E}(\frac{b}{t_f})^2 + 0.15 \frac{F_{yw}}{E}(\frac{d}{t_w})^2} \qquad (5a.J8)$$

F_{cr} = the smaller value of ${}_fF_{cr}$ and ${}_wF_{cr}$

where

b = half of the width of flange plate

d = depth of web plate

F_{yf} = yield stress of flanges

F_{yw} = yield stress of webs

$${}_fF_{cr} = (67 - \frac{b}{t_f}\sqrt{F_y}\,) \frac{F_y}{50} \qquad (5a.J9)$$

$${}_wF_{cr} = (540 - \frac{d}{t_w}\sqrt{F_y}\,) \frac{F_y}{440} \qquad (5a.J10)$$

t_f = thickness of flange plate

t_w = thickness of web plate

(B) Hot-formed, welded rectangular hollow sections

$$P_{ux} = \frac{P_y}{0.71 + 0.17\frac{F_y}{E}(\frac{B}{t})^2} \qquad (5a.J11)$$

$$F_{cr} = (124.5 - \frac{B}{t}\sqrt{F_y}) \frac{F_y}{32.5} \qquad (5a.J12)$$

Biaxial Bending

For H-shaped section members

$$(\frac{M_x}{M_{px}})^2 + (\frac{M_y}{M_{py}}) \leq 1 \qquad (5a.J13)$$

where

M_{px} = plastic moment under axial force and bending with respect to the strong axis

M_{py} = plastic moment under axial force and bending with respect to the weak axis

M_x = applied moment with respect to the strong axis

M_y = applied moment with respect to the weak axis

JRA 90

Biaxial Bending

$$ {}_{t}f_{a} + {}_{t}f_{bx} + {}_{t}f_{by} \leq {}_{t}F_{a} \tag{5a.J14}$$

$$ - {}_{t}f_{a} + {}_{c}f_{bx} + {}_{c}f_{by} \leq {}_{p}F_{a} \tag{5a.J15}$$

where

$_{p}F_{a}$ = permissible compressive stress for local buckling
$_{t}F_{a}$ = permissible tensile stress
$_{c}f_{bx}, {}_{c}f_{by}$ = applied compressive stress due to bending about the strong/weak axis
$_{t}f_{a}$ = applied axial tensile stress
$_{t}f_{bx}, {}_{t}f_{by}$ = applied tensile stress due to bending about the strong/weak axis

NORTH AMERICA

CHAPTER EDITOR
Wai-Fah Chen, Purdue University

The design of beam-columns in North American is facilitated by using interaction equations. Both linear and nonlinear forms of design interaction equations are used in AISC and CSA Specifications.

The linear interaction equation has the general form

$$\frac{P}{P_u} + \frac{M_{max}}{M_u} = 1.0 \tag{5a.N1}$$

Depending on the design philosophy, the ultimate axial capacity of the member P_u can be represented by the ASD curve, PD curve, LRFD curve and the CSA curve. The quantity M_{max} is the second-order moment or the amplified moment in the member and is represented by

$$M_{max} = A_F M_o \tag{5a.N2}$$

where

A_F = P-δ moment amplification factor

$$= \frac{C_m}{1 - \frac{P}{P_{ek}}} \tag{5a.N3}$$

C_m = equivalent uniform moment factor
M_o = maximum first-order moment
P_{ek} = Euler buckling load considering the end restrained condition.

The general interaction equation for the stability limit loads of the beam-column has the form

$$\frac{P}{P_u} + \frac{C_m M_o}{M_u(1 - P/P_{ek})} = 1.0 \tag{5a.N4}$$

Cross-Section Capacity

If the slenderness of the member is very low or zero, no instability will occur and the nonlinear interaction equation of the form

$$(\frac{P}{P_y})^a + \frac{M_o}{bM_p} = 1.0 \tag{5a.N5}$$

is used for the ultimate strength interaction curve of various cross-sections. The constants a and b define the shape of the interaction curve.

Member Resistance to Bi-axial Bending

When bending occurs in both axes, the interaction equation takes the form

$$\frac{P}{P_u} + \frac{C_{mx} M_{ox}}{M_{ux}(1-P/P_{ex})} + \frac{C_{my} M_{oy}}{M_{uy}(1-P/P_{ey})} = 1.0 \tag{5a.N6}$$

in which the subscripts x and y refer to the bending action about the x and y axes of the member, respectively.

AASHTO 87

Allowable stress design is essentially the same as the linear interaction equations used in the AISC ASD 78.

For LRFD design the section strength is given by

$$\frac{P}{0.85A_sF_y} + \frac{M}{M_p} \leq 1.0 \qquad (5a.N7)$$

where A_s is the effective column cross section.

AISC ASD 78

The AISC ASD format is a stress-based design format. As a result, allowable stresses and service loads rather than ultimate strengths and maximum loads are used as the basis for design. The linear interaction equations are converted to stresses and a factor of safety applies to them to bring them into the service load range.

If yielding of material controls, the interaction equation is

$$\frac{f_a}{0.60F_y} + \frac{f_{bx}}{F_{bx}} + \frac{f_{by}}{F_{by}} \leq 1.0 \qquad (5a.N8)$$

This equation pertains to cases in which yielding rather than instability controls the design. This happens at support locations in braced frames and for members with low slenderness ratios in unbraced frames. The number 0.6 (=1/1.67) in the denominator of the first term that reflects the axial force effect is a safety factor applied to the CRC curve in order to obtain the AISC ASD curve at KL/r = 0.

The AISC PD format provides two interaction equations for the design of beam-columns, one for yielding and the other for instability. If yielding controls,

$$\frac{P}{P_y} + \frac{M}{1.18M_p} \leq 1.0 \qquad (5a.N9)$$

where

M_p = full plastic moment capacity of the section
P_y = yield load of the section

P and M are the factored axial force and moment (service loads time load factor). The load factor is 1.7 if only gravity loads are acting and is 1.3 if wind or earthquake loads are acting with gravity loads.

AISC LRFD 86

The bi-linear interaction equations given by Eqs. 5d.N6 and 5d.N7 can also be used for a section capacity check. For I-shaped sections the following nonlinear interaction equations are provided in the Appendix of the AISC LRFD 86 as alternatives

If yielding controls

$$\left(\frac{M_{ox}}{\phi_b M'_{px}}\right)^{\zeta} + \left(\frac{M_{oy}}{\phi_b M'_{py}}\right)^{\zeta} \leq 1.0 \qquad (5a.N10)$$

where

$$M'_{px} = 1.2\ M_{px}[1-(P/P_y)] \leq M_{px}$$

$$M'_{py} = 1.2\ M_{py}[1-(P/P_y)^2] \leq M_{py}$$

$$\zeta = 1.6 - \frac{P/P_y}{2[\ln(P/P_y)]} \qquad \text{for} \quad 0.5 \leq b_f/d \leq 1.0$$

b_f = flange width
d = member depth

AREA 87

Section Strength

$$\frac{f_a}{20{,}000} + \frac{f_{b1}}{F_{b1}} + \frac{f_{b2}}{F_{b2}} \leq 1.0 \qquad (5a.N11)$$

where

F_a, F_{b1}, F_{b2} = allowable axial stress and bending stresses about axes 1-1 and 2-2, respectively
f_a, f_{b1}, f_{b2} = computed axial stress and bending stresses about axes 1-1 and 2-2, respectively

CSA S16.1 M89 84b

The general CSA format is a limit states design format. Two linear interaction equations are given in the design of beam-columns, one for yielding and the other for instability. If yielding controls

$$\frac{P}{P_y} + \frac{M_{fx}}{M_{rx}} + \frac{M_{fy}}{M_{ry}} \leq 1.0 \qquad (5a.N12)$$

For I–shaped members, if yielding controls, the following two equations are checked.

$$\frac{M_{fx}}{M_{px}} + \frac{M_{fy}}{M_{py}} \leq 1.0 \qquad (5a.N13)$$

$$\frac{P}{P_y} + \frac{0.85 M_{fx}}{M_{px}} + \frac{0.60 M_{fy}}{M_{py}} \leq 1.0 \qquad (5a.N14)$$

Sections are classified as follows:

(a) Class 1 sections (plastic design sections) will permit attainment of the plastic moment and subsequent redistribution of bending moment;

(b) Class 2 sections (compact sections) will permit attainment of the plastic moment, but need not allow for subsequent moment redistribution;

(c) Class 3 sections (noncompact sections) will permit attainment of the yield moment; and

(d) Class 4 sections (slender sections) will generally have local buckling of elements in compression as the limit state of structural capacity.

Member adequacy is assumed by application of the dual requirements, ultimate capacity of the cross section at member ends and stability of the member as a whole.

Linear Interaction Equation

In essence, any technique that would demonstrate the cross section's capacity to maintain equilibrium under the action of the applied force and moments is acceptable.

For Class 1 and Class 2 sections of I-shaped members, it is required that

$$\frac{M_{fx}}{M_{rx}} + \frac{M_{fy}}{M_{ry}} \leq 1.0 \tag{5a.N15}$$

and

$$\frac{C_f}{C_r} + 0.85\,\frac{M_{fx}}{M_{rx}} + 0.60\,\frac{M_{fy}}{M_{ry}} \leq 1.0 \tag{5a.N16}$$

$$C_r = \phi A F_y \tag{5a.N17}$$

where

$M_{rx} = M_{px}$ for class 1 and 2 sections
$M_{ry} = M_{py}$ for class 1 and 2 sections

For partially stress-relieved square hollow structural sections ($F_r < 0.3\ F_y$), the following inequality shall be satisfied

$$\frac{M_{fx}}{M_{rx}} + 0.5\,\frac{M_{fy}}{M_{ry}} \leq 1.0 \tag{5a.N18}$$

For Class 3 and 4 sections:

$$\frac{C_f}{C_r} + \frac{M_{fx}}{M_{rx}} + \frac{M_{fy}}{M_{ry}} \leq 1.0 \tag{5a.N19}$$

Nonlinear Interaction Equation

For more refined application, the following expressions may be used for these sections:

$$\left(\frac{M_{fx}}{M_{rcx}}\right)^{\zeta} + \left(\frac{M_{fy}}{M_{rcy}}\right)^{\zeta} \leq 1.0 \tag{5a.N20}$$

where

$$M_{rcx} = 1.18\, M_{rx}\, (1 - C_f/C_y) \leq M_{rx}$$

$$M_{rcy} = 1.19\, M_{ry}\, [1 - (C_f/C_y)^2] \leq M_{ry}$$

$$\zeta = 1.6 - \frac{C_f/C_y}{2[\ln (C_f/C_y)]}$$

Section Strength

$$f_a + f_b \leq 0.6\, F_y \qquad (5a.N21)$$

where the terms used have the same meaning as in section 5d.

OHBDC 83

Compact Sections, H-Shapes

$$\frac{M_{fx}}{M_{rx}} + \frac{M_{fy}}{M_{ry}} \leq 1.0 \qquad (5a.N22)$$

where M_r is $\phi Z F_y = \phi M_p$ for compression flange that meets the following requirements:

for $\frac{M_1}{M_2} > 0.5$ $\qquad L \leq 500\, r_y/\sqrt{F_y}$

or (5a.N23)

for $\frac{M_1}{M_2} \leq 0.5$ $\qquad L \leq 980\, r_y/\sqrt{F_y}$

where $\frac{M_1}{M_2}$ is the ratio of the smaller moment to the larger moment at opposite ends of the unbraced length, at ultimate limit state in the plane of bending considered.

Other Singly and Doubly Braced Symmetrical Sections

$$\frac{C_f}{C_r} + \frac{M_{fx}}{M_{rx}} + \frac{M_{fy}}{M_{ry}} \leq 1.0 \qquad (5a.N24)$$

where

$C_r = \phi A F_y$

$M_r = \phi S F_y = \phi M_y$ for compression flange that meets the following requirements

$$L \leq \frac{140{,}000 A_f}{F_y d}$$

or (5a.N25)

$$L \leq \sqrt{\frac{1{,}700{,}000\, r_t^2}{F_y}}$$

whichever is greater.

WEST EUROPE

CHAPTER EDITOR
David A. Nethercot, The University of Nottingham

BS 5400 P.3 82

Both uniaxial and biaxial bending are covered by the same expression for members that meet the necessary limits on overall slenderness and where individual plate elements satisfy the rules for compactness such that

$$\lambda_{lt} \leq 45\sqrt{355/\sigma_y}$$

and $\ell_e/\mathrm{r} \leq 15\sqrt{355/\sigma_y}$

and $\mathrm{b/t} \leq 24\sqrt{355/\sigma_y}$ for elements supported on both edges

or $\mathrm{b/t} \leq 7\sqrt{355/\sigma_y}$ for outstands (5a.W1)

Cross-sectional capacity may be based on any assumed stress distribution that satisfies equilibrium and nowhere exceeds $\sigma_y/\gamma_m\ \gamma_{f3}$; that is, fully plastic cross sectional behavior may be used.

For all classes of cross section an elastic limit given by

$$\frac{\mathrm{P}}{\mathrm{A}} \pm \frac{\mathrm{M}_x}{\mathrm{Z}_x} \pm \frac{\mathrm{M}_y}{\mathrm{Z}_y} \leq \sigma_y/\gamma_m\ \gamma_{f3} \qquad (5a.W2)$$

may be used.

where

γ_{f3} = 1.1 for ultimate limit states
γ_m = $\gamma_{m1}\ \gamma_{m2}$
γ_{m1} = partial factor on characteristic yield stress
γ_{m2} = partial factor for modeling uncertainties

BS 5950 P.1 85

Uniaxial Bending

For Class 1 or Class 2 sections, that is, those able to at least attain the full plastic moment capacity, the reduced moment capacity M_r in the presence of axial load may be obtained based on full cross-sectional plasticity. No explicit formulae are given, although these are provided in the supporting literature.

For class 3 or 4 sections, that is, those able to attain the yield moment or insufficiently compact to reach even this level of moment capacity, elastic limit is used

$$\frac{\mathrm{F}}{\mathrm{A\,p}_y} + \frac{\mathrm{M}}{\mathrm{M}_c} \leq 1 \qquad (5a.W3)$$

Biaxial Bending

Class 1 or 2 sections must satisfy

$$\left(\frac{\mathrm{M}_x}{\mathrm{M}_{rx}}\right)^{\mathrm{Z1}} + \left(\frac{\mathrm{M}_y}{\mathrm{M}_{ry}}\right)^{\mathrm{Z2}} \leq 1 \qquad (5a.W4)$$

where

Z1 = 2.0 for H and I sections, 5/3 for solid and closed hollow sections and 1.0 for all other cases

Z2 = 5/3 for solid and closed hollow sections and 1.0 for all other cases.

For Class 3 and Class 4 sections, and for Class 1 and Class 2 sections if a simpler alternative is required, the elastic limit is extended to

$$\frac{F}{A\,p_y} + \frac{M_x}{M_{cx}} + \frac{M_y}{M_{cy}} \leq 1 \qquad (5a.W5)$$

CNR 10111 85

A simple elastic limit is used in all cases and is given by

$$\sigma_n + \sigma_m \leq f_d \qquad (5a.W6)$$

where σ_m is $\left(\frac{M_x}{W_x} + \frac{M_y}{W_y}\right)\frac{1}{\phi}$.

CTICM 80

Uniaxial Bending

General formulas for strong axis and weak axis bending are given as

$$M_{ux} = M_{px}\left(\frac{1 - N/N_p}{1 - c}\right) \qquad (5a.W7)$$

where c is the lesser of $\frac{1}{2}\frac{A_w}{A}$ and 0.25

$$M_{uy} = M_{py}\left(\frac{1 - N/N_p}{1 - c}\right) \qquad (5a.W8)$$

where c is the lesser of $\frac{A_f}{A}$ and 0.25

Specialized versions of these formulae are also provided for IPE and HE sections.

Biaxial Bending

All cases are governed by

$$\left(\frac{M_x}{M_{ux}}\right)^{\alpha} + \left(\frac{M_y}{M_{uy}}\right)^{\beta} \leq 1 \qquad (5a.W9)$$

for $0 \leq N/N_p \leq 0.2$ and I sections: $\alpha = 2,\ \beta = 1$

for $0.2 \leq N/N_p \leq 1$ and I sections: $\alpha = 2,\ \beta = 5N/N_p$

EC 3 84

Uniaxial Bending

Three different approaches are given depending upon the classification of the cross section based on the extent to which local buckling will affect the ability of each plate element to attain and/or maintain its full capacity.

For class 1 and 2 sections full plasticity may be attained and for I and H sections this is governed by

H-sections, strong axis
for $N > 0.1\ N_{pl}$

$$M_{red} = 1.11\ M_{pl}\ (1-N/N_{pl}) \tag{5a.W10}$$

I-sections, strong axis
for $N > 0.18\ N_{pl}$

$$M_{red} = 1.22\ M_{pl}\ (1-N/N_{pl}) \tag{5a.W11}$$

H-section, weak axis
for $N > 0.2\ N_{pl}$

$$M_{red} = M_{pl}\left(1-\left(\frac{N/N_{pl}-0.2}{0.8}\right)^2\right) \tag{5a.W12}$$

I-sections, weak axis
for $N > 0.36\ N_{pl}$

$$M_{red} = M_{pl}\left(1-\left(\frac{N/N_{pl}-0.36}{0.64}\right)^2\right) \tag{5a.W13}$$

For class 3 sections a simple elastic limit is used

$$\frac{N}{A} + \frac{M}{W} \leq f_y \tag{5a.W14}$$

For class 4 sections the full cross-sectional area should be used but stresses must be kept within the buckling limits.

Biaxial Bending

For class 1 and 2 sections biaxial bending is covered by

$$\left(\frac{M_x}{M_{xred}}\right)^{\alpha} + \left(\frac{M_y}{M_{yred}}\right)^{\alpha} \leq 1 \tag{5a.W15}$$

where

M_{xred}, M_{yred} = uniaxial capacities in the presence of the axial load

α = $1.6 - (N/N_{pl})\ /\ (2\ \ln\ (N/N_{pl}))$ for I and H sections with $B/D > 0.3$

α = 1.0 for all other cases

For class 3 sections the elastic limit formula is extended to

$$\frac{N}{A} + \frac{M_x}{W_x} + \frac{M_y}{W_y} \leq f_y \tag{5a.W16}$$

NBN B51002 88

All types of cross section are governed by the following expression, valid for both uniaxial and biaxial bending

$$\frac{P}{A} + \frac{M_x}{(I/v)_x} + \frac{M_y}{(I/v)_y} \leq f_y \tag{5a.W17}$$

NEN 3851 72

Uniaxial Bending

Bending about either the strong or weak axis is covered by

$$\frac{F}{A} + \frac{M}{W} \leq \sigma_e \tag{5a.W18}$$

For plastic design this is replaced (strong axis bending only) by

$$\frac{M_x}{M_p} \leq 1.18\ (1 - \frac{F}{A\ \sigma_e}) \tag{5a.W19}$$

SIA 161 79

Uniaxial Bending

For doubly symmetrical I-sections capacity is given by

(A) Strong axis bending

$$M_{pcx} = 1.12\ M_{px}(1 - N^*/N_p) \leq M_{px} \tag{5a.W20}$$

(B) Weak axis bending

for $N^*/N_p \leq A_w/A$

$$M_{ply} = M_{py} \tag{5a.W21}$$

for $N^*/N_p > A_w/A$

$$M_{ply} = M_{py}\left\{1 - \left(\frac{\frac{N^*}{N_p} - \frac{A_w}{A}}{1 - \frac{A_w}{A}}\right)^2\right\} \tag{5a.W22}$$

Biaxial Bending

This is governed by

$$\left(\frac{M_x^*}{M_{pcx}}\right) + \left(\frac{M_y^*}{M_{pcy}}\right)^{\alpha} \leq 1 \tag{5a.W23}$$

where α is $5N^*/N_p \geq 1.1$ for $N^*/N_p \leq 0.9$

StBK N1 70

Uniaxial Bending

Cross sectional capacity should be checked using

for strong axis bending

$$\left(\frac{N_{sd}}{N_{Rd}}\right)^{\alpha_o} + \frac{M_{sxd}}{M_{Rxd}} \leq 1.0 \tag{5a.W24}$$

for weak axis bending

$$\left(\frac{N_{sd}}{N_{Rd}}\right)^{\alpha} + \left(\frac{M_{syd}}{M_{Ryd}}\right)^{\alpha_o} \leq 1.0 \tag{5a.W25}$$

Biaxial Bending

The second equation given above should be replaced by

$$\left(\frac{N_{sd}}{N_{Rd}}\right)^{\alpha_o} + \left(\frac{M_{sxd}}{M_{Rxd}}\right)^{\beta_o} + \left(\frac{M_{syd}}{M_{Ryd}}\right)^{\gamma_o} \leq 1.0 \tag{5a.W26}$$

Note: International Comparisons is at the end of Chapter 5. (following page 5c(W))

5b. In-Plane Member Capacity

SPECIFICATION PROVISIONS & EXPLANATION

AUSTRALIA

CHAPTER EDITOR
Mark A. Bradford, University of New South Wales

AS DR87164 87

For a compact doubly symmetric I-section member, the nominal member capacity M_i for in-plane bending of a member without transverse loads is given by

$$M_i = M_s\left[\left\{1-\left(\frac{1+\beta_m}{2}\right)^3\right\}\left(1-\frac{N^*}{\phi N_c}\right)+1.18\left(\frac{1+\beta_m}{2}\right)^3\left(1-\frac{N^*}{\phi N_c}\right)^{1/2}\right] \quad (5b.A1)$$

where

M_s = section moment capacity in the plane of bending,
N_c = axial load capacity for the member buckling in the same plane,
β_m = ratio of the end moments.

For members with transverse loads, this expression may still be used with a modified value of β_m. For any other member, the nominal member capacity for in-plane bending is given by

$$M_i = M_s\,(1 - N^*/\phi N_c) \quad (5b.A2)$$

For a member in a frame analyzed by the simple rigid plastic method, there are additional requirements that restrict the cross-section shape and slenderness, the spacing of lateral bracing, and the in-plane member slenderness.

CHINA

CHAPTER EDITORS

Shao-Fan Chen, Xian Institute of Metallurgy & Construction Engineering

Bao Kang He, Xian Institute of Metallurgy & Construction Engineering

Zu-Yan Shen, Tong-Ji University

GBJ 17 88

The in-plane stability-checking formula of beam-columns with one axis of symmetry subject to bending moments acting in the plane of symmetry (about x-axis) is

$$\frac{N}{\phi_x A} + \frac{\beta_{mx} M_x}{\gamma_x W_{1x}(1 - 0.8\,\frac{N}{N_{Ex}})} \le f \qquad (5b.C1)$$

For T, channel, and unsymmetrical I-sections, when the larger flange is subjected to the maximum compressive stress, the following additional equation must be satisfied

$$\left| \frac{N}{A} - \frac{\beta_{mx} M_x}{\gamma_x W_{2x}(1 - 1.25\,\frac{N}{N_{Ex}})} \right| \le f \qquad (5b.C2)$$

where

- f = design normal strength of steel
- N_{Ex} = Euler buckling load about the x-axis (strong axis)
- W_{1x} = section modulus about the x-axis with respect to the fiber subjected to the maximum compressive stress
- W_{2x} = section modulus about the x-axis with respect to the extreme fiber of the smaller flange
- β_{mx} = equivalent moment factor
- γ_x = plastic adaptation coefficient
- ϕ_x = strength reduction factor of centrally compressed members bent about x-axis (see Chapter 1)

For columns of unbraced frames and cantilevers

$\beta_{mx} = 1$

For columns of braced frames and ends-supported members

(i) β_{mx} = 0.65 + 0.35(M_2/M_1) when the member is not subjected to transverse loads between supports, where (M_2/M_1) is the ratio of the smaller to larger moments by absolute value at opposite ends of the unbraced length. M_2/M_1 is positive where the member is bent in single curvature and negative when it is bent in reverse curvature.

(ii) β_{mx} = 1.0 when the member is subjected simultaneously to end moments and transverse loads, the curvatures produced by which have the same signs.

β_{mx} = 0.85 when the member is subjected simultaneously to end moments and transverse loads, the curvatures produced by which have the opposite signs.

(iii) β_{mx} = $1 - 0.2(N/N_{Ex})$ when the member is subjected only to a concentrated transverse load at mid-span;

β_{mx} = 1.0 when the member is subjected to other kinds of transverse loads.

For plastic design the in-plane checking formula for beam-columns with symmetrical I-sections or box sections subjected to bending moments acting in one plane of symmetry is

$$\frac{N}{\phi_x A f_p} + \frac{\beta_{mx} M_x}{W_{px} f_p (1 - 0.8 \frac{N}{N_{Ex}})} \leq 1 \tag{5b.C3}$$

where

f_p = design strength of steel for plastic design which equals 0.9f.

W_{px} = plastic section modulus about the x-axis (x-axis is the strong axis of I-sections)

TBJ 2 85

Use Eq. 5c.C3 by making $\phi_2 = 1$ and ϕ_1 related to in-plane buckling.

COMMENTARY

The in-plane interaction equation of a beam-column derived from the initial yield criterion is

$$\frac{N}{A f_y} + \frac{M_x + N e_o}{W_{1x} f_y (1 - \frac{N}{N_{Ex}})} = 1 \tag{5b.C4}$$

For T, channel, and unsymmetrical I-sections, when the larger flange is subjected to the maximum compressive stress, the interaction equation may be

$$-\frac{N}{A f_y} + \frac{M_x + N e_o}{W_{2x} f_y (1 - \frac{N}{N_{Ex}})} = 1 \tag{5b.C5}$$

where

e_o = equivalent eccentricity of initial imperfections

f_y = specified minimum yield stress

By eliminating e_o and taking into account the plastic behavior of beam-columns and introducing the equivalent moment factor, these equations become

$$\frac{N}{\phi_x A f_y} + \frac{\beta_{mx} M_x}{\gamma_x W_{1x} f_y (1 - K_1 \frac{N}{N_{Ex}})} = 1 \tag{5b.C6}$$

and

$$\left| \frac{N}{A f_y} - \frac{\beta_{mx} M_x}{\gamma_{2x} W_{2x} f_y (1 - K_2 \frac{N}{N_{Ex}})} \right| = 1 \qquad (5b.C7)$$

To determine the coefficients K_1 and K_2 in Eqs. 5a.C6 and 5a.C7, the ultimate strength of beam-columns of eleven widely used sections with slenderness ratio λ_x=20, 40, 60, 80, 100, 120, 160, 200 and relative eccentricity $\frac{MA}{NW}$ = 0.2, 0.6, 1.0, 2.0, 4.0, 10.0, 20.0 are computed by numerical calculation method. The major initial imperfections including residual stresses and initial crookedness of 1/1000 are taken into account. Through comparison the values $K_1 = 0.8$ and $K_2 = 1.25$ appear to be most suitable.

Calculations have also been made on the ultimate strength of beam-columns subjected to end moments whose ratio is 1.0, 0.5, 0, −0.5, −1.0 with slenderness ratio λ_x = 40, 60, 80, 100, 120. Based on these calculated results, the equivalent moment factor $\beta_{mx} = 0.65 + 0.35\, M_2/M_1$ is adopted. It considers the effect of plastic behavior of beam-columns and is slightly different from that derived for elastic beam-columns.

EAST EUROPE

CHAPTER EDITORS
Ioan A. Caraba, Polytechnic Institute of Timisoara
Miklos Ivanyi, Technical University of Budapest
Dan Mateescu, Polytechnic Institute of Timisoara

CMEA 83

Members Subject to Axial Compression and Monoaxial Bending

These members are checked by interaction equations taking into account the shape of the moment diagram. In cases where the lateral deformation is prevented, the following equation is used

$$\frac{N}{AR_{fl}} + \frac{M_e}{R_{fll}W_{xp}(1 - \frac{N}{AR_{cr}})} \leq \phi \tag{5b.E1}$$

where

M_e = design equivalent bending moment, which produces the same effect as the actual loading

R_{cr} = buckling critical strength of the ideal bar having the same support conditions as the bar being checked

R_{fl} = minimum buckling design strength

R_{fll} = lateral buckling design strength defined by the relationship, $R_{fll} = R_{cr}^m/\gamma_{fll}$; R_{cr}^m represents the critical strength in the elastic and plastic range for the value of the maximum critical moment depending on the modified slenderness ratio λ_{tr} of the compression flange and on the steel grade; λ_{fll} is the safety coefficient of the member for lateral buckling

W_{xp} = plastic strength modulus of the cross section for the most compressed fiber

ϕ = coefficient whose value is generally 1 or 1.1 depending on the cross section type

The equivalent bending moment M_e for symmetrical sections can be computed as a function of the magnitudes of the member end moments taken with their signs, thus

$$M_e = 0.6M_1 + 0.4M_2 \geq 0.4M_1 \qquad M_1 > M_2 \tag{5b.E2}$$

The bending moment for transverse loads is computed by the tabulated equivalence coefficients.

MSZ 15024/1 85

For members with full section, with one or two symmetry axes subjected to eccentric compression or to centric compression with bending moment, whose vector is on the maximum inertia axis of the cross section x-x, stability is checked by

$$\frac{N}{\phi A} + \frac{\beta_k M_{max}^{II}}{\phi_L \overline{W}_c} \leq \sigma_H \tag{5b.E3}$$

where

A = area of the cross section

M_{min}, M_{max} = values of the bending moments on the neighboring supports, which are introduced with their signs

M^{I}_{max} = maximum first-order bending moment, which is calculated for the unstrained member

M^{II}_{max} = maximum second-order bending moment, which is determined by

$$M^{II}_{max} = \psi M^{I}_{max} \qquad (5b.E5)$$

R_y = yield stress of steel

$\overline{W}_c$ = strength modulus calculated for the most compressed fiber

α = a coefficient depending on buckling curves

β $= \dfrac{1 + \alpha(\overline{\lambda} - 0.2) + \overline{\lambda}^2}{2\overline{\lambda}^2}$

β_k = reduction coefficient depending on the variation of the bending moment along the member axis. For the linear variation of the moments diagram along the member, the β_k coefficient is calculated by

$$\beta_k = 0.6 + 0.4\,\frac{M_{min}}{M_{max}} \quad \text{but} \quad \beta_k \geq 0.4 \qquad (5b.E4)$$

λ = ℓ_f/i, the slenderness ratio with respect to either the x-x axis ($\lambda_x = \ell_{fx}/i_x$) or the y-y axis ($\lambda_y = \ell_{fy}/i_y$)

λ_E $= \pi\sqrt{E/R_y}$

λ_k $= \pi\sqrt{\dfrac{1.1\,E}{\sigma_{ki}}}$

λ_k = torsional slenderness of the member

$\overline{\lambda}$ $= \dfrac{\lambda}{\pi}\sqrt{\dfrac{R_y}{E}}$ = reduced slenderness ratio

$\overline{\lambda}_k$ $= \lambda_k/\lambda_E$

σ_H = limit strength depending on the steel grade and the thickness of the element

σ_{ki} = torsional critical unitary stress for a straight member, made from a perfectly elastic material

ϕ $= \beta - \sqrt{\beta^2 - \dfrac{1}{\overline{\lambda}^2}}$

ϕ_k = coefficient depending on the influence of torsion upon bending buckling and is calculated by the formula

$$\phi_k = \left(\frac{1}{1 + \overline{\lambda}_k^5}\right)^{0.4} \qquad (5b.E7)$$

ψ = coefficient depending on the influence of eccentricity upon the value of the bending moment that has the form

$$\psi = \frac{1}{1 - \dfrac{N}{A\sigma_H\left(\dfrac{\lambda}{\lambda_E}\right)^2}} \qquad (5b.E6)$$

PN 80/B-03200 87

For members with full section subjected to compression and bending in the symmetry plane the cross section (Fig. 5b.E1), buckling stress is checked as follows

for $e_c \geq e_r$

$$\sigma = \frac{Pm_w}{F_{br}} + \frac{M}{W_{c,br}} \leq 1.05R \tag{5b.E8}$$

for $e_c < e_r$

$$\left.\begin{aligned} \sigma &= \frac{Pm_w}{F_{br}} + \frac{M}{W_{c,br}} \leq 1.05R \\ \sigma &= \frac{Pm_w}{F_{br}} + \frac{300 + 2\lambda}{1000}\frac{M}{W_{r,br}} \leq R \end{aligned}\right\} \tag{5b.E9}$$

where

F_{br} = gross area of the cross section
m_w = buckling multiplier determined from tables as a function of the ratio λ/λ_p
R = design strength
$W_{c,br}$, $W_{r,br}$ = gross strength moduli calculated with respect to the x-x axis, for the compressed and stretched fibers, respectively.
λ = $\lambda_y = l_{fy}/i_y$ (Fig. 5b.E1)
λ_p = $1675/\sqrt{R}$

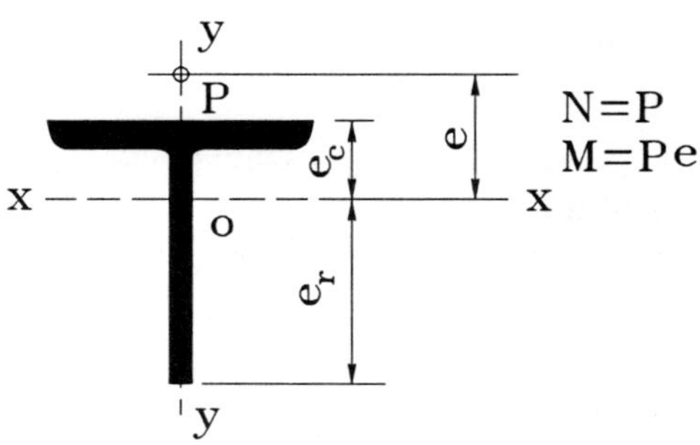

Fig. 5b.E1 Definitions of terms

SNiP II-23-81 82

Stability of steel members is checked separately in the plane of the bending moment and in the plane normal to it.

In-plane Stability Check

When members are compressed eccentrically or subject to compression and bending with constant section, the following formula is used

$$\frac{N}{\phi_e A} \leq R_y \gamma_c \tag{5b.E10}$$

where

R_y = yield stress
γ_c = coefficient of service conditions
ϕ_e = coefficient of yield stress reduction by eccentric compression

The coefficient of yield stress reduction, ϕ_e, is determined in relation to non-dimensional slenderness $\overline{\lambda}$

$$\overline{\lambda} = \lambda\sqrt{R_y/E} \tag{5b.E11}$$

and modified relative eccentricity

$$m_{ef} = \eta m \tag{5b.E12}$$

where

e = M/N
m = eA/W_c, relative eccentricity
W_c = section modulus of a member corresponding to the most compressed fiber
λ = greatest slenderness of a member
η = coefficient in relation to section shape

The calculated values for axial load N and for bending moment M will be determined for the same load assumption. These values will be obtained through a first-order analysis considering the elastic behavior of steel.

The values of the coefficient of yield stress reduction ϕ are given in a table in the specification:

- for full-web members in relation to non-dimensional slenderness $\overline{\lambda}$, as well as to modified eccentricity, m_{ef}
- for lattice members joined with lattice bars of plates positioned in the plane of bending in relation to the modified non-dimensional slenderness $\overline{\lambda}_{ef}$ as well as to the specific relative eccentricity m, determined by the formula

$$m = \frac{eAa}{I} \tag{5b.E13}$$

where a is the distance from the main axis of the section, perpendicular to the bending plane, to the most compressed fiber

For the computation of eccentricity e, magnitudes of M and N (calculated for the same combination of loads) will be taken into consideration.

If for full-web members $m_{ef} > 20$ and for lattice members $m > 20$, a stability check is not needed.

Here, a is the distance between the most compressed branch and the major axis of section.

Out-plane Stability Check

Checking the stability of eccentrically compressed members bended in the plane of minimum stiffness ($I_y < I_x$ and $e_y \neq 0$), if $\lambda_x > \lambda_y$ the relation given in Eq. 5b.E14 will be used for out-plane stability check.

$$\frac{N}{\phi_x A} \leq R_y \gamma_c \qquad (5b.E14)$$

where the coefficient ϕ_x is the same as ϕ_y above.

If $\lambda_x \leq \lambda_y$ stability checking outside the plane of the moment is not required.

STAS 763/1 71

Members with full section subjected to axial compression and uniaxial bending, are checked with the formulas

$$\frac{N}{\varphi A} + \frac{M}{W} \leq \sigma_a \qquad (5b.E15)$$

where

φ = the buckling coefficient
σ_a = the allowable stress

STAS 10108/0 78

The stability of I-section members about the strong axis check is given by

$$\frac{N}{\varphi A} - \frac{c_x M_x}{\varphi_g (1 + \frac{\sigma}{\sigma_E}) W_x} \leq R \qquad (5b.E16)$$

where

c_x = coefficient correcting the value of moment M_x in relation to distribution manner

$$c_x = \sqrt{0.3(1 + \frac{M_2^2}{M_1^2}) + 0.4\frac{M_2}{M_1}} \qquad \text{with } M_1 \geq M_2 \qquad (5b.E17)$$

I_D = the inertia moment for free torsion of the member
ℓ = the length of member between supports
R = the design strength of steel
μ = coefficient that multiplies the length of the member depending on the degree of end restraint of the member
σ = N/A, the normal unitary stress produced by the compressive axial force
σ_E = $(\pi^2 E)/\lambda^2$
φ = the minimum value of the buckling coefficient, determined as a function of section group, steel grade and maximum slenderness ratio (considering λ_x, λ_y, or λ_{tr})
φ_g = coefficient obtained from buckling curves, in relation to section shape, steel grade and modified slenderness ratio

$$\lambda_{tr} = \gamma \frac{\mu \ell}{i_y} \qquad (5b.E18)$$

in which γ depends on

$$\frac{\ell^2 I_D}{h^2 I_y} \qquad (5b.E19)$$

For bars with full section with one symmetry axis, which are subjected to compression and bending, stability is checked by Eq. 5b.E16, where φ_g is taken from tables providing the values of buckling coefficients, depending on section group, steel grade, and the modified slenderness ratio λ_{tr} obtained by the relation

$$\lambda_{tr} = \gamma_1 \frac{\mu \ell}{i_j} \tag{5b.E20}$$

For members with one symmetry axis

$$\lambda_{tr} = \gamma_1 \frac{\mu \ell}{i_y} \tag{5b.E21}$$

where

$$c^2 = \frac{I_\omega (\mu \ell)^2 (\mu_o \ell_o)^2 + 0.039 \mu^2 \ell^2 I_D}{I_y}$$

e = the distance between the central axis x-x of the cross section and the axis of the most compressed flange

I_ω = the sectional inertia moment

ℓ_o = the length of member between end points ensured against overturning

$r_x = \int y(x^2 + y^2)\, dA / I_x$

v = the distance from the level of application of load in the member axis

y_c = the distance between the center of gravity 0 and the center of torsion C

$$\gamma_1 = \sqrt{\frac{I_x}{eA\left(\sqrt{\left(0.506 v \mu^2 + \frac{r_x}{3} - y_c\right)^2 \pm c^2} - \left(0.506 v \mu^2 + \frac{r_x}{3} - y_c\right)\right)}}$$

μ_o = coefficient that takes into account the degree of preventing the deformation of the member plane. The values of coefficients μ and μ_o are:

= 1.00 for members at ends with prevented overturning and free warping deformation of the cross section plane

= 0.50 for members restrained at both ends with prevented overturning and warping deformation

= 0.70 for members restrained at one end and articulated at the other for bending, respectively the warping deformation is prevented at one end and free at the other. Currently, in practice $\mu = \mu_o$

If $N/\varphi A \leq 0.15R$, the stability of the member is checked with the relation

$$\frac{N}{\varphi A} + \frac{M_x}{\varphi_g W_x} \leq R \tag{5b.E22}$$

For compound-section members stiffened with plates or latticed bars subjected to compression and bending in plane x-x

$$\frac{N}{\phi A} + \frac{c_y\, M_y\, e}{(1 - \frac{\sigma}{\sigma_E})\, I_y} \leq R \tag{5b.E23}$$

where e is the distance between the y-y axis and the most compressed fiber.

If bending occurs in plane y-y

$$\frac{N}{\mu A} + \frac{c_x\, M_x\, h_1}{(1 - \frac{\sigma}{\sigma_e})I_x} \leq R \qquad (5b.E24)$$

where h_1 is the distance between the x-x axis and the most compressed fiber.

TGL 13503 82

Monosymmetric Section Subjected to Compression and Bending
Buckling in the plane of moment

$$\sigma_c(1.0 + \mu_N\, f_N) + \sigma_{bc}\, f_M \leq \sigma_{all} \qquad (5b.E25)$$

$$\sigma_c(-1.0 + \mu_N f_N) + \sigma_{bz}\, f_M \leq \sigma_{all} \qquad (5b.E26)$$

where

$\sigma_{bc} = M/W_d$
$\sigma_{bz} = M/W_z$
$\sigma_c = N/A$

A strength modulus can be introduced

$$W_T = (W_{el} + W_{pl})/2 \leq 1.2\, W_{el} \qquad (5b.E27)$$

where

f_N, f_M = factors of amplification of bending moments according to second order theory
σ_{all} = allowable stress in relation to ultimate stress = σ_F/v_{kr}

$$\mu_N = \frac{93\overline{\lambda} - c_1}{c_2} \geq 0 \qquad (5b.E28)$$

where c_1, c_2 are the imperfections in relation to buckling curve.

Loss of Stability Due to Torsion of the Member Subjected to Bending

$$M \leq \frac{M_{cr}}{\nu_{cr}} \quad \text{or} \quad \sigma_b(1 + \mu_M f_{oM}) \leq \sigma_{all} \qquad (5b.E29)$$

where

$f_{oM} = M_{ki}/(M_{ki} - \nu_{cr}M)$
i_G = radius of gyration of upper flange (flange plus 2/5 of the compressed area in the web)
l_k = distance between supporting points of upper flange
$M_F = W\sigma_F$
$M_T = (M_F + M_{pl})/2 \leq 1.2M_F$
$M_{cr} = M_T\phi_M$ - critical moment with bending only
M_{ki} = ideal moment of loss of stability

$$M_{pl} = W_{pl}\sigma_F$$
$$\overline{\lambda}_M = \sqrt{M_T/M_{ki}}$$
$$\overline{\lambda}_M = \theta\, l_k/\lambda_E\, i_G \qquad (5b.E30)$$

$\theta = 1.25$ with respect to upper flange, welded sections
$\theta = 1.00$ - rolled sections
$\theta = 0.80$ with respect to lower flange

$\mu_M = (93\,\overline{\lambda}_M - c_1)/c_2$
ν_{cr} = safety coefficient of the member for lateral buckling
ϕ_M = factor of loss of stability in relation to $\overline{\lambda}_M$

c_1 and c_2 for the four curves a, b, c and d with respect to area technology and dimensions h_s/s (web height as compared to web thickness)

Checking is not required when

for sections (curve a)	$\overline{\lambda}_M \leq 0.60$
for sections (curve b)	$\overline{\lambda}_M \leq 0.40$
for sections (curve c)	$\overline{\lambda}_M \leq 0.28$

where curve “a” is the maximum strength column curve for tubes, curve “b” for box sections and rolled I-shapes, and curve “c” for wide flange I-sections and welded I-shapes with weak axis bending.

Loss of Stability Due to Torsion with Axial Force

$$\frac{N}{N_{all}} + \left(\frac{M}{M_{all}}\right)^n \leq 1 \qquad (5b.E31)$$

where

$M_{all} = M_{kr}/V_{cr}$
$N_{all} = A\sigma_{all}\phi$
n = Exponent in relation to the shape of the moment diagram (Fig. 5b.E2)

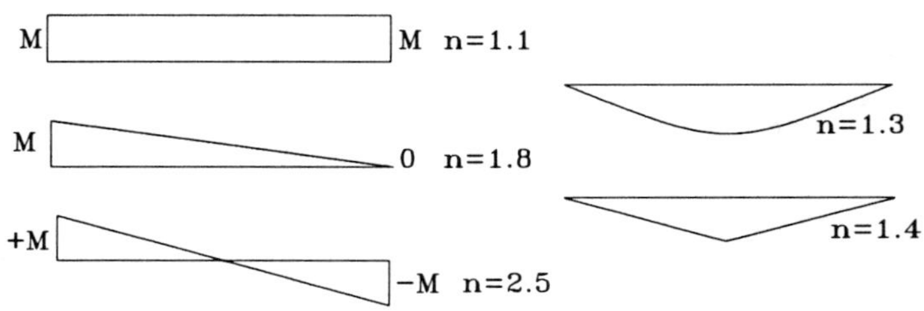

Fig. 5b.E2 The Effect of the Shape of the Moment Diagram

JAPAN

CHAPTER EDITORS

Takeshi Nakamura, Kyoto Institute of Technology

Koichi Takanashi, The University of Tokyo

AIJ 90

In-plane stability is checked by

$$\frac{P}{\phi_c P_{ux}} \leq 1.0 \qquad (5b.J1)$$

$$\frac{P}{\phi_p P_y} + 0.85\psi_c \frac{M}{\phi_p M_p} \leq 1.0 \qquad (5b.J2)$$

where

M_p = plastic moment
M_1 = larger end moment
M_2 = smaller end moment
P_{ex} = Euler buckling load in plane of bending
P_{ux} = axial compressive capacity for buckling in plane of bending
P_y = axial compressive capacity for yielding
β = M_2/M_1, end moment ratio
λ_c = $\sqrt{P_y/P_{ex}}$, modified slenderness ratio
ϕ_c = resistance factor for compression
ϕ_p = resistance factor for tension
ψ_c = 1.0 for $(P/P_y)\,\lambda_c^2 < 0.25\,(1+\beta)^2$
ψ_c = $[1 - 0.5(1+\beta)\sqrt{P/P_{ex}}]/(1 - P/P_{ex})$ for $(P/P_y)\lambda_c^2 \geq 0.25(1+\beta)^2$

JRA 90

In-plane stability is checked by

$$_cf_a + \frac{_cf_{bx}}{(1 - \frac{_cf_a}{F_{ex}})} + \frac{_cf_{by}}{(1 - \frac{_cf_a}{F_{ey}})} \leq {_pF_a} \qquad (5b.J3)$$

where

F_{ex}, F_{ey} = Euler buckling stress with respect to the strong and weak axis, respectively
$_pF_a$ = permissible compressive stress for local buckling
$_cf_a$ = applied axial compressive stress
$_cf_{bx}$, $_cf_{by}$ = applied compressive stress due to bending about the strong and weak axis, respectively

JSCE 83

Allowable stress design approach, in which the equations giving the allowable stress in the tension and compression flanges are

for axial tension force and biaxial bending

$$\frac{P}{A_n} + \frac{M_x}{I_x} y_t \frac{A_{fg}}{A_{fn}} + \frac{M_y}{I_y} y'_t \frac{A'_{fg}}{A'_{fn}} \leq \sigma_{ta} \tag{5b.J4}$$

$$-\frac{P}{A_g} + \frac{M_x}{I_x} y_c + \frac{M_y}{I_y} y'_c \leq \sigma_{cao} \tag{5b.J5}$$

for axial compression force and biaxial bending

$$-\frac{P}{A_g} + \frac{M_x}{I_x} y_t \frac{A_{fg}}{A_{fn}} + \frac{M_y}{I_y} y'_t \frac{A'_{fg}}{A_{fn}} \leq \sigma_{ta} \tag{5b.J6}$$

$$\frac{P}{A_g} + \frac{M_x}{I_x} y_c + \frac{M_y}{I_y} y'_c \leq \sigma_{cao} \tag{5b.J7}$$

where

- A_{fn}, A_{fg} = net flange area and gross flange area, respectively, with respect to the strong axis
- A'_{fn}, A'_{fg} = net flange area and gross flange area, respectively, with respect to the weak axis
- A_n, A_g = net section area and gross section area, respectively
- M_x, M_y = bending moments with respect to the strong axis and the weak axis, respectively. For out-of-plane member resistance, M_x and M_y must be the equivalent bending moment M_{eq}, which is evaluated by $M_{eq} = 0.6M_1 - 0.4M_2$ but not less than $0.4M_1$, where M_1 is the greater end moment whereas M_2 the smaller
- y_t, y_c = distance from the strong axis to the extreme fibers in the tension and compression sides of the section, respectively
- y'_t, y'_c = distance from the weak axis to the extreme fibers in the tension and compression sides of the section, respectively
- σ_{cao} = permissible stress in compression
- σ_{ta} = permissible stress in tension

COMMENTARY

These formulas are simple linear equations for a combination of an axial force and biaxial bending moment. For a compressive axial force, the P - δ effect must be considered. But it is expected to be so small that amplification factors in the formulas are neglected.

NORTH AMERICA

CHAPTER EDITOR
Wai-Fah Chen, Purdue University

AASHTO 87

Member Stability

$$\frac{P}{P_u} + \frac{MC}{M_u(1 - \frac{P}{A_s F_e})} \leq 1.0 \qquad (5b.N1)$$

where

A_s = effective column cross-section
C = equivalent moment factor
= $0.6 + 0.4\,(M_2/M_1) > 0.4$ (same as the ASD formula in AISC ASD 78)
F_{cr} = same as the ASD Column curve in AISC ASD 78 where the effective length factor K is determined by
(a) For members having lateral support in both directions at its ends:
$K = 0.75$ for riveted, bolted, or welded end connections
$K = 0.875$ for pinned ends
(b) Others by a rational procedure
F_e = Euler buckling stress considering the end restrained condition
M_u = beam strength
P_u = $0.85\ A_s\ F_{cr}$

AISC ASD 78

In-plane and Out-of-plane Bending Member Stability (also see Eqs. 5d.N4 and 5d.N5). If stability controls,

$$\frac{P}{P_{uk}} + \frac{C_m M_o}{M_m(1-P/P_{ek})} \leq 1 \qquad (5b.N2)$$

where

C_m = same as in AISC ASD 78
M_m = the maximum resisting moment in the absence of axial force, taken as M_p if the member is braced against lateral torsional buckling and taken as

$$M_m = \left[1.07 - \frac{(L/r_y)\sqrt{F_y}}{3{,}160}\right] M_p \leq M_p \qquad (5b.N3)$$

if the member fails by lateral torsional buckling. In this equation, the units are inches and ksi
P_{ek} = $\pi^2\ EI/(KL)^2$
P_{uk} = 1.7 times AISC ASD 78 column curve using the effective length of the column KL

Again, in actual design, both the strength interaction equation and the stability interaction equation are used to check the adequacy of the section.

AISC LRFD 86 (See Eqs. 5d.N6 and 5d.N7)

WEST EUROPE

CHAPTER EDITOR
David A. Nethercot, The University of Nottingham

BS 5400 82

In-plane stability is checked by

$$\frac{F}{R_c} + \frac{M}{M_c} \leq 1 \qquad (5b.W1)$$

in which F and M are the maximum compression and moment within the middle third of the member length between points of restraint. M should be determined taking into account the amplification effects where appropriate.

BS 5950 P.1 85

Two alternative interaction formulas are provided, both within the context of the full biaxial problem. The governing condition for in-plane failure is

$$\frac{F}{P_c} + \frac{1}{2}\frac{F}{P_c}\frac{m\,M}{M_c} \leq 1 \qquad (5b.W2)$$

$$\frac{F}{P_{cy}} + \frac{m_y\,M}{p_y\,Z_y} \leq 1 \qquad (5b.W3)$$

where Z_y is the elastic modulus about the minor axis.

COMMENTARY

For Eq. 5b.W2 the basis is the ECCS e* formula but with the amplification effect included in a simpler approximate way. The same formula is used for in-plane failure about either the major or minor axes.

For the simpler Eq. 5b.W3 only weak axis behavior is considered. It is presented in the Code as part of the biaxial equation and this does not reduce to cover in-plane failure in the absence of minor axis bending.

CNR 10111 85

$$w\,\frac{N}{A} + \frac{M_{eq}}{\phi W(1 - N/N_{cr})} \leq f_d \qquad (5b.W4)$$

COMMENTARY

Use of ϕ limits the amount of residual plastic deformation due to partial plastification of the cross-section. Numerical values are not provided, only strain limits for various span-to-depth ratios and steel grades.

CTICM 80

In-plane buckling is governed by

$$k_o \frac{N}{N_p} + k_f \frac{M_m}{M_p} \leq 1.0 \qquad (5b.W5)$$

where

C_m = equivalent uniform moment factor

$k_f = \dfrac{C_m}{1 - \overline{\lambda}^2 \frac{N}{N_p}}$

k_o = coefficient to allow for flexural buckling due to compression

$\overline{\lambda} = \frac{\lambda}{x}\sqrt{\frac{\sigma_e}{E}}$

DIN 18800 P.2 86

For uniaxial bending with no possibility of lateral-torsional buckling the design condition is

$$\frac{F}{\chi \, A \, p_y} + \beta_m \frac{M}{M_{pl}} + \Delta n \leq 1 \qquad (5b.W6)$$

where

M_{pl} = plastic moment capacity
p_y = material strength
β_m = a coefficient depending on the pattern of moment, see Table 5b.W1

$\Delta n = \frac{F}{\chi A p_y}(1 - \frac{F}{\chi A p_y})\chi^2 \, \overline{\lambda}^2 \leq 0.1$

$\overline{\lambda}$ = non-dimensional column slenderness
χ = column buckling coefficient

If F/Ap_y is greater than 0.2, M_{pl} may be replaced by 1.1 M_{pl}.

EC 3 88

For uniaxial bending about a principal axis overall buckling in the plane of the applied moments should be checked using

$$\frac{F}{\chi \, p_y \, A} + \frac{m \, M}{M_c} \leq 1 - \Delta n \qquad (5b.W7)$$

where

M_c = moment capacity depending on section classification, 1.1 M_c may be used for doubly symmetrical rolled sections or comparable welded section with $A_{web} > 0.1A$ providing $F/p_y \, A > 0.1$
m = coefficient to allow for moment gradient, see Table 5b.W1
p_y = material yield stress

$$\Delta n = \frac{F}{\chi\, p_y\, A} \cdot \left(1 - \frac{F}{\chi\, p_y\, A}\right) \chi^2\, \overline{\lambda}^2 \text{ or}$$

$\Delta n = 0.25\, \chi^2\, \overline{\lambda}^2$ for $0.1 < F/p_y A$ and $0.1 \leq \beta M/M_c$
$\Delta n = 0$ for $0.1 > F/p_y A$ and $0.1 > \beta M/M_c$
χ = reduction factor for column strength depending on slenderness $\overline{\lambda}$

COMMENTARY

Use of the limitation on m given in Table 5b.W1 automatically ensures that the cross section has sufficient capacity locally at all points to resist coincident axial load and moment without the need for a separate capacity check.

Equation 5b.W7 is based on a rearrangement of the original ECCS e* formula.

Table 5b.W1 Equivalent uniform moment factors for use with Eq. 5b.W7

Moment distribution	m	
M 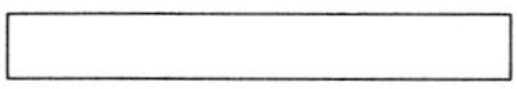	1.10	
M 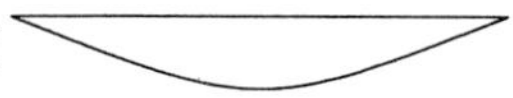	1.00	
M 	1.00	$\geq \left(1 - \frac{N}{N_{k1}}\right)^*$
M 	0.66	
M 	0.44	
M  ψM, $-1 < \psi < 1$	$0.66 + 0.44\psi > 0.44$	

NBN B51002 88

In-plane stability is checked by

$$\frac{F}{A\chi} + \frac{mM}{Z(1 - P/P_{cr})} = p_y \tag{5b.W8}$$

where

$$m = \left(0.3\left(1 + \left(\frac{M_{min}}{M_{max}}\right)^2\right) + 0.4\,\frac{M_{min}}{M_{max}}\right)^{0.5} \text{ if sidesway is prevented}$$

$m = 0.85$ if sidesway is possible.

COMMENTARY

Equation 5b.W8 is valid for both major and minor axis behavior.

NEN 3815 72

For bending about the x-axis only, when lateral-torsional buckling is not a possible failure mode, the design must satisfy

$$\chi_x \frac{F}{A} + m_x \frac{P_{crx}/F}{P_{crx}/F - 1} \frac{M_x}{Z_x} \leq p_y \tag{5b.W9}$$

$$\chi_y \frac{F}{A} \leq p_y \tag{5b.W10}$$

$$\frac{F}{A} + \frac{M_x}{Z_x} \leq p_y \tag{5b.W11}$$

where

$m_x = 0.6 + 0.4\beta_x$
$\beta_x \geq 0.4$ if the actual member length is used when determining χ_x
$\beta_x \geq 0.6$ if the effective member length is used when determining χ_x
$\chi_x = P_c/P_s$ is the column buckling coefficient

For bending about the y-axis only, Eqs. 5b.W9 - 5b.W11 should also be used but with the appropriate change of subscripts.

Modified versions of Eqs. 5b.W9 - 5b.W11 are provided for use when sidesway is possible; in the absence of transverse loads these take $m_x = m_y = 0.85$. For members containing a plastic hinge at one or both ends, when sidesway is prevented, Eqs. 5b.W9 - 5b.W10 should be modified to

$$\chi_x \frac{F}{A} + m_x \left(\frac{P_{crx}/F}{P_{crx}/F - 1} \right) \frac{M_x}{S_x} \leq p_y \tag{5b.W12}$$

In addition the presence of adequate rotation capacity must either be checked by rigorous analysis or by satisfying

$$\frac{F}{A\, p_y} + \frac{\lambda\chi}{120} \leq 1 \tag{5b.W13}$$

When sidesway is possible m_x is Eq. 5b.W12 must be taken as 0.85, the figure of 120 in Eq. 5b.W13 should be reduced to 100.

COMMENTARY

Use of the rules is assisted by the provision of an integral explanation.

In the proposed revision to this code the alternative of using second order analysis with a cross-sectional strength interaction is also permitted.

NS 3472 84

In-plane stability is checked by

$$\frac{F}{P_c} + \frac{mM}{(1 - \frac{F}{P_{cr}} \frac{P_{cx}}{P_s})M_c} \leq 1 \tag{5b.W14}$$

where

M_c = M_p for class 1 and 2 sections, M_y otherwise
m = $0.6 + 0.4\beta \geq 0.4$ if sidesway is prevented
m = 1.0 if sidesway is permitted

ÖNORM B4600 78

It is assumed that the section is symmetrical about the axis normal to the plane of bending. The in-plane stability is checked by

$$\frac{\sigma N}{\sigma_k F} + 0.9 \frac{M}{W_d} \leq \sigma \tag{5b.W15}$$

where

M = $[0.3(M_1^2 + M_2^2) + 0.4M_1M_2]^{0.5}$
W_d = elastic section modulus, compression side

For monosymmetric sections it is also necessary to satisfy

$$\frac{\sigma N}{\sigma_k F} + \frac{300 + 2\lambda}{1000} \frac{M}{W_z} = \sigma \tag{5b.W16}$$

where W_z is the elastic section modulus, tension side. If $W_z \leq W_d$, Eq. 5b.W16 is not needed.

SIA 161 79

Overall buckling should be checked using

$$\frac{N^*}{A} + \frac{1}{1 - N^*/N_{cr}} \frac{w M^*_{max} + N^* e_o}{W} \leq \sigma_f \tag{5b.W17}$$

where

e_o = $l_k/300$ for elastic design
e_o = $l_k/250$ for plastic design
N_{cr} = $\pi^2 EA/\lambda_k^2$ is the critical load about the axis under consideration, using the appropriate effective length.
N^*, M^*_{max} = maximum internal forces and moments obtained from a first order calculation
W = elastic or plastic section modulus
w = $0.6 + 0.4(M^*_{min}/M^*_{max}) \leq 0.4$ for linear variation of moment between the smaller end moment M^*_{min} and the larger M^*_{max}
w = 1.0 for all other cases

COMMENTARY

Equation 5b.W17 corresponds to the ECCS e* concept.

StBK N1 70

Depending on the axis about which the moment M_s is applied.

for flexure about the major-axis of an I-section

$$\left(\frac{N_{sd}}{N_{Rxcd}}\right)^{\lambda_{xc}} + \frac{M_{sxd}}{M_{Rxd}} \leq 1.0 \qquad (5b.W18)$$

for flexure about the minor-axis of an I-section

$$\left(\frac{N_{sd}}{N_{Rycd}}\right)^{\alpha_c} + \left(\frac{M_{syd}}{M_{Ryd}}\right)^{\gamma_{yc}} \leq 1.0 \qquad (5b.W19)$$

where

f_{yd}	= material strength
M_{Rxd}, M_{Ryd}	= moment capacity $\eta W f_{yd}$
M_{sxd}, M_{syd}	= moment determined by first order theory
N_{Rxcd}, N_{Rycd}	= column strengths
N_{sd}	= axial load
W	= elastic modulus
α_c	= $\alpha_o W_{yc} \geq 0.8$
α_o^2	= $\eta_x^2 \eta_y^2$ but ≥ 1 and ≤ 2
γ_o	= η_x^2 but ≥ 1
γ_x	= $\gamma_o W_{xc} \geq 0.8$
γ_{yc}	= $\gamma_o W_{yc} \geq 0.8$
η, η_x, η_y	= shape factors

COMMENTARY

Separate strength and stability checks are required because M_{sxd} and M_{syd} in Eqs. 5b.W18 and 5b.W19 may be obtained using the equivalent uniform moment concept.

Modifications to the above formulae are included for use with other forms of cross-section.

Note: International Comparisons is at the end of Chapter 5. (following page 5c(W)

5c. Out-of-Plane Member Capacity

SPECIFICATION PROVISIONS & EXPLANATION

AUSTRALIA

CHAPTER EDITOR
Mark A. Bradford, University of New South Wales

AS DR87164 87

For a compact doubly symmetric I-section member the nominal out-of-plane moment capacity M_{ox} for a member bent about its principal axis (for example, the x axis), which may buckle laterally is given by

$$M_{ox} = C_{bc}\, M_{bxo}\left[\left(1 - \frac{N^*}{\phi N_{cy}}\right)\left(1 - \frac{N^*}{N_z}\right)\right]^{1/2} \tag{5c.A1}$$

$$\frac{1}{C_{bc}} = \frac{1-\beta_m}{2} + \left(\frac{1+\beta_m}{2}\right)^3\left(0.4 - 0.23\,\frac{N^*}{\phi N_{cy}}\right) \tag{5c.A2}$$

$$N_z = (GJ + \pi^2 EI_w/L^2)\, A/(I_x + I_y) \tag{5c.A3}$$

where

M_{bxo} = the lateral buckling resistance of the member as a beam in uniform bending

N_{cy} = the axial load capacity for the member buckling out of the plane

For any other member, the nominal out-of-plane moment capacity for a member bent about its major (x) principal axis is given by

$$M_{ox} = M_{bx}\,(1 - N^*/\phi N_{cy}) \tag{5c.A4}$$

where M_{bx} is the lateral buckling resistance of the member as a beam.

CHINA

CHAPTER EDITORS
Shao-Fan Chen, Xian Institute of Metallurgy & Construction Engineering
Bao Kang He, Xian Institute of Metallurgy & Construction Engineering
Zu-Yan Shen, Tong-Ji University

GBJ 17 88

Lateral-torsional buckling checking formula is

$$\frac{N}{\phi_y A} + \frac{\beta_{tx} M_x}{\phi_b W_{1x}} \le f \qquad (5c.C1)$$

where

α_b = $I_1/(I_1 + I_2)$, see section 3a(C1)

T-sections, when the flange is compressed under the bending moment

Section built up with two angles $\phi_b = 1 - 0.0017 \cdot \lambda_y \sqrt{f_y/235}$

Section welded with two plates $\phi_b = 1 - 0.0022 \cdot \lambda_y \sqrt{f_y/235}$

T-sections when the flange is stretched under the bending moment

$\phi_b = 1.0$

β_{tx} = equivalent moment factor

1. For members having lateral supports
   (i) $\beta_{tx} = 0.65 + 0.35(M_2/M_1)$, when the member portion under consideration is not subjected to transverse loads. β_{tx} shall not be less than 0.4.
   (ii) $\beta_{tx} = 1.0$, when the member portion under consideration is subjected simultaneously to end moments and transverse loads, the curvatures produced by which have the same signs
   (iii) $\beta_{tx} = 0.85$, when the member portion under consideration is subjected simultaneously to end moments and transverse loads, the curvatures produced by which have the opposite signs.
2. For cantilevers $\beta_{tx} = 1.0$

ϕ_b = strength reduction factor of laterally unsupported beams subjected to uniform bending

For box section $\phi_b = 1.0$

For T- and I-sections, the following approximate formulas can be used to calculate ϕ_b:

Doubly symmetric I sections $\phi_b = 1.07 - \frac{\lambda_y^2}{44{,}000} \cdot \frac{f_y}{235}$

Monosymmetric I sections $\phi_b = 1.07 - \frac{W_{1x}}{(2\alpha_b + 0.1)A_h} \cdot \frac{\lambda_y^2}{14{,}000} \cdot \frac{f_y}{235}$

ϕ_y = strength reduction factor of centrally compressed members bending about the weak axis

For plastic design, the following formula is used

$$\frac{N}{\phi_y A f_p} + \frac{\beta_{tx} M_x}{\phi_b W_{px} f_p} \le 1 \qquad (5c.C2)$$

TBJ 2 85

The design formula for beam-columns loaded in one of its principal planes is

$$\frac{N}{\phi_1 A_m} + \frac{M}{\mu \phi_2 W_m} \leq [\sigma] \tag{5c.C3}$$

where

A_m, W_m = gross sectional area and gross section modulus

M = non-factored maximum bending moment within the middle third portion of the member

N = non-factored compressive axial force

n_1 = factor of safety = 1.7 for load combination of primary forces and 1.4 for load combination of primary and secondary forces.

μ = coefficient of bending moment magnification due to the presence of axial compression

for $\frac{N}{A_m} \leq 0.15\phi_1[\sigma]$ $\quad \mu=1$

for $\frac{N}{A_m} > 0.15\phi_1[\sigma]$ $\quad \mu=(1 - n_1 \frac{N}{N_e})$

$[\sigma]$ = allowable normal stress of steel

ϕ_1 = strength reduction factor of centrally compressed members

ϕ_2 = strength reduction factor of beams due to the presence of lateral-torsional buckling
For box section, $\phi_2 = 1$
For other cases, ϕ_2 is given in Chapter 3

COMMENTARY

The out-of-plane interaction equation of a beam-column in the elastic range is as follows

$$(1 - \frac{N}{N_y})(1 - \frac{N}{N_\omega}) - (\frac{M}{M_o})^2 = 0 \tag{5c.C4}$$

where

M_o = buckling moment of a uniformly bent beam

N_ω = torsional buckling load

N_y = Euler buckling load about the y-axis

Since $(N/N_y)-(M/M_o)$ curve plotted according to Eq. 5c.C4 is a convex curve, it can be conservatively replaced by a straight line. The interaction equation then becomes

$$\frac{N}{N_y} + \frac{M}{M_o} = 1 \tag{5c.C5}$$

Analysis has shown that for members of $\lambda_y = 60-105$ buckling in the inelastic range, the sum $\left(\frac{N}{N_y} + \frac{M}{M_o}\right)$ is always slightly larger than 1.0, varying from 1.01 to 1.07.

Equation 5c.C5 is also valid for T- and monosymmetric I-sections if N_y is replaced by flexural-torsional buckling load of an axially compressed member. Experimental results of 29 doubly symmetric I-sections, 4 monosymmetric I-sections and 13 T-sections have shown that Eq. 5c.C5 is always on the safe side. By introducing $N=\phi_y Af$, $M_o=\phi_b W_{1x} f$ and the equivalent moment factor β_{tx}, Eq. 5c.C1 is reached.

Equation 5c.C3 is different from 5a.C1 because a magnification coefficient μ is introduced to the moment term and it is in allowable stress format.

EAST EUROPE

CHAPTER EDITORS
Ioan A. Caraba, Polytechnic Institute of Timisoara
Miklos Ivanyi, Technical University of Budapest
Dan Mateescu, Polytechnic Institute of Timisoara

CMEA 83

In case where the lateral deformation is not prevented, the following relation is used

along the x-x direction

$$\frac{N}{AR_{fl}} + \frac{M_e}{R_{fll}W_{xp}(1 - \frac{N}{AR_{cr}})} \leq \phi \tag{5c.E1}$$

along the y-y direction

$$\frac{N}{AR_{fl}} + \frac{M_e}{R_{fll}W_{yp}(1 - \frac{N}{AR_{cr}})} \leq \phi \tag{5c.E2}$$

where

M_e = design equivalent bending moment, which produces the same effect as the actual loading

R_{cr} = buckling critical strength of the ideal bar having the same support conditions as the bar being checked.

R_{fl} = minimum buckling design strength, strength for λ_x, λ_y, for flexural buckling, and for λ_{tr}, for flexural torsional buckling, whose values are tabulated.

R_{fll} = lateral buckling design strength defined by the relation: $R_{fll} = R^m_{cr}/\gamma_{fll}$; R^m_{cr} represents the critical strength in the elastic and plastic range for the value of the maximum critical moment depending on the modified slenderness ratio λ_{tr} of the compression flange and on the steel grade; λ_{fll} is the safety coefficient of the member for lateral buckling.

W_{xp}, W_{yp} = plastic strength moduli of the cross section for the most compressed fiber.

ϕ = coefficient whose value is generally 1 or 1.1 depending on the cross section type.

SNiP II-23-81 82

If the moment of inertia of the section corresponding to this plane is smaller than the moment of inertia corresponding to the plane of the bending moment, ($I_y < I_x$) stability is checked by the relation

$$\frac{N}{c\varphi_y A} \leq R_y\varphi_c \tag{5c.E3}$$

where c is the coefficient that is determined depending on the eccentricity magnitude m_x, by means of the formulas.

for $m_x = M_x/N \leq 5$

$$c = \frac{\beta}{1 + \alpha m_x}$$

for $5 < m_x < 10$

$$c = c_5\,(2 - 0.2m_x) + c_{10}(0.2m_x - 1)$$

for $m_x \geq 10$

$$c = \frac{1}{1 + m_x \varphi_y/\varphi_b}$$

α, β = coefficients that have calculation expressions depending on the type of section, the magnitude of eccentricity m_x and the slenderness ratio

$$c_5 = \frac{\beta}{1 + 5\alpha} \quad \text{and} \quad c_{10} = \frac{1}{1 + 10\ \varphi_y/\varphi_b}$$

φ_b = lateral buckling coefficient of the most compressed flange
φ_y = the buckling coefficient is a function of non-dimensional slenderness $\overline{\lambda}$; it has the following value:

for $0 < \overline{\lambda} \leq 2.5$

$$\varphi_y = 1 - (0.073 - 5.530\ R_y/E)\ \overline{\lambda}\ \sqrt{\overline{\lambda}}$$

for $2.5 < \overline{\lambda} \leq 4.5$

$$\varphi_y = 1.47 - 13.00R_y/E - (0.371 - 27.300R_y/E)\ \overline{\lambda} + (0.0275 - 5.5300R_y/E)\ \overline{\lambda}$$

for $\overline{\lambda} > 4.5$

$$\varphi_y = 332/\overline{\lambda}^2\ (51 - \overline{\lambda})$$

TGL 13503 82

The moment acts about x-axis and buckling occurs about y-axis.

$$\sigma_c(1.0 + \mu_N\, f_N) + \sigma_{bcx}\, f_x + \sigma_{bcy}\, f_{yi} \leq \sigma_{all} \tag{5c.E4a}$$
$$\sigma_c(-1.0 + \mu_N\, f_N) + \sigma_{bzx}\, f_x + \sigma_{bzy}\, f_{yi} \leq \sigma_{all} \tag{5c.E4b}$$

where

$$c = \sqrt{\frac{I_\omega\,(\beta/\beta_o l_o)^2 + 0.039(\beta l)^2\, I_D}{I_y}}$$

f_N, μ_N = amplification and imperfection factor with a maximum slenderness ratio between λ_x and λ_{yi}
f_x = amplification factor calculated with λ_x
f_{yi} = amplification factor calculated with λ_{yi}
I_D = moment of inertia upon twisting
I_ω = sectorial moment of inertia
$i_m = \sqrt{i_p^2 + y_m^2}$
$i_p = \sqrt{i_x^2 + i_y^2}$
l, l_o = lengths upon bending and twisting

y_m = the distance between center of shearing and center of gravity

β, β_o = supporting factors upon bending and twisting ($\beta = \beta_o = 1$ - hinged supports; $\beta = \beta_o = 0.5$ - braced supports)

λ_x = slenderness ratio about x-axis

λ_{yi} = slenderness ratio for buckling with bending-torsion about symmetry y-axis

$$\lambda_{yi} = \frac{\beta l}{i_y}\sqrt{\frac{c^2 + i_m^2}{2c^2}\left(1 \pm \sqrt{1 - \frac{4c^2(i_p^2 + 0.093\ \beta/\beta_o - 1)y_m^2)}{(c^2 + i_m^2)^2}}\right.} \tag{5c.E5}$$

$\sigma_{bcx} = M_x/W_{dx}$

$\sigma_{bcy} = M_y/W_{dy}$

$\sigma_{bzx} = M_x/W_{zx}$

$\sigma_{bzy} = M_y/W_{zy}$

For point sections and double symmetrical sections the buckling is decisive when $\lambda_{yi} > \lambda_y$ and $\lambda_{yi} > \lambda_x$.

The slenderness ratio results

$$\lambda_{yi} = \frac{\beta l}{i_y} \tag{5c.E6}$$

$$\frac{i_p}{c} = \sqrt{\frac{I_x + I_y}{I_\omega/(\beta_o l_o)^2 + 0.039\ I_D}} \tag{5c.E7}$$

JAPAN

CHAPTER EDITORS
Takeshi Nakamura, Kyoto Institute of Technology
Koichi Takanashi, The University of Tokyo

AIJ 70

The allowable stress design approach is used, in which stresses of a beam-column subjected to combined axial compression and bending must satisfy the following equations

$$\frac{f_a}{{}_cF_a} + \frac{{}_cf_{bx,y}}{F_{bx,y}} \leq 1 \tag{5c.J1}$$

$$\frac{{}_tf_{bx,y} - f_a}{{}_tF_a} \leq 1 \tag{5c.J2}$$

where

$F_{bx,y}$ = permissible bending stress about strong, weak axis
${}_cF_a$ = permissible axial stress in compression
${}_tF_a$ = permissible axial stress in tension
f_a = applied axial stress
${}_cf_{bx,y}$ = compressive applied bending stress about strong, weak axis
${}_tf_{bx,y}$ = tensile applied bending stress about strong, weak axis

Stresses of beam-columns subjected to combined axial tension and bending must satisfy the following equations

$$\frac{f_a + {}_tf_{bx,y}}{{}_tF_a} \leq 1 \tag{5c.J3}$$

$$\frac{{}_cf_{bx,y} - f_a}{F_{bx,y}} \leq 1 \tag{5c.J4}$$

AIJ 90

$$\frac{M}{\phi_b M_{uy}} \leq 1.0 \tag{5c.J5}$$

$$\frac{P}{\phi_c P_{uy}} + 0.85\frac{M}{\phi_b M_{uy}} \leq 1.0 \tag{5c.J6}$$

where

M_{uy} = moment capacity with respect to the weak axis
P_{uy} = axial compression capacity for buckling out of the plane of bending
ϕ_b = resistance factor for bending
ϕ_c = resistance factor for compression

JRA 80

$$-\frac{{}_t f_a}{{}_t F_a} + \frac{{}_t f_{bx}}{F_{bx}} + \frac{{}_t f_{by}}{F_{bo}} \leq 1 \qquad (5c.J7)$$

$$\frac{{}_c f_a}{{}_c F_{ay}} + \frac{{}_c f_{bx}}{F_{bx}(1 - \frac{{}_c f_a}{F_{ex}})} + \frac{{}_c f_{by}}{F_{bo}(1 - \frac{{}_c f_a}{F_{ey}})} \leq 1 \qquad (5c.J8)$$

where

F_{bo} = the upper bound of permissible stress for bending about the strong axis without local buckling
F_{bx} = permissible stress for bending about the strong axis without local buckling
F_{ex} = Euler buckling stress with respect to the strong axis
F_{ey} = Euler buckling stress with respect to the weak axis
${}_c F_{ay}$ = permissible compressive stress for buckling with respect to the weak axis
${}_p F_a$ = permissible compressive stress for local buckling
${}_t F_a$ = permissible tensile stress
${}_c f_a$ = applied axial compressive stress
${}_c f_{bx} / {}_c f_{by}$ = applied compressive stress due to bending about the strong axis/the weak axis
${}_t f_a$ = applied axial tensile stress
${}_t f_{bx} / {}_t f_{by}$ = applied tensile stress due to bending about the strong axis/the weak axis

COMMENTARY

Allowable stress design approach is adopted. When non-uniform bending is applied, F_{bx} can be increased by being multiplied (M/M_{eq})

where

$M_{eq} = 0.6\ M_1 + 0.4\ M_2 \geq 0.4M_2$
M_1 = larger end moment
M_2 = smaller end moment

JSCE 83

$$\frac{P}{P_{ca}} + \frac{M_x}{M_{xa}} + \frac{M_y}{M_{yao}} \leq 1.0 \qquad (5c.J9)$$

in which P = 0, in case the axial force is in tension.

where

M_{xa} = permissible lateral buckling bending moment with respect to the strong axis
M_{yao} = permissible bending moment with respect to the weak axis
P_{ca} = the smaller out of the permissible buckling loads with respect to the strong and the weak axis

NORTH AMERICA

CHAPTER EDITOR
Wai-Fah Chen, Purdue University

AISC 78

See Eqs. 5b.N2, 5d.N4, and 5d.N5

AISC LRFD 86

See Eqs. 5d.N6 and 5d.N7

WEST EUROPE

CHAPTER EDITOR
David A. Nethercot, The University of Nottingham

BS 5400 P.3 82

$$\frac{F}{P_{cy}} + \frac{M_x}{M_b} = 1 \tag{5c.W1}$$

where F, M_x is the maximum applied compression and moment within the middle third of the member length between points of restraint.

COMMENTARY

M_x should be determined taking amplification effects into account where appropriate.

BS 5950 P.1 85

For out-of-plane buckling under compression and major-axis moment the two alternative Code provisions, presented in the context of the full biaxial problem, both reduce to

$$\frac{F}{P_{cy}} + \frac{mM_x}{M_b} = 1 \tag{5c.W2}$$

CNR 10111 85

$$\frac{wN}{A} + \frac{M_{xeq}\, f_y/\sigma_D}{\phi_x\, W_x(1 - N/N_{cr,x})} = f_d \tag{5c.W3}$$

where

f_y = material yield strength
w = (material yield strength)/(column strength)
σ_D = limiting stress for lateral-torsional buckling
ϕ_x = plastic adaptation factor

CTICM 80

Out-of-plane buckling is controlled by

$$k_o \frac{N}{N_p} + \frac{k_{fx}}{k_D}\frac{M_{mx}}{M_{px}} \leq 1.0 \tag{5c.W4}$$

where k_D is the coefficient to allow for lateral torsional buckling

DIN 18800 P.2 86

For checking lateral-torsional buckling under compression and strong-axis bending the condition is

$$\frac{F}{\chi_y\, A\, p_y} + \frac{M_x}{\chi_M\, M_{plx}}\, k_x \le 1 \qquad (5c.W5)$$

where

$k_x = 1 - \frac{F}{\chi_y\, A\, p_y}[0.15\overline{\lambda}_y\beta_M - 0.15]$

$k_y \le 1$ with the term in square brackets in $k_y \le 0.9$

χ_M = lateral-torsional buckling coefficient

EC 3 84

Buckling out of the plane of the applied moments should be checked using

$$\frac{F}{\chi_y\, p_y\, A} + \left(\frac{M}{\chi_M\, M_{cx}}\right)^{\beta_M} \le 1 \qquad (5c.W6)$$

where

β_M = coefficient to allow for moment distribution, see Table 5c.W1

χ_M = reduction coefficient for lateral-torsional buckling

Table 5c.W1 Coefficient β_M

Moment Distribution	β_M
M_{max}  ψM	$1.8 - 0.7\psi$
M_{max} 	1.3
M_{max} 	1.4

NBN B51002 85

$$\frac{F}{A\chi} + \frac{m_x\, M_x\, \chi_m}{Z_x(1 - F/P_{crx})} = p_y \qquad (5c.W7)$$

where $m_x = [0.3(1 + (\frac{M_{min}}{M_{max}})^2) + 0.4\frac{M_{min}}{M_{max}}]^{0.5}$

In cases where $F/A\chi$ is less than 0.15, m_x and $(1 - F/P_{crx})$ should be taken as unity.

NEN 3851 72

$$\frac{F}{\chi_y\, A} + \chi_m m_x \left[\frac{P_{crx}/F}{P_{crx}/F - 1}\right] \left|\frac{m_x\, M_{xz} + M_{Dx}}{Z_x}\right| = p_y \qquad (5c.W8)$$

$$\frac{F}{A} + \chi_m \left|\frac{M_{xz} + M_{Dx}}{Z_x}\right| = p_y \qquad (5c.W9)$$

where

M_{Dx} = maximum "free" moment with member length, taken as zero if $|M_x| \leq |2M_z|$

M_{xz} = larger end moment

p_b = lateral-torsional buckling design stress

$\chi_m = \dfrac{5p_y}{p_b(8-3\beta_x)}$

COMMENTARY

Equations 5c.W8 and 5c.W9 are based on the full ECCS e* concept, generalized to include the effects of intermediate transverse loads. Interpretation of M_{Dx} for specific cases is facilitated by the provision of an explanatory table. The basis for its inclusion is a simple addition of the maximum mid-span moment caused by transverse loads in the statically determinate system to the mid-span moments due to top and bottom moments.

NS 3472 84

$$\frac{N}{N_{kyd}} + \frac{\overline{M}_x}{M_{vd}}\frac{1}{K_E}\frac{1}{1-\dfrac{N}{N_{Eyd}}\dfrac{N_{kyd}}{N_d}} = 1 \qquad (5c.W10)$$

where

$K_E = (1 - N/N_{exd})/(1 - N/N_{eyd})$

M_{vd} = moment capacity allowing for lateral-torsional buckling

$\overline{M}_x = mM_{xmax}$ = equivalent uniform moment

$m = 0.6 + 0.4\beta \geq 0.4$ if sidesway is prevented

$m = 1.0$ if sidesway is permitted

N_{Eyd} = weak axis column strength based on effective column length

N_d = squash load of member

N_{kyd} = weak axis column strength

StBK 70

For out-of-plane buckling of doubly symmetrical I-sections the following formulas should be used

$$\left(\frac{N_{sd}}{N_{Rxcd}}\right)^{\gamma_{xc}} + \frac{M_{sxd}}{M_{Rxd}} \leq 1.0 \qquad (5c.W11)$$

$$\left(\frac{N_{sd}}{N_{Rycd}}\right)^{\alpha_c} + \left(\frac{M_{sxd}}{M_{Rxcd}}\right)^{\beta_c} \leq 1.0 \qquad (5c.W12)$$

where M_{Rxcd} is the moment capacity allowing for lateral-torsional buckling

COMMENTARY

In determining N_{Rycd} both flexural buckling about the weak axis and flexural-torsional buckling should be considered. For all other types of cross-section the user is directed to the technical literature.

5d. Biaxial Bending

SPECIFICATION PROVISIONS & EXPLANATION

AUSTRALIA

CHAPTER EDITOR
Mark A. Bradford, University of New South Wales

AS DR87164 87

A member subject to biaxial bending is required to satisfy

$$\left(\frac{M_x^*}{\phi M_{ux}}\right)^{\gamma_m} + \left(\frac{M_y^*}{\phi M_{iy}}\right)^{\gamma_m} \leq 1 \qquad (5d.A1)$$

where

M_{ux} = lesser of the in-plane moment capacity M_{ix} and the out-of-plane moment capacity M_{ox}

γ_m = 1.4

COMMENTARY

Linear interaction equations are given for the section and moment capacities, which may be used for any type of cross-section. These equations provide linear interpolations between the capacities of beams and compression members for the effects of yielding, local buckling, and member buckling. They are easy to use, but are often conservative.

Non-linear interaction equations may be used for doubly symmetric compact I-section members. These equations are more difficult to use than the linear equations, but are often more economical, especially for members with high moment gradients.

Compact I-section beam-columns are defined as those that both satisfy the compact section requirements for beams (Chapter 3), which have fully effective sections as compression members (Chapter 2). This definition makes it unnecessary to consider the actual variation in the stresses across the section of the beam-column and their effect on local buckling.

The basis for determining the uniaxial section capacities for compact I-sections is well-established (Trahair, 1988). The non-dimensional interaction equation for the section capacity in biaxial bending incorporates an index γ_s. The value of γ_s is an approximation to the research findings (Tebedge, 1974).

The expression for the in-plane moment capacity M_i is a cubic approximation to the more accurate analytical interaction equations (Trahair, 1986). For members with transverse loads, a modified value of β_m is used in this expression (Bridge, 1987).

The expression for out-of-plane moment capacity M_{ox} has been developed from both analytical and experimental studies (Cuk, 1986).

Finally, the non-dimensional interaction equation for member capacity incorporates in index γ_m. The value of 1.4 for γ_m is a lower bound to the more complex equation based on research findings (Tebedge, 1974).

CHINA

CHAPTER EDITORS
Shao-Fan Chen, Xian Institute of Metallurgy & Construction Engineering
Bao Kang He, Xian Institute of Metallurgy & Construction Engineering
Zu-Yan Shen, Tong-Ji University

GBJ 17 88

The design formulas for bi-axially loaded beam-columns with symmetrical I-sections or box sections are given by

$$\frac{N}{\phi_x A} + \frac{\beta_{mx} M_x}{\gamma_x W_{1x}(1 - 0.8 \frac{N}{N_{Ex}})} + \frac{\beta_{ty} M_y}{\phi_{by} W_{1y}} \leq f \qquad (5d.C1)$$

$$\frac{N}{\phi_y A} + \frac{\beta_{tx} M_x}{\phi_{bx} W_{1x}} + \frac{\beta_{my} M_y}{\gamma_y W_{1y}(1 - 0.8 \frac{N}{N_{Ey}})} \leq f \qquad (5d.C2)$$

All the notations used in Eqs. 5d.C1 and 5d.C2 are the same as that in Eqs. 5a.C1, 5b.C1, and 5c.C1, and the subscripts y represents the y-axis.

No formula is given for plastic design.

TBJ 2 85

No provision is given.

COMMENTARY

For biaxially loaded beam-columns with symmetrical I-sections or box sections, simplified linear interaction equations are adopted. They are established to have the form that is compatible with the conditions of centrally loaded columns, uniaxially loaded beam-columns, and beams of uniaxial bending and biaxial bending.

As between Eqs. 5d.C1 and 5d.C2, the latter is usually the governing one in design. This equation has been checked by theoretical strengths computed on a welded I-section of h/b=2 and proves to be satisfactory. Comparison with three groups of experimental data (Klöppel, Winkelmann, Chuvikin, Massonnet) has shown a remarkable margin of safety. This is because the specimens tested were all strong in resisting torsion.

EAST EUROPE

CHAPTER EDITORS
Ioan A. Caraba, Polytechnic Institute of Timisoara
Miklos Ivanyi, Technical University of Budapest
Dan Mateescu, Polytechnic Institute of Timisoara

CMEA

Members subject to compression and biaxial bending are checked by the following relation

$$\frac{N}{AR_{fl}} + \frac{M_e^x}{R_{fll}W_{xp}(1 - \frac{N}{AR_{cr,x}})} + \frac{M_e^y}{R_{fll}W_{yp}(1 - \frac{N}{AR_{cr,y}})} \leq \phi \qquad (5d.E1)$$

The terms in Eq. (5d.E1) have the same significance as the terms in Eqs. 5b.E1 - 5b.E2.

PN 80/B-03200 87

For members with full section under compression and biaxial bending, the bending buckling is checked by the relations:

$$\sigma = \frac{P\ m_w}{k\ F_{br}} + \frac{M_x}{I_{x,br}} y + \frac{M_y}{I_{y,br}} x \leq 1.1R \qquad (5d.E2a)$$

$$\sigma = \frac{Pm_w}{F_{br}} + \frac{300 + 2\lambda}{1000} + \frac{M_x}{I_{x,br}} y_r + \frac{300 + 2\lambda}{1000} \frac{M_y}{I_{y,br}} x_r \leq 1.1R \quad (5d.E2b)$$

where

x, y = coordinates of the point where σ is calculated with respect to the principal axes of inertia

x_r, y_r = the coordinates of the stretched or less compressed fiber from the cross sectional plane, with respect to the principal axes of inertia.

The torsion-bending buckling is checked by the relations Eq. 5b.E8, 5b.E9, 5d.E3 where m_w is calculated as a function of the slenderness ratio λ_j, determined for monosymmetrical sections, by the relation

$$\lambda_j = \gamma_j \lambda_y \qquad (5d.E3)$$

where

$c^2 = (\frac{\mu l_t}{\mu_o l_o})^2 \frac{I_\omega}{I_y} + \frac{0.04(\mu l_t) I_D}{I_y}$

$e = M/P$ (Fig. 5b.E2)

$i_p^2 = i_x^2 + i_y^2$

$i_s^2 = i_p^2 + y_s^2$

l_t, l_o = length of the member respectively, the distance between two consecutive joints along the member

$$r_x = \int \frac{y(x^2 + y^2)}{I_x} dA$$

y_s = distance measured along the y-y axis on the cross section between the center of gravity and the center of torsion.

$$\gamma_j = \sqrt{\frac{c^2 + i_s^2 + e(r_x - 2y_s)}{2c^2}} \left\{ 1 \pm \sqrt{\frac{4c^2[i_p^2 + e(r_x - e) + 0.1\,(\frac{\mu^2}{\mu_o^2} - 1)\,(e - y_s)^2]}{[c_2^2 + i_s^2 + e(r_x - 2y_s)^2]}} \right.$$

$$\lambda_y = \frac{\mu l_t}{i_y}$$

μ, μ_o = multiplication coefficients of the lengths l and l_o whose values range between 1 and 0.5

If the eccentric compression force P is applied in the center of torsion ($e = y_s$) then the slenderness ratio λ_j is calculated by the relation

$$\lambda_j = \gamma_k \lambda_y \tag{5d.E4}$$

where γ_k is equal to $\sqrt{\dfrac{i_s^2 + y_s\,(\gamma_x - 2y_s)}{c^2}}$

If $\gamma_k \leq 1$, then m_w must be determined as a function of $\lambda_j = \lambda_y$.

SNiP II-23-81 82

To check the stability of members with full section subjected to compression and bending along two directions the following formula is applied

$$\frac{N}{\phi_{exy} A} \leq R_y \gamma_c \tag{5d.E5}$$

where

c = is determined depending on the magnitude of eccentricity m_y, from Eq. 5c.E3.

ϕ_{exy} = $\phi_{ey}(0.6\ c^{1/3} + 0.4\ c^{1/4})$

ϕ_{ey} = is determined depending on $\overline{\lambda}_y$ and $m_{ef,y}$ by replacing λ with $\lambda_y = \lambda_{fy}/_{iy}$ in Eq. 5b.E10 and m with $m_y = M_y/N$ in Eq. 5b.E12.

If $m_{ef,y} < m_x$, then, besides the check of the member stability with Eq. 5d.E5, it is necessary to carry two further checks by Eq. 5b.E10 and 5c.E3 where $e_y = 0$. In this case, the magnitude of relative eccentricities is determined by the relations

$$m_x = e_x \frac{A}{W_{cx}} \tag{5d.E6}$$

$$m_y = e_y \frac{A}{W_{cy}} \tag{5d.E7}$$

where W_{cx}, W_{cy} are the strength moduli of the cross section calculated for the most compressed fibers with respect to the x-x and y-y axes.

If $\lambda_x > \lambda_y$, then in addition to the check of the member stability with Eq. 5d.E5, it is necessary to carry out a supplementary checking by Eq. 5b.E10, where $e_y = 0$

In case the symmetry plane of the cross section does not coincide with the maximum stiffness plane ($I_x > I_y$) then the calculated value of eccentricity is increased by 25%.

STAS 763/1 71

$$\frac{N}{\varphi A} + \frac{M_x}{W_x} + \frac{M_y}{W_y} \leq 1.10\sigma_a \tag{5d.E8}$$

where φ is the buckling coefficient that is determined from tables depending on maximum slenderness, between λ_x and λ_y and on steel grade.

STAS 1911 75

Members with unitary section, under centric or eccentric compression, are checked for buckling by the equation

$$\sigma = \frac{N}{\varphi A} + 0.9\,\frac{M_x}{W_x} + 0.9\,\frac{M_y}{W_y} \leq \sigma_a \tag{5d.E9}$$

where φ is the buckling coefficient determined as a function of section group, steel grade, and slenderness ratio λ.

STAS 10108/0 78

For bars subjected to compression and bending in two directions, stability checking is obtained as

$$\frac{N}{\varphi A} + \frac{c_x M_x}{\varphi_g(1 - \frac{\varphi}{\varphi_E})\, W_x} + \frac{c_y M_y}{(1 - \frac{\varphi}{\varphi_E})\, W_y} \leq 1.1R \tag{5d.E10}$$

where

M_x,M_y = the maximum respectively different bending moments, M_x=maximum, M_y=different and converse

φ = the buckling coefficient corresponding to the buckling plane

φ_g = lateral buckling coefficient

φ_E = Euler buckling coefficient correpsonding to the buckling plane

JAPAN

CHAPTER EDITORS
Takeshi Nakamura, Kyoto Institute of Technology
Koichi Takanashi, The University of Tokyo

AIJ 90

For members without buckling

$$\frac{P}{\phi_p P_y} + 0.85 \frac{M_{1x}}{\phi_p M_{px}} + 0.85 \frac{M_{1y}}{\phi_p M_{py}} \leq 1 \quad (5d.J1)$$

For members with buckling

$$\frac{P}{\phi_c P_{uy}} + 0.85 \frac{M_{1x}}{\phi_b M_{ux}} + 0.85 \frac{M_{1y}}{\phi_p M_{py}} \leq 1 \quad (5d.J2)$$

where

M_{px}, M_{py} = plastic moment with respect to the strong and the weak axis
M_{ux} = moment capacity with respect to the strong axis
M_{1x}, M_{1y} = applied bending moment with respect to the strong and the weak axis, respectively
P_{uy} = axial compressive capacity for buckling out of plane of bending
P_y = axial compressive capacity for yielding
ϕ_b = resistance factor for bending
ϕ_c = resistance factor for compression
ϕ_p = resistance factor for tension

NORTH AMERICA

CHAPTER EDITOR
Wai-Fah Chen, Purdue University

AA 86

Bending moment at center $\geq$ 0.9 of maximum bending moment in span

$$\frac{f_a}{F_a} + \frac{f_b}{F_b(1-f_a/F_{ec})} \leq 1.0 \tag{5d.N1}$$

where

F_a and F_b = allowable axial and bending stresses
F_{ec} = $\pi^2 E/[\eta_u (L/r)^2]$ where η_u is factor of safety on yield strength
f_a and f_b = applied axial and bending stresses

Bending moment at center $\leq$ 0.5 of maximum bending moment in span

$$\frac{f_a}{F_a} + \frac{f_b}{F_b} \leq 1.0 \tag{5d.N2}$$

Bending moment at center between 0.5 and 0.9 of maximum bending moment in span

$$\frac{f_a}{F_a} + \frac{f_b}{F_b\left[1-\left(\frac{2M_c}{M_m} - 1\right)\frac{f_a}{F_{ec}}\right]} \leq 1.0 \tag{5d.N3}$$

where

M_c = moment at center of span
M_m = maximum bending moment in span

AISC ASD 78

If stability controls, the interaction equation is,

$$\frac{f_a}{F_a} + \frac{C_{mx} f_{bx}}{F_{bx}(1 - f_a/F'_{ex})} + \frac{C_{my} f_{by}}{F_{by}(1 - f_a/F'_{ey})} \leq 1.0 \tag{5d.N4}$$

where

F_a, F_{bx}, F_{by} = allowable axial compressive stress and allowable flexural stresses, respectively
f_a, f_{bx}, f_{by} = first-order elastic axial stress, and flexural stresses at service load, respectively

For members braced against joint translation and without transverse loading between supports, C_m is referred to as the equivalent moment factor and is defined and $C_m = 0.6 - 0.4\, M_1/M_2 \geq 0.4$ where M_1/M_2 is the ratio of the smaller to larger end moments. It is negative if the member is bent in single curvature and is positive if the member is bent in reverse curvature.

For members braced against joint translation with transverse loading between supports, C_m is referred to as the moment reduction factor. It is defined as $C_m = 1 + \psi\ P/P_e$ and it is an integral part of the moment magnifier. However, for simplicity, the AISC ASD 78 suggests the use of $C_m = 0.85$ for members with restrained ends and $C_m = 1.0$ for members with unrestrained ends.

For members not braced against joint translation, C_m is considered 0.85.

F'_{ex}, F'_{ey} are the critical elastic buckling stresses about the x and y axes, respectively, divided by a factor of safety of 23/12 and evaluated using the effective length of the member.

In actual design, both the stability and yield interaction equations should be checked. However, if the axial force in the member is small, that is, $f_a/F_a \leq 0.15$ the specification allows the use of the following interaction equation, instead of the other two equations.

$$\frac{f_a}{F_a} + \frac{f_{bx}}{F_{bx}} + \frac{f_{by}}{F_{by}} \leq 1.0 \qquad (5d.N5)$$

Thus, in the design, the value f_a/F_a is first evaluated. If f_a/F_a is less than or equal to 0.15, this equation is used to check the adequacy of the section.

If f_a/F_a is greater than 0.15, then the first two equations are used to check the adequacy of the section.

Biaxial bending of beam-columns is not permitted in plastic design.

AISC LRFD 86

The AISC LRFD format is based on the exact inelastic solutions of 82 beam-columns and uses the following bi-linear interaction equation for sway and non-sway beam-columns. It combines the section capacity check with the in-plane and out-of-plane bending capacity checks into one bi-axial bending capacity equation of the simple form.

for $P/\phi_c P_u \geq 0.2$

$$\frac{P}{\phi_c P_u} + \frac{8}{9}\left(\frac{M_{ax}}{\phi_b M_{ux}} + \frac{M_{ay}}{\phi_b M_{uy}}\right) \leq 1.0 \qquad (5d.N6)$$

for $P/\phi_c P_u < 0.2$

$$\frac{P}{2\phi_c P_u} + \frac{M_{ax}}{\phi_b M_{ux}} + \frac{M_{ay}}{\phi_b M_{uy}} \leq 1.0 \qquad (5d.N7)$$

where

M_{ax}, M_{ay} = second-order design moment for the section bent about the x and y axes, respectively, calculated either by a direct second-order elastic frame analysis or by a first-order elastic frame analysis using the following amplification factors

$$M_a = B_1 M_{nt} + B_2 M_{lt} \qquad (5d.N8)$$

M_{ux}, M_{uy} = ultimate moment resisting capacity of the member bent about the x and y axes, respectively, according to the beam strength curve

P_u = axial compressive strength in accordance with column strength curve

ϕ_c, ϕ_b = column resistance factor (=0.85) and the beam resistance factor (=0.90) respectively

where

B_1 = P-δ moment amplification factor (designated as A_F in the previous section, see Eq. 5a.N3)

$$B_1 = \frac{C_m}{1 - \dfrac{P}{P_{ek}}} \geq 1.0 \qquad \text{(5d.N9)}$$

B_2 = P-Δ moment amplification factor, evaluated by

$$B_2 = \frac{1}{1 - \dfrac{\Sigma P \Delta_o}{\Sigma HL}} \qquad \text{(5d.N10)}$$

or alternatively

$$B_2 = \frac{C_m}{1 - \dfrac{\Sigma P}{\Sigma P_{ek}}} \qquad \text{(5d.N11)}$$

M_{lt} = moment in member as a result of lateral translation of the frame only calculated by using first-order elastic analysis;

M_{nt} = moment in member assuming there is no lateral translation in the frame calculated by using first-order elastic analysis;

where

C_m = $0.6 - 0.4M_1/M_2$ same as defined previously in ASD, except that the limit condition $C_m > 0.4$ has been removed in LRFD. This limit was found to be too conservative for $M_1/M_2 = 0.5$ to 1.0 when compared with the elastic C_m factor used in ASD with the exact elastic-plastic computer solutions.

L = story height

Δ_o = first-order translational deflection of the story under consideration

ΣH = sum of all story horizontal forces producing Δ_o

ΣP = axial loads on all columns in a story

Alternate Interaction Equations

For biaxially loaded I-shaped members used in braced frames only, nonlinear biaxial interaction equations for section capacity and member stability capacity are provided in the appendix of the specification.

Non-linear Biaxial Bending

These equations are given as follows based on the research findings reported in (Tebedge and Chen, 1974) and summarized in the two-volume book (Chen, 1976-1977).

If stability controls

$$\left(\frac{C_{mx}\,M_{ox}}{\phi_b\,M'_{nx}}\right)^{\eta} + \left(\frac{C_{my}\,M_{oy}}{\phi_b\,M'_{ny}}\right)^{\eta} \leq 1.0 \tag{5d.N12}$$

where

b_f = flange width
d = member depth
$M'_{nx} = M_{nx}[1-(P/\phi_c P_u)](1-P/P_{ex})$ (5d.N13)
$M'_{ny} = M_{ny}[1-(P/\phi_c P_u)](1-P/P_{ey})$
$\eta = 0.4+P/P_y+b_f/d \geq 1.0$ for $b_f/d \geq 0.3$
$\eta = 1.0$ for $b_f/d < 0.3$

COMMENTARY

The P-Δ moment amplification for the first B_2 factor was developed based on the story stiffness concept, that is, (i) each story can behave independently of other stories, and (ii) the additional moments in the columns caused by the P-Δ effect is equivalent to that caused by a lateral force of $\Sigma P\Delta/L$.

The alternate expression for B_2 factor was developed based on the multiple column buckling concept, that is, when instability is to occur in a story, all columns in that story will become unstable simultaneously.

The use of this method will sometimes overestimate the maximum moment in the member because the amplified moments resulting from the P-δ and P-Δ effects do not necessarily coincide at the same location. For elastic behavior, the P-Δ effect usually magnifies the end moments. However, because of the assumptions involved in developing the P-δ and P-Δ amplification factors, as well as the difficulties involved in locating the exact location of each of the magnified moments in the member, it is justifiable to estimate the second-order design moment for the member by simply summing up the two effects together.

Note that, unlike the ASD and PD interaction equations in which both the yielding and stability interaction equations are needed in the design process, only one interaction equation is needed if the LRFD approach is used. The applicable equation is determined by the term $P/\phi_c P_u$. In the ASD or PD approach, if the member is subjected to sway, the factor C_m is taken as 0.85, therefore, the moment magnification factor is $0.85/(1-P/P_{ek})$ and this moment magnification factor is applied to the total first-order moment of the member despite whether it is caused by gravity load (M_{nt}) or lateral loads (M_{lt}).

AREA 87

Member Stability

for $f_a/F_a \leq 0.15$

$$\frac{f_a}{F_a} + \frac{f_{b1}}{F_{b1}} + \frac{f_{b2}}{F_{b2}} \leq 1.0 \tag{5d.N14}$$

for $f_a/F_a > 0.15$

$$\frac{f_a}{F_a} + \frac{f_{b1}}{F_{b1}\left[1 - \frac{f_a}{200x10^6}\left(\frac{k_1 l_1}{r_1}\right)^2\right]} + \frac{f_{b2}}{F_{b2}\left[1 - \frac{f_a}{200x10^6}\left(\frac{k_2 l_2}{r_2}\right)^2\right]} \leq 1.0 \quad (5d.N15)$$

where

F_a, F_{b1}, and F_{b2} = allowable axial stress and bending stresses about axes 1-1 and 2-2 respectively

f_a, f_{b1}, and f_{b2} = computed axial stress and bending stresses about axes 1-1 and 2-2 respectively

$\frac{k_1 l_1}{r_1}$ and $\frac{k_2 l_2}{r_2}$ = effective length ratio

CSA CAN3-S6-M78

Member Stability

$$\frac{f_a}{F_a} + \frac{f_b}{F_b} \leq 1.0 \quad (5d.N16)$$

where

F_a and F_b = allowable axial and bending stresses
f_a and f_b = calculated axial and bending stresses

in which f_b shall be determined from the bending moment value not less than

(a) $\sqrt{0.3(M_1^2 + M_2^2) + 0.4 M_1 M_2}$ if the ends of the member are restrained against translation

(b) $0.85M_1$ if the ends of the member are restrained.

where

M_1 = larger end moment
M_2 = lesser end moment

The sign of M_2 is negative with respect to M_1 if M_2 causes member curvature opposite to M_1.

CSA S16.1 M89 84b

Linear Interaction Equation

The following expression is applied identically to all members, despite class or shape

$$\frac{C_f}{C_r} + \frac{\omega_x M_{fx}}{M_{rx}(1 - \frac{C_f}{C_{ex}})} + \frac{\omega_y M_{fy}}{M_{ry}(1 - \frac{C_f}{C_{ey}})} \leq 1.0 \quad (5d.N17)$$

where

C_r = the capacity of the member under axial load only, normally based on SSRC column curve #2 (Galambos, 1988), but curve #1 for partially stress relieved hollow structural sections (to which roller straightened I-shapes are being added), and curve #3 for very heavy sections

M_{rx}, M_{ry} = maximum uniform moment capacities of the member about the respective axes, having regard for lateral torsional buckling

ω = equivalent moment factor, taken as:
$\omega = 0.6 + 0.4M_1/M_2$ for single curvature
$\omega = 0.6 - 0.4M_1/M_2$ but not less than 0.4 for double curvature

For I-shapes, the maximum value of M_{rx} will be M_{px} for class 1 & 2 sections, $M_{yield\ x}$ for class 3 sections. M_{ry} will be M_{py} for class 1 & 2 sections and $M_{yield\ y}$ for class 3 sections.

Nonlinear Interaction Equation

More refined expressions for class 1 & 2 sections are given in an appendix.

$$\left(\frac{\omega_x M_{fx}}{M_{ox}}\right)^{\eta} + \left(\frac{\omega_y M_{fy}}{M_{oy}}\right)^{\eta} \leq 1.0 \qquad (5d.N18)$$

where M_{ox}, M_{oy} are the maximum single curvature factored moment resistances of the column in the presence of axial load, having regard for lateral torsional buckling, but in the absence of the other orthogonal moment, which may be taken as:

$$M_{ox} = M_{rx}(1 - C_f/C_r)(1 - C_f/C_{ex}) \qquad (5d.N19)$$
$$M_{oy} = M_{ry}(1 - C_f/C_r)(1 - C_f/C_{ey})$$

for $b/d \geq 0.3$

$$\eta = 0.4 + C_f/C_y + b/d$$

for $b/d < 0.3$

$$\eta = 1.0$$

where

b = width of I-shapes

d = depth of I-shapes

For hot rolled partially stress relieved hollow square structural sections, the sum of the moment terms in the interaction equation, mentioned above, may be reduced by the coefficient ν where

$$\nu = \frac{[(\omega_x M_{fx})^2 + (\omega_y M_{fy})^2]^{1/2}}{\omega_x M_{fx} + \omega_y M_{fy}} \qquad (5d.N20)$$

COMMENTARY

Design is based generally on elastic analysis of the structural framework. Plastic analysis is admissible subjected to the customary provisions relating to lateral bracing, material thicknesses and material properties. A second-order inelastic analysis is required if the influence of inelastic deformation on the strength of the structure is significant.

In the normally used elastic analysis, instability effects are taken into account by the requirement for a second-order elastic analysis. Thus, the effective length of beam-columns normally is considered unity.

In common with almost all codes, members are selected based on their ultimate inelastic strength. This incongruity appears accepted almost universally at this time.

Stability constraints currently being introduced are:

– despite lateral wind, seismic or equipment lateral forces, the minimum lateral design force shall be 0.5 percent of the dead and live loads.
– that elastic second-order effects shall not exceed a prescribed fraction of the first-order effects unless an inelastic second-order analysis is done.

OHBDC 83

Compact Sections, H-Shapes

$$\frac{C_f}{C_r} + 0.85\,\frac{M_{fx}}{M_{rx}} + 0.60\,\frac{M_{fy}}{M_{ry}} \leq 1.0 \qquad (5d.N21)$$

Other Singly- and Doubly-symmetrical Sections

$$\frac{C_f}{C_r} + \frac{\omega_x M_{fx}}{M_{rx}\left(1-\frac{C_f}{C_{ex}}\right)} + \frac{\omega_y M_{fy}}{M_{ry}\left(1-\frac{C_f}{C_{ey}}\right)} \leq 1.0 \qquad (5d.N22)$$

WEST EUROPE

CHAPTER EDITOR
David A. Nethercot, The University of Nottingham

BS 5400 P.3 82

Overall buckling should be checked using

$$\frac{F}{P_c} + \frac{M_x}{M_{cx}} + \frac{M_y}{M_{cy}} \leq 1 \qquad (5d.W1)$$

where F, M_x, M_y are the maximum applied compression and moments respectively, within the middle third of the member length between points of restraint.

BS 5950 P.5 85

Overall buckling should be checked using

$$\frac{m_x M_x}{M_{ax}} + \frac{m_y M_y}{M_{ay}} \leq 1 \qquad (5d.W2)$$

where

M_{ax} = maximum buckling moment about the x-axis in the presence of axial load F, taken as the lesser of

$$M_{ax} = \frac{1 - F/P_{cx}}{(1 + 0.5\, F/P_{cx})} M_{cx} \qquad (5d.W3)$$

or

$$M_{ax} = (1 - F/P_{cy})\, M_b \qquad (5d.W4)$$

M_{ay} = maximum buckling moment about the y-axis in the presence of axial load F, taken as

$$M_{ay} = \frac{1 - F/P_{cy}}{(1 + 0.5\, F/P_{cy})} M_{cy} \qquad (5d.W5)$$

M_b = lateral torsional buckling resistance moment
M_{cx}, M_{cy} = moment capacity about x and y-axes, respectively, calculated as $p_y S$
m_x, m_y = equivalent uniform moment factors
P_{cx}, P_{cy} = compressive resistance about x and y-axes, respectively*

As an alternative the following simplified overall buckling check may be used

$$\frac{F}{A\, p_c} + \frac{m_x M_x}{M_b} + \frac{m_y M_y}{p_y Z_y} \leq 1 \qquad (5d.W6)$$

where

p_c = compressive strength*
Z_y = elastic modulus about the minor axis

* See Chapter 1 – Compression Members

COMMENTARY

Equation 5d.W2 follows the so-called "Chen concept" (Galambos, 1988) with powers of unity on both terms. For uniaxial bending the appropriate component Eqs. 5d.W3–W5 should be satisfied. Of these Eq. 5d.W3 and 5d.W5 represents a simplification of the original ECCS e* concept for in-plane failure ECCS (1978).

The alternative, simplified approach of Eq. 5d.W6 presumes that strength due to the action of F and M_x will be controlled by lateral-torsional buckling, that is, Eq. 5d.W1 rather than Eq. 5d.W6 is critical; it is thus less easy to interpret in situations where this is not the case for example, for closed sections. Due to the nature of the approximations involved it also gives higher results than the more exact method in certain instances.

CNR 10111 85

For compression plus biaxial bending it is necessary to satisfy

$$\frac{F}{\chi\, A\, p_y} + \frac{m_x\, M_x}{\chi\, M_{ux}(1 - F/P_{crx})} + \frac{m_y\, M_y}{M_{uy}(1 - F/P_{cry})} \leq p_y \qquad (5d.W7)$$

where m is $0.6 + 0.4\,\dfrac{M_{min}}{M_{max}}$

CTICM 80

Biaxial bending is governed by

$$k_o\,\frac{N}{N_p} + \frac{k_{fx}}{k_d}\,\frac{M_{mx}}{M_{px}} + k_{fy}\,\frac{M_{my}}{M_{py}} \leq 1.0 \qquad (5d.W8)$$

DIN 18800 P.2 86

For biaxial bending with no possibility of lateral-torsional buckling the condition is

$$\frac{F}{\chi\, A\, p_y} + \frac{M_x\,\beta_{mx}}{M_{plx}}\,k_x + \frac{M_y\,\beta_{my}}{M_{ply}}\,k_y + \Delta \leq 1 \qquad (5d.W9)$$

where

$$c_y = \frac{1}{c_x} = \frac{1 - \dfrac{F}{A\,p_y}\bar{\lambda}\,x^2}{1 - \dfrac{F}{A\,p_y}\bar{\lambda}\,y^2}$$

$$k_x = 1 \quad \text{and} \quad k_y = c_y \qquad \text{for } \chi_x < \chi_y$$
$$k_x = c_x \quad \text{and} \quad k_y = 1 \qquad \text{for } \chi_y < \chi_x$$

when lateral-torsional buckling is possible Eq. 5d.W9 should be replaced by

$$\frac{F}{\chi\, A\, p_y} + \frac{M_x}{\chi_M\, M_{plx}}\,k_x + \frac{M_y}{M_{ply}}\,k_y \leq 1 \qquad (5d.W10)$$

where $k_y = 1 - \dfrac{F}{\chi\, A\, p_y}\left[\bar{\lambda}\,y(2\beta_M - 4) + 0.5\right] \leq 1.5,$

with the term in square brackets ≤ 0.8

COMMENTARY

The basis for Eq. 5d.W10 is a substantial comparison against all available test data undertaken by Lindner and Gietzelt (1983). Its simplification for the special case where lateral-torsional buckling effects are negligible in Eq. 5d.W9 either because of the cross-sectional shape or because the beam slenderness λ_D is low has been discussed by Lindner and Gietzelt (1985).

EC 3 84

When moments about both x and y axes are present overall buckling of members of doubly symmetrical cross-section should be checked using

$$\left(\frac{M_y}{M_{ay}}\right)^{Z_1} + \left(\frac{M_x}{M_{ax}}\right)^{Z_2} \leq 1 \qquad (5d.W11)$$

where

M_{ax} = smaller value of M obtained from Eq. 5b.W7 or Eq. 5c.W6
M_{ay} = value of M obtained from Eq. 5b.W7
Z_1, Z_2 = may conservatively be taken as unity

As an alternative verification for I-sections considering only the flanges under axial force using Eq. 5b.W7 is also permitted.

NBN B51002 88

For members subject to compression plus biaxial bending the governing condition is

$$\frac{F}{A\chi} + \frac{m_x M_x \chi_M}{Z_x(1 - F/P_{crx})} + \frac{m_y M_y}{Z_y(1 - F/P_{cry})} \leq p_y \qquad (5d.W12)$$

where

$$m_x, m_y = \left[0.3\left(1 + \left(\frac{M_{min}}{M_{max}}\right)^2\right) + 0.4\,\frac{M_{min}}{M_{max}}\right]^{0.5} \text{ if sidesway is prevented}$$

m_x, m_y = 0.85 if sidesway is possible

In cases where the value of $F/A\chi$ is less than 0.15 m_x,m_y, $(1 - F/P_{crx})$ and $(1 - F/P_{cry})$ may all be taken as unity.

NEN 3851 72

$$\frac{F}{\chi_{min} A} + \chi_M\, m_x\left[\frac{P_{crx}/F}{P_{crx}/F - 1}\right]\left|\frac{m_x M_{xz}}{Z_x}\right| + m_y\left[\frac{P_{cry}/F}{P_{cry}/F - 1}\right]\left|\frac{m_y M_{yz} + M_{zy}}{Z_y}\right| \leq p_y \qquad (5d.W13)$$

$$\frac{F}{A} + \chi_M\left|\frac{M_{xz} + M_{Dx}}{Z_x}\right| + \left|\frac{M_{yz} + M_{Dy}}{Z_y}\right| \leq p_y \qquad (5d.W14)$$

where

M_{Dx}, M_{Dy} = maximum "free" moment within member length, taken as zero if $|M_d| \leq |2M_z|$

M_{xz}, M_{yz} = larger end moment about x and y axis respectively

p_b = lateral-torsional buckling design stress

$\chi_M = \dfrac{5p_y}{p_b(8 - 3\beta_x)}$

χ_{min} = smaller of χ_x and χ_y

COMMENTARY

Equations 5d.W13 and 5d.W14 are based on the full ECCS e* concept, generalized to include the effects of intermediate transverse loads. Interpretation of M_{Dx} and M_{Dy} for specific cases is facilitated by the provision of an explanatory table. The basis for its inclusion is a simple addition of the maximum mid-span moment caused by transverse loads in the statically determinate system to the mid-span moments due to top and bottom moments.

NS 3472 84

For members under compression and biaxial bending the following should be satisfied

$$\frac{F}{A\,p_y\,\chi_y} + \left(\frac{m_x\,M_x}{\chi\,M_{ux}}\frac{1}{k_E} + \frac{m_y\,M_y}{M_{uy}}\right)\frac{1}{1 - \dfrac{F}{P_{cry}}\dfrac{\chi_y A p_y}{p_y A}} \leq 1.0 \qquad (5d.W15)$$

where

$k_E = \dfrac{1 - F/P_{crx}}{1 - F/P_{cry}}$

M_{ux}, M_{uy} = moment capacities, elastic or plastic depending on section classification

m_x, m_y = $0.6 + 0.4M_{min}/M_{max} \leq 0.4$ for non-sway cases
= 1.0 for sway cases

χ = lateral-torsional buckling coefficient, obtained from an ECCS type buckling curve

StBK N1 70

For biaxial bending of I-sections stability is controlled by Eq. 5c.W12 used with

$$\left(\frac{N_{sd}}{N_{Rycd}}\right)^{\alpha_c} + \left(\frac{M_{sxd}}{M_{Rxcd}}\right)^{\beta_c} + \left(\frac{M_{syd}}{M_{Ryd}}\right)^{\gamma_{yc}} \leq 1.0 \qquad (5d.W16)$$

Similarly strength is controlled by Eq. 5a.W24 used with

$$\left(\frac{N_{sd}}{N_{Rd}}\right)^{\alpha_o} + \left(\frac{M_{sxd}}{M_{Rxd}}\right)^{\beta_o} + \left(\frac{M_{syd}}{M_{Ryd}}\right)^{\gamma_o} \leq 1.0 \qquad (5d.W17)$$

INTERNATIONAL COMPARISONS & COMMENT

WHAT ARE THE DIFFERENCES?

The common theme in each of the 33 codes examined is the use of interaction equations that utilize the strength of the member (or cross-section) under the component load cases of compression or bending about one principal axis. Thus the interactive nature of the problem is reflected in the use of a design approach that provides a smooth transition between the two extremes of the column and the beam. The exact type of interaction formulas used reflects, at least in part, the general philosophy of the particular code. Thus while most are expressed in terms of loads, some still operate with stresses. Similarly the end points of the formulas, since they depend directly on the particular code provisions for columns (Chapter 1) and beams (Chapter 3) will vary in the manner already discussed in those chapters. Any comparison between provisions should therefore recognize the important role of these contributing items. For this reason, as well as because of the large number of interacting variables, it is not possible to provide quantitative comparisons of the type used elsewhere for example to illustrate differences in the approach for centrally loaded columns, beams, or b/t limits for local buckling.

Generally speaking, the more modern codes provide separate treatments of the strength and stability aspects of beam-column design. In this context strength is taken to mean the local strength of the cross-section under combined loading, whilst stability is taken to mean the load carrying capacity of the member as a whole allowing for the effects of slenderness. Separate treatments are not required if the stability check always assumes the most severe loading arrangement for bending of uniform single curvature moments.

Similarly, the three different forms of beam-column behavior: in-plane response, out-of-plane buckling and biaxial bending are identified and covered separately in recent codes, whereas older documents tend to provide only a single formula that in many instances does not cover all cases. One specialized aspects of this concerns the treatment of tubular members for which out-of-plane buckling does not need to be considered on account of their high torsional stiffness. A few codes recognize this and provide a separate, specialized treatment; more details of this may be found in Chapter 9.

Since interaction equations are essentially a means for checking the suitability of an individual member within a frame, some method is necessary for representing the influence of other parts of the frame, for the purposes of both the loads transferred and the restraint provided. The traditional approach to this has been to use first order analysis to determine the end forces on the member, to allow for amplification of these within the member length and to make an approximate allowance for end restraint conditions - including the important distinction between members in non-sway and sway frames - via the effective length factor. Some recent codes have modified a part of this process by permitting the use of second-order moments directly without the need for approximate sway amplification or enhanced effective length treatments. Since this topic is closely related to design approaches for frames, interested readers should also consult Chapter 6.

The basis of many formulas is the SSRC equation. More recently the approach first proposed by ECCS as the e* concept has also been adopted and subsequently adapted for use in several specifications. Both approaches may be used in either an allowance stress or an ultimate strength form. Another often used general concept is the so-called Chen format for combining formulas for the two uniaxial cases to cover biaxial bending.

WHY ARE THERE DIFFERENCES?

Member forces, amplification factors and effective length factors

Only the most recent codes of these regions where structures likely to be sensitive to second order effects, e.g., tall unbraced frames are common, specifically allow for the possibility that member forces will be determined by anything other than a first order analysis. Although tall buildings are often constructed in Japan, the dominant effect of the seismic load case results in stocky construction so that a more precise treatment of frame instability effects is considered unnecessary. In China, GBJ 17 88 is intended principally for low rise buildings and industrial structures not prone to second order effects. Thus it is the North American codes

Table 5.1 Concept Matrix Comparison

REGION		A	C		J	
TOPICS	CODES	AS DR 87164	GBJ 17 88	TBJ 2 85	AIJ 89b	JRA 80
Member forces	first order analysis	X	X		X	X
	second order analysis	X				
Amplification factors	due to member deformation(P-δ)	X	X	X	X	X
	due to member deformation(P-Δ)	X				
Effective length factors	to determine column strength	X	X	X		X
	when using amplification factors	X	X	X		X
Cross-sections	provides for other than doubly symmetric cross-sections		X	X		
	special provisions for tubular members		X	X	X	
Type of interaction factor	linear	X	X	X	X	X
	nonlinear	X				
Cases covered	cross-section (a)	X	X	X	X	X
	in-plane (b)	X	X		X	X
	out-of-plane (c)	X	X		X	
	biaxial (d)	X			X	
Basis for formula	SSRC					
	ECCS e*		X			
	other	X				
Moment capacity	fully plastic	X			X	
	first yield	X				X
	other eg plastic adaption		X	X		
	reduced for slender sections	X				
Covers plastic design		X	X		X	
Use of equivalent moments		X	X		X	X

that pay most attention to overall frame stability effects, not just permitting the use of second order analysis but also in their concern for the use of amplification effects and effective length factors. Thus the level of attention given to these three factors that each assume a greater significance as the effect of isolating the beam-column from its parent frame becomes more questionable depends directly on the expected usage of the code provisions.

Type of interaction formula and basis

The type of interaction equation(s) used depends principally on the date and/or level of sophistication of the code. Whereas recent documents favor nonlinear formulas - frequently based on the e* and Chen concepts - the more traditional SSRC type of interaction equation is the basis for several of the other codes.

Similarly explicit recognition of the different cases (b), (c), and (d), appears to be largely a feature of the more modern codes, several of the older documents not covering the biaxial case at all.

Table 5.1 (Cont'd) Concept Matrix Comparison

REGION		J	N	W		
TOPICS	CODES	JSCE 83	AISC LRFD86	BS5950 P.1 85	BS5400 P.3 82	EC 3 85
Member forces	first order analysis	X		X	X	X
	second order analysis		X			
Amplification factors	due to member deformation(P-δ)		X	X	X	X
	due to member deformation(P-Δ)		X	X		X
Effective length factors	to determine column strength	X	X	X	X	X
	when using amplification factors		X			X
Cross-sections	provides for other than doubly symmetric cross-sections		X	X	X	X
	special provisions for tubular members					X
Type of interaction factor	linear	X	X	X	X	
	nonlinear		X	X		X
Cases covered	cross-section (a)		X	X	X	X
	in-plane (b)	X	X	X	X	X
	out-of-plane (c)		X	X	X	X
	biaxial (d)		X	X	X	X
Basis for formula	SSRC		X			
	ECCS e*			X		X
	other		X		X	
Moment capacity	fully plastic		X	X	X	X
	first yield	X	X	X	X	X
	other eg plastic adaption					
	reduced for slender sections		X	X	X	X
Covers plastic design			X	X		X
Use if equivalent moments		X	X	X		X

Inclusion of a separate cross-section check depends directly on whether the code uses the equivalent uniform moment concept since the requirement for such a check only exists if the design approach permits the use of less than the full values.

Most codes use equations expressed in terms of loads. Some East European specifications contain formulas in a stress format. In some cases these appear to have been derived with particular forms of lattice construction in mind - rather than the more general building frame basis of most of the other equations. The basis of several of these expressions does not appear to be either of the conventional SSRC or e^* approaches.

Where moment capacity is used, this usually reflects the class of cross-section as governed by local buckling limitations. The exception is China where the plastic adaption concept is employed.

Symbol List for Chapter 5 -- Beam-Columns

A = gross area of cross section (5a.C), (5b.E), (5b.W) & (5d.E)
A_F = moment amplification factor (5a.N)
A_f = area of flange (5a.N) & (5a.W)
A_{fg} = gross flange area with respect to the strong axis (5b.J)
A'_{fg} = gross flange area with respect to the weak axis (5b.J)
A_{fn} = net flange area with respect to the strong axis (5b.J)
A'_{fn} = net flange area with respect to the weak axis (5b.J)
A_g = gross section area (5b.J)
A_m = gross sectional area (5c.C)
A_n = net sectional area (5a.C) & (5b.J)
A_s = effective column cross-section (5b.N)
A_w = area of the web (5a.W)
A_{web} = area of the web (5b.W)
a = distance from the main axis of the section, perpendicular on the bending plane, to the most compressed fiber (5b.E)
a = constant to define the shape of interaction curve (5a.N)
B_1 = P-δ moment amplification factor (5d.N)
B_2 = P-Δ moment amplification factor (5d.N)
b = constant to define the shape of interaction curve (5a.N)
b = outstanding width of flange (5a.C)
b = half of the width of flange plate (5a.J)
b′ = outstanding width of the flange (5a.E)
b_f = flange width (5a.N) & (5d.N)
b_o = width of flanges of box-section members between the two webs (5a.C)
C = magnification factor for allowable stress for biaxial bending (5a.C)
C_{bc} = moment gradient modifier, given by Eq. 5c.A2 (5c.A)
C_{ex} = Euler buckling load about x-axis (5d.N)
C_{ey} = Euler buckling load about y-axis (5d.N)
C_f = axial force (5a.N) & (5d.N)
C_m = equivalent uniform moment factor (5a.N) & (5b.W)
C_m = sectorial moment of inertia (5c.E)
C_{mx} = equivalent uniform moment factor about x-axis (5a.N) & (5d.N)
C_{my} = equivalent uniform moment factor about y-axis (5a.N) & (5d.N)
C_r = axial compressive strength in accordance with column strength curve (5a.N) & (5d.N)
c = the coefficient that is determined depending on the eccentricity magnitude m_x, by means of the formulas (5c.E) & (5d.E)
c_x = coefficient correcting the value of moment M_x in relation to distribution manner (5b.E) & (5d.E)
c_5 = coefficient to determine c, given in Eq. 5c.E3 (5c.E)
c_{10} = coefficient to determine c, given in Eq. 5c.E3 (5c.E)
d = member depth (5a.N) & (5d.N)
d = depth of web plate (5a.J)
E = Young's modulus (5b.E)
e = eccentricity (5b.E)
e = distance between the central axis x-x of the cross section and the axis of the most compressed flange (5b.E)

e = distance between the y-y axis and the most compressed fiber (5b.E)
e_c = the distance from the centroid to the extreme compressive fiber (see Fig. 5b.E1) (5b.E)
e_o = equivalent eccentricity of initial imperfections (5b.C) & (5b.W)
e_r = the distance from the centroid to the extreme tensile fiber (see Fig. 5b.E1) (5b.E)
e_y = eccentricity outside the plane of bending (5b.E)
F = axial load (5b.W) & (5d.W)
F = maximum applied compression within the middle third of the member length between points of restraint (5c.W)
F_a = allowable axial stress (5d.N)
F_b = allowable bending stress (5d.N)
F_{bo} = the upper bound of permissible stress for bending about the strong axis without local buckling (5c.J)
F_{br} = gross area of the cross section (5b.E) & (5d.E)
F_{bx} = allowable bending stresses about x-axis (5a.N), (5c.J) & (5d.N)
$F_{bx,y}$ = permissible bending stress about strong, weak axis (5c.J)
F_{by} = allowable bending stresses about y-axis (5a.N) & (5d.N)
F_{b1} = allowable bending stress about axis 1-1 (5a.N) & (5d.N)
F_{b2} = allowable bending stress about axis 2-2 (5a.N) & (5d.N)
F_{cr} = allowable axial strength of column (5b.N)
F_e = Euler buckling stress (5b.N)
F_{ex} = Euler buckling stress with respect to the strong axis (5b.J) & (5c.J)
F'_{ex} = critical elastic buckling stress about x-axis, divided by a factor of safety of 23/12 and evaluated using the effective length of member (5d.N)
F_{ey} = Euler buckling stress with respect to the weak axis (5b.J) & (5c.J)
F'_{ey} = critical elastic buckling stress about y-axis, divided by a factor of safety of 23/12 and evaluated using the effective length of member (5d.N)
F_y = yield stress (5a.J) & (5b.N)
F_{yf} = yield stress of flanges (5a.J)
F_{yw} = yield stress of webs (5a.J)
${}_cF_a$ = permissible axial stress in compression (5c.J)
${}_pF_a$ = permissible compressive stress for local buckling (5a.J), (5b.J) & (5c.J)
${}_tF_a$ = permissible tensile stress (5a.J) & (5c.J)
${}_cF_{ay}$ = permissible compressive stress for buckling with respect to the weak axis (5b.C) & (5c.J)
f = design normal strength of steel (5a.C), (5c.C), & (5d.C)
f_a = applied axial stress (5a.N), (5c.J), & (5d.N)
f_b = applied bending stress (5a.N) & (5d.N)
f_{bx} = applied bending stress about the x-axis (5a.N) & (5d.N)
f_{by} = applied bending stress about the y-axis (5a.N) & (5d.N)
f_{b1} = computed bending stress about axis 1-1 (5a.N) & (5d.N)
f_{b2} = computed bending stress about axis 2-2 (5a.N) & (5d.N)
f_d = design strength (5a.W), (5b.W), & (5c.W)
f_M = factor of amplification of bending moments according to second order theory (5b.E)
f_N = factor of amplification of bending moments according to second order theory (5b.E)

f_N = amplification factor with a maximum slenderness ratio between λ_x and λ_{yi} (5c.E)
f_p = design normal strength of steel for plastic design (5a.C), (5b.C), & (5c.C)
f_x = amplification factor calculated with λ_x (5c.E)
f_y = material yield strength (5a.C), (5a.W), (5b.C), (5c.C) & (5c.W)
f_{yd} = material strength (5b.W)
f_{yi} = amplification factor calculated with λ_{yi} (5c.E)
${}_cf_a$ = applied axial compressive stress (5b.J) & (5c.J)
${}_tf_a$ = applied axial tensile stress (5c.J)
${}_cf_{bx}$ = applied compressive stress due to bending about the strong axis (5a.J), (5b.J) & (5c.J)
${}_tf_{bx}$ = applied tensile stress due to bending about the strong axis (5a.J) & (5c.J)
${}_cf_{by}$ = applied compressive stress due to bending about the weak axis (5a.J), (5b.J) & (5c.J)
${}_tf_{by}$ = applied tensile stress due to bending about the weak axis (5a.J) & (5c.J)
h_o = depth of the web of I- or box section (5a.C) & (5a.E)
h_1 = distance between the x-x axis and the most compressed fiber (5b.E)
I = gross moment of inertia of the cross section (5a.W) & (5b.E)
I_D = moment of inertia upon twisting (5c.E)
I_w = moment of inertia for warping (5c.A)
I_x = moment of inertia with respect to the strong axis (5b.J) & (5b.E)
$I_{x,br}$ = inertia moment of cross section with respect to x-axis (5d.E)
I_y = moment of inertia with respect to the weak axis (5b.E) & (5b.J)
$I_{y,br}$ = moment of inertia of cross section with respect to y-axis (5d.E)
I_ω = sectional inertia moment (5b.E) & (5c.E)
i_G = radius of gyration of upper flange (5b.E)
i_m = torsional radius of inertia about the shear center (5c.E)
i_p = torsional radius of inertia about the centroid (5c.E)
i_s = torsional radius of inertia about the shear center (5d.E)
i_x = inertia radius of the member section with respect to the x-x axis (5b.E)
i_y = inertia radius of the member section with respect to the y-y axis (5b.E)
J = torsional moment of inertia (5c.A)
K_1 = coefficient in Eq. 5b.C6 (5b.C)
K_2 = coefficient in Eq. 5b.C7 (5b.C)
k_D = coefficient to allow for lateral torsional buckling (5c.W)
k_{fx} = coefficient to allow for flexural buckling about x-axis (5c.W) & (5d.W)
k_{fy} = coefficient to allow for flexural buckling about y-axis (5d.W)
k_o = coefficient to allow for flexural buckling due to compression (5b.W), (5c.W), & (5d.W)
L = story height (5a.N) & (5d.N)
l = length of member between supports (5b.E) & (5c.E)
l_k = distance between supporting points of upper flange (5b.E)
l_o = length of member between end points ensured against overturning (5b.E) & (5d.E)
l_t = length of the member between two consecutive joints along the member against torsion (5d.E)
ℓ_e = effective length of plate elements (5a.W)
ℓ_f = distance between supporting points (5b.E)
ℓ_{fx} = distance between in-plane supporting points (5b.E)
ℓ_{fy} = distance between lateral supporting points (5b.E)

M = bending moment (5a.J), (5b.E), (5b.J), (5b.W), (5c.C), (5c.J) & (5d.E)
M_A = smaller end moment of members (5d.N)
M_{ax} = second-order design moment for the section bent about the x-axis (5d.N)
M_{ax} = maximum buckling moment about the x-axis in the presence of axis load F (5d.W)
M_{ay} = second-order design moment for the section bent about the y-axis (5d.N)
M_{ay} = maximum buckling moment about the y-axis in the presence of axial load F (5d.W)
M_B = larger end moment of members (5d.N)
M_b = lateral torsional buckling resistance moment (5c.W) & (5d.W)
M_{bx} = the lateral buckling resistance of the member as a beam (5c.A)
M_{bxo} = the lateral buckling resistance of the member as a beam in uniform bending (5c.A)
M_c = moment capacity depending on section classification (5b.W)
M_c = moment at center of span (5d.N)
M_c = ultimate design bending moment (5a.J) & (5b.W)
M_{cr} = critical moment (5b.E)
M_{cx} = moment capacity about x-axis (5d.W)
M_{cy} = moment capacity about y-axis (5d.W)
M_{Dx} = maximum "free" moment within member length (5c.W) & (5d.W)
M_{Dy} = maximum "free" moment within member length (5d.W)
M_{Eyd} = weak axis slenderness based on effective column length (5c.W)
M_e = design equivalent bending moment (5b.E) & (5c.E)
M_{eq} = design equivalent bending moment (5b.W)
M_F = yield moment of members (5b.E)
M_{fx} = applied flexural stress about the x-axis (5a.N) & (5d.N)
M_{fy} = applied flexural stress about the y-axis (5a.N) & (5d.N)
M_i = nominal member capacity for in-plane bending of a member without transverse loads (5b.A)
M_{ix} = the in-plane moment capacity of members (5d.A)
M_{ki} = ideal moment of loss of stability (5b.E)
M_{lt} = moment in member as a result of lateral translation of the frame only calculated by using first-order elastic analysis (5d.N)
M_m = maximum resisting moment in the absence of axial force, taken as M_p if the member is braced against lateral torsional buckling (5b.N)
M_m = maximum bending moment in span (5b.W) & (5d.N)
M_{max} = maximum second-order moment or the amplified moment in the member (5a.N)
M_{max} = larger moment at the two ends of beams (5b.E)
M_{min} = smaller moment at the two ends of beams (5b.E)
M_{mx} = maximum bending moment about x-axis (5c.W) & (5d.W)
M_{my} = maximum bending moment about y-axis (5d.W)
M_{nt} = moment in member assuming there is no lateral translation in the frame calculated by using first-order elastic analysis (5b.N) & (5d.N)
M_{nx} = ultimate moment resisting capacity about x-axis (5d.N)
M_{ny} = ultimate moment resisting capacity about y-axis (5d.N)
M_o = maximum first-order moment (5a.N) & (5b.N)
M_o = buckling moment of a uniformly bent beam (5c.C)
M_{ox} = the maximum single curvature factored moment resistance of the column in the presence of axial load (5a.N) & (5d.N)

M_{ox} = nominal out-of-plane capacity (5c.A)
M_{oy} = the maximum single curvature factored moment resistance of the column in the presence of axial load (5a.N) & (5d.N)
M_p = plastic moment (5a.J), (5a.N), (5b.J), & (5b.N)
M_{pcx}= plastic moment under axial force and bending with respect to x-axis(5a.W)
M_{pcy}= plastic moment under axial force and bending with respect to y-axis(5a.W)
M_{pl} = plastic moment capacity (5b.E) & (5b.W)
M_{plx}= plastic moment about x-axis (5c.W)
M_{px} = plastic moment under axial force and bending with respect to the strong axis (5a.J), (5a.N), (5a.W), & (5d.J)
M_{py} = plastic moment under axial force and bending with respect to the weak axis (5a.J), (5a.N), (5a.W), & (5d.J)
$M_{\mathrm{R}xcd}$ = moment capacity allowing for lateral-torsional buckling (5c.W)
$M_{\mathrm{R}xd}$ = moment capacity for strong axis bending (5a.W), (5b.W), & (5c.W)
$M_{\mathrm{R}yd}$ = moment capacity for weak axis bending (5a.W), (5b.W), & (5c.W)
M_r = nominal section plastic moment capacity reduced for axial force (5a.A)&(5a.N)
M_{rcx} = moment capacity allowing for lateral-torsional buckling about strong axis (5a.N)
M_{rcy} = moment capacity allowing for lateral-torsional buckling about weak axis (5a.N)
M_{red} = reduced moment capacity due to axial force (5a.W)
M_{rx} = nominal section plastic moment capacity about x-axis (5a.A), (5a.N), (5a.W), & (5d.W)
M_{ry} = nominal section plastic moment capacity about y-axis (5a.A), (5a.N), (5a.W), & (5d.W)
M_{ry} = maximum uniform moment capacity of the member about the y-axis, having regard for lateral-torsional buckling (5a.N) & (5d.N)
M_s = section moment capacity in the plane of bending (5b.A)
M_{sx} = nominal beam section capacity about x-axis (5a.A)
M_{sxd} = applied moment about x-axis determined by first order theory (5a.W) & (5b.W)
M_{sy} = nominal beam section capacity about y-axis (5a.A)
M_{syd} = applied moment about y-axis determined by first order theory (5c.W) & (5d.W)
M_u = section capacity, the in-plane member capacity or the out-of-plane member capacity (5a.A)
M_{ux} = moment capacity with respect to the strong axis (5a.N), (5d.A), (5d.J), (5d.N)&(5d.W)
M_{uy} = moment capacity with respect to the weak axis(5a.N), (5c.J), (5d.N)&(5d.W)
M_{vd} = moment capacity allowing for lateral-torsional buckling (5c.W)
M_x = factored maximum bending moment about the x-axis (5a.C), (5b.C)&(5c.C)
M_x = applied moment with respect to the strong axis (5a.J), (5b.J), (5b.E)&(5d.E)
M_x = maximum applied moment about the x-axis within the middle third of the member length between points of restraint (5d.W)
$\overline{M}_x$ = equivalent uniform moment (5c.W)
M_{xa} = permissible lateral buckling bending moment with respect to the strong axis (5b.J) & (5c.J)
M_{xmax} = maximum bending moment about x-axis (5c.W)
M_{xred} = uniaxial capacity in the presence of the axial load (5a.W)

M_{xz} = larger end moment about the x-axis (5c.W) & (5d.W)
M_y = factored maximum bending moment about the y-axis (5a.C), (5b.C)&(5c.C)
M_y = bending moment with respect to the weak axis (5b.J) & (5d.E)
M_y = maximum applied moment about the y-axis within the middle third of the member length between points of restraint (5d.W)
M_{yao} = permissible bending moment with respect to the weak axis (5b.J) & (5c.J)
M_{yred} = uniaxial capacity in the presence of the axial load (5a.W)
M_{yz} = larger end moment about the y-axis (5d.W)
M^{I}_{max} = maximum first-order bending moment, which is calculated for the unstrained member (5b.E)
M^{II}_{max} = maximum second-order bending moment (5b.E)
M_1 = larger end moment (5b.E) & (5d.N)
M_{1x} = applied bending moment with respect to the strong axis (5d.J)
M_{1y} = applied bending moment with respect to the weak axis (5d.J)
M_2 = smaller end moment (5b.E), (5b.J) & (5d.N)
M^x_e = design equivalent bending moment about x-axis (5d.E)
M^y_e = design equivalent bending moment about y-axis (5d.E)
M^* = bending moment (5a.A)
M^*_{max} = maximum bending moment obtained from a first order calculation (5a.A), (5b.A), (5b.W) & (5c.A)
M^*_x = bending moment about x-axis (5a.A) & (5d.A)
M^*_y = bending moment about y-axis (5a.A) & (5d.A)
m = relative eccentricity (5b.E)
m = coefficient to allow for moment gradient, see Table 5b.W1 (5b.W) & (5c.W)
m_{ef} = modified relative eccentricity (5b.E)
$m_{ef,y}$ = modified relative eccentricity with respect to y-axis (5d.E)
m_w = buckling multiplier determined from tables (5b.E)
m_x = equivalent uniform moment factor (5c.W) & (5d.W)
m_x = equivalent relative eccentricity with respect to x-axis (5d.E)
m_y = equivalent relative eccentricity with respect to y-axis (5d.E)
m_y = equivalent uniform moment factor (5d.W)
N = design axial force (5a.C), (5b.E), (5c.C) & (5c.E)
N_{Rxcd} = column strength (5b.W)
N_{Rycd} = column strength (5b.W)
N_c = axial load capacity for the member buckling in the same plane (5b.A)
N_{cr} = critical load about the axis under consideration, using the appropriate effective length (5b.W)
$N_{cr,x}$ = critical load about x-axis (5c.W)
N_{cy} = axial load capacity for out-of-plane buckling (5c.A)
N_d = squash load of member (5c.W)
N_{Ex} = Euler buckling load about the x-axis (strong axis) (5b.C) & (5d.C)
N_{Exd} = strong axis column strength based on effective column length (5c.W)
N_{Ey} = Euler buckling load about y-axis (5d.C)
N_{Eyd} = weak axis column strength based on effective column length (5c.W)
N_{ki} = column strength (5b.W)
N_{kyd} = weak axis column strength (5c.W)

N_{Rd} = column strength (5a.W)
N_{Rxcd} = column strength about x-axis (5c.W)
N_{Rycd} = column strength about y-axis (5c.W)
N_p = yield capacity of axially loaded members (5a.W), (5b.W) & (5c.W)
N_{pl} = yield capacity of axially loaded members (5a.W)
N_s = nominal column section capacity (5a.A)
N_{sd} = axial load (5a.W), (5b.W), (5c.W) & (5d.W)
N_y = Euler buckling load about the y-axis (5c.C)
N_z = torsional buckling load of column (5c.A)
N_ω = torsional buckling load (5c.C)
N^* = maximum axial force obtained from a first order calculation (5a.A), (5b.A), (5b.W) & (5c.A)
n = exponent in relation to the rate of the moment diagram (5b.E)
n_1 = factor of safety (5c.C)
P = axial compression force (5a.J), (5b.J), (5c.J), & (5d.J)
P_{ay} = axial compressive capacity for buckling out of plane of bending (5d.J)
P_c = compressive resistance of column (5b.W) & (5d.W)
P_{ca} = smaller among the permissible buckling loads with respect to the strong and the weak axis (5b.J) & (5c.J)
P_{cr} = Euler buckling load (5b.W)
P_{crx} = Euler buckling load about x-axis (5b.W), (5c.W) & (5d.W)
P_{cry} = Euler buckling load about y-axis (5d.W)
P_{cx} = compressive resistance about x-axis (5d.W)
P_{cy} = compressive resistance about y-axis (5b.W) & (5c.W)
P_e = Euler buckling load (5d.N)
P_{ek} = Euler buckling load considering the end restrained condition (5a.N) & (5b.N)
P_{ex} = Euler buckling load in plane of bending (5a.N) & (5b.J)
P_{ey} = Euler buckling load out-of-bending plane (5a.N) & (5d.N)
P_u = ultimate axial capacity of the member (5a.N), (5b.N) & (5d.N)
P_{uk} = axial capacity of member considering end restrained condition (5b.N)
P_{ux} = axial compressive capacity for buckling in plane of bending (5a.J) & (5b.J)
P_{uy} = axial compression capacity for buckling out of the plane of bending (5c.J)
P_y = yield load of the section (5a.J), (5a.N), (5b.J) & (5d.J)
p_b = lateral-torsional buckling design stress (5c.W) & (5d.W)
p_c = compressive strength (5d.W)
p_y = material yield stress (5b.W), (5c.W) & (5d.W)
Q_a = shearing force from external loads (5b.E)
Q_i = ideal shearing force (5b.E)
R_c = column design strength (5b.W)
R_{cr} = buckling critical strength of the ideal bar having the same support conditions as the bar being checked (5b.E) & (5c.E)
$R_{cr,x}$ = buckling critical strength of ideal bar having the same support conditions as the bar being checked about x-axis (5d.E)
$R_{cr,y}$ = buckling critical strength of ideal bar having the same support conditions as the bar being checked about y-axis (5d.E)
R_{fl} = minimum buckling design strength (5b.E) & (5c.E)
R_{fll} = lateral buckling design strength (5b.E) & (5c.E)
R_o = design strength (5a.E)
R_y = yield stress (5b.E) & (5c.E)

R^m_{cr} = lateral buckling strength (5b.E)
r = radius of gyration of plate elements (5a.W)
r_t = radius of gyration of compression flange, plus one-third of the compression portion of the web taken about an axis in the plane (5a.N)
r_y = the radius of gyration of section about y-axis (5a.N) & (5b.N)
s_x = section modulus with respect to x-axis (5b.W)
t = thickness of flanges of I- or box section (5a.C)
t_f = thickness of flanges (5a.E)
t_w = thickness of web (5a.C) & (5a.E)
${}_t t_a$ = applied axial tensile stress (5a.J)
V_{cr} = safety factor (5b.E)
v = distance from the level of application of load in the member axis (5b.E)
v = distance from the principle axis to the extreme stress point (5a.W)
W = strength modulus of the gross cross section (5b.E) & (5d.E)
W = elastic or plastic section modulus (5b.W)
W_c = section modulus of a member corresponding to the most compressed fiber (5b.E)
$\overline{W}_c$ = strength modulus calculated for the most compressed fiber (5b.E)
$W_{c,br}$ = gross strength moduli calculated with respect to the x-x axis, for the compressed fit (5b.E)
W_d = elastic section modulus, compression side (5b.E) & (5b.W)
W_{el} = elastic modulus (5b.E)
W_m = gross section modulus (5c.C)
W_{nx} = net section modulus about x-axis (5a.C)
W_{ny} = net section modulus about y-axis (5a.C)
W_{pl} = plastic modulus (5b.E)
W_{pnx} = net plastic section modulus about the x-axis (5a.C)
W_{px} = plastic section modulus about the x-axis (x-axis is the strong axis of I-sections) (5b.C) & (5c.C)
$W_{r,br}$ = gross strength moduli calculated with respect to the x-x axis, for the stretched fit (5b.E)
W_T = strength modulus, given by Eq. 5b.E27 (5b.E)
W_x = strength modulus of the gross cross section (5b.E) & (5d.E)
W_{xp} = plastic strength modulus of the cross section for the most compressed fiber (5b.E) & (5c.E)
W_y = strength modulus of the gross cross section (5b.E) & (5d.E)
W_{yp} = plastic strength modulus of the cross section for the most compressed fiber (5b.E) & (5c.E)
W_z = the elastic section modulus, tension side (5b.W)
W_{1x} = section modulus about the x-axis with respect to the fiber subjected to the maximum compressive stress (5b.C), (5c.C), & (5d.C)
W_{1y} = section modulus about y-axis with respect to the fiber subjected to the maximum compressive stress (5b.C) & (5d.C)
W_{2x} = section modulus about the x-axis with respect to the extreme fiber of the smaller flange (5b.C)
w = (material yield strength)/(column strength) (5c.W)
w_{cx} = strength modulus of cross section calculated for the most compressed fiber with respect to the x-axis (5d.E)
w_{cy} = strength modulus of cross section calculated for the most compressed fiber with respect to the y-axis (5d.E)

y_c = distance between the center of gravity and the center of torsion (5b.E)
y_c = distance from the strong axis to the extreme fiber in the compression side of the section (5b.J)
y'_c = distance from the weak axis to the extreme fiber in the compression side of the section (5b.J)
y_m = distance between center of shearing and center of gravity (5c.E)
y_s = distance between the shear center and the center of gravity (5d.E)
y_t = distance from the strong axis to the extreme fiber in the tension side of the section (5b.J)
y'_t = distance from the weak axis to the extreme fiber in the tension side of the section (5b.J)
Z = elastic section modulus (5b.W)
Z_e = elastic section modulus (5a.J)
Z_x = section modulus about x-axis (5a.W) & (5c.W)
Z_y = elastic modulus about the minor axis (5a.W), (5b.W) & (5d.W)
Z_1 = exponent to define the shape of interaction curve (5a.W)
Z_2 = exponent to define the shape of interaction curve (5a.W)
α = coefficient depending on buckling curves (5b.E)
β = M_2/M_1, end moment ratio (5b.J)
β = supporting factor upon bending (5c.E)
β_k = reduction coefficient depending on the variation of the bending moment along the member axis (5b.E)
β_M = coefficient to allow for moment distribution, see Table 5c.W1 (5c.W)
β_m = ratio of the end moments (5b.A) & (5c.A)
β_m = a coefficient depending on the pattern of moment, see Table 5b.W1 (5b.W)
β_{mx} = equivalent moment factor for in-plane stability (5b.C)
β_{my} = equivalent moment factor for in-plane stability (5d.C)
β_o = supporting factor upon twisting (5c.E)
β_{tx} = equivalent moment factor for lateral-torsional buckling (5c.C) & (5d.C)
β_{ty} = equivalent moment factor for lateral-torsional buckling (5d.C)
β_x = coefficient to determine equivalent uniform moment (5b.W) & (5c.W)
γ_c = coefficient of service conditions (5b.E)
γ_{fu} = safety coefficient of the member for lateral buckling (5b.E)
γ_{f3} = factor for axial compression (5a.W)
γ_j = coefficient, given by Eq. 5d.E3 (5d.E)
γ_k = coefficient, given by Eq. 5d.E4 (5d.E)
γ_m = resistance factor for bending (5a.W) & (5d.A)
γ_s = exponent in Eq. 5a.A5 (5a.A)
γ_x = plastic adaptation coefficient (5a.C), (5b.C), & (5d.C)
γ_{xc} = γ_o W_{xc}, coefficient to define the shape of interaction relation (5b.W) & (5c.W)
γ_y = plastic adaptation coefficient (5a.C), (5b.C), & (5d.C)
γ_{yc} = γ_o W_{yc}, coefficient to define the shape of interaction relation (5b.W) & (5d.W)
Δ_o = first-order translational deflection of the story under consideration (5d.N)
ζ = exponent given in Eq. 5a.N10 (5a.N)
η = coefficient in relation to section shape (5b.E) & (5b.W)
η_u = factor of safety on yield strength (5d.N)
η_x = shape factor (5b.W)
η_y = shape factor (5b.W)

θ = coefficient whose value in generally 1.0 or 1.1 depending on the cross section type (5b.E) & (5d.E)
λ = slenderness ratio of members, which is in the plane of bending moment (5a.C)
λ = greatest slenderness of a member (5b.E) & (5b.W)
$\overline{\lambda}$ = non-dimensional column slenderness (5b.E), (5b.W), (5c.E) & (5d.W)
λ_c = modified slenderness ratio (5b.J)
λ_E = limiting slenderness (5b.E)

$\overline{\lambda}_{ef}$ = modified non-dimensional slenderness (5b.E)
λ_j = slenderness ratio, given by Eq. 5d.E3 (5d.E)
λ_k = torsional slenderness of the member (5b.E)
$\overline{\lambda}_k$ = non-dimensional slenderness ratio (5b.E)
λ_{lt} = slenderness ratio of plate elements (5a.W)
λ_{tr} = modified slenderness ratio (5b.E)
λ_x = slenderness ratio about x-axis (5b.E) & (5c.E)
λ_y = slenderness ratio about the y-axis (5b.E) & (5c.C)
λ_{yi} = slenderness ratio for buckling with bending-torsion about symmetry y-axis (5c.E)
μ = coefficient of bending moment magnification due to the presence of axial compression (5c.C)
μ = multiplication coefficient of the length l (5b.E) & (5d.E)
μ_e = coefficient of yield stress reduction by eccentric compression (5b.E)
μ_{ey} = coefficient of yield stress reduction by eccentric compression with respect to y-axis (5d.E)
μ_k = coefficient depending on the influence of torsion upon bending buckling (5b.E)
μ_N = imperfection factor with a maximum slenderness ratio between λ_x and λ_y (5c.E)
μ_o = multiplication coefficient of the length l_o (5b.E) & (5d.E)
ν = reduction coefficient for moment, given by Eq. 5d.N20 (5d.N)
ν_{cr} = safety coefficient of the member for lateral buckling (5b.E)
σ = normal unitary stress produced by the compressive axial force (5b.E)
σ = material yield strength (5b.W)
σ_a = allowable stress (5b.E) & (5d.E)
σ_{all} = permissible strength in relation to ultimate stress (5b.E)
σ_b = bending stress (5b.E)
σ_{bc} = bending stress (5b.E)
σ_{bcx} = elastic bending compressive stress about x-axis (5c.E)
σ_{bcy} = elastic bending compressive stress about y-axis (5c.E)
σ_{bz} = bending stress (5b.E)
σ_{bzx} = plastic bending compressive stress about x-axis (5c.E)
σ_{bzy} = plastic bending compressive stress about y-axis (5c.E)
σ_c = axial stress due to axial force (5b.E)
σ_{cao} = permissible stress in compression (5b.J)
σ_D = limiting stress for lateral-torsional buckling (5c.W)
σ_E = buckling critical normal unitary stress, corresponding to the x-x plane, or y-y plane, respectively where bending occurs (5b.E)
σ_F = ultimate stress (5b.E)
σ_H = limit strength depending on the steel grade and the thickness of the element (5b.E)

σ_k = critical stress of column (5b.W)
σ_{ki} = torsional critical unitary stress for a straight member, made from a perfectly elastic material (5b.E)
σ_m = given by Eq. 5a.W6 (5a.W)
σ_{max} = maximum compressive edge stress of web (5a.C)
σ_{min} = stress at other side of web, taking compression in positive sign and tension in negative (5a.C)
σ_{m1} = larger stress due to bending moment M_x, M_y (5a.C)
σ_{m2} = smaller stress due to bending moment M_x, M_y (5a.C)
σ_n = axial design stress due to axial force (5a.W)
σ_{ta} = permissible stress in tension (5b.J)
σ_y = nominal yield stress of material (5a.W)

$[\sigma]$ = allowable normal stress of steel (5a.C) & (5c.C)

ϕ = resistance factor (5a.A), (5a.N), (5a.W), (5b.A), (5b.E), (5c.A)&(5d.A)
ϕ = coefficient to limit the amount of residual plastic deformation due to partial plastification of the cross section (5b.W)
ϕ_b = resistance factor for bending (5a.J), (5a.N), (5b.J), (5c.J), (5d.J) & (5d.N)
ϕ_b = strength reduction factor of laterally unsupported beams subjected to uniform bending (5c.C)
ϕ_{bx} = strength reduction factor of laterally unsupported members subjected to uniform bending about the x-axis (5d.C)
ϕ_{by} = strength reduction factor of laterally unsupported members subjected to uniform bending about the y-axis (5d.C)
ϕ_c = resistance factor for compression (5a.J), (5b.J), (5c.J), (5d.J)&(5d.N)
ϕ_e = coefficient of yield stress reduction by eccentric compression (5b.E)
ϕ_M = factor of loss of stability in relation to $\overline{\lambda}_M$ (5b.E)
ϕ_p = resistance factor for tension (5a.J), (5b.J) & (5d.J)
ϕ_x = strength reduction factor of centrally compressed members bent about x-axis (5b.C)
ϕ_x = plastic adaptation factor (5c.W)
ϕ_x = coefficient of yield stress reduction by eccentric compression (5b.E)
ϕ_y = strength reduction factor of centrally compressed members bent about the y-axis (weak axis) (5c.C)
ϕ_1 = strength reduction factor of centrally compressed members (5c.C)
ϕ_2 = strength reduction factor of beams due to the presence of lateral-torsional buckling (5c.C)
φ = buckling coefficient (5b.E) & (5d.E)
φ_b = lateral buckling coefficient of the most compressed flange (5c.E)
φ_E = Euler buckling coefficient corresponding to the buckling plane (5d.E)
φ_g = coefficient obtained from buckling curves, in relation to section shape, steel brand and slenderness ratio transformed (5b.E) & (5d.E)
φ_y = buckling coefficient (5c.E)
χ = column buckling coefficient (5b.W)
χ = lateral-torsional buckling coefficient, obtained from an ECCS type buckling curve (5c.W) & (5d.W)
χ_M = reduction coefficient for lateral-torsional buckling (5c.W) & (5d.W)
χ_{min} = smaller of χ_x and χ_y (5d.W)
χ_x = column buckling coefficient about x-axis (5b.W)
χ_y = column buckling coefficient about y-axis (5b.W), (5c.W) & (5d.W)

ψ = coefficient depending on the influence of eccentricity upon the value of the bending moment (5b.E)
ψ_c = coefficient (5b.J)
ω = equivalent moment factor (5d.N)
ω_x = equivalent moment factor with respect to x-axis (5d.N)
ω_y = equivalent moment factor with respect to y-axis (5d.N)

Chapter 6

Frames

TOPICAL EDITORS

Le-Wu Lu, Lehigh University
John Springfield, Carruthers & Wallace Limited

CHAPTER EDITORS

Russell Q. Bridge, The University of Sydney
Miklos Ivanyi, Technical University of Budapest
Le-Wu Lu, Lehigh University
Shosuke Morino, Mie University
John Springfield, Carruthers & Wallace Limi^t^ed
Huibi Wu, Chongqing Architectural Institute

SPECIFICATION PROVISIONS & EXPLANATION

AUSTRALIA

CHAPTER EDITOR

Russell Q. Bridge, The University of Sydney

AS 1250 81

Frames are analyzed to determine the maximum values of the action effects (bending moment, shear and axial force) at each end of the component members and the distribution of these action effects along the length of each member. The component members are then designed in accordance with the requirements of separate sections of the Standard appropriate to the action or combination of actions (beams, columns, or beam-columns).

Second-order effects are included in the determination of the maximum values of the action effects. This is done either directly using a second-order analysis or indirectly using a first-order analysis together with moment magnifiers based on a buckling analysis of the member or the frame as appropriate.

First-Order Elastic Analysis – Braced Members

For a member with an axial force N_d and maximum moment M_m along the length of the member as determined by a first-order analysis, the maximum design bending moment M_d is given by

$$M_d = \delta_b M_m \tag{6.A1}$$

where

$$\delta_b = \frac{C_m}{1 - N_d/N_{omb}} \tag{6.A2}$$

and N_{omb} is the buckling load for the braced member for the particular conditions of end restraint as provided by the surrounding structure.

For members subject to end moments only, the factor C_m is given by

$$C_m = 0.6 - 0.4\,\beta \tag{6.A3}$$

where β is the ratio of the smaller to the larger moment at the ends of the member, taken as positive when the member is bent in reverse curvature.

First-Order Elastic Analysis – Sway Members

In the simple method, the maximum design moment M_d is given by

$$M_d = \delta_m\, M_m \tag{6.A4}$$

where δ_m is the larger of δ_b (Eq. 6.A2) or δ_s (Eq. 6.A5)

$$\delta_s = \frac{1}{1 - N_d/N_{oms}} \tag{6.A5}$$

and N_{oms} is the buckling load for the sway member for the particular conditions of end restraint provided by the surrounding structure.

In the alternative method, the end-moment M_f obtained from a first-order sway analysis is separated into two components

$$M_f = M_{fb} + M_{fs} \tag{6.A6}$$

where

M_{fb} = end moment obtained from a first-order analysis of the frame with sway prevented

M_{fs} = end moment with sway permitted

The maximum design moment M_d is given by

$$M_d = \delta_b\,(M_{fb} + \delta_s\, M_{fs}) \tag{6.A7}$$

For columns in a story of a rectangular frame, δ_s is given by

$$\delta_s = \frac{1}{1 - \Sigma(N_d/L)/\Sigma(N_{oms}/L)} \tag{6.A8}$$

and the summations include all columns within the story.

For non-rectangular, irregular or asymmetric structures, δ_s is taken as a constant for all members where

$$\delta_s = \frac{1}{1 - 1/\lambda_c} \tag{6.A9}$$

and λ_c is the ratio of the elastic critical buckling load of the entire to the design load for the particular axial load combination under consideration.

Second-Order Elastic Analysis – Linear Stiffness

Although changes in geometry under load are accounted for, the reduction in elastic stiffness of the members due to axial compressive force is neglected. This assumption enables a direct solution to be obtained without iteration, For frames where the second-order effects increase the end-moments by more than 25%, this assumption can lead to inaccuracies and a non-linear stiffness model should be used.

The maximum design moment M_d is obtained by multiplying the second-order end-moment by δ_b.

Second-Order Elastic Analysis – Non-Linear Stiffness

Both changes in geometry and reduction in the elastic stiffness with axial compressive force are accounted for in the analysis. The maximum design moment M_d is obtained by multiplying the second-order end moment by δ_b.

Plastic Analysis

Plastic analysis is restricted to frames containing compact doubly symmetric I-sections with ductile steels. The elastic buckling load must be more than three times the plastic collapse load. Regions of plastic hinges have special requirements for lateral restraint. A second-order analysis is used for frames over three stories in height. Connections must have the required rotation capacity to allow the assumed collapse mechanism to develop.

Rigorous Analysis

The analysis must take into account the relevant material properties, geometric properties, three-dimensional effects and interaction with the foundations. Modified action effects are automatically generated by such an analysis.

Buckling Analysis

Two types of analysis are carried out: a member buckling analysis and a frame buckling analysis.

The buckling load N_{om} of a member for the particular conditions of end restraint provided by the surrounding structure is calculated from

$$N_{om} = \pi^2 \, EI/(k_m \, L)^2 \tag{6.A10}$$

where k_m is a member effective length factor determined from design charts in the specification. The influence of axial force on the stiffness of the restraining members is taken into account.

The axial force N_{of} in a member at elastic critical buckling of the frame for the particular set of design axial loads N_d under consideration is calculated from

$$\begin{aligned} N_{of} &= \lambda_c \, N_d \\ &= \pi^2 \, EI/L_e^2 \end{aligned} \tag{6.A11}$$

where λ_c is the elastic critical buckling load factor of the entire frame from which the effective length L_e can be determined. This effective length is used in the determination of the axial load capacity of compression members.

EAST EUROPE

CHAPTER EDITOR

Miklos Ivanyi, Technical University of Budapest

MSZ 15024/1 85

The Hungarian code of practice offers three different procedures for the design of plane frameworks:

(i) Spatial model – Considering both in- and out-of-plane imperfections.
(ii) Considering in-plane imperfections only.
(iii) Effective length concept without initial imperfections.

In cases (i) and (ii) generally a second-order analysis must be carried out whereas for case (iii) this can be replaced by using the product of first-order internal forces and the magnification factor.

Initial Imperfections
(A) Braced system
The maximum value of initial out-of-straightness of the form of a half sine-wave or second-order parabola as shown below is used.

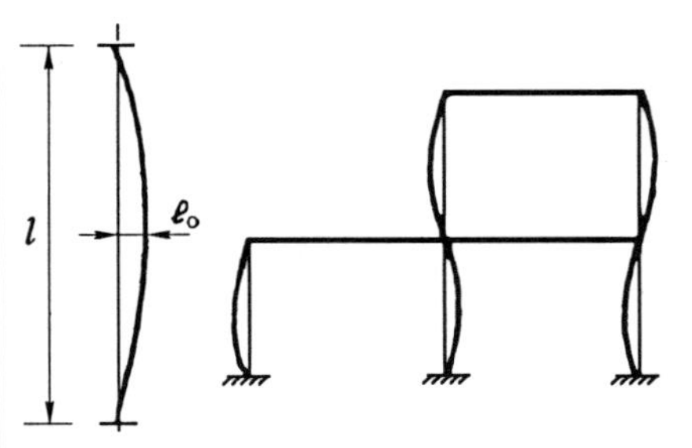

Buckling curve	Initial imperfection e_o
a	$\frac{l}{500}$
b	$\frac{l}{250}$
c	$\frac{l}{200}$

(B) Sway frames
The value of initial out-of-plumb as shown below is used.

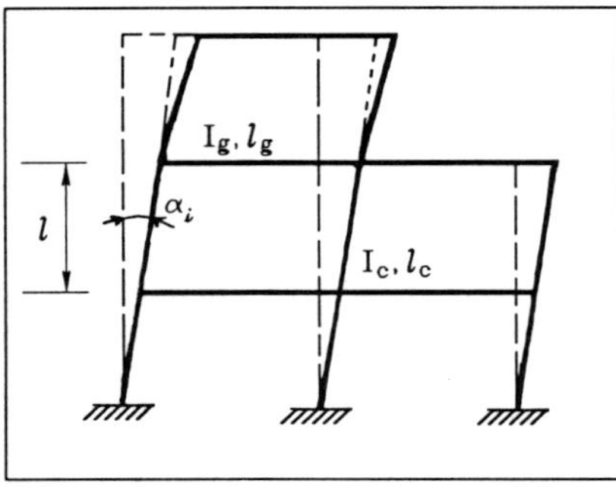

Range of application:

$$\epsilon_i = l_i \sqrt{\frac{N_i}{E\, I_i}} \leq 1.6$$

$$V_j = \frac{\Sigma I_g/l_g}{\Sigma I_c/l_c}$$

$$\alpha_i = \frac{1}{500}\,\gamma \qquad \gamma = \frac{1}{2}\left(1 + \frac{1}{n-1}\right)$$

n=number of columns in one story in the plane

For members not satisfying any of the two conditions listed above, a more refined analysis is required. The second-order analysis may be simplified for sway frames consisting of relatively rigid members (all eccentricity loaded columns satisfy $\epsilon_i \leq 1.6$).

Case 1:

$$\text{if} \quad \frac{\Delta H_j}{Q_j} \leq 0.1$$

first-order computation of internal forces is sufficiently accurate

Case 2:

$$\text{if} \quad 0.1 < \frac{\Delta H_j}{Q_j} \leq 0.25$$

first-order computation with increased values of horizontal forces, Q'_j may be adopted, where

$$Q'_j = \frac{1}{1 - \dfrac{\Delta H_j}{Q_j}} Q_j$$

where

ΔH_j = $P_j \, \Delta \alpha_j$, increment of horizontal force at level j due to column chord rotation $\Delta \alpha_j$ from the initial rotation α_j

Q_j = $\Sigma H + \alpha_j P_j$, sum of original horizontal force above level j, $\Delta \alpha_j$ is the rotation due to applied loads, ΣH is the cumulative horizontal force above level j, and P_j the total axial force in columns between levels j−1 and j

JAPAN

CHAPTER EDITOR
Shosuke Morino, Mie University

AIJ 88

The effective length approach is used. Because of the large earthquake forces that buildings are designed to resist, the effect of frame instability tends to be less important. AIJ states:

> "In braced frames, the effective length of columns shall be taken as the distance between panel points unless the analysis shows that a smaller value may be used. In unbraced frames, it shall be determined by a rational method and shall not be less than the distance between panel points."

Axial force ratio n_y and normalized slenderness ratio ${}_f\lambda_c$ of a column in a frame shall satisfy the following inequalities:

(A) For columns in a frame which is designed on the basis of plastic analysis

$$n_y \leq 0.75 \tag{6.J1}$$

$$n_y \cdot {}_f\lambda_c^2 \leq 0.25 \tag{6.J2}$$

(B) For columns with plastic hinges in a frame which is designed by plastic analysis

$$\left.\begin{array}{ll} \text{for } -0.5 < \kappa \leq 1.0 & n_y \cdot {}_f\lambda_c^2 \leq 0.10\,(1+\kappa) \\ \text{for } -1.0 \leq \kappa \leq -0.5 & n_y \cdot {}_f\lambda_c^2 \leq 0.05 \end{array}\right\} \tag{6.J3}$$

(C) For columns in a frame which is designed on the basis of the elastic analysis

$$n_c \leq 1.0 \tag{6.J4}$$

where

N_e $= \pi^2\, EI/(L/r)^2$
N_y = yield axial force
${}_fN_c$ = critical strength of the centrally loaded column determined according to effective length KL
${}_fN_e$ $= \pi^2\, EI/(KL/r)^2$
n_c $= N/{}_fN_c$
n_y $= N/N_y$
κ $= M_1/M_2$
λ_c $= \sqrt{N_y/N_e}$
${}_f\lambda_c$ $= \sqrt{N_y/{}_fN_e}$

Requirements in (A) above are to guarantee attainment of the stability strength limit of a frame, particularly a tall frame, in which a number of plastic hinges form at the ultimate stage, and the columns are subjected to large axial forces. The normalized slenderness ratio of a column shall be determined based on the effective length KL. In addition to these requirements, it is necessary to secure sufficient rotation capacity at the plastic hinges in columns, which is considered in (B). The normalized slenderness ratio of a column shall be determined for the real column length taken equal to the distance between panel points in this case. Item (C) is to prevent buckling of a frame in which the plastic hinges are not allowed to form.

Column Axial Force and Slenderness Ratio

Axial force N and slenderness ratio KL/r of a beam column subjected to combined compression and bending shall satisfy the following inequalities

$$\left.\begin{array}{lll} \text{for } N/N_y \le 0.15 & KL/r \le 150 & \\ \text{for } N/N_y > 0.15 & N/N_y + (KL/r)120 \le 1.0 & \text{for SS41 steel} \\ & N/N_y + (KL/r)100 \le 1.0 & \text{for SM50 steel} \end{array}\right\} \quad (6.J5)$$

These limits are to insure sufficient rotation capacity at the plastic hinges.

For a frame subjected to combined vertical and horizontal loads caused by earthquakes, the P-Δ effect should be considered in determining design shear force at each story. The story drift index may be taken as equal to 0.015 to 0.020, if the frame is designed under the condition that the design seismic coefficient is 0.2, and the load factor for the earthquake load is 1.5.

AIJ 88 requires second-order plastic analysis, but AIJ 75 requires only first-order plastic analysis, since the assumed P-Δ moment is included in the design horizontal shear force.

NORTH AMERICA

CHAPTER EDITORS
Le-Wu Lu, Lehigh University
John Springfield, Carruthers & Wallace Limited

AISC ASD 89

In trusses and in those frames where lateral stability is provided by adequate attachment to diagonal bracing, to shear walls, to an adjacent structure having adequate lateral stability, or to floor slabs or roof decks secured horizontally by walls or bracing systems parallel to the plane of the frame, the effective length factor, K, for the compression members shall be taken as unity, unless analysis shows that a smaller value is permitted.

In frames where lateral stability is dependent upon the bending stiffness of rigidly connected beams and columns, the effective length KL of compression members shall be determined by analysis and shall not be less than the actual unbraced length.

Allowable Stress Design
Members subjected to both axial compression and bending stresses shall be proportioned to satisfy the following requirements

$$\frac{f_a}{F_a} + \frac{C_{mx} f_{bx}}{\left(1 - \frac{f_a}{F'_{ex}}\right) F_{bx}} + \frac{C_{my} f_{by}}{\left(1 - \frac{f_a}{F'_{ey}}\right) F_{by}} \leq 1.0 \qquad (6.N1)$$

$$\frac{f_a}{0.60\, F_y} + \frac{f_{bx}}{F_{bx}} + \frac{f_{by}}{F_{by}} \leq 1.0 \qquad (6.N2)$$

When $f_a/F_a \leq 0.15$, Eq. 6.N3 is permitted in lieu of Eqs. 6.N1 and 6.N2

$$\frac{f_a}{F_a} + \frac{f_{bx}}{F_{bx}} + \frac{f_{by}}{F_{by}} \leq 1.0 \qquad (6.N3)$$

In Eqs. 6.N1 to 6.N3, the subscripts x and y combined with subscripts b, m, and e, indicate the axis of bending about which a particular stress or design property applies, and

where

F_a = axial compressive stress that would be permitted if axial force alone existed

F_b = compressive bending stress that would be permitted if bending moment alone existed

$F'_e = \dfrac{12\pi^2 E}{23(Kl_b/r_b)^2}$, Euler stress divided by a factor of safety (In the expression for F'_e, l_b is the actual unbraced length in the plane of bending, and r_b is the corresponding radius of gyration. K is the effective length factor in the plane of bending.) F'_e may be increased by 1/3 in the case of combined gravity and lateral loads.

f_a = computed axial stress

f_b = computed compressive bending stress at the point under consideration

C_m = coefficient whose value shall be taken as follows:

a. For compression members in frames subject to joint translation (sidesway) = 0.85

b. For restrained compression members in frames braced against joint translation and not subject to transverse loading between their supports in the plane of bending

$$C_m = 0.6 - 0.4\ (M_1/M_2)$$

where M_1/M_2 is the ratio of the smaller to larger moments at the ends of that portion of the member unbraced in the plane of bending under consideration. M_1/M_2 is positive when the member is bent in reverse curvature, negative when bent in single curvature.

c. For compression members in frames braced against joint translation in the plane of loading and subjected to transverse loading between their supports, the value of C_m may be determined by an analysis. However, in lieu of such analysis, the following values may be used:

i. For members whose ends are restrained against rotation in the plane of bending $C_m = 0.85$

ii. For members whose ends are unrestrained against rotation in the plane of bending $C_m = 1.0$

Plastic Design

For one- or two-story frames, the maximum strength is permitted to be determined by a routine plastic analysis procedure and ignores the frame instability effect (P-Δ). For braced multi-story frames, provisions shall be made to include the frame instability effect in the design of the bracing system and frame members. For unbraced multi-story frames, the frame instability effect shall be included directly in the calculations for maximum strength.

(A) Stability of Braced Frames

The vertical bracing system for a plastically designed braced multi-story frame shall be adequate, as determined by an analysis to

1. prevent buckling of the structure under factored gravity loads
2. maintain the lateral stability of the structure, including the overturning effects of drift, under factored gravity plus factored horizontal loads.

It is permitted to consider that the vertical bracing system functions together with in-plane shear-resisting exterior and interior walls, floor slabs and roof decks, if these walls, slabs and decks are secured to the structural frames. The columns, girders, beams and diagonal members, when used as the vertical bracing system, could be considered to comprise a vertical-cantilever, simply connected truss in the analyses for frame buckling and lateral stability. Axial deformation of all members in the vertical bracing system shall be included in the lateral stability analysis. The axial force in these members caused by factored gravity plus factored horizontal loads shall not exceed $0.85P_y$, where P_y is the product of yield stress times the profile area of the member.

Girders and beams included in the vertical bracing system of a braced multi-story frame shall be proportioned for axial force and moment caused by the concurrent factored horizontal and gravity loads, in accordance with Eq. 6.N4, with P_{cr} taken as the maximum axial strength of the beam, based on the actual slenderness ratio between braced points in the plane of bending.

(B) Stability of Unbraced Frames
The strength of an unbraced multi-story frame shall be determined by an analysis which includes the effect of frame instability and column axial deformation. Such a frame shall be designed to be stable under (1) factored gravity loads and (2) factored gravity loads plus factored horizontal loads. The axial force in the columns at factored load levels shall not exceed $0.75\,P_y$.

Members subject to combined axial load and bending moment shall be proportioned to satisfy the following interaction formulas

$$\frac{P}{P_{cr}} + \frac{C_m M}{\left(1 - \frac{P}{P_e}\right) M_m} \leq 1.0 \qquad (6.N4)$$

$$\frac{P}{P_y} + \frac{M}{1.18\,M_p} \leq 1.0; \qquad M \leq M_p \qquad (6.N5)$$

where

C_m = coefficient as defined above
M = maximum factored moment
M_m = maximum moment that can be resisted by the member in the absence of axial load
M_p = plastic moment
P = factored axial load
P_{cr} = maximum compressive load that can be resisted by the member in the absence of bending moment (based on applicable slenderness ratio)
P_e = Euler buckling load

AISC LRFD 86

The interaction of flexure and compression in symmetric shapes shall be limited by

for $\frac{P_u}{\phi_c P_n} \geq 0.2$

$$\frac{P_u}{\phi_c P_n} + \frac{8}{9}\left(\frac{M_{ux}}{\phi_b M_{nx}} + \frac{M_{uy}}{\phi_b M_{ny}}\right) \leq 1.0 \qquad (6.N6)$$

for $\frac{P_u}{\phi_c P_n} < 0.2$

$$\frac{P_u}{2\phi_c P_n} + \left(\frac{M_{ux}}{\phi_b M_{nx}} + \frac{M_{uy}}{\phi_b M_{ny}}\right) \leq 1.0 \qquad (6.N7)$$

where

M_n = nominal flexural strength
M_u = required flexural strength
P_n = nominal compressive strength
P_u = required compressive strength
ϕ_b = resistance factor for flexure = 0.90
ϕ_c = resistance factor for compression = 0.85

Determination of M_u

In structures designed on the basis of elastic analysis, M_u may be determined from a second-order elastic analysis using factored loads. In structures designed on the basis of plastic analysis, M_u shall be determined from a plastic analysis which includes the effect of frame instability. In structures designed on the basis of elastic first-order analysis the following procedure for the determination of M_u may be used in lieu of a second-order analysis.

$$M_u = B_1 M_{nt} + B_2 M_{\ell t} \tag{6.N8}$$

where

$$B_1 = \frac{C_m}{(1-P_u/P_e)} \geq 1 \tag{6.N9}$$

$$B_2 = \frac{1}{1-\Sigma P_u\left(\frac{\Delta_{oh}}{\Sigma HL}\right)} \tag{6.N10}$$

or

$$B_2 = \frac{1}{1-\frac{\Sigma P_u}{\Sigma P_e}} \tag{6.N11}$$

C_m = a coefficient whose value shall be taken as follows:

i. For restrained compression members in frames braced against joint translation and not subject to transverse loading between their supports

$$C_m = 0.6-0.4\,(M_1/M_2) \tag{6.N12}$$

where M_1/M_2 is the ratio of the smaller to larger moments at the ends of that portion of the member unbraced in the plane of bending under consideration. M_1/M_2 is positive when the member is bent in reverse curvature, negative when bent in single curvature.

ii. For compression members in frames braced against joint translation in the plane of loading and subjected to transverse loading between their supports, the value of C_m can be determined by rational analysis. In lieu of such analysis, the following values may be used:

for members whose ends are restrained $C_m = 0.85$
for members whose ends are unrestrained $C_m = 1.0$

L = story height
$M_{\ell t}$ = required flexural strength in member as a result of lateral translation of the frame only

M_{nt} = required flexural strength in member assuming there is no lateral translation of the frame
P_e = Euler load based on effective length factor K in the plane of bending (shall not be less than unity)
Δ_{oh} = translation deflection of the story under consideration
ΣH = sum of all story horizontal forces producing Δ_{oh}
ΣP_u = required axial load strength of all columns in a story

CSA S16.1 M89 84

Stability effects are prescribed in the Canadian National Standard CSA S16.1 M89, Limit States Design of Steel Structures. It is stipulated that included in the analysis of a structure shall be the sway effects in each story produced by the vertical loads acting on the structure in its displaced configuration. By loads is meant factored loads. Analysis refers to either elastic or plastic, whichever is adopted. In the latter case, the inelastic deformation evidently is intended, a severe but necessary handicap on plastic analysis. It is stated that these second-order effects due to relative translational displacement (sway) of the ends of a member shall preferably be determined from a second-order analysis. In common terminology, the frame P-Δ effects are referred to, not the member P-δ effects, which are treated by amplification of the bending moment in an interaction equation.

By second-order analysis is meant a stiffness analysis in which the stiffness matrix is modified to account for sway effects. Alternatively, one of two other techniques may be employed. These are iterative analysis in which equivalent story horizontal shear forces are introduced to simulate sway effects, or the amplification of first-order translational moments by the factor

$$U_2 = \frac{1}{1 - \left[\frac{\Sigma C_f \Delta_f}{\Sigma V_f h}\right]} \tag{6.N13}$$

Thus, $M_f = M_{fg} + U_2 M_{ft}$

where

C_f = factored compressive force in column
h = story height
M_{fg} = member first-order moment due to gravity loads
M_{ft} = member first-order translational moment
V_f = member shear
Δ_f = first-order story drift under factored load

The summation in Eq. 6.N13 is over the entire framework, but any corresponding values of Δ and ΣV will suffice.

It will be noted that U_2 is identical to B_2 in the AISC LRFD 86 specification.

If second-order effects exceed 40% of the primary effects, that is, $U_2 > 1.4$, it is required that frame stiffness be increased or a second-order elasto-plastic analysis be performed. The 40% limit is to minimize the probability of inelastic response actually occurring under factored loads in frameworks analyzed elastically, yet the members are still selected on the basis of their ultimate capacity which almost certainly involves inelastic response. In fact, examination suggests that for laterally unbraced W-Sections, response would remain elastic about the major axis but if the primary bending were about the minor axis, this may not be the case.

Note that the treatment of frame stability by effective length methods is excluded from the standard. The effect of this is to force designers to determine the sway stiffness of frameworks from a lateral load analysis; it is necessary to determine story sway Δ under a set of horizontal loads. The value of U_2 assumes great significance since it is a function not only of sway stiffness but also the total vertical load which is to stabilized.

Under gravity load only, it is required that initial lateral story shears, which comprise the starting point for a second-order analysis, be 0.005 of the factored story gravity load. This shear is greater than other codes, the effect of which appeared trivial in practice.

Sections are to be evaluated using the following interaction equation

$$\frac{C_f}{C_r} + \frac{0.85\,U_{1x}M_{1x}}{M_{rx}} + \frac{0.60\,U_{1y}M_{1y}}{M_{ry}} \leq 1.0 \qquad (6.N14)$$

The coefficient 0.85 and 0.60 apply only to sections capable of plastic moment redistribution, being increased to 1.0 for sections which are only compact.

C_f and M_{fx} are corresponding values, that is major axis buckling capacity with major axis bending capacity and minor axis buckling capacity with lateral torsional bending capacity.

M_{fx} and M_{fy} are maximum factored moments in the member about respective axes.

U_1 accounts for second-order effects due to deformation between member ends, taken as

$$U_1 = \frac{\omega_1}{1 - (C_f/C_e)} \qquad (6.N15)$$

where

C_e = Euler buckling load of member about axis under consideration, with effective length of unity

κ = ratio of minimum to maximum member moment

$\omega_1 = 0.6 - 0.4\kappa \geq 0.4$

WEST EUROPE

CHAPTER EDITOR

John Springfield, Carruthers & Wallace Limited

BS 5950 85

While the British Standard BS 5950 P.1 85 retains the classification of multi-story frames as sway or non-sway, it employs the rational classification of lateral stiffness to do so. Under notional horizontal loads of 0.5% of factored gravity loads applied at the roof and each floor, the following criteria are applied to classify frames as non-sway:

For clad frames in which the stiffening effect of cladding is not taken into account in determining lateral deflection,

$$\frac{\Delta}{h} \leq \frac{1}{2000} \tag{6.W1}$$

For unclad frames or clad frames in which the stiffening effect of cladding is taken into account in determining lateral deflection,

$$\frac{\Delta}{h} \leq \frac{1}{4000} \tag{6.W2}$$

Where Δ is the story drift (the standard uses δ) and h the story height. The maximum value of the ratio Δ/h is to be used.

Stability of sway frames is assessed by one of two methods, termed "extended simple design" and "amplified sway method."

Extended simple design, which presumably is applied to regular frames rather than irregular frames with leaning columns or widely varying column stiffnesses, is an effective length approach. The charts which are provided for the determination of effective length are similar to those adopted by Australia, but different in form to the American Julian & Lawrence alignment charts.

The amplified sway method of increasing moments due to horizontal loads is more akin to the Canadian method of using an amplification term, U_2. It is dependent on the elastic critical load factor λ_{cr}, which is the ratio by which the critical elastic buckling load of the framework exceeds the total factored applied load. The effective length in the plane of the frame is taken as 1.0. Under gravity loads only, the sway moments to be amplified are those due to the notional horizontal loads defined above. The factor is $\lambda_{cr}/(\lambda_{cr}-1.0)$.

The method may be applied to virtually any framework to which the principle of superposition applies. The critical buckling load may be determined either by the refined approaches available today utilizing computer programs or by an approximate method as follows

$$\lambda_{cr} = \frac{1}{200\,(\Delta/h)\,\max} \tag{6.W3}$$

in which the maximum sway index is due to the notional horizontal forces.

For frames designed by the rigid-plastic mechanism method, it is required that λ_{cr} should exceed 4.6 and that, depending on the range of values within which λ_{cr} lies, the value of the rigid-plastic load factor should exceed certain values.

BS 5950 P.1 85 supplements the above requirements with guidance to designers which is unusual in North American codes. This includes load patterns to be used in deriving column moments, differences in treatment of beams with or without concrete slabs, permissible moment redistribution and two aspects to which BS 5950 P.1 85 paid much attention, namely, effects of cladding and ridge roof portal frames.

COMMENTARY

It is interesting to compare the criterion, Δ/h, with the American term B_2 or the Canadian U_2. Both contain the story drift Δ/h and lateral load which is related to the total vertical load for which stability is being assessed.

The Merchant-Rankine load factor, λ_F, determined from

$$\lambda_F = \frac{\lambda_p}{0.9 + \lambda_p/\lambda_{cr}} \qquad (6.W4)$$

is not used explicitly in BS 5950 P.1 85, although the limitations on λ_{cr} and λ_p discussed above appear to be related to the Merchant-Rankine expression.

INTERNATIONAL COMPARISONS & COMMENT

WHAT ARE THE DIFFERENCES?

Approaches to stability may be classified into the following broad categories according to the basis adopted:

- elastic critical buckling
- second-order elastic analysis
- second-order elastic-plastic analysis

Despite the lack of compatibility, the first two generally rely on elastic analysis of the frame but accept ultimate member strength for member selection, which implies inelastic member response. This is considered by many to be justified provided the framework capacity is limited at the formation of the first hinge.

Elastic Critical Buckling

The critical elastic buckling load of a framework, which reflects only the effects of gravity loads acting in the columns, is useful for the classification of frames but is of little use in actual design. Stevens (1967) and Horne (1981) suggested that if the ratio of elastic critical buckling load to service load exceeded 10, frame stability would not be a problem. Accordingly, the German specification, DIN 18 800 Part 2 permits first-order elastic analysis with no modifications for frame stability for structures in which the ratio of elastic buckling load to factored service load exceeds 10.

The elastic critical buckling method most widely used is the column effective length method. This has its origins in Euler elastic flexural buckling of single columns. Through the effective length factor, K, the actual column elastic critical buckling load, which is dependent on end restraint of both rotation and translation, is related to that of a pin-ended column. Structures are classified into two distinct categories, sidesway permitted and sidesway prevented. Various techniques have been developed to take account of the complexities that develop in real structures versus isolated columns, such as variation in end restraint of adjacent-columns in a floor from zero to infinity, permitting weaker columns to "lean" on the stronger.

Despite the fact that effective length is a concept applicable only to centrally loaded columns, it is extended to interaction equations which are almost universally used for the design of beam-columns. The equation is usually of the form

$$\frac{P}{P_u} + \frac{C_m\,M}{M_u(1-P/P_e)} < 1.0 \tag{6.1}$$

where

C_m = factor to transform M into an equivalent uniform moment
M = maximum factored end moment
M_u = bending resistance of member, taking account of reduction due to lateral torsional buckling
P = factored axial load
P_e = Euler buckling capacity of column of length KL in plane of bending
P_u = centrally loaded resistance of columns of least unbraced length KL

Account is taken of column instability effects through the quantity KL in the term P_u. Allowance for the increase in primary bending in the column due to the axial load acting on the flexurally deformed member is made through the moment amplification factor $1/(1-P/P_e)$.

Until about 20 years ago, this so-called "effective length" approach was in almost universal use and it remains in wide use even today. The ease with which the elastic critical buckling load of a framework may be determined using computer software has led to the development of another form of moment magnifier, based on the ratio of the total factored load to the elastic critical buckling load of the frame.

Eurocode No. 3 permits the amplification of first-order elastic sway moments by the ratio

$$\frac{0.9\ V/V_{cr}}{(V_{cr}/V) - 1} \quad \text{but} \geq 1.0 \tag{6.2}$$

in which V_{cr} is the story critical elastic buckling load and V is the story vertical load. This expression provides a continuous modifier when the ratio is less than 10, to the lower limit of applicability of 4. This provision was originally adopted in the British code (BS 5950).

In essence, the foregoing two approaches form the basis of all the elastic critical buckling load methods in use today. Although the methods can be shown to give reliable predictions for sub-assemblages, and appear well suited to frameworks subject to gravity load only, they defy strict theoretical derivation. By that is meant that there is no theoretical reason to suppose that the capacity determined through interaction equations is, in fact, the collapse load.

Criticisms that may be levied against the effective length approach to frame stability are that column moments are magnified but beam moments are not and that the division between sidesway permitted and sidesway prevented is not clear cut in tall buildings. The AISC LRFD 86 has addressed the first criticism by requiring a second-order analysis. The second criticism is less widely appreciated but Springfield and Adams (Springfield, 1972) reported finding that the second-order effects in a 57-story building were greater in the plane of bracing than in the rigidity framed direction. The British Standard has addressed this problem by defining the limiting deflection/height ratio under lateral loading for which the structure may be considered sidesway prevented: the prescribed lateral load is a fraction of the gravity load, not the applied wind or seismic load.

Second-Order Elastic Analysis

The mode of failure in which a structure suddenly lurches sideways under the action of gravity load only is well understood. However, if the structure is acted upon by combined lateral and gravity loading, no such sudden failure occurs. Although it is elastic, under increasing load, the structure deflects laterally at an increasing rate due to the effect of the gravity load acting on the displaced structure. Second-order elastic analysis takes account of this lateral sway.

The isolated column in Fig. 6.1, displaced laterally by an amount Δ, requires an additional stabilizing force to maintain equilibrium

$$Q = \frac{P\Delta}{h} \tag{6.3}$$

Thus, the destabilizing effects sometimes are termed the P-Δ effects.

In an early application of second-order elastic analysis (1970), Springfield and Adams (1972) used an iterative procedure in which the story PΔ/h shears were added to the frame lateral forces in a first-order analysis. It was quickly realized that iteration could be avoided by use of the terminal value of the progression.

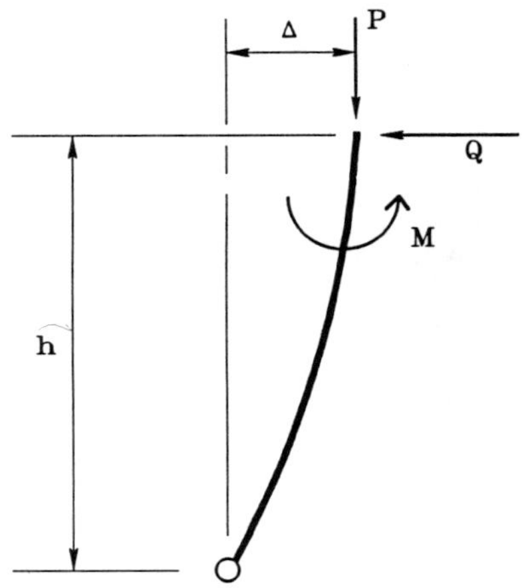

Fig. 6.1 P-Δ Effects in Frames

This is derived as the product of the first-order drift and the amplification factor

$$\frac{1}{1 - \frac{P\Delta}{Qh}} \tag{6.4}$$

This factor may be derived from any pair of corresponding values of Q and Δ. The amplification factor is a fundamental characteristic of the framework and its gravity loads, including the effect of leaning columns. The Australian, Hungarian, and Canadian standards, AS 1250 81, MSZ 85, and CSA S16.1 M89 84, limit the applicability of second-order elastic analysis by limiting the following ratios:

2nd order Δ / 1st order Δ	$\leq$ 0.25	(AS 1250 81)
2nd order Δ / (1st + 2nd order Δ)	$\leq$ 0.25	(MSZ 85)
2nd order Δ / 1st order Δ	$\leq$ 0.4	(CSA S16.1 M89 84)

Under the Hungarian standard, the limiting ratio of second-order to first-order sway would be 0.33. It is interesting to note that, by application of Horne's approximate method of determining the critical elastic buckling load of a frame and setting at 4 the ratio of elastic critical buckling load to factored load, the ratio of second-order to first-order sway can be shown to be 0.38 (Stevens, 1967 and Horne, 1981). Thus, for the few standards which apply limits, there is a measure of agreement.

The amplification factor given above does not include the effect of the axial load resultant acting on the curved member, the so-called P-δ effect. This can be shown to increase the bending in the member by a maximum of 20% in most practical situations. As a consequence, when the amplification factor is used as a simple approximation, DIN 18800 requires its value to be increased by 20%.

The effects that a rigorous second-order elastic analysis should include, but may not be significant in all cases, are

- The effects of gravity loads acting on the laterally displaced frame.
- The effects of the member loads acting on the deformed members.

The advantages of the second-order elastic analysis procedure are its simple concept and the ease with which complex frameworks can be analyzed using standard programs. The disadvantages are that limits are thought to be required to guard against the onset of significant inelastic response, that under gravity load only, articificial lateral loads must be applied and, in common with the elastic critical buckling approach, the collapse load is not determined. The method seeks to ensure that a specific loading can be sustained safely by the deformed frame, a subtle but significant difference.

Second-Order Inelastic Analysis

A second-order inelastic analysis represents most closely the actual response of the structure to the point of maximum load and beyond. In general, techniques and computer programs for non-linear second-order inelastic analysis are not available in design offices, although it is anticipated that such capabilities will become available within the not too distant future.

Several standards, such as AS 1250 81 and CSA S16.1, refer the user to such analyses when the limits of applicability of other simpler analyses are exceeded. The reasons are two-fold. First, it directs the user in a positive rather than negative direction. Secondly, the user who does in fact invoke such a method can demonstrate readily to a regulatory official that the standard is being followed.

The simplest form of inelastic analysis is first-order, the method that has been used in the plastic analysis of low-rise frames for many years. Material is assumed to be elastic, rigid plastic, precipitating the formation of a finite number of concentrated hinges. Moments are redistributed but the effect of gravity loads acting on the displaced frame are not considered. The justification for this assumption stems from the low level of axial loads in columns; the columns are primarily bending members.

Several degrees of complexity may be adopted in a second-order inelastic analysis. The simplest approach is to assume the formation of concentrated hinges, the member segments being assumed straight between hinges and nodes. Accounting for distributed plasticity, such as is likely in columns under heavy axial load, increases the complexity of the situation considerably; the effect of residual stress in permitting partial plastification of the section would need to be included also in this condition.

Merchant-Rankine

The Merchant -Rankine expression for the actual load factor at failure, λ_{F}, retains some favor in Europe, especially when simple plastic analysis is used. The original and modified expressions are

$$\frac{1}{\lambda_{\mathrm{F}}} = \frac{1}{\lambda_p} + \frac{1}{\lambda_{cr}} \tag{6.5}$$

$$\lambda = \frac{\lambda_p}{0.9 + \lambda_p / \lambda_{cr}} \tag{6.6}$$

in which λ_p is the rigid-plastic load factor and λ_{cr} is the critical elastic buckling load factor.

Until approximately 1970, elastic critical buckling was the only approach in use for the treatment of frame stability. Since that time, the second-order analysis approach has been adopted in more and more standards. Fundamental to even an approximate second-order analysis is a reliable first-order prediction of lateral displacement. With the ever increasing availability of structural analysis computer programs, some of which will perform second-order analyses directly, the popularity of the second-order analysis approach is increasing. There appears to be little reason to believe that it will not eventually replace elastic critical buckling completely.

The most marked differences of interest are between the AISC LRFD 86 requirements and DIN 18800 P.2 86. The former requires second-order analysis but requires in addition the use of effective length factors in the design of the columns, in both the axial load term and the moment magnifier term. Admittedly there are no other constraints. DIN 18800 P.2 86 limits $L\sqrt{N/EI}$ only, but artificially increases the sway shear to account for local bending, the P-δ effect.

Since buckling out-of-plane frequently is critical for W-columns, the effective length requirement retained by the AISC LRFD 86 is not as severe as it might appear. Therefore the differences are not all that marked.

Comparison

A comparative study of frames designed by the following five codes has been conducted: AISC ASD 89, AISC LRFD 86, BS 5950 P.1 85, CSA S16.1 M89 84, and MSZ 15024/1 85. The frames included in the study and their service loads are shown in Fig. 6.2. They are one-bay frames with a bay width of 56 ft and a constant story height of 14 ft. Frame A has ten stories and Frames B and C have fourteen stories. Frames A and B are designed in A36 steel (yield stress=36 ksi) and for "usual" lateral loads. Frame C is designed in A572(50) steel (yield stress=50 ksi) and for a more severe lateral load condition. The gravity loads are the same for all the frames. In order to make the comparison consistent the member sizes are selected for all the designs based on the allowable stress procedure AISC ASD 89, but the frame stability provisions of the various codes are incorporated as prescribed.

Comparisons of the member sizes of the frames were made and the strength of the frames was determined by second-order elastic-plastic analyses which account for the P-Δ effect. Using the AISC ASD 89 design as the reference, the relationship of the member sizes of the other four designs may be summarized as follows: the AISC LRFD 86 design is heavier, the British is slightly heavier but very close, the Canadian is between the AISC ASD 86 and AISC LRFD 86, and the Hungarian is the same as AISC ASD 89. This implies that although the five codes have different frame stability provisions (at least in appearance) their net effect on member size selection is insignificant. The ultimate lateral load factors achieved by the frames under a gravity load factor of 1.3 are summarized in Table 6.1.

Table 6.1 Calculated Load Factors for Selected Frames

	AISC (ASD)	AISC (LRFD)	British	Canadian	Hungarian
Frame A	3.0	3.5	*	3.0	3.0
Frame B	2.5	2.5	*	2.5	2.5
Frame C	2.0	2.0	2.0	2.0	2.0

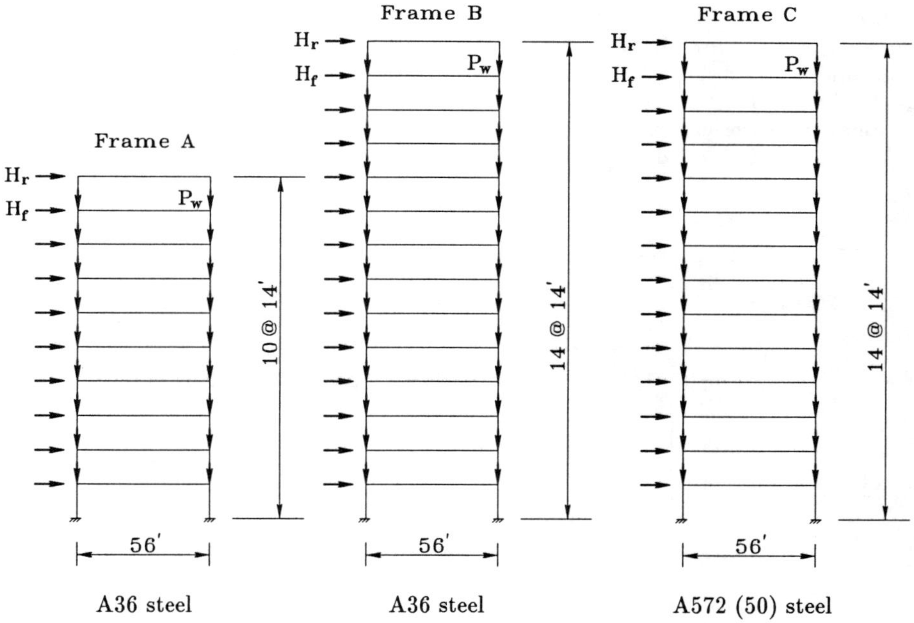

Lateral Loads (service loads):

$H_r = 4.2$ kips	$H_r = 4.2$ kips	$H_r = 6.3$ kips
$H_f = 8.4$ kips	$H_f = 8.4$ kips	$H_f = 12.6$ kips

Gravity Loads (service loads, the same for all frames):

Roof:	Floor:	Wall:
$w_r = 21.$ kips/ft	$w_f = 3.75$ kips/ft	$P_w = 18$ kips

Fig. 6.2 Frames Selected in Design Study

Figure 6.3 shows a typical lateral load-drift relationship of the frames studied. It is shown in terms of lateral load-factor vs. drift index in order that the presentation can be non-dimensional.

WHY ARE THERE DIFFERENCES?

The answer to why are there differences seems be that we are in a period of transition as standards are revised. There is a trend towards adoption of a second-order approach.

The differences among specifications are not all that marked. Such differences as remain appear to stem from the comparisons that have been made by the individual researchers in the respective countries. All of the comparisons have not been available readily to the standard committees, so decisions have not been based on the same data. Furthermore, tradition and use are bound to influence committee decisions, for example, the British preoccupation with the effect of cladding and with ridged portal frames. The only real reason for the differences seem to be the degree of caution exercised by the standard committees and how this caution is applied.

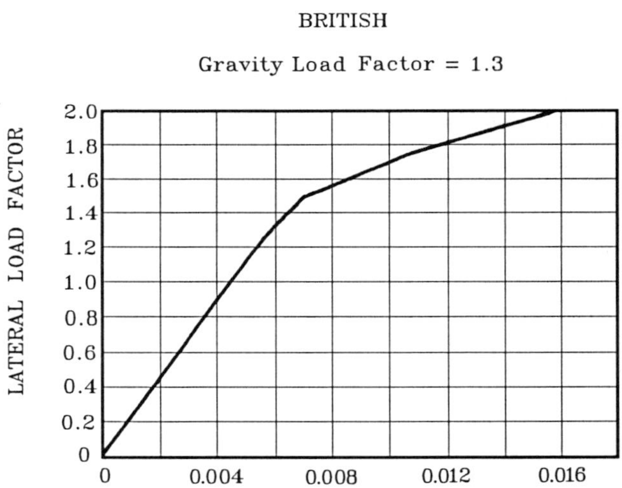

Fig. 6.3 Drift Index — Delta/Height

Symbol List for Chapter 6 -- Frames

B_1 = coefficient to determine the required flexural strength (6.N)
B_2 = coefficient to determine the required flexural strength (6.N)
C_e = Euler buckling load of member (6.N)
C_f = factored compressive force in columns (6.N)
C_m = bending coefficient dependent upon moment gradient (6.A), (6.N) & (6)
C_{mx} = bending coefficient dependent upon moment gradient about x-axis (6.N)
C_{my} = bending coefficient dependent upon moment gradient about y-axis (6.N)
C_r = critical strength of centrally loaded columns (6.N)
e_o = initial imperfection (6.E)
F_a = axial compressive stress that would be permitted if axial force alone exists (6.N)
F_b = compressive bending stress that would be permitted if bending moment alone exists (6.N)
F_{bx} = compressive bending stress about x-axis that would be permitted if bending moment alone exists (6.N)
F_{by} = compressive bending stress about y-axis that would be permitted if bending moment alone exists (6.N)
F'_e = Euler stress divided by a factor of safety (6.N)
F'_{ex} = Euler stress about x-axis divided by a factor of safety (6.N)
F'_{ey} = Euler stress about y-axis divided by a factor of safety (6.N)
F_y = yield stress (6.N)
f_a = computed axial stress (6.N)
f_b = computed compressive bending stress at the point under consideration (6.N)
f_{bx} = computed compressive bending stress due to the moment about x-axis at the point under consideration (6.N)
f_{by} = computed compressive bending stress due to the moment about y-axis at the point under consideration (6.N)
H = story horizontal force (6.E) & (6.N)
H_f = lateral load applied at floor level (6)
H_r = lateral load applied at level of roof (6)
h = story height (6.N) & (6.W)
I_c = moment of inertia of columns (6.E)
I_g = moment of inertia of beams (6.E)
K = effective length factor (6.N)
k_m = effective length factor (6.A)
L = story height (6.A) & (6.N)
L_e = effective length (6.A)
l = length of member (6.E)
l_b = actual unbraced length in plane of bending (6.N)
l_c = height of columns (6.E)
l_i = story height at level i (6.E)
l_g = span of beams (6.E)
M = maximum factored moment (6.N) & (6)
M_d = maximum design moment (6.A)
M_f = end-moment obtained from a first-order sway analysis (6.A)
M_f = required flexural strength (6.N)

M_{fb} = end-moment obtained from a first-order analysis of a frame with sway prevented (6.A)
M_{fg} = member moment due to gravity loads (6.N)
M_{fs} = end moment with sway permitted
M_{ft} = member translational moment (6.N)
$M_{\ell t}$ = required flexural strength in a member as a result of lateral translation of a frame only (6.N)
M_m = maximum bending moment (6.A)
M_m = maximum moment that can be resisted by a member in the absence of axial load (6.N)
M_{nt} = required flexural strength in a member assuming there is no lateral translation of a frame (6.N)
M_{nx} = nominal flexural strength about x-axis (6.N)
M_{ny} = nominal flexural strength about y-axis (6.N)
M_p = plastic moment (6.N)
M_{rx} = critical bending strength about x-axis (6.N)
M_{ry} = critical bending strength about y-axis (6.N)
M_u = required flexural strength (6.N)
M_u = bending resistance of a member, taking into account the reduction due to lateral torsional buckling (6)
M_{ux} = required flexural strength about x-axis (6.N)
M_{uy} = required flexural strength about y-axis (6.N)
M_1 = smaller end moment (6.J) & (6.N)
M_{1x} = maximum factored moment in a member about x-axis (6.N)
M_{1y} = maximum factored moment in a member about y-axis (6.N)
M_2 = larger end moment (6.J) & (6.N)
N_d = axial force (6.A)
N_e = Euler's buckling load (6.J)
N_i = axial force at level i (6.E)
N_{of} = axial force in a member at elastic critical buckling of a frame (6.A)
N_{om} = buckling load of a member (6.A)
N_{omb} = buckling load for braced member (6.A)
N_{oms} = buckling load for sway members (6.A)
N_y = yield axial force (6.J)
${}_fN_c$ = critical strength of centrally loaded column (6.J)
${}_fN_e$ = Euler's buckling load determined according to effective length (6.J)
n = number of columns in one story in the plane (6.E)
n_c = ratio of axial force to critical strength (6.J)
n_y = ratio of axial force to yield axial force (6.J)
P = factored axial load (6.N)
P_{cr} = maximum compressive load that can be resisted by a member in the absence of bending moment (based on applicable slenderness ratio) (6.N)
P_e = Euler load based on effective length factor K in the plane of bending (6.N) & (6)
P_j = total axial force in columns between levels j−1 and j (6.E)
P_n = nominal compressive strength (6.N)
P_u = required compressive strength (6.N)
P_u = required axial load strength of a column in a story (6.N)
P_u = resistance of centrally loaded columns with unbraced length KL (6)
P_w = vertical loads applied to walls (6)

P_y = product of yield stress times the profile area of a member (6.N)
Q = shear force (6)
Q_j = sum of original horizontal force above level j (6.E)
Q'_j = modified sum of original horizontal force above the level j (6.E)
r_b = radius of gyration in plane of bending (6.N)
U_1 = coefficient to account for second-order effects (6.N)
U_{1x} = moment amplification factor to account for second-order effect (6.N)
U_{1y} = moment amplification factor to account for second-order effect (6.N)
U_2 = coefficient to determine the required flexural strength (6.N)
V = story vertical load (6)
V_{cr} = story critical elastic buckling load (6)
V_f = member shear force (6.N)
V_j = ratio of the sum of line stiffnesses of beams to columns at one joint (6.E)
w_f = gravity loads applied at floor level (6)
w_r = gravity loads applied at roof level (6)
α_i = initial rotation (6.E)
α_j = initial rotation (6.E)
β = ratio of the smaller to larger moment at the ends of a member (6.A)
γ = coefficient used to determine initial rotation (6.E)
Δ = story drift (6.W)
ΔH_j = increment of horizontal force at level j (6.E)
Δ_f = story drift under factored loads (6.N)
Δ_{oh} = translation deflection of the story under consideration (6.N)
δ_b = coefficient to determine maximum design bending moment (6.A)
δ_m = coefficient to determine the maximum design bending moment (6.A)
δ_s = coefficient to determine the maximum design bending moment (6.A)
κ = ratio of smaller to larger moment at the ends of a member (6.J)
κ = ratio of minimum to maximum member moment (6.N)
λ_c = ratio of the elastic critical buckling load of the entire to the design load for the particular axial load combination under consideration (6.A)
λ_c = elastic critical buckling load factor (6.A)
λ_c = normalized slenderness ratio (6.J)
λ_{cr} = elastic critical load factor (6.W) & (6)
λ_F = Merchant-Rankine load factor (6.W)
λ_p = rigid-plastic load factor (6)
${}_f\lambda_c$ = normalized slenderness ratio (6.J)
ϕ_b = resistance factor for flexure (6.N)
ϕ_c = resistance factor for compression = 0.85 (6.N)
ω_1 = bending coefficient dependent upon moment gradient (6.N)

Chapter 7

Arches

TOPICAL EDITOR
Shigeru Kuranishi, Tohoku University

CHAPTER EDITORS
Lajos Kollar, Technical University of Budapest
Shigeru Kuranishi, Tohoku University
Nicholas S. Trahair, The University of Sydney
Sriramulu Vinnakota, Marquette University

7a. Inplane Stability of Ribs

SPECIFICATION PROVISIONS & EXPLANATION

EAST EUROPE

CHAPTER EDITOR
Lajos Kollar, Technical University of Budapest

ČSN 73 1401 84

The linear elastic buckling should be checked.
For stress analysis, the amplification factor of bending moment is specified as

$$M_{II,\,L/4} = \left[\frac{1}{1 - \dfrac{f_{I,L/4}\, H_I}{M_{I,L/4}}} \right] M_{I,\,L/4} \tag{7a.E1}$$

where

$f_{I,L/4}$ = deflection at the quarter span due to the loading according to first-order theory
H_I = horizontal force acting on the arch according to first-order theory
$M_{I,L/4}$ = bending moment at the quarter of the span according to first-order theory
$M_{II,L/4}$ = bending moment at the quarter of the span according to second-order theory

MSZ 15024/1 85

The strength is verified by linear elastic buckling and 2nd order elastic analysis.

TGL 13503/01 82

The strength is verified by linear elastic buckling and 2nd order elastic analysis.

REGIONAL COMPARISON

The common feature of the above codes is that they contain no special provisions for arches. They have very short chapters on buckling of the arch in its own plane. They deal mainly with the buckling length of the arch in its own plane, which is given in tables based on the well-known results of the elastic theory.

The most detailed code is that of Czechoslovakia, where tables are included that contain factors for variable moment of inertia. The other codes give only simple rules to transform the variable moment of inertia into a constant one: either by the requirement of identical deflections of two simply supported beams, one having variable and the other constant moment of inertia (GDR and Hungarian code), or simply taking the moment of inertia at the quarter of the span as the constant one (Soviet code).

For the calculation of the compressive and bending stresses, ČSN 73 1401 84, MSZ 15024/1 85 and TGL 13503/01 82 take the effect of the deflections into account. ČSN 73 1401 84 checks the stress at the quarter point by using the deflection at the point obtained by first-order analysis.

JAPAN

CHAPTER EDITOR

Shigeru Kuranishi, Tohoku University

JRA 89

Arch bridges are classified into two categories depending on their stiffness and dead load using the following equations

$$w = \frac{8\,\alpha}{\gamma}\,\frac{E\,I}{L^3}\,\frac{f}{L} \tag{7a.J1}$$

where

- E = elastic modulus material (t/m^2)
- I = mean value of inplane moment of inertia of an arch along the length. In case of stiffened arch, it shall be taken as the sum of those of the arch and stiffened girder (m^4)
- w = load intensity per one main structure of arch to check the necessity of the second order elastic analysis for stress calculation (t/m)
- α, γ = coefficients concerning inplane buckling and relative stiffness of arch and stiffening girder given by Table 7a.J1

Table 7a.J1 Buckling coefficient α (buckling in the arch plane) and correction factor γ

Type of structure				f/L					γ
				0	0.10	0.15	0.20	0.30	
				α					
Arch without stiffening girder	2 hinged arch			39.5	36.0	32.0	28.0	20.0	9
	Fixed arch			81.0	76.0	69.5	63.0	48.0	
2 hinged stiffened arch without axial forces in stiffening girder	Without side span			39.5	36.0	32.0	28.0	20.0	12
	With side span	λ	0	81.0	76.0	69.5	63.0	48.0	
			0.25	63.0	58.5	52.5	47.0	34.5	
			0.50	55.5	51.5	46.5	41.5	30.5	
			0.75	51.5	48.0	43.0	38.5	28.5	
			1.00	49.0	45.5	41.0	36.5	27.0	
			2.00	45.0	41.0	36.5	32.0	22.5	

(a)

$$\lambda = \frac{a}{L}\left(1 + \frac{I_A}{I_G}\right)$$

where

- a = Length of a side span of the stiffening girder (m)
- f = rise of the arch(m)
- I_A = Mean value of in-plane moment of inertia along the single arch member(m^4)
- I_G = Mean value of moment of inertia along the single stiffening girder(m^4)
- L = Span length of the arch(m)

(b) Coefficient α may be obtained by linear interpolation for intermediate values of f/L and/or λ.

When the arch is subjected to greater dead load than the value w given by Eq. 7a.J1, then second-order elastic analysis must be applied. The allowable stress is obtained by the beam-column formula in which the effective length is taken equal to the panel length. The ultimate strength of arches shall be checked by calculating the stress σ_u produced by 1.7 times live plus impact load and 1.7 times dead load using the following equations

for tensile stress

$$\sigma_u \leq 1.7\sigma_{ta} \tag{7a.J2}$$

for compressive stress

$$\sigma_u \leq 1.7\sigma_{cal}$$

where

σ_{cal} = allowable compressive stress for local buckling (kg/cm^2)
σ_{ta} = allowable tensile stress (kg/cm^2)

For stiffened arches with slender arch ribs (Langer or Stab bogen type of stiffened arch), the arch member is designed by the column formula if Eq.7a.J3 is satisfied and the ribs are straight and not deeper than 1/10 of the distance between panel points.

$$\beta \frac{\sigma^A_{ca}\, h^G}{\sigma_{ta}\, h^A} > 1 \tag{7a.J3}$$

where

h^A = mean value of depth along the arch member (cm)
h^G = mean value of depth along the stiffening girder (cm)
σ^A_{ca} = mean value of allowable compressive stress for arch member (kg/cm^2)
σ_{ta} = mean value of allowable stress for lower flange of stiffening girder (kg/cm^2)
β = $0.004 + 0.004\ l/r$ for the stiffening girder with axial force
β = $1.75(0.04+0.004\ l/r)$ for the stiffening girder with axial force

The effect of curvature and eccentricity of arch rib between adjacent panel points shall be considered.

NORTH AMERICA

CHAPTER EDITOR
Sriramulu Vinnakota, Marquette University

AASHTO 89

This specification gives two alternative design methods, namely, Service Load Design Method (SLDM) and Load Factor Design Method (LFDM) for solid rib arches. In what follows, the design equations for use with SLDM are indicated by (.a) and those for LFDM are indicated by (.b) in the equation number.

Buckling in the Vertical Plane
Allowable axial stress is given by

$$F_a = \frac{F_y}{2.12}\left[1 - \frac{(\frac{KL}{r})^2 F_y}{4\pi^2 E}\right] \tag{7a.N1.a}$$

$$F_a = \frac{F_y}{1.18}\left[1 - \frac{(\frac{KL}{r})^2 F_y}{4\pi^2 E}\right] \tag{7a.N1.b}$$

where

F_a = allowable axial unit stress
K = effective length factor
L = one-half the length of arch rib

Effective length factors for three-hinged, two-hinged and fixed arches for different rise-to-span ratios are given in the following Table 7a.N1

Combined Compression and Bending
For this case, the amplification factor method is used, namely, the arch rib is to be proportioned to satisfy the following requirement:

$$\frac{f_a}{F_a} + \frac{f_b}{F_b} \leq 1 \tag{7a.N2}$$

Table 7a.N1: K values for use in calculating F_a and F_e

Rise to span ratio	3-hinged arch	2-hinged arch	Fixed arch
0.1 - 0.2	1.16	1.04	0.70
0.2 - 0.3	1.13	1.10	0.70
0.3 - 0.4	1.16	1.16	0.72

Live load plus impact moments that are determined by an analysis that neglects arch rib deflection is to be increased by an amplification factor A_F given as

$$A_F = \frac{1}{1 - \dfrac{1.70T}{AF_e}} \qquad (7a.N3.a)$$

$$A_F = \frac{1}{1 - \dfrac{1.18T}{A\,F_e}} \qquad (7a.N3.b)$$

where

A = area of cross-section of the rib
A_F = moment amplification factor
F_b = allowable bending stress
F_e = Euler buckling stress
f_a = computed axial stress
f_b = calculated bending stress, including moment amplification at the extreme fiber
I_s = moment of inertia of a stiffener
T = arch rib thrust at quarter point from dead plus live plus impact loading

WEST EUROPE

CHAPTER EDITOR

Shigeru Kuranishi, Tohoku University

DIN 18800 P.2 86

Arch with Constant Cross Section

The verification for the compressive strength of a funicular curved arch (centrally compressed) uses the compressed column formula and the effective buckling length factor as given by Table 7a.W1.

Table 7a.W1 Factor of Effective Buckling Length

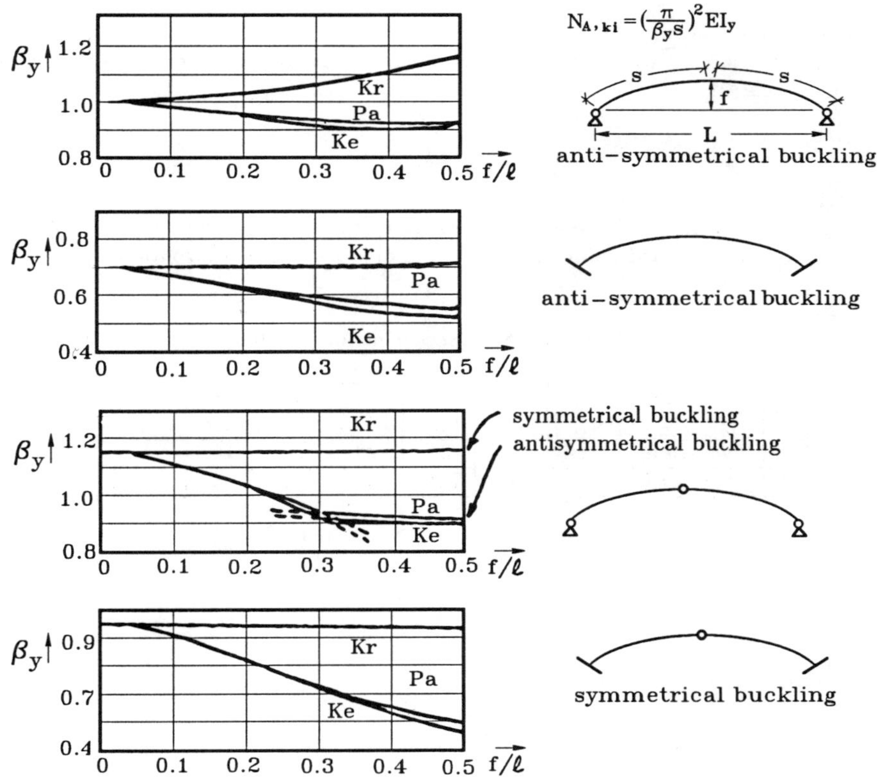

Pa: parabolic, Ke: catenary, Kr: circular. The direction of loads for parabolic and catenary arches are assumed to be vertical; for circular arches they are assumed to be perpendicular to the axis of the arch.

The effective buckling length factors β given for various cases of structural systems and support are illustrated in Table 7a.W1. The effective buckling length (s_k) is given by

$$s_k = s\ \beta_y \tag{7a.W1}$$

where

s = one half of the arch length
s_k = the effective length of arch

As for N, the axial force at the crown can be used as an approximation.

Arch with Tie
The compressive strength of arches connected with a tie by hangers should be verified for the whole structure.

Snap-Through Buckling
For shallow arches with constant cross section, symmetrical snap-through buckling might occur, if

$$\ell \sqrt{\frac{A}{12\ I_y}} > k \tag{7a.W2}$$

where

A = cross sectional area and the value of k is taken from Table 7a.W2
I_y = standard moment of inertia of arch rib about horizontal axis

Table 7a.W2 Supplementary Values

	1	2	3	4	5	6	7
1		f/l	0.05	0.075	0.10	0.15	0.20
2	Two–hinged arch		35	23	17	10	8
3	Fixed arch	k	319	97	42	13	6

Arch with Variable Cross Section

The compressive strength is verified by a 2nd order elastic analysis using the equivalent imperfections given by Table 7a.W3.

Combined Compression and Bending

For the strength of arches analyzed by the treatment recommended by (DIN 18800 P.2 86), the equivalent imperfections shown in Table 7a.W3 are used in the most unfavorable way.

Strength of arches that fulfill the provisions for inplane strength can be verified according to first-order theory without equivalent imperfections if

$$\beta_y \, s \sqrt{\frac{N}{E \, I_y}} < 1 \qquad (7a.W3)$$

where

β_y = value of β for inplane buckling
s = one-half of the arch length

Table 7a.W3 Equivalent In-Plane Geometrical Imperfection

	1	2	3			
	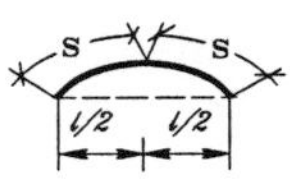	distribution of equivalent geo. imperfection (sine of parabolic form)	w_o imperfection factor for cross sections of buckling curve			
			a	b	c	d
1	three-hinged arch for symmetrical buckling	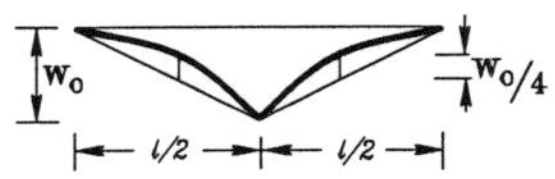	$\frac{s}{300}$	$\frac{s}{250}$	$\frac{s}{200}$	$\frac{s}{150}$
2	two-hinged arch, fixed arch, three-hinged arch in anti-symmetrical buckling	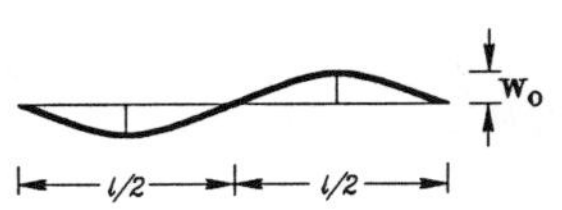	$\frac{s}{600}$	$\frac{s}{500}$	$\frac{s}{400}$	$\frac{s}{300}$

INTERNATIONAL COMPARISONS & COMMENT

The arch is defined as a vertically curved structure with end constraints. The compressive axial force produced by gravity loads in its ribs causes not only inplane but also spatial buckling. However, if arches have enough out-of-plane strength and rigidity, their inplane strength can be evaluated independent of the spatial behavior. Arches are used in various configurations in structures, for example, as roofs of buildings and main structures of bridges with stiffening girders and bracing systems and with various degrees of constraint. Consequently, the code provisions for arch strength must cover all aspects of their configurations. Currently, rational and detailed design rules have not been established.

The most unfavorable loading for arches is the asymmetric one, and the cross section becomes critical at the quarter point of the span of two-hinged arches. Therefore, some provisions indicate implicitly to check the stress according to the critical value applicable at the quarter point only.

In JRA 89, the safety factor (expressed by a load factor), adopted to assure that the ultimate strength, is taken equal to 1.7 for a combination of live load, impact, and dead load.

The inplane strength of arches is assured chiefly by one of the following:

1. Linear buckling strength for funicular curved arch or effective buckling length
2. Effective buckling length and beam-column formula
3. Amplification factor for bending moment calculated by first-order elastic analysis caused by live loads or finite displacement effect (second-order analysis) and allowable stress or ultimate strength.

WHAT ARE THE DIFFERENCES?

The most significant difference is that some specifications include provisions concerning arches but others have no provision at all.

Nearly all specifications that do have provisions for arches require assurance of linear buckling strength, using an effective buckling length or buckling coefficient. The compressive strength so calculated yields almost the same value, and there is no substantial difference between the codes of the various countries.

According to JRA 89, it is not required to check the linear buckling, but the buckling coefficient is specified. The coefficient and dead load intensity are used to estimate the effect of deflections.

Two specifications, ČSN 73 1401 84 and AASHTO 89, employ an amplification factor for the stresses calculated by first-order theory. The former uses the deflection calculated by first-order theory, and the latter uses the beam-column formula for the estimation of the effect of deflection. Most of the others verify the strength and safety using 2nd order theory, but they do not indicate a definite characteristic value. Among them, only JRA 89 specifies a characteristic value. The value 1.7 for the factor of safety is adopted to adjust the proportion of stiffening girder designed by this code to one designed by the conventional design method according to first order theory and effective length. A higher safety factor for live load than for dead load yields a larger stiffening girder than a conventionally designed one. Since the stiffening girder is not subjected to bending moment under dead load.

DIN 18800 P.2 86 allows for the design of arches by various analytical methods and design concepts. AASHTO 89 gives two alternative design methods, namely SLDM and LFDM.

Almost every specification referred to has general provisions related to the linear buckling of arch ribs with constant cross section. A few specifications also contain provisions for variable cross-sections. JRA 89 and DIN 18800 P.2 86 require a check of the compressive strength of the arch rib between adjacent posts or hangers, considering the influence of its curvature.

The concept matrix if Table 7a.1 presents a more detailed picture of the differences.

Table 7a.1 Concept Matrix, International Comparison of Specified Items

Region	Specification	Items														
		1								2		3		4		
		a	b	c	d	e	f	g	h	a	b	a	b	a	b	c
E	ČSN 73 1401 84	o		o		o	o			o						o
	MSZ 15024/1 85	o		o	o	o					o					
	TGL 13503/01 82	o		o	o	o		o			o					
J	JRA 89*		o					o	o		o	o	o	o	o	
N	AASHTO 89	o								o				o	o	
W	DIN 18800 P.2 86*	o		o				o	o		o			o	o	

* JRA 89 has a provision concerning local buckling of arch rib between adjacent hangers or posts.

* DIN 18800 P.2 86 has provisions concerning equivalent initial imperfections and snap through buckling.

Explanation of "Items" in table 7a.1

1 COMPRESSIVE STRENGTH
- (1a) Application of column strength formula
- (1b) Buckling coefficient
- (1c) Effective buckling length
- (1d) Constant cross section
- (1f) Variable cross section
- (1g) Effect of stiffening girder
- (1h) Effect of supporting conditions

2 STRESS VERIFICATION
  (2a) Use of amplification factor
  (2b) Calculation according to 2nd order elastic theory

3 STRENGTH VERIFICATION
  (3a) Application of beam-column formula
  (3b) Characteristic value for stresses by 2nd order elastic theory

4 DESIGN METHOD
  (4a) Service Load Design Method
  (4b) Load Factor Design Method
  (4c) Limit States Design Method

WHY ARE THERE DIFFERENCES?

The lack of provisions for arches in some specifications may arise because arch structures are complex and should be designed by selected engineers using their own judgment and responsibility. There are no rational design rules covering all aspects of arch strength presently available.

The adoption of the amplification factor for bending stresses results in a simpler and easier analysis. ČSN 73 1401 84 uses deflections calculated by first-order theory and AASHTO 89 adopts the beam-column formulas that account for the interaction effect of the deflection with the axial thrust to the critical thrust by the approximate linear buckling.

The second-order theory can cope with a variety of arch configurations. However, it involves a more complicated analytical process, and computer programs that consider second-order effects are not always available. In the future, a combination of second-order analysis and characteristic values will be used widely.

Note that the ultimate strength of highly redundant structures such as fixed arches cannot be estimated using 2nd order analysis which obtains a lower value than the actual one. Further study is needed.

7b. System Stability

SPECIFICATION PROVISIONS & EXPLANATION

EAST EUROPE

CHAPTER EDITOR
Lajos Kollar, Technical University of Budapest

ČSN 73 1401 84 and TGL 13503/01 82

The effective buckling length for lateral buckling is given by

$$\ell_0 = \alpha\, \beta\, L/2 \tag{7b.E1}$$

where

L = span of the arch
ℓ_0 = buckling half wave length
α = factor considering the behavior of the load during buckling
for the load remains vertical

$$\alpha = 1$$

for a part q_1 of the load q is suspended (Fig.7b.E1-a)

$$\alpha = 1 - 0.35\,\frac{q_1}{q}$$

for a part q_2 of the load q is transmitted by columns (Fig.7b.E1-b)

$$\alpha = 1 + 0.45\,\frac{q_2}{q} \tag{7b.E2}$$

where

q = total load on rib
q_1 = part of q transmitted by hangers
q_2 = part of q transmitted by columns

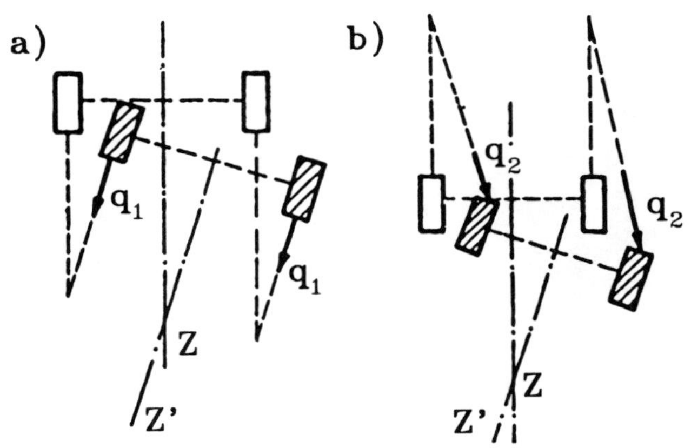

Fig. 7b.E1 Load Transmission during Buckling

REGIONAL COMPARISON

As far as the buckling length in lateral buckling is concerned, the Czechoslovakia and GDR codes give two factors that consider the variation of the moment of inertia and the behavior of the loads during buckling, namely whether they are transmitted by hangers or by columns (which become inclined), or whether they act directly on the arch, i.e., they remain vertical. The other codes contain no provisions for lateral buckling or system strength.

Table 7b.E1 β: buckling length factor

f/l	I_x=const.	$I_x \cos\phi$=const.
0.05	1.00	1.00
0.10	1.08	1.04
0.20	1.30	1.18
0.30	1.64	1.42
0.40	2.14	1.72

where

I_x = the moment of inertia with respect to the axis in the plane of arch
ϕ = the inclination of the axis to the horizontal line

JAPAN

CHAPTER EDITOR

Shigeru Kuranishi, Tohoku University

JRA 89

The lateral buckling strength of an arch with specified lateral bracing under full loading with dead and live distributed loads may be checked by the following equation

$$H/A_g < 0.85\sigma_{ca} \tag{7b.J1}$$

where

A_g = mean value of gross sectional area of arch of a single arch member along the length (m^2)
b = distance between axes of twin arch rib (m)
H = the horizontal thrust of each rib (t)
I_z = mean value of the moment of inertia of cross section of arch member about the z axis along the length
k = factor that indicates live load p and dead load w carried by the hangers or posts and expressed by (p+w) k
L = span length of arch (m)
l = $\phi\beta_z$ L
r = the radius of gyration

$$r = \sqrt{\frac{I_z + A_g(\frac{b}{2})^2}{A_g}} \tag{7b.J2}$$

σ_{ca} = the allowable axial compressive stress

β_z is given in Table 7b.J1. Linear interpolation for the intermediate values is allowed. ϕ is specified as follows

for through type stiffened arches

ϕ=1− 0.35k

for deck type stiffened arches

ϕ=1− 0.45k

for half through type of arches

ϕ=1

Table 7b.J1 Coefficient β_z

Rise to span ratio	0.05	0.10	0.20	0.30	0.40
I_z=constant	0.05	0.54	0.65	0.82	1.07
$I_z(x)=I_{z,c}/\cos\phi_x$	0.50	0.52	0.59	0.71	0.86

NORTH AMERICA

CHAPTER EDITOR
Sriramulu Vinnakota, Marquette University

AASHTO 89

This specification does not cover lateral buckling of individual ribs, or spatial stability of the system.

COMMENTARY

AASHTO 89 does not give specific guidance on lateral buckling or flexural torsional buckling of individual ribs, nor spatial stability of the system composed of the arch ribs, lateral bracing and hangers or columns.

A majority of the arch bridges constructed in the U.S.A. are composed of box section ribs, widely spaced and braced by lateral members. Therefore, lateral stability and spatial stability are not, generally, considered to be critical. Studies confirming this assumption are not available.

WEST EUROPE

CHAPTER EDITOR

Shigeru Kuranishi, Tohoku University

DIN 18800 P.2 86

The buckling perpendicular to arch-plane for arches not supported laterally between end bearings may be checked approximately according to the following equations:

For parabolic arches, the buckling strength is verified by the column strength formula using the following slenderness parameter $\overline{\lambda}_k$:

$$\overline{\lambda}_k = \beta_1 \beta_2 \frac{\ell}{I_z \lambda_s} \qquad (7b.W1)$$

Table 7b.W1 Buckling Length Factor β_1

	1	2	3	4	5	6	7
1	f/ℓ	0.05	0.1	0.2	0.3	0.4	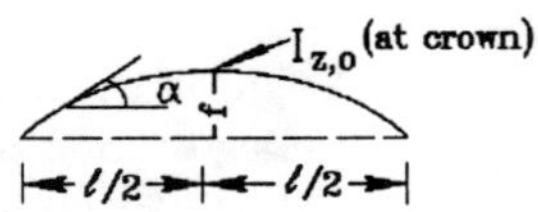
2	I_z=const.	0.50	0.54	0.65	0.82	1.07	
3	$I_z = \frac{I_{z,0}}{\cos\alpha}$	0.50	0.52	0.59	0.71	0.86	

where

I_z = moment of inertia of arch section out-of-plane
$I_{z,0}$ = moment of inertia of arch section at crown
$\lambda_s = \pi\sqrt{E/F_y}$ e.g. 92.9 for St 37 for F_y=240 N/mm^2

Table 7b.W2 Buckling Length Factor β_2

	1	2	3
	loading	β_2	
1	unchanged direction	1	q total load
2	by hangers	$1-0.35\ q_h/q$	q_h load component taken by hangers
3	by posts*	$1+0.45\ q_{st}/q$	q_{st} load component taken by posts
4	* floor slab is rigidly connected with arch crown laterally		

for circular arches

$$\overline{\lambda}_k = \frac{N_{pl}\ \gamma_m}{N_{ki,kr}} \qquad (7b.W2)$$

where

I_t = torsional constant of rib cross section

$$k = \frac{EI_z}{GI_t} \tag{7b.W3}$$

$$N_{ki} = \frac{EI_z(\pi^2 - \alpha^2)^2}{r^2\alpha^2(\pi^2 + \alpha^2 k)} \tag{7b.W4}$$

$N_{ki,kr}$ = the least bifurcation load of fork-supported circular arch with constant, double symmetrical cross section and uniformly distributed, readily unchanged direction loading,

r = radius of the circular arch

α = central angle of the circular arch

Arch with Wind Bracing and Portal Frames

For out-of-plane buckling of twin arch ribs with wind bracing and end portal frames, the buckling may be estimated approximately considering end portal frame columns only. For the verification of the portal frame columns, the column strength formula is used with $\bar{\lambda}_k$ where:

$$\bar{\lambda}_k = \frac{\beta h}{i_z \lambda_s} \tag{7b.W5}$$

where

h = height of the portal frame in the frame plane

i_z = radius of moment of inertia of the portal column about the z axis. For lateral loads, an additional verification of the portal columns has to be made considering the bending moment using effective buckling length factor given by Eq.7b.W5

β = effective buckling length factor

$\lambda_s = \pi\sqrt{E/F_y}$

COMMENTARY

DIN 18800 P.2 86 has provisions concerning lateral buckling of fixed or fork-supported arches subjected to end forces in the chord direction.

INTERNATIONAL COMPARISONS & COMMENT

Arches buckle in the lateral direction under inplane loading or a combination of inplane and out-of-plane loading.

In general, lateral buckling of arch ribs is similar to that of a column with the same supporting conditions, in case of a sufficient torsional rigidity and of properly checked dimensions.

Concerning bridge structures, arches are spatial structures and generally consist of two or more arch ribs, their bracing systems include the portal frames and/or stiffening girders. Therefore, the system stability of the arch is affected by the inplane and out-of-plane flexural rigidity and torsional rigidity of each arch rib, shear rigidity of the structure as a whole, number of arch ribs, the bracing system, location of the stiffening girder, relative stiffness of the girder to the arch rib, and inclination of their hangers or posts during buckling.

Consequently, it is very difficult to cover all aspects of system stability of archs in terms of code provisions.

A significant feature of DIN 18800 P.2 86 is the specification of the initial imperfection to be used in cases where a second-order analysis is required.

WHAT ARE THE DIFFERENCES?

ČSN 73 1401 84, TGL 13503/01 82, JRA 89 and DIN 18800 P.2 86 include the effect of the inclination of posts and hangers during buckling, following DIN 4114. Consequently, the final values are almost the same.

The provisions of DIN 18800 P.2 86 can deal with a wide range of variables and provide detailed values mainly according to the results from an elastic analysis and effective length concept. The specifications of the other countries seem to be too conservative and don't reflect the results obtained from recent research.

A detailed statement of the differences is given in Table 7b.1.

Table 7b.1 Concept Matrix, International Comparison of Specified Items

Region	Specification	Items					
		(1a)	(1b)	(1c)	(1d)	(1e)	(2)
E	ČSN 73 1401 84	o	o	o			
	MSZ 15024/1 85						
	TGL 13503/01 82	o	o	o			
J	JRA 89	o	o	o			
N	AASHTO 89						
W	DIN 18800 P.2 86*	o	o	o	o	o	o

* DIN 18800 P.2 86 has a provision concerning equivalent initial imperfections.

Explanation of "Items" in Table 7b.1

1 COMPRESSIVE STRENGTH (UNDER GRAVITY LOADS)
- (1a) Application of column strength formula
- (1b) Effective buckling length
- (1c) Effect of the inclination of hangers or posts during buckling
- (1d) Effect of bracing systems
- (1e) Effect of end portal frames

2 COMPRESSION AND BENDING STRENGTH (UNDER GRAVITY AND LATERAL LOADS)
- (2) Effect of load combination

Symbol List for Chapter 7 -- Arches

A = area of cross-section of the rib (7a.N)
A = cross sectional area (7a.W)
A_F = moment amplification factor (7a.N)
A_g = mean value of gross sectional area of arch of a single arch member along the length (7b.J)
a = span length of a side span connected to the stiffening girder in the main span (7a.J)
b = distance between axes of twin arch rib (7b.J)
E = elastic modulus of material (7a.J) & (7a.N)
F_a = allowable axial unit stress (7a.N)
F_b = allowable bending stress (7a.N)
F_e = Euler buckling stress (7a.N)
F_y = yield stress of material (7a.N) & (7a.W)
f = rise of the arch (7a.J), (7a.W), & (7b.E)
$f_{I,\,L/4}$ = deflection at the quarter point of arch span (7a.E)
f_a = computed axial stress (7a.N)
f_b = calculated bending stress, including moment amplification at the extreme fiber (7a.N)
H = horizontal component of axial force acting on a single arch rib due to loading (7b.J)
H_I = horizontal force on the arch (1st order theory) (7a.E)
h = height of the portal frame in the plane of frame (7b.W)
h^A = mean value of depth of an arch member along the length (7a.J)
h^G = mean value of depth of a stiffening girder along the length (7a.J)
I = mean value of inplane moment of inertia of an arch along the length. In case of stiffened arch, it shall be taken as the sum of those of the arch and stiffened girder (7a.J)
I = standard moment of inertia (7a.W)
I_s = moment of inertia of a stiffener (7a.N)
I_t = torsional constant of rib cross section (7b.W)
I_x = moment of inertia with respect to the axis in the arch plane (7b.E)
I_y = moment of inertia of arch rib about horizontal axis (7a.W)
I_z = moment of inertia of arch section out-of-plane (7b.W)
I_z = mean value of the moment of inertia of cross section of arch member about the z axis along the length (7b.J)
I_{z0} = moment of inertia of arch section at crown (7b.W)
I_A = mean value of the inplane moment of inertia of arch along the length (7a.J)
I_G = mean value of the inplane moment of inertia of a stiffening girder along the length (7a.J)
i_z = radius of gyration of portal frame column (7b.W)
K = effective length factor (7a.N)
k = factor that indicates live and dead load carried by the hangers or posts and expressed by $(P+W)\,k$ (7a.J)
k = ratio of bending stiffness to torsional stiffness (7b.W)
L = span length of arch (7b.E) & (7a.J)

L = one-half the length of arch rib (7a.N)
l = span length of arch (7a.W)
ℓ = effective buckling length of arch (7b.J)
ℓ_0 = effective buckling length for lateral buckling (7b.E)
$M_{I, L/4}$ = bending moment at the quarter point (1st order theory) (7a.E)
$M_{II, L/4}$ = bending moment at the quarter point (2nd order theory) (7a.E)
N = thrust at the crown in case of arch (7a.W)
N_A = thrust at the arch springing (7.W)
$N_{ki,kr}$ = the least bifurcation load of fork-supported circular arch with constant, double symmetrical cross section and constant readily unchanged direction loading (7b.W)
N_{pl} = the yield capacity of circular arch section (7b.W)
p = uniformly distributed live load intensity (7a.J)
q = total load on rib (7b.E) & (7b.W)
q_h = load component taken by arch hangers (7b.W)
q_{st} = load component taken by the arch posts (7b.W)
q_1 = part of q transmitted by hangers (7b.E)
q_2 = part of q transmitted by columns (7b.E)
r = radius of circular arch (7b.W)
r = radius of gyration of column section (7a.N)
s = one half the arch length (7a.W)
s_k = effective length of arch for inplane buckling (7a.W)
T = arch rib thrust at quarter point from dead plus live plus impact loading (7a.N)
w = load intensity per one main structure of arch to check the necessity of the application of second order elastic analysis for stress calculation (7a.J)
w_o = equivalent inplane imperfection (7.W)
$\overline{w}$ = dead load intensity (7a.N)
α = buckling coefficient (buckling in the arch plane) (7a.J)
α = factor considering the behavior of loading (7b.E)
α = central angle of circular arch (7b.W)
β = coefficient to determine the design method of arch member as an axially compressed member (7a.J)
β = specified buckling length factor for uniform distributed load, $=s_k/s$ (7a.W)
β = effective buckling length factor (7.E)
β_y = value of β for inplane buckling (7a.W)
β_z = effective lateral buckling length coefficient (7b.J)
β_1 = lateral buckling length factor (7b.W)
β_2 = lateral buckling length factor depending on load transmission (7b.W)
γ = coefficient concerning inplane buckling and relative stiffness of arch and stiffening girder given by Table 7a.J1 (7a.J)
γ_m = resistance factor for bending (7b.W)
λ = correction factor for arch plane buckling coefficient concerning the effect of side spans (7a.J)
λ = effective slenderness, $=s_k/i$ (7a.W)
$\overline{\lambda}_k$ = slenderness parameter (7b.W)
λ_s = $\pi\sqrt{E/F_y}$ e.g. 92.9 for St 37 for F_y=240 N/mm^2 (7b.W)
$\overline{\lambda}$ = λ/λ_s (7b.W)

ϕ = inclination of the arch axis to the horizontal line (7b.E)
σ_{ca} = mean value of allowable axial compression stress (7a.J)
σ_{cal} = allowable compressive stress for local buckling (7a.J)
σ_{ca}^{A} = mean value of allowable axial compressive stress for arch member (7a.J)
σ_{ta} = allowable tensile stress (7a.J)
σ_{ta} = mean value of allowable stress for lower flange of stiffening girder (7a.J)
σ_u = 1.7 σ_{ta} for tensile stress or, (7a.J)
= 1.7 σ_{cal} for compressive stress (7a.J)

Chapter 8

Triangulated Structures

TOPICAL EDITOR
Carlo Urbano, Technical University of Milan

CHAPTER EDITORS
Hiroshi Akiyama, University of Tokyo
Pierre Dubas, Swiss Federal Institute of Technology
Zaitian Guo, Xian Institute of Metallurgy & Construction Engineering
Miklos Ivanyi, Technical University of Budapest
Nicholas F. Morris, Manhattan College
Lewis Schmidt, University of Wollongong
Carlo Urbano, Technical University of Milan

8a. Effective Length

SPECIFICATION PROVISIONS & EXPLANATION

AUSTRALIA

CHAPTER EDITOR
Lewis Schmidt, University of Wollongong

AS 1250 81

Axial forces are found by assuming pin connected members. Stresses arising from eccentricity are kept within prescribed limits.

The effective length of compression members is the distance between braced panel points ℓ when it is considered that all members are buckling or yielding at the same load level acting on the truss. In other cases a rational elastic buckling analysis is permitted.

Alternatively when $f_y \leq 280$ MPa, and when members adjacent to the member under consideration are stressed to less than 0.75 times their permissible capacities, the effective length is not less than 0.85ℓ.

For battened members or for discontinuous members connected at the ends by single connectors or to gusset plates having no significant stiffness in the plane of buckling, the effective length is not less than ℓ.

SEC 81

Design of towers, as carried out by State Electricity Commission, Victoria, identify three types of element. The legs of the tower have an effective length equal to thepanel point length, ℓ. Where the rib members are cross-braced, the compressive element has an effective length equal to the distance between point of crossing to the panel point. Single angle members are designed for eccentricity and torsional buckling.

CHINA

CHAPTER EDITOR
Zaitian Guo, Xian Institute of Metallurgy & Construction Engineering

GBJ 17 88

For in-plane buckling of trusses, the effective length ℓ_{ef} of chord members and end posts should be taken equal to the distance between panel points. For the other web members, ℓ_{ef} should be taken equal to 0.8ℓ.

For out-of-plane buckling, the effective length of web members should be taken equal to their geometrical length ℓ whereas that of chord members should be taken equal to the distance between the laterally supported panel points.

For those members whose principal axes do not lie in the plane of the truss, such as web members consisting of a single angle or of two angles built-up in cruciform cross-section, the effective length should be taken equal to 0.9ℓ. This clause does not cover end posts.

If the distance between laterally supported points of compression chords is equal to twice the panel length of the truss and the two panels are subjected to unequal compressive force (Fig. 8a.C1), the effective length for out-of-plane buckling should be taken as

$$\ell_{ef}=\ell_1(0.75+0.25\,\frac{N_2}{N_1}) \quad \text{but} \quad \ell_{ef} \geq 0.5\ell_1 \tag{8a.C1}$$

where N_1 and N_2 are the greater and smaller compression force, respectively.

For double diagonals of trusses or cross bracings, the effective length for in-plane buckling should be taken equal to the distance from panel point to the crossing point. For out-of-plane buckling, ℓ_{ef} should be taken as follows:

Tension in one member, when both members are continuous, 0.5ℓ
Tension in one member, when one of the members is interrupted and connected by gusset plate, 0.7ℓ
All cases not mentioned above, ℓ.

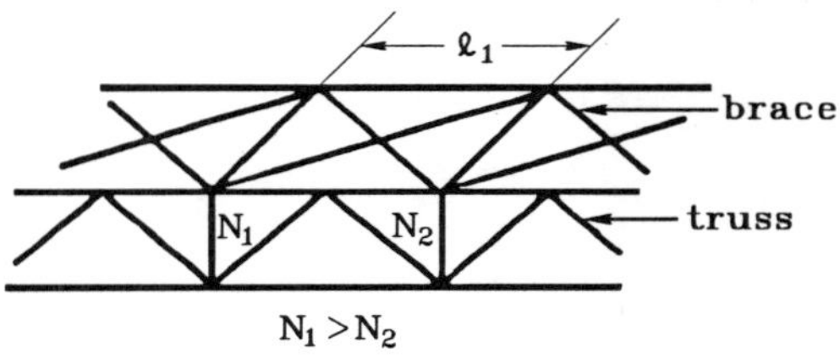

Fig. 8a.C1 Lattice Elements

Notes

1) ℓ = distance between panel points, but the crossing point is not considered as a panel point.
2) When both members are in compression, interruption of either one at the crossing point is not recommended.
3) When determining the slenderness ratio of single angle compression members in their principal plane, the effective length should be taken equal to the distance between panel point and the crossing point.

COMMENTARY

Eq. 8a.C1 is a well-known formula obtained by stability analysis.

The reduced factor 0.8 for in-plane buckling of web members is due to the elastic restraining effect at panel points.

The factor 0.9 for single-angle and cruciform sections is an average of the factors 1.0 and 0.8, and has been confirmed by elastic buckling analysis.

GBJ 18 87

The effective length of members are specified similarly to the Specification for Design of Steel Structures (GBJ 17 88), except that no restraining effect from adjacent members is considered for web members.

HRS 87

The slenderness ratio λ of chord members, diagonal members, transverse and diaphragm members for tower and mast structures should be calculated according to Table 8a.C1, Table 8a.C2 and Table 8a.C3, respectively.

TBJ 2 85

The effective length of all members of main trusses are specified the same as in the Specification for Design of Steel Structures (GBJ 17 88), except that for in-plane buckling of end posts ℓ_{ef} could be taken as 0.9 ℓ .

Table 8a.C1 Slenderness ratio of chord members for towers and masts

Types of Chord members	Panel points in adjacent faces staggered	Panel points in adjacent faces coincident
Figures	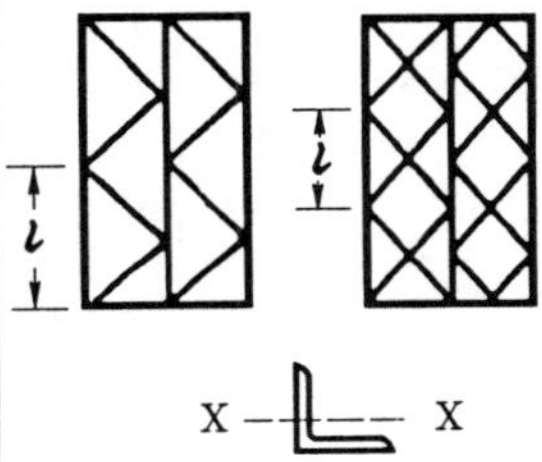	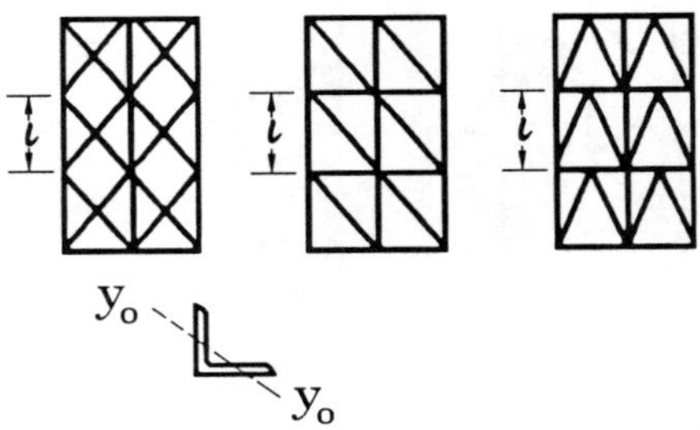
Slenderness ratios	$\lambda = \frac{1.2\ \ell}{\mathrm{i}_x}$	$\lambda = \frac{\ell}{\mathrm{i}_{yo}}$

Table 8a.C2 Slenderness ratio of diagonal members for towers and masts

Types of diagonal members	Single diagonal system	Double diagonal system	Double diagonals with redundant members	
Figures	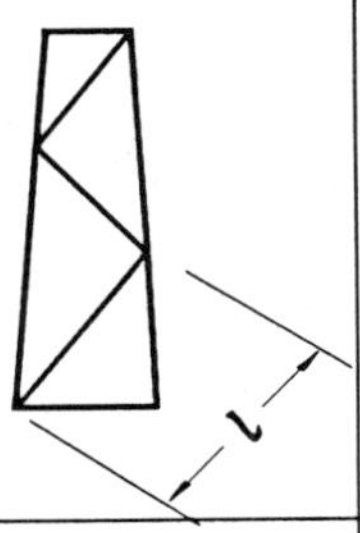	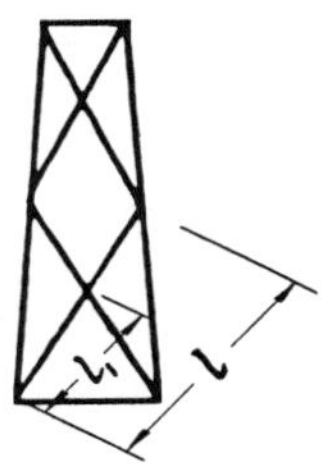	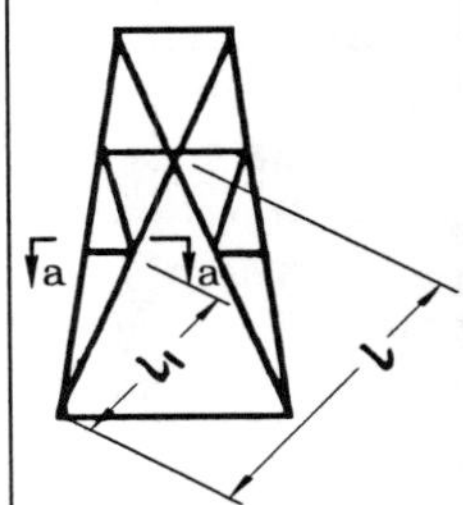	
Slender-ness ratios	$\lambda = \frac{\ell}{i_{yo}}$	both members are continuous $\lambda = \frac{\ell_1}{i_{yo}}$ with one member dis-continuous $\lambda = \frac{0.7\ell}{i_{yo}}$	with reliable brace in a-a plane $\lambda = \frac{\ell_1}{i_{yo}}$ no brace in a-a plane $\lambda = \frac{\ell}{i_x}$	the diagonal members are continuous $\lambda = \frac{\ell_1}{i_x}$

Note: In Table 8a.C2, "reliable brace" means that there is no distortion of cross section in the horizontal plane through a-a.

Table 8a.C3 Calculation of slenderness ratio of transverse and diaphragm members for towers and masts

Figure	Types of cross section	Transverse	Diaphragm
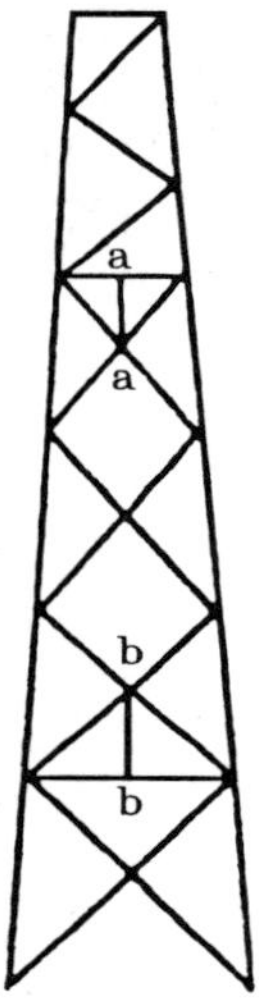	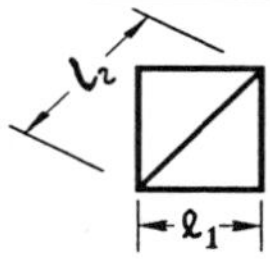	with tie rod a-a $\lambda=\frac{\ell_1}{i_x}$ without tie rod a-a $\lambda=\frac{\ell_1}{i_{yo}}$	$\lambda=\frac{\ell_2}{i_{yo}}$
	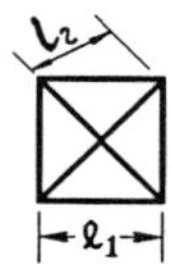	with tie rod a-a $\lambda=\frac{\ell_1}{i_x}$ without tie rod a-a $\lambda=\frac{\ell_1}{i_{yo}}$	when one of the diagonals is discontinuous and joined by gusset plate $\lambda=\frac{1.4\ell_2}{i_{yo}}$
	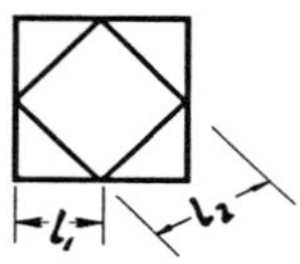	with tie rod a-a $\lambda=\frac{\ell_1}{i_{yo}}$ without tie rod a-a $\lambda=\frac{2\ell_1}{i_x}$	$\lambda=\frac{\ell_2}{i_{yo}}$
	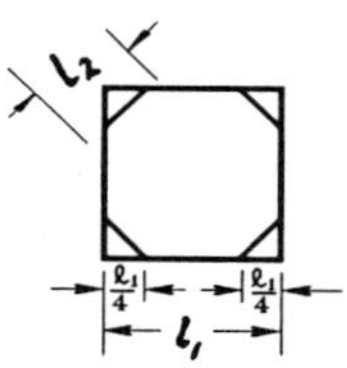	with tie rod a-a $\lambda=\frac{\ell_1}{2\,i_{yo}}$ without tie rod a-a $\lambda=\frac{\ell_1}{i_x}$	$\lambda=\frac{\ell_2}{i_{yo}}$

Note: The expressions for determining the slenderness ratio of chords, diagonals, transverses and diaphragm members are based on a series of experimental investigations for some commonly used types of web member system in steel transmission towers.

EAST EUROPE

CHAPTER EDITOR
Miklos Ivanyi, Technical University of Budapest

CMEA 72 and CMEA 76

The effective length of compression chords for both in-plane and out-of-plane buckling is equal to the distance between supported points (Fig. 8a.E1-a and 8a.E1-b)

Elastic restraint can be taken into account in case of in-plane buckling of diagonals by $\ell_k = 0.8\ \ell$.

For non-uniform axial force (Fig. 8a.E1-c)

$$\ell_k = \ell\,(0.75 + 0.25\frac{N_2}{N_1}); \qquad N_2 < N_1 \qquad (8a.E1)$$

For out-of-plane buckling of compressed diagonals in Fig. 8a.E1-d
If diagonals are continuous:

compression in both diagonals: $\ell_k = \ell_1$ (8a.E2)
tension in one diagonal: $\ell_k = \ell$ (8a.E3)
no axial force in one diagonal: $\ell_k = 0.7\,\ell_1$ (8a.E4)

If diagonal is non-continuous but connection provides flexural rigidity:

compression or no axial force in one diagonal: $\ell_k = \ell_1$ (8a.E5)
tension in one diagonal: $\ell_k = 0.7\,\ell_1$ (8a.E6)

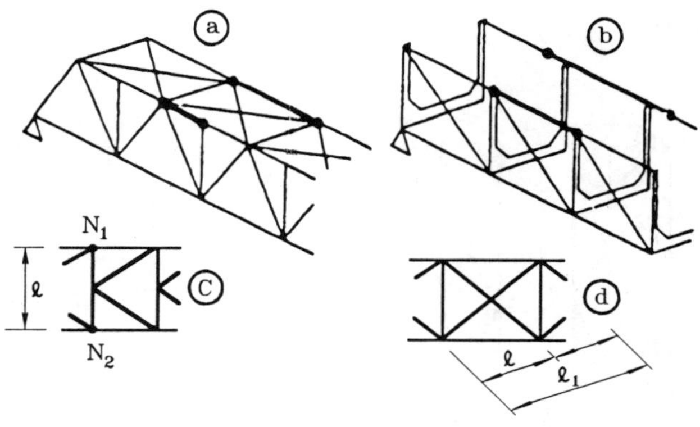

Fig. 8a.E1 Support Points

SNIP II-23-81 82

$$\lambda = \overline{\mu}\, \ell / i_x \qquad (8a.E7)$$

where

i_x = radius of gyration of angle section about the x-axis
J = moment of inertia of the chord (Fig. 8a.E2)
J_d = moment of inertia of the web member (Fig. 8a.E2)
$\overline{\mu}$ = given in Fig. 8a.E2

COMMENTARY

Triangulated structures are composed of linear members. They are characterized by small displacements of the panel points and high axial forces. This is why it is acceptable to investigate separately the stability of each compression member, but taking into account its connection to the rest of the structure by means of the effective length concept. This concept may be used for triangulated structures for the following reasons:

(i) Members are mainly subjected to axial forces; bending moments have a secondary effect, in contrast with the case of frames.
(ii) The buckling of a truss member has a local character, being mainly influenced by the adjacent members from their both ends, while the instability of a frame (especially if unbraced) is influenced by the whole structure.

The next stage is to define the effective length factor. At first glance, this appears to be a very simple task (if the effective length is defined as the distance between two inflection points), but it proves to be a very intricate problem, which has not been satisfactory solved yet. If the only question would be to estimate the favorable effect of partial restraint, no major difficulties would arise.

Approximate formulas and diagrams for effective length are given in most specifications.

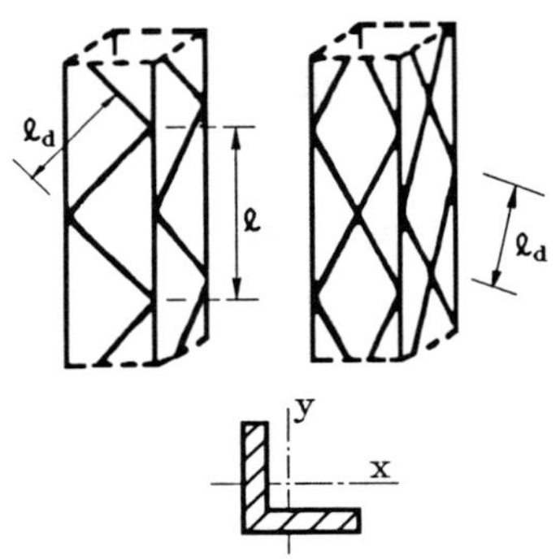

$\frac{J\,\ell_d}{J_d\,\ell}$	$\overline{\mu}$
1.0	0.98
1.25	1.00
2.50	1.03
5.0	1.08
10.0	1.13

Fig. 8a.E2 Triangulated Elements

JAPAN

CHAPTER EDITOR
Hiroshi Akiyama, University of Tokyo

AIJ 70

The effective length of compression members in trusses is taken as follows according to geometric and stress conditions.

The effective length of chord members is taken as the distance between panel points (panel-point distance) for in-plane buckling, and as the distance between laterally stiffened supports for out-of-plane buckling.

For in-plane buckling of web members, the effective length is taken as the distance between panel points. Where members are rigidly fixed at the ends, the effective length can be considered as the distance between the centers of gravity of groups of fasteners at connections. For out-of-plane buckling of web members, the effective length is taken as the distance between panel points.

When unequal compression forces act on the halves of a member, which is restrained against translation at both ends, the effective length is obtained from Eq. 8a.J1 based on the larger compression force.

$$\left.\begin{aligned} \ell_k &= \ell\left(0.75+0.25\,\frac{N_2}{N_1}\right) \\ \text{and}\quad \ell_k &\geq 0.5\,\ell \end{aligned}\right\} \qquad (8a.J1)$$

where

ℓ = length of member
N_1= larger compression
N_2= smaller compression with a negative value for the tensile force

Where web members of a truss are composed of pairs of bars intersecting each other at the centers, the effective length of such web elements for out-of-plane buckling is determined by considering axial force acting on intersecting elements and the condition of continuity at the intersecting points as follows.

for Fig. 8a.J1

$$\ell_k=\ell_1\sqrt{1-0.75\,\frac{N_2\ell_1}{N_1\ell_2}} \geq 0.5\,\ell_1 \qquad (8a.J2)$$

for Fig. 8a.J2

$$\ell_k=\ell_1\sqrt{1-\frac{N_2\ell_1}{N_1\ell_2}\left(0.75+\frac{\pi^2 E\,I_2}{\ell_2^2\,N_2}\right)} \geq 0.5\,\ell_1 \qquad (8a.J3)$$

The effective length of chord members can be modified by the stress gradient as follows.

for Fig. 8a.J3
$$\ell_k = \ell \sqrt{\frac{1+0.83 \cdot N_2/N_1}{1.88}} \quad \text{and} \quad \ell_k \geq 0.66\ \ell \qquad (8a.J4)$$

for Fig. 8a.J4
$$\ell_k = \ell \sqrt{\frac{1+2.18 \cdot N_2/N_1}{3.18}} \quad \text{and} \quad \ell_k \geq 0.42\ \ell \qquad (8a.J5)$$

for Fig. 8a.J5
$$\ell_k = \ell \sqrt{\frac{1+1.09 \cdot N_2/N_1}{2.09}} \quad \text{and} \quad \ell_k \geq 0.62\ \ell \qquad (8a.J6)$$

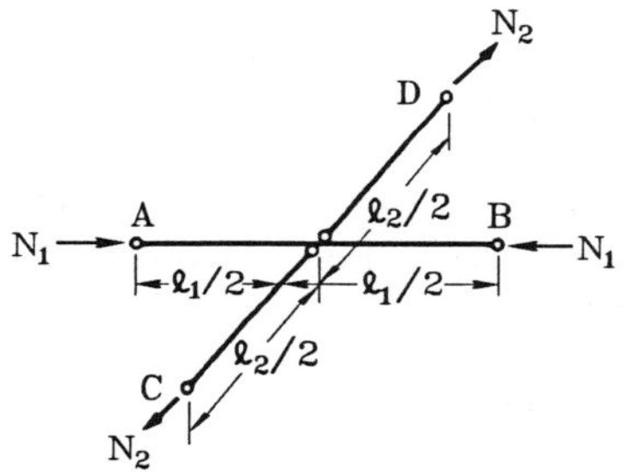

Fig. 8a.J1 Forces and Gradients

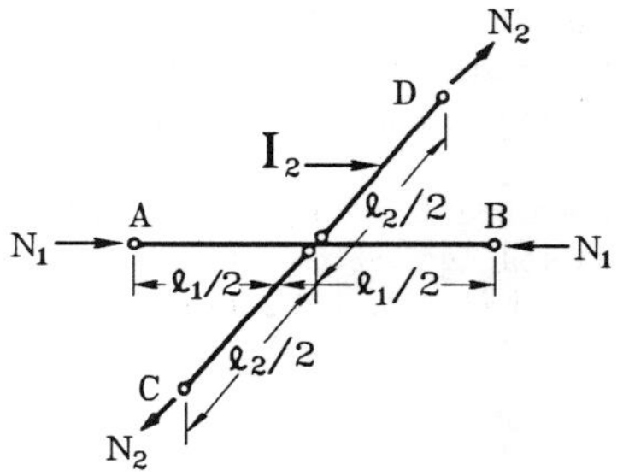

Fig. 8a.J2 Forces and Gradients

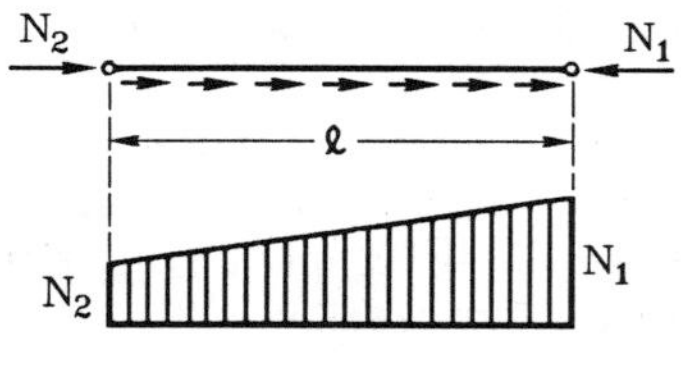

Fig. 8a.J3 Forces and Gradients

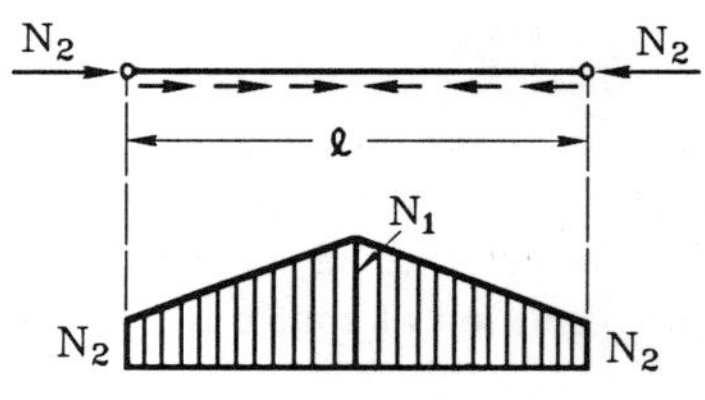

Fig. 8a.J4 Forces and Gradients

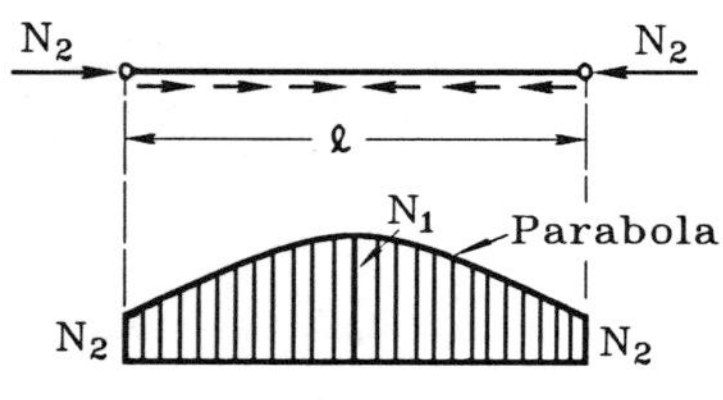

Fig. 8a.J5 Forces and Gradients

JEC 127 79

For compression members with no intermediate supports, the effective length is taken as the panel-point distance with reference to the smallest radius of gyration of member as shown in Fig. 8a.J6(a).

For compression members, which are restrained by subsidiary members, the effective length is given by the distance between the panel point and the support or two adjacent supports, referring to the smallest radius of gyration of member as shown in Fig. 8a.J6(d).

When the support with subsidiary members provides only uni-directional restraint as shown in Fig. 8a.J6(c), the effective length is given by the panel-point distance with reference to the radius of gyration of member about the axis along the restrained direction. In this case, buckling of the supported length, ℓ' must be also checked with reference to the minimum radius of gyration of member.

The effective length of crossed diagonal members is given by the distance between the panel point and the intersecting point of diagonal members as shown in Fig. 8a.J6(b) under the condition that the intersecting point is joined with bolts.

When the diagonal members in adjacent planes do not meet on one point of main column member, the effective length for the main column is given by the panel-point distance ℓ as shown in Fig. 8a.J6(e), (f) with reference to the radius of gyration of the column about the axis along the restrained direction. In this case, buckling of the supported length, ℓ' must be also checked with reference to the smallest radius of gyration.

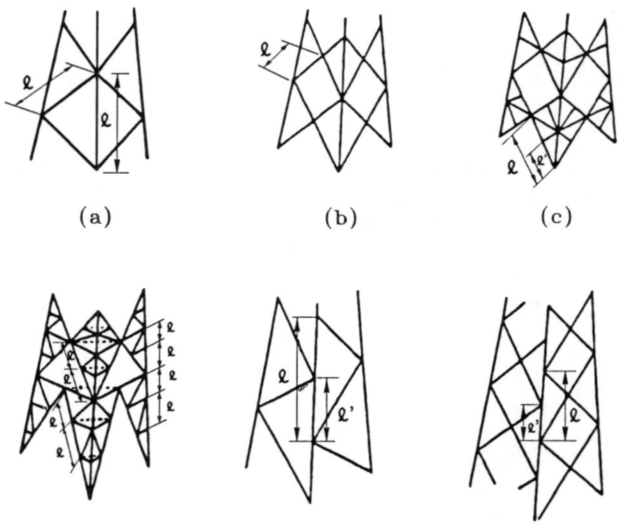

(a) (b) (c) (d) (e) (f)

Fig. 8a.J6 Member Lengths

JRA 80

In-plane Buckling

The effective length of compression chords is taken as the distance between panel points.

The effective length of web members connected to chords with gusset plates is taken as the distance between the centers of gravity of groups of fasteners, but should not be less than 0.8 times the panel-point distance.

For members connected with single gusset plates, the effective length is 0.9 times the panel-point distance.

When a compression member is effectively supported at an intermediate support by another member, the effective length can be taken as the distance between the panel-point and the support. In this case, the strength of the support must be greater than one-fourth of the strength of the joint at the panel point.

Out-of-plane Buckling

The effective length is taken as the distance between panel points.

When the chord members are sufficiently supported by restraining systems, the effective length can be taken as the distance between supports.

When unequal compression forces act on the halves of members without supports in the direction of out-of-plane, the effective length is given by Eq. 8a.J1.

JSCE 83

For chord members, the effective length is taken as the distance between panel points.

For out-of-plane buckling of web members, the effective length is taken as the panel-point distance. For in-plane buckling of web members, the effective length can be reduced to 0.9 times the panel-point distance.

NORTH AMERICA

CHAPTER EDITOR
Nicholas F. Morris, Manhattan College

AASHTO 83

Actual unbraced length, ℓ, shall be assumed as follows: for the top chords of half-through trusses, the length between panel points laterally supported; for other main members, the length between panel point intersections or centers of braced points or centers of end connections; for secondary members, the length between the centers of the end connections of such members or centers of braced points.

Bracing of Towers

Towers shall be braced, both transversely and longitudinally, with stiff members having either welded, high-strength bolted or riveted connections. The sections of members of longitudinal bracing in each panel shall not be less than those of the members in corresponding panels of the transverse bracing.

The bracing of long columns shall be designed to fix the column about both axes at or near the same point.

Horizontal diagonal bracing shall be placed in all towers having more than two vertical panels, at alternate intermediate panel points.

AISC ASD 78

$k = 1.00$ unless analysis shows a smaller value could be used.

At points of support, beam, girders and trusses shall be restrained against rotation about their longitudinal axis.

AISC LRFD 86

In trusses and frames where lateral stability is provided by diagonal bracing, shear walls, or equivalent means, the effective length factor k for compression members shall be taken as unity, unless structural analysis shows that a smaller value may be used.

Bracing requirement is identical with AISC ASD 78.

AREA 85

$k\ell$ is the effective length of the compression member. Under usual conditions:

k = 7/8 for members with pin-ended connections

k = 3/4 for members with riveted, bolted or welded connections.

CSA S6 M89 84a

The effective length factor k for compression members shall not be less than 1.00 except as otherwise indicated by rational analysis, or otherwise accepted as sound engineering practice.

Bracing - Trusses

Through truss spans, deck truss spans, and spandrel-braced arch spans shall have top and bottom lateral bracing systems.

The lateral bracing of compression chords preferably shall be as deep as the chords and effectively connected to both flanges of the chords.

Through truss spans shall have portal bracing, preferably of the double, plane or box type, rigidly connected to the end post and the top-chord flanges, and as deep as the clearance will allow. If single-plane portal bracing is used, it shall be located preferably in the central transverse plane of the end posts, with diaphragms between the end post webs to provide for stress distribution. Portal bracing shall be designed to take the full end reaction of the top chord lateral system and the end posts shall be designed to transfer this reaction to the truss bearings. Consideration shall be given to sway bracing and, if required, sway bracing shall be located at such panel points as necessary and shall be kept as shallow as practicable and at a depth not less than that of the top chord.

Deck truss spans shall have sway bracing in the plane of the end posts and at other locations where required. This bracing shall extend the full depth of the trusses below the floor system. The end sway bracing shall be proportioned to carry the entire upper lateral stress to the supports through the end posts of the truss.

CSA S16.1 M89 84b

Unless otherwise specified in this standard or unless analysis shows that a smaller value is applicable, the effective length factor, k, shall be taken as 1.00 for compression members in trusses.

Bracing members assumed to provide lateral support to the compression flange of beams and girders or to the compression chords of trusses, and the connections of such bracing, shall be proportioned to resist a force equal to 1 percent of the force in the compression flange or chord at the point of support.

CSA S136 M89 84c

The factored resistance of braces shall be at least 2 percent of either the factored compressive force in the compression member at the braced location or the factored compressive force in the compressive flange of the member in bending. When more than one brace acts at a common location and the nature of the braces is such that combined action is possible, the bracing force may be shared proportionately. The slenderness ratio of compressive braces shall not exceed 200.

OHBDC 83

$k = 1.00$ unless analysis shows that a smaller value is applicable.

Through truss spans, deck truss spans, and spandrel braced arch spans shall have top and bottom lateral bracing systems.

The lateral bracing of compression chords preferably shall be as deep as the chords and effectively connected to both flanges of the chords.

WEST EUROPE

CHAPTER EDITORS
Pierre Dubas, Swiss Federal Institute of Technology
Carlo Urbano, Technical University of Milan

BS 5400 P.3 82

Effective Length of Compression Members

The effective length ℓ_k of a compression member either should be obtained from Table 8a.W1 or should be determined by an elastic critical buckling analysis of the truss. In applying Table 8a.W1, the end raker or end posts of a truss should be considered as a web member.

Lateral Restraint by Deck to Compression Chord

A compression chord, continuously supporting a steel or reinforced concrete deck, may be deemed to be effectively restrained laterally throughout its length if the frictional or other connection of the deck to the chord is capable of resisting a lateral force, distributed uniformly along its length, of 2.5% of the maximum force in the chord. The effective length ℓ_k of such a compression chord should be taken as zero where friction provides an adequate restraint, or as equal to the spacing of discrete connections where these are provided.

Table 8a.W1 Effective length ℓ_k for compression members in trusses

Members		Effective length ℓ_k		
		Buckling in plane of truss	Buckling normal to plane of truss when:	
			Compression chord is effectively braced by internal system	Compression chord is unbraced
Chord		0.85 x distance between intersections with web members	0.85 x distance between intersection with lateral bracing members or rigidly connected cross beams	See page 8b(W1)
Web Member	Single triangulated system	0.70 x distance between intersections with chords	0.85 x distance between intersections with chords	Distance between intersections with chords
Web Member	Multiple intersection system with adequate connections at all points of intersection	0.85 x greatest distance between any two successive intersections	0.70 x distance between intersections with chords	0.85 x distance between intersection with chords

Lateral Bracing

Sufficient bracing should be provided between main trusses to ensure that all external and stabilizing loads and load effects can be transmitted to the supporting structures, and that restraint is provided at all intersection points where such restraint is assumed in determining the effective length of compression members, and at each point where a compressive force is applied to a web member owing to change of direction of a chord (whether the chord is in tension or compression).

Bracing members and their connections to compression chords, or to U-frames restraining compression chords, should be designed to resist the forces given below.

Forces on bracing. Lateral restraint should be provided to compression chords in such a way that at all transverse sections of the bridge the following stabilizing lateral shear force can be resisted:

(a) $\Sigma P_c/80$ when the load combination includes wind and other laterally applied forces;
(b) $\Sigma P_c/40$ when the load combination excludes wind and other laterally applied forces;

where ΣP_c is the sum of the greatest coexistent axial forces in any two chords at the section under consideration.

CNR 10011 85

Chords

For buckling in the plane of the truss, the effective length is taken equal to the theoretical distance between the points of intersection of member center lines. For buckling perpendicular to the truss plane, the effective length is taken equal to the distance between laterally braced nodes.

Web members

For buckling in the plane of the truss, the effective length is taken equal to the distance between the centers of the end connections. The effective length shall not be less than 80% of the distance between the points of intersection of member center lines. In double intersection systems with adequate connections at all points of intersection, the effective length may be reduced, for buckling in the plane of the truss, up to 1/2 of the system length of the web-member (Fig. 8a.W1).

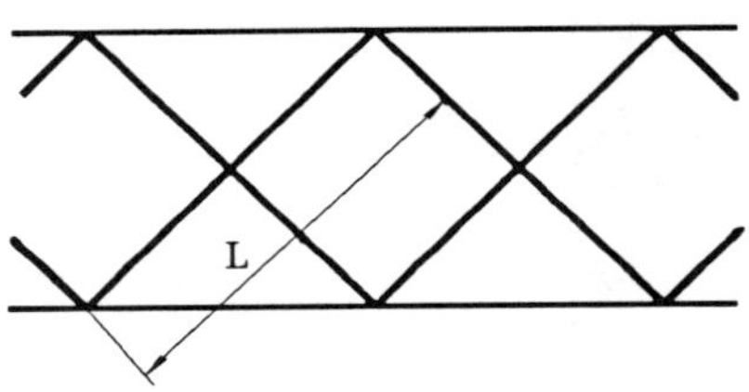

Fig. 8a.W1 Web Member Elements

DIN 18800 P.2 86

Laced Built-Up Compression Members
The effective length ℓ_k of the chords to be adopted for the buckling check is shown in Table 8a.W2.

Trusses
Effective length of truss members under axial compression if the nodes are fixed normal to the plane of the truss.

> In the case of struts and posts of trusses (except for the end struts of trapezoidal beams), $\ell_k = 0.9\ \ell$ may be assumed as buckling length for buckling in the plane of the truss; for buckling perpendicular to the plane of the truss $\ell_k = \ell$.

For struts and posts, which are held or elastically restrained by cross beams or cross bars at the nodes perpendicular to the plane of the truss, the effective length for buckling perpendicular to the plane of the truss is given in Fig. 8a.W2 and 8a.W3.

Effective length of truss members under axial compression, the nodes of which are restrained normal to the plane of the truss and which are centrally intersected by another truss member.

Table 8a.W2 Effective Length ℓ_k

		1	2	3	4	5
1		Gitterstäbe				
		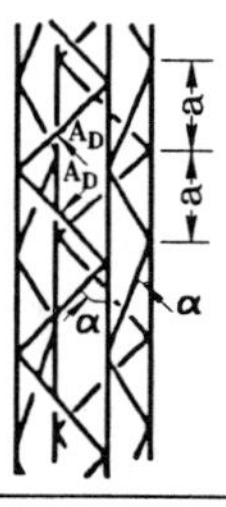	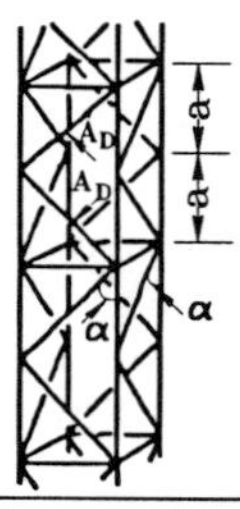	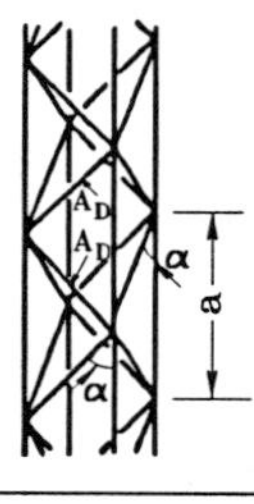	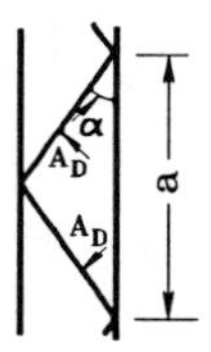	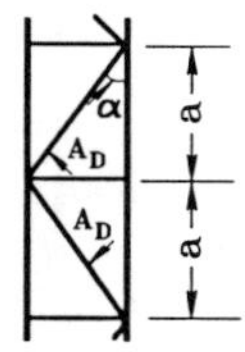
		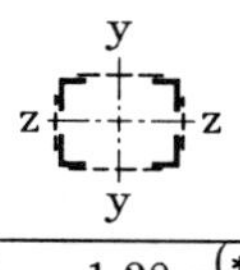			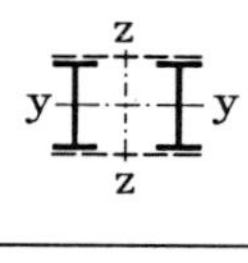	
2	ℓ_k	1.52 a(*)	1.28 a(*)	a(*)	a	

(*) For chord angles the effective length ℓ_k is valid only if the slenderness is calculated with reference to the smallest radius of gyration i_1.

For buckling in the plane of the truss, the distance to the point of central intersection may be adopted as effective length for the above members.

For buckling perpendicular to the truss plane, the effective lengths are given in Table 8a.W3. At the point of intersection, both members have to be connected, either directly or by a gusset plate, in order to enable the transfer of one fourth of the maximum value of the compression force of the two members.

Effective length of web members with elastic lateral restraint at the middle section.

For buckling perpendicular to the plane of the truss, the effective length is given by

$$\ell_k = \frac{\ell}{2a + 1.5} \left(1 + \sqrt{0.25 - a}\right) \qquad (8a.W1)$$

where (see Fig. 8a.W4)

$a = (C\,\ell) / (16N) \leq 0.25$ (8a.W2)

C = stiffness of the restraining frame

ℓ = system length of the member

N = maximum compression force (N_1 or N_2)

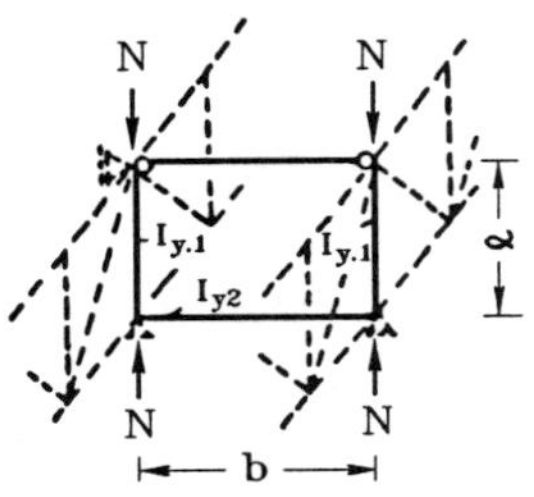

$$\ell_{ky} = 0.8\,\ell$$

$$\frac{\ell \cdot I_{y.2}}{b \cdot I_{y.1}} \geqslant 2.64$$

Fig. 8a.W2

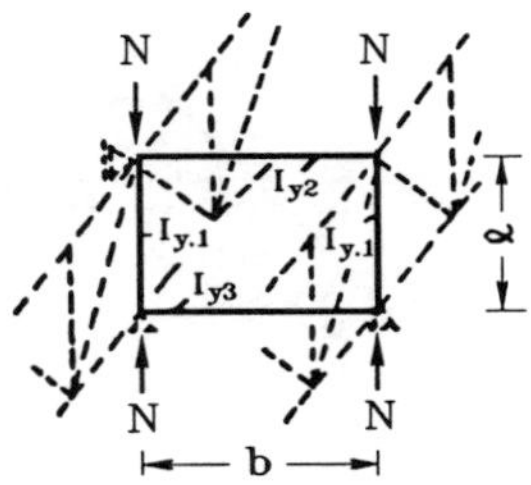

$$\ell_{ky} = 0.7\,\ell$$

$$\frac{I_{y.2}}{I_{y.1}} >= \frac{3.23 - 0.0781\frac{I_{y.3}}{I_{y.1}}\frac{\ell}{b}}{\frac{I_{y.3}}{I_{y.1}}\frac{\ell^2}{b^2} + 0.0781\frac{\ell}{b}} \qquad (E502)$$

Fig. 8a.W3

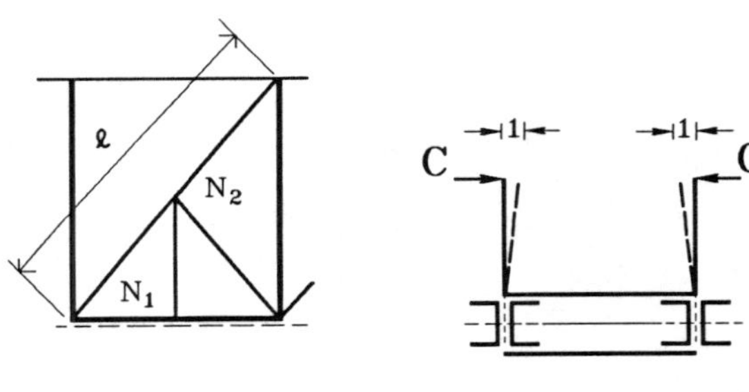

Fig. 8a.W4

Centrally compressed web members consisting of one single angle.

For compressed truss web members or web members of lattice masts consisting of one single angle, connected eccentrically but rigidly with one leg to the chords (for example according to Fig. 8a.W5), the eccentricity may be neglected and buckling strength may be verified with reference to the modified slenderness $\overline{\lambda}'$ quoted in Table 8a.W4.

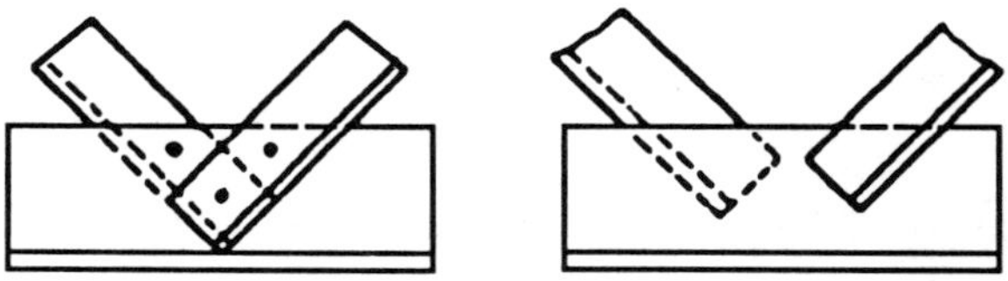

Fig. 8a.W5

Table 8a.W3 Effective Length for Out-of-Plane Bending of Wide-Flange Members

1	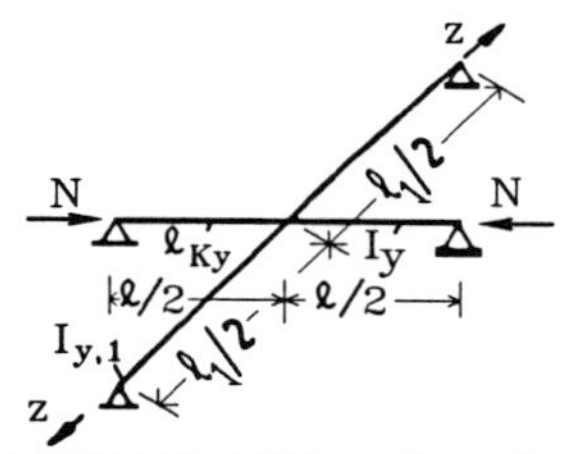	$\ell_{Ky} = \ell \sqrt{\dfrac{1 - \frac{3}{4}\frac{Z\,\ell}{N\,\ell_1}}{1 + \dfrac{I_{y,1}\,\ell^3}{I_y\,\ell_1^3}}}$ but $\ell_{Ky} \geq 0.5\,\ell$	
2	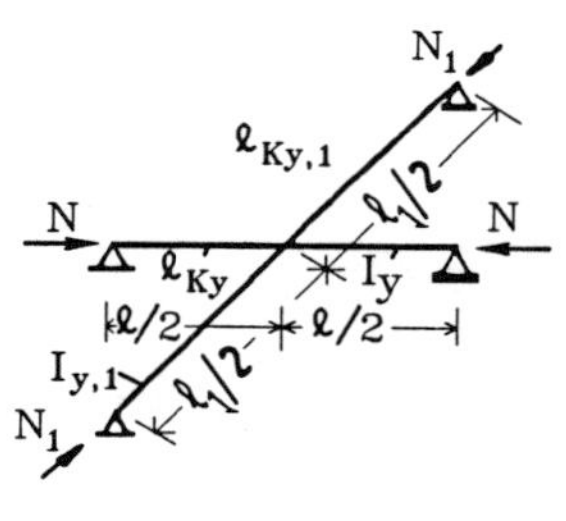	$\ell_{Ky} = \ell \sqrt{\dfrac{1 + \frac{N_1\,\ell}{N\,\ell_1}}{1 + \dfrac{I_{y,1}\,\ell^3}{I_y\,\ell_1^3}}}$ but $\ell_{Ky} \geq 0.5\,\ell$	$\ell_{Ky_1} = \ell_1 \sqrt{\dfrac{1 + \frac{N\,\ell_1}{N_1\,\ell}}{1 + \dfrac{I_y\,\ell_1^3}{I_{y,1}\,\ell^3}}}$ but $\ell_{Ky_1} \geq 0.5\,\ell_1$
3	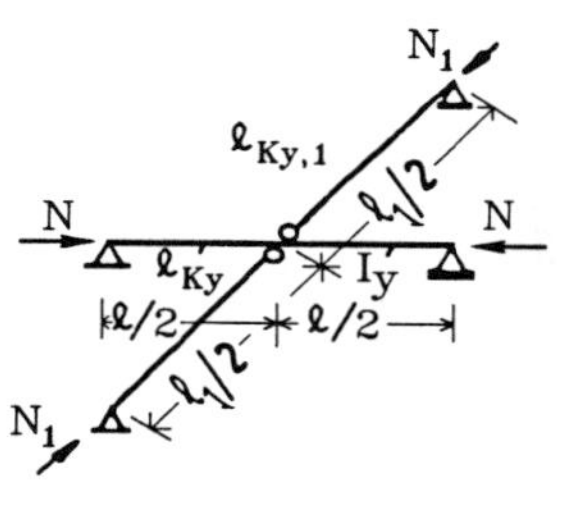	continuous compression members $\ell_{Ky} = \ell \sqrt{1 + \frac{\pi^2}{12}\frac{N_1\,\ell}{N\,\ell_1}}$ but $\ell_{Ky} \geq 0.5\,\ell$	pin-connnected compression members $\ell_{Ky_1} = 0.5\,\ell_1$ if $EI_y \geq \frac{N_1\,\ell^3}{\pi^2\,\ell_1}\left(\frac{\pi^2}{12} + \frac{N\,\ell_1}{N_1\,\ell}\right)$

Table 8a.W3 Effective Length for Out-of-Plane Bending of Wide-Flange Members (Cont.)

4	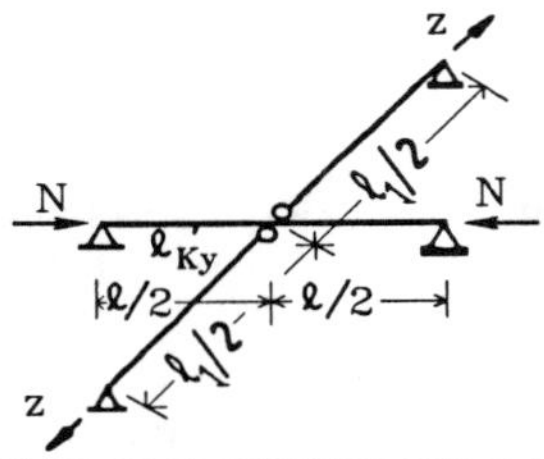	$\ell_{Ky} = \ell\sqrt{1-0.75\,\frac{Z\,\ell}{N\,\ell_1}}$ but $\ell_{Ky} \geq 0.5\,\ell$	
5	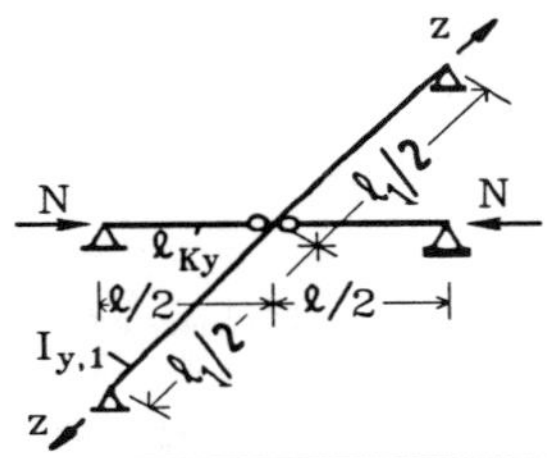	$\ell_{Ky} = 0.5\,\ell$ if $\frac{N\,\ell_1}{Z\,\ell} \leq 1$ or if $EI_{y,1} \geq \frac{3\,Z\,\ell_1^2}{4\,\pi^2}\left(\frac{N\,\ell_1}{Z\,\ell} - 1\right)$	
6	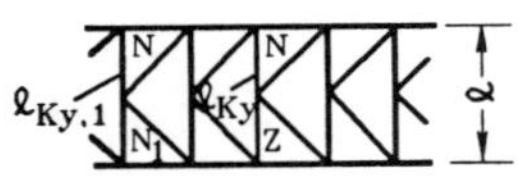	$\ell_{Ky}=\ell\left(0.75-0.25\left\|\frac{Z}{N}\right\|\right)$ but $\ell_{Ky} \geq 0.5\,\ell$	$\ell_{Ky,1}=\ell\left(0.75+0.25\,\frac{N_1}{N}\right)$ $N_1 < N$
7	ℓ, ℓ_1 design length of members N, N_1 axial force (compression) Z axial force (tension)	cross section axis y may be in the plane of truss	

Table 8a.W4 Modified Slenderness $\overline{\lambda}'$

	1	2
1	$0 < \overline{\lambda} \leq \sqrt{2}$	$\overline{\lambda}' = 0.35 + 0.7525\overline{\lambda}$
2	$\sqrt{2} \leq \overline{\lambda} \leq 3.0$	$\overline{\lambda}' = 0.50 + 0.6464\overline{\lambda}$
	$\lambda = \ell/i_1$ slenderness of web-member $\overline{\lambda} = \lambda/\lambda_p$ ℓ = system length of web-member i_1 = smallest radius of gyration of angle cross-section λ_p = limiting slenderness ratio = $\pi\sqrt{E/\sigma_e}$ σ_e = design value of the yiels stress	

EC 3 88

Triangulated Structures

Truss chords and towers may be analyzed assuming that there are hinges at the member ends.

Compression members in such structures may be designed as axially compressed members, as bending and compression members or as centrally compressed built-up members, as the case may be.

The effective length for web members may be determined as follows:
Web members may be designed for in-plane failure with an effective length smaller than the geometrical length (the length between the points of intersection of member center-lines) provided the chords give good end restraint and at least two bolts or equivalent connections are present. For sections other than angle sections, the factor k to be applied to the geometrical length is 0.9.

Angles as Web Members

If the chords give good end restraint to the web members and at least two bolts or rivets (or the equivalent in welds) are present at the end connections of the web members or if the ends are continuous, the eccentricities may be neglected and end fixities allowed for x (see Fig. 8a.W6); the effective slenderness ratio is modified as follows:

for buckling about V-V

$$\overline{\lambda}_{eff} = 0.35 + 0.7\,\overline{\lambda} \tag{8a.W3}$$

for buckling about X-X

$$\overline{\lambda}_{eff} = 0.50 + 0.7\,\overline{\lambda} \tag{8a.W4}$$

The modified slenderness ratio $\overline{\lambda}_{eff}$ is then used to obtain the buckling stress.

Care should be taken when only single bolts are used at the ends of very slender members or when the connection has poor stiffness. In such cases the eccentricity must be taken into account with no allowance for end fixities.

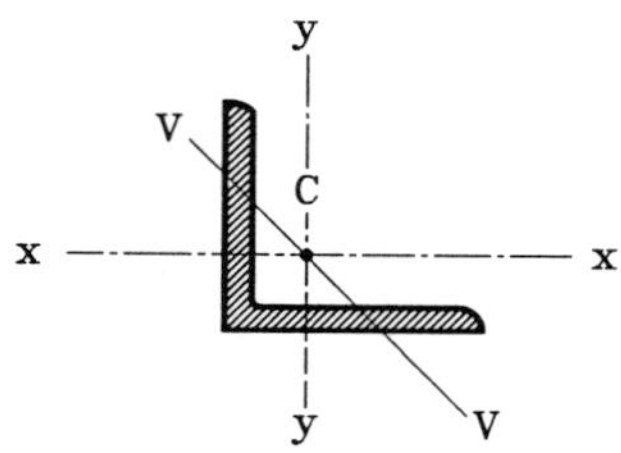

Fig. 8a.W6

NEN 3851 72

Compression Members in Plane Lattice Systems

In the case of built-up members special provisions are given. For determining the effective length in buckling, a distinction is drawn between chord members and web members and between buckling within the plane of the lattice structure and buckling out of that plane.

If a more accurate analysis is dispensed with, the following rules may be applied.

Chord Members

a. The effective length of chord members in the case of buckling within the plane of the lattice structure is equal to the system length.

b. 1. The effective length of a chord member in the case of buckling out of the plane of the lattice structure is equal to the distance between the points where the member is rigidly supported against displacement out of that plane.

   2. If the member is elastically supported, see page 8b(W5).

   3. If the joints A and B of the continuous member AB in Fig. 8a.W7 are rigidly supported, if joint C is not supported, and if the compressive forces in AC and BC differ in magnitude (N_1 and N_2 respectively, $N_1 > N_2$), then the member AB may be designed with a compressive force N_1 and an effective length:

$$\ell_k = 2\,\ell\,(0.75 + 0.25\frac{N_2}{N_1}) \qquad \text{(8a.W5)}$$

c. If the chord members of a truss are loaded also at intermediate points between the joints, the effect of bending upon stability with regard to buckling within the plane of the truss may be taken into account by performing the following check

$$\omega\frac{N}{A} + \frac{2}{3}\frac{M}{W} \leq \sigma_e \qquad \text{(8a.W6)}$$

where

- M = bending moment in the member due to the transverse load acting between the joints in the design state, calculated as though the chord member were pin-jointed at the lattice joints
- σ_e = design value of the yield stress
- ω = buckling coefficient within the plane of the truss, determined with the system length

Web Members

a. The effective length of a web member in the case of buckling within the plane of the lattice structure is equal to the system length. If the member is fully connected, a reduced value may be adopted for the effective length.

b. The effective length of a web member in the case of buckling out of the plane of the lattice structure is equal to the system length.

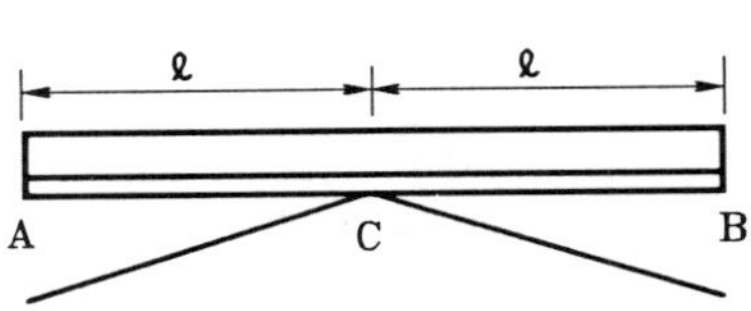

Fig. 8a.W7 Support of Continuous Member

c. If a web member is composed of a single angle section and is therefore eccentrically connected, its stability should be checked by verifying that both of the following conditions are satisfied:

$$\omega_x \frac{N}{A} + \frac{Ne_x}{W_x} \leq \sigma_e \qquad (8a.W7)$$

$$\omega_{max} \frac{N}{A} \leq \sigma_e \qquad (8a.W8)$$

where

e_x = eccentricity as indicated in Fig. 8a.W8

ω_{max} = buckling coefficient determined with ℓ_k = the system length and i_η about the η–axis in Fig. 8a.W8

ω_x = buckling coefficient determined with ℓ_{kx} = 0.7 times the system length and i_x about the x-axis in Fig. 8a.W8

COMMENTARY

By fully connected it is understood that the connection of the member to the joints should, within the plane of the lattice, be able completely to transmit the plastic moment M_p of the member.

Equation 8a.W7 is applicable for low slenderness ratios, Eq. 8a.W8 is applicable for high slenderness ratios. The borderline is located at a value of λ between 115 and 130, depending on the section of the member and on the thickness of the gusset plate.

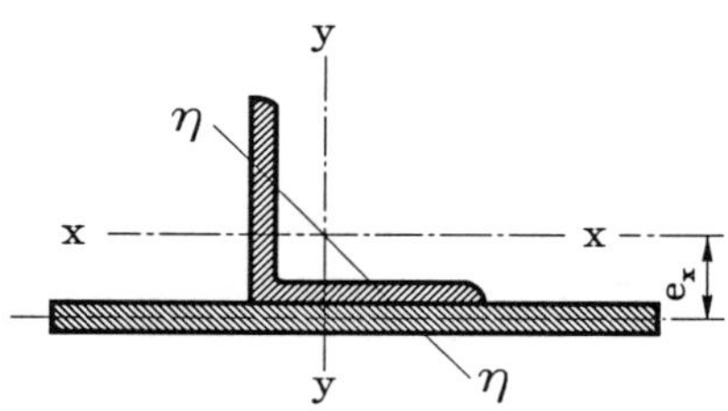

Fig. 8a.W8 Section Axes

SIA 161 79

The effective length ℓ_k shall usually be equal to the theoretical length ℓ of the member, measured between nodal points.

A reduction of the effective length ℓ_k is allowable if the maximum compression loads do not act simultaneously on the adjacent members and if the end restraints justify such a reduction.

If these conditions are satisfied, the effective lengths given in Table 8a.W5 may be employed for common trusses.

Members consisting of a single eccentric angle, or of angles forming a cross, shall, in addition, be checked for buckling about the smallest axis of inertia, with $\ell_k = 0.9\ell$.

Table 8a.W5 Usual effective lengths

Member	Truss	Buckling Mode in plane	Buckling Mode out of plane
Chords		0.9ℓ	0.9ℓ
Diagonals	Single	0.8ℓ	1.0ℓ
	Double	0.4ℓ	$0.5\ell^{o}$
	Multiple	1.0m	1.5m
	K shaped	0.9ℓ	1.0ℓ
Struts	Single or Double	0.8ℓ	1.0ℓ
	K shaped	0.4ℓ	$0.5\ell^{o}$

m = truss mesh size

^{o} if the tension and compression forces are equal

Should the ends of a member not be fixed, the effective length ℓ_k can exceed its theoretical length ℓ.

REGIONAL COMPARISON

Effective Length

Nearly all European Codes include simplified rules for the effective length of compression members in plane trusses. The prescriptions are generally in good agreement and recent numerical simulations (Dubas, 1986) in the elastoplastic range (taking into account second order effects, geometrical imperfections and residual stresses) have shown that at least for typical industrial trusses the simplified rules adopted in West Europe guarantee a safe and economic design.

Chords

As far as buckling in the plane of the truss is considered, the effective length of the chords may be obtained as $\ell_k = 0.85/1$ times the system length ℓ (the distance between intersections with web members). The lowest figure $\ell_k = 0.85\ \ell$ is adopted in BS 5400 P.3 82; the intermediate figure $\ell_k = 0.9\ \ell$ is adopted in DIN 18800 P.2 86, SIA 161 79; the highest figure $\ell_k = \ell$ is adopted in NEN 3851 72 and in CNR 10011 85.

For buckling normal to the plane of the truss, the effective length ℓ_k adopted in most codes is equal to the system length (the distance between the intersections with lateral bracing members); in BS 5400 P.3 82 a reduction of the effective length to 0.85ℓ is allowed for chords buckling normal to truss plane.

Web Members

The effective length of web members buckling in the plane of the truss may be obtained as $\ell_k = 0.7/0.9$ times the distance ℓ between intersections with the chords. $\ell_k = 0.7\ell$ suggested in BS 5400 P.3 82; SIA 161 79 and CNR 10011 85 adopt $\ell_k = 0.8\ell$; DIN 18800 P.2 86 (FRG) adopts $\ell_k = 0.9\ \ell$.

For buckling normal to the plane of truss the effective length ℓ_k is generally adopted equal to the system length ℓ. Moreover in the German Code DIN 18800 P.2 86 cross bracings and K bracings are considered in greater detail.

INTERNATIONAL COMPARISONS & COMMENT

WHAT ARE THE DIFFERENCES?

In the United States and Canada, excluding bridges, the effective length is taken as the member length unless analysis shows a smaller effective length can be used. Bridges in the United States and other international codes allow some reduction in effective length, particularly for web members.

WHY ARE THERE DIFFERENCES?

When all members in a truss are buckling or yielding at the same load level acting on the structure, the effective length ℓ_k of compression members is taken equal to the theoretical length ℓ, measured between fixed nodal points.

In other cases it is permitted to take into account the favorable effects of partial end restraint. All over the world specifications and codes present approximate formulas, tables or diagrams for the effective length of compression members in common trusses.

If one compares in detail the approximate formulas of the codes, it will be observed that there is some scattering in the figures. The discrepancies arise from the differences in the elastic models adopted for the simplified analysis of the effects of the end restraints. This is true both for chords and for web members.

8b. Unbraced Compression Chords

SPECIFICATION PROVISIONS & EXPLANATION

AUSTRALIA

CHAPTER EDITOR
Lewis Schmidt, University of Wollongong

AS 1250 81

Lateral restraining systems for reduction of effective lengths are covered. Guidelines are given for the determination of stiffness and the force induced in the restraining systems.

COMMENTARY

The problem of the ultimate load of both planar and space trusses is being studied. Experimental and theoretical studies continue on the influence of chord continuity and eccentricity of connection of webs and chords. Effective length determination in the inelastic range needs further investigation.

CHINA

CHAPTER EDITOR

Zaitian Guo, Xian Institute of Metallurgy & Construction Engineering

TBJ 2 85

For out-of-plane buckling of compression chords of pony truss where the panel length $d \leq L/(3m)$, the effective length is equal to αL, with α shown in Table 8b.C1.

Table 8b.C1 Factors α and m

<table>
<tr><td>β</td><td>0</td><td>5</td><td>10</td><td>19</td><td>20</td><td>30</td><td>50</td><td>100</td><td>150</td></tr>
<tr><td>α</td><td>0.696</td><td>0.524</td><td>0.443</td><td>0.370</td><td>0.367</td><td>0.353</td><td>0.330</td><td>0.290</td><td>0.264</td></tr>
<tr><td>m</td><td colspan="4">1</td><td colspan="5">2</td></tr>
<tr><td>β</td><td>200</td><td>242</td><td>243</td><td>300</td><td>500</td><td>700</td><td>1000</td><td>1060</td><td>1061</td></tr>
<tr><td>α</td><td>0.246</td><td>0.234</td><td>0.234</td><td>0.225</td><td>0.204</td><td>0.189</td><td>0.174</td><td>0.171</td><td>0.171</td></tr>
<tr><td>m</td><td colspan="2">2</td><td colspan="6">3</td><td>4</td></tr>
<tr><td>β</td><td>1500</td><td>2000</td><td>2500</td><td>3000</td><td>3082</td><td>3083</td><td>4000</td><td>5000</td><td>6000</td></tr>
<tr><td>α</td><td>0.160</td><td>0.149</td><td>0.142</td><td>0.136</td><td>0.135</td><td>0.135</td><td>0.127</td><td>0.121</td><td>0.116</td></tr>
<tr><td>m</td><td colspan="5">4</td><td colspan="4">5</td></tr>
<tr><td>β</td><td>7129</td><td>7130</td><td>8000</td><td>10000</td><td colspan="5" rowspan="3"></td></tr>
<tr><td>α</td><td>0.112</td><td>0.112</td><td>0.109</td><td>0.103</td></tr>
<tr><td>m</td><td>5</td><td colspan="3">6</td></tr>
</table>

where

- B = distance between the two main trusses
- d = panel length
- h = height from the centroid of cross section of compression chord to the upper surface of floor beam
- I = moment inertia of gross section of compression chord about the vertical axis (average value within the total length)
- I_b = moment of inertia of the floor beam section
- I_c = moment of inertia of the gross section of the verticals for out-of-plane buckling
- L = span length of trusses
- m = number of half-waves at the onset of chord buckling

$$\beta = \frac{L^4}{16 d \delta E I} \quad (8b.C1)$$

- δ = the maximum deflection caused by unit horizontal force (P=1) acting on the upper panel point of one chord of an intermediate semi-frame

$$\delta = \frac{h^3}{3EI_c} + \frac{Bh^2}{2EI_b} \quad (8b.C2)$$

COMMENTARY

Some restraining effects from tension chord are considered for end posts. In Table 8b.C1, the factor α provides for the elastic restraining effect from the semi-rigid frame.

JAPAN

CHAPTER EDITOR
Hiroshi Akiyama, University of Tokyo

AIJ 70 and JRA 80

For the purpose of reducing effective lengths of compression chords, lateral restraining systems are provided. The lateral restraining systems must be strong and stiff enough to resist concentrated lateral forces (applied at every supporting point) which are not less than 1% (JRA 80) or 2% (AIJ 70) of the axial compression forces in the compression chords.

NORTH AMERICA

CHAPTER EDITOR
Nicholas F. Morris, Manhattan College

AASHTO 83

The vertical truss members and the floor beams and their connections in half-through truss spans shall be proportioned to resist a lateral force of not less than 300 pounds per linear foot applied at the top chord panel of each truss.

The top chord of such spans shall be considered as a column with elastic lateral supports at the panel points. The critical buckling force of the column, so determined, shall exceed the maximum force from the dead load, live load, and impact in any panel of the top chord by not less the 50 percent.

AREA 85

The lateral bracing of compression chords or flanges of trusses and deck girders, and between the posts of viaduct towers shall be proportioned for a transverse shear equal to 2.5% of the total axial stress in both members in that panel, in addition to the shear from the specified lateral forces.

CSA S6 M89 84a

In half-through or pony truss spans the vertical truss members, floor beams and connections between them shall be proportioned to resist a lateral force of not less than 4.4 kN/m, applied at the top chord panel points of each truss. This requirement does not preclude meeting other requirements as may be necessary and established practice in similar structures. The top chord shall be considered as a column having elastic lateral supports at each panel point. The critical buckling load of the column, so determined, shall be at least 1.5 times the maximum force in any panel of the top chord resulting from dead load, live load, and impact.

WEST EUROPE

CHAPTER EDITORS

Pierre Dubas, Swiss Federal Institute of Technology
Carlo Urbano, Technical University of Milan

BS 5400 P.3 82

Effective Length

Where a compression chord is not provided with a system of lateral bracing, but is restrained laterally by U-frames comprising cross members and truss web members (see Figure 8b.W2), the effective length ℓ_k of the chord member may be taken as

$$\ell_k = 2.5k_3(EI_c a\delta)^{0.25} \quad \text{but} \geq a \tag{8b.W1}$$

where

- a = distance between U-frames
- I_c = maximum second moment of area of the chord about the Y-Y axis shown in Fig.8b.W2
- k_3 = 1.0, except that when the compression chord is restrained against bending in plane at its section over the supports of the truss, a lower value of k_3 may be obtained from Fig. 8b.W1
- δ = lateral deflection that would occur in the U-frame, at the level of the centroid of the chord being considered, when a unit force acts laterally to the U-frame only at this point and simultaneously at each corresponding point on the other chord or chords connecting to the same U-frame. The direction of each unit force should be such as to produce the maximum aggregate value of δ. The U-frame should be taken as fixed in position at each point of intersection between the cross member and an upright, and as free and unconnected at all other points.

In cases of symmetrical U-frames, where cross members and verticals are each of constant moment of inertia throughout their own length, it may be assumed that:

$$\delta = \frac{{d_1}^3}{3EI_1} + \frac{u\,s\,d_2^2}{EI_2} + fd_2^2 \tag{8b.W2}$$

where

- d_1 = distance from the centroid of the compression chord to the nearer face of the cross member of the U-frame
- d_2 = distance from the centroid of the compression chord to the centroidal axis of the cross member of the U-frame
- f = flexibility of the joint between the cross member and the verticals of the U-frame, expressed in radians per unit moment; f may be taken as:

  0.5×10^{-10} rad/Nmm, when the cross member is bolted or riveted through unstiffened end plates or cleats (see Fig. 8b.W3, type (a)), or 0.2×10^{-10} rad/Nmm, when the cross member is bolted or riveted through stiffened end plates (see Fig. 8b.W3, type (b)), or

0.1 x 10^{-10} rad/Nmm, when the cross member is welded right round its cross section or the connection is by bolting or riveting between stiffened end-plates on the cross member and a stiffened part of the vertical or a stiffened section of the chord (see Fig. 8b.W3, type (c))

I_1 = second moment of area of the web member forming an arm of the U-frame in its plane of bending

I_2 = second moment of area of the cross member in its plane of bending

s = distance between centers of main girders or trusses connected by the U-frame

u = 0.5 for an outer beam, or 0.33 for an inner beam, if there are three or more beams interconnected by U-frames

In the case of modified Warren or similar trusses (Fig. 8b.W2), the end raker should be considered as a chord for the purposes of this clause.

Effect of Loading on a Cross Member

When the compression chord is held in position over the supports by means of end members in the plane of the truss, the lateral flexure of a compression chord due to loading on a cross member should be considered. In the absence of a rigorous analysis of the interaction of main girders and cross members, the maximum value M_y of the lateral bending moment in a compression chord may be taken as

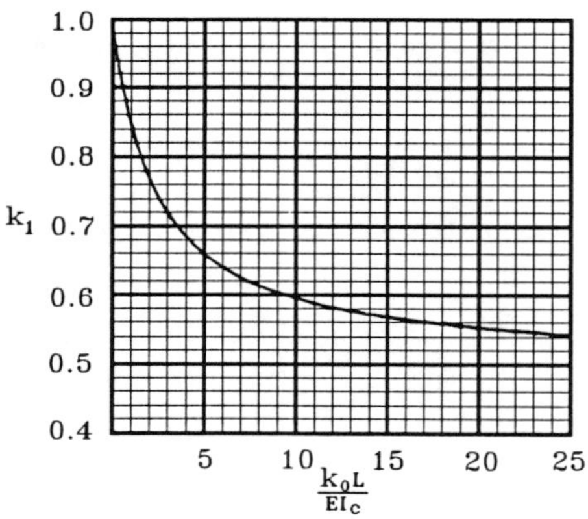

(a) Effect of rotational end restraint

Note 1. k_o= smaller value, at either end of the rotational stiffness to lateral bending of the compression flange, chord or strut.

I_c defined in Eq. 8b.W1

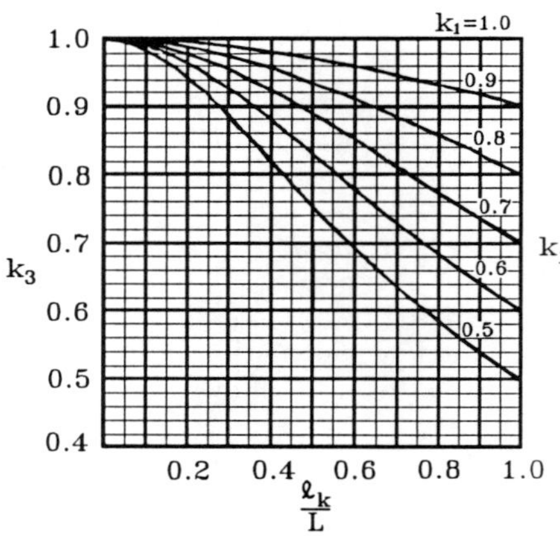

(b) Effect of bending restraint

L = span of the beam or truss or length between the ends of a compression member effectively held in positon.

ℓ_k= obtained from Eq. 8b.W1 with k_3=1.0.

Fig. 8b.W1 Influence on effective length of compression flange restraint

$$M_y = \frac{5EI_c\theta d_2}{L\ell_k\left(1 - \frac{P_c}{P_E}\right)}\left(1 + \frac{\frac{L}{\ell_k} - 1.25}{2.8 + 3.5\left(\frac{P_c}{P_e}\right)^2}\right) \tag{8b.W3}$$

provided that each main girder is in a vertical plane, and that the two chords of a main girder are parallel

where

I_c, d_2 = defined in Eq. 8b.W1

L = span of the main girder being considered

ℓ_k = defined in Eq. 8b.W1

P_c = maximum force in the compression chord in the span being considered

P_E = taken as follows

(a) $$P_E = \frac{\pi^2 EI_c}{\ell_k^2} \tag{8b.W4}$$

if ℓ_k is less than three times the spacing of U-frames

(b) $$P_E = \frac{1.25\pi^2 EI_c}{\ell_k^2} \tag{8b.W5}$$

if ℓ_k is more than four times the spacing of U-frames

(c) P_E is obtained by linear interpolation, for intermediate values of ℓ_k.

θ = rotation (in rad) of the cross member at its junction with the main girder under consideration, under the loading used when calculating P_c. θ may be calculated neglecting any interaction between the cross member and the main girders. If, because of non-uniform loading, θ varies between cross members, the average value of θ for those cross members within the loaded portion of the span should be used

For uniformly distributed loading when placed over the whole span, the maximum moment M_y derived above should be assumed to act anywhere within a horizontal distance ℓ_k from each bearing support of the beam. Elsewhere the bending moment should be assumed to be $0.5M_y$.

For all other loading cases it should be assumed that M_y acts anywhere within the span.

U-Frames

Each intermediate U-frame and its connections should be designed to resist all of the following:

(a) Wind and other applied forces

(b) Horizontal forces F_u acting normal to the compression chord at the level of its centroid are given by

$$F_u = \left(\frac{P_c}{P_E - P_c}\right)\frac{\ell_k}{667\delta} \tag{8b.W6}$$

but not greater than $$\left(\frac{P_c}{P_E - P_c}\right)\frac{EI_c}{16.7a^2} \qquad (8b.W7)$$

where P_c, P_E, ℓ_k, δ, I_c and a are as defined above.

In the case of several interconnected trusses, two such forces F_u should be applied in the same or opposite directions in such a way as to produce the most severe effect in the part being considered.

(c) Horizontal forces F_c applied to the U-frame at the same points and in the same manner as in (b) for F_u. The forces F_c result from the interaction between the bending of the cross member and the vertical members of the U-frame, and, in the absence of a more rigorous analysis, should be assumed to be given by:

$$F_c = \frac{3EI_1\theta}{{d_2}^2} \qquad (8b.W8)$$

where θ, I_1 and d_2 are as defined above.

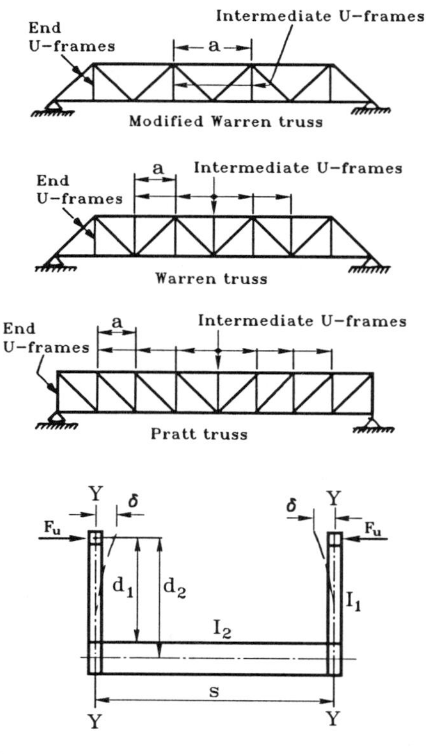

Fig. 8b.W2 Lateral restraint by U-frames

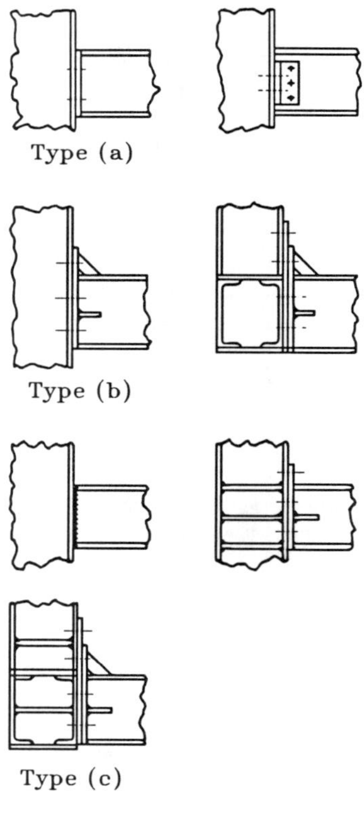

Fig. 8b.W3 U-frame joints

End U-Frames

End U-frames of trusses restrained laterally by a system of U-frames should be designed to resist all applied loads, and in addition, lateral forces each equal to $2(F_u + F_c)$, where the magnitudes of F_u and F_c and their manner of application are given above.

In the case of Warren-type trusses, where the extreme node of the compression chord is laterally restrained by U-frames both in the plane of the end diagonals and in the plane of the extreme verticals, the total lateral forces given above may be assumed to be shared by the two sets of U-frames.

NEN 3851 72

Axially Compressed Elastically Supported Members

In lieu of a more rigorous analysis, the following approximate method may be used.

- Determine the lesser stiffness of the two end supports c_e expressed as the force necessary to displace the support a distance equal to the unit of length
- Determine the stiffness of the intermediate supports c_1 to c_n, the least of these stiffnesses being c_{min}.
- Determine for each sub-length an equivalent buckling coefficient ω_i with the aid of the formula

$$\omega_i = \frac{A\,\sigma_e}{N} \tag{8b.W9}$$

where

A = cross-sectional area of the member

N = largest compressive force in the sub-length of the member due to the design load

σ_e = design value for the steel used

- Determine for each sub-length the equivalent slenderness ratio λ_i corresponding to ω_i with the aid of the table of the buckling coefficients in the specification.
- Determine for each sub-length the ratio λ^* between the equivalent effective length ℓ_{ki} and the sub-length ℓ:

$$\lambda^* = \frac{\ell_{ki}}{\ell} = \frac{\lambda_i}{\ell}\cdot i \quad \text{but } \lambda^* \geq 1 \tag{8b.W10}$$

where

i = radius of gyration in the direction of elastic support.

- Determine the arithmetical mean of all the values λ^*, i.e., λ^*_m.
- Determine the stiffness $c_o = \dfrac{4\,N_{max}}{\lambda^{*\,2}_m\,\ell_{min}}$

where

ℓ_{min} = shortest sub-length

N_{max} = largest value of all the largest compressive forces N

In the case of rigid end supports ($c_e \approx \infty$) it must be shown that $c_{min} \geq c_o$.

In the case of elastic end supports, determine the values

$$\xi = \frac{c_{min}}{c_o} \qquad (8b.W11)$$

and then the coefficient

$$\xi^* = \frac{1 + 0.6\xi\lambda_m^*}{2}\left(1 + \sqrt{1 - \frac{1.44\xi\lambda_m^*}{(1 + 0.6\xi\lambda_m^*)^2}}\right) \qquad (8b.W12)$$

It must be shown that $c_{min} \geq \xi^* c_o$

SIA 161 79

The effective length ℓ_k of unbraced compression chords is given by

$$\ell_k = \pi\left(\sqrt[4]{0.25\ \mathrm{EIav}}\right) > a \qquad (8b.W13)$$

where

a = distance between the supporting frames (elastic supports)
EI = relevant flexural stiffness of the chord
v = lateral deflection that would occur in the supporting frame, when a unit force acts laterally

REGIONAL COMPARISON

Starting from the results of the elastic buckling analysis of compression members elastically supported (the well-known Engesser formula), two equivalent design approaches are derived in the European codes. The Swiss and the English codes express the effective length ℓ_k of the chord as a function of the stiffnesses both of the chord itself and of the supports. The Italian and the Dutch codes express the required stiffness for the supports as a function of the imposed buckling strength of the chord.

INTERNATIONAL COMPARISONS & COMMENT

WHAT ARE THE DIFFERENCES?

In many codes, guidelines are given for the design of unbraced compression chords and their lateral restraining systems.

In some codes, approximate formulas are given in order to derive the stiffness of the supporting frames as a function of the buckling strength of the chord. In other codes, on the contrary, the effective length of the chord is given by approximate formulas as a function of the stiffness of the supporting frames.

The effective length (ℓ_k) of the chords is usually written in the form

$$\ell_k = k\ \ell \tag{8b.1}$$

where

EI = flexural stiffness of the chord
k = $f(\ell^3/wEI)$
ℓ = distance between the supporting frames
w = lateral deflection of the supporting frame, when a unit force acts laterally

In Fig. 8b.1 a comparison is shown of the approximate formulas for $f(\ell^3/wEI)$ adopted in the English (GB) code BS 5400 82, in the Swiss code (CH) SIA 161 79 and in the Chinese "Specification for the Design of Railway Bridges." (In the Chinese Specifications the number n of panels is also considered: n = 10 for the upper curve in Fig. 8b.1, n = 5 for the lower one). A significant variation of the values is shown.

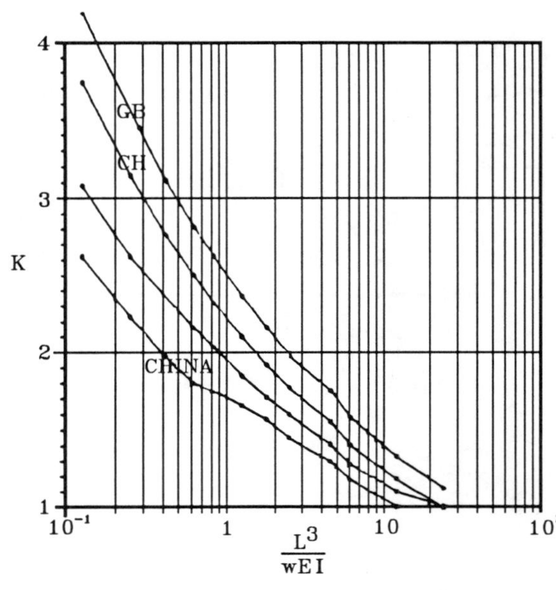

Fig. 8b.1 Approximate Formulas

The stiffness requirements for the elastic supports (of frames) laid down by the codes of A, ND, I, E, H, CS, P . . . are in the following

$$C = \frac{1}{w} \simeq \frac{4F_{max}}{k^2 \ell} \tag{8b.2}$$

where

F_{max} = the design compression force of the chord
k = effective length factor derived from the relevant buckling curve as denoted in the code
ℓ = already defined

Fig. 8b.2 indicates the procedures used in applying the above equations.

United States and Canadian codes specify the types of analysis to be carried out, but provide no details for doing so.

WHY ARE THERE DIFFERENCES?

The well-known Engesser buckling analysis of elastically supported compression members is the theoretical basis for both of the above approximate approaches. The differences in the various approximations take into account some favorable effects (varying compression forces, spans, moments of inertia, spring constants).

In the British and Swiss approaches the effective length ℓ_k is derived assuming elastic behavior. In the other codes mentioned, the effective length ℓ_k is derived taking into account elastic-plastic behavior implicitly through the buckling curve.

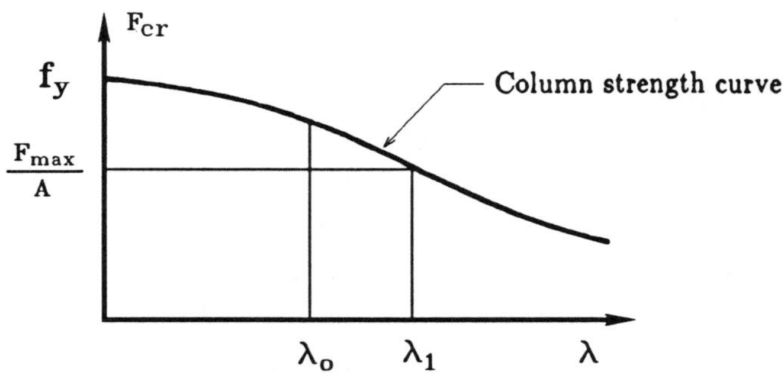

$\lambda_o = \frac{\ell}{r} \quad \lambda_1 = k\frac{\ell}{r} \quad k = \lambda_1/\lambda_o$

Fig. 8b.2

8c. Maximum Slenderness Ratio

SPECIFICATION PROVISIONS & EXPLANATION

CHINA

CHAPTER EDITOR
Zaitian Guo, Xian Institute of Metallurgy & Construction Engineering

GBJ 17 88, GBJ 18 87, and TBJ 2 85

The maximum slenderness ratios for compression members in triangulated structures are given in Tables 8c.C1 and 8c.C2.

Table 8c.C1 Maximum Slenderness Ratios

GBJ 17 88	GBJ 18 87	HRS 87	Maximum slenderness ratios
Columns, Truss members, Column lacings, Column braces under crane runway girders	Columns, Chords and End posts of trusses	Chords, Diagonals, Transverses	150
Column braces (except those under crane runway girders), Braces used to reduce the slenderness ratio of compression members	Other members and Braces	Diaphragms, Other secondary members	200

Table 8c.C2 Maximum Slenderness Ratios

Specification for the design of railway bridges and culverts TBJ 2 85		
Main members	Chords, Members subject to compression or reversal loading	100
	Web members (not subject to live load)	150
Bracing systems	Longitudinal bracing systems, Transverse bracing systems at supports	130
	Intermediate transverse bracing systems	150

JAPAN

CHAPTER EDITOR
Hiroshi Akiyama, University of Tokyo

AIJ 70

The maximum slenderness ratio of compression members is 250, and that of columns is 200.

JEC 127

The maximum slenderness ratios of compression members of electrical transmission towers are

200 for main column members,
220 for compression members other than main column members,
250 for secondary stiffening members.

JSCE 83 and JRA 83

The maximum slenderness ratios of compression members are

120 for primary members
150 for secondary members

NORTH AMERICA

CHAPTER EDITOR

Nicholas F. Morris, Manhattan College

AASHTO 83

For compression members, the slenderness ratio shall not exceed 120 for main members, or those in which the major stresses result from dead or live load, or both; and shall not exceed 140 for secondary members, or those whose primary purpose is to brace the structure against lateral or longitudinal force, or to brace or reduce the unbraced length of other members, main or secondary.

AISC ASD 78

The maximum slenderness ratio of compression members shall not exceed 200.

AISC LRFD 86

For members whose design is based on compressive force, the slenderness ratio preferably shall not exceed 200.

AREA 85

The maximum slenderness ratios are

100 for main compression members
120 for wind and sway bracing in compression
140 for single lacing
200 for double lacing
200 for tension members

CSA S6 M89 84a

The slenderness ratio shall not exceed

a) 120 for main compression member.
b) 160 for bracing of main compression members.
c) 200 for all other bracing unless otherwise specified or approved by the engineer.

CSA S16.1 M89 84b

The slenderness ratio of a compression member shall not exceed 200.

CSA S136 M89 84c

The effective slenderness ratio of members in compression shall not exceed 200.

OHBDC 83

The slenderness ratio shall not exceed 120 for main compression members or 160 for secondary and bracing members.

REGIONAL COMPARISON

American and Canadian codes recommend that truss members be designed as simply supported compression members ($k=1$). As could be expected, the maximum slenderness ratio permitted for buildings is larger than that permitted for bridges. Bracing, where specified, must usually be designed to resist some percent of the compression forces in the braced members. However, there is no consensus concerning the percentage to be resisted.

WEST EUROPE

CHAPTER EDITORS
Pierre Dubas, Swiss Federal Institute of Technology
Carlo Urbano, Technical University of Milan

BS 5400 P.3 82

The value of the slenderness ratio for compression members should not exceed the following

180 for members resisting loads other than wind loads
250 for members resisting self weight and wind loads only
350 for any member normally acting as a tie, but subject to reversal of stress resulting from the action of wind

CNR 10011 85

The following limits for the maximum slenderness ratio of compression members are adopted

200 for main members
250 for bracing members

For bridges and other dynamically loaded structures the above limits are reduced to

150 for main members
200 for bracing members

SIA 161 79

The limits for the maximum slenderness ratio of compression members are

250 for bracing and secondary members
200 for main members
160 for dynamically loaded members

REGIONAL COMPARISON

West European codes are in good agreement on the following limits for the maximum slenderness ratios of compression members

For main members the slenderness should not exceed 200
For secondary members the slenderness should not exceed 250

In the case of bridges and other dynamically loaded structures the codes of some countries (Switzerland, Spain, Italy) limit the maximum slenderness ratio of compression members in the range 150 to 175.

INTERNATIONAL COMPARISONS & COMMENT

WHAT ARE THE DIFFERENCES?

There is a general agreement on the upper limit for the maximum slenderness ratio of compression members in building structures. This limit is 200. This is not the case with bridge structures. In the United States and Canada, the maximum slenderness ratio is 120 for the main compression members. Other codes allow higher slenderness ratios, typically in the range 140 to 175.

WHY ARE THERE DIFFERENCES?

The upper limit for building structures is chosen more by practical considerations than by strength requirements. A very slender member may easily undergo inadvertent damage during erection and in use. Bridge specifications are also controlled by considerations outside of stability requirements; i.e., fatigue, corrosion, maintenance, etc.

Table 8c.1 Comparison of Provisions

ITEM		Australia		China			E. Europe	Japan	
		SEC	SAA	GBJ	TBJ	HRS	CMEA	AIJ	JSCE
Effective length factor K	Chords	1	< 1	< 1	< 1	< 1	< 1	1	1
	Web Member	1	< 1	< 1	< 1	< 1	< 1	1	1
Maximum slenderness ratio	Building	180	180	200			200	250	
	Bridge		120	150	100/150		120		150
Unbraced compression chord eff length, E=elastic, EP=elastic plastic			E		E		EP	E	E

Table 8c.1 (Cont'd) Comparison of Provisions

ITEM		North America							
		AASHTO	AISC 83	AISC 86	AREA	CMTC	CSA.1	CSA.2	CSA.3
Effective length factor	Chords	1	1	1	0.75	1	1	1	1
	Web Member	1	1	1	0.75	1	1	1	1
Maximum slenderness ratio	Building		200	200				200	200
	Bridge	120			100	120	120		
Unbraced compression chord effective length, E=elastic, EP=elastic plastic									

Table 8c.1 (Cont'd) Comparison of Provisions

ITEM		Western Europe						
		EC3	SIA 78	BS 82	DIN 86	CNR.1 85	NEN	BS 85
Effective length factor	Chords	< 1	< 1	< 1	< 1	< 1	< 1	< 1
	Web Member	< 1	< 1	< 1	< 1	< 1	< 1	< 1
Maximum slenderness ratio	Building	none	200		none	200	none	180
	Bridge	none	160	none	none	150	none	
Unbraced compression chord effective length, E=elastic, EP=elastic plastic		EP	E	E	EP	E	EP	

Symbol List for Chapter 8 -- Triangulated Structures

A = cross-sectional area of the member (8b.W)
A_D = section area of diagonal members (8a.W)
a = distance between the supporting frames (elastic supports) (8b.W)
a = length of member, defined in Table 8a.W2 (8a.W)
B = distance between the two main trusses (8b.C)
C = stiffness of the restraining frame (8a.W)
c_e = less stiffness of the two end supports (8b.W)
c_{min} = least of the stiffness of the intermediate supports (8b.W)
c_o = required stiffness of supports (8b.W)
d = panel length (8b.C)
d_1 = distance from the centroid of the compression chord to the nearer face of the cross member of the U-frame (8b.W)
d_2 = distance from the centroid of the compression chord to the centroidal axis of the cross member of the U-frame (8b.W)
E = modulus of elasticity (8b.C)
e_x = eccentricity as indicated in Fig. 8a.W8 (8a.W)
F_c = horizontal forces applied to the U-frame resulting from the interaction between the bending of the cross member and the vertical members (8b.W)
F_{max} = design compression force of the chord (8b.1)
F_u = horizontal forces acting normal to the compression chord at the level of its centroid (8b.W)
f = flexibility of the joint between the cross member and the verticals of the U-frame (8b.W)
f_y = yield stress of material (8a.A)
h = height from the centroid of cross section of compression chord to the upper surface of floor beam (8b.C)
I = moment inertia of gross section of compression chord about the vertical axis (average value within the total length) (8b.C)
I_b = moment of inertia of the floor beam section (8b.C)
I_c = moment of inertia of the gross section of the verticals for out-of-plane buckling (8b.C)
I_c = maximum second moment of area of the chord about the Y-Y axis shown in Fig.8b.W2 (8b.W)
I_y = moment of inertia, defined in Table 8a.W3 (8a.W)
$I_{y,1}$ = moment of inertia, defined in Fig. 8a.W4 (8a.W)
$I_{y,2}$ = moment of inertia, defined in Fig. 8a.W4 (8a.W)
$I_{y,3}$ = moment of inertia, defined in Fig. 8a.W4 (8a.W)
I_1 = second moment of area of the web member forming an arm of the U-frame in its plane of bending (8b.W)
I_2 = second moment of area of the cross member in its plane of bending (8b.W)
I_2 = moment of inertia of the member subjected to smaller axial force, see Fig. 8a.J2 (8a.J)
i = radius of gyration in the direction of elastic support (8b.W)
i_x = radius of gyration of angle section about x-axis (8a.C)
i_{yo} = smallest radius of gyration of angle section about the principle axis (8a.C)
i_1 = smallest radius of gyration of angle cross-section (8a.W)

J = moment of inertia of the chord (8a.C)
J_d = moment of inertia of the web member (8a.C)
K = effective length factor (8c.1)
k = effective length factor (8a.N), (8b.N) & (8b.1)
k_o = smaller value, at either end of the rotational stiffness to lateral bending of the compression flange, chord or strut (8b.W)
k_3 = modification coefficient of effective length (8b.W)
L = span of the beam or truss or length between the ends of a compression member effectively held in position (8b.C) & (8b.W)
L_f = effective length for chords of laced members with four angles as chords (8a.W)
ℓ = distance between braced panel points (8a.A), (8a.C), (8a.J), (8a.N)&(8a.W)
ℓ = distance between the supporting frames (8b.1)
ℓ_d = length of diagonal members, defined in Fig. 8a.E2 (8a.C)
ℓ_{ef} = effective length of chord members and end posts (8a.C)
ℓ_k = effective length of a compression member (8a.E), (8a.J), (8a.W), & (8b.W)
ℓ_{ky} = effective length of members, defined in Table 8a.W5 (8a.W)
$\ell_{ky,1}$ = effective length of members, defined in Table 8a.W5 (8a.W)
ℓ_{min} = shortest sub-length (8b.W)
ℓ_1 = distance betwwen laterally supported points of compression chords, see Fig. 8a.C1 (8a.C)
ℓ_1 = length of diagonal, transverse or diaphram members for towers and masts, defined in Tables 8a.C2 and 8a.C3 (8a.C)
ℓ_1 = length of members, defined in Fig. 8a.J1 (8a.J)
ℓ_2 = length of members, defined in Fig. 8a.J2 (8a.J)
M = bending moment in the member due to the transverse load acting between the joints in the design state (8a.W)
M_p = plastic moment of the member (8a.W)
M_y = maximum moment (8b.W)
m = truss mesh size (8a.W)
m = number of half-waves at the onset of chord buckling (8b.C)
N = maximum compression force (8a.W) & (8b.W)
N_{max} = largest value of all the largest compressive forces N (8b.W)
N_1 = greater compression force (8a.E), (8a.C) & (8a.J)
N_2 = smaller compression force, with a negative value for the tensile force (8a.E), (8a.C) & (8a.J)
P_c = maximum force in the compression chord in the span being considered (8b.W)
P_E = buckling load (8b.W)
s = distance between centers of consecutive main girders connected by the U-frame (8b.W)
v = lateral deflection that would occur in the supporting frame, when a unit force acts laterally (8b.W)
W = section modulus within the plane of the truss (8a.W)
W_x = section modulus of angle section about x-axis (8a.W)
w = lateral deflection of the supporting frame, when a unit force acts laterally (8b.1)
z = axial force (8a.W)
α = effective length factor (8b.C)
α = angle between chords and diagonal member (8a.W)

δ = maximum deflection caused by unit horizontal force acting on the upper panel point of one chord of an intermediate semi-frame (8b.C)
δ = lateral deflection that would occur in the U-frame (8b.W)
θ = rotation (in rad) of the cross member at its junction with the main girder (8b.W)
λ = slenderness ratio of chord members, diagonal members, transverse and diaphragm members (8a.C) & (8a.W)
$\overline{\lambda}_{eff}$ = modified slenderness ratio (8a.W)
λ_i = equivalent slenderness ratio (8b.W)
λ_o = slenderness ratio (8b.2)
λ_p = limiting slenderness ratio (8a.W)
λ_1 = effective slenderness ratio (8b.2)
λ^* = ratio between the equivalent effective length ℓ_{ki} and the sub-length ℓ (8b.W)
λ^*_m = arithmetical mean of all the values of λ^* (8b.W)
$\overline{\lambda}'$ = modified slenderness (8a.W)
$\overline{\mu}$ = modification factor of slenderness ratio (8a.E)
ξ = stiffness ratio in the case of elastic end supports (8b.W)
ξ^* = required stiffness ratio (8b.W)
σ_e = design value of the yield stress (8b.W)
ω = buckling coefficient within the plane of the truss (8a.W)
ω_i = equivalent buckling coefficient (8b.W)
ω_{max} = buckling coefficient (8a.W)
ω_x = buckling coefficient (8a.W)

Chapter 9

Tubular Structures*

TOPICAL EDITORS
Ben Kato, Toyo University
Donald R. Sherman, University of Wisconsin-Milwaukee

CHAPTER EDITORS
Greg J. Hancock, The University of Sydney
Ben Kato, Toyo University
Dan Mateescu, Polytechnic Institute of Timisoara
Jacques Rondal, University of Liege
Donald R. Sherman, University of Wisconsin-Milwaukee

9a. Column Strength

SPECIFICATION PROVISIONS & EXPLANATION

AUSTRALIA

CHAPTER EDITOR
Greg J. Hancock, The University of Sydney

AS 1250 81

Allowable stress design with the factor of safety is used. The design column curve is given by

$$f_c = \frac{\sigma_y \xi}{\gamma}\left\{1-\sqrt{1-\frac{1}{\gamma^2\xi^2}}\right\} \tag{9a.A1}$$

where

$$\gamma = 1.67$$

$$\eta = 0.00003\left(\frac{\ell}{i}\right)^2$$

$$\xi = \frac{\lambda^2+1+\eta}{2\lambda^2}$$

*The notation CHS refers to circular hollow sections and SHS refers to square and rectangular hollow sections. Most of the equation in this Chapter are presented in a nondimensional form. The SI system was used for any equations which inherently include units through a substitution of a numerical value of the modulus of elasticity, E.

AS DR87164 87

The ultimate limit state design approach is used. The design ultimate column strength is given by

$$N_f = \phi N = \alpha_d N_y \leq N_y \qquad \text{with } \phi = 0.9 \tag{9a.A2}$$

where

$$\alpha_a = \frac{90(\lambda - 0.15)}{3.86\lambda^2 - 0.656\lambda + 0.976} \tag{9a.A3}$$

$$\alpha_d = \xi \left\{ 1 - \sqrt{1 - \frac{1}{\lambda_m^2 \xi^2}} \right\} \tag{9a.A4}$$

$$\eta = 0.293(\lambda_m - 0.15) \geq 0, \tag{9a.A5}$$

$$\lambda_m = \lambda + \frac{\alpha_a \alpha_b}{90} \tag{9a.A6}$$

$$\xi = \frac{\lambda_m^2 + 1 + \eta}{2\lambda_m^2} \tag{9a.A7}$$

where

α_b = −1.0 for hot-formed SHS and CHS and cold-formed (stress-relieved) SHS and CHS (This curve is approximately equivalent to ECCS a_o and SSRC1)
= −0.5 for cold-formed (non stress-relieved) SHS and CHS (This curve is approximately equivalent to ECCS a)
= 0.0 for welded box sections (This curve is approximately equivalent to ECCS curve b and SSRC curve 2)

COMMENTARY

Both specifications summarized in AS 1250 81 and AS DR87164 87 use the Perry formula as a column strength curve. However AS 1250 81 has a single central curve whereas AS DR87164 87 uses the multiple column curve concept. The justification for the choice of the value of α_b for cold-formed tubes was given by Key and Hancock (1988).

EAST EUROPE

CHAPTER EDITOR
Dan Mateescu, Polytechnic Institute of Timisoara

STAS 10108/0 78

The limit states design approach is used. The design criterion for compression members with cross sectional area A is

$$N_f \leq \psi \, A \, R \qquad (9a.E1)$$

where the design yield stress R is defined by

$$R = \sigma_y / \gamma_m \qquad (9a.E2)$$

where γ_m is the material homogeneity factor which is in the range of 1.05 to 1.10.

The buckling coefficient ψ is given by

$$\psi = \frac{\alpha + \beta\lambda^2}{2\lambda^2} - \frac{1}{2\lambda^2}\sqrt{(\alpha+\beta\lambda^2)^2 - 4\lambda^2} \qquad (9a.E3)$$

where

$\alpha = 1.258$
$\beta = 1.293$

COMMENTARY

The column curve is approximately equivalent to the old version of the ECCS curve a, and coefficients α and β have been evaluated accordingly.

JAPAN

CHAPTER EDITOR
Ben Kato, Toyo University

AIJ 80

Allowable stress design with the factor of safety is used. The column design curves are given by Eqs. 9a.J1 and 9a.J2.

for $\lambda \leq 1.291$

$$f_c = \frac{(1-0.24\lambda^2)\sigma_y}{\gamma} \quad (9a.J1)$$

for $1.291 < \lambda$

$$f_c = \frac{0.462\sigma_y}{\lambda^2} \quad (9a.J2)$$

$\gamma = 1.5+0.4\lambda^2 \leq 2.167 =$ factor of safety

where

f_c = allowable compressive stress (MPa)

λ = $\left(\frac{k\ell}{i}\right)\frac{1}{\pi}\sqrt{\frac{\sigma_y}{E}}$ is the normalized slenderness ratio

σ_y = minimum specified yield stress (MPa)

COMMENTARY

In inelastic region, column strength is given in the form of CRC (Column Research Council) parabola.

AIJ 89b

The ultimate limit states design approach is used. The column design curves are given by Eqs. 9a.J3 to 9a.J5.

for $\lambda \leq \lambda_p$

$$N = N_y \quad \text{with } \phi = 0.9 \quad (9a.J3)$$

for $\lambda_p < \lambda \leq \lambda_e$

$$N = \left(1-0.5\frac{\lambda-\lambda_p}{\lambda_e-\lambda_p}\right) N_y$$

$$\text{with } \phi = 0.90-0.05\left(\frac{\lambda-\lambda_p}{\lambda_e-\lambda_p}\right) \quad (9a.J4)$$

for $\lambda_e < \lambda$

$$N = \left(\frac{1}{1.2\lambda^2}\right) N_y$$

$$\text{with } \phi = 0.85 \quad (9a.J5)$$

where

N = flexural buckling strength
N_y = $A\sigma_y$, axial yield strength
λ = normalized slenderness ratio
λ_e = $1/\sqrt{0.6}$, elastic limit slenderness ratio
λ_p = 0.15, plastic limit slenderness ratio
ϕ = resistance factor

COMMENTARY

The formulas have been derived statistically from the experimental data.

NORTH AMERICA

CHAPTER EDITOR
Donald Sherman, University of Wisconsin-Milwaukee

In North America, there are six steel specifications or guidelines that contain criteria for tubular sections used as structural members. These are in either allowable stress design or ultimate limit states design formats, and are referred to ASD and LRFD respectively in each specification. The yield strength is determined from a test coupon taken from the wall of finished CHS and SHS manufacture tubes and from the virgin plate of fabricated CHS.

AISC ASD 78

for $\lambda < \sqrt{2}$

$$f_c = \frac{(1 - \frac{1}{4}\lambda^2)\sigma_y}{\frac{5}{3} + \frac{3}{8\sqrt{2}}\lambda - \frac{1}{16\sqrt{2}}\lambda^3} \tag{9a.N1a}$$

for $\lambda > \sqrt{2}$

$$f_c = \frac{12\sigma_y}{23\lambda^2} \tag{9a.N1b}$$

AISC LRFD 86

The limit states design format is given by Eq. 9a.N2a and 9a.N2b.

$N_f = \phi N$ with $\phi = 0.85$ (9a.N2a)

for $\lambda < 1.5$

$$N = (0.658^{\lambda^2})\, N_y \tag{9a.N2b}$$

for $\lambda > 1.5$

$$N = \frac{0.877}{\lambda^2} N_y \tag{9a.N2c}$$

AISI 76

Equations 9a.N1a and 9a.N1b for CHS and SHS hot-formed or cold-formed stress-relieved tubes.

For SHS cold-formed tubes

for $\lambda < \sqrt{3}$

$$f_c = \frac{(1 - \frac{2}{3\sqrt{3}}\lambda)\sigma_y}{\frac{5}{3} + \frac{1}{4\sqrt{3}}\lambda}$$

for $\lambda > \sqrt{3}$

$$f_c = \frac{12\,\sigma_y}{23\,\lambda^2}$$

(9a.N3)

AISI 86

Equations 9a.N1a and 9a.N1b

API RP2A 86

Equations 9a.N1a and 9a.N1b

CSA S16.1 M89 84b

The ultimate strength is given as following, depending on the manufacturing processes, with ϕ=0.90.

For hot-formed and cold-formed stress relieved CHS and SHS tubes

for $\lambda < 0.15$
$$N=N_y \tag{9a.N4a}$$
for $0.15 < \lambda < 1.2$
$$N=(0.990+0.122\lambda - 0.367\lambda^2)N_y \tag{9a.N4b}$$
for $1.2 < \lambda < 1.8$
$$N=(0.051+\frac{0.801}{\lambda^2})N_y \tag{9a.N4c}$$
for $1.8 < \lambda < 2.8$
$$N=(0.008+\frac{0.942}{\lambda^2})N_y \tag{9a.N4d}$$
for $\lambda > 2.8$
$$N=\frac{N_y}{\lambda^2} \tag{9a.N4e}$$

For cold-formed CHS and SHS tubes

for $\lambda < 0.15$
$$N=N_y \tag{9a.N5a}$$
for $0.15 < \lambda < 1.0$
$$N=(1.035-0.202\lambda-0.222\lambda^2)N_y \tag{9a.N5b}$$
for $1.0 < \lambda < 2.0$
$$N=(-0.111+\frac{0.636}{\lambda}+\frac{0.087}{\lambda^2})N_y \tag{9a.N5c}$$
for $2.0 < \lambda < 3.6$
$$N=(0.009+\frac{0.877}{\lambda^2})N_y \tag{9a.N5d}$$
for $\lambda > 3.6$
$$N=\frac{N_y}{\lambda^2} \tag{9a.N5e}$$

REGIONAL COMPARISON

The ultimate strength implied in the various North American criteria are compared in Fig. 9a.N1. In this figure, equations from ASD specifications have been multiplied by 1.67, which is the stated factor of safety for members. Equations for the LRFD specifications have been plotted directly so that the resistance factor does not appear.

Specification writers in the United States have expressed a clear preference for one column curve applicable to all sections. For inelastic column buckling, this curve is the Column Research Council (now SSRC) parabola with the variable factor of safety adopted in AISC ASD 78. For elastic buckling it is the Euler equation. Therefore, in AISC ASD 78, AISI 86 and API RP2A 86 criteria for tubular members are automatically included in the same column curve as other sections. AISC LRFD 86 continues the practice of including all sections under one column curve. Although the equation appears in an entirely different form, it gives essentially the same implied ultimate strengths as AISC ASD 78.

AISI 76 was written specifically for tubular members. It recognizes that residual stresses cause a distinct difference in column strengths between cold-formed and hot-formed SHS. (Both analytical investigations and experiments show this.) The experimental data base was largely from the CIDECT programs although some North American data was also available.

CSA has accepted the multiple column curve concept in design and has adopted the SSRC equations. Hot-formed tubes have been assigned to SSRC Curve 1, which is somewhat higher than the AISC equation and recognizes the low level of residual stresses in these products. All cold-formed tubes, both CHS and SHS, are assigned to SSRC Curve 2.

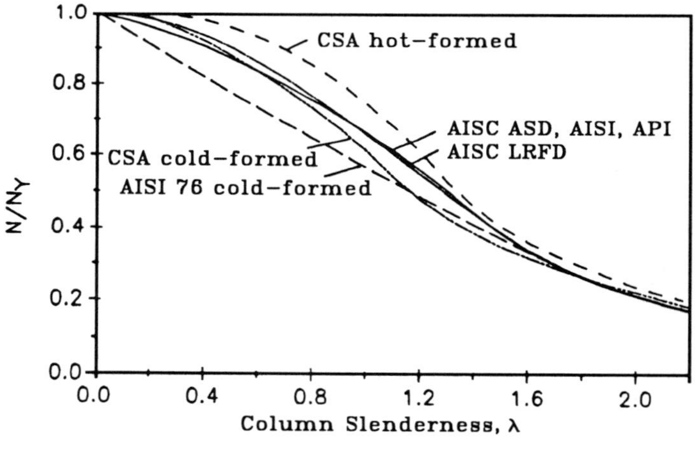

Fig. 9a.N1 Column Buckling Curves

WEST EUROPE

CHAPTER EDITOR
Jacques Rondal, University of Liege

Since there are many West European national standards, only the rules specified by EC 3 88 are presented. Generally, the West European national standards are more or less in line with the rules given in EC 3 88.

EC 3 88

The ultimate limit states design approach is used. For column strength, the buckling curve shall be determined from Table 9a.W1. Hot rolled sections are checked using buckling ECCS curve a.

Cold formed structural hollow sections are checked using one of the following:

(a) The average strength f_{ya} of the member after cold-forming, determined in conformity with the definition given in Table 9a.W2, with buckling curve c.
(b) The basic yield strength f_{yb} of the flat sheet material out of which the member is made by cold-forming, with buckling curve b.

The column strength is given by the relation

$$N_f = \frac{\chi\, N_y}{\gamma_m} \tag{9a.W2}$$

where

N_y = the plastic limit load
γ_m = the partial safety factor equal to 1.1
χ = a reduction factor determined from

Table 9a.W1 Selection of buckling curve for a cross section

<table>
<tr><th colspan="2">Cross section</th><th>buckling perpendicular to axis</th><th>ECCS buckling curve</th></tr>
<tr><td rowspan="3">hollow sections 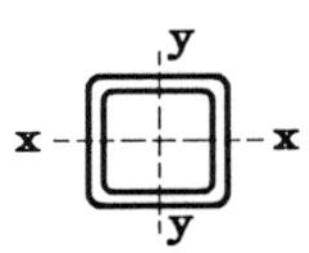 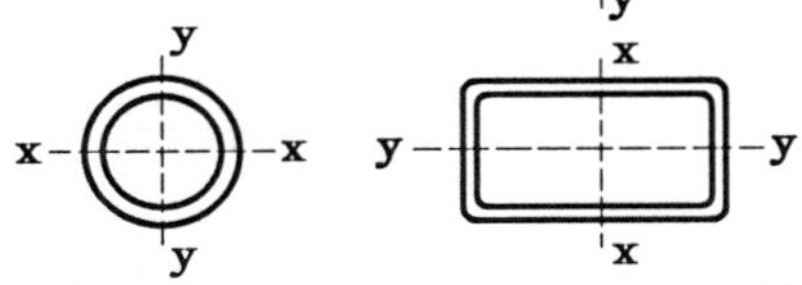</td><td>hot rolled</td><td>any</td><td>a</td></tr>
<tr><td>cold formed using f_{yb}</td><td>any</td><td>b</td></tr>
<tr><td>cold formed using f_{ya}</td><td>any</td><td>c</td></tr>
</table>

$$\chi = \frac{1}{\phi + \sqrt{\phi^2 - \lambda^2}} \leq 1 \qquad (9a.W3)$$

where

$$\lambda = \frac{k\ell}{\pi i}\sqrt{\frac{f_y}{E}} \qquad (9a.W5)$$

$$\phi = 0.5\left[1+\alpha(\lambda - 0.2) + \lambda^2\right] \qquad (9a.W4)$$

The imperfection factor α corresponding to the appropriate buckling curve shall be obtained from Table 9a.W3.

Table 9a.W2 Average yield strength f_{ya} of cold formed structural hollow sections

Average Yield Strength

The average yield strength (f_{ya}) may be determined from tests of full size sections or from Eq. 9a.W1

$$f_{ya} = f_{yb} + (4Ct^2/A)\,(f_u - f_{yb}) \qquad (9a.W1)$$

where

- A = gross cross sectional area
- C = coefficient as a function of the type of forming
  = 7 for rolled material
  = 5 for other methods of forming
- f_{yb}, f_u = tensile yield strength and tensile ultimate strength of the basic material
- t = material thickness

f_{ya} should not exceed f_u or 1.2 f_{yb}

The increase in yield strength due to cold working should not be utilized for members which are galvanized or subject to heat treatment after forming which may produce softening.

Basic Material

Basic material is the flat sheet material out of which sections are made by cold-forming.

Table 9a.W3 Imperfection factors

Buckling curve	a	b	c
Imperfection factor α	0.21	0.34	0.49

INTERNATIONAL COMPARISONS & COMMENT

WHAT ARE THE DIFFERENCES?

Both allowable stress design formats and ultimate limit state formats are used. In allowable stress design formats, traditional formulas such as Perry and CRC (Column Research Council parabola) are generally used. On the other hand, almost all ultimate limit state formats are based on statistical evaluations of experimental data.

Some countries adopt multiple column curves distinguishing hot-formed sections and cold-formed section. The mechanical properties of the latter are affected by work-hardening and the resulting residual stresses. Some countries adopt a single column curve.

There are several ways in which the yield strength is defined. In some cases yield is determined from tension coupons taken from the wall of the finished tubular product whereas in others, the coupon is from the virgin plate material before the tube is formed. In still other cases, yield is a measured or calculated average value for the full cross section.

Beam-columns have not been discussed since in all codes it is acceptable to use the same interaction equations that are used for other sections. There are a few codes that permit the use of alternate interaction equations for certain restricted types of tubes.

The differences in factors of safety and of resistance factors are not substantial between the various countries.

Keeping these observations in mind, a comparison of design concepts is made in Table 9a.1. Column curves for allowable stress design are compared in Fig. 9a.1 removing the factor of safety, and column curves for ultimate limit state design are compared in Fig. 9a.2 removing the resistance factor. A steel with the nominal yield stress of 320 MPa is assumed when it is necessary.

Table 9a.1 Concept Matrix – Allowable Stress Design (ASD)

Codes		Type of formula	Single or multiple curve*	Factor of safety
Australia	AS 1250 81	Perry	single	1.67
Japan	AIJ 80	CRC	single	$1.5+0.4\lambda^2 < 2.167$
USA	AISC ASD 78 AISI 86 API RP2A 86	CRC	single	1.67–1.92
	AISI 76	CRC & empirical	multiple	1.67–1.92

WHY ARE THERE DIFFERENCES?

Resistance factors are closely related to load factors. For load combinations, weighting among loads combined may be different from region to region. If a load factor is very large, resistance factor may be led to unity. A similar discussion could be made to the factor of safety for allowable stress design. From these points of view, the difference of resistance factors and factors of safety among regions as shown in Table 9a.1 seems to be quite natural.

For ultimate limit state design, multiple curves are adopted in Australia, Canada and Eurocode, which consider the manufacturing of tubes. Other standards adopt a single column curve, presumably due to the fact that (a)emphasis is placed upon simplicity or (b)one manufacturing process dominates in that particular country.

As for the classification of multiple column curves, the following comments are made. Cold-formed steel tubes contain the residual stresses which give an adverse

Table 9a.1 Concept Matrix (Cont.) – Limit States Design (LSD)

Codes		Type of formula	Single or multiple curve*	Resistance factor
Australia	AS DR87164 87	Perry	multiple	0.9
Romania	STAS 10108/0 81	Perry	single	0.95 ~ 0.87
Japan	AIJ 89b	empirical	single	0.9 ~ 0.85
USA	AISC LRFD 86	empirical	multiple	0.85
Canada	CSA S16.1 M89 84b	empirical	multiple	0.9
West Europe	EC 3 88	Perry	multiple	0.9

* multiple curves are classified according to the manufacturing processes of tubes.

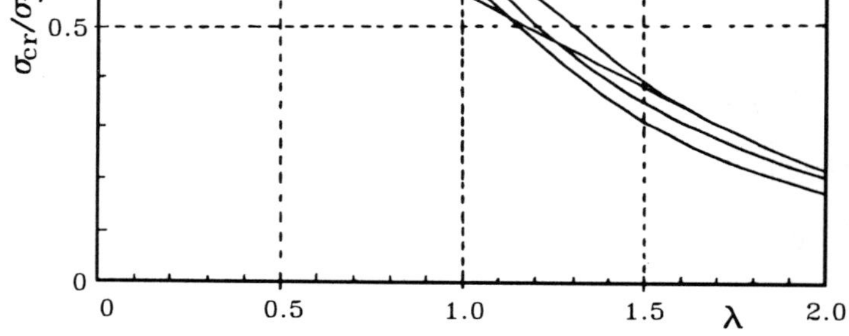

Fig. 9a.1 Column Curves For Allowable Stress Design (1), AS 1250 81, (2), AIJ 80, (3), AISC ASD 78, (4), AISI 76

effect on stiffness. On the other hand, the yield stress increases due to the work-hardening. These two effects partially compensate each other in the inelastic region. Therefore, if the design yield stress for cold-formed hollow sections is to be determined on the basis of finished products, the column curves for cold-formed CHS and SHS should be assigned to the lower strength class. However, if the design yield stress is that of their base material, the upper class curves could be assigned. Eurocode 3 is adopting this idea.

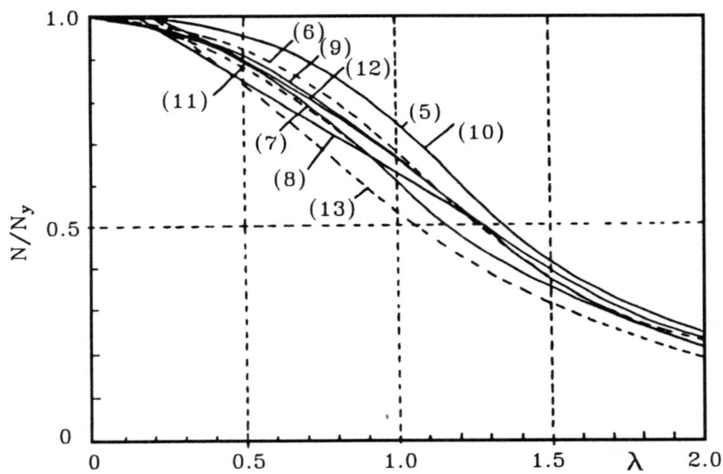

Fig. 9a.2 Column Curves For Ultimate Limit States Design
(5), AS. Hot-formed, (6), AS. Cold-formed, (7), STAS 10108 81, (8), AIJ 89b, (9), AISC LRFD 86, (10), CSA S16.1. Hot-formed, (11), CSA S16.1. Cold-formed, (12), EC 3 88. Hot-formed, (13), EC 3 88 Cold-formed

9b. D/t Limitations, Local Buckling and Column Strength Interaction for CHS

SPECIFICATION PROVISIONS & EXPLANATION

AUSTRALIA

CHAPTER EDITOR
Greg J. Hancock, University of Sydney

Axial Compression

AS 1250 81

If a circular hollow section is used as a strut, the effective geometrical properties of the section shall be calculated using the specified outside diameter, D, and a value for wall thickness, t, which shall be the lesser of

$$\frac{23000\ t}{\left(\frac{D}{t}\right)\sigma_y} \quad \text{and} \quad t \qquad (9b.A1)$$

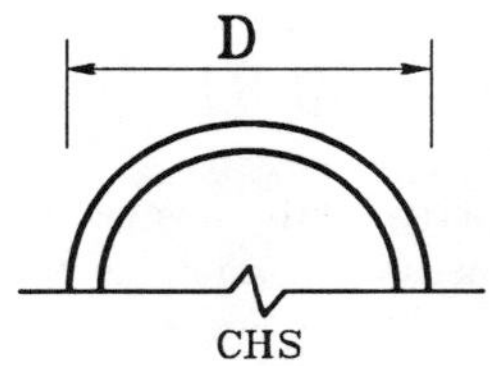

Fig. 9b.A1 Outside Diameter

AS DR87164 87

In limit states design, the following effective diameters are imposed

$$\text{for } \frac{D_i}{t} \leq \frac{22750}{\sigma_y} \qquad D_e = D_i \qquad (9b.A2)$$

$$\text{for } \frac{D_i}{t} > \frac{22750}{\sigma_y} \qquad D_e = \frac{22750}{\sigma_y}\,t \qquad (9b.A3)$$

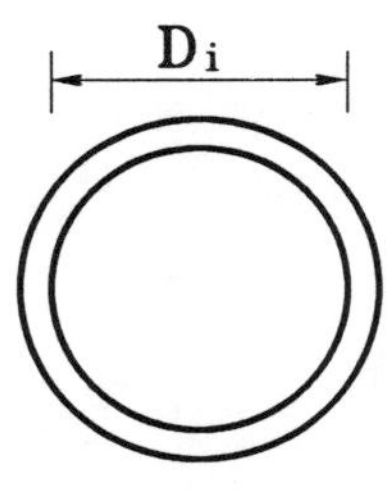

Fig. 9b.A2 Inside Diameter

where D_i is the inside diameter of the section.

For interaction between local and flexural buckling, AS 1250 81 uses the effective section concept based on an effective thickness whereas AS DR87164 87 uses the effective yield concept based on a Q factor. The Perry formulation has been modified in AS DR87164 87 by incorporation of $\sqrt{Q}$ in the normalized slenderness, λ, as well as a reduced stub column strength equal to $\alpha_d Q N_y$.

To account for the interaction between local and flexural buckling the following equations are used.

$$Q = A_e/A \tag{9b.A4}$$

$$\lambda_Q = \sqrt{Q} \cdot \lambda \tag{9b.A5}$$

$$\lambda_m = \lambda_Q + \frac{\alpha_a \alpha_b}{90} \tag{9b.A6}$$

$$\alpha_a = \frac{90\ (\lambda_Q - 0.15)}{3.86\lambda_Q^2 - 0.656\lambda_Q + 0.976} \tag{9b.A7}$$

The value of A_e in Eq. 9b.A4 is based on the effective diameter defined in Eqs. 9b.A2 and 9b.A3.

Bending

AS 1250 81

If a circular hollow section is used as a beam, the effective geometrical properties of the section shall be calculated using the specified outside diameter and a value for wall thickness which shall be the lesser of

$$\frac{23000\ t}{\left(\frac{D}{t}\right)\sigma_y} \quad \text{and} \quad t \tag{9b.A8}$$

AS DR87164 87

for $\frac{D_i}{t} < \frac{17500}{\sigma_y}$

$$M_{ult} = M_p \tag{9b.A9}$$

for $\frac{17500}{\sigma_y} \leq \frac{D_i}{t} \leq \frac{30000}{\sigma_y}$

$$M_{ult} = M_y + \left[\frac{\frac{30000}{\sigma_y} - \frac{D_i}{t}}{\frac{17500}{\sigma_y}}\right](M_p - M_y) \tag{9b.A10}$$

for $\frac{D_i}{t} > \frac{30000}{\sigma_y}$

M_{ult} = lesser of

$$M_y\sqrt{\frac{30000}{\sigma_y}\frac{t}{D_i}} \quad \text{or} \quad M_y\left[\frac{30000}{\sigma_y}\frac{2t}{D_i}\right]^2 \tag{9b.A11}$$

EAST EUROPE

CHAPTER EDITOR
Dan Mateescu, Polytechnic Institute of Timisoara

STAS 10108/0 78

For CHS, the D/t ratio is limited to

$$\frac{D}{t} \leq 120 \left(\frac{230}{\sigma_y}\right) \tag{9b.E1}$$

where σ_y is the yield stress.

For higher D/t ratios, the compression stress shall not exceed the following limit

$$\sigma \leq 0.14E \frac{t}{D} \tag{9b.E2}$$

JAPAN

CHAPTER EDITOR

Ben Kato, Toyo University

AIJ 80

In allowable stress design the following width-to-thickness ratio limitations is imposed

$$\frac{D}{t} \leq \frac{23520}{\sqrt{\sigma_y}} \qquad (9b.J1)$$

where σ_y is the minimum specified yield stress (MPa).

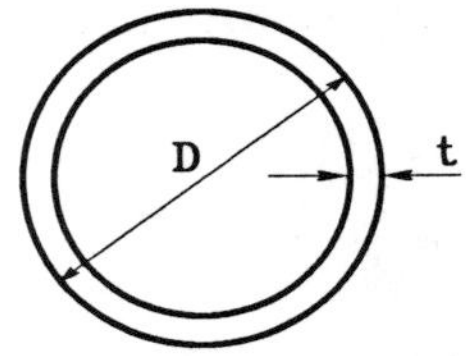

Fig. 9b.J1 Tube Dimensions

AIJ 89b

The minimum D/t ratio is given, for compression by

$$D/t \leq 90 \qquad (9b.J2)$$

and for bending by

$$D/t \leq 92 \qquad (9b.J3)$$

If in compression the D/t exceeds this limit, the local buckling strength can be calculated for cold-formed sections by means of the formula

$$N_l = \frac{N_y}{0.777 + 2.12\left(\frac{\sigma_y}{E}\right)\left(\frac{D}{t}\right)} \qquad (9b.J4)$$

with a ϕ factor equal to 0.9

This formula has been derived statistically from experimental data. The design column strength should be the smaller of the column strength calculated with flexural buckling or with local buckling.

NORTH AMERICA

CHAPTER EDITOR

Donald R. Sherman, University of Wisconsin-Milwaukee

Axial Compression

The various specifications and guidelines use different formats for expressing the criteria for thin CHS under axial compression. In the following summary, some rearrangement has been made so that the equations are in terms of the nondimensional wall slenderness parameter $\alpha_c = \frac{E}{\sigma_y}\left(\frac{t}{D}\right)$ whereever possible.

AISC ASD 78

for $8.79 > \alpha_c > 2.23$

$$f_c = (0.0226\alpha_c + 0.4)\sigma_y \qquad (9b.N1)$$

Comment: The allowable stress is the smaller of Eqs. 9a.N1 and 9b.N1. No interaction is considered.

AISC LRFD 86

for $8.79 > \alpha_c > 2.23$

$$N_{cr} = (0.0379\alpha_c + 2/3)N_y \qquad (9b.N2)$$

Comment: The inelastic local buckling load, N_{cr}, is substituted for N_y in Eq. 9a.N3b to account for interaction and the column slenderness is multiplied by $\sqrt{N_{cr}/N_y}$.

AISI 76

Equation 9b.N1 is used for local buckling.

Comment: The allowable stress is the smaller of Eqs. 9a.N1 and 9b.N1. No interaction is considered.

AISI 86

for $9.0 > \alpha_c > 2.27$ and $\lambda < \sqrt{2}$

$$A_e/A = 1-(1-\lambda^2/2)(0.333 - 0.037\alpha_c) \qquad (9b.N3)$$

Comment: Equation 9a.N1a is multiplied by A_e/A to account for interaction. For elastic buckling, Eq. 9a.N1b is used and no local buckling interaction is considered.

API RP2A 86

for $60 < D/t < 300$

$$\sigma_{cr} = (1.64 - 0.23(D/t)^{0.25})\sigma_y < \frac{0.6E}{D/t} \qquad (9b.N4)$$

Comment: The inelastic local buckling stress, σ_{cr} is substituted for σ_y in Eq. 9a.N1 to account for interaction.

CSA S16.1 M89 84b

Comment: Local buckling is not considered and α_c must be greater than 8.70.

REGIONAL COMPARISON

There are three fundamental differences in the local buckling criteria just summarized.

1. The equation used for local buckling.
2. The wall slenderness parameter used.
3. Whether an interaction between local and column buckling is considered.

Fig. 9b.N1 shows the critical stress for local buckling plotted against the nondimensional wall slenderness parameter, α_c. Factors of safety have been removed from the ASD criteria, and the AISI 86 criteria is not included due to its different form. It is evident that all but API RP2A 86 are based on the same critical stress criteria. API RP2A 86 expresses the slenderness of a CHS wall simply as the D/t ratio. The other criteria all use a modification of the nondimensional parameter.

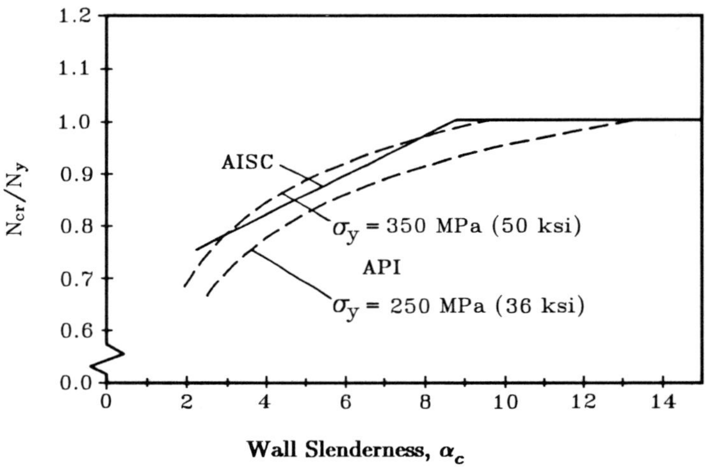

Fig. 9b.N1 Local Buckling Of CHS Under Axial Compression

The older criteria, AISC ASD 78 and AISI 76, do not consider any interaction between local and flexural buckling and use the smaller of two to obtain an allowable stress for axial compression. Interaction in the other criteria is accounted for by replacing the yield load in the column buckling equations by the inelastic local buckling load. In API RP2A 86 and AISC LRFD 86, the replacement is made in all terms where the yield strength appears, including the nondimensional column slenderness. AISI 86 modifies only the strength and not the column slenderness so that the limit between inelastic and elastic flexural buckling does not change.

Bending

AISC ASD 78

for $\alpha_c > 8.79$

$$f = 0.66\sigma_y \tag{9b.N5a}$$

for $8.79 > \alpha_c > 2.23$

$$f = (0.0226\alpha_c + 0.4)\sigma_y \tag{9b.N5b}$$

AISC LRFD 86

for $\alpha_c > 14.0$

$$M_{ult} = M_p \tag{9b.N6a}$$

for $14.0 > \alpha_c > 3.23$

$$M_{ult} = (0.0207\alpha_c + 1)\sigma_y Z \tag{9b.N6b}$$

for $3.23 > \alpha_c > 2.23$

$$M_{ult} = 0.330\alpha_c\sigma_y Z \tag{9b.N6c}$$

AISI 76

for $\alpha_c > 8.79$

$$f = 0.72\sigma_y \tag{9b.N7}$$

Eq. 9b.N5b for smaller α_c

AISI 86

for $\alpha_c > 14.3$

$$f = 0.75\sigma_y \tag{9b.N8a}$$

for $14.3 > \alpha_c > 3.13$

$$f = (0.581 + 0.0120\alpha_c)\sigma_y \tag{9b.N8b}$$

for $3.13 > \alpha_c > 2.27$

$$f = 0.196\alpha_c\sigma_y \tag{9b.N8c}$$

API RP2A 86

for D/t < 60

$$f = 0.66\sigma_y \tag{9b.N9a}$$

for 60 < D/t < 300

$$f = 0.66[1.64-0.23(D/t)^{0.25}]\sigma_y < 0.396E/(D/t) \tag{9b.N9b}$$

CSA S16.1 M89 84b

for $\alpha_c > 11.1$

$$M_{ult} = M_p \tag{9b.N10a}$$

for $11.1 > \alpha_c > 8.70$

$$M_{ult} = M_y \tag{9b.N10b}$$

Comment: Local buckling is not considered and α_c must be greater than 8.70.

REGIONAL COMPARISON

The various criteria are compared in Fig. 9b.N2 in a plot of M_{ult}/M_p vs. the wall slenderness α_c. The ASD provisions have been multiplied by 1.67 for the comparison. AISC ASD 78, API RP2A 86 and AISI 76 use the same local buckling stress as they use for axial compression. The ASD criteria do not take full advantage of the high shape factor in CHS and permit increased allowable stresses ranging from the 10% used for W shapes to 25% used for solid rectangles.

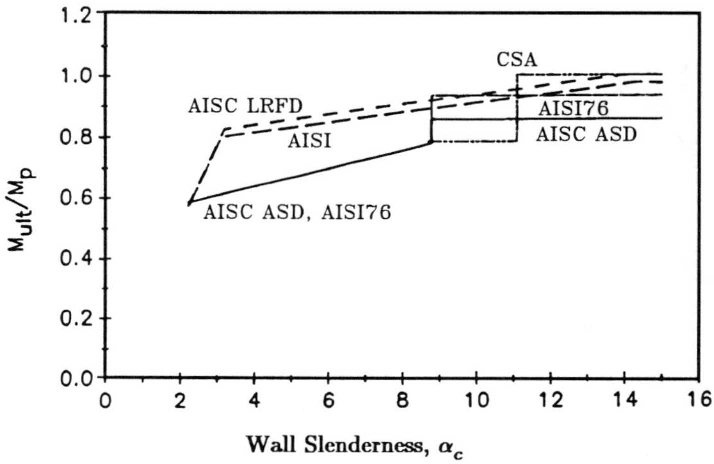

Fig. 9b.N2 Local Buckling Of CHS Under Bending

WEST EUROPE

CHAPTER EDITOR
Jacques Rondal, University of Liege

Since there are many West European national standards, only the rules specified by EC 3 88 are presented. Generally, the West European national standards are more or less in line with the rules given in EC 3 88.

EC 3 88

The maximum D/t ratio is given, for compression or bending, by the relation

$$D/t \le 90\,\epsilon^2 \qquad (9b.W1)$$

where

$$\epsilon = \sqrt{235/f_y} \qquad (9b.W2)$$

f_y = yield strength (MPa)

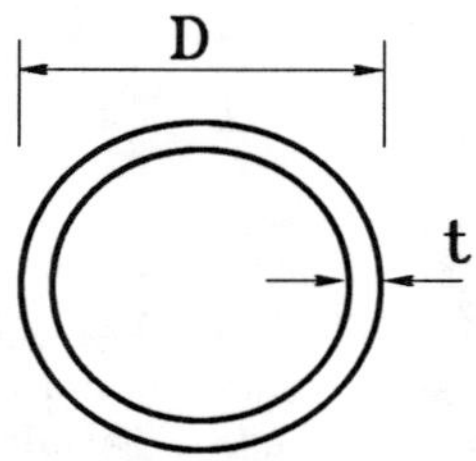

Fig. 9b.W1 Tube Dimension

For bending, the maximum D/t ratio is given by the relations:

1) Plastic cross section (plastic analysis):

$$D/t \le 50\,\epsilon^2 \qquad (9b.W3)$$

2) Compact cross section (plastic moment, elastic analysis):

$$D/t \le 70\,\epsilon^2 \qquad (9b.W4)$$

3) Semi-compact cross section (elastic moment, elastic analysis):

$$D/t \le 90\,\epsilon^2 \qquad (9b.W5)$$

INTERNATIONAL COMPARISONS & COMMENT

WHAT ARE THE DIFFERENCES?

Diameter-to-thickness ratio limitations provided by each country (region) are compared in Table 9b.1. All specifications were adjusted to conform to the same format, with σ_y given in MPa.

Definitions of diameter are somewhat different by specifications as shown in Fig. 9b.1, and they are identified by letters (a, b) in the table.

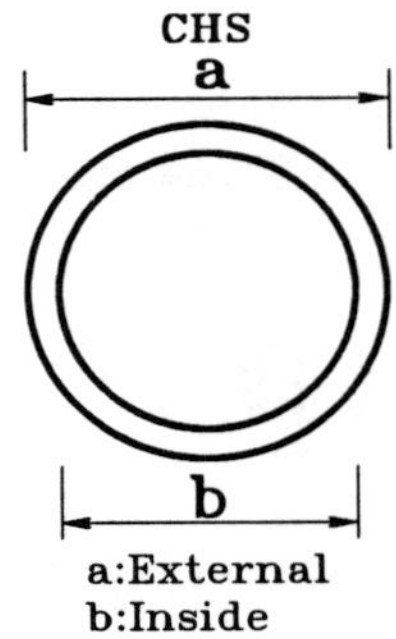

Fig. 9b.1 Diameter Definitions

If the D/t limit has been exceeded there are considerable differences in the standards. Some do not permit the use of these thin tubes while others use various strength criteria. Standards that permit the use of thin tubes also differ as to whether an interaction with local buckling is considered or whether the strength is the lower of the column and local buckling strength.

WHY ARE THERE DIFFERENCES?

There are essentially no differences in the D/t limits for axial compression because the criteria are based on a well-established data base that has been accepted in most of the world. Older codes used the same criteria for compressive stresses in bending. Two newer North American codes (AISC LRFD 86 and AISI 86) use a newer data base (Sherman, 1986).

The differences in strength criteria when the D/t limit has been exceeded is the result of two viewpoints of code writers: (a)the limited application of thin tubes for structures covered in the scope of the code, (b)degree of confidence in the data base due to the scatter resulting from the imperfection sensitivity of cylinders under compression.

Table 9b.1 Comparison of D/t Limitations (CHS)

Code	Axial compression	Bending		Diameter definition
		Plastic limit	Yield limit	
AS 1250 81	$\frac{23000}{\sigma_y}$	–	$\frac{23000}{\sigma_y}$	a
AS DR87164 87	$\frac{22750}{\sigma_y}$	$\frac{17500}{\sigma_y}$	$\frac{30000}{\sigma_y}$	b
STAS 10108/0 78	$\frac{27600}{\sigma_y}$	–	$\frac{27600}{\sigma_y}$	a
AIJ 80	$\frac{23520}{\sigma_y}$	–	$\frac{23520}{\sigma_y}$	a
AIJ 89b	90	–	92	a
AISC ASD 78	$\frac{22750}{\sigma_y}$	$\frac{22750}{\sigma_y}$	$\frac{22750}{\sigma_y}$	a
AISI 86	$\frac{22600}{\sigma_y}$	$\frac{13980}{\sigma_y}$	$\frac{63880}{\sigma_y}$	a
API RP2A 86	60	60	60	a
AISI 76	$\frac{22750}{\sigma_y}$	$\frac{22750}{\sigma_y}$	$\frac{22750}{\sigma_y}$	a
AISC LRFD 86	$\frac{22750}{\sigma_y}$	$\frac{14530}{\sigma_y}$	$\frac{62970}{\sigma_y}$	a
CSA S16.1 M89 84b	$\frac{23000}{\sigma_y}$	$\frac{18020}{\sigma_y}$	$\frac{23000}{\sigma_y}$	a
EC 3 88	$\frac{21150}{\sigma_y}$	$\frac{16950}{\sigma_y}$	$\frac{21150}{\sigma_y}$	a

9c. b/t Limitations, Local Buckling and Column Strength Interaction for SHS

SPECIFICATION PROVISIONS & EXPLANATION

AUSTRALIA

CHAPTER EDITOR
Greg J. Hancock, University of Sydney

Axial Compression

AS 1250 81

In allowable stress design, the following effective widths are imposed

for $\frac{b}{t} \leq \frac{635}{\sqrt{\sigma_y}}$ $$b_e = b \qquad (9c.A1)$$

for $\frac{b}{t} > \frac{635}{\sqrt{\sigma_y}}$ $$b_e = \frac{635\,t}{\sqrt{\sigma_y}} \qquad (9c.A2)$$

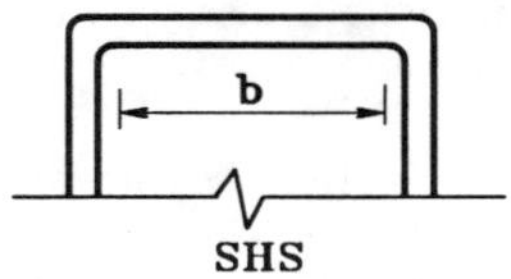

Fig. 9c.A1 Flat Width

For non-fully effective sections, the effective section must be used in design when calculating both the section area and second moment of area for use in the determination of λ and i.

AS DR87164 87

In limit states design, the following effective widths are imposed

Cold-formed non-stress relieved

for $\frac{b}{t} \leq \frac{632}{\sqrt{\sigma_y}}$ $$b_e = b \qquad (9c.A3)$$

for $\frac{b}{t} > \frac{632}{\sqrt{\sigma_y}}$ $$b_e = \frac{632\,t}{\sqrt{\sigma_y}} \qquad (9c.A4)$$

Hot-formed and cold-formed stress relieved

for $\frac{b}{t} \leq \frac{712}{\sqrt{\sigma_y}}$ $$b_e = b \qquad (9c.A5)$$

for $\frac{b}{t} > \frac{712}{\sqrt{\sigma_y}}$ $$b_e = \frac{712\,t}{\sqrt{\sigma_y}} \qquad (9c.A6)$$

For interaction between local and flexural buckling, AS 1250 81 uses the effective section concept whereas AS DR87164 87 uses the effective yield concept based on a

Q factor. The Perry formulation has been modified in AS DR87164 87 by the incorporation of $\sqrt{Q}$ in the normalized slenderness, λ, as well as a reduced column strength equal to $\alpha_d Q N_y$.

To account for the interaction between local and flexural buckling the following equations are used

$$Q = A_e/A \tag{9c.A7}$$

$$\lambda_Q = \sqrt{Q} \cdot \lambda \tag{9c.A8}$$

$$\lambda_m = \lambda_Q + \frac{\alpha_a \alpha_b}{90} \tag{9c.A9}$$

$$\alpha_a = \frac{90(\lambda_Q - 0.15)}{3.86\lambda_Q^2 - 0.656\lambda_Q + 0.976} \tag{9c.A10}$$

For all cold-formed tubes, the yield strength is defined in AS 1163 81 to be the tensile strength of a longitudinal specimen taken from the flat faces after forming of the tube.

The justification for the width-to-thickness limit for cold-formed rectangular and square hollow sections is given in Key and Hancock (1988).

Bending

AS 1250 81

for $\frac{b}{t} < \frac{318}{\sqrt{\sigma_y}}$

$$f = 0.66\sigma_y \tag{9c.A11}$$

for $\frac{318}{\sqrt{\sigma_y}} \leq \frac{b}{t} \leq \frac{635}{\sqrt{\sigma_y}}$

$$f = \left(0.72 - \frac{0.12}{635}\frac{b}{t}\sqrt{\sigma_y}\right)\sigma_y \tag{9c.A12}$$

for $\frac{b}{t} > \frac{635}{\sqrt{\sigma_y}}$

$$f = \frac{381}{\sqrt{\sigma_y}}\frac{t}{b}\sigma_y \tag{9c.A13}$$

AS DR87164 87

for $\frac{b}{t} < \frac{474}{\sqrt{\sigma_y}}$

$$M_{ult} = M_p \tag{9c.A14}$$

for $\frac{474}{\sqrt{\sigma_y}} \leq \frac{b}{t} \leq \frac{632^*}{\sqrt{\sigma_y}}$

$$M_{ult} = M_y + \left[\frac{\frac{632}{\sqrt{\sigma_y}} - \frac{b}{t}}{\frac{158}{\sqrt{\sigma_y}}}\right](M_p - M_y) \tag{9c.A15}$$

for $\frac{b}{t} > \frac{632^*}{\sqrt{\sigma_y}}$

$$M_{ult} = M_y\left(\frac{632}{\sqrt{\sigma_y}}\frac{t}{b}\right) \tag{9c.A16}$$

*For hot-formed and cold-formed stress relieved sections, replace 632 by 712.

EAST EUROPE

CHAPTER EDITOR
Dan Mateescu, Polytechnic Institute of Timisoara

STAS 10108/0 78

For SHS, if the width and the height of RHS are denoted by b and h respectively, the limitations are given by

$$\frac{h}{t} \leq 20\sqrt{\left[4+3\left(1-\frac{b}{h}\right)\right]\frac{230}{\sigma_y}} + 0.2\Lambda \qquad (9c.E1)$$

$$\frac{b}{t} \leq 20\sqrt{\left[4+3\left(1-\frac{h}{b}\right)\right]\frac{230}{\sigma_y}} + 0.2\Lambda \qquad (9c.E2)$$

where

$\Lambda = \frac{k\ell}{i}$ is the slenderness ratio.

σ_y = the yield stress

If $\sigma = \frac{N}{A} \leq \psi R$, these limit ratios can be increased by multiplying them by the factor $\sqrt{\frac{\psi R}{\sigma}}$. However the increase must not exceed 25%.

where

R = design yield stress (Eq. 9a.E2)

ψ = buckling coefficient (Eq. 9a.E3)

JAPAN

CHAPTER EDITOR
Ben Kato, Toyo University

AIJ 80

In allowable stress design, the following width-to-thickness ratio limitation is imposed for uniform compression

$$\frac{b}{t} \leq \frac{733}{\sqrt{\sigma_y}} \quad (9c.J1)$$

where

b = flat part of the flange (Fig. 9c.J1)
σ_y = minimum specified yield stress (MPa)

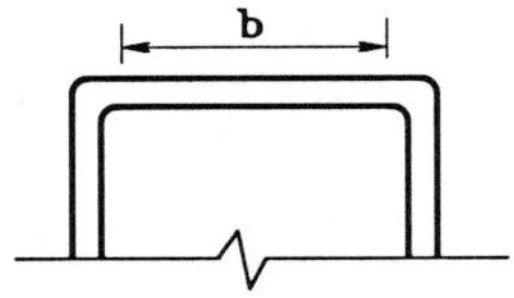

Fig. 9c.J1 Flat Width

AIJ 89b

The maximum b/t ratio is given, for compression, by

Hot-formed sections: $b/t \leq \frac{733}{\sqrt{\sigma_y}}$ (9c.J2)

Cold-formed sections: $b/t \leq 36$ (9c.J3)

where σ_y is given in MPa.

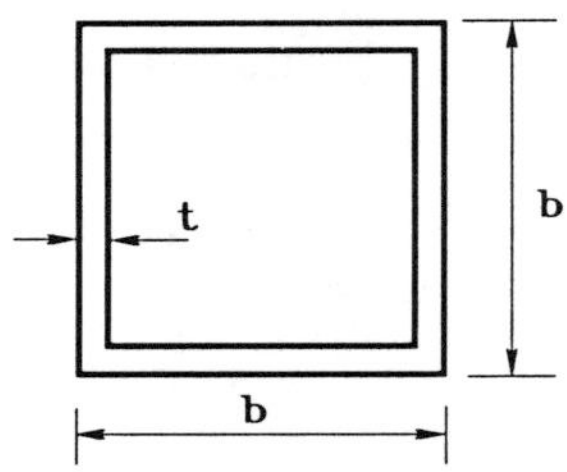

Fig 9c.J2 Width and Tickness

For bending, the maximum b/t ratio is given by

Hot-formed sections: $b/t \leq \frac{911}{\sqrt{\sigma_y}}$ (9c.J4)

Cold-formed sections: $b/t \leq 43$ (9c.J5)

The local buckling strength must be calculated by means of the following

Hot-formed and hot-finished sections:

$$N_l = \frac{N_y}{0.710 + 0.17\left(\frac{\sigma_y}{E}\right)\left(\frac{b}{t}\right)^2} \quad (9c.J6)$$

Cold-formed sections:

$$N_l = \frac{N_y}{0.778 + 0.20\left(\frac{\sigma_y}{E}\right)\left(\frac{b}{t}\right)^2} \tag{9c.J7}$$

In these formulas, a resistant factor ϕ=0.9 is used. All these formulas have been based on statistically evaluated experimental data.

The design column strength should be the smaller of the flexural column strength and the local buckling strength.

NORTH AMERICA

CHAPTER EDITOR
Donald R. Sherman, University of Wisconsin-Milwaukee

Axial Compression

All of the criteria use a b/t ratios based on the flat width of the side of a tube to determine if local buckling must be considered. Effective widths b_e are determined for the sides that exceed the b/t limit and an effective area A_e of the cross section is determined. Column equations are then modified by factors that include the effective area. A nondimensional wall slenderness parameter $\alpha_s=\sqrt{E/\sigma_y}\ (\frac{t}{b})$ is used.

AISC ASD 78

for $\alpha_s < 0.716$

$$b_e/b = \frac{1.49\alpha_s}{\sqrt{\sigma/\sigma_y}}\left(1 - \frac{0.295\alpha_s}{\sqrt{\sigma/\sigma_y}}\right) < 1 \qquad (9c.N1)$$

Comment: σ is the average axial allowable stress on the gross area. $\sigma_y A_e/A$ is substituted for σ_y in Eq. 9a.N1.

AISC LRFD 86

for $\alpha_s < 0.716\sqrt{1-\sigma_r/\sigma_y}$

$$b_e/b = \frac{1.91\alpha_s}{\sqrt{\sigma_f/\sigma_y}}\left(1 - \frac{0.381\alpha_s}{\sqrt{\sigma_f/\sigma_y}}\right) < 1 \qquad (9c.N2)$$

Comment: σ_f is the average axial stress at factored loads on the gross area and σ_r is the residual stress. A footnote states that σ_r is 10 ksi for hot-rolled shapes and 16.5 ksi for welded shapes. No advice is given for cold-formed SHS. However, since the residual stresses are tension in the middle of a wall of an SHS, it is probably satisfactory to use σ_r=0. $\sigma_y A_e/A$ is substituted for σ_y in Eq. 9a.N2.

AISI 76

Identical to AISC ASD 78 (Eq. 9c.N1).

AISI 86

for $\alpha_s < 0.781\sqrt{\sigma_n/\sigma_y}$

$$b_e/b = \frac{1.90\alpha_s}{\sqrt{\sigma_n/\sigma_y}}\left(1 - \frac{0.418\alpha_s}{\sqrt{\sigma_n/\sigma_y}}\right) \qquad (9c.N3)$$

Comment: for $\lambda > \sqrt{2}$ $\sigma_n = \frac{\sigma_y}{\lambda^2}$

for $\lambda < \sqrt{2}$ $\sigma_n = \sigma_y\left(1 - \frac{\lambda^2}{4}\right)$

Eq. 9a.N1 is multiplied by A_e/A

CSA S16.1 M89 84b

Eq. 9c.N3 (with different definition of σ_n) (9c.N4)

Comment: σ_n is the average axial stress computed on the basis of the effective area. Eqs. 9a.N4 or 9a.N5 are multiplied by A_e/A.

REGIONAL COMPARISON

Aside from the different column equations used in the various specifications, there are two differences in the local buckling criteria. These are the equation used for b_e and the manner in which the column equations are modified.

There is really no difference between Eqs. 9c.N1 and 9c.N2 if σ_r is taken as zero. In AISC ASD 78, σ is at service loads whereas in AISC LRFD 86 it is at factored loads. If the stress in AISC ASD 78 is multiplied by 1.67 to obtain the ultimate condition, the equation is the same as AISC LRFD 86. In AISC ASD 78 and AISC LRFD 86, the effective width is used to modify both the column strength equation and the column slenderness ratio.

In AISI 86 and CSA S16.1 M89 84b, it modifies only the strength equation. This does not produce significant differences for short columns, but is more of a factor as the column length increases.

Bending

The various specifications and guidelines use different formats for expressing the criteria for bending strengths of SHS. In the following summary, the criteria have been rearranged as allowable moments M_a in the ASD formats and ultimate moments without resistance factors M_{ult} in LRFD formats. The criteria are also expressed in terms of the wall slenderness parameter, α_s.

AISC ASD 78

for $\alpha_s > 0.896$ $M_a = 0.66Z\sigma_y$ (9c.N5a)

for $0.896 > \alpha_s > 0.716$ $M_a = 0.6Z\sigma_y$ (9c.N5b)

for $\alpha_s < 0.716$ $M_a = 0.6Z_e\sigma_y$ (9c.N5c)

Comment: Z_e is the effective elastic section modulus using Eq. 9c.N1 for the compression flange.

AISC LRFD 86

for $\alpha_s > 0.896$

$$M_{ult} = M_p \tag{9c.N6a}$$

for $0.896 > \alpha_s > 0.716$

$$M_{ult} = M_p-(M_p-M_r)(\frac{3.55}{\alpha_s}-3.96) \tag{9c.N6b}$$

where $M_r = Z\sigma_y$

for $\alpha_s < 0.716$

$$M_{ult} = Z_e\sigma_y \tag{9c.N6c}$$

Comment: Z_e is the effective elastic section modulus using Eq. 9c.N2 for the compression flange.

AISI 76

for $\alpha_s > 0.811$

$$M_a = CZ\sigma_y \tag{9c.N7a}$$

where

C = 0.72 for major axis bending of rectangular tubes
= 0.69 for bending of square tubes
= 0.66 for minor axis bending of rectangular tubes

for $0.811 > \alpha_s > 0.695$

$$M_a = 0.60Z\sigma_y \tag{9c.N7b}$$

for $\alpha_s < 0.695$

$$M_a = 0.60Z_e\sigma_y \tag{9c.N7c}$$

Comment: Z_e is the effective elastic section modulus using Eq. 9c.N1 for the compression flange.

AISI 86

for $\alpha_s > 0.782$

$$M_a = 0.60Z\sigma_y \tag{9c.N8a}$$

for $\alpha_s < 0.782$

$$M_a = 0.60Z_e\sigma_y \tag{9c.N8b}$$

Comment: Z_e is the effective elastic section modulus using Eq. 9c.N3 for the compression flange. An alternate procedure which accounts for inelastic reserve capacity may be used.

CSA S16.1 M89 84b

for $\alpha_s > 0.852$

$$M_{ult} = M_p \tag{9c.N9a}$$

for $0.852 > \alpha_s > 0.668$

$$M_{ult} = M_y \tag{9c.N9b}$$

for $\alpha_s < 0.668$

$$M_{ult} = Z_e \sigma_y \tag{9c.N9c}$$

Comment: Z_e is the effective elastic section modulus using $b_e/b = 1.50\ \alpha_s$ for the compression flange. An alternate procedure for Z_e based on Eq. 9c.N4 is provided.

REGIONAL COMPARISON

The moment capacities for rectangular compact and noncompact sections bending about the major axis are compared in Fig. 9c.N1. Allowable moments, M_a, in the ASD criteria have been multiplied by 1.67 and a shape factor of 1.20 is used. The comparison has not been extended to thin walled sections since this comparison is more dependent on section properties.

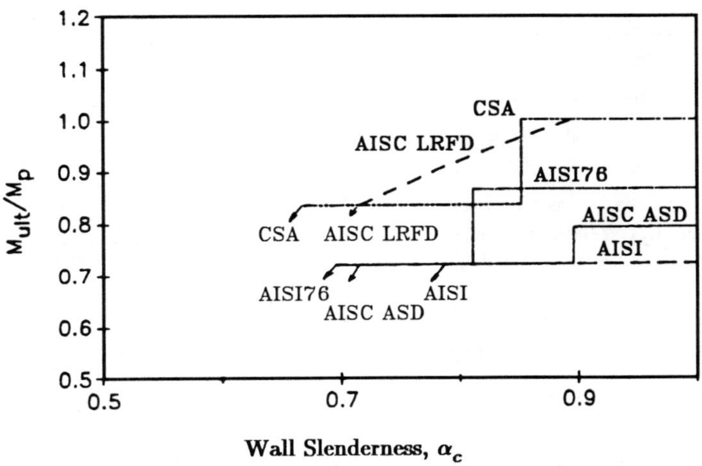

Fig. 9c.N1 Inelastic Local Buckling of SHS Under Bending

WEST EUROPE

CHAPTER EDITOR
Jacques Rondal, University of Liege

Since there are many West European national standards, only the rules specified by EC 3 88 are presented. Generally, the West European national standards are more or less in line with the rules given in EC 3 88.

EC 3 88

The maximum b/t ratio is given, for compression, by the relation

$$b/t \leq 45\,\epsilon \qquad (9c.W1)$$

where

f_y = the yield strength (MPa)

$$\epsilon = \sqrt{235/f_y} \qquad (9c.W2)$$

For bending, the maximum b/t ratio is given by

1) Plastic cross section (plastic analysis):

$$b/t \leq 33\,\epsilon \qquad (9c.W3)$$

2) Compact cross section (plastic moment, elastic analysis):

$$b/t \leq 38\,\epsilon^2 \qquad (9c.W4)$$

3) Semi-compact cross section (elastic moment, elastic analysis):

$$b/t \leq 45\,\epsilon^2 \qquad (9c.W5)$$

If the b/t ratio is greater than the maximum b/t ratio, the effective cross-section method must be used.

Table 9c.W1 gives the equations for the calculation of the effective width b_e.

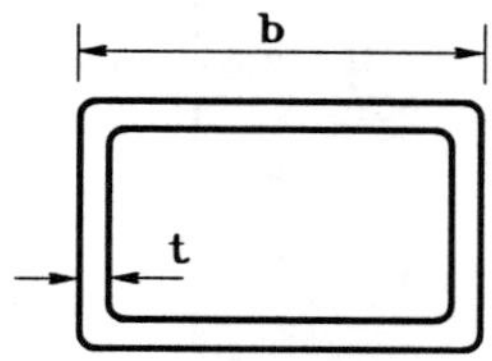

Fig. 9c.W1 Width and Thickness

The reduction factor ρ is obtained from the following equations

for $\lambda_p \leq 0.673$

$$\rho = 1 \qquad (9c.W6)$$

for $\lambda_p > 0.673$

$$\rho = (\lambda_p - 0.22)/\lambda_p^2 \qquad (9c.W7)$$

where λ_p is the plate slenderness given by

$$\lambda_p = \sqrt{f_y/f_e} = \frac{b/t}{28.4\ E\ \sqrt{k_\sigma}} = \frac{1.056}{\sqrt{k_r}}\sqrt{\frac{f_y}{E}}\left(\frac{b}{t}\right) \qquad (9c.W8)$$

where k_r is the buckling factor corresponding to the stress ratio ψ from Table 9c.W1 as appropriate.

The interaction between local buckling and flexural column strength is performed by means of the Q factor method where the effective width is used to modify both the column strength equation and the column slenderness ratio. In this respect f_y is multiplied by A_e/A in Eq. 9a.W5 and N_y is also multiplied by A_e/A in Eq. 9a.W2.

Table 9c.W1 Effective Width and Buckling Factors

Internal compression elements	
Stress distribution(compression positive)	Effective width b_e
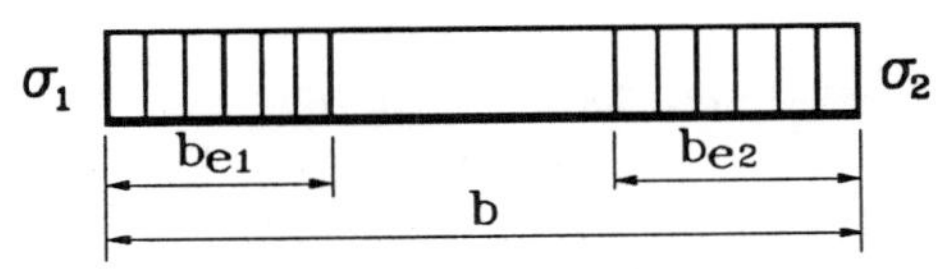	$b_e = \rho\, b$ $b_{e1} = 0.5\, b_e$ $b_{e2} = 0.5\, b_e$
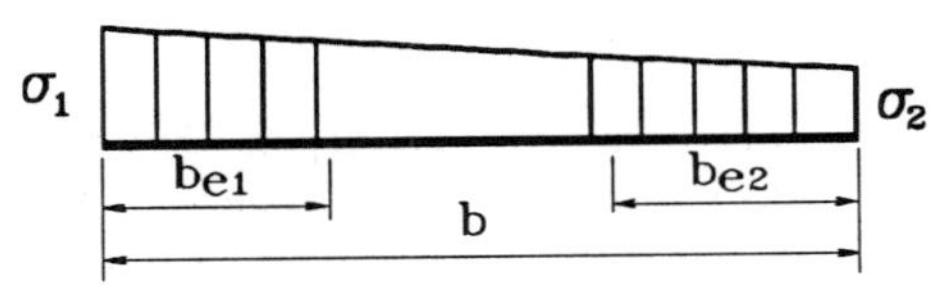	$b_e = \rho\, b$ $b_{e1} = \frac{2\, b_e}{5-\psi}$ $b_{e2} = b_e - b_{e1}$ $\psi = \frac{\sigma_2}{\sigma_1}$
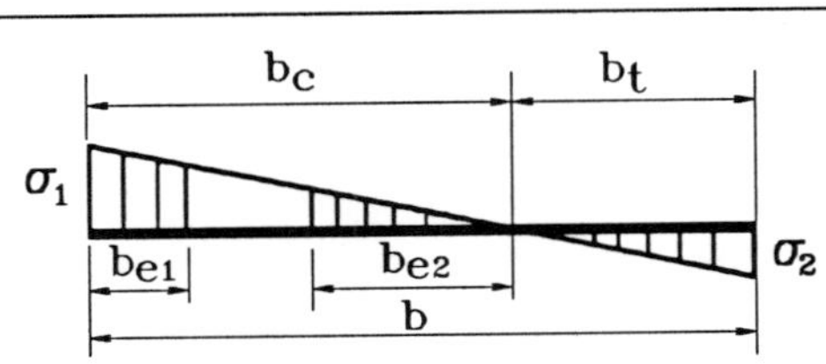	$b_e = \rho\, b_c$ $b_{e1} = 0.4\, b_e$ $b_{e2} = 0.6\, b_e$

$\psi=\sigma_2/\sigma_1$	+1	$1 > \psi > 0$	0	$0 > \psi > -1$	−1
Buckling factor k_r	4.0	$\frac{8.2}{1.05+\psi}$	7.81	$7.81 - 6.29\psi + 9.78\psi^2$	23.9

Alternatively:

$$k_r = \frac{16}{\sqrt{(1+\psi)^2 + 0.112(1-\psi)^2} + (1+\psi)}$$

INTERNATIONAL COMPARISONS & COMMENT

WHAT ARE THE DIFFERENCES?

Width-to-thickness ratio limitations provided by each country (region) are compared in Table 9b.1. All specifications were adjusted to conform to the same format, where σ_y is given in MPa.

Definitions of width are somewhat different by specifications as shown in Fig. 9c.1 and they are identified by letters (a, b, c) in the table.

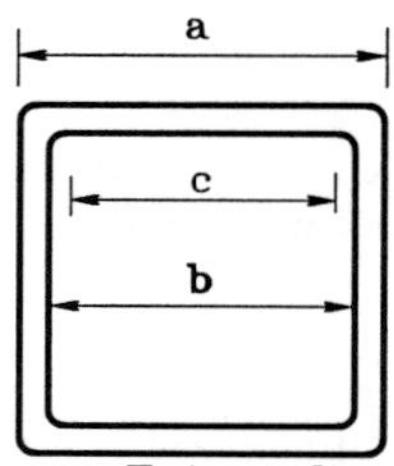

a: External
b: Internal c: Flat

Fig. 9c.1 Width Definitons

In all the specifications considered, other than the Japanese, the interaction between local buckling and flexural column buckling has been included for the postbuckling range of local buckling. In the Japanese code, the axial load carrying capacity of an SHS member is the smaller of the maximum plate strength and the flexural buckling load.

For those codes which consider interaction between local buckling and flexural column buckling, an effective width formula is used to compute an effective area. However, some codes calculate the effective area at yield whilst others calculate the effective area at the stress level pertaining in the column.

WHY ARE THERE DIFFERENCES?

Various standards have elected to use equations for effective widths or effective areas in one of several different forms (von Kármán, Winter). Combined with somewhat different data bases, these result in slight variations in the b/t limits for local buckling. Further differences in how the interaction with column buckling are considered are the result of a desire for simplicity and the limited availability or use of very thin tubes in some regions.

Table 9c.1 Comparison of b/t Limitations (SHS)

Code	Axial compression	Bending		Plate width definition
		Plastic limit	Elastic limit	
AS 1250 81	$\frac{635}{\sqrt{\sigma_y}}$	$\frac{318}{\sqrt{\sigma_y}}$	$\frac{635}{\sqrt{\sigma_y}}$	b
AS DR87164 87	$632/\sqrt{\sigma_y}^{*}$ $712/\sqrt{\sigma_y}^{**}$	$\frac{474}{\sqrt{\sigma_y}}$	$632/\sqrt{\sigma_y}^{*}$ $712/\sqrt{\sigma_y}^{**}$	b
STAS 10108/0 78	–	–	–	–
AIJ 80	$\frac{733}{\sqrt{\sigma_y}}$	–	$\frac{733}{\sqrt{\sigma_y}}$	c
AIJ 89b	36* $733/\sqrt{\sigma_y}^{**}$	35* $743/\sqrt{\sigma_y}^{**}$	43* $911/\sqrt{\sigma_y}^{**}$	a
AISC ASD 78	$\frac{625}{\sqrt{\sigma_y}}$	$\frac{499}{\sqrt{\sigma_y}}$	$\frac{625}{\sqrt{\sigma_y}}$	c
AISI 86	$\frac{577}{\sqrt{\sigma_y}}$	–	$\frac{568}{\sqrt{\sigma_y}}$	c
AISI 76	$\frac{625}{\sqrt{\sigma_y}}$	$\frac{551}{\sqrt{\sigma_y}}$	$\frac{643}{\sqrt{\sigma_y}}$	c
AISC LRFD 86	$\frac{625}{\sqrt{\sigma_y}}$	$\frac{499}{\sqrt{\sigma_y}}$	$\frac{625}{\sqrt{\sigma_y}}$	c
CSA S16.1 M89 84b	$\frac{577}{\sqrt{\sigma_y}}$	$\frac{525}{\sqrt{\sigma_y}}$	$\frac{669}{\sqrt{\sigma_y}}$	c
EC 3 88	$\frac{690}{\sqrt{\sigma_y}}$	$\frac{583}{\sqrt{\sigma_y}}$	$\frac{690}{\sqrt{\sigma_y}}$	a

* for cold-formed non-stress relieved hollow sections

** for hot-formed and cold-formed stress relieved hollow sections

** for square hollow sections only

Symbol List for Chapter 9 -- Tubular Structures

A = cross sectional area (9a.E), (9a.W), (9b.A), (9b.N), (9c.A), (9c.E), & (9c.N)
A_e = effective area (9b.A), (9b.N), (9c.A), & (9c.N)
a = external diameter (9b.1)
a = external width of square tube (9c.1)
b = width of a plate or section (9c.A), (9c.E), (9c.J), (9c.N), & (9c.W)
b = inside diameter (9b.1)
b = inside width of square tube (9c.1)
b_c = compressive width of internal compression elements (9c.W)
b_e = effective width (9c.A), (9c.N), & (9c.W)
b_{e1} = effective width, defined in Table 9c.W1 (9c.W)
b_{e2} = effective width, defined in Table 9c.W1 (9c.W)
b_t = tensile width of internal compression elements (9c.W)
C = coefficient as a function of the type of forming, defined in Table 9a.W2 (9a.W)
c = coefficient for axis of bending, defined in Eq. 9c.N7a (9c.N)
c = flat width of square tube (9c.1)
D = external diameter (9b.A), (9b.E), (9b.J), (9b.N), & (9b.W)
D_e = effective diameter (9b.A)
D_i = inside diameter (9b.A)
E = Young's modulus (longitudinal modulus of elasticity) (9a.J), (9a.W), (9b.E), (9b.J), (9b.N), (9c.J), (9c.N), & (9c.W)
f = allowable stress (9b.N)
f_c = allowable column stress (9a.A), (9a.J), & (9a.N)
f_u = ultimate tensile strength of virgin material (9a.W)
f_y = yield strength (9a.W), (9b.W), & (9c.W)
f_{ya} = average yield strength of the member after cold-forming (9a.W)
f_{yb} = yield strength of virgin material (9a.W)
h = height (9c.E)
i = radius of gyration (9a.A), (9a.J), (9a.W), (9c.A), & (9c.E)
k = effective length factor (9a.J), (9a.W), & (9c.E)
k_r = buckling factor (9c.W)
k_σ = local buckling factor (9c.W)
ℓ = span, length of a member (9a.A), (9a.J), (9a.W), & (9c.E)
M_a = allowable bending moment (9c.N)
M_p = plastic bending moment (9b.A), (9b.N), (9c.A), & (9c.N)
M_r = limiting elastic bending moment with residual stress (9c.N)
M_{ult} = ultimate unfactored bending moment (9b.A), (9b.N), (9c.A), & (9c.N)
M_y = yield moment (9b.A), (9b.N), (9c.A), & (9c.N)
N = ultimate unfactored column axial load (9a.A), (9a.J), (9a.N), & (9c.E)
N_{cr} = inelastic local buckling load (9b.N)
N_f = factored column axial force=ϕN (9a.A), (9a.E), (9a.N), & (9a.W)
N_ℓ = local buckling load of cold-formed section (9b.J) & (9c.J)
N_y = column yield load (9a.A), (9a.J), (9a.N), (9a.W), (9b.A), (9b.J), (9b.N)&(9c.J)
Q = factor, defined in Eq. 9b.A3 (9b.A) & (9c.A)
R = design yield stress (9a.E) & (9c.E)
t = thickness (9a.W), (9b.A), (9b.E), (9b.J), (9b.N), (9b.W), (9c.A), (9c.E), (9c.J), (9c.N), & (9c,W)
Z = elastic section modulus (9b.N) & (9c.N)

Z_e = effective elastic section modulus (9c.N)
α = imperfection factor (9a.W)
α = constant, given in Eq. 9a.E3 (9a.E)
α_a = compression member factor (9a.A), (9b.A), & (9c.A)
α_b = compression member section constant (9a.A), (9b.A), & (9c.A)
α_c = nondimensional local buckling parameter for CHS$=\frac{E}{\sigma_y}\left(\frac{t}{D}\right)$ (9b.N)
α_d = compression member slenderness reduction factor (9a.A) & (9b.A)
α_s = nondimensional local buckling parameter for SHS$=\sqrt{\frac{E}{\sigma_y}}\left(\frac{t}{b}\right)$ (9c.N)
β = constant, given in Eq. 9a.E3 (9a.E)
γ = factor of safety (9a.A) & (9a.J)
γ_m = material homogeneity factor (9a.E)
γ_m = partial safety factor (9a.W)
ϵ = factor, defined in Eq. 9b.W2 & 9c.W2 (9b.W) & (9c.W)
η = coefficient, defined in Eq. 9a.A1 (9a.A)
Λ = slenderness ratio (9c.E)
λ = normalized slenderness (9a.A), (9a.E), (9a.J), (9a.N), (9a.W), (9b.N), (9c.A), & (9c.N)
λ_e = elastic limit normalized slenderness (9a.J)
λ_m = modified normalized slenderness ratio defined by Eq. 9a.A6 (9a.A)
λ_m, λ_Q = modified normalized slenderness (9b.A) & (9c.A)
λ_p = plastic limit normalized slenderness (9a.J)
λ_p = plate slenderness, defined in Eq. 9c.W8 (9c.W)
ξ = coefficient, defined in Eq. 9a.A1 (9a.A)
ρ = reduction factor of effective width (9c.W)
σ = average axial stress on the gross area (9c.E)
σ = average axial allowable stress on the gross area (9c.N)
σ_{cr} = inelastic local buckling stress for CHS (9b.N)
σ_{cr} = unfactored column buckling stress (9a.1)
σ_f = average axial stress at factored loads on gross area (9c.N)
σ_f = average axial stress computed on the basis of the effective area (9c.N)
σ_n = column buckling stress (9c.N)
σ_r = residual stress (9c.N)
σ_y = yield stress (9a.A), (9a.E), (9a.J), (9a.N), (9b.A), (9b.E), (9b.J), (9b.N), (9c.A), (9c.E), (9c.J), & (9c.N)
σ_1 = stress on the internal compression elements, defined in Table 9c.W1 (9c.W)
σ_2 = stress on the internal compression elements, defined in Table 9c.W1 (9c.W)
ϕ = resistance factor (9a.A), (9a.J), (9a.N), (9a.W), & (9b.J)
ϕ = coefficient, defined in Eq. 9a.W4 (9a.W)
χ = reduction factor; column buckling coefficient (9a.W)
ψ = buckling coefficient (9a.E) & (9c.E)
ψ = stress ratio, defined in Table 9c.W1 (9c.W)

Abbreviations

ASD : Allowable Stress Design
CHS : Circular hollow section
CRC : Column Research Council
LRFD : Load and resistance factor design
LSD : Limit States Design
SHS : Square and rectangular hollow section

Chapter 10

Shells

TOPICAL EDITOR

Miklos Ivanyi, Technical University of Budapest

CHAPTER EDITORS

Miklos Ivanyi, Technical University of Budapest
Zbigniew K. Mendera, Politechnika Krakowska
Nicholas F. Morris, Manhattan College
J. Michael Rotter, The University of Sydney
Daniel Vandepitte, Ghent University

10a. Unstiffened Cylinders Subjected to Meridional Compression

SPECIFICATION PROVISIONS & EXPLANATION

AUSTRALIA

CHAPTER EDITOR

J. Michael Rotter, The University of Sydney

AISC 83

The average stress σ_u at failure can be expressed in the form

for $\beta \frac{r}{t} < 55$

$$\frac{\sigma_u}{f_y} = 1.0 \tag{10a.A1}$$

for $55 < \beta \frac{r}{t} < 1030$

$$\frac{\sigma_u}{f_y} = \frac{110}{\beta \frac{r}{t}} - \frac{3025}{\left(\beta \frac{r}{t}\right)^2} \tag{10a.A2}$$

for $\beta \frac{r}{t} > 1030$

$$\frac{\sigma_u}{f_y} = \frac{484}{\beta \frac{r}{t}} \left\{1 - 0.9 \left(1 - e^{-0.0625\sqrt{\beta \frac{r}{t}}}\right)\right\} \tag{10a.A3}$$

where

e = exponential function
f_y = specified yield stress
r = radius of the cylinder
t = thickness of the cylindrical shell
β = $\frac{f_y}{250}$, a correction factor that allows for steels with differing values specified yield stress f_y (in MPa)

Values of σ_u/f_y are plotted in Fig. 10a.A1.

COMMENTARY

For cylinders with high r/t values, virtually all of the experimental data available refer to aeronautical cylinders. An approximate lower bound strength based on aerospace industry experiments has been proposed in (Weingarten, 1968).

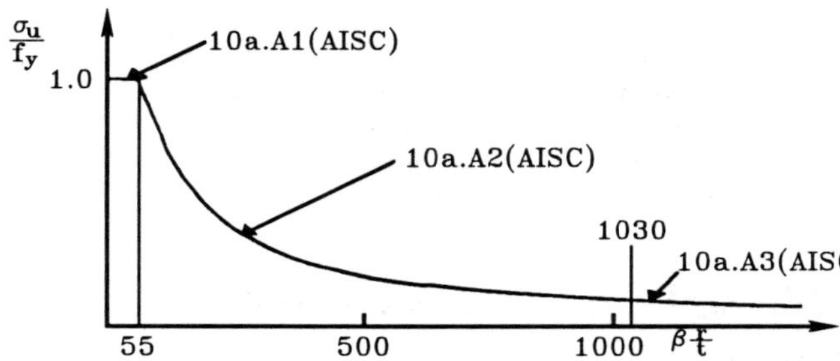

Fig. 10a.A1 Buckling curve for axially loaded cylindrical shells by AISC 83

EAST EUROPE

CHAPTER EDITORS
Miklos Ivanyi, Technical University of Budapest
Zbigniew K. Mendera, Politechnika Krakowska

ČSN 73 1401 84

Similar methods are adopted with concise formulas for all steel grades, using the notation

$$\beta_1 = \sqrt{\frac{210}{R_d}} \qquad (R_d \text{ and } f_y \text{ in MPa})$$

for $\frac{r}{t} \leq 50\ (\beta_1)^{4/3}$

$$\sigma_u = \frac{f_y}{\gamma_m} = R_d \tag{10a.E1}$$

for $\frac{r}{t} \geq 180\ (\beta_1)^{4/3}$

$$\sigma_u = \left(\frac{155\ t}{r}\right)^{3/2} (\beta_1)^2\ R_d \tag{10a.E2}$$

with linear interpolation between the two limits (Fig. 10a.E1),

where

- f_y = specified yield stress
- R_d = design yield stress
- r = radius of the cylindrical shell
- t = thickness of the shell
- β_1 = correction factor
- γ_m = material coefficient

When a member is subjected to eccentrically compressed force and uniform bending, the limit stress is multiplied by a modification factor α as follows

$$\alpha = 1 + 0.1 \left(1 - \frac{\sigma'_x}{\sigma_x}\right) \tag{10a.E3}$$

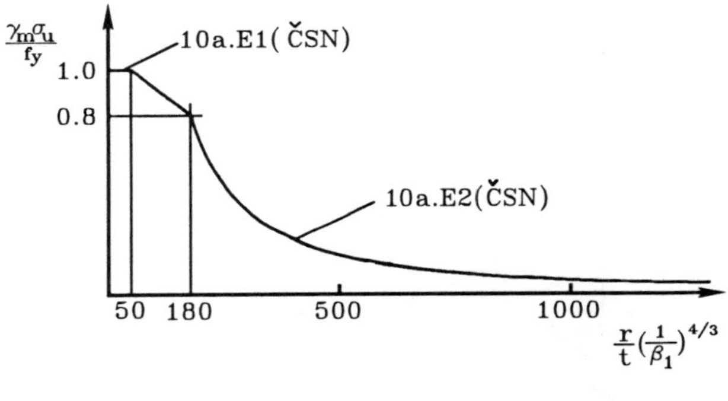

Fig. 10a.E1 Buckling curve for axially loaded cylindrical shells by CSN 73 1401 84

where

σ'_x = the smallest meridional stress (tension stress is negative)
σ_x = the highest meridional stress

The modified limit stress ($\alpha \sigma_u \leq R_d$) can be considered together with the maximum shear stress τ_{max} (in the location of the maximum bending stress) if

$$\tau_{max} \leq 126 \left(\frac{24\ t}{r}\right)^{3/2} \tag{10a.E4}$$

MI 04.184 78

The limit stress σ_u for axially compressed shells is given by

$$\sigma_u = \phi \frac{f_y}{\gamma_m} \tag{10a.E5}$$

where

$a = 1 + b + 0.257 \sqrt{\frac{r}{t}}$

$b = \frac{\sigma_{cr}}{f_y} = \frac{0.605\ E\ t}{r\ f_y}$

r = radius of the cylindrical shell
t = thickness of the shell
γ_m = material coefficient
σ_{cr} = critical (buckling) stress
σ_y = specified yield stress

$\phi = \frac{1}{2}\left(a - \sqrt{a^2 - 4\ b}\right)$

The relationship between $\frac{\sigma_u}{f_y}$ and $\frac{r}{t}$ is represented graphically in Fig. 10a.E2.

COMMENTARY

In the Hungarian specifications the limit stresses of axially compressed cylindrical shells are calculated from a wide-column formula. Fig. 10a.E3 shows the effect of the initial imperfections (Alekseev, 1956).

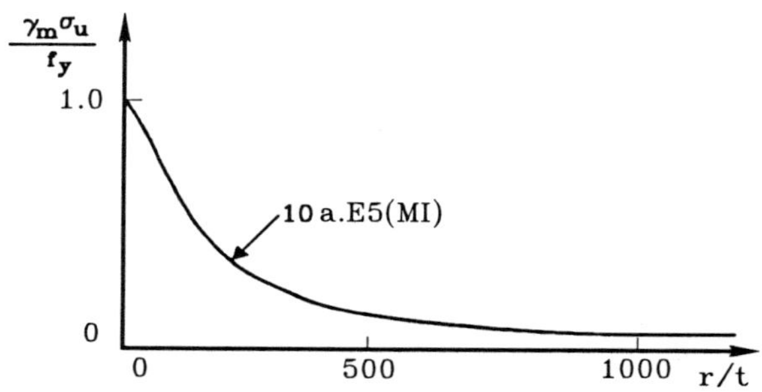

Fig. 10a.E2 Buckling curve for axially loaded cylindrical shells by MI 04.184 78

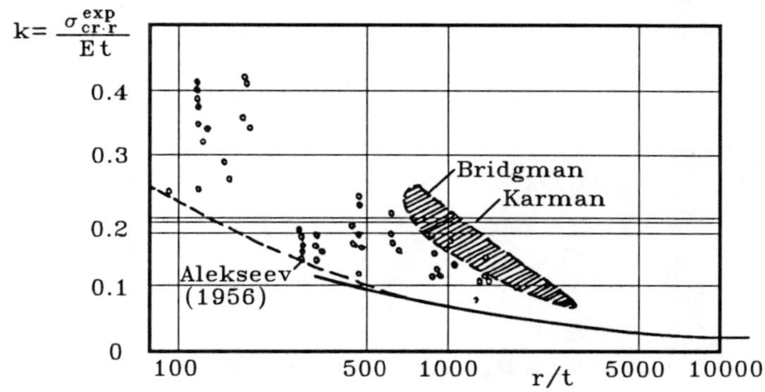

Fig. 10a.E3 Test results for buckling of cylinders under axial compression and recommended reduction factor by Alekseev

SNiP II-23-81 82

Axially Compressed Cylindrical Shells

In the elastic range, the limit stress is given by

$$\sigma_u = \sigma_{cr} = c\ \frac{E\ t}{r} \qquad (10a.E6)$$

where

c = an empirical factor based on experiments given in Table 10a.E1
r = radius of the shell
t = thickness of the shell

In the inelastic range

$$\sigma_u = \psi\ \frac{f_y}{\gamma_m} \qquad (10a.E7)$$

where $\psi = 0.97 - \left(0.00025 + 0.95\,\frac{f_y}{E}\right)\frac{r}{t}$ for $0 < \frac{r}{t} \leq 300$.

The relationship between $\frac{\gamma_m \sigma_u}{f_y}$ and $\frac{r}{t}$ is represented graphically in Fig. 10a.E4.

Bending of the Cylinder in the Meridional Direction or Axial Compression Load Combined with Bending

A modified limit stress σ'_u is given by

$$\sigma'_u = \sigma_u\left(1.1 - 0.1\,\frac{\sigma'_x}{\sigma_x}\right) \qquad (10a.E8)$$

where

σ'_x = the smallest meridional stress (tension stress is negative)
σ_x = the highest meridional stress

Table 10a.E1

r/t	100	200	300	400	600	800	1000	1500	2500
c	0.22	0.18	0.16	0.14	0.11	0.09	0.08	0.07	0.06

The modified limit stress σ'_u can be considered together with the maximum shear stress τ_{max} in the location of maximum bending stress if

$$\tau_{max} < 0.07 \ E \left(\frac{t}{r}\right)^{3/2} \qquad (10a.E9)$$

Tubular Columns

Results are given as a limit stress with the following slenderness ratio

$$\bar{\lambda} = \lambda \sqrt{\frac{f_y}{E}} \geq 0.65 \qquad (10a.E10)$$

and

$$\frac{r}{t} \leq 3.14 \sqrt{\frac{E}{f_y}} \qquad (10a.E11)$$

The buckling of a welded tubular column is not treated, when

$$\frac{r}{t} \leq 1.57 \sqrt{\frac{E}{f_y}} \qquad (10a.E12)$$

These formulas can be used only if tolerances in fabrication and erection given in the corresponding codes are not surpassed.

COMMENTARY

In the inelastic range a lower limit is given according to Broude's theory (1963), reducing the actual problem to the analysis of an eccentricity compressed beam-column on elastic foundations of effective length $\ell_x = \ell_o \sqrt{2}$ (Fig. 10a.E4-b), and initial eccentricity a_o as represented by Fig. 10a.E4.

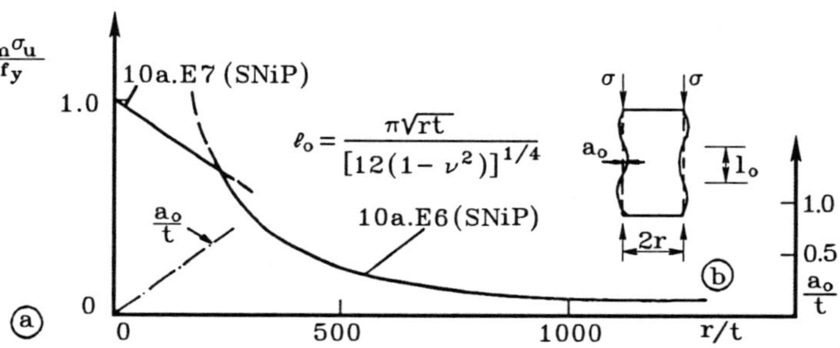

Fig. 10a.E4 Buckling curve for axially loaded cylindrical shells by SNiP II-23-81 81

TGL 13503 82

The limit stress σ_u of axially compressed cylindrical shells is given as follows

for $c \cdot \sigma_{cr} \leq 0.6 \ f_y$

$$\sigma_u = c \cdot \sigma_{cr} \qquad (10a.E13)$$

for $c \cdot \sigma_{cr} > 0.6 \ f_y$

$$\sigma_u = f_y \left(1.1 - 0.3 \ \frac{f_y}{c \cdot \sigma_{cr}} \right) \leq f_y \qquad (10a.E14)$$

where

c = a constant as given in Table 10a.E2

f_y = specified yield stress

$$\sigma_{cr} = 0.605 \text{ k E } \frac{t}{r} \tag{10a.E15}$$

Values of the factor k:

- Pure axial compression load: k = 1
- Pure bending of the cylinder in the meridional direction: k = 1.2

with linear interpolation between the two cases.

The relationship between $\frac{\sigma_u}{f_y}$ and $\frac{r}{t}$ is represented graphically in Fig. 10a.E5.

The maximum amplitude $\overline{w}$ of the initial imperfections is measured as shown in Fig. 10a.E3.

Table 10a.E2 Effect of imperfections

Class of imperfections	Technology of cylindrical shell production	Max amplitude $\overline{w}$ of the initial imperfections	c
a	Rolled shell or welded in a spiral way	0.3 t	0.33
b	as above	t	0.25
c	as above	2.0 t	0.19
	longitudinally welded shell	t	
d	as above	2.0 t	0.14

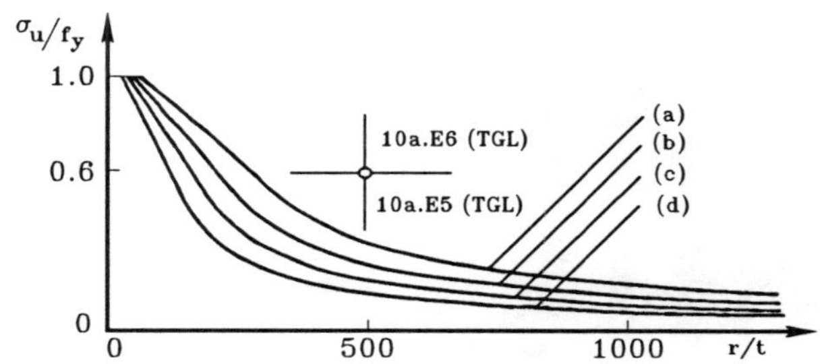

Fig. 10a.E5 Buckling curves for axially loaded cylindrical shells by TGL 13503 82

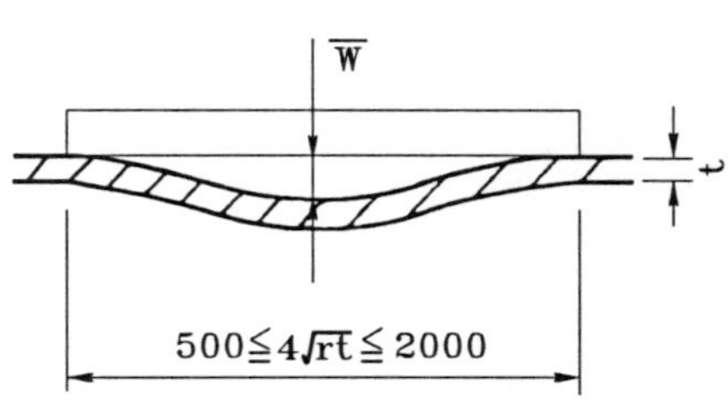

Fig. 10a.E6 Allowable imperfections in cylindrical shells by TGL 13503 82

$$\text{If } \frac{r}{t} \leq \frac{42300 \text{ N/mm}^2}{f_y} c \qquad \sigma_u = \frac{f_y}{\gamma_m} \tag{10a.E16}$$

and Table 10a.E3 shows the maximum $\left(\frac{r}{t}\right)$.

COMMENTARY

The TGL 13503 82 standard takes into account the technological imperfections that appear in the elastic-plastic range. The inclusion of the latter ones is important because of various technologies and precision of manufacture of shells.

Table 10a.E3 Maximum (r/t)

f_y N/mm^2	c			
	0.33	0.25	0.19	0.14
240	58	44	33	25
300	47	35	27	20
360	39	29	22	17
450	31	24	18	13

NORTH AMERICA

CHAPTER EDITOR
Nicholas F. Morris, Manhattan College

API RP2A 89

The allowable buckling stress σ_{ui} is given by

$$\sigma_{ui} = \alpha_o \, \eta \, \sigma_{cri} \qquad (10a.N1)$$

where

α_o = the imperfection reduction factor
η = the plasticity reduction factor
σ_{cri} = elastic buckling stress in failure mode "i"

The elastic buckling stress for unstiffened cylinders under axial compression is as follows. (Note that API RP2A 89 includes an imperfection reduction factor in its buckling stress equations.)

$$\sigma_{x,l} = \alpha_o \frac{c_x \, Et}{r} \qquad (10a.N2)$$

for $\delta_x \leq 1.5$ $\qquad c_x = 0.630$

for $1.5 < \delta_x < 1.73$ $\qquad c_x = \frac{0.940}{\delta_x^2} + 0.1013\delta_x^2$

for $\delta_x \geq 1.73$ $\qquad c_x = 0.605$

The imperfection reduction factors are

for $r/t \geq 600$ $\qquad \alpha_o = 0.207$

for $r/t < 610$ $\qquad \alpha_o = \frac{169\,\bar{c}}{195 + r/t} < 0.90$

for $\delta_x \geq 1.5$ $\qquad \bar{c} = 2.65$

for $1.5 < \delta_x < 15$ $\qquad \bar{c} = \frac{3.13}{\delta_x^{0.42}}$

for $\delta_x \geq 15$ $\qquad \bar{c} = 1.0$

(10a.N3)

The plasticity reduction factors are

$$\left.\begin{array}{ll} \text{for } r/t < 300 & \eta = \dfrac{233 f_y}{166 + r/t} \leq f_y \\ \text{for } r/t > 300 & \eta = 0.50\, f_y \end{array}\right\} \quad (10a.N4)$$

where f_y is the yield stress

Note that

$$\delta_x = \frac{\ell}{\sqrt{r\,t}} \quad (10a.N5)$$

where ℓ is the distance between stringers or supports points.

ASME N284 80

The basic design equations of this code case are similar to API RP2A 89 for cylindrical shells but are of an earlier date. It is recommended that the API RP2A 89 be followed.

WEST EUROPE

CHAPTER EDITOR

Daniel Vandepitte, Ghent University

ECCS No. 29 88

The strength requirement is $\sigma_d \leq \sigma_u$.

σ_u is given by the equations

for $\alpha\sigma_{cr} \leq \frac{1}{2}f_y$ $$\sigma_u = \frac{\alpha\sigma_{cr}}{\gamma} \quad (10a.W1)$$

for $\alpha\sigma_{cr} > \frac{1}{2}f_y$ $$\sigma_u = \left(1-0.4123\left(\frac{f_y}{\alpha\,\sigma_{cr}}\right)^{0.6}\right)f_y \quad (10a.W2)$$

where

f_y = specified yield stress
α = reduction factor
$\gamma = \frac{4}{3}$
$\sigma_{cr} = 0.605E\frac{t}{r}$

With the parameter $\bar{\lambda}$ defined by $\bar{\lambda} = \sqrt{\frac{f_y}{\alpha\sigma_{cr}}}$ Eqs. 10a.W1 and 10a.W2 may be written

for $\bar{\lambda} \geq \sqrt{2}$ $$\frac{\sigma_u}{f_y} = \frac{0.75}{\bar{\lambda}^2} \quad (10a.W3)$$

for $\bar{\lambda} \leq \sqrt{2}$ $$\frac{\sigma_u}{f_y} = 1-0.4123\,\bar{\lambda}^{1.2} \quad (10a.W4)$$

The relationship between $\frac{\sigma_u}{f_y}$ and $\bar{\lambda}$ is represented graphically in Fig. 10a.W1.

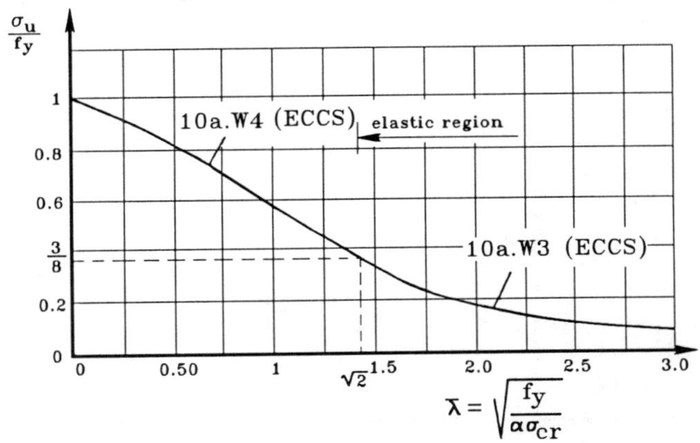

Fig. 10a.W1 Buckling curve for axially loaded cylindrical shells by ECCS 88

Values of the Reduction Factor α

(A) Pure Axial Compression Load (See Fig. 10a.W2)

for r/t < 212

$$\alpha = \alpha_o \text{ and } \alpha_o = \frac{0.83}{\sqrt{1+0.01\ r/t}} \tag{10a.W5}$$

for r/t > 212

$$\alpha_o = \frac{0.70}{\sqrt{0.1+0.01\ r/t}}$$

(B) Pure Bending of the Cylinder in the Meridional Direction (Fig. 10a.W2)

$$\alpha = \alpha_b$$
$$\alpha_b = 0.1887+0.8113\ \alpha_o \tag{10a.W6}$$

and α_o given by Eq. 10a.W5.
α_b has been experimentally verified for $\frac{r}{t} < 1500$.

(C) Axial Compression Load Combined with Bending

$$\alpha = \frac{\alpha_o\ \sigma_o+\alpha_b\ \sigma_b}{\sigma_o+\sigma_b} \tag{10a.W7}$$

with

α_o given by Eqs. 10a.W5 and
α_b given by Eq. 10a.W6

where

σ_b = the maximum compression stress due to the design bending moment.
σ_o = the uniform compression stress due to the design axial load.

(D) Axial Compression Load Combined with Internal Pressure (See Fig. 10a.W3)

$$\alpha = \alpha_p = \alpha_o+(1-\alpha_o)\frac{\rho}{\rho+0.007} \tag{10a.W8}$$

where $\rho = \frac{p\ r\ \sqrt{r}}{E\ t\ \sqrt{t}}$ when an axial load acts together with the internal pressure p.

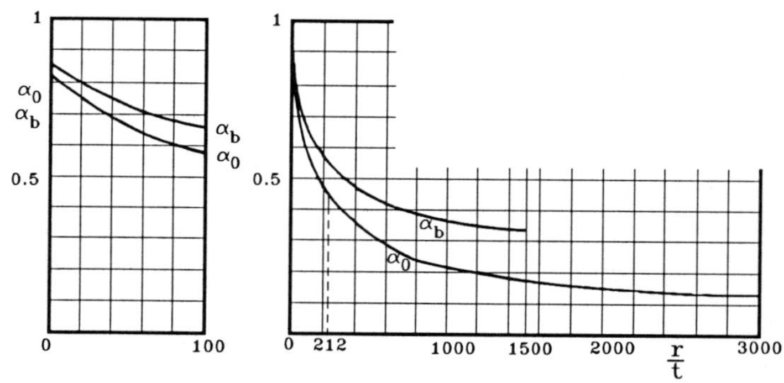

Fig. 10a.W2 Evaluation of α_o and α_b

COMMENTARY

The range of applicability of the above recommendation is limited as follows:

- The boundary conditions have to be such that the displacements in the radial and the circumferential direction are zero along both edges of the cylinder.
- The length ℓ of the cylinder is limited by the requirement $\frac{\ell}{r} \leq 0.95 \sqrt{\frac{r}{t}}$.
- Since axially loaded cylindrical shells are very imperfection sensitive, the amplitude $\overline{w}$ of the imperfections is measured from a straight rod (Fig. 10a.W4–a) and from a circular template (Fig. 10a.W4–b), held anywhere between welds respectively against any meridian and against any parallel circle. The length of the rod and of the template is $\ell_r = 4\sqrt{rt}$, but not greater than 95% of the distance between welds. A rod of length $\ell_r = 25t$ is used across circular welds (Fig. 10a.W4–c).

  When $\overline{w} \leq 0.01\ \ell_r$, the reduction factors α_o apply.
  When $\overline{w} = 0.02\ \ell_r$, the values of α_o must be halved.
  When $0.01\ \ell_r \leq \overline{w} < 0.02\ \ell_r$ linear interpolation between α_o and $\alpha_o/2$ provides the reduction factor to be applied.

- Although a slight, practically unavoidable unevenness of the supports of the shell is covered by the reduction factor α_o, care must be taken to introduce the compressive forces uniformly into the cylinder and to avoid edge disturbances.

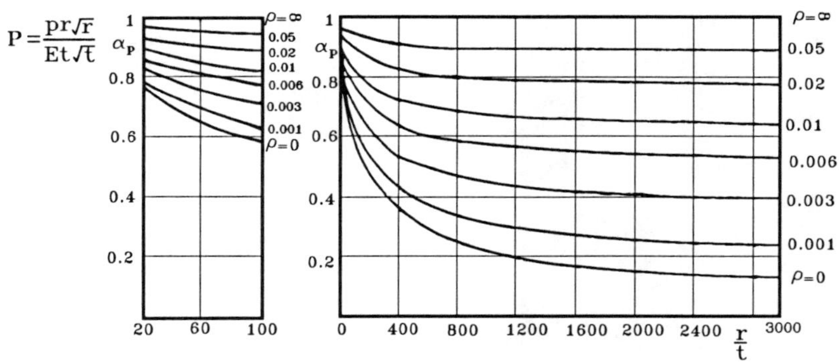

Fig. 10a.W3 Evaluation of α_p

The imperfection factor α accounts for reduction due to accidental imperfections. The incremental safety factor γ accounts for the pronounced imperfection sensitivity of meridionally compressed cylinders. Plasticity reduction Eq. 10a.W2 allows σ_u to approach f_y for very stocky cylinders, thus allowing the incremental safety factor in effect to approach 1 gradually. The values of $\alpha\sigma_{cr}$ given by Eq. 10a.W1 and of σ_u given by Eq. 10a.W2 represent reasonably well lower bounds of scatter bands of experimental results.

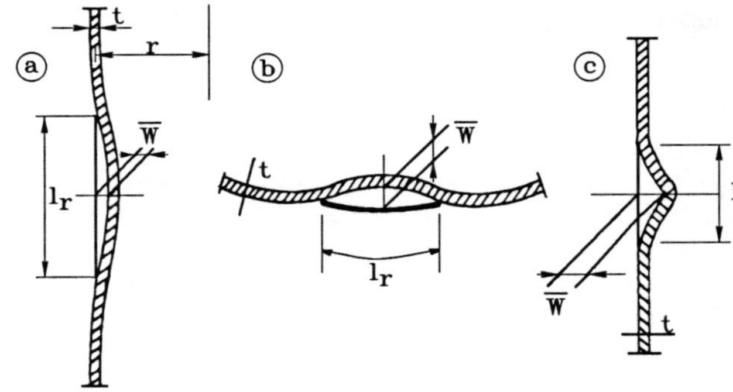

Fig. 10a.W4 Allowable imperfections in cylindrical shells by ECCS 88

INTERNATIONAL COMPARISONS & COMMENT

WHAT ARE THE DIFFERENCES?

The basic shell stress formulas specified in the specifications are compared graphically in Fig. 10a.1.

In all formulas, the shell stress σ_u is normalized by the yield stress f_y and the shell radius to thickness ratio $\frac{r}{t}$ is given.

The following provisions for cylinders subjected to meridional compression are included in the recommendations, specifications, or codes: (Table 10a.1)-(the designation **0** means that it applies)

Table 10a.1 Comparison of specifications for design of unstiffened cylinders subjected to meridional compression

Region	Specification	Item (1)	(2)	(3)	(4)	(5)	(6)	(7)	(8)	(9)	(10)	(11)
AUSTRALIA												
	AISC 83	0		0			0		0			0
EAST EUROPE												
Czechoslovakia	ČSN 73 1401 84	0		0					0	0		0
GDR	TGL 13503 82		0	0		0	0	0	0	0		
Hungary	MI 04.184 78	0			0				0			
Soviet Union	SNIP II-23-81 82	0		0					0	0	0	
U.S.A.												
	ASME N284 80	0		0				0	0	0		0
	AWWA D100 73	0			0				0		0	
	API RP2A 89	0		0	0			0	0	0	0	0
	AISI 82	0		0	0				0		0	
WEST EUROPE												
	ECCS No. 29 88	0		0			0	0	0	0	0	

(1) single shell curve developed
(2) multiple shell curves developed
(3) shell stress formula constructed from two or more curves
(4) continuous stress formula
(5) classification of strength curve due to technological imperfections
(6) describe the magnitudes of imperfections
(7) describe the techniques to be used to measure imperfections
(8) axially compressed cylinder
(9) cylinders subjected to meridional compression
(10) effect of shell length
(11) formulas for concerning the effect of steel grades

A major difficulty arises here in the design of civil engineering shells, as it is not easy for the designer to predict the level of imperfections that may be found in his structure, or for the fabricator to control the magnitude and shape of the imperfections he may produce.

Consequently, most design guides are based on an empirically-derived lower bound of experimental strengths.

Two approaches are used in developing design rules:

- the first is simply to draw a lower bound curve on all available appropriate test results;
- the second attempts to make use of an expression to categorize test results into groups of specimens of different quality. Empirical relationships are then proposed for the maximum imperfection amplitude to be expected in a shell of given quality (Koiter, 1945).

In the first approach four proposals using simple lower bounds are shown in Fig. 10a.2.

(Donnell, 1934)'s proposal is still quite widely used in the USA, but it is based on old and limited data, and suffers the disadvantage of predicting zero strength for $\frac{r}{t} > 2450$.

The proposal of (Weingarten, Morgan and Seide, 1965) is only for elastic buckling failures and was derived from a large data base gathered chiefly for aerospace applications. (Baker et al., 1971)

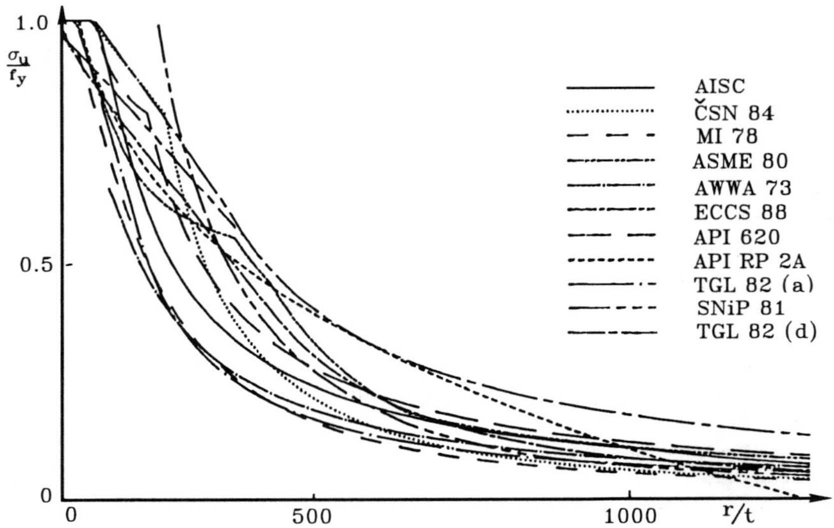

Fig. 10a.1 Basic buckling curves in comparison with the referred specifications for axially loaded cylinders

Most recent curves of Steinhardt and Schulz, V. (1971) and Schulz, V. (1981) are based on a complete set of data ($70 < \frac{r}{t} \leq 2800$), and provides a 90 percentile lower bound on experimental results.

Most design guides are based on empirically-derived lower bounds on experimental strength: (ČSN 73 1401 84, TGL 13503 82, SNiP II-23-81 82, AWWA D100 73, ASME N284 80, API RP2A 89, AISI 82).

In the second approach, the strength is directly related to a defined level of imperfection.

Design recommendations in this group include those of ECCS No. 29 88, AISC 83, MI 04.184 78.

Hoff and Soong (1967) seem to have been first to suggest that design strength could be based on Koiter's equation (1945) and that different quality cylinders could be separated for design purposes according to their imperfection levels. The alternative cylinders covered by (Hoff and Soong, 1967), (ESDU, 1974), (Odland, 1978), ECCS No. 29 88, MI 04.184 78 and AISC 83 are shown in Fig. 10a.3.

In Fig. 10a.4 the basic shell stress formulas specified in the referred specifications and the four proposals using simple lower bounds are compared graphically.

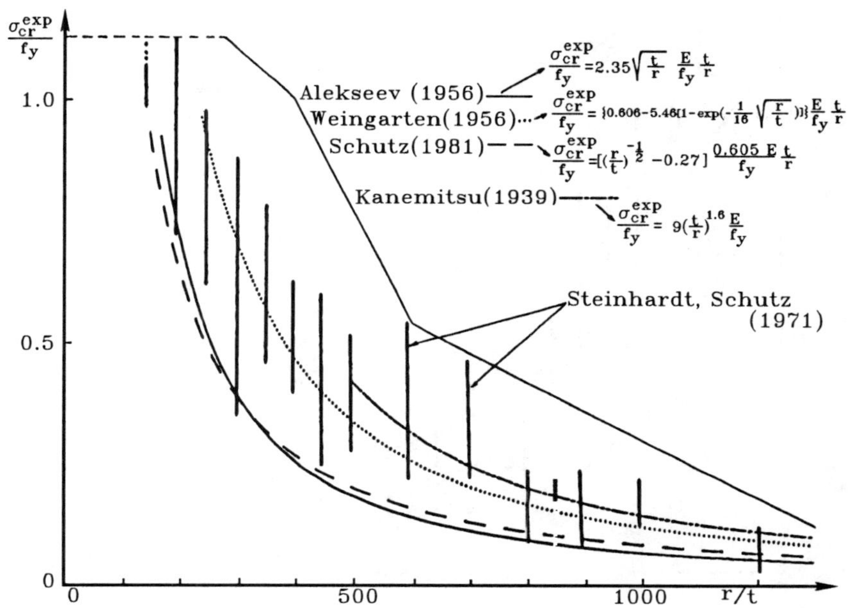

Fig. 10a.2 Test data and empirically-derived lower bound design curves (Individual results are not shown; instead the range for each r/t group is indicated by a solid line)

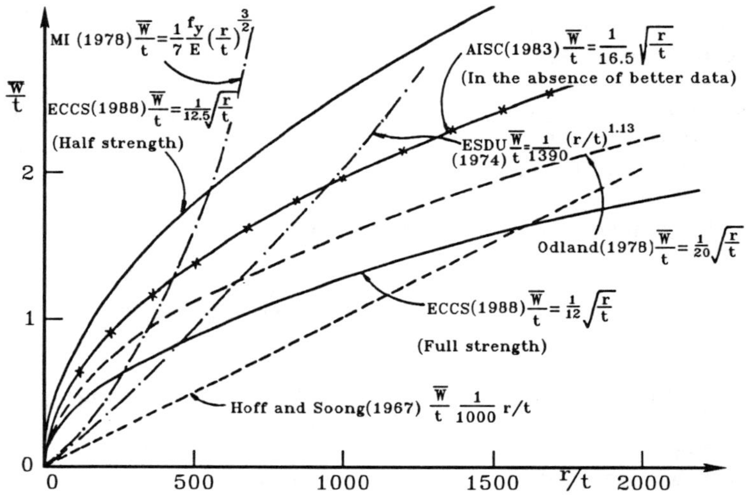

Fig. 10a.3 Maximum allowable imperfections in poor quality cylinders

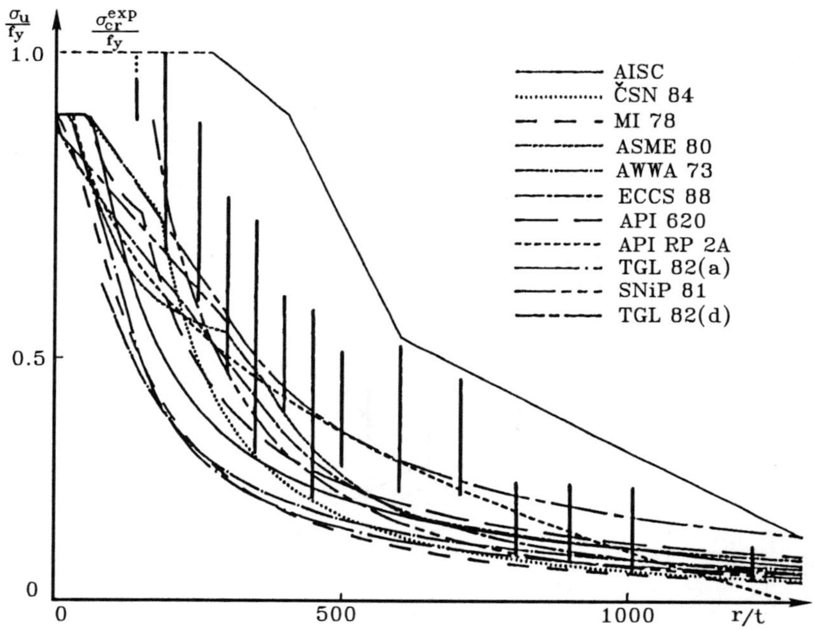

Fig. 10a.4 Test data and the referred specfications for axially loaded cylinders

WHY ARE THERE DIFFERENCES?

The empirical data base of imperfections is different for each code considered. Therefore, different numerical factors are used.

In making comparisons between the above proposals the following observations provide useful principles:

(1) The most important parameter affecting the buckling strength is the amplitude of imperfections. It is desirable that the designer or fabricator be aware of the physical dimensions of the imperfections on which the design has been based.

(2) Some of the existing design proposals are based on old, limited or inappropriate data: since these rules are empirical, the data-base should be as wide and as recent as possible.

(3) If the strength is to be related to the imperfection amplitude, since there is insufficient experimental data to establish the relationship experimentally, a simple soundly-based conservative relation should be adopted. Arbocz (1981) and Singer (1982) collected data for an imperfection data bank that is intended to provide more accurate information for prediction of the carrying capacity of shells in the future.

(4) Finally, it may be observed that development of computer codes for stability problems, with regard to bifurcation, collapse, and imperfection sensitivity is in progress.

Most of the codes address problems that may be characterized as "elementary"; they give rules for the design of simple shells and shell elements under (uniform) loading conditions. The ASME N284 80 touches on the problem of how to interpret buckling load predictions obtained by use of finite element analysis (Bushnell, 1985).

10b. Stiffened Cylinders Subjected To Meridional Compression

SPECIFICATION PROVISIONS & EXPLANATION

AUSTRALIA

CHAPTER EDITOR
J. Michael Rotter, The University of Sydney

AISC 83

Four modes of buckling can be limiting: stiffener buckling, local buckling of the shell panels, panel buckling with stiffener rotation, and general instability as an orthotropic cylinder.

Where a limited number of stiffeners is used on bin walls, these generally fall into the class of broad-paneled stiffened cylinders. The buckling strength may then be satisfactorily assessed by determining the buckling load carried by the wall as an isotropic cylinder ($2\pi rt\sigma_u$), where σ_u is the axial membrane stress at buckling without internal pressure, and adding the loads carried by the stiffeners, making proper allowance for axial strain compatibility between the two elements.

When stiffeners are more closely spaced than

$$b = 2.44\sqrt{rt} \tag{10b.A1}$$

narrow-panel buckling occurs and the shell buckling stress should be taken instead as the simply-supported plate buckling stress.

The buckling of the wall as an orthotropic shell is examined by a simple treatment, that a unit width of shell between panel centers b is treated as a prismatic member. The effective inertia of this section EI_{eff} is then used to assess the buckling strength of a perfect orthotropic shell as

$$\sigma_{u,p} = \frac{Et}{r}\,\frac{2\sqrt{EI_{eff}\,b/t}}{b\,t + A_s} \tag{10b.A2}$$

where A_s is the stiffener area.

Local torsional buckling of the stiffener is generally covered by simple rules:

- a flat bar stiffener may be treated as an outstanding flange and the buckling stress determined as

$$\sigma_1 = 0.39E\left(\frac{t}{b}\right)^2 \tag{10b.A3}$$

- a T section stiffener may be conservatively treated as a long column buckling in torsion.

$$\sigma_1 = \frac{GC_s}{I_f + I_s} \tag{10b.A4}$$

where

C_s = St. Venant torsional constant

I_f = flange inertia about the web centerline

I_s = T section inertia about a circumferential axis through the point of attachment to the shell wall

COMMENTARY

Stringer stiffeners are quite often used on bins and silos. There appear to have been fewer failures of stiffened bins than of unstiffened bins, but it is difficult to be certain on this point. A considerable amount of research work on the buckling of stringer-stiffened cylinders has been undertaken in recent years, chiefly in relation to aerospace and offshore construction. Many of the shells in these studies involved thicker cylinders than are used in silos.

EAST EUROPE

CHAPTER EDITORS
Miklos Ivanyi, Technical University of Budapest
Zbigniew K. Mendera, Politechnika Krakowska

ČSN 73 1401 84

Limit Stress of Cylindrical Panels under Uniform Meridional Compression

If $\frac{b^2}{rt} < 4$, the cylindrical panel is treated as a plate.

If $4 < \frac{b^2}{rt} < 20$: for $\frac{b}{t} \leq \frac{37}{\sqrt{1 + \frac{1}{2\beta_1{}^2}}}$

$$\sigma_{us} \leq \frac{f_y}{\gamma_m} \qquad (10b.E1)$$

for $\frac{b}{t} \geq 58\beta_1$

$$\sigma_{us} \leq (\frac{52t}{b})^2 \beta_1{}^2 (\frac{f_y}{\gamma_m}) \qquad (10b.E2)$$

with linear interpolation.

$$\beta_1 = \sqrt{\frac{210}{f_y}} \qquad \text{(in MPa)}$$

If $20 < \frac{b^2}{rt}$, Eqs. 10b.E1 and 10b.E2 are to be applied.

Circular Cylindrical Shells with Longitudinal Stiffeners and Subjected to Axial Compression

The circumferential spacing of longitudinal stiffeners is "b". The longitudinal stiffeners are designed as columns. The effective width of a circular panel is

$$b_e = b\sqrt{\frac{\sigma_u}{\sigma_1}} \qquad (10b.E3)$$

where

σ_u = limit stress (Eq. 10b.E1 and 10b.E2)
σ_1 = axial stress on the cross section A

MI 04.184 78

Cylindrically Stiffened Cylindrical Shells

Limit stress is given by Eq. 10b.E1, but the effective thickness of shells t_{eff} is used instead of the thickness of unstiffened shell t.

for $N_y \geq \frac{2E}{r}\sqrt{(t + \frac{A_{so}}{b})I_x}$

$$t_{eff}{}^2 = 3.3\sqrt{(t + \frac{A_{so}}{b})I_x} \qquad (10b.E4)$$

for $N_y < \frac{2E}{r}\sqrt{(t + \frac{A_{so}}{b})I_x}$

$$t_{eff}^2 = 3.3\sqrt{\frac{t + \frac{A_{so}}{b}}{\frac{1}{I_x} + \frac{1}{I_y}}} + 1.65\left(1 - \sqrt{\frac{1}{1 + \frac{I_x}{I_y}}}\right)\frac{rN_y}{E} \qquad (10b.E5)$$

where

$$A_{so} = \min\begin{cases} A_{sc}\text{—area of ring stiffener} \\ A_{sl}\text{—area of longitudinal stiffener}\end{cases}$$

$$I_x = \frac{I_{bx}}{b} + \frac{t^3}{12}$$

$$I_y = \frac{I_{by}}{l_l} + \frac{t^3}{12}$$

N_y = circumferential stress resultant (force per unit circumference)

The notations used are defined in Fig. 10b.E1.

Local Shell Instability of Cylindrical Panel

The design value of the local shell instability of cylindrical panel, i.e. buckling of shell panels between stiffeners is higher than the design value of the buckling mode, in which both stringers and ring participate,

if both

$$b = \frac{1}{\sqrt{\frac{1}{\pi l_l}\sqrt{\frac{3.3}{t\,r}} + \frac{1}{l_l^2}}} \qquad (10b.E6)$$

and

$$b \leq 1.484t\left(t^2r^2\left(\frac{1}{I_{bx}} + \frac{l_l/b}{I_{by}}\right)\right)^{1/3} \qquad (10b.E7)$$

where

$b = 3.46\sqrt{rt}$ (circumferential)

$l_l = 1.73\sqrt{rt}$ (longitudinal)

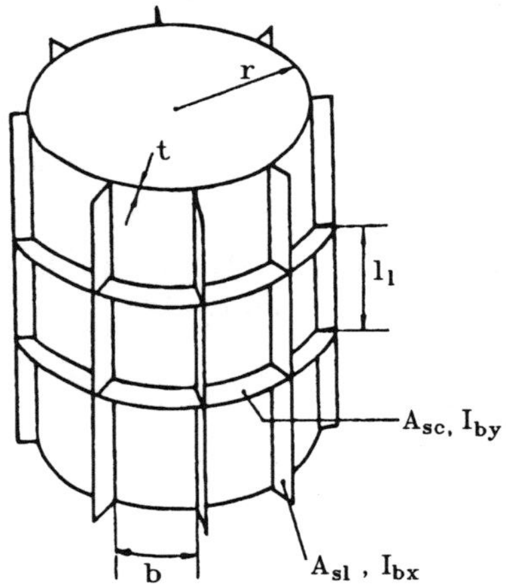

Fig. 10b.E1 Notations of the stiffened cylindrical shells

when only stringers are in the shells

$$b \leq 2.4\sqrt{rt} \qquad (10b.E8)$$

COMMENTARY

In Hungary the limit stress of a circular stiffened shell can be calculated also from a wide flange column formula (the effective thickness of the shell is used). The effective thickness of the shell was determined on the basis of the buckling of an orthogonally stiffened cylinder.

SNiP II-23-81 82

The limit stress of cylindrical panels under uniform meridional compression is given by

If $b^2/rt \leq 20$, the cylindrical panel is treated as a plate

for $\frac{b}{t} \leq \frac{37}{\sqrt{1 + 500\frac{f_y}{E}}}$

$$\sigma_{uc} = \frac{f_y}{\gamma_m} \qquad (10b.E9)$$

for $\frac{b}{t} \geq 7.9\sqrt{\frac{E}{0.8\,f_y}}$

$$\sigma_{uc} \leq \frac{f_y}{\gamma_m}\frac{\pi^2 E}{3(1-\nu)^2}\left(\frac{t}{b}\right)^2 \qquad (10b.E10)$$

for $\frac{37}{\sqrt{1 + 500\frac{f_y}{E}}} < \frac{b}{t} < 7.9\sqrt{\frac{E}{0.8\,f_y}}$ Linear interpolation is used See Fig. 10b.E2

If $b^2/rt > 20$, the limit stress σ_u is given by Eqs. 10a.E6 and 10a.E7.

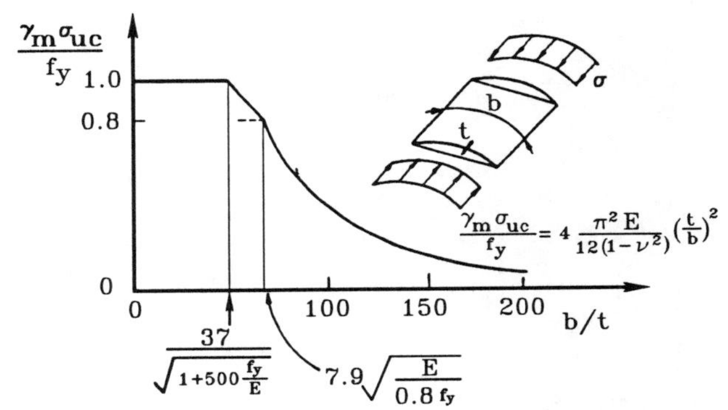

Fig. 10b.E2 Buckling curve for axially loaded cylindrical panels by SNiP II-23-81 82

TGL 13503 82

The limit stress of circular cylindrical shells with longitudinal stiffeners is given by Eqs. 10a.E13 and 10a.E14, but

$$\sigma_{cr} = 0.605\mathrm{kE}\left(\frac{\mathrm{t}}{\mathrm{r}} + 6\left(\frac{\mathrm{t}}{\mathrm{b}}\right)^2\right) \quad (10b.E11)$$

where

b = the circumferential spacing of longitudinal stiffeners (Fig. 10b.E3)
k = a factor specified in Eq. 10a.E15

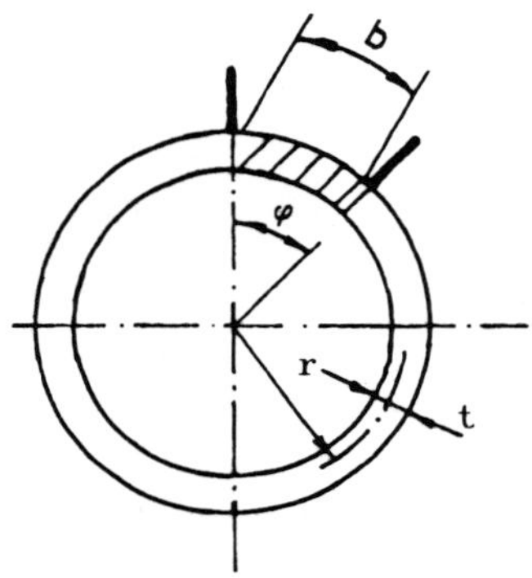

Fig. 10b.E3 Notations of the cylindrical shells with longitudinal stiffeners

NORTH AMERICA

CHAPTER EDITOR

Nicholas F. Morris, Manhattan College

API RP2A 89

Five failure modes must be investigated for any cylindrical shell stiffened by either stringers or stringers and rings.

1) Local shell buckling, ℓ (buckling of the shell plate between stiffeners)
2) Bay instability, B (buckling of the stringers together with the attached shell plate between rings or the ends of the cylinders if there are no ring stiffeners)
3) General instability, g (buckling of one or more rings together with the attached shell plus stringers for ring and stringer stiffened cylinders)
4) Local stiffener buckling (buckling of the stiffener elements)
5) Column buckling (buckling of the cylinder as a column)

The elastic buckling stress for local buckling is

$$\sigma_{x\ell} = \alpha_{o\ell}\, C_x\, E\, t/r \tag{10b.N1}$$

for $1.5 < \delta_\theta < 3.46$

$$\alpha_{o\ell}\, C_x = 3.254 + 0.0253\, \alpha_o\, \delta_\theta^{\,2}$$

for $\delta_\theta > 15.0$

$$\alpha_{o\ell}\, C_x = 0.605 \tag{10b.N2}$$

where

$\delta_\theta = \dfrac{b}{\sqrt{rt}}$

b = circumferential spacing of stringers (longitudinal stiffeners)

α_o = imperfection reduction factor (see Eq. 10a.N3)

$\alpha_{o\ell}\, C_x$ is found from linear interpolation in the range $3.46 < \delta_\theta < 15$. If $\delta_\theta < 1.50$ use $\delta_\theta = 1.50$ in Eq. 10b.N2.

The plastic reduction factors η are

for $\Delta \le 0.55$

$$\eta = 1$$

for $0.55 < \Delta < 1.6$

$$\eta = \frac{0.45}{\Delta} + 0.18 \tag{10b.N3}$$

for $1.6 < \Delta \le 6.25$

$$\eta = \frac{1.31}{1 + 1.15\,\Delta}$$

for $\Delta > 6.25$

$$\eta = \frac{1}{\Delta}$$

where

$\Delta = \dfrac{\sigma_u}{f_y}$

η = applied according to Eq. 10a.N4

Note that the equation for η is general, so σ_u represents any of the five types of buckling critical stresses under investigation.

The equation for general buckling of ring-stiffened cylinders is

$$\sigma_{xg} = 0.605\ \alpha_{xg}\ \frac{\mathrm{E\ t}}{\mathrm{r}} \sqrt{1 + \overline{\mathrm{A}}_r} \qquad (10b.N4)$$

where

$$\overline{\mathrm{A}}_r = \frac{\mathrm{A}_r}{\ell_h\ \mathrm{t}}$$

A_r = cross-sectional area of the ring-stiffener
ℓ_h = distance between the ring-stiffeners

The imperfection factors α_{xg} for the case of general buckling are

for $\overline{\mathrm{A}}_r \geq 0.2$ $\qquad \alpha_{xg} = 0.72$

for $0.06 < \overline{\mathrm{A}}_r < 0.2$ $\qquad \alpha_{xg} = (3.6 - 5.0\ \alpha_o)\ \overline{\mathrm{A}}_r + \alpha_o \qquad$ (10b.N5)

for $\overline{\mathrm{A}}_r \leq 0.06$ $\qquad \alpha_{xg} = \alpha_o$

Bay instability and general instability of stringer-stiffened cylinders require the solution of orthotropic shell equations using member properties provided in the bulletin. The equations must be solved repeatedly to compute the wave numbers corresponding to the minimum buckling load.

Imperfection factors applied to the theoretical buckling stresses found from the orthotropic shell equations for bay instability are

for $\overline{\mathrm{A}}_s \geq 0.06$ $\qquad \alpha_{ob} = 0.65$

for $\overline{\mathrm{A}}_s < 0.06$ $\qquad \alpha_{ob} = \alpha_o \qquad$ (10b.N6)

where

$$\overline{\mathrm{A}}_s = \frac{\mathrm{A}_s}{\mathrm{b\,t}}$$

A_s = area of the stiffener

The cylinder must also satisfy column buckling equations when considered as a circular tube

for $\frac{\mathrm{K}\ell}{\mathrm{r}} > 0.5 \sqrt{\frac{\mathrm{E}}{\sigma_{xl}}}$

$$\sigma_{xc} = \frac{\alpha_{xc}\ \pi^2\ \mathrm{E}}{(\mathrm{K}\ \ell/\mathrm{r})^2} \qquad (10b.N7)$$

where α_{xc} is 0.87

Stiffener design requirements are presented in Section 10d.

ASME N284 80

The equations in ASME N284 80 are superseded by those of the API RP2A 89.

WEST EUROPE

CHAPTER EDITOR

Daniel Vandepitte, University of Ghent

ECCS No. 29 88

Rules given are intended to preclude local stringer instability. The strength requirements are $\sigma_d \leq \sigma_{ul}$, $\sigma_d \leq \sigma_{up}$ and $1.2\sigma_d \leq \sigma_{ug}$. The third requirement does not apply when there are no ring stiffeners.

where

σ_d = design value of acting stress, i.e. stress produced by the factored loads

σ_{ug} = acting stress which an imperfect stiffened cylinder will withstand before general instability occurs

σ_{ul} = acting stress which an imperfect stiffened cylinder will withstand before local shell buckling occurs

σ_{up} = acting stress which an imperfect stiffened cylinder will withstand before panel buckling occurs

Local Shell Buckling

σ_{ul} is the higher of the stresses

$$\sigma_{ul1} = \frac{3}{4}\,\alpha_0\,0.605\mathrm{E}\,\frac{\mathrm{t}}{\mathrm{r}} \tag{10b.W1}$$

$$\sigma_{ul2} = 0.83 \times 3.6\mathrm{E}\,\left(\frac{\mathrm{t}}{\mathrm{b}}\right)^2 \tag{10b.W2}$$

where b is the stringer spacing

If neither $0.605\,\alpha_0\,\mathrm{E}\frac{\mathrm{t}}{\mathrm{r}}$ nor $0.83 \times 3.6\mathrm{E}(\frac{\mathrm{t}}{\mathrm{b}})^2$ exceeds $0.5\mathrm{f}_y$, α_0 is given by Eq. 10a.W5 for "good" panels ($\overline{\mathrm{w}} \leq 0.01l_r$). The value given by Eq. 10a.W5 has to be halved for "poor" panels ($\overline{\mathrm{w}} = 0.02l_r$) and α_0 is obtained by linear interpolation for intermediate values of $\overline{\mathrm{w}}$. A cylindrical panel ($\mathrm{b}/\sqrt{\mathrm{rt}} > 2.44$) is imperfection sensitive, and σ_{ul1} is used, while a supposedly flat panel ($\mathrm{b}/\sqrt{\mathrm{rt}} < 2.44$) is not, and σ_{ul2} is used.

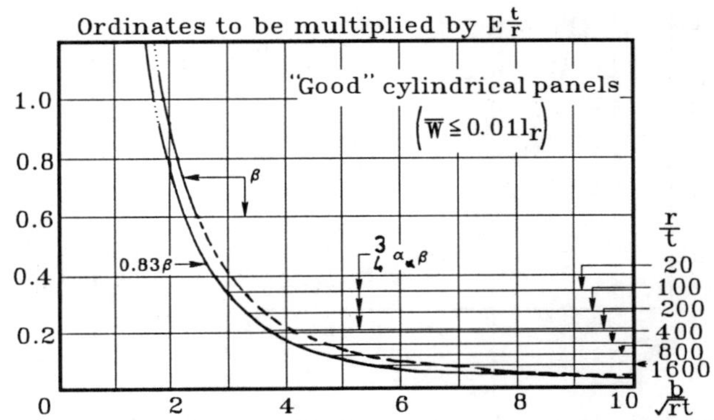

Fig. 10b.W1 Diagram for determination of local shell buckling stress for "good" cylindrical panels

Fig. 10b.W1 is applicable for "good" panels, and Fig. 10b.W2 is applicable for "poor" panels.

If either $0.605\ \alpha_o\ E\ \frac{t}{r}$ or $0.83 \times 3.6\ E\ (\frac{t}{b})^2$ exceeds $0.5\ f_y$, σ_{ul} is the higher of the stresses

$$\sigma_{ul1} = \left(1 - 0.4123\left(\frac{rf_y}{0.605\alpha_0 Et}\right)^{0.6}\right) f_y \tag{10b.W3}$$

and

$$\sigma_{ul2} = \left(1 - 0.25\ \frac{b^2 f_y}{0.83 \times 3.6Et^2}\right) f_y \tag{10b.W4}$$

Stiffened Panel Buckling

for $\alpha_{sp}\sigma_{cr,p} \leq 0.5f_y$

$$\sigma_{up} = \frac{3}{4}\ \alpha_{sp}\sigma_{cr,p}$$

where

A_s = cross-sectional area of one stringer stiffener
α_{sp} = 0.65 when $A_s > 0.2bt$
α_{sp} = α_0 (α_0 given by Eq. 10a.W5) when $A_s < 0.06bt$ and also when $r < 60t$
α_{sp} = linear interpolation for intermediate values of A_s/bt
$\sigma_{cr,p}$ = theoretical panel buckling stress
σ_{up} = panel buckling stress

$\sigma_{cr,p}$ is obtained by minimizing the complex expression (44) in ECCS No.29 84, considered as a function of the integers n and m.

If $\alpha_{sp}\ \sigma_{cr,p} > 0.5\ f_y$, then σ_{up} is given by

$$\sigma_{up} = \left(1 - 0.4123\left(\frac{f_y}{\alpha_{sp}\sigma_{cr,p}}\right)^{0.6}\right) f_y \tag{10b.W5}$$

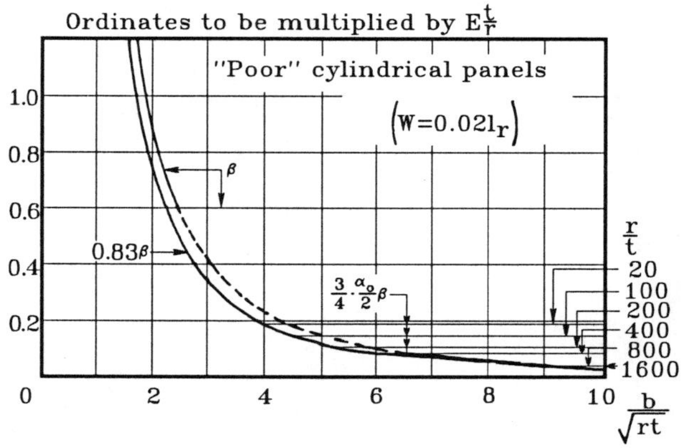

Fig. 10b.W2 Diagram for determination of local shell buckling stress for "poor" cylindrical panels

General Instability

$\sigma_{cr,g}$ is obtained by minimizing the expression (58) in ECCS No. 29 88, considered as a function of the integers n and m.

for $\alpha_g\sigma_{cr,g} \leq 0.5f_y$ $\quad \sigma_{ug} = \frac{3}{4}\alpha_g\sigma_{cr,g}$ (10b.W6)

for $A_s > 0.2bt$ $\quad \alpha_g = 0.65$

for $A_s < 0.06bt$ and also when $r < 60t$ $\quad \alpha_g = \alpha_o$

where

α_g = obtained by linear interpolation for intermediate values of A_s/bt
α_o = given by Eqs. 10a.W5
$\sigma_{cr,g}$ = stress causing general instability of a perfect elastic ring and stiffened cylinder

If $\alpha_g\sigma_{cr,g} > 0.5f_y$, then σ_{ug} is given by

$$\sigma_{ug} = [1-0.4123 \{\frac{f_y}{\alpha_g\sigma_{cr,g}}\}^{0.6}]f_y \qquad (10b.W7)$$

Limitation of Imperfections

The above procedure is valid when all the following conditions are satisfied: $A_s < 2bt$, $I_s < 15bt^3$ and $GC_s < 10bD$, $\overline{w} \leq 0.01l_r$ for the cylindrical panels between stringers, $\overline{w} \leq 0.0015l_G$ when $A_s \geq 0.06bt$, where $\overline{w}$ is the radial lack of straightness of the stringer and l_G is either equal to the cylinder length l or to the critical meridional half wave length (as determined from minimizing expression (44) or expression (58) in ECCS No.29 84, when rings are absent or present, respectively), whichever is smaller, $\overline{w} \leq 0.0015l_G$ and $\overline{w} \leq 0.01l_r$ for the stringers when $A_s < 0.06bt$; $\overline{v}$ is the lateral misalignment of the stiffener attachment to the shell (Fig. 10b.W4) and $\overline{v}$ is limited in the same way as $\overline{w}$. The initial tilt of the stringer web and of the stringer flange, as defined in Fig. 10b.W3, shall be limited by

$$\frac{\overline{v}_1}{h_w} \leq 0.008 \quad \text{and} \quad \frac{\overline{v}_2}{b_f/2} \leq 0.008 \qquad (10b.W8)$$

$n_s \geq 3.5n$, where n is the integer minimizing expression (44) or expression (58) in ECCS No.29 84, when rings are absent or present, respectively.
$n_r \geq 2m$ (that implies $n_r \geq 2$), where m is the integer minimizing expression (58) in ECCS No. 29 88, when there are intermediate rings.
(n_s and n_r are the number of stringers and of intermediate rings, respectively)

COMMENTARY

Complying with the conditions $\sigma_d \leq \sigma_{ul}$ and $\sigma_d \leq \sigma_{up}$ and $1.2\sigma_d \leq \sigma_{ug}$ should preclude local shell instability, stiffened panel buckling, and general instability, respectively.

σ_{ul1} and σ_{ul2} are calculated as if a shell panel between the stringers were either part of an imperfect cylinder or an imperfectly flat plate, respectively. The plasticity reduction, when applicable, results from Eq. 10b.W3 or Eq. 10b.W4.

When starting to calculate $\sigma_{cr,p}$ or $\sigma_{cr,g}$ one does not know the effective circumferential shell panel width b_e appearing in expressions (44) and (58) in ECCS No. 29 88. The term b_e must be determined through iteration, by repeatedly minimizing expression (44) or (58) until convergence is achieved. Charts in ECCS No. 29 88 are intended for preliminary sizing of outside or inside flat bar stiffeners.

ECCS No. 29 88 also provides a simplified, non iterative method of stiffened panel buckling analysis.

Comparison with known test results have shown that the procedure described above yields safe predictions.

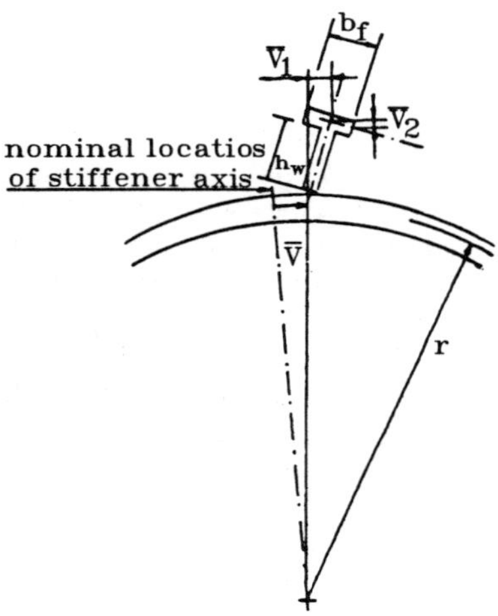

Fig. 10b.W3 The initial tilt of the stringer web and of the stringer flange

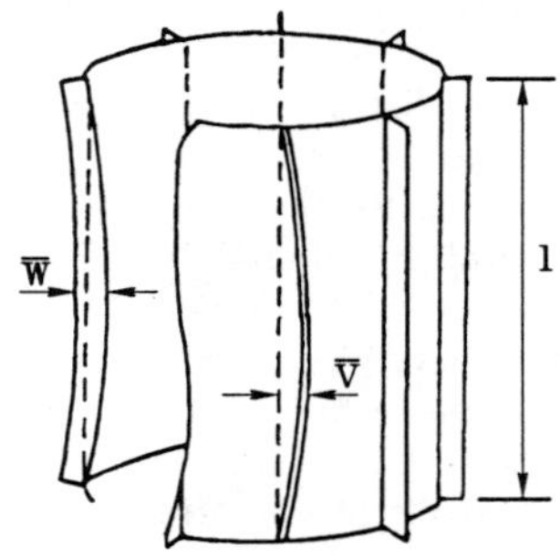

Fig. 10b.W4 The lateral misaligment of the attachment to the shell

INTERNATIONAL COMPARISONS & COMMENT

WHAT ARE THE DIFFERENCES?

The following provisions for stringer stiffened cylinders subjected to meridional compression are included in the recommendations, specifications or codes: (Table 10b.1)

Table 10b.1 Comparison of specifications for design of stiffened cylinders subjected to meridional compression

Region	Specification	item (1) a	b	c	d	(2) a	b	c	d	(3)	(4)	(5)	(6)	(7)
AUSTRALIA														
	AISC 83	0				0					0			
EAST EUROPE														
Czechoslovakia	ČSN 73 1401 84	0		0							0			0
GDR	TGL 13503 82		0	0								0		
Hungary	MI 04.184 78	0			0	0			0		0		0	
Soviet Union	SNiP II-23-81 82	0		0										0
NORTH AMERICA														
	ASME N284 80	0		0	0	0		0	0	0	0		0	0
	API RP2A 89	0		0	0	0		0	0	0	0		0	0
WEST EUROPE														
	ECCS No. 29 88	0		0		0		0		0	0	0	0	

(1) Local shell buckling
  a. single shell curve developed
  b. multiple shell curve developed
  c. shell stress formula constructed
  d. continuous stress formula
(2) Stiffened panel buckling
  a. single shell curve developed
  b. multiple shell curve developed
  c. shell stress formula constructed
  d. continuous stress formula
(3) General instability of stiffened shell
(4) Local stiffener tripping
(5) Describe the magnitude of imperfections
(6) Effect of shell length
(7) Formula for considering the effect of steel grades

The different modes of instability that may occur are:

Local Shell Buckling

This mode is characterized by the panel skin plating between stiffeners buckling locally such that the panel/stiffener junctions remain essentially straight.

Generally, for low slenderness ratio (many stiffeners) the panel can be expected to exhibit a stable initial post-buckling behavior. For high slenderness ratios, the post-buckling behavior is unstable. There is a crucial slenderness ratio.

Stiffened Panel Buckling

This form of buckling is characterized by the stiffener/panel junction deflecting between the end support or between the ring stiffeners in orthogonally stiffened shells.

The development of "smeared" orthotropic analysis has been specifically pursued for this shell form, and this type of analysis can give reasonable estimates for the initial buckling loads. (Ellinas and Croll, 1983)

An alternative method of analysis of shell, in which the stringers may be quite stiff is to consider an "equivalent column" approach. Essentially, this consists of treating a stiffener and portions of continuous plating as a column on a pseudo-elastic foundation (Walker and Sridharan, 1980).

In Hungarian code, allowance is made for the column-like behavior of the stringers with their associated shell plating, taking into account the shell curvature contribution to strength (Hunyadi and Ivanyi, 1977).

General Instability

This involves buckling of both stringer and ring stiffeners together with the shell plating. It can have catastrophic consequences, and for this reason it is usual to suppress this mode by placing limitations on the second moment of area of the ring stiffeners.

There is a dearth of experimental information for the general instability of axially-compressed orthogonally-stiffened shells.

Local Stiffener Tripping

In circumstances, in which the torsional stiffness or the stiffeners is low and the shell skin thickness-to-diameter ratio is relatively high, the stiffener can suffer torsional instability at a stress lower than that required for local or orthotropic buckling.

When the stiffener buckles, it loses a large part of its effectiveness in maintaining a straight generator for the shell. This mode of failure is suppressed in specifications by placing restrictions on the geometry of the stiffeners.

Comparison of Test Results

Comparisons between test results and code recommendations are presented in Fig. 10b.1 for local buckling of ring stiffened cylinders under axial compression. (Samuelson, Vandepitte and Paridaens, 1987) (Ellinas and Williams, 1987) (Kinra and Miller, 1987)

In Fig. 10b.1(a) comparison of Hungarian test data with code recommendations for stiffened panel buckling is presented for MI 04.184 78.

WHY ARE THERE DIFFERENCES?

There are many physical parameters that control the stability of shells; especially the ring- and stringer-stiffened cylindrical shells.

The codes which were developed in various countries usually arose in response to a particular need, therefore many national codes cover only a limited number of load cases and shell geometries. This tends to make the equations for each code appear to be very different. These codes are based on limited experimental data.

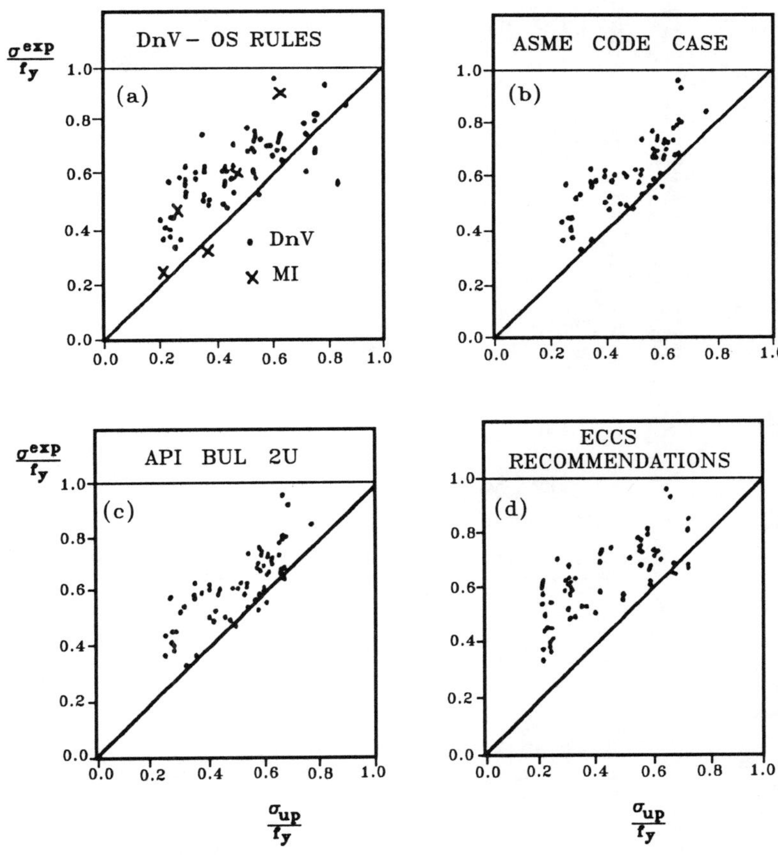

Fig. 10b.1 Code validation through comparison with experimental results

10c. Cylindrical Shells Under Uniform External Pressure

SPECIFICATION PROVISIONS & EXPLANATION

EAST EUROPE

CHAPTER EDITORS

Miklos Ivanyi, Technical University of Budapest

Zbigniew K. Mendera, Politechnika Krakowska

ČSN 73 1401 84

Expressions for the limit stress σ_{ur} of a cylindrical shell of length ℓ under uniform external radial pressure p_e are given by the following.

If $0.5 \le \frac{\ell}{r} \le 10$, for $\frac{\ell}{r}\left(\frac{r}{t}\right)^{\frac{3}{2}} \le 330\ \beta_1^2$

$$\sigma_{ur}=\frac{f_y}{\gamma_m} \tag{10c.E1}$$

for $\frac{\ell}{r}\left(\frac{r}{t}\right)^{\frac{3}{2}} \ge 510\ \beta_1^2$

$$\sigma_{ur}=408\,\frac{r}{\ell}\left(\frac{t}{r}\right)^{\frac{3}{2}}\frac{1}{\beta_1^2}\frac{f_y}{\gamma_m} \tag{10c.E2}$$

If $\frac{\ell}{r} \ge 20$, for $\frac{r}{t} \le 9\beta_1$

$$\sigma_{ur}=\frac{f_y}{\gamma_m} \tag{10c.E3}$$

for $\frac{r}{t} \ge 13\beta_1$

$$\sigma_{ur}=126\left(\frac{t}{r}\right)^2\beta_1^2\,\frac{f_y}{\gamma_m} \tag{10c.E4}$$

where β_1 is equal to $\sqrt{\frac{210}{R_d}}$,

Linear interpolation is used when $10 < \frac{\ell}{r} < 20$.

SNiP II-23-81 82

The limit stress $\sigma_{ur} = \frac{p_e r}{t}$ of a cylindrical shell of length ℓ under uniform external radial pressure p_e is given by the following.

for $0.5 \le \frac{\ell}{r} \le 10$

$$\sigma_{ur} = 0.6\,\frac{\pi\sqrt{2}}{3\sqrt{3}}\,\frac{E}{(1-\nu^2)^{3/4}}\,\frac{r}{\ell}\left(\frac{t}{r}\right)^{3/2}=0.55\,\frac{r}{\ell}\left(\frac{t}{r}\right)^{3/2}E \tag{10c.E5}$$

(Mushtari, 1954)

for $\frac{\ell}{r} \ge 20$

$$\sigma_{ur} = 0.6\,\frac{E}{(1-\nu^2)}\left(\frac{t}{r}\right)^2 = 0.17E\left(\frac{t}{r}\right)^2 \tag{10c.E6}$$

COMMENTARY

The effect of initial imperfections is taken into account in both equations by the factor 0.6.

NORTH AMERICA

CHAPTER EDITOR
Nicholas F. Morris, Manhattan College

API RP2A 89

The failure modes are the same as those considered in Section 10b.
The local buckling stress is

$$\sigma_{pl} = \alpha_{pl} \frac{p_{el}\, r\, K_{pl}}{t} \tag{10c.N1}$$

where K_{pl} is 1.0 for $\delta_x \geq 3.42$.

The value of δ_x will be greater than 3.42 for unstiffened cylinders (see Eq. 10d.N4).

$$\delta_x = \ell_r/\sqrt{rt}$$

for $\delta_x > 1.5$ and $\overline{A} < 2.5$

$$p_{el} = \frac{1.27}{\overline{A}^{1.18} + 0.50}\, E\, (t/r)^2$$

for $2.5 < \overline{A} < 0.208\ r/t$

$$p_{el} = \frac{0.92\, E\, (t/r)^2}{\overline{A}} \tag{10c.N2}$$

for $0.208 < C_p < 2.85$

$$p_{el} = 0.836\, C_p^{1.061} E\, (t/r)^3$$

for $C_p > 2.85$

$$p_{el} = 0.275\, E\, (t/r)^3$$

where

$\overline{A} = \delta_x - 1.17 + 1.068k$

$C_p = \dfrac{\overline{A}}{r/t}$

k = 0 for radial pressure and 0.50 for hydrostatic pressure

ℓ_r = distance between stiffening rings

r = shell radius

t = shell thickness

The local buckling imperfection factor, α_{pl}, is fixed at 0.80.

General instability of a ring stiffened cylinder is governed by the following equations

$$\sigma_{pg} = \frac{\alpha_{pg}\, p_{eg}\, R_o\, K_{pg}}{t} \tag{10c.N3}$$

where K_{pg} is $\dfrac{(1 - 0.3k)\, \ell_e t}{A_r + \ell_e t}$, if $\delta_x > 3.42$ (10c.N4)

A more complicated equation must be used for K_{pg} if $\delta_x < 3.42$ (see Eq. 10d.N3).

$$p_{eg} = \frac{E(t/r)\overline{\tau}_g^4}{(n^2 + k\overline{\tau}_g^2 - 1)\,(n^2 + \overline{\tau}_g^2)^2} + \frac{EI_{er}(n^2 - 1)}{\ell_e R_c^2 R_o} \qquad \text{(10c.N5)}$$

where

A_r = section area of ring
b = stiffener spacing

$I_{er} = I_r + \dfrac{A_r\, z_r^2\, \ell_h t}{A_r + \ell_h t} + \dfrac{\ell_h t^3}{12}$

$\ell_e = 1.56\sqrt{rt} + t_s$
ℓ_h = ring spacing
R_c = radius to the centroid of the effective section
R_o = radius to the outside of the shell
t_s = stiffener thickness
z_r = distance from the centerline of the shell to the centroid of the stiffener ring
$\overline{\tau}_g = \dfrac{\pi r}{b}$

The correct value of n will be that which gives a minimum for p_{eg}.
The imperfection factor α_{pg} is also 0.80 for this buckling mode.

Bay instability of stringer-stiffened or ring- and stringer-stiffened cylinders and general instability of ring- and stringer-stiffened cylinders requires the application of the orthotropic shell equations. These equations must be solved by trial and error to determine the wave numbers corresponding to the lowest buckling pressures.

Stiffener design is considered in section 10d.

ASME N284 89

The equations herein are similar to those of API RP2A 89.

WEST EUROPE

CHAPTER EDITOR
Daniel Vandepitte, Ghent University

ECCS No. 29 88

Note: The rules apply if the out-of-roundness does not exceed 0.5% of the radius.

Unstiffened Cylinders
The strength requirement is $p_d \leq p_u$.

$p_y = tf_y/r$, $p_{cr} = Et\beta_{min}/r$ and $\bar{\lambda} = \sqrt{\frac{p_y}{p_{cr}}}$ must be calculated. β_{min} is a function of $\frac{t}{r}$ and $\frac{\ell}{r}$ (ℓ: unsupported length of the cylinder); its value can be either read from diagram 25 in ECCS No. 29 88 or obtained by minimizing Eq. 10c.W1 as a function of the integer n (≥ 2).

$$\beta = \frac{1}{n^2-1+\frac{1}{2}\left(\frac{\pi r}{\ell}\right)^2}\left\{\frac{1}{\left(\left(\frac{n\ell}{\pi r}\right)^2+1\right)^2}+\frac{t^2}{12r^2(1-\nu^2)}\left(n^2-1+\left(\frac{\pi r}{\ell}\right)^2\right)^2\right\} \quad (10c.W1)$$

When $\bar{\lambda} \geq 1$, $p_u = \frac{1}{2}\, p_{cr}$.

When $0 \leq \bar{\lambda} \leq 1$, $\frac{p_u}{p_y}$ is obtained directly from the upper curve in Fig. 10c.W1.

Ring-Stiffened Cylinders
Local stiffener instability can be avoided by dimensioning the rings according to requirements given in ECCS No.29 84.

It is further required that $p_d \leq p_u$. The value of p_u is obtained as previously under unstiffened cylinders, except that p_y is now found from $(1-\omega\zeta)\, p_y = tf_y/r$, ζ

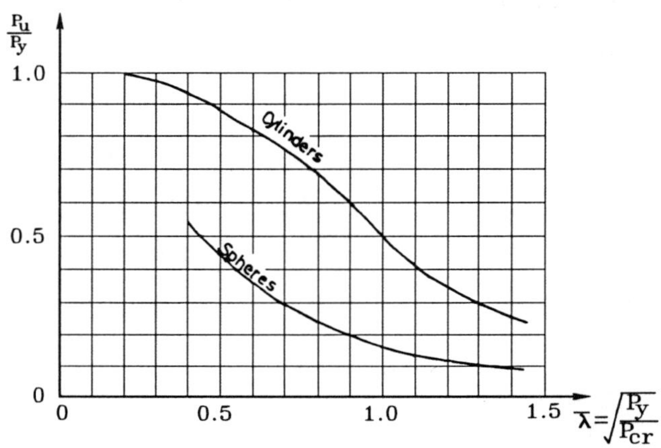

Fig. 10c.W1 Evaluation of p_u/p_y

(and η) depending on $\delta = 1.285\ \ell/\sqrt{rt}$ and ω depending on δ and on shell-stiffener geometry.

This p_u is also used to check the additional requirements

$$\sigma_r = \frac{p\, r^2 \left(1-\frac{\nu}{2}\right)}{\left(t + \dfrac{A^*}{b + \dfrac{2\,\eta\,\ell}{\delta}}\right) r_f} + 0.005\, \frac{(n^2-1)\, p}{p_n - p}\, \frac{E_r\, h_2}{r} \le f_{yr} \qquad (10c.W2)$$

and $p \le p_n$ (10c.W3)

where

$A^* = \dfrac{r^2}{r_r^2} A_r$

A_r = area of the stiffener cross-section
b = depends on the cylinder-ring geometry
E_r = modulus of elasticity of ring stiffener material
h_2 = centroid dimension of the stiffeners
p = $4\, p_u/3$ for cold-formed stiffening rings
p = $3.6\, p_u/3$ for fabricated or hot-formed stiffening rings
r_f = maximum radius of circle of shell
r_r = radius of circle of stiffness centroid

$$\zeta = 2\, \frac{\mathrm{sh}\left(\frac{\delta}{2}\right) \cos\left(\frac{\delta}{2}\right) + \mathrm{ch}\left(\frac{\delta}{2}\right) \sin\left(\frac{\delta}{2}\right)}{\mathrm{sh}\ \delta + \sin \delta}$$

$$\eta = \frac{\mathrm{ch}\ \delta - \cos \delta}{\mathrm{sh}\ \delta + \sin \delta}$$

ν = Poisson's ratio

$$\omega = \frac{A\left(1 - \frac{\nu}{2}\right)}{A + bt + 2\, \frac{\nu t \ell}{\delta}}$$

$$p_n = \frac{E\, t}{\left(n^2 - 1 + \frac{1}{2}\left(\frac{\pi r}{\ell_h}\right)^2\right) \cdot \left(\left(\frac{n \ell_h}{\pi r}\right)^2 + 1\right)^2 r} + (n^2-1)\, \frac{E_r\, I}{r^3\, \ell_l} \qquad (10c.W4)$$

where

I = second moment of area of a light stiffener together with the effective length ℓ_{ef} of the cylinder wall
ℓ_h = distance between end diaphragms or between heavy ring stiffeners
ℓ_l = distance between intermediate light stiffeners

Equations 10c.W2 and 10c.W3 must be fulfilled also for heavy stiffeners, if any. However, Eq. 10c.W4 is then replaced by $p_n = \dfrac{3\, E_r\, I}{r^3\, \ell_h}$, n=2 is set in Eq. 10c.W2, and I, h_2, etc., relate to the heavy stiffeners.

COMMENTARY

The upper curve in Fig. 10c.W1 is a lower bound curve of a cloud of about 700 experimental points.

The values of the integer n to be considered when checking light stiffeners with Eqs. 10c.W2, 10c.W3, and 10c.W4 are n $\geq$ 2 but it is usually sufficient to assume n = 2, 3, 4, 5 and 6. ℓ_{ef} and, hence, I depends on n. Complying with Eq. 10c.W2 and 10c.W3 ensures that the pressure p would not cause yielding of the flange of the stiffening ring assumed to have an initial radial imperfection described by n sine waves of amplitude 0.005 r along its circumference. ECCS No. 29 88 also contains a simplified design method.

INTERNATIONAL COMPARISONS & COMMENT

WHAT ARE THE DIFFERENCES?

The following provisions for cylinders subjected to external pressure are included in the recommendations, specifications and codes: (Table 10c.1)

Table 10c.1 Comparison of specifications for design of cylindrical shells under uniform external pressure

Region	Specification	Item (1)	(2)	(3)	(4)	(5)	(6)
EAST EUROPE							
Czechoslovakia	ČSN 73 1401 84	0	0		0	0	
Soviet Union	SNIP II -23-81 82	0	0			0	
NORTH AMERICA							
	ASME N284 80	0	0	0	0	0	0
	AWWA D100 73						
	API RP2A 89	0	0	0	0	0	0
	AISI 83	0					
WEST EUROPE							
	ECCS N0.29 88	0	0	0		0	0

(1) single shell curve developed
(2) shell stress formula constructed from two or more curves
(3) describe the magnitude of imperfections
(4) formulas for considering the effect of steel grades
(5) shell stress formula for unstiffened cylinders
(6) shell stress formula for stiffened cylinders

The strength of a shell under external pressure depends on its $\frac{\ell}{2r}$ and $\frac{2r}{t}$ ratios and upon the physical properties of the material. It also depends on the amount of deviation of the shell from a true circular form.

Stringer stiffeners are not generally used to increase the critical external pressure of a cylinder. Ring stiffeners are preferred for pressure loading but in cases when the cylinder must withstand several loading conditions, stringer stiffeners may also be present. If stringers are spaced more closely than the buckling half wave length of the shell, they will increase the critical pressure.

WHY ARE THERE DIFFERENCES?

The empirical database of imperfections is different for each code considered. Therefore, different parameters are used in each code. There are some differences because each design code uses different theoretical approaches. In the various countries, the codes were developed arising from particular needs and are based on limited local experimetnal data.

10d. Interaction with Meridional and Radial Compression

SPECIFICATION PROVISIONS & EXPLANATION

AUSTRALIA

CHAPTER EDITOR
J. Michael Rotter, The University of Sydney

AISC 83

Cylinders under Combined Axial Load and Internal Pressure
The strength requirement is given by

$$\sigma^{p}_{ud} = \sigma_{cr} \sqrt{1 - \left(1 - \frac{\overline{p}}{5}\right)^2 \left(1 - \left\{\frac{\sigma_u}{\sigma_{cr}}\right\}^2\right)} \qquad (10d.A1)$$

where

$\overline{p} = \dfrac{p}{\sigma_{cr} \frac{t}{r}}$

$\sigma_{cr} = 0.605 \frac{E\,t}{r}$

σ_u = buckling stress of unpressured cylinder Eq. 10a.A1, 10a.A2, and 10a.A3

COMMENTARY

As the circumferential tension approach about $\overline{p} \simeq 5$, the axial buckling stress approaches the classical elastic critical value σ_{cr}.

EAST EUROPE

CHAPTER EDITORS
Miklos Ivanyi, Technical University of Budapest
Zbigniew K. Mendera, Politechnika Krakowska

ČSN 73 1401 84

Cylinders under Combined Axial Load and External Pressure

Interaction of meridional and radial compression is expressed by the linear interaction formula (Dunkerley's equation).

Cylinders under Combined Axial Load and Internal Pressure

In case of interaction of axial compression σ_d and internal pressure p_i , limit stress as σ'_{ud} and σ''_{ud} are as follows:

If $\sigma'_r = p_i \frac{r}{t} \geq 1.33\ \sigma'_{ud}$

for $\frac{r}{t} \leq 360\ \beta_1^2$

$$\sigma'_{ud} = \frac{f_y}{\gamma_m} \tag{10d.E1}$$

for $\frac{r}{t} \geq 560\ \beta_1^2$

$$\sigma'_{ud} = 450\ \frac{t}{r}\ \beta_1^2\ \frac{f_y}{\gamma_m} \tag{10d.E2}$$

If $\sigma'_r < 1.33\ \sigma'_{ud}$

$$\sigma''_{ud} = \sigma_u + (\sigma'_{ud} - \sigma_u)\ \sqrt{\frac{\sigma'_r}{2\sigma'_{ud}}} \tag{10d.E3}$$

MI 04.184 78

Cylinders under Combined Axial Load and External Pressure

Interaction of meridional and radial compression is expressed by the linear interaction formula (Dunkerley's equation).

Cylinders under Combined Axial Load and Internal Pressure

The limit stress is given by

$$\sigma^p_{ud} = \sigma_{cr} \left(\frac{1-3.884\ \sqrt{t/r}}{\overline{p} + 1.653}\ \overline{p} + 3.884\ \sqrt{t/r} \right) \tag{10d.E4}$$

where

$$\overline{p} = \frac{\sigma_y}{\sigma_{cr}} = \frac{p\ r}{t\sigma_{cr}}$$

$\sigma_{cr} = 0.605\ E\ \frac{t}{r}$

σ_y = hoop-tension stress

COMMENTARY

Cylinders with initial imperfections were investigated by Hutchinson (1965). He found that the internal pressure increases the axial critical stress in the case of axisymmetric imperfections to a much smaller extent. The result of Hutchinson's calculations for axisymmetric imperfections are shown in Fig. 10d.E1.

The interaction equation is

$$\sigma_{ud}^{p} = \frac{\sigma_{cr} - \sigma_{cr}^{exp}}{\sigma_y + \overline{\sigma}_o}\,\sigma_y + \sigma_{cr}^{exp} \qquad (10d.E5)$$

where

$\sigma_{cr}^{exp} = 2.35\sqrt{\frac{t}{r}}\;E\,\frac{t}{r}$ [Alekseev 1956]

$\overline{\sigma}_o = E\,\frac{t}{r}$

The interaction equation is shown in Fig. 10d.E1.

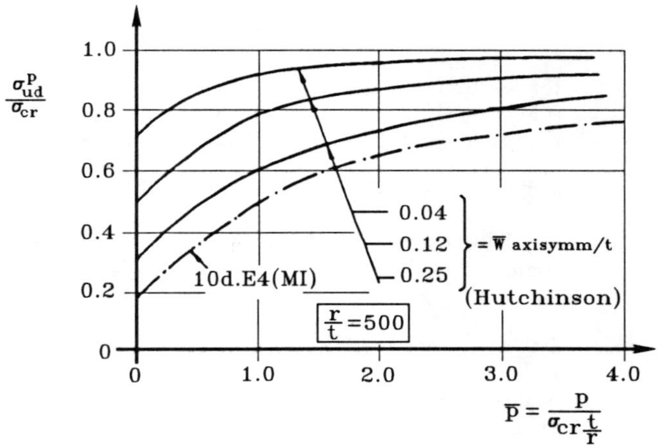

Fig. 10d.E1 Comparison between Hutchinson's results and the design curve

SNiP II-23-81 82

Cylinders under Combined Axial Load and External Pressure

Interaction of meridional and radial compression is expressed by the linear interaction formula (Dunkerley's equation).

$$\frac{1}{\sigma_{cri}} = \frac{1}{\sigma_{\phi i}} + \frac{1}{\sigma_{\theta i}}$$

NORTH AMERICA

CHAPTER EDITOR
Nicholas F. Morris, Manhattan College

API RP2A 89

Interaction equations are given for determining the failure stresses for cylinders subjected to combined loads. The stresses due to bending moment are treated as equivalent axial stresses. The interaction equations are applicable to all modes of failure and to both elastic and inelastic buckling stresses. Each mode must be checked independently.

Axial Tension, Bending and Hoop Compression

The failure stresses are the lower of the following two values; the longitudinal stress is the sum of the axial and bending stresses

$$\sigma_{\theta cj} = \sigma_{pj}\,(1 - 0.25\,\sigma_{\phi cj}/f_y) \qquad (10d.N1)$$

$$\sigma_{\theta cj} + \sigma_{\phi cj} = f_y \qquad (10d.N2)$$

where

j = the failure mode
σ_p = p r
σ_{pj} = the predicted failure stress for radial pressure
$\overline{\sigma}_x$ = $\dfrac{P}{2\pi r} + \dfrac{M}{\pi r^2}$, meridional compression stress
$\sigma_{\theta cj}$ = the failure stress in the hoop direction corresponding to the combined loading
$\sigma_{\phi cj}$ = $\sigma_{\theta cj}\,k\,K_{pj}/K_{xj}$ (10d.N3)
K_{pj}, K_{xj} = given in Table 10d.N1
k = $\overline{\sigma}_x/\sigma_p$
M = bending moment
P = axial tension
p = hoop compression

Table 1d.N1 Stress Factors

K_{ij}	Unstiffened	Ring Stiffened	Stringer Stiffened	Ring and Stringer Stiffened
K_{xl}	1.0	1.0	1.0	1.0
K_{xb}			t/t_s	t/t_s
K_{xg}		1.0		t/t_s
K_{pl}	1.0	Eq. 10d.N4	Eq. 10d.N4	Eq. 10d.N4
K_{pb}			same as K_{pl}	same as K_{pl}
K_{pg}		Eq. 10c.N4		Eq. 10c.N4

$$\left.\begin{array}{ll} \text{for } \delta_x \geq 3.42 & K_{pl} = 1.0 \\ \text{for } \delta_x < 3.42 & K_{pl} = 1 - \omega_o \psi_o \end{array}\right\} \qquad (10d.N4)$$

where

$$A_m = A_r\,(r/R_o)^2$$

$$\omega_o = \frac{1 - 0.3k}{1 + \ell_e t/A_m}$$

$$\begin{array}{ll} \text{for } \delta_x \leq 1.26 & \psi_o = 1.0 \\ \text{for } 1.26 < \delta_x < 3.42 & \psi_o = 1.58 - 0.46\,\delta_x \\ \text{for } \delta_x \geq 3.42 & \psi_o = 0 \end{array}$$

where

A_r = cross-sectional area of ring stiffener
ℓ = distance between strings or support points
$\ell_e = 1.56\sqrt{rt} + t_s$
R_o = maximum radius of circle
$\delta_x = \dfrac{\ell}{\sqrt{rt}}$

Axial Compression, Bending, and Hoop Compression

The longitudinal stress is the sum of the axial and bending stresses. If the bending stress exceeds the axial stress, Eqs. 10d.N1 and 10d.N2 apply. The following equation applies in all cases.

$$R_a^2 - c_o R_a R_h + R_h^2 = 1.00 \qquad (10d.N5)$$

where

$R_a = \sigma_{\phi cj}/\sigma_{xj}$
$R_h = \sigma_{\theta cj}/\sigma_{pj}$

The equations for c are dependent upon the cylinder geometry.

(a) Unstiffened and ring and stringer stiffened cylinders (all modes)

$$c_o = \frac{\sigma_{xj} + \sigma_{pj}}{f_y} \leq 1.0 \qquad (10d.N6)$$

(b) Stringer-stiffened and ring- and stringer-stiffened cylinders (local buckling)

$$c_o = \frac{0.40(\sigma_{xj} + \sigma_{pj})}{f_y} - 0.80 \qquad (10d.N7)$$

(c) Stringer-stiffened and ring- and stringer-stiffened cylinders (bay and general instability)

$$c_o = \frac{1.5(\sigma_{xj} + \sigma_{pj})}{f_y} - 2.0 \qquad (10d.N8)$$

Design of Stiffeners

The sizes of the stringers are determined from the bay instability mode equations and the sizes of the rings are determined from the general instability mode equations.

COMMENTARY

The shell buckling stress equations were developed based on no interaction between the buckling modes. The buckling, however, may be reduced if the predicted buckling stress for either bay instability or general instability is approximately equal to the predicted local buckling stress. Similarly, if the predicted general instability stress is approximately equal to the bay instability stress, the actual stress for either of these modes may be less than predicted.

Mode interaction can be avoided by applying a factor β to the strains corresponding to the buckling stresses. It is desirable to provide a hierarchy for failure with general instability preceded by bay instability and bay instability preceded by local shell buckling. A minimum factor of $\beta = 1.2$ is recommended for both the bay and instability modes. The designer may elect to select a higher β value for the general instability mode.

To preclude stiffener buckling prior to shell buckling, the local stiffener buckling stress must be greater than the shell buckling stresses given in the previous sections. The local stiffener buckling stress can be assumed equal to the yield stress for stiffeners that satisfy compact section requirements.

The stringer properties are determined by equating the elastic design stresses for local buckling and bay instability and the ring properties are determined by equating the elastic design stresses for local buckling and general instability.

WEST EUROPE

CHAPTER EDITOR

Daniel Vandepitte, Ghent University

ECCS No. 29 88

Cylinders under Combined Axial Load and External Pressure

Two conditions should be satisfied:

$\sigma_d \leq \sigma_u$

$$\text{for } \sigma_d \leq \frac{rp_u}{2t} \qquad \frac{p_d}{p_u} \leq 1$$

$$\text{for } \sigma_d > \frac{rp_u}{2t} \qquad \frac{p_d}{p_u} + \frac{\sigma_d - \frac{rp_u}{2t}}{\sigma_u - \frac{rp_u}{2t}} \leq 1 \tag{10d.W1}$$

where

p_d = the factored acting pressure
σ_d = the factored meridional acting stress

σ_u and p_u are evaluated as explained in Sections 10a.W and 10c.W.

Cylinders under Combined Axial Load and Internal Pressure p_i

The strength requirement is given by

$$\sigma_{eff,d} = \sqrt{\sigma_{xd}^2 + \sigma_{yd}^2 - \sigma_{xd}\sigma_{yd}} \leq \sigma_{eff,u} \tag{10d.W2}$$

The effective stress at buckling $\sigma_{eff,u}$ is defined by

$$\sigma_{eff,u} = \sqrt{\sigma_{xu}^2 + \sigma_{yu}^2 - \sigma_{xu}\sigma_{yu}} \tag{10d.W3}$$

and it is evaluated by means of the equations

$$\text{for } \overline{\lambda}_p \geq \sqrt{2} \qquad \frac{\sigma_{eff,u}}{f_y} = \frac{0.75}{\overline{\lambda}_p^2} \tag{10d.W4}$$

$$\text{for } \overline{\lambda}_p \leq \sqrt{2} \qquad \frac{\sigma_{eff,u}}{f_y} = 1\text{–}0.4123\ \overline{\lambda}_p^{1.2}$$

The parameter $\overline{\lambda}_p$ is defined by

$$\overline{\lambda}_p = \sqrt{\frac{\overline{\beta} f_y}{\alpha_p \sigma_{cr}}} \tag{10d.W5}$$

where

$\overline{\beta}$ = the ratio of the meridional membrane stress σ_{xu} to the effective stress $\sigma_{eff,u}$ just before buckling occurs

$$\overline{\beta} = \frac{\sigma_{xu}}{\sigma_{eff,u}} = \frac{\sigma_{xu}}{\sqrt{\sigma_{xu}^2 + \sigma_{yu}^2 - \sigma_{xu}\sigma_{yu}}} \tag{10d.W6}$$

and where the numerical value of α_p is given by Eq. 10a.W8. In this context the compressive meridional membrane stress σ_{xu} is to be treated as a positive quantity and the tensile circumferential membrane stress $\sigma_{yu} = -\frac{r}{t}\, p_i$ as a negative quantity.

COMMENTARY

Experimental evidence shows that the Eq. 10d.W1 is conservative.

If the acting meridional compression and the acting internal pressure are proportional, the number $\overline{\beta}$ appearing in Eq. 10d.W5 and defined by Eq. 10d.W6 is equal to the ratio of the design value σ_{xd} of the meridional membrane stress to the design value $\sigma_{eff,d}$ of the effective stress, and the evaluation of $\overline{\beta}$ is then straightforward. If the meridional compression and the internal pressure are physically independent (the internal pressure may, for example, be constant), it is necessary to apply an iteration procedure until convergence shows that the correct values of $\overline{\beta}$ and of $\sigma_{eff,u}$ at buckling have been obtained. Section C1.5.4 of ECCS No. 29 88 contains a numerical example of each case.

INTERNATIONAL COMPARISONS & COMMENT

WHAT ARE THE DIFFERENCES?

Cylinders under Combined Axial Load and External Pressure

In ČSN 73 1401 84, MI 04.184 78, SNiP II-23-81 82, and ECCS No. 29 88 the interaction of meridional and radial compression is expressed by Dunkerley's equation.

However, the ECCS No. 29 88 interaction equation accounts for the fact that only that part of the axial load that exceeds the load acting on the end closures (in the case of hydrostatic pressure) causes p_d to decrease below the p_u evaluated according to section 10c.W.

Cylinders under Combined Axial Load and Internal Pressure

An internal pressure p_i does not raise the critical axial load computed for a perfect elastic cylinder. But a high p_i does alleviate the detrimental effect of imperfections so thoroughly that the reduction factor approaches unity.

AISC 83, ČSN 73 1401 84, and MI 04.184 78 assess the strength increase to be dependent on the parameter

$$\overline{p} = \frac{p_i\, r}{t\, \sigma_{cr}} \tag{10d.1}$$

that is supported by Hutchinson (1965) from a theoretical standpoint.

ECCS No. 29 88 uses instead the parameter

$$\rho = \frac{p_i}{E}\left(\frac{r}{t}\right)^{1.5} \tag{10d.2}$$

The above differences make comparisons a little difficult, especially as each source uses a different assessment of the unpressurized strength. As a result, only a sample comparison, for $\frac{r}{t} = 500$ is given in Fig. 10d.1. The dimensionless strength

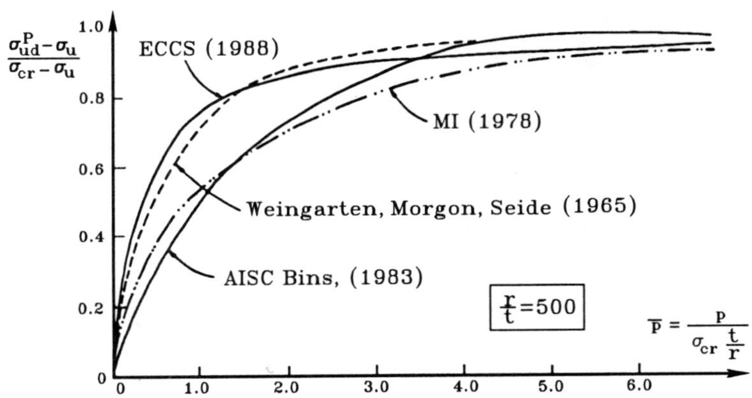

Fig. 10d.1 Comparison of rules for pressurisation strength gain

gain predicted by each of the three above descriptions is compared with the original lower bound curve of (Weingarten, Morgan and Seide, 1965).

The ECCS No. 29 88 curve provides a close match to the experimental data. The AISC 83 curve is quite conservative at low pressures.

The design rules are based on the very comprehensive study of (Weingarten et al., 1965). Unfortunately, the authors did not propose an algebraic relation to describe their results. As a consequence, some considerable differences have arisen when their experimental observations were approximated by later workers.

More recent careful experiments on steel cylinders have been put forward to support the ECCS No. 29 88 equations (Saal et al., 1979). The validity of high strength gains has been argued by Esslinger, (1977) especially as the experimental scatter decreases with increasing internal pressure.

WHY ARE THERE DIFFERENCES?

Each code uses different assessments of unpressurized strength. In addition, the codes are based on different sets of experiments.

10e. Conical Shells

SPECIFICATION PROVISIONS & EXPLANATION

AUSTRALIA

CHAPTER EDITOR

J. Michael Rotter, The University of Sydney

AISC 83

The failure stress σ_u at the smallest radius may be approximated by multiplying the corresponding cylinder failure stress (Eq. 10a.A1 - 10a.A4) by cos β. (β is the apex half angle.)

EAST EUROPE

CHAPTER EDITORS
Miklos Ivanyi, Technical University of Budapest
Zbigniew K. Mendera, Politechnika Krakowska

ČSN 73 1401 84

Axially Compressed Shells
The limit load is given by Eq. 10e.E2, σ_u being the limit stress of a cylindrical shell of radius r_e.

$$r_e = \frac{0.9r_1+0.1r_2}{\cos\beta} \tag{10e.E1}$$

$$N \le N_u = 2\pi r_e t\sigma_u \cos^2\beta \tag{10e.E2}$$

where

$r_1 > r_2$
$\beta \le 60°$

Conical Shells with Axial Compression and External Pressure p
The interaction formula is given by

$$\frac{\sigma_n}{\sigma_u} + \frac{\sigma_r}{\sigma_{ur}} = \sqrt{1+3\,\frac{\sigma_n\sigma_r}{\left(\frac{f_y}{\gamma_m}\right)^2}} \tag{10e.E3}$$

where

$A_e = 2\pi r_e t\cos^2\beta$
$\sigma_n = \frac{N}{A_e}$
$\sigma_r = p\frac{r_e}{t}$
σ_u = (see Eq. 10a.E2)
σ_{ur} = (see Eqs. 10c.E1 – 10c.E4, but r_e is used)

SNiP II-23-81 82

The limit load of a conical shell is given by Eq. 10e.E5, σ_u being the limit stress of a cylindrical shell of radius r_e.

$$r_e = \frac{0.9r_1+0.1r_2}{\cos\beta} \tag{10e.E4}$$

$$N \le N_u = 2\pi r_e t\sigma_u \cos^2\beta \tag{10e.E5}$$

In case of external pressure, Eq. 10c.E5 can be used, replacing r by r_e (Fig. 10e.E1).

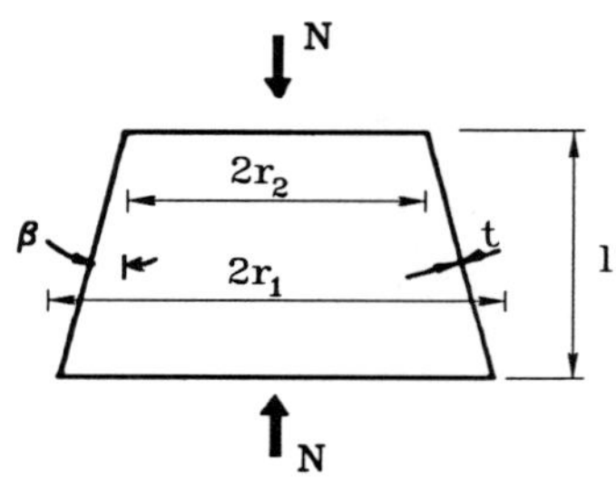

Fig. 10e.E1 Notations of the conical shells

NORTH AMERICA

CHAPTER EDITOR
Nicholas F. Morris, Manhattan College

ASME N284 80

Conical shells are not explicitly discussed in ASME N284 80. However, the code encourages one to carry out the bifurcation analysis of shells of revolution and to apply imperfection reduction factors and plasticity factors to the buckling values obtained. The reduction factors would be those for a cylindrical shell.

WEST EUROPE

CHAPTER EDITOR

Daniel Vandepitte, Ghent University

ECCS No. 29 88

Liquid-filled Conical Shells with Simply Supported Edge (Fig. 10e.W1)
The conical shell contains a liquid of weight γ'

The quantities σ_x (> 0) and $\sigma_y = -\dfrac{\gamma' \, hr_2}{t \cos \beta_o}$ (< 0) denote meridional and circumferential membrane stresses, respectively, in the cone wall in the vicinity of the supported edge. The additional subscripts u or d denote that the stresses are taken at the buckling stage or are generated by the factored acting loads, respectively. The design value $\sigma_{eff,u}$ of the effective stress at buckling is then defined by

$$\sigma_{eff,u} = \sqrt{\sigma_{xu}^2 + \sigma_{yu}^2 - \sigma_{xu}\sigma_{yu}} \qquad (10e.W1)$$

and the general strength requirement is

$$\sigma_{eff,d} = \sqrt{\sigma_{xd}^2 + \sigma_{yd}^2 - \sigma_{xd}\sigma_{yd}} \le \sigma_{eff,u} \qquad (10e.W2)$$

(A) Elastic buckling of cones with $15^\circ \le \beta_o < 65^\circ$
The strength condition 10e.W2 may be simplified into

$$\sigma_d \le \sigma_u \qquad (10e.W3)$$

where σ_d has the same meaning as σ_{xd} in the preceding paragraph, and where σ_u is given by the following

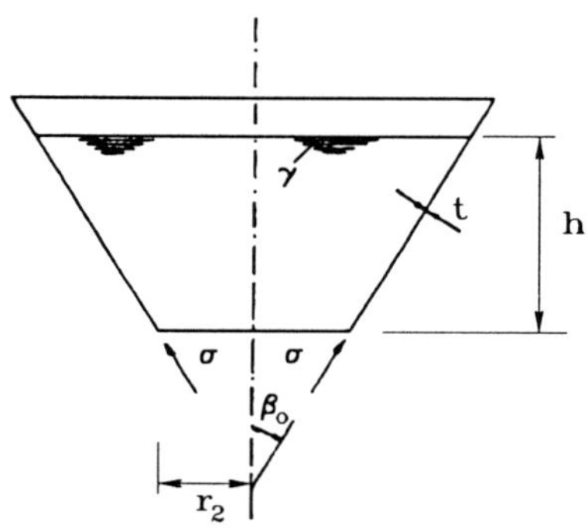

Fig. 10e.W1 Notations of the liquid-filled conical shells

for $\overline{w} \leq 0.008\ell_r$

$$\sigma_u = \frac{44{,}620}{1.1}\,\underline{\omega}^{-1.724}\,\frac{\gamma' r^2}{t\cos\beta} \qquad (10e.W4)$$

for $0.008\ell_r < \overline{w} \leq 0.02\ell_r$

$$\sigma_u = \frac{39{,}030}{1.1}\,\underline{\omega}^{-1.713}\,\frac{\gamma' r^2}{t\cos\beta}$$

where

$$\ell_r = 3.6\sqrt{\frac{r_2 t}{\cos\beta_o}}$$

$$\underline{\omega} = \frac{1000\,r}{t\cos\beta_o}\sqrt{\frac{\gamma' r_2}{E}\sqrt{1-\nu^2}} \qquad (10e.W5)$$

$\overline{w}$ (Fig. 10b.W4) is the largest imperfection measured from a straight template of length ℓ_r held anywhere against any meridian in the region of width $20\sqrt{r_2 t/\cos\beta_o}$ adjoining the supported edge. Eq. 10e.W4 is applicable provided that the following conditions hold: At least 80° of σ_d is due to the weight of the liquid in the conical vessel, $15° \leq \beta_o < 65°$, and $1.1\sigma_u \leq 0.5f_y$.

(B) Elastic-plastic buckling of cones with $15° \leq \beta_o < 65°$
When $1.1\sigma_u$ calculated according to Eq. 10e.W4 turns out to exceed $0.5f_y$, the full strength condition of Eq. 10e.W2 must be used and $\sigma_{eff,u}$ is evaluated by means of the following

for $\overline{\lambda} \leq \sqrt{2}$

$$\frac{\sigma_{eff,u}}{f_y} = 1-0.3061\,\overline{\lambda}^{1.6666} \qquad (10e.W6)$$

The parameter $\overline{\lambda}$ is defined by

$$\overline{\lambda} = \sqrt{\frac{\beta' f_y}{1.1\sigma_u}} \qquad (10e.W7)$$

where

β' = the ratio of the meridional stress σ_{xu} to the effective stress $\sigma_{eff,u}$ at buckling

$$= \frac{\sigma_{xu}}{\sigma_{eff,u}} = \frac{\sigma_{xu}}{\sqrt{\sigma_{xu}^2 + \sigma_{yu}^2 - \sigma_{xu}\sigma_{yu}}} \qquad (10e.W8)$$

σ_u = value obtained from one of the Eq. 10e.W4

If the meridional and the circumferential stresses increase proportionally between the working condition and the buckling stage, β' is simply equal to the known ratio of σ_{xd} to $\sigma_{eff,d}$. If the meridional compression and the circumferential tension may not be assumed to vary proportionally when the load factors increase, the correct values of β' and $\sigma_{eff,u}$ result from an iteration procedure similar to the one in Section 10d(W), section (B).

(C) Elastic buckling of cones with $\beta_o < 15°$
The value of σ_{u1} is calculated as the stress σ_u given by Eq. 10e.W4, in which β is not set equal to its real value, but to 15°. Next σ_{u2} is calculated for an axially

loaded cylinder of radius r_2 and thickness t, with an internal pressure p_i equal to the liquid pressure existing at the bottom of the actual conical shell under working conditions

$$\sigma_{u2} = \frac{\alpha_p \sigma_{cr}}{\gamma} \qquad \text{with } \gamma = \frac{4}{3} \text{ and } \sigma_{cr} = 0.605 \text{ E} \frac{t}{r_2} \tag{10e.W9}$$

and with α_p given by Eqs. 10a.W8 and 10a.W5 when $\overline{w} \leq 0.01 \ \ell_r$. α_p must be reduced to 0.5 α_p when $\overline{w} = 0.02 \ \ell_r$. When $0.01 \ \ell_r < \overline{w} \leq 0.02 \ \ell_r$, linear interpolation between α_p and 0.5 α_p provides the factor to be used in Eq. 10e.W9 when calculating σ_{u2}. σ_u for the liquid-filled conical shell with $\beta_o < 15°$ may be then evaluated by interpolation between σ_{u2} and σ_{u1}

$$\sigma_u = \sigma_{u2} + \frac{\beta_o}{15}(\sigma_{u1} - \sigma_{u2}) \tag{10e.W10}$$

provided that neither 1.1 σ_{u1} nor $\frac{4}{3}\sigma_{u2}$ exceeds $\frac{f_y}{2}$.

(D) Elastic-plastic buckling of cones with $\beta_o < 15°$
When either $1.1\sigma_{u1}$ or $4\sigma_{u2}/3$ exceeds $f_y/2$, with σ_{u1} and σ_{u2} evaluated according to section (C), then the strength requirements of Eqs. 10e.W2 and 10e.W11 are applicable.

$\sigma_{eff,u1}$ is the effective buckling stress $\sigma_{eff,u}$ obtained in accordance with section (B) for a fictitious cone having the same r_2 and t as the actual cone, but having generatrices that form an angle of 15° with the axis of revolution.

$\sigma_{eff,u2}$ is the effective buckling stress $\sigma_{eff,u}$ (obtained by means of Eqs. 10d.W4, 10d.W5, and 10d.W6 and of the procedure referred to in the Commentary at the end of Section 10d(W)) for an axially loaded fictitious cylinder of radius r_2 and thickness t, with an internal pressure p equal to the liquid pressure existing at the bottom of the actual conical shell under working conditions.

$\sigma_{eff,u}$ for the actual liquid-filled conical shell with $\beta_o < 15°$ then may be evaluated with the equation

$$\sigma_{eff,u} = \sigma_{eff,u2} + \frac{\beta_o}{15}\left[\sigma_{eff,u1} - \sigma_{eff,u2}\right] \tag{10e.W11}$$

COMMENTARY

Checking the imperfections is especially important in the part of the cone adjoining the supported edge, say in a region of width $20\sqrt{r_2 t/\cos \beta_o}$, measured along the generatrices.

The values of $1.1\sigma_u$ given by Eq. 10e.W4 very nearly represent lower bounds of stresses obtained from hundreds of tests on models that buckled primarily elastically. 1.1 is an incremental safety factor accounting for the moderate imperfection sensitivity of liquid-filled cones. The above recommendations may also be applied to vessels in the shape of other shells of revolution, such as a sphere or a torus, provided that the meridian is everywhere either straight or concave towards the vertical axis of revolution.

Continuity between cones and cylinders is achieved by using the interpolation formulas given in (C) and (D) when $\beta_o < 15°$.

Conical Shells under Meridional Compression, Possibly Combined with Uniform Internal Pressure

The boundary conditions must be such that u sin β_o + w cos $\beta_o = 0$ along both edges of the cone (Fig. 10e.W2). Moreover, β_o must be less than 65°. The imperfections must remain within the limits defined in Section 10a(W). They are measured in the way described in Section 10a(W), the length of the rod and of the circular template used at each specific location on the shell being determined based on the principal radius of curvature of the cone at the midpoint of the rod or template. $\overline{w}$ is measured normal to the shell.

$$\sigma_d t \cos \beta_o = \frac{Q_d}{2\pi r} + \frac{M_d}{\pi r^2} - \frac{r p_d}{2} \qquad (10e.W12)$$

Equation 10e.W12 gives the extreme design value of the stresses acting along a given parallel circle of radius r (Fig. 10e.W2) and generated by the design values of the various loads Q, M and p.

The strength condition $\sigma_d \leq \sigma_u$ applies to any point on the shell wall. The calculation of the buckling stress σ_u by means of the procedure described in Section 10a(W) for circular cylinders, including the evaluation of the knock-down factor α, shall be based on the local principal radius of curvature $r_e = r/\cos \beta_o$. In cases with no internal pressure the small radius $r=r_2$ shall be used. When internal pressure is present, calculations have to be carried out for different values of r and the strength condition must be fulfilled for all of them.

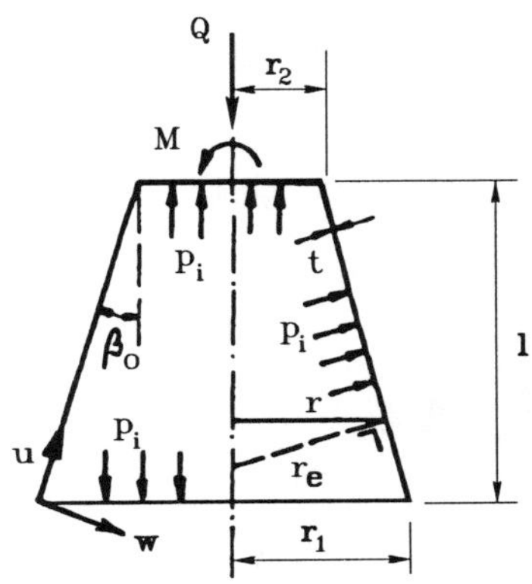

Fig. 10e.W2 Notations of the conical shells under meridional compression, combined with internal pressure

Conical Shells under Uniform Hydrostatic External Pressure

The boundary conditions must be such that u sin β_o + w cos β_o = 0 (Fig. 10e.W2) along both edges, and β_o must be less than 65°. The out-of-roundness may not exceed 0.5% of the radius.

The strength requirement is $p_d \leq p_u$. p_u is evaluated by means of the procedure described in Section 10c(W) for unstiffened circular cylinders. However, the pressure p_y is obtained from $p_y = \frac{t \cos \beta_o}{r_1} f_y$ (Fig.10e.W3) and in the calculation of p_{cr}, the following parameters shall be used:

$$\text{radius } \bar{r}_e = \frac{r_1 + r_2}{2 \cos \beta_o} \quad \text{and} \quad \ell_e = \frac{\ell}{\cos \beta_o} \tag{10e.W13}$$

As in Section 10c(W), p_u may be derived from p_y, p_{cr} and $\bar{\lambda}$ by means of the upper curve in Fig. 10c.W1.

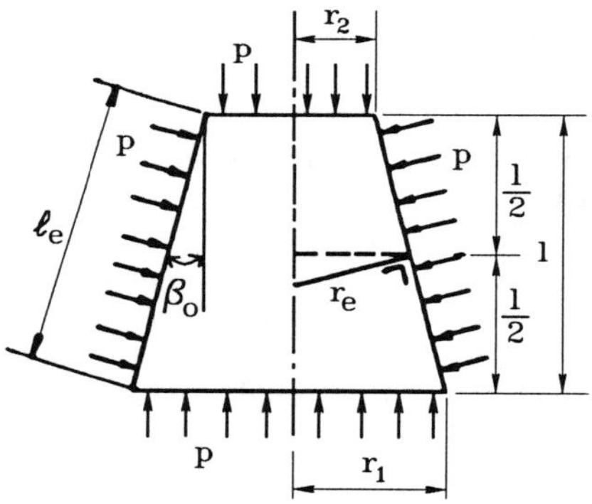

Fig. 10e.W3 Notations of conical shells under external pressure

INTERNATIONAL COMPARISONS & COMMENT

WHAT ARE THE DIFFERENCES?

The following provisions for conical shells are included in the various recommendations, specifications or codes. (Table 10e.1)

The buckling problem of a conical shell was solved based on the linear theory by (Seide, 1956). The main result of this investigation is that the cone can be substituted for by a cylinder with the radius r/cos β_o equal to the actual principal radius of curvature of the cone. Therefore the computation of conical shells becomes completely identical to that of cylinders. (AISC 83, ČSN 73 1401 84, SNiP II-23-81 82, AISI 79, ECCS No. 29 88).

WHY ARE THERE DIFFERENCES?

Design codes for conical shells are based on a small number of experiments. Therefore, interpretation of the empirical data can differ.

ECCS No. 29 88 seems to be the only set of recommendations considering liquid-filled conical vessels. In the past water-towers have been known to collapse due to buckling of the bottom shell.

Table 10c.1 Comparsion of specifications for design of conical shells

Region	Specification	Item (1)	(2)	(3)	(4)	(5)	(6)	(7)	(8)
AUSTRALIA									
	AISC 83	0	0				0		
EAST EUROPE									
Czechoslovakia	ČSN 73 1401 84	0	0	0			0		
Soviet Union	SNiP II -23-81 82	0	0				0		
NORTH AMERICA									
	AISI 79	0							
WEST EUROPE									
	ECCS No. 29 88	0	0	0	0	0	0	0	

(1) use the radius of equivalent cylinder
(2) axially compressed cone
(3) other type of loading
(4) describe the magnitude of imperfections
(5) the field of elastic buckling itself
(6) the field of elastic and plastic buckling
(7) use the partial safety factor
(8) stiffened and unstiffened cones

10f. Spherical Shells Under Uniform External Radial Pressure

SPECIFICATION PROVISIONS & EXPLANATION

EAST EUROPE

CHAPTER EDITORS
Miklos Ivanyi, Technical University of Budapest
Zbigniew K. Mendera, Politechnika Krakowska

ČSN 73 1401 84

Limit stress $\sigma_{usp} = \frac{p_u R}{2t}$ of a spherical under uniform external radial pressure p_u is given by

for $\frac{R}{t} \leq 750$

$$\sigma_{usp} = \frac{80t}{R}\,\beta_1\,\frac{f_y}{\gamma_m} \tag{10f.E1}$$

where R is the radius of spherical shell.

SNiP II-23-81 82

Limit stress $\sigma_{usp} = \frac{p_u R}{2t}$ of a spherical shell under uniform external radial pressure p_u is given by

for $\frac{R}{t} \leq 750$

$$\sigma_{usp} = 0.1\mathrm{E}\,\frac{t}{R} \tag{10f.E2}$$

replacing factor $1/\sqrt{3(1-\nu^2)}=0.606$ in theoretical formula by 0.1 because of effect of imperfections. (R=radius of spherical shell)

NORTH AMERICA

CHAPTER EDITOR
Nicholas F. Morris, Manhattan College

ASME N284 80

The local elastic buckling stress is

$$\sigma_{\phi\ell} = \alpha_\phi \, C_\phi \, E \, t/R \qquad (10f.N1)$$

for $1.5 < \delta_\phi < 1.73$

$$C_\phi = \frac{0.904}{\delta_\phi^{\,2}} + 0.1013 \, \delta_\phi^{\,2}$$

for $\delta_\phi \geq 1.73$

$$C_\phi = 0.605 \qquad (10f.N3)$$

for $1.5 < \delta_\phi < 1.73$

$$\alpha_\phi = 0.837 - 0.014 \, \delta_\phi$$

for $1.73 < \delta_\phi < 23.$

$$\alpha_\phi = \frac{0.826}{\delta_\phi^{\,0.46}} \qquad (10f.N2)$$

for $\delta_\phi > 23.6$

$$\alpha_\phi = 0.124$$

$$\delta_\phi = \frac{\ell_s}{\sqrt{R \, t}}$$

ℓ_s = arc length of support in circumferential direction

The plasticity reduction factor is the same as that for a cylinder under axial compression.

When stiffeners are used in both directions, the general buckling stress is

$$\sigma_{\phi G} = \frac{0.583 \, E \, t_s^{\,0.25}}{R \, t_r^{\,0.75}} \, \alpha_\phi \left(\frac{I_s}{b}\right)^{1/3} \left(\frac{I_r}{\ell_r}\right)^{1/6} \qquad (10f.N4)$$

where t_r and t_s are equivalent shell thicknesses taking into account the stiffeners and rings.

For two-way orthogonal stiffeners

$$\sigma_{\phi G} \geq \alpha_\phi \frac{4.81 \, E \, t}{\delta_s^{\,0.40} \, R} \qquad (10f.N5)$$

where δ_s is the smaller of δ_ϕ and δ_θ.

The stiffeners must satisfy the relation

$$I_s = \frac{62.4 \, B \, t^3}{\delta_s^{\,1.8}} \left(\frac{t}{t_s}\right)^{3/4} \qquad (10f.N6)$$

Note that toroidal and ellipsoidal shells shall be analyzed as equivalent spheres by substituting R_1 or R_2 for R in the spherical shell equations. See Fig. 10f.N1 for R_1 and R_2.

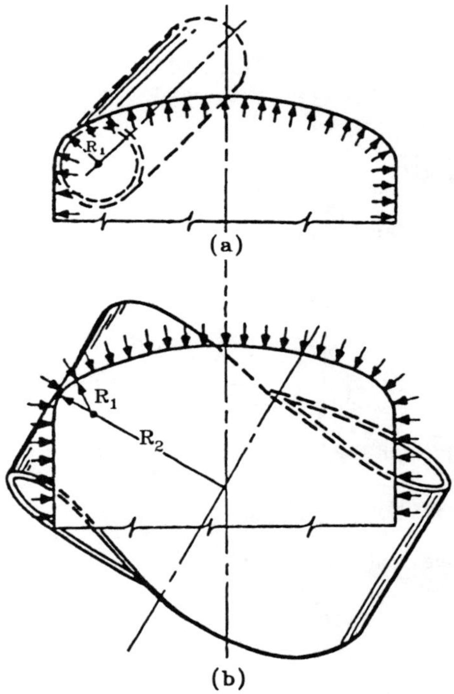

Fig. 10f.N1 Correlation of ellipsoidal head and equivalent cylinders

WEST EUROPE

CHAPTER EDITOR

Daniel Vandepitte, Ghent University

ECCS No. 29 88

The strength requirement is $p_d \leq p_u$.

$$p_y = 2t\frac{f_y}{R} \qquad (10f.W1)$$

$$p_{cr} = 1.21\ E\left(\frac{t}{R}\right)^2 \text{ and} \qquad (10f.W2)$$

$$\bar{\lambda} = \sqrt{p_y/p_{cr}} \qquad (10f.W3)$$

p_u/p_y is read directly from the lower curve in Fig. 10c.W1.

$$p_u = 0.18\ p_{cr} \text{ holds for } \bar{\lambda} \geq \sqrt{2}.$$

This procedure applies only to spheres that are spherical to within 1% on the radius and whose local radius of curvature, measured on an arc length of $2.4\sqrt{Rt}$, does not exceed the nominal value by more than 30%.

COMMENTARY

The ordinates of the curve in Fig. 10c.W1 are those of a lower bound of test points, divided by the partial safety factor $\gamma = \frac{4}{3}$ that accounts for the notorious imperfection sensitivity of externally pressurized spheres.

Chapter 10 of ECCS No. 29 88 treats torispherical end closures subject to uniform internal pressure.

INTERNATIONAL COMPARISONS & COMMENT

WHAT ARE THE DIFFERENCES?

The following provisions for spherical shells included in the specifications or codes: (Table 10f.1)

It is a known fact (Volmir, 1967) that for an ideal spherical shell under hydrostatic pressure in the elastic range the critical value is determined exactly the same way as for the cylindrical shell under axial compression. The non-linearity of the problem and effect of imperfections on the stability of the spherical shell are analogous to the cylindrical shell. However, we have far less experimental information than in the case of cylindrical shells.

WHY ARE THERE DIFFERENCES?

The various codes are based on different sets of experiments.

Table 10f.1 Comparison of specifications for design of spherical shells under uniform external radial pressure

Region	Specification	Item (1)	(2)	(3)	(4)
EAST EUROPE					
Czechoslovakia	ČSN 73 1401 84		0		
Soviet Union	SNIP II-23-81 82		0		
NORTH AMERICA					
	ASME N284 80		0	0	
WEST EUROPE					
	ECCS No. 29 88	0		0	0

(1) shell stress formula constructed from two curves
(2) continuous stress formula
(3) describe the magnitudes of imperfections
(4) use the partial safety factor

10g. Cylindrical Shells under Torsion and/or Shear Stress

SPECIFICATION PROVISIONS & EXPLANATION

NORTH AMERICA

CHAPTER EDITOR

Nicholas F. Morris, Manhattan College

ASME N284 80

The local elastic buckling stress is

$$\tau_l = \alpha_{\theta\phi\ell}\, E(t/r)\, C_\tau \qquad (10g.N1)$$

where

$$\alpha_{\theta\phi\ell} = 0.80 \qquad r/t \le 250$$

$$\alpha_{\theta\phi\ell} = 1.323 - 0.218 \log(r/t) \qquad 250 < r/t < 1000 \qquad (10g.N2)$$

for $1.5 \le \delta_\phi < 26$

$$C_\tau = \frac{4.84}{\delta_\phi^2}\sqrt{1 + 0.0239\,\delta_\phi^3}$$

for $26 \le \delta_\phi < 8.69\,(r/t)$

$$C_\tau = \frac{0.746}{\sqrt{\delta_\phi}} \qquad (10g.N3)$$

for $\delta_\phi > 8.69\,(r/t)$

$$C_\tau = 0.253\sqrt{t/r}$$

The buckling factor for stringer stiffened shells is

for $\delta < 26$ and $\frac{\ell_h}{b} < 3.0$

$$C_\tau = \frac{1}{\delta^2}[4.82\sqrt{1+0.0239\,\delta^3} + 2.62\,(\frac{b}{\ell_h})^2] \qquad (10g.N4)$$

where δ is equal to $b/\sqrt{rt}$

If $b > \ell_h$ their ratio is reversed in the equation for C_τ .

The general elastic buckling stress is

$$\tau_g = \frac{\alpha_{\theta\phi\ell}\, 3.46\, E\, t_s^{0.375}}{\ell_h^{0.5}\, r^{0.75}\, t_s}\left(\frac{I_s}{b}\right)^{0.625} \qquad (10g.N5)$$

subject to the restriction that $\tau_g > 1.2\,\tau_\ell$.

The stiffeners must satisfy the requirements

$$I_s = 0.184\ C_\tau\ \delta^{0.8}\ t^3\ b \qquad (10g.N6)$$

The plasticity reduction factors are

for $\dfrac{\tau\ \gamma}{f_y} \leq 0.48$

$$\eta_\tau = 1.0$$

for $0.48 < \dfrac{\tau\ \gamma}{f_y} \leq 0.60$

$$\eta_\tau = \frac{0.10}{1 - \dfrac{0.43\ f_y}{\tau\ \gamma}} \qquad (10g.N7)$$

When a shear force V is acting, it is also generates normal stresses in the meridional direction. In the case the interaction formula

$$\frac{\sigma_d}{\sigma_{ui}} + \left(\frac{\tau_d}{\eta_\tau\ \tau_l}\right)^2 \leq 1 \qquad (10g.N8)$$

must be used, where σ_d and τ_d are stresses generated at the same point by factored loads and where σ_{ui} is the buckling stress under meridional compression (Eq. 10a.N1). τ_l is the local elastic buckling stress (Eq. 10g.N1). η_τ is the plasticity reduction factor for shear (Eq. 10g.N7).

WEST EUROPE

CHAPTER EDITOR

Daniel Vandepitte, Ghent University

ECCS No. 29 88

The strength requirement is

$$\tau_d \leq \tau_u$$

$$\tau_d = \frac{V_d}{\pi rt} + \frac{T_d}{2\pi\ r^2\ t} \qquad (10g.W1)$$

where

T_d = acting torque due to factored loads
V_d = acting shear force due to factored loads

In order to evaluate τ_u, the quantity τ_{cr} is first calculated

for $\frac{\ell}{r} \leq 8.9\sqrt{\frac{r}{t}}$

$$\tau_{cr} = 5.34\sqrt{1 + 0.0257\ Z^{1.5}}\ \frac{\pi^2\ E}{12(1-\nu^2)}\ (\frac{t}{\ell})^2 \qquad (10g.W2)$$

for $\frac{\ell}{r} > 8.9\sqrt{\frac{r}{t}}$

$$\tau_{cr} = 0.25\ E\ (\frac{t}{r})^{1.5} \qquad (10g.W3)$$

where Z is $\sqrt{1-\nu^2}\ \frac{\ell^2}{rt}$

τ_u is then given by

for $0.65\ \tau_{cr} \leq 0.5\ \frac{f_y}{\sqrt{3}}$

$$\tau_u = 0.65\ \tau_{cr} \qquad (10g.W4)$$

for $0.65\ \tau_{cr} \geq 0.5\ \frac{f_y}{\sqrt{3}}$

$$\tau_u = (1-0.25\ \frac{f_y}{0.65\sqrt{3}\ \tau_{cr}})\ \frac{f_y}{\sqrt{3}} \qquad (10g.W5)$$

The procedure is applicable provided that the deviations from the nominal cylindrical geometry are limited to

$$\frac{r - r_{nom}}{r_{nom}} \leq 0.01 \qquad (10g.W6)$$

When a shear force V is acting, it also generates normal stresses in the meridional direction. In that case the interaction formula

$$\frac{\sigma_d}{\sigma_u} + \frac{\tau_d}{\tau_u} \leq 1.0 \qquad (10g.W7)$$

must be used, where σ_d and τ_d are stresses generated at the same point by factored loads and where σ_u and τ_u are the buckling stresses under meridional compression acting alone and shear acting alone, respectively.

COMMENTARY

A torque T is assumed to produce a uniform shear stress around the circumference, while a shear force V is assumed to produce a shear stress distribution which varies harmonically. The maximum value of the shear stress is used in the analysis.

A cylinder subjected to shear loading and torsion is moderately imperfection sensitive. Experiments show approximately the same sensitivity as in the case of external pressure.

Since cylinders under shear loading are only moderately imperfection sensitive, the partial safety factor γ may be taken equal to 1.0.

The experimental results that are available suggest, that the checking procedure consisting in the application of Eqs. 10g.W2, 10g.W4, and 10g.W5 and of strength condition 10g.W1 is conservative.

INTERNATIONAL COMPARISONS & COMMENT

WHAT ARE THE DIFFERENCES?

The following provisions for cylindrical shells under torsion and/or shear stress are included in the specifications

WHY ARE THERE DIFFERENCES?

There are some differences because each design code uses different experimental data and different theoretical approach.

The problem of simultaneous action of twist and axial compression was investigated by Kromm (1940), on the basis of the linear theory. He showed that the interaction formulas for combined loading depend on the geometry of the cylindrical shell.

When $\omega = 1$, the interaction formula was parabolic, and
When $\omega = 1300$, the interaction formula was linear,

where $\omega = \dfrac{12\,(1-\nu^2)}{\pi^2}\,\dfrac{\ell^4}{r^2\,t^2}$

Table 10g.1 Comparison of specifications for design of cylindrical shells under torsion and/or shear stress

Region	Specification	Item (1)	(2)	(3)	(4)	(5)	(6)	(7)
AUSTRALIA								
CHINA								
EAST EUROPE								
JAPAN								
NORTH AMERICA								
	ASME 86		0	0	0	0		0
WEST EUROPE								
	ECCS 88	0		0	0		0	

(1) shell stress formula constructed from two curves
(2) continous stress formula
(3) describe the magnitudes of imperfections
(4) shell stress formula for unstiffened cylinders
(5) shell stress formula for stiffened cylinders
(6) linear interaction formula for combined loading
(7) parabolic interaction formula for combined loading

Chapter 10 – SUMMARY

Code comparisons are made in the discussion of each characteristic topic. Nonetheless, a brief overview of why there are code differences is in order. There are few codes available for the buckling investigation of steel shells. Those which were developed in the various countries usually arose in response to a particular research need and are based on limited local experimental data. This has tended to make the equations for each code appear to be very different. In addition, many national codes cover only a limited number of load cases and shell geometries.

It should be emphasized that there are many physical parameters which control the stability of shells; a good example is ring- and stringer-stiffened cylindrical shells. There are few codes available for investigating the buckling of steel shells. This makes it very difficult to compare experimental results to design rules. A much greater number of experiments is needed to make such comparisons. Hence, the need for more research data is very great in this field.

Most shell buckling codes are based on empirical data obtained in the aeronautical engineering industry. There are, however, great differences between the shells employed in aeronautical applications and in civil engineering applications. Aerospace applications use very thin shells made of aluminum alloys. There are no residual stresses in these shells and buckling is elastic. Imperfections are important because the shells are so thin. Shells used in civil engineering applications are made of steel and are much thicker. Hence residual stresses and plasticity are important. Imperfections are important in these applications as well, but their effect may not be as pronounced.

It does appear that there is a great deal of parallelism in the recent general shell buckling codes such as ECCS and the ASME Code Case. The buckling modes investigated, the use of knock-down factors, and the use of plasticity reduction factors are similar in many codes. Since the imperfections existing in shells are part of the fabrication process, which may differ from country to country, this is probably the best that one can hope for.

Symbol List for Chapter 10 -- Shells

A_e = equivalent cylindrical area (10e.E)
A_m = modified area of ring stiffener, defined in Eq. 10d.N4 (10d.N)
A_r = area of a ring stiffener
A_s = cross-sectional area of one stringer stiffener (10b.E)
A_{sc} = area of ring stiffener (10b.E)
A_{sl} = area of longitudinal stiffener (10b.E)
A_{so} = minimum area of ring stiffener and longitudinal stiffener (10b.E)
$\overline{A}$ = coefficient given in Eq. 10c.N2 (10c.N)
$A^* = \frac{\gamma^2}{\gamma_r^2} A_r$, modified area of ring stiffener (10c.W)
a = coefficient given in Eq. 10a.E5 (10a.E)
a_o = initial deflection (10a.E)
b = circumferential spacing of longitudinal stiffeners (10b.E) & (10b.W)
b = ratio of critical stress to yield stress (10a.E)
b = width of flange of stiffener (10c.W)
b_e = effective width of circular panel (10b.E)
C_p = coefficient given in Eq. 10c.N2 (10c.N)
C_s = St. Venant torsional constant (10b.A)
C_x = coefficient given in Eq. 10b.N2 (10b.N)
C_τ = buckling factor for cylindrical shell under torsion and/or shear stress (10g.N)
C_ϕ = coefficient used to determine local elastic buckling stress of a spherical shell (10f.N)
c = coefficient given in Table 10a.E2 (10a.E)
c = reduction factor of limit stress given in Table 10a.E1 (10a.E)
$\bar{c}$ = coefficient given in Eq. 10a.N3 (10a.N)
c_x = coefficient given in Eq. 10a.N2 (10a.N)
E_r = modulus of elasticity of ring stiffener material (10c.W)
f_{yr} = yield stress of ring stiffener material (10c.W)
f_y = yield strength of material (10a.A)
h = height of a liquid in a conical shell (10e.W)
h_w = height of web (10b.W)
h_2 = dimension of the stiffener centroid, defined in Eq. 10c.W2 (10c.W)
I_{bx} = moment of inertia of longitudinal stiffener (10b.E)
I_{by} = moment of inertia of ring stiffeners (10b.E)
I_f = flange inertia about the web centerline (10b.A)
I_r = moment of inertia of a ring stiffener (10c.N)
I_s = T section inertia about a circumferential axis through the point of attachment to the shell wall (10b.A)
I_x = equivalent moment of inertia per unit length (10b.E)
I_y = equivalent moment of inertia per unit length (10b.E)
K = effective length factor (10b.N)
K_{pb} = modification coefficient for buckling stress given in Table 10d.N1 (10d.N)
K_{pg} = modification coefficient for buckling stress given in Eq. 10c.N4 (10c.N)
K_{pg} = modification coefficient for buckling stress given in Table 10d.N1 (10d.N)
K_{pj} = modification coefficient for buckling stress given in Table 10d.N1 (10d.N)
K_{pl} = modification coefficient for buckling stress given in Eq. 10c.N1 (10c.N)

K_{pl} = modification coefficient for buckling stress given in Table 10d.N1 (10d.N)
K_{xb} = modification coefficient for buckling stress given in Table 10d.N1 (10d.N)
K_{xg} = modification coefficient for buckling stress given in Table 10d.N1 (10d.N)
K_{xj} = modification coefficient for buckling stress given in Table 10d.N1 (10d.N)
K_{xl} = modification coefficient for buckling stress given in Table 10d.N1 (10d.N)
k = factor concerning loading type (10a.E)
k = ratio of axial stress to the stress in the hoop direction (10d.N)
ℓ = length or unsupported length of a cylinder (10a.W)
ℓ_{ef} = effective length of cylindrical shell, acting with a stiffener (10c.W)
ℓ_G = length of a template for measuring imperfections of stringer stiffener (10b.W)
ℓ_h = distance between end diaphragms or between heavy ring stiffeners (10c.W)
ℓ_l = distance between intermediate light ring stiffeners (10b.E) & (10c.W)
ℓ_r = length of a template for measuring imperfections of shell wall (10a.W)
ℓ_s = arc length of support in circumferential direction (10f.N)
M = bending moment applied at the top of cones (10e.W)
M_d = design value of the bending moment (10e.W)
m = number of half buckling waves in the longitudinal direction (for a cylinder) (10b.W)
N = axial compression load (10e.E)
N_u = axial limit load of a conical shell (10e.E)
n = number of longitudinal stiffeners (10b.E)
n = number of full circumferential buckling waves (10c.W)
n_r = number of intermediate rings (10b.W)
n_s = number of stringer stiffeners (10b.W)
p = internal pressure (10d.W)
p = external pressure (10e.E)
$\overline{p}$ = ratio of hoop-tension stress to critical stress (10d.A) & (10d.E)
p_{cr} = critical elastic (bifurcation) pressure for a perfect shell (10e.W) & (10f.W)
p_d = factored acting pressure (10d.W)
p_e = limit external radial pressure (10c.E)
p_{eg} = external radial pressure at global buckling (10c.N)
p_{el} = external radial pressure at local buckling (10c.N)
p_i = internal pressure (10d.E)
p_u = minimum pressure which the shell will withstand (10f.W)
p_y = pressure causing yield of the steel (10f.W)
Q = axial load of cylindrical or conical shell (10e.W)
Q_d = design value of the acting axial load (10e.W)
R = radius of spherical shell (10f.E)
R_c = radius to the centroid of the effective section (10c.N)
R_o = radius to the outside of the shell (10c.N)
R_1 = equivalent spherical radius of toroidal shell (10f.N)
R_2 = equivalent spherical radius of ellipsoidal shell (10f.N)
r = radius of unstiffened cylinders (10a.A)
r = smaller radius of a conical shell (10e.E)
r_e = equivalent radius of conical shells, defined by Eq. 10e.E1 (10e.E)
r_e = local principal radius of curvature (10e.W)
r_f = maximum radius of circle of shell (10c.W)
r_{nom} = nominal radius of cylindrical shell (10g.W)
r_1 = larger radius of a conical shell (10e.E)

r_1 = smaller radius of a conical shell (10e.W)
r_2 = larger radius of a conical shell (10e.W)
T_d = acting torque due to factored loads (10g.W)
t = thickness of unstiffened cylinder (10a.A) & (10b.E)
t_{eff} = effective thickness of shells (10b.E)
t_r = equivalent shell thicknesses taking into account the stiffeners and rings (10f.N)
t_s = thickness of cylinder with smeared-out stringer stiffener (10c.N)
V_d = acting shear force due to factored loads (10g.W)
$\overline{v}$ = circumferential initial out of straightness of a stringer (10b.W)
$\overline{v}_1$ = parameter defining tilt of stringer web (10b.W)
$\overline{v}_2$ = parameter defining tilt of stringer flange (10b.W)
$\overline{w}$ = maximum amplitude of the initial imperfection (10a.E) & (10e.W)
α_b = reduction factor for limit strength under pure bending (10a.W)
α_g = imperfection reduction factor for global buckling (10b.W)
α_o = reduction factor for cylinder under axial compression (10a.N)
α_{ob} = imperfection reduction factors applied to the theoretical buckling stress found from the orthotropic shell equations (10b.N)
$\alpha_{o\ell}$ = reduction factor for stiffened cylinders under meridional compression (10b.N)
α_{pg} = imperfection reduction factor global buckling under external radial pressure (10c.N)
α_{pl} = imperfection reduction factor for local buckling under external radial pressure (10c.N)
α_{sp} = reduction factor for stiffened panel buckling of a stiffened cylinder (10b.W)
α_{xc} = imperfection reduction factor for axial buckling as a column (10b.N)
α_{xg} = imperfection reduction factor for general buckling (10b.N)
$\alpha_{\theta\phi l}$ = imperfection reduction factor for local buckling of a cylindrical shell under torsion and/or shear stress (10g.N)
α_ϕ = modification reduction factor for local buckling of a spherical shell (10f.N)
β = ratio of the meridional membrane stress to the effective stress (10d.W)
β = taper of a conical shell (10e.E)
β' = ratio of the meridional stress to the effective stress at buckling (10e.W)
β_{min} = coefficient to determine critical stress of cylinder stress (10c.W)
γ = resistance factor (10a.W)
γ_m = resistance factor of material (10a.E)
γ' = weight of a liquid (10e.W)
Δ = ratio of the buckling critical stress to the yield stress (10b.N)
δ = geometric parameter used to determine the buckling stress (10g.N)
δ_s = geometric parameter used to determine the buckling stress (10f.N) & (10g.N)
δ_x = geometric parameter used to determine the buckling stress (10a.N)
δ_θ = geometric parameter used to determine the buckling stress (10b.N)
δ_ϕ = geometric parameter used to determine local elastic buckling stress (10f.N)
ζ = coefficient depending on $\delta = 1.285\,\ell/\sqrt{rt}$ (10c.W)
η = plasticity reduction factor (10a.N)
$\overline{\lambda}$ = slenderness parameter (10a.E), (10c.W) & (10e.W)
$\overline{\lambda}_p$ = slenderness parameter, defined by Eq. 10d.W5 (10d.W)
ν = Poisson's ratio (10c.E) & (10c.W)
σ_b = maximum compression stress due to the design bending moment (10a.W)

σ_{cr} = elastic critical stress for a perfect shell (10a.E) & (10d.A)
$\sigma_{cr,g}$ = elastic buckling stress for global buckling (10b.W)
$\sigma_{cr,p}$ = stress causing stiffened panel buckling of a perfect elastic stiffened cylinder (10b.W)
σ_{cri} = elastic buckling stress in failure mode "i" (10a.N)
σ_d = factored meridional acting stress (10d.W)
$\sigma_{eff,d}$ = effective design stress under combined loads (10d.W)
$\sigma_{eff,u}$ = effective stress at buckling under combined loads (10d.W)
$\sigma_{eff,u1}$ = effective buckling stress (10e.W)
$\sigma_{eff,u2}$ = effective buckling stress (10e.W)
σ_n = axial compressive stress (10e.E)
σ_o = uniform compression stress due to the design axial load (10a.W)
σ_p = radial compressive stress (10d.N)
σ_{pg} = global buckling stress of a cylindrical shell under external radial pressure (10c.N)
σ_{pj} = elastic buckling stress of cylindrical shell under uniform external pressure in failure mode "j" (10d.N)
σ_{pl} = local buckling stress of cylindrical shell under uniform external pressure (10c.N)
σ_r = total circumferential stress in ring stiffener (10c.W)
σ_r' = circumferential stress due to internal pressure (10d.E)
σ_u = uniform compressive stress (in cylinder) due to design bending moment (10a.A)
σ_u = limit stress (10b.E)
σ_u' = modified limit stress (10a.E)
σ_{ud}^p = buckling stress for cylinders under combined axial load and internal pressure (10d.A)
σ_{ud}' = limit stress for cylinders under combined axial load and internal pressure (10d.E)
σ_{ud}'' = limit stress for cylinders under combined axial load and internal pressure (10d.E)
σ_{ug} = limit stress for global buckling (10b.W)
σ_{ui} = allowable buckling stress (10a.N)
σ_{ul} = local buckling stress of a stiffened cylinder (10b.W)
σ_{ul1} = local buckling stress of a stiffened cylinder (10b.W)
σ_{ul2} = local buckling stress of a stiffened cylinder (10b.W)
σ_{up} = buckling stress for stiffened panel buckling (10b.W)
σ_{ur} = limit stress of a cylindrical shell under uniform external radial pressure (10c.E)
σ_{us} = limit stress of cylindrical panels under uniform meridional compression (10b.E)
σ_{usp} = limit stress of a spherical shell under uniform external radial pressure (10f.E)
σ_{u1} = elastic buckling stress of cones (10e.W)
σ_{u2} = elastic buckling stress (10e.W)
σ_x = the highest meridional stress (10a.E)
σ_x = meridional stress (10d.N)
σ_x' = the smallest meridional stress (tension stress is negative) (10a.E)

$\sigma_{x,\ell}$ = elastic buckling stress for unstiffened cylinders under axial compression (10a.N)
σ_{xc} = column buckling stress when the cylinder is considered as a circular tube (10b.N)
σ_{xd} = design stress in the meridional direction under combined axial load and internal pressure (10d.W)
σ_{xg} = elastic buckling stress for general buckling of ring-stiffened cylinders (10b.N)
$\sigma_{x\ell}$ = elastic buckling stress for local buckling (10b.N)
σ_{xu} = meridional membrane stress at buckling under combined axial load and internal pressure (10d.W)
σ_y = hoop-tension stress (10d.E)
σ_{yd} = design stress in the circumferential direction under combined axial load and internal pressure (10d.W)
σ_{yu} = circumferential membrane stress at buckling under combined axial load and internal pressure (10d.W)
σ_1 = axial stress on the cross section (10b.E)
$\sigma_{\theta cj}$ = the failure stress in the hoop direction corresponding to the combined loading (10d.N)
$\sigma_{\phi cj}$ = the failure stress in the axial direction corresponding to the combined loading (10d.N)
$\sigma_{\phi g}$ = general buckling stress of a spherical shell under uniform external radial pressure (10f.N)
$\sigma_{\phi\ell}$ = local elastic buckling stress of a spherical shell under uniform external radius pressure (10f.N)
τ = shear stress (10g.N)
τ_{cr} = elastic buckling stress of cylindrical shell under torsion and/or shear stress (10g.W)
τ_g = general elastic buckling stress of cylindrical shell under torsion and/or shear stress (10g.N)
τ_i = local elastic buckling stress of cylindrical shell under torsion and/or shear stress (10g.N)
τ_{max} = maximum shear stress in the place of the maximum bending stress (10a.E)
τ_u = buckling stresses under shear acting alone (10g.W)
ϕ = buckling modification factor, given in Eq. 10a.E9 (10a.E)
ψ = buckling modification factor in elastic range (10a.E)
ψ = coefficient given in Eq. 10d.N3 (10d.N)
ω = coefficient used to determine the elastic buckling stress of a conical shell containing liquid (10e.W)

Chapter 11

Cold-Formed Members

TOPICAL EDITOR
Ton Toma, TNO Institute for Building Materials and Structures

CHAPTER EDITORS
Victor Gioncu, Polytechnic Institute of Timisoara
Greg J. Hancock, The University of Sydney
Joachim Lindner, Technische Universitat Berlin
Jindrich J. Melcher, Technical University of Brno
Tetsuro Ono, Nagoya Institute of Technology
Teoman Pekoz, Cornell University
Toshiro Suzuki, Tokyo Institute of Technology
Ton Toma, TNO Institute for Building Materials and Structures
Shi-Ji Wang, Hunan University
Zhong-Quan Zhang, Zhongnan Building Design Institute

11a. Plate Elements

SPECIFICATION PROVISIONS & EXPLANATION

AUSTRALIA

CHAPTER EDITOR
Greg J. Hancock, The University of Sydney

AS 1538 88

Properties of Stiffened and Unstiffened Compression Elements

In calculating the section properties of a flexural member, or values of Q for compression members, the flat width, b, of any stiffened or unstiffened compression element shall be taken as the effective design width, b_e, determined as given below. The portion of the total width, which is considered removed in order to determine the effective design width, shall be located at and symmetrically about the center-line of the element in stiffened elements and shall be located adjacent to the free edge in an unstiffened element.

(A) Flat Elements with low b/t:
For flat compression elements where the flat-width ratio, b/t, is less than $(b/t)_{lim}$ given below, b_e shall be taken as the flat width b

for strength calculations

$$(b/t)_{lim} = 225\sqrt{(k/\sigma)} \tag{11a.A1}$$

for deflection calculations

$$(b/t)_{lim} = 290\sqrt{(k/\sigma)} \tag{11a.A2}$$

where σ is the compression stress caused by the unfactored load (in MPa).

The plate buckling coefficient, k, may be assumed to be 4.0 for stiffened elements and 0.5 for unstiffened elements. Alternatively, the plate buckling coefficient for each flat element shall be determined from a rational elastic buckling analysis of the whole section as a plate assemblage subjected to the longitudinal stress distribution in the section prior to buckling.

(B) Flat Elements with Medium b/t:
For flat elements where b/t exceeds the appropriate value of $(b/t)_{lim}$ above, b_e shall be given by

for strength calculations

$$\frac{b_e}{t} = \frac{332\sqrt{k}}{\sqrt{\sigma}}\left(1 - \frac{72.5\sqrt{k}}{(b/t)\sqrt{\sigma}}\right) \tag{11a.A3}$$

for deflection calculations

$$\frac{b_e}{t} = \frac{428\sqrt{k}}{\sqrt{\sigma}}\left(1 - \frac{93.5\sqrt{k}}{(b/t)\sqrt{\sigma}}\right) \tag{11a.A4}$$

CHINA

CHAPTER EDITORS
Shi-Ji Wang, Hunan University
Zhong-Quan Zhang, Zhongnan Building Design Institute

GBJ 18 87

Local Buckling of Component Plate Elements

Two design approaches consider the local buckling of component plate elements, the effective width approach and the permitted width approach. The effective width approach is valid for stiffened plate elements under uniform compression or a combination of compression and bending, as well as for edge-stiffened elements under uniform compression. The values of effective width-to-thickness ratios b_e/t for various edge-supporting and loading conditions are given in the form of tables.

The effective width b_e is assumed to be divided into two equal portions as shown in Fig. 11a.C1 for both uniformly and non-uniformly compressed plate elements.

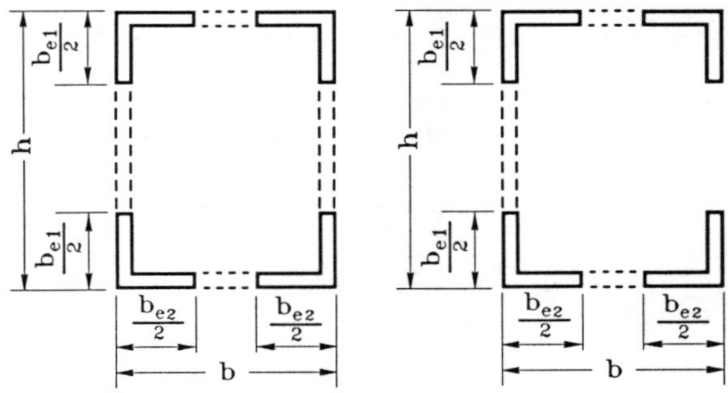

Fig. 11a.C1 Effective widths in two cross sections

The permitted width approach is chosen for cases other than those mentioned above. The width-to-thickness ratio of such plate elements is controlled by the following formula

$$\frac{b}{t} \leq 100\sqrt{\frac{\xi}{\sigma_{max}}} \qquad (11a.C1)$$

where

α	$= (\sigma_{max} - \sigma_{min})/\sigma_{max}$
ξ	= coefficient depending on the supporting conditions of the unloaded edges, the location of σ_{max}, and the value of α
σ_{max}, σ_{min}	= maximum compressive stress caused by the factored loads (N/mm^2) at one of the unloaded edges and the corresponding stress at another edge, respectively

Values of ξ for unstiffened plate elements under uniform compression, unstiffened and edge-stiffened plate elements with σ_{max} at the stiffened edge or at the free/lipped edge are tabulated.

Adequate values for width-to-thickness ratio of edge stiffeners corresponding to width-to-thickness ratio of edge-stiffened plate elements are also tabulated.

Effective Width under Compression and Bending Forces

In order to determine the effective width of stiffened plate elements under a combination of compression and bending, the actual post-buckling stress is replaced by a linearly varying fictitious stress distributed over a reduced effective width b_e as shown in Fig. 11a.C2(b).

The effective width is determined on the assumption that the ultimate state of the plate element is reached when the maximum fictitious stress σ_1 is equal to f_y. Two equilibrium equations (refer to Fig. 11a.C2) can thus be obtained.

$$\sigma_{max} = \frac{N_u}{A} + \frac{M_u}{W} = \frac{N_u}{A}(1 + \frac{M_u}{N_u} \cdot \frac{A}{W}) = \frac{N_u}{A}(1 + \frac{\alpha}{2-\alpha}) = \sigma_u \qquad (11a.C2)$$

$$\sigma_1 = \frac{N_u}{A_e}(1 + \frac{M_u}{N_u} \cdot \frac{A_e}{W_e}) = \frac{N_u}{A_e}(1 + \frac{\alpha}{2-\alpha} \cdot \frac{W}{A} \cdot \frac{A_e}{W_e}) = f_y \qquad (11a.C3)$$

where

- A = gross area
- A_e = effective area
- N_u, M_u = axial force and bending moment corresponding to the ultimate state
- W = section modulus of gross area for bending
- W_e = effective section modulus for bending
- σ_u = nominal ultimate stress obtained from the conversion of ultimate load governed by strength or deformation.

Dividing Eq. 11a.C3 by 11a.C2 we obtain a cubic equation

$$\left(\frac{b_e}{b}\right)^3 - A_o\left(\frac{b_e}{b}\right)^2 + B_o\left(\frac{b_e}{b}\right) - C_o = 0 \qquad (11a.C4)$$

where

- $A_o = (1 + 3\psi)/\psi$
- $B_o = 3(1 + \psi)/\psi$
- $C_o = (3 + \chi)/\psi$
- $\chi = \alpha/(2-\alpha)$
- $\psi = (1 + \chi)f_y/\sigma_u$

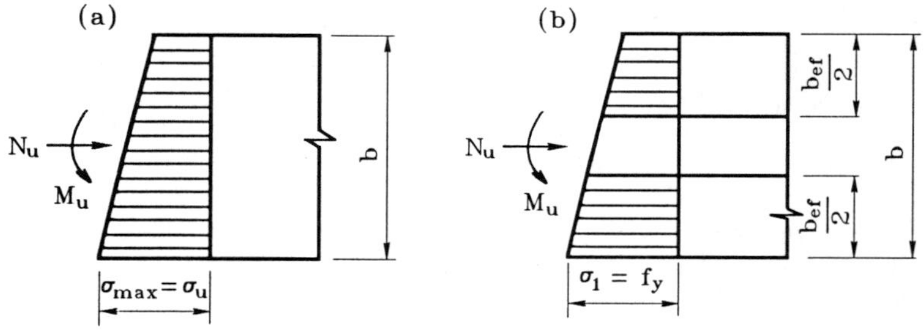

Fig. 11a.C2 Stress distribution caused by axial force and bending moment and the corresponding effective width

Finally, the value of b_e/b can be determined by solving Eq. 11a.C4.

The permitted width approach is based on the linear theory of plate buckling. The coefficient ξ in Eq. 11a.C1 is derived from the interaction formula for combination of compression and shear. Plastic reduction is taken into account.

COMMENTARY

The effective width approach is based on the large deflection theory of thin plates, taking into account the post-buckling strength and the influence of initial imperfections. For stiffened and edge-stiffened plate elements under uniform compression, the effective width is determined by the equation

$$\frac{b_e}{t}\sqrt{\frac{\sigma_d}{E}} = K_1 \frac{b}{t}\sqrt{\frac{\sigma_d}{E}} + K_2 \frac{\sqrt{\tau}}{\frac{b}{t}\sqrt{\frac{\sigma_d}{E}}} \qquad \text{(11a.C5)}$$

where

f_p = proportional limit taken as $0.8f_y$ in the specification

f_y = yield strength

K_1,K_2 = coefficients determined by experiments, and evaluated in the specification as follows:

for stiffened elements

$K_1 = 0.25 \qquad K_2 = 2.35$

for edge-stiffened elements

$K_1 = 0.125 \qquad K_2 = 1.06$

σ_{cr} = elastic critical stress of the plate element

σ_d = governing nominal stress in design

τ = $\dfrac{(f_y - \sigma_{cr})\sigma_{cr}}{(f_y - f_p)\, f_p}$, plastic reduction factor

EAST EUROPE

CHAPTER EDITORS

Victor Gioncu, Polytechnic Institute of Timisoara
Jindrich J. Melcher, Technical University of Brno

ČSN 73 1402 88

The uniform formula for the effective width of the compression web is

$$b_e = \rho \cdot b \tag{11a.E1}$$

where

b = the width of the straight part of the web
ρ = the reduction factor defined as

$$\rho = \frac{0.95}{\beta}\sqrt{\frac{kE}{\sigma_{max}}}\left(1 - \frac{0.209}{\beta}\sqrt{\frac{kE}{\sigma_{max}}}\right) \tag{11a.E2}$$

If we define E = 210,000 MPa and the design stress $\sigma_{max} = R_d = 210$ MPa (Steel grade with ultimate tensile strength of 370 MPa), ρ is directly given by

$$\rho = \frac{30.04}{\beta}\sqrt{k} - \frac{198.53k}{\beta^2} \tag{11a.E3}$$

where

R_d = see Eq. 1a.E2 and the explanation to code ČSN 73 1401 84 in Chapter 1 (MPa)
t = web thickness
β = b/t

The plate buckling coefficient k in Eq. 11a.E2 and Eq. 11a.E3 shall be derived for the ideal web according to the appropriate load and boundary conditions.

For example, at the uniformly compressed unstiffened elements of angle and T section webs (Fig. 11a.E1b), the plate buckling coefficient k = 0.425. For uniformly compressed stiffened elements (Fig. 11a.E1a), the basic value is k = 4.0. Also more detailed data are given for webs with stress gradient.

The edge stiffener can be regarded as a support element of the stiffened web if the minimum moment of inertia I_{st} of the stiffener about its own centroidal axis parallel to the element to be stiffened is

for $\beta \leq 40\sqrt{210/R_d}$

$$I_{st} = \frac{6\beta\sqrt{R_d/210} - 120\sqrt{k}}{2 - \sqrt{k}}\ t^4 \tag{11a.E4}$$

for $\beta > 40\sqrt{210/R_d}$

$$I_{st} = (2.875\beta\sqrt{R_d/210} + 5)\ t^4 \tag{11a.E5}$$

where

k = the plate buckling factor of the free (unstiffened) web to be stiffened
t = thickness of stiffener

For edge stiffeners the round corner between the stiffener and the element to be stiffened shall not be considered as a part of the stiffener.

The minimum moment of inertia of the intermediate stiffener about its own centroidal axis shall be

for $20\sqrt{210/R_d} < \beta \leq 60\sqrt{210/R_d}$

$$I_{st} = (2.5\ \beta\sqrt{R_d/210} - 5)\ t^4 \qquad (11a.E6)$$

for $\beta > 60\sqrt{210/R_d}$

$$I_{st} = (6.4\ \beta\sqrt{R_d/210} - 285)\ t^4 \qquad (11a.E7)$$

The design provisions for the flexible stiffeners that do not meet the requirement of Eq. 11a.E4 to Eq. 11a.E7 are also given in ČSN 73 1402 88.

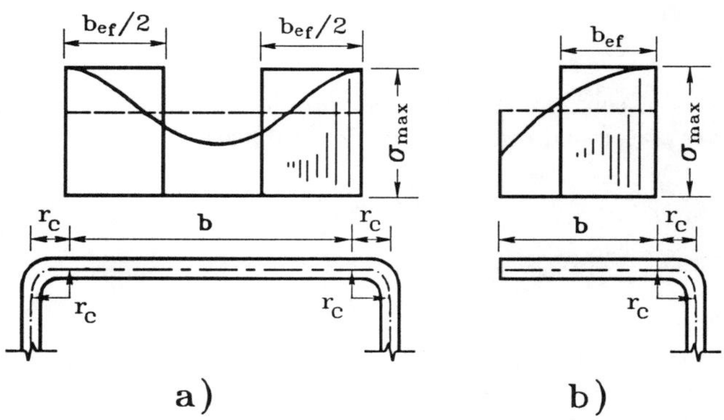

Fig. 11a.E1 Effective widths for double and single longitudinally supported plates

JAPAN

CHAPTER EDITORS
Tetsuro Ono, Nagoya Institute of Technology
Toshiro Suzuki, Tokyo Institute of Technology

AIJ 85

Width-Thickness Ratio and Effective Width

(A) Width-thickness ratios of plate elements of sections that are subject to compressive force by axial compression or bending or shear force shall be as follows

(1) Single-angle struts and double-angle struts with separators

$$b/t \leq 13 \tag{11a.J1}$$

(2) Flanges of columns and other compression members, plates and angles projecting from compression portions of beams, compression flanges of beams, stems of tees, and stiffeners subject to concentrated load

$$b/t \leq 16 \tag{11a.J2}$$

(3) Web plates of columns and other compressions members, flanges of box-type columns, cover plates, and compressed stiffened flanges

$$b/t \leq 48 \tag{11a.J3}$$

(4) Web plates of beams

$$b/t \leq 71 \tag{11a.J4}$$

(B) When the actual width-to-thickness ratio of a particular plate element exceeds the prescribed value in (A), the portion of plate exceeding the prescribed value may be regarded as being absent for the purpose of stress analysis. Effective width concepts are shown in Fig. 11a.J1.

(C) Effective width, b_e, of a single-angle strut subject to compression and not stiffened against torsion shall be determined by the following equation

$$b_e/b = 250(t/b)^2 \leq 1 \tag{11a.J5}$$

(D) When the span of a beam subject to concentrated load is equal to or less than 30 times the flange protrusion, b_f, the flange effective width is subject to the following limitations in terms of both compression and tension.

$$b_e/b_f = 0.3 + 0.13\sqrt{L/b_f} \leq 1 \tag{11a.J6}$$

where

b_f = Width of flange for I-shaped section, width of flange divided by 2 for box or V-shaped section, and width of flange plus height of lip for section with lip (see Fig. 11a.J2)

L = Equivalent span, that is, the entire length for simple beam, the distance between points of inflection for continuous beam, and the entire length multiplied by 2 for cantilever beam

(E) For a lip considered as an edge stiffener of a compression plate element or for any other stiffener, the moment of inertia about the gravity axis shall not be less than the value obtained by the following equation

$$I_{min} = 1.9t^4\sqrt{(b/t)^2 - 150} \geq 8.3t^4 \tag{11a.J7}$$

where t is steel core thickness of plate element

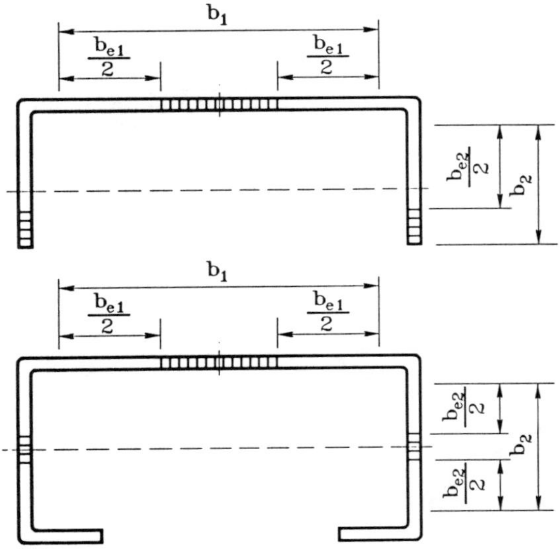

Fig. 11a.J1 Effective widths for two cross sections

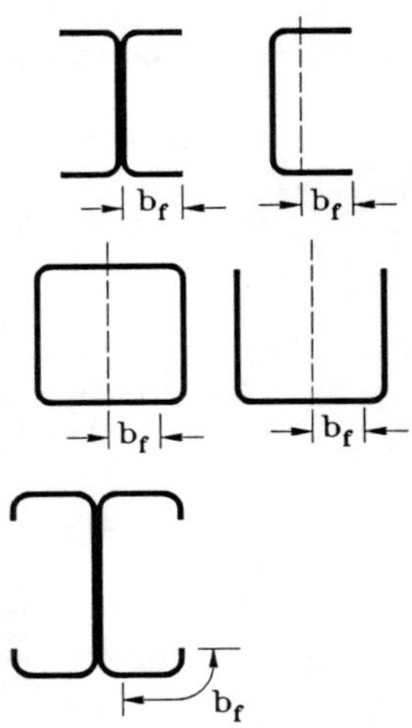

Fig. 11a.J2 Width of flanges b_f as used in Eq. 11a.J6

NORTH AMERICA

CHAPTER EDITOR
Teoman Pekoz, Cornell University

AISI 86

Behavior of Plate Elements

The post-buckling behavior of plate elements is represented by a generalization of the Winter (1947) effective width equation. The Winter equation was derived for plate elements supported adequately along the two longitudinal edges. This concept has been extended to

- unstiffened elements (elements supported only on one longitudinal edge)
- perforated elements
- elements with an edge stiffener or an intermediate stiffener of any size
- elements with stress gradients such as webs

Effective Width for Determination of Load-Carrying Capacity

The following are the provisions for determining the effective width b_e of elements containing compressive stresses

for $\lambda \leq 0.673$
$$b_e = b$$
for $\lambda > 0.673$ (11a.N1)
$$b_e = \rho b$$

where

b = flat width

F_{cr} = plate buckling stress calculated as

$$= \frac{\pi^2 k\ E}{12(1-\upsilon^2)\ (b/t)^2}$$

t = thickness of element

λ = plate slenderness ratio

$$\lambda = \sqrt{\frac{\sigma_d}{F_{cr}}}$$

ρ = $(1 - 0.22/\lambda)/\lambda$

σ_d = stress in the element caused by the factored load, calculated on the basis of the effective section

Thus

for $\nu = 0.3$

$$\lambda = \frac{1.052}{\sqrt{k}} \cdot \frac{b}{t} \cdot \sqrt{\frac{\sigma_d}{E}}$$

The unified approach of AISI 86 involves the determination of the effective widths of elements with all edge conditions and stress gradients with the above equations.

Special conditions of these cases are accounted for by using the appropriate plate buckling coefficient k. For uniform compressed unstiffened and stiffened elements, the plate buckling coefficients are 0.43 and 4, respectively. The buckling coefficients belonging to stress gradient cases have been given in AISI 86.

Plate buckling coefficients are also given for plate elements with an edge stiffener or one intermediate stiffener. Stiffeners themselves present two problems to be considered. First is evaluating the adequacy and effectiveness of the stiffener in restraining the element to be stiffened. Second is how the stiffener itself should be considered in calculating the overall section properties such as the moment of inertia, area, and the section modulus.

Effective Width for Determination of Deflection

The effective width for deflection calculations can be determined using the above expressions by substituting for σ_d the stress at design load and replacing the section with the effective one in calculating the section properties. This approach has been shown to be quite conservative. A more accurate approach is provided in AISI 86 for stiffened elements.

Dimensional Limits

AISI 86 stipulates the limits on the width to thickness ratios. Elements with width-to-thickness ratios larger than those given may safely carry loads but may show visible local buckles.

WEST EUROPE

CHAPTER EDITOR

Ton Toma, TNO Institute for Building Materials and Structures

ECCS 87

Behavior of Plate Elements

For stiffened elements under uniform compression the effective width (for load bearing capacity and deflection) is given by the following expressions.

$$b_e = \rho b_p \tag{11a.W1}$$

b_p is not the flat width but depends on the corner radius

where

f_y = design value of yield stress

k = buckling coefficient; if there is no special investigation, k = 4.0 may be used

$$\lambda = 1.052 \frac{b_p}{t}\sqrt{\frac{\sigma_d}{Ek}}$$

$$\overline{\lambda}_p = 1.052 \frac{b_p}{t}\sqrt{\frac{f_y}{Ek}}$$

ρ = reduction coefficient given by

for $\lambda \leq 0.673$: $\rho = 1$

for $\lambda > 0.673$; $\rho \leq 1$: $\rho = (1-0.22/\lambda)/\lambda + 0.18 \dfrac{\overline{\lambda}_p - \lambda}{\overline{\lambda}_p - 0.6}$

for $\sigma_d = f_y$: $\rho = (1 - 0.22/\overline{\lambda}_p)/\overline{\lambda}_p$

σ_d = actual compressive stress caused by design load

In the European recommendations of 1987, formulas have also been given to determine the effective width for the following structures

- stiffened elements under stress gradient
- unstiffened elements under stress gradient

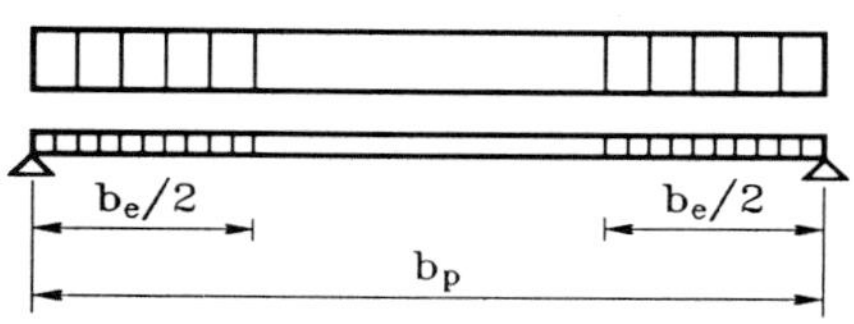

Fig. 11a.W1 Effective width for double support and uniform compression

For all of the cases, values or formulas have been given for the plate buckling coefficient k.

Furthermore, elements with edge or intermediate stiffeners have been treated. This has been done by distinguishing different cases of stiffener adequacy. For every case effective widths and effective stiffener areas have been defined.

INTERNATIONAL COMPARISONS & COMMENT

WHAT ARE THE DIFFERENCES?

The referenced specifications are all based on the effective width procedure for taking into account the postbuckling behavior. To compare the different specifications, a plate element supported at two longitudinal edges under uniform compression with maximum stress equal to the yield stress (f_y) will be taken as an example. For purposes of the comparison, the formulas in the referenced specifications have been generalized using the following definitions

$$\lambda = 1.052\,\frac{b}{t}\sqrt{\frac{f_y}{Ek}} \qquad (11a.1)$$

$$b_e = \rho \cdot b \qquad (11a.2)$$

See page 11a(W1) for definition of terms.

For the different regions this leads to:

AUSTRALIA

b = flat width

allowable stress: $\sigma = 0.6f_y$

for $\lambda \le 0.667$

$$\rho = 1$$

for $\lambda > 0.667$

$$\rho = \frac{0.98}{\lambda}\left(1 - \frac{0.21}{\lambda}\right) \qquad (11a.3)$$

So b_e has been calculated at the level of allowable stress.

CHINA

b = total width

$$\rho = 0.25 + 0.65\,\frac{\sqrt{\tau}}{\lambda^2} \qquad \tau = \frac{\lambda^2 - 1}{0.16\lambda^4}$$

$$\rho = 0.25 + \frac{1.625}{\lambda^4}\sqrt{\lambda^2 - 1} \qquad (11a.4)$$

EAST EUROPE

b = flat width

$$\rho = \frac{1}{\lambda}\left(1 - \frac{0.22}{\lambda}\right) \qquad (11a.5)$$

JAPAN

b = flat width

for $\lambda \le 0.0551\sqrt{f_y}$

$$\rho = 1$$

for $\lambda > 0.0551\sqrt{f_y}$

$$\rho = \frac{0.0551\sqrt{f_y}}{\lambda} \qquad (11a.6)$$

In case of $f_y = 330\ \mathrm{N/mm^2}$

for $\lambda \le 1$

$$\rho = 1$$

for $\lambda > 1$

$$\rho = \frac{1}{\lambda} \qquad (11a.7)$$

NORTH AMERICA

b = flat width

for $\lambda \leq 0.673$

$$\rho = 1$$

for $\lambda > 0.673$

$$\rho = \frac{1}{\lambda}\left(1 - \frac{0.22}{\lambda}\right) \tag{11a.8}$$

WEST EUROPE

$b = b_p$

with b_p larger than flat width, in most cases

with r: inner bending radius of corner

t: steel core thickness

b_p = flat width + $2(r + \frac{1}{2}t)$

for $\lambda \leq 0.673$

$$\rho = 1$$

for $\lambda > 0.673$

$$\rho = \frac{1}{\lambda}\left(1 - \frac{0.22}{\lambda}\right) \tag{11a.9}$$

Remark: In all specifications except in North American the yield stress f_y does not include the influence of cold forming for compressed elements.

Table 11a.1 shows the different treatments by the regions of the items effective design width and stiffeners.

Table 11a.1 Comparison of the treatment by the various regions of the items effective design width and stiffeners

Region	Effective design width	Stiffeners
A	I	VI
C	II	IV
E	I	V
J	III	IV
N	I	V
W	I	V

Notes

I. Effective width based on Winter formula

II. Effective width concept and permitted width concept depending on edge conditions of the plate element and the stress situation. This approach is based on the large deflection theory of thin plates.

III. Width-thickness ratios for certain cross sections.
Widths exceeding these ratios may be taken into account for the portion that will result in the ratio criteria being satisfied. For cross sections with an effective width concept, the value of the effective width only depends on geometrical parameters and not on the actual or yield stress.

IV. Edge or intermediate stiffeners can be regarded as a supporting element when the stiffener has a specified adequate stiffness.

V. Same as IV with additional provisions given for any size of edge or intermediate stiffener.

VI. Same as IV with a rational analysis permitted when the stiffener does not have the specified adequate stiffness.

WHY ARE THERE DIFFERENCES?

Australia has almost the same formula for ρ as East Europe, North America and West Europe. This means that in Australia the value of b_e at allowable stress is equal to b_e for the other countries in the ultimate limit state. For the allowable force in the treated plate element this means

– Australia: allowable force $= b_e \cdot 0.6f_y$ (11a.10)

– Other countries: allowable force $= b_e \cdot f_y \frac{1}{\gamma}$ (11a.11)
with γ as the load factor

Although the formula for ρ of the West Europe seems to be the same as in other countries, there is a small difference. West Europe does not use the flat width, but a width that is somewhat larger.

China has a different formula for ρ and a different definition for b. The difference of ρ with the other countries depends on the value of λ (see Fig. 11a.1) and a limitation caused by τ.

From Fig. 11a.1 the wrong conclusion could be drawn that in Japan the effective width depends on the yield stress. This is only caused by reformulating the Japan calculation rules for this second World View. In reality the effective width in Japan is constant namely $b_e = 48t$.

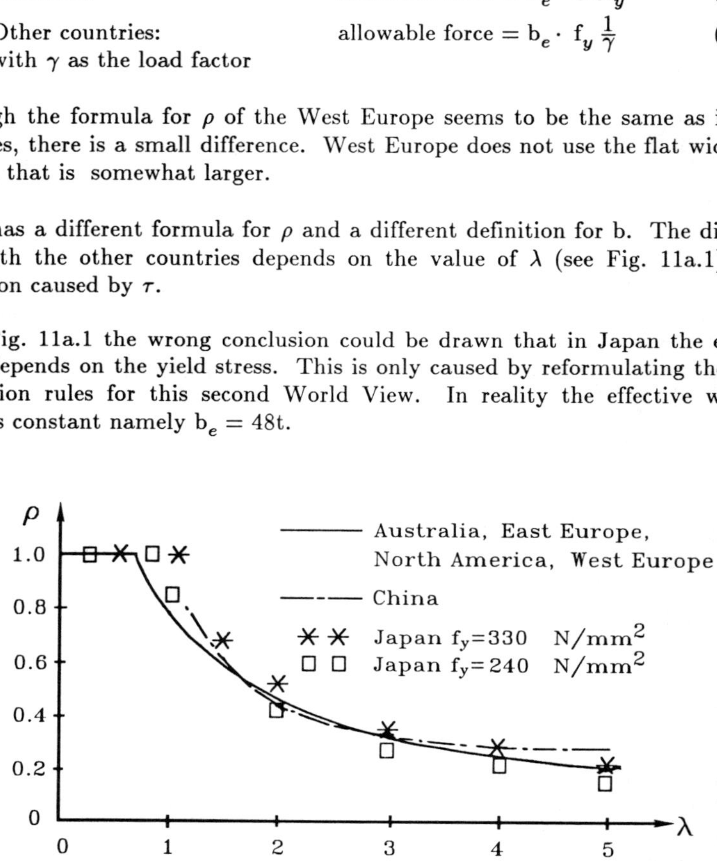

Fig. 11a.1 Relation Between ρ and λ for the Referred Specifications

11b. Columns

SPECIFICATION PROVISIONS & EXPLANATION

AUSTRALIA

CHAPTER EDITOR
Greg J. Hancock, The University of Sydney

AS 1538 88

Axially Loaded Compression Members
The column curve is based on the Perry formula as follows

The maximum permissible average axial stress F_a in a compression member, calculated on the full cross-sectional area of the member, shall be

$$F_a = \frac{QF_y}{\Omega}\left[\frac{1+(1+\eta)\ (F_{oc}/QF_y)}{2} \quad \sqrt{[\frac{1+(1+\eta)\ (F_{oc}/QF_y)}{2}]^2 - \frac{F_{oc}}{QF_y}}\right] \qquad (11b.A1)$$

where

F_{oc} = elastic buckling stress
F_y = yield stress

$$Q = \frac{\text{Effective Design Area}}{\text{Full Area of the Cross-Section}}$$

η = imperfection parameter = $(1.25-Q)QF_y/F_{oc}$
Ω = load factor = 1.67

The effective design area is determined from the effective widths of the stiffened and unstiffened elements.

The elastic buckling stress F_{oc} is the least value of the flexural or flexural-torsional buckling stress.

The term (1.25−Q) in the imperfection parameter lowers the column curve as Q is reduced below 1.0 in order to allow for the interaction of local and global buckling.

CHINA

CHAPTER EDITORS
Shi-Ji Wang, Hunan University
Zhong-Quan Zhang, Zhongnan Building Design Institute

GBJ 18 87

Compression Members

Flexural and flexural-torsional buckling of centrally loaded compression members are checked by the following formula

$$\frac{N}{\phi A_e} \leq f \tag{11b.C1}$$

where

A_e = effective area
f = design strength
N = design value of axial load
ϕ = a strength reduction factor depending upon the slenderness ratio and steel grade, whose value is determined from the basic column curves and tabulated in the appendix of the specification

For singly symmetric open sections with or without battens flexural-torsional buckling should be checked in addition to flexural buckling by introducing an equivalent slenderness ratio λ_ω into the basic column curve.

$$\lambda_\omega = \lambda_x \sqrt{\frac{s^2+i_o^2}{2s^2} + \sqrt{\left(\frac{s^2+i_o^2}{2s^2}\right)^2 - \frac{i_o^2-\alpha e_o^2}{s^2}}} \tag{11b.C2}$$

where

A = gross area
e_o = abscissa of the shear center
I_t = St. Venant torsion constant
$I\omega$ = warping constant
i_x, i_y = radii of gyration about the symmetric and non-symmetric principal axes, respectively
i_o^2 = $e_o^2 + i_x^2 + i_y^2$ (11b.C3)
$L\omega$ = effective length of warping torsion, evaluated in such a manner: for members without intermediate batten plate
$L\omega$ = effective length of flexural buckling about the symmetric axis; for members, without intermediate batten plate
L_ω = largest center to center spacing of batten plates for battened members

s^2 = $\frac{\lambda_x^2}{A}\left(\frac{\beta I_\omega}{L_\omega^2}+0.039I_t\right)$ (11b.C4)

α and β = coefficients reflecting supporting conditions at the ends of the member, whose values are given in Table 11b.C1
λ_x = slenderness ratio for flexural buckling about the x-axis (symmetrical axis)

COMMENTARY

The basic column curves shown in Fig. 11b.C1 are based on experimental results of over 160 axially loaded columns with various types of cold formed steel shapes. Statistical analysis of the testing results shows that the single column curve is adequate for cold formed steel members. The strength reduction factor ϕ is expressed in Perry-Robertson model as follows

Maximum permitted slenderness ratio is 150 for main members, and 200 for secondary members.

$$\phi = \frac{1}{2}\{1+\frac{1}{\lambda_o^2}(1+\epsilon_o) - \sqrt{[1+\frac{1}{\lambda_o^2}(1+\epsilon_o)]^2-\frac{4}{\lambda_o^2}}\} \qquad (11b.C5)$$

Table 11b.C1 Coefficients α and β

Supporting Conditions	Unbattened members		Battened members	
	α	β	α	β
Both ends hinged and free to warp	1.00	1.00	—	—
Both ends fixed	1.00	1.00	0.80	1.00
Both ends hinged but warping prevented	0.72	4.00	0.80	1.00

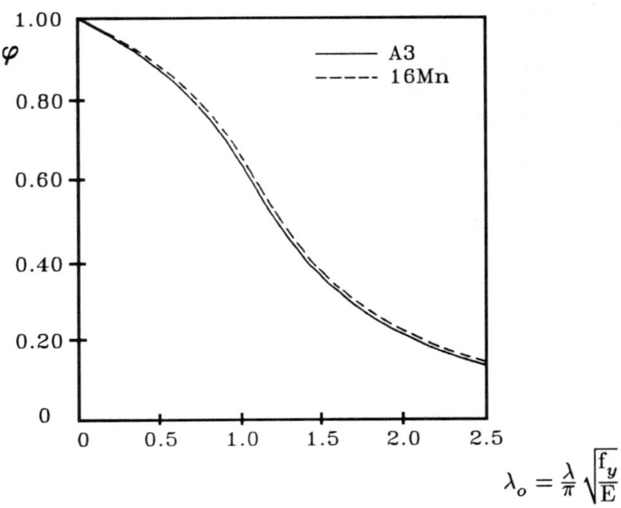

Fig. 11b.C1 Relation between ϕ and λ_o according to Eq. 11b.C5 for steels A3 and 16 Mn

where

E = modulus of elasticity

f_y = specified minimum yield stress

ϵ_o = equivalent initial eccentricity ratio considering the effect of imperfections, whose value is taken as follows:

for A3 steel:

for $\lambda_o \leq 0.5$

$$\epsilon_o = 0.25\lambda_o$$

for $0.5 < \lambda_o \leq 1.0$

$$\epsilon_o = 0.05 + 0.15\lambda_o$$

for $1.0 < \lambda_o$

$$\epsilon_o = 0.05 + 0.15\lambda_o^2$$

for 16Mn steel:

for $\lambda_o \leq 0.5$

$$\epsilon_o = 0.23\lambda_o$$

for $0.5 < \lambda_o \leq 1.3$

$$\epsilon_o = 0.05 + 0.13\lambda_o$$

for $1.3 < \lambda_o$

$$\epsilon_o = 0.05 + 0.10\lambda_o^2$$

λ = effective slenderness ratio of the member

λ_o = $(\lambda/\pi)\sqrt{f_y/E}$

Eq. 11b.C2 is derived from the theory of elastic flexural-torsional buckling. For thin-walled open sections the flexural-torsional buckling load can be remarkably increased if a few batten plates are attached to the open side of the member. It is possible to transform the mode of failure from flexural-torsional buckling to Euler buckling if a sufficient number of batten plates are provided. Analysis of the buckling behavior for partially closed sections, i.e. open sections with battens, is based on the assumption that the warping deformation at battened cross-sections of the member is fully restrained, in other words, the rate of change of twisting angle at these sections vanishes. Analysis is reduced to that of flexural buckling by introducing an equivalent slenderness ratio as mentioned before, which is valid for both battened and unbattened members.

EAST EUROPE

CHAPTER EDITORS

Victor Gioncu, Polytechnic Institute of Timisoara
Jindrich J. Melcher, Technical University of Brno

ČSN 73 1402 88

The design strength of centrally loaded compression member is checked by the following equation

$$N \leq \varphi \cdot A_{ef} \cdot R_d \qquad (11b.E1)$$

where

A = area of gross cross section
A_{ef} = area of effective cross section at uniform compression
C = $1 + m_o + \overline{\lambda}^2_{ef}$
f_y = yield stress
m_o = imperfection parameter depending on the type of cross section (see Chapter 1 - Compression members)
R_d = design strength of material used (see Chapter 1 - Compression Members)
λ = slenderness ratio for the full unreduced section
λ_{ef} = $\lambda(A_{ef}/A)^{1/3}$
$\overline{\lambda}_{ef}$ = $\frac{\lambda_{ef}}{\pi}\sqrt{f_y/E}$
φ = strength reduction factor depending upon the slenderness ratio and steel grade according to the expression:

$$\varphi = \frac{1}{2\overline{\lambda}^2_{ef}}\left(C - \sqrt{C^2 - 4\,\overline{\lambda}^2_{ef}}\right) \qquad (11b.E2)$$

The effective area is determined using the effective widths of the stiffened and unstiffened elements.

In the case of flexural-torsional buckling the slenderness ratio λ is derived from the critical stress of the full unreduced section.

JAPAN

CHAPTER EDITORS
Tetsuro Ono, Nagoya Institute of Technology
Toshiro Suzuki, Tokyo Institute of Technology

AIJ 85

Compression Members

(A) Allowable compressive stress

Allowable compressive stress of a compression member shall be calculated as follows

$$\lambda \leq \Lambda \qquad f_c = \frac{\{1-0.4(\frac{\lambda}{\Lambda})^2\}}{\nu}F \tag{11b.J1}$$

$$\lambda > \Lambda \qquad f_c = \frac{0.277}{(\frac{\lambda}{\Lambda})^2}F$$

$$\Lambda = \sqrt{\frac{\pi^2 E}{0.6F}} \tag{11b.J2}$$

where

- E = modulus of elasticity of steel
- F = yield stress
- f_c = allowable compressive stress
- Λ = critical slenderness ratio
- λ = slenderness ratio of compression member
- ν = $\frac{3}{2} + \frac{2}{3}(\lambda/\Lambda)^2$

(B) For single-angle struts, etc., in which compressive stress is preceded by flexural torsional buckling, the allowable compressive stress shall be obtained by substituting QF for F in Eq. 11b.J1

$$Q = A_{eff}/A \tag{11b.J3}$$

where

- A = total section
- A_{eff} = net section based on effective width described in Eq. 11a.J5

(C) Slenderness ratio of a compression member shall not exceed the following values

Main members	200
Secondary members	250

(D) The effective length of a compression member subject to buckling varies significantly according to the rotation and the restraint of displacement of member ends. In designing compression members, therefore, it is necessary to determine the effective length of buckling based on thorough examinations of the individual conditions affecting it.

Built-up Compression Members

(A) Slenderness ratio of built-up compression members

(1) For the solid web axis of a built-up compression member, the slenderness ratio shall be determined in the same way as that of a compression member composed of identical elements.

(2) For the open web axis of a built-up compression member, the slenderness ratio shall be determined taking into account the effective shear deformation due to buckling.

(B) Stiffeners of built-up compression member

(1) When stiffeners are used to reinforce a plate, the rivet, or weld pitch ρ_1 in the direction of member axis shall not exceed the value obtained by the following equations

$$\rho_1 = 50t \qquad (11b.J4)$$

and

$$\rho_1 = \frac{b}{2} \qquad (11b.J5)$$

where

- b = width of stiffened plate, that is, the distance between fillet tips for bent stiffener, and the distance between rows of welds for riveted or welded stiffeners
- t = thickness of stiffened plate
- ρ_1 = rivet or weld pitch

(2) When a stiffener is provided (includes folding) at the edge of a plate, the moment of inertia of the stiffener protrusion shall not be less than the value shown in Eq. 11a.J7.

(3) The moment of inertia of intermediate stiffeners shall not be less than twice the value shown in Eq. 11a.J7.

(C) Design considerations for built-up compression members. Lacings or tie plates for built-up compression members shall be designed taking into account the shear force due to external force and shear force due to buckling. In addition, the following rules shall be followed

(1) As far as possible, the principal axes of chords and bracing shall converge on a single point. If this is impossible the effect of eccentricity shall be reflected in the design.

(2) For a compression member fabricated by welding two channels with or without lip back to back or into a box, the spacing of separators l_1 shall conform to the provisions given in Eq. 11b.J6.

$$l_1 \le \frac{50g\left(1 + \frac{i_1^2}{g^2}\right)R_s}{f_c A} \qquad (11b.J6)$$

where

- A = sum of sectional areas of components
- f_c = allowable compression stress
- g = distance from y–axis to center of gravity
- i_1 = radius of gyration of component about 1–1 axis
- l_1 = spacing of separators
- R_s = allowable shear force of welds, spot welds, rivets, or bolts

(3) Detailed requirements for built-up members
- Slenderness ratio λ_1 of components shall not exceed 50.
- The number of sections divided by welds or lacings shall not be less than 3.
- Member ends shall be provided with a gusset plate, plate, etc. to secure sufficient rigidity.
- In the case of (a) and (b) in Fig. 11b.J1, the value of c shall not be less than 30 mm.
- Slenderness ratio of lattices shall not be more than 160.

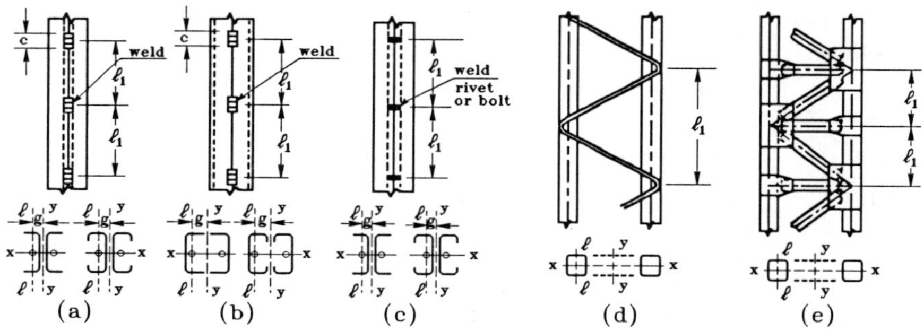

Fig. 11b.J1 Notations for built-up compression members

NORTH AMERICA

CHAPTER EDITOR

Teoman Pekoz, Cornell University

AISI 86

For locally buckled channel and other singly symmetrical sections, concentric axial loading with respect to the centroid of the effective section is an exceptional situation. The centroid of the effective section depends on the magnitude of loading. The location of the centroid moves as the load is increased. However, the allowable concentric loading is important as a parameter in the interaction equation.

AISI 86 treats all overall buckling modes in an unified manner. First the least buckling stress is determined checking for flexural, torsional-flexural and torsional buckling. The least of these three buckling stresses is used to determine the nominal buckling stress for the full unreduced cross-section.

Interaction of Local and Flexural Column Buckling

The approach consists of the following steps. First the elastic flexural buckling stress, F_e, is calculated for the full unreduced section

$$F_e = \pi^2 E/(KL/r)^2 \qquad (11b.N1)$$

Then the failure stress, F_n, is determined

for $F_e \leq F_y/2$ $\quad F_n = F_e \qquad$ (11b.N2)

for $F_e > F_y/2$ $\quad F_n = F_y(1 - F_y/4F_e)$

and the ultimate column load is calculated as

$$P_n = A_e^* F_n \qquad (11b.N3)$$

where A_e^* is the effective area computed at stress F_n.

Interaction of Local and Torsional-Flexural Column Buckling

The approach for the torsional-flexural buckling of locally buckled columns is exactly the same as that for columns subject to flexural buckling, except in this case F_e is determined according to the torsional flexural buckling theory for the full unreduced section.

Evaluation of Stub Column Test Results

The column design approach of AISI 86 necessitates an expression for the effective area A_e as a function of the stress f on the effective area. The stress f is taken as F_n in calculating column strength.

When A_e cannot be calculated, such as when the column has dimensions or geometry outside the range of applicability of the generalized effective width equations, a functional relation between f and A_e can be obtained by stub column tests. A stub column is a short column that is long enough to reflect the local buckling behavior but short enough so that the behavior is not affected by the overall buckling. The effective area A_{eu} at the ultimate load P_u is

$$A_{eu} = P_u/F_y \tag{11b.N4}$$

The effective area at any stress on the effective area f can be calculated as follows

$$A_e = A - (A - A_{eu})\,(f/F_y)^{(A_{eu}/A)} \tag{11b.N5}$$

where A is the full unreduced area of the section.

WEST EUROPE

CHAPTER EDITORS
Joachim Lindner, Technische Universitat Berlin
Ton Toma, TNO Institute for Building Materials and Structures

DASt 016 88

Flexural Buckling due to Concentric Axial Load

The following equation should be satisfied

$$\frac{N}{N_u} \leq 1 \tag{11b.W1}$$

where N_u is the smallest value of N_u^D or N_u^B below

$$N_u^D = A_{ef}^D \cdot f_y \tag{11b.W2}$$

$$N_u^B = \kappa^1 \cdot A_{ef}^B \cdot f_y \tag{11b.W3}$$

where

A_{ef}^B = area of the effective cross section due to bending taken into account different imperfection directions. For simplification constant compressive stresses may be assumed

A_{ef}^D = area of the effective cross section assuming uniform compression at yield stress

i_{ef} = $\sqrt{\frac{I_{ef}}{A_{ef}}}$

k' = $\frac{1}{2}\left[1 + \alpha' + (\bar{\lambda}_k' - 0.2) - (\bar{\lambda}_k')^2 + \frac{\Delta W_o \cdot r_{d,ef}}{i_{ef}^2}\right]$

N = calculated axial force due to design load

N_u^B = buckling load based on an effective area belonging to bending taking into account different imperfection parameters

N_u^D = buckling load based on an effective area belonging to uniform compression at yield stress

r_d, $r_{d,ef}$ = distance of the compression flange to the centroid of the gross respectively effective cross section

α = reduction factor of the European buckling curves, see Table 1a.W1

α' = $\alpha \frac{i \cdot r_{d,ef}}{i_{ef} \cdot r_d} \geq \alpha$

ΔW_o = distance between the gross and the effective section centroids for uniform compression at yield stress

κ^1 = reduction factor as defined by Eq. 11b.W4

$$\kappa^1 = \frac{1}{k' + \sqrt{(k')^2 - \bar{\lambda}_k^2}} \leq 1 \tag{11b.W4}$$

$\lambda_a \quad = \lambda_{pl}$, see Chapter 1a(W1)

$\overline{\lambda}'_k \quad = \dfrac{L}{i_{ef} \cdot \lambda_a}$

ECCS 87

Flexural Buckling Due To Concentric Axial Force

The design strength N_d is to be determined from the following expressions

$$N_d = \kappa \, f_y \, A_g \tag{11b.W5}$$

$$\kappa = F - \sqrt{F^2 - Q/\overline{\lambda}^2} \quad \text{but not greater than Q} \tag{11b.W6}$$

$$F = \frac{1}{2}\,[Q + (1 + \eta(\overline{\lambda} - 0.2))/\overline{\lambda}^2] \tag{11b.W7}$$

where

A_e = area of the effective cross section
A_g = area of gross cross section
f_y = design value of yield stress
i_g = radius of gyration of gross cross section
l = effective length of member
$Q = A_e/A_g$
α = coefficient depending on type of cross section and plane of buckling
$\eta = \alpha(4 - 3Q) \;\; \leq \;\; 0.76$

$$\overline{\lambda} = \sqrt{(f_y/\sigma_E)} = \frac{l}{i_g}\,\frac{1}{\pi}\sqrt{\frac{f_y}{E}} \tag{11b.W8}$$

σ_E = Euler stress

COMMENTARY

The coefficient κ reflects the reduction in strength of a compressed member due to local and overall buckling. For short members (stub columns) local buckling is governing but for long columns overall buckling (e.g., Euler mode) is predominant. This behavior can also be seen from the expression for κ

$$\text{for} \quad \overline{\lambda} \leq 0.2 \qquad \kappa = Q = A_e/A_g$$

$$\text{for} \quad \overline{\lambda} \rightarrow \infty \qquad \kappa \rightarrow 1/\overline{\lambda}^2 \tag{11b.W9}$$

Between these two extreme cases there exists some transition curve where interaction between local and overall buckling is present. The shape of this curve depends mainly on the geometry of the cross section. Formally this has been taken into account by the coefficient $\eta = \alpha(4 - 3Q)$. When Q=1 (no local buckling) $\eta=\alpha$. Depending on type of cross section and overall buckling direction each section is to be referred to some cases. Both theory and test indicate that when local buckling has a significant influence, an increased value of η is generally required. Therefore, α is adjusted by the factor (4–3Q), which gives a reasonable agreement with test results.

INTERNATIONAL COMPARISONS & COMMENT

WHAT ARE THE DIFFERENCES?

To make possible comparisons of the interaction on local and flexural column buckling, the formulas of the reference specifications have been generalized by using following definitions

$$\lambda = \sqrt{\frac{f_y}{\sigma_E}} = \frac{l}{i_g}\frac{1}{\pi}\sqrt{\frac{f_y}{E}} \tag{11b.1}$$

where

A_e = area of effective cross section at yield stress level
A_g = area of gross cross section
f_y = design yield stress
i_g = radius of gyration of gross cross section
l = effective length
N_d = axial load on member caused by design load
Q = A_e/A_g
σ_E = critical Euler stress based on gross cross section

For the different countries this leads to:

AUSTRALIA

$$\frac{N_d}{A_g f_y} = \frac{Q}{2}\left[\{1+\frac{1+\eta_A}{Q\lambda^2}\} - \sqrt{\{1+\frac{1+\eta_A}{Q\lambda^2}\}^2 - \frac{4}{Q\lambda^2}}\right] \tag{11b.2}$$

$$\eta_A = (1.25 - Q)\, Q\lambda^2 \tag{11b.3}$$

CHINA

$$\frac{N_d}{A_g f_y} = \frac{Q}{2}\left[\{1+\frac{1+\epsilon_o}{\lambda^2}\} - \sqrt{\{1+\frac{1+\epsilon_o}{\lambda^2}\}^2 - \frac{4}{\lambda^2}}\right] \tag{11b.4}$$

ϵ_o = a coefficient that is a function of λ and depends on the type of steel.

EAST EUROPE

$$\frac{N_d}{A_g f_y} = \frac{Q}{2}\left[\{1+\frac{1+m_o}{\lambda^2 Q^{2/3}}\} - \sqrt{\{1+\frac{1+m_o}{\lambda^2 Q^{2/3}}\}^2 - \frac{4}{\lambda^2 Q^{2/3}}}\right] \tag{11b.5}$$

JAPAN

for $\lambda \leq 1.29$

$$\frac{N_d}{A_g f_y} = \frac{1-0.24\lambda^2}{0.9 + 0.24\,\lambda^2}\, Q$$

for $\lambda > 1.29$ (11b.6)

$$\frac{N_d}{A_g f_y} = \frac{0.77}{\lambda^2}\, Q$$

NORTH AMERICA

for $\lambda < \sqrt{2}$

$$\frac{N_d}{A_g f_y} = (1 - \frac{\lambda^2}{4})Q^*$$

for $\lambda \geq 2$ (11b.7)

$$\frac{N_d}{A_g f_y} = \frac{Q^*}{\lambda^2}$$

$Q^* = \frac{A_e^*}{A_g}$ in which A_e^* is determined at the failure stress F_n (Eq. 11b.N2)

Remark: In all specifications except in North American the yield stress f_y does not include the influence of cold forming for compressed elements.

WEST EUROPE

$$\frac{N_d}{A_g f_y} = \frac{Q}{2}\left[\left\{1+\frac{1+\eta_\omega(\lambda-0.2)}{Q\lambda^2}\right\} - \sqrt{\left\{1+\frac{1+\eta_\omega(\lambda-0.2)}{Q\lambda^2}\right\}^2 - \frac{4}{Q\lambda^2}}\right] \leq Q \quad (11b.8)$$

where

α = a coefficient depending on the type of cross-section

$\eta_\omega = \alpha(4-3Q)$

Table 11b.1 shows a survey of how the different regions have treated the failure modes flexural buckling and torsional buckling. Furthermore, information has been provided if a region has a procedure to evaluate stub column tests. This survey in the table also gives an indication of "Why are there differences?"

Table 11b.1 Comparison of the treament by the various regions of the items flexural and torsional buckling

Region		Flexural Buckling		Torsional Buckling		Procedure for Evaluation Stub Column Test
		Locally Stable	Locally Unstable	Locally Stable	Locally Unstable	
A		I, VI	II	I, VI	II	–
C		I, VI	III	I, VI, VII, X	III, X	–
E		IX	III	IX, VII	III	–
J		V	II	?	?	–
N	USA	IV	III	V, VII	III	VIII
	CSA	IV	III	IV, VII	III	–
W		I, VI	II	I, VI	II	VIII

Notes

- I. Column curve accounting for out-of-straightness explicitly
- II. Q-method of accounting for the interaction of local and overall buckling
- III. Determining overall buckling stress for full section and multiplying it by the effective area at that stress to get column failure load
- IV. CRC column curve
- V. Modified column curve that gives lower results
- VI. Believed to be an improvement over the CRC column curve on the basis of test results
- VII. Explicit account of torsional-flexural buckling is taken
- VIII. Procedure given
- IX. Single column curve in U.S.S.R., in others E lower, multiple column curve concept
- X. The effect of batten plates on the torsional-flexural buckling load of singly symmetrical open sections can be taken into account.
- – No procedure given
- ? No information available or uncertain about information

WHY ARE THERE DIFFERENCES?

1. From Fig. 11b.1 it appears that AISI 86 gives for Q = 1 the most optimistic values. The West European values seem to be the conservatives ones.

2. The formulas from Australia, China, East Europe and West Europe have the format of the Perry-Robertson formula. These have only a difference in defining the Perry-coefficient.

3. For the West European specifications the values of α are defined as follows:
   $\alpha = 0.21$ for square tube
   $\alpha = 0.34$ for channel with lips

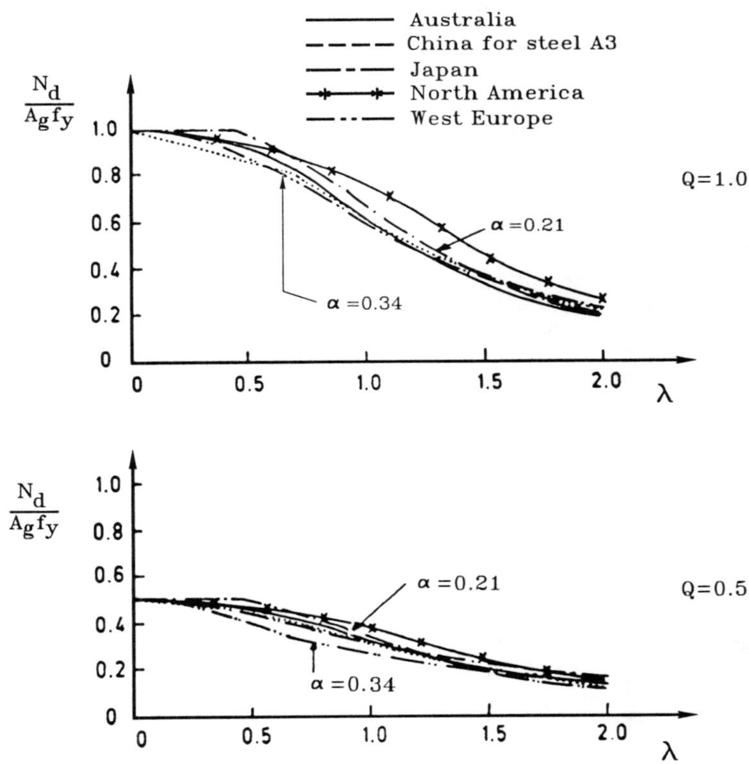

Fig. 11b.1 Relation between $\frac{N_d}{A_g f_y}$ and λ concerning columns for the referenced specifications.

11c. Flexural-Members

SPECIFICATION PROVISIONS & EXPLANATION

AUSTRALIA

CHAPTER EDITOR
Greg J. Hancock, The University of Sydney

AS 1538 88

Laterally Unbraced Beams

Bending stresses are computed on the effective section (see page 11a(A1)). The formulas for the maximum permissible stress in bending (F_b) are the same as in AS 1250 81 and are defined as follows

$$\begin{aligned}
&\text{for } F_o \le F_y && F_b = (0.55 - 0.10\, F_o/F_y)\, F_o \\
&\text{for } 2.04\, F_y \ge F_o > F_y && F_b = (0.95 - 0.50\sqrt{F_y/F_o}\,)\, F_y \\
&\text{for } F_o > 2.04\, F_y && F_b = 0.60\, F_y
\end{aligned} \tag{11c.A1}$$

where

F_o = elastic critical stress for flexural-torsional buckling
F_y = yield stress

The formulas for elastic critical stress for flexural-torsional buckling have been revised to distinguish between I, T, Z and Channel beams as follows

Doubly Symmetric I and Monosymmetric I and T Beams

The elastic critical stress, F_o (tensile or compressive) of a simply supported monosymmetric I-beam or T-beam with equal and opposite end moments is equal to

$$F_o = \frac{\pi\sqrt{(E\, I_y\, G J)}}{Z_{xmin}\, L}\left[\frac{\pi\,\delta}{2} + \sqrt{\left\{\frac{\pi\delta}{2}\right\}^2 + \left\{1 + \frac{\pi^2 E\, I_w}{G J\, L^2}\right\}}\,\right] \tag{11c.A2}$$

where

β_x = monosymmetry section constant

$\delta = \pm\dfrac{\beta_x}{L}\sqrt{\dfrac{E\, I_y}{G\, J}}$

Channel and Z-Beams

The elastic critical stress F_o of simply supported channels and Z-beams with unequal end moments may be approximated by

$$F_o = \left\{\frac{1}{C_m}\right\} \left\{\frac{\pi^2 E \sqrt{I_y I_w}}{L^2}\right\} \left\{\frac{1}{Z'_{xmin}}\right\} \qquad (11c.A3)$$

where

I_y = the value computed about the inclined minor principal axis for Z-beams. Alternatively, for Z-beams restrained against lateral movement by sheeting effectively bracing the tension flange, the values of I_y and I_w shall be those for an equivalent channel where the direction of the flange of the Z-beam attached to the sheeting is reversed

Z'_{xmin} = the smaller value of the section modulus computed about the x, axis perpendicular to the web

For a channel or Z-beam that is intermediately braced, the bracing interval *a* shall be used instead of the length L in the formula for F_o. Values of the bracing interval *a* and coefficient C_m for uniformly distributed loads applied within the span of intermediately braced simply supported beams are given in Table 11c.A1 below.

Table 11c.A1 C_m Factors for Simply Supported Beams Uniformly Loaded Within The Span

Loading position	No bracing a=L	One central brace a=0.5 L	Third point bracing a=0.33 L
Tension flange	0.52	0.63	0.68
Shear center	0.82	0.73	0.73
Compression flange	1.30	0.84	0.78

CHINA

CHAPTER EDITORS

Shi-Ji Wang, Hunan University
Zhong-Quan Zhang, Zhongnan Building Design Institute

GBJ 18 87

The lateral buckling of beams subjected to major axis bending is checked by the formula

$$\frac{M_{max}}{\phi_b W_{ef}} + \frac{B}{W_\omega} \leq f \qquad (11c.C1)$$

where

B	= bimoment at the section corresponding to M_{max}
d	= distance between shear center and point of application of the transverse load, being positive/negative when load is applied below/above the shear center
f_y	= yield strength (MPa)
h	= height of the cross-section
I_y	= moment of inertia about the minor axis
L_o	= lateral unsupported length of the beam
M_{max}	= maximum bending moment
U_y	$= \int_A x(x^2 + y^2)\, dA \qquad (11c.C2)$
W_{ef}	= effective section modulus for bending
W_x	= section modulus about the major axis
W_ω	= section modulus for warping torsion
β_x	$= \frac{U_y}{2I_y} - e_o \qquad (11c.C3)$
ζ	$= \frac{4I_\omega}{h^2 I_y} + \frac{0.156 I_t}{I_y}(\frac{L_o}{h}) \qquad (11c.C4)$
η	$= 2(\xi_2 d + \xi_3 \beta_x)/h \qquad (11c.C5)$
ξ_1, ξ_2, ξ_3	= coefficients according to the loading and supporting conditions of the beam, whose values are tabulated in the appendix of the specification
ϕ_b	= strength reduction factor for beams determined by the following equation

$$\phi_b = \frac{4320Ah}{\lambda_y^2 W_x} \xi_1 (\sqrt{\eta^2+\zeta} + \eta) \frac{240}{f_y} \qquad (11c.C6)$$

If ϕ_b obtained from Eq. 11c.C6 exceeds 0.7, it should be replaced by a modified factor ϕ_b'

$$\phi_b' = 1.091 - 0.274/\phi_b \leq 1 \qquad (11c.C7)$$

COMMENTARY

Equation 11c.C6 is based on the elastic lateral-torsional buckling theory. Various loading conditions such as uniformly distributed load, concentrated load(s), equal or unequal end moments; and various lateral-supporting conditions are taken into consideration. In the inelastic domain ϕ_b is modified by making use of tangent modulus theory. The proportional limit is assumed to be $0.7f_y$, and the analytic values have been adjusted so that $\phi_b' = 1.0$ when $\phi_b = 3.0$, and $\phi_b' = 0.7$ when $\phi_b = 0.7$.

EAST EUROPE

CHAPTER EDITORS
Victor Gioncu, Polytechnic Institute of Timisoara
Jindrich J. Melcher, Technical University of Brno

ČSN 73 1402 88

The design strength of flexural members subjected to major axis bending shall be checked by the equation

$$M \leq \lambda_{lat} W_{ef} R_d \tag{11c.E1}$$

where

- M = maximum bending moment caused by the design load
- R_d = design strength of material used (see Chapter 1 - Compression Members)
- W_{ef} = section modulus of the effective cross section for the compression fibers
- λ_{lat} = lateral buckling reduction factor for the critical beam slenderness ratio determined for the full unreduced section (see Chapter 3 - Beams, East Europe)

Calculations of effective widths are based on the Eqs. 11a.E1 and 11a.E2. The values of the web buckling coefficient k for the example of single-channel section is presented, in Fig. 11c.E1 together with the distribution of the effective web portion.

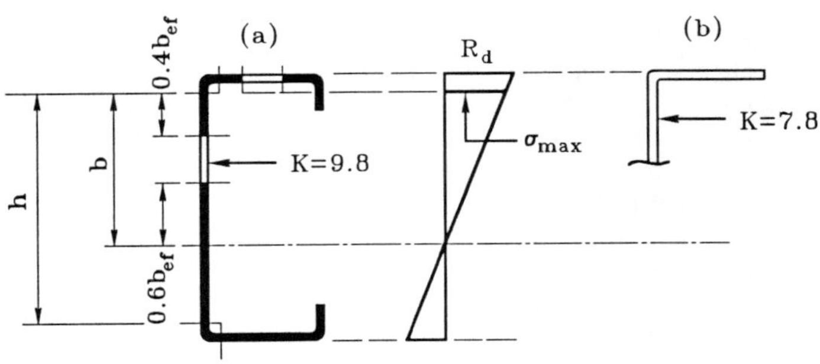

Fig. 11c.E1 Effective web portion in case of flexural members

JAPAN

CHAPTER EDITORS

Tetsuro Ono, Nagoya Institute of Technology
Toshiro Suzuki, Tokyo Institute of Technology

AIJ 85

Allowable Bending Stress

(A) When a laterally unsupported straight flexural member is subjected to bending stress about its strong axis, the allowable bending stress for long sustained loading shall be as follows

(1) For I–shaped section (including built-up)

for $\lambda_y \leq 120\sqrt{C_b}$

$$f_b = (1.1 - 0.3\,\frac{F}{\pi^2\,E\,C_b}\lambda_y^2)\,f_t \leq f_t \qquad (11c.J1)$$

for $\lambda_y > 120\sqrt{C_b}$

$$f_b = \frac{2}{3}\,\frac{\pi^2 E}{\lambda_y^2}\,C_b \qquad (11c.J2)$$

(2) For single-channel section, point symmetrical Z shape section

for $\lambda_y \leq 85\sqrt{C_b}$

$$f_b = (1.1 - 0.6\,\frac{F}{\pi^2\,E\,C_b}\lambda_y^2)\,f_t \leq f_t \qquad (11c.J3)$$

for $\lambda_y > 85\sqrt{C_b}$

$$f_b = \frac{1}{3}\,\frac{\pi^2 E}{\lambda_y^2}\,C_b \qquad (11c.J4)$$

where

C_b = $1.75 - 1.05(M_2/M_1) + 0.3(M_2/M_1)^2 \leq 2.3$
M_2 is the smaller bending moment about the strong axis at the end of section subject to buckling, and M_1 is the larger bending moment about the strong axis at the same end. M_2/M_1 is positive for single curvature , and negative for double curvature. When the moment at the middle of the section is greater than M_1, the value of C is assumed to be 1

F = yield stress

f_b = allowable bending stress

f_t = allowable tensile stress

i = radius of gyration in the plane of web of tee section consisting of compression flange and 1/6 of the depth of beam; for single channel section or point symmetrical Z shape section, the radius of gyration of the entire section about the axis parallel with the web

l_b = Unsupported length of compression flange (cm)

λ_y = l_b/i

(B) When the flexural member is subject to a bending stress about the web axis or free from lateral buckling, the allowable bending stress shall be equivalent to the allowable tension stress.

NORTH AMERICA

CHAPTER EDITOR
Teoman Pekoz, Cornell University

AISI 86

In the design of flexural-members three types of failure modes are considered.

Reaching the Section Strength
The section strength can be designed in two different ways.
(A) Reaching the yield stress on the effective section.
(B) Reaching the compressive failure yield strain in the section. (Several conditions must be satisfied in order to be able to use this criterion of failure.)

Lateral Buckling Strength
The approach for this case is consistent with the one used for columns.
(A) First the elastic lateral buckling moment M_e is calculated on the basis of torsional-flexural buckling theory assuming the section to be fully effective. For certain types of sections simplified lateral buckling equations may be used.

The lateral buckling moment M_c can be either the elastic lateral buckling moment M_e or the inelastic buckling moment depending on the ratio of M_e to the yield moment M_y for the full section. It is determined as follows

$$\text{for} \quad M_e \le M_y/2 \qquad M_c = M_e \tag{11c.N1}$$

$$\text{for} \quad M_e > M_y/2 \qquad M_c = M_y(1 - M_y/4M_e)$$

(B) The lateral buckling stress is determined by dividing the lateral buckling moment M_c by the section modulus S_f of the full unreduced section for the compressive fiber. The nominal buckling moment M_n is calculated by multiplying the lateral buckling stress determined in b above by the section modulus of the effective section S_f. The effective section modulus is calculated for the extreme compression fiber and for the compressive stress in that fiber equal to the lateral buckling stress determined in b above. The allowable moment M_a for flexural members determined by dividing the nominal bending moment capacity M_n by a factor of safety equal to 1.67. The nominal bending moment capacity is the smaller of the ultimate moment capacity determined by the section strength buckling moment.

Other Failure Modes
Provisions are provided to safeguard against failure due to web crippling or shear buckling strength. Interaction equations are provided for the cases of combined bending and shear and bending and web crippling.

WEST EUROPE

CHAPTER EDITORS
Joachim Lindner, Technische Universitat Berlin
Ton Toma, TNO Institute for Building Materials and Structures

DASt 016 88

Lateral Torsional Buckling
The following equation should be satisfied

$$\frac{M_x}{M_{u,\,B}} \leq 1 \tag{11c.W1}$$

where

k = plate buckling coefficient
M_{ki} = elastic critical moment for lateral torsional buckling of the gross section
$M_{ki,\,PB}$ = elastic critical plate buckling moment
= $k \cdot \sigma_e \cdot S_{x,\,ef}$
$M_{ki,\,ef}$ = elastic critical moment for lateral torsional buckling of the effective cross section

$$M_{ki,\,ef} = M_{ki}\sqrt{\frac{1}{1+\left(\frac{M_{ki}}{M_{ki,\,PB}}\right)^2}}$$

$M_{u,\,B}$ = $\kappa_m \cdot S_{x,\,ef} \cdot f_y$
$S_{x,\,ef}$ = effective section modulus with the outer fiber at actual stress or yield stress
κ_m = reduction factor

for $\bar{\lambda}'_m \leq 0.4$ $\quad \kappa_m = 1.0$

for $\bar{\lambda}'_m > 0.4$ $\quad \kappa_m = \left(\frac{1}{1+(\bar{\lambda}'_m)^5}\right)^{0.4}$

$$\bar{\lambda}'_m = \sqrt{\frac{f_y}{\sigma_{ki,\,ef,\,m}}}$$

$$\sigma_e = \frac{\pi^2 E}{12(1-\mu^2)(\frac{b}{t})^2}$$

$$\sigma_{ki,\,ef,\,m} = \frac{M_{ki,\,ef}}{S_{x,\,ef}}$$

ECCS 87

Members not Subject to Lateral Buckling
Calculation of section properties, such as cross-sectional area and section modulus shall in general be based on the effective cross section.

When yielding first occurs at the compression edge of the section, the design strength M_d is determined by

$$M_d = f_{ty} W_{ef} \tag{11c.W2}$$

where

f_{ty} = design value of yield stress
W_{ef} = section modulus of the effective cross section. In order to avoid an iterative procedure the effective portions of the web may be based on Ψ obtained by assuming the compression flange reduced but the web being fully effective

When yielding first occurs on the tension side, the plastic reserve of the tension zone can be utilized until the compressive stress reaches f_{ty}. This will normally lead to iterative calculations.

A conservative value of M_d in the latter case can however be obtained by basing the effective cross section on the compressive stress $= f_{ty}$

Note: Channel- or Z-sections having both flanges braced and used as beams, loaded in the plane of the web can be designed according to the above rules. However due consideration must be taken to the design of bracings.

Lateral Torsional Buckling
The design strength M_d can be calculated from the following expressions

$$M_d = \kappa_M f_{ty} W_g \tag{11c.W3}$$

$$\kappa_M = F - \sqrt{F^2 - \frac{W_{ef}}{W_g}\frac{1}{\bar{\lambda}_M^2}} \tag{11c.W4}$$

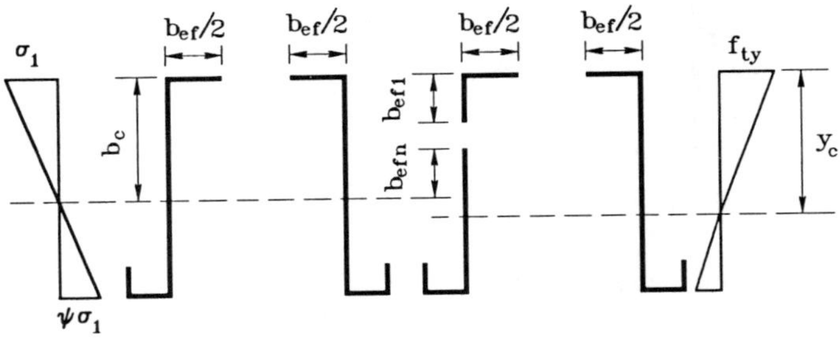

Fig. 11c.W1 Example of determining effective cross-section of a thin-walled flexural member

$$F = \frac{1}{2}\left[\frac{W_{ef}}{W_g} + \frac{1}{\overline{\lambda}_M^2}\left(1 + \eta(\overline{\lambda}_M - 0.2)\right)\right] \qquad (11c.W5)$$

where

W_{ef} = section modulus of the effective cross-section = I_{ef}/y_c
W_g = section modulus of gross cross section
η = 0.85 $(1 - (\sigma_{crlb}\, W_g)/(f_{ty}\, W_{ef}))$ but not less than 0.1
$\overline{\lambda}_M$ = $(f_{ty}/\sigma_{crM})^{1/2}$
σ_{crM} = theoretical critical stress for elastic lateral buckling based on the gross cross-section
σ_{crlb} = elastic critical stress for local buckling of the compression flange

COMMENTARY

The reduction factor κ_M with respect to lateral buckling has been given the same format as for the column curves. The expression for the parameter η is partly based on test results. (Davies and Thomasson, 1983)

Further Design Aspects Concerning Flexural Members

Besides the moment capacity for flexural members following design aspects have been treated in the European recommendations:

- effects of shear lag
- shear strength
- web crippling
- combination of bending moment and shear force
- combination of bending moment and support reaction

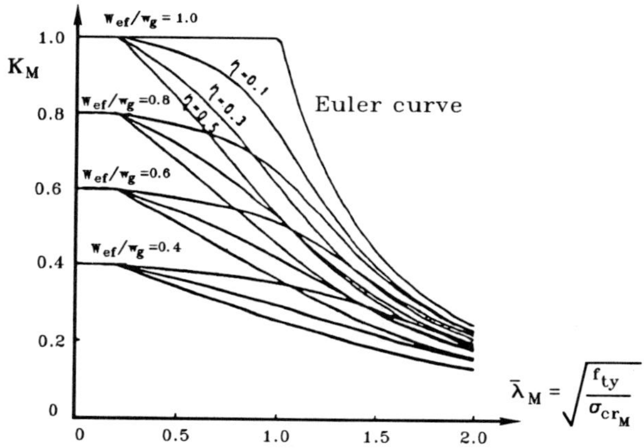

Fig. 11c.W2 Relation between reduction factor for lateral torsional buckling (κ_M) and the relative slenderness ($\overline{\lambda}_M$)

INTERNATIONAL COMPARISONS & COMMENT

WHAT ARE THE DIFFERENCES?

For comparison the most important case will be treated, that for which lateral buckling can occur. For convenience the formulas of the reference specifications have been reformulated using the following definitions

f_c = elastic critical stress
f_y = design yield stress
M_d = maximum bending moment caused by the design load
W_{ef} = section modulus of the effective cross section
(Definition of the stresses at which the effective section is to be determined are given in the corresponding subchapter.)
W_g = section modulus of the gross cross section

For the different countries this leads to the following comparison.

AUSTRALIA

for $f_c \le f_y$
$$\frac{M_d}{f_y W_g} = (0.92 - 0.17 \frac{f_c}{f_y}) \frac{f_c}{f_y} \frac{W_{ef}}{W_g}$$

for $f_y < f_c \le 2.04\, f_y$
$$\frac{M_d}{f_y W_g} = (1.58 - 0.83\sqrt{\frac{f_y}{f_c}}) \frac{W_{ef}}{W_g}$$

for $f_c > 2.04\, f_y$
$$\frac{M_d}{f_y W_g} = \frac{W_{ef}}{W_g} \qquad (11c.1)$$

For f_c formulas have been given depending on the type of section:
- doubly symmetric I, monosymmetric I and T-beams
- channel and Z-beams

CHINA
$$\frac{M_d}{f_y W_g} = \phi_b \frac{W_{ef}}{W_g} (1 - \frac{B}{W_\omega f_y}) \qquad (11c.2)$$

Although it cannot be seen directly, the methods in China have been based on the theoretical elastic critical stress. They are related to the effective cross section.

EAST EUROPE
$$\frac{M_d}{R_d W_g} = \lambda_{lat} \frac{W_{ef}}{W_g} \qquad (11c.3)$$

JAPAN
$$\frac{M_d}{f_y W_g} = \frac{f_b}{0.6 f_y} \qquad (11c.4)$$

f_b has been defined and expressed in formulas, which depend on:
- type of the section
- distribution of moments in the member

No information has been provided to take into account the lack of full effectiveness of cross sections.

NORTH AMERICA

for $f_c \leq 0.5f_y$

$$\frac{M_d}{f_y W_g} = \frac{f_c}{f_y} \frac{W_{ef}}{W_g}$$

for $f_c > 0.5f_y$

$$\frac{M_d}{f_y W_g} = (1 - \frac{f_y}{4 f_c}) \frac{W_{ef}}{W_g} \tag{11c.5}$$

f_c is based on the torsional-flexural buckling theory assuming the section to be fully effective.

Table 11c.1 Comparison of the treatment by the various regions of the moment capacity for flexural members in case of lateral buckling prevented or not

Region	Lateral buckling	
	Prevented	Not prevented
A	I, II	III, V
C	I	VIII
E	I	I, III
J	VI	VII
N	I, II	III, V
W	I, II	III, IV

Notes

I Moment capacity determined by multiplying yield stress and section modulus of the effective cross section.

II Plastic reserve may be used under certain conditions.

III Method based on theoretical critical stress for elastic lateral buckling derived from the gross cross section

IV A Q method is used for taking into account the interaction of local and lateral instability.

V The ultimate lateral buckling stress will be multiplied by the effective section modulus belonging to that stress.

VI Bending stress equal to tension stress. For influence of local instability no information has been provided.

VII The allowable bending stress depends on
- type of section
- slenderness of the beam
- distribution of moments in the member

VIII In the checking formula the following have been taken into account
- a reduction factor based on the elastic lateral-torsional buckling theory
- bi-moment at the section corresponding to the maximum moment
- section modulus for warping
- effective section modulus for bending

WEST EUROPE

$$\frac{M_d}{f_y\, W_g} = K \qquad (11c.6)$$

K has been defined and expressed in formulas, which depend on:

- effectiveness of cross section
- theoretical critical stress for elastic lateral buckling based on the gross cross section.

Remark: In all specifications except in North American the yield stress f_y does not include the influence of cold forming for compressed elements.

Table 11c.1 compares the basis for the treament of flexural-members by the different regions. There have been distinguished cases in which lateral buckling has been prevented and cases in which it has not been prevented.

WHY ARE THERE DIFFERENCES?

1. The specifications of Japan do not take into account the reduction of the cross section caused by local buckling.
2. The theoretical elastic lateral buckling stress is not defined in all the specifications.
3. In the different specifications various parameters are taken into consideration in different ways. An example is the moment distribution.
4. The definition of W_{ef} is different according to different specifications.

11d. Beam-Columns

SPECIFICATION PROVISIONS & EXPLANATION

AUSTRALIA

CHAPTER EDITOR

Greg J. Hancock, The University of Sydney

AS 1538 88

Combined Axial and Bending Stresses

The rules for monosymmetric beam-columns have been expanded to include sections bent both in a plane of symmetry and about a plane of symmetry as follows:

Monosymmetric Sections

Sections Bent in a Plane of Symmetry

Sections that are symmetric about the y–axis and bent in the plane of symmetry shall satisfy the requirements of a normal linear interaction formula for beam-columns except that the values of F_b (the maximum permissible stress in bending, see Eq. 11c.A1, shall be calculated using the elastic flexural buckling stress F_o given by

$$F_o = \frac{A\ F_{oy}\ r_{o1}}{2\ Z_{xmin}} \left[\pm \frac{\beta_x}{r_{o1}} + \sqrt{\left[\frac{\beta_x}{r_{o1}}\right]^2 + \frac{4\,F_{oz}}{F_{oy}}} \right] \tag{11d.A1}$$

Sections Bent about a Plane of Symmetry

Sections that are symmetric about the x-axis and bent about the plane of symmetry shall satisfy the requirements of the value of F_b (the maximum permissible stress in bending, see Eq. 11c.A1, shall be calculated using the elastic flexural-torsional buckling stress F_o given by

$$F_o = \frac{A\ r_{o1}\sqrt{F_{oy}\ F_{oz}}}{Z_{xmin}} \tag{11d.A2}$$

The terms F_{oy} and F_{oz} are the y-axis flexural and torsional buckling stresses, respectively.

The term β_x is the monosymmetry section constant and the term r_{o1} is the polar radius of gyration about the shear center.

CHINA

CHAPTER EDITORS
Shi-Ji Wang, Hunan University
Zhong-Quan Zhang, Zhongnan Building Design Institute

GBJ 18 87

For Beam-Columns with Doubly Symmetric Cross-Sections

The following formulas should be satisfied

$$\frac{N}{\phi_x A_{ef}} + \frac{\beta_{mx} M_x}{(1 - \frac{N}{N_{Ex}}\phi_x) W_{efx}} + \frac{M_y}{\phi_{by} W_{efy}} \leq f \tag{11d.C1}$$

$$\frac{N}{\phi_y A_{ef}} + \frac{\beta_{my} M_y}{(1 - \frac{N}{N_{Ey}}\phi_y) W_{efy}} + \frac{M_x}{\phi_{bx} W_{efx}} \leq f \tag{11d.C2}$$

where

M_x, M_y = design values of bending moments
N_{Ex}, N_{Ey} = Euler buckling loads
W_{efx}, W_{efy} = effective section moduli
β_{mx}, β_{my} = equivalent uniform moment factors, the subscripts x and y represent the x- and y-axis, respectively
ϕ_x, ϕ_y = strength reduction factors for columns
ϕ_{bx}, ϕ_{by} = strength reduction factors for beams

M and β_m are evaluated in the following manner:

(A) For braced members without transverse load $\beta_m = 0.6 + 0.4M_1/M_2 \geq 0.4$, $M = M_2$; in which M_1 and M_2 are end moments, M_2 being the numerically larger one; the ratio M_1/M_2 is positive/negative when the member is bent in single/reverse curvature.

(B) For braced members with transverse load(s) $\beta_m = 1.0$, M = maximum moment within the mid-third portion of the member, but not less than one half of the maximum moment within the whole length.

(C) For unbraced member $\beta_m = 1.0$, M = maximum moment within the whole length of the member.

For Beam-Columns with Singly Symmetric Open Sections

When the beam is loaded in the plane of symmetry as shown in Fig. 11d.C1 the in-plane instability is checked by Eq. 11d.C1 or 11d.C2. However, the flexural-torsional buckling is checked by Eq. 11d.C1, in which the strength reduction factor ϕ is obtained by introducing an equivalent slenderness ratio λ_ω into the basic column curve, where

$$\lambda_\omega = \lambda_x \sqrt{\frac{s^2+a^2}{2s^2} + \sqrt{(\frac{s^2+a^2}{2s^2})^2 - \frac{a^2 - \alpha(e_o-e_x)^2}{s^2}}} \qquad (11d.C3)$$

where

$$a^2 = e_o^2 + i_x^2 + i_y^2 + e_x(U_y/I_y - 2e_o) \qquad (11d.C4)$$

$e_x = \beta_m M/N$, being positive/negative when the open side is on the concave/convex surface of the deformed member, s^2 has been defined in Eq. 11b.C4; other notations used here have the same meaning as that in Eq. 11b.C4 and Table 11b.C1.

COMMENTARY

According to the initial yield criterion the ultimate state for in-plane instability of uniaxially loaded beam-columns can be expressed in the form of the interaction equation as follows

$$\frac{N}{A_{ef}\, f_y} + \frac{M + N\, e^*}{(1 - \frac{N}{N_E})W_{ef}\, f_y} = 1 \qquad (11d.C5)$$

where e^* is equal to equivalent eccentricity of initial imperfections.

In case $M = 0$, $N = N_o = \phi\, A_{ef}\, f_y$, Eq. 11d.C5 becomes

$$\frac{N_o}{A_{ef}\, f_y} + \frac{N_o\, e^*}{(1 - \frac{N_o}{N_E})W_{ef}\, f_y} = 1 \qquad (11d.C6)$$

Eliminating e^* in Eqs. 11d.C5 and 11d.C6, we obtain the in-plane interaction equation as follows

$$\frac{N}{\phi\, A_{ef}\, f_y} + \frac{M}{(1-\phi \frac{N}{N_E})W_{ef}\, f_y} = 1 \qquad (11d.C7)$$

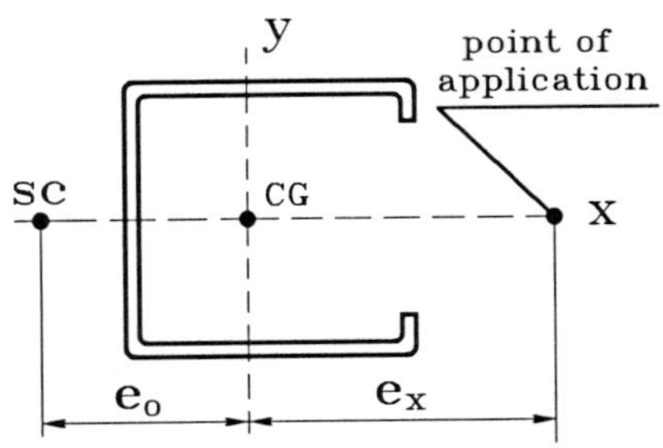

Fig. 11d.C1 Beam-column of singly symmetric open section loaded in-plane of symmetry

The ultimate state for out-of-plane instability of beam-columns bent about the strong axis can be expressed as

$$\frac{N}{\phi A_{ef} f_y} + \frac{M}{\phi_b W_{ef} f_y} = 1 \qquad (11d.C8)$$

which is compatible with the conditions for both centrally loaded columns and beams. Based on Eqs. 11d.C7 and 11d.C8 a pair of semi-empirical formulas, Eqs. 11d.C1 and 11d.C2, are adopted for checking the stability of biaxial loaded beam-columns with doubly symmetric sections. However for beam-columns of singly symmetric open sections loaded in the plane of symmetry, the out-of-plane stability is checked by the equivalent slenderness ratio approach instead of the interaction equation because the former is more accurate and valid for both battened and unbattened members. This problem has been studied theoretically and experimentally. A total of 213 eccentrically loaded thin-walled columns have been tested with various slenderness ratios, positive or negative eccentricities and number of battens (or without batten plate). Experimental results are in satisfactory agreement with the theory.

EAST EUROPE

CHAPTER EDITORS
Victor Gioncu, Polytechnic Institute of Timisoara
Jindrich J. Melcher, Technical University of Brno

ČSN 73 1402 88

Major Axis Bending and Compression

The following check condition for the compression fibers should be satisfied

$$\frac{N}{\lambda\,A_{ef}} + \frac{M_y}{\lambda_{lat}\,W_{ef,\,y}} \leq R_d \tag{11d.E1}$$

where

R_d = design strength of material used

λ, A_{ef} = strength reduction factor and area of the effective cross section, respectively, for compression members (see Section 11b(E))

λ_{lat}, $W_{ef,\,y}$ = lateral buckling reduction factor and section modulus of the effective cross section, respectively, for flexural members (see Section 11c(E))

In Eq. 11d.E1 M and N are the maximum bending moment and axial force respectively, within the whole length of the member. For members without transverse load the concept of equivalent uniform moment shall be used (similar to GBJ 18 87, Section 11d(C)).

Biaxial Bending, Compression and Torsion

It shall be verified that

$$\frac{N}{\lambda\,A_{ef}} + \frac{M_y}{\lambda_{lat}\,W_{ef,\,y}} + \frac{M}{W_{ef,\,z}} + \frac{B}{W_w} \leq R_d \tag{11d.E2}$$

where

B = bimoment

M_z, $W_{ef,z}$ = the minor axis bending moment and section modulus, respectively

W_w = warping section modulus for the full unreduced cross section.

JAPAN

CHAPTER EDITORS
Tetsuro Ono, Nagoya Institute of Technology
Toshiro Suzuki, Tokyo Institute of Technology

AIJ 85

A beam-column subjected to axial compression and bending stress simultaneously shall be designed using the following equations.

$$\frac{\sigma_c}{f_c} + \frac{c\,\sigma_b}{f_b} \leq 1 \qquad (11d.J1)$$

$$\frac{t\,\sigma_b - \sigma_c}{f_t} \leq 1 \qquad (11d.J2)$$

f_c and f_b are as defined for columns and flexural members, respectively

where

f_t = allowable tensile stress
$c\,\sigma_b$ = compression caused by bending
$t\,\sigma_b$ = tension caused by bending
σ_c = compression stress caused by axial load

Slenderness ratio of beam-columns must not exceed the following values
a) Main beam-column : 120
b) Studs, etc. : 160

NORTH AMERICA

CHAPTER EDITOR

Teoman Pekoz, Cornell University

AISI 86

Locally Stable Open Section Beam-Columns

For sections with fully effective plate elements the following interaction equation is used

$$\frac{P}{P_o} + \frac{M_x\, C_{mx}}{M_{xo}(1-P/P_x)} + \frac{M_y\, C_{my}}{M_{yo}(1-P/P_y)} \leq 1 \qquad (11d.N1)$$

where

C_{mx}, C_{my} = corrections to reflect the moment gradient in the member
M_{xo} = failure moment for bending about the x-axis in the absence of an axial load or bending about the y-axis
M_{yo} = failure moment for bending about the y-axis in the absence of an axial load or bending about the x-axis
P, M_x, M_y = axial force and the moments about the x- and y-axis, respectively, due to the applied loading
P_o = failure load in the absence of any moment
P_x, P_y = flexural buckling loads about the x- and y-axes, respectively

P_o, M_{xo} and M_{yo} are determined considering both flexural and torsional flexural buckling.

Biaxially Loaded Locally Unstable Beam-Columns

Some of the parameters in the above interaction are redefined for singly or doubly symmetric open sections and closed tubes. P_o is determined as described above for locally unstable columns. It may be governed by flexural or torsional flexural buckling. M_{xo} and M_{yo} are determined by the approach described above for lateral buckling. All eccentricities are taken with respect to the centroid of the effective section for the axial load alone. The parameters P_x and P_y are the elastic buckling loads for the full unreduced section.

WEST EUROPE

CHAPTER EDITORS
Joachim Lindner, Technische Universitat Berlin
Ton Toma, TNO Institute for Building Materials and Structures

DASt 016 88

Members in Flexure and Compression
(A) Flexure in the plane of symmetry, lateral torsional buckling is prevented

Eq. 11d.W1 or 11d.W2 should be satisfied

$$\frac{N}{A_{ef}} + \frac{M_x^{II}}{S_{x,ef}} \leq f_y \qquad (11d.W1)$$

$$\frac{N}{N_u} + \beta \frac{M_x^{I}}{M_u} + \Delta n \leq 1 \qquad (11d.W2)$$

where

M_u = $S_{x,ef} \cdot f_y$ (11d.W3)

M_x^I = calculated moment about the x-axis according to the first order theory

M_x^{II} = calculated moment about the x-axis, taking into account second order effects, especially initial imperfections and the shift of the centroid between the gross and the effective cross section

N_u = see Eq. 11b.W2 or 11b.W3

$$\Delta n = \frac{N}{N_u}\left(1 - \frac{N}{N_u}\right)(\kappa')^2 \cdot (\overline{\lambda}')^2 \leq 0.1 \qquad (11d.W4)$$

κ', $\overline{\lambda}'_k$ = see Section 11b(W)

(B) Lateral torsional buckling not prevented

The following equation should be satisfied

$$\frac{N}{N_u} + \frac{M_x}{M_{u,b}} \cdot k_x \leq 1 \qquad (11d.W5)$$

where

$$a_y = 0.15\,\overline{\lambda}'_{ky} \cdot \beta_m - 0.15 \leq 0.9 \qquad (11d.W7)$$

$$k_x = \left(1 - \frac{N}{N_u}\right) a_y \leq 1 \qquad (11d.W6)$$

$M_{u,b}$ = see Section 11c(W), Eq. 11c.W1

M_x = calculated moment about the x-axis. End moments must be calculated according to the second order theory if appropriate

N_u = flexural buckling load according to Section 11b(W) with regard to flexure about the y-axis or pure torsional buckling

ECCS 87

Members in Flexure and Compression, With Mono- or Bisymmetric Cross Sections

(A) Flexure in the plane of symmetry. Lateral-torsional buckling prevented. Constant moment. It should be verified that

$$\frac{N}{N_{xd}} + \frac{1}{1 - \dfrac{N(\kappa\overline{\lambda}_x)^2}{Q N_{xd} R_x}} \frac{M_x}{M_{xd}} \leq 1.0 \qquad (11d.W8)$$

where

M_x = calculated moment about the x–axis. For an eccentric axial force, M_x is the product of the axial force and its eccentricity with respect to the centroid of the effective cross section at uniform compression, see Fig. 11d.W1

M_{xd} = design value of moment about the x–axis in pure flexure according to 11c.W

N = calculated axial force due to design load

N_{xd} = design value of concentric axial compressive force according to Eq. 11b.W5 (buckling about the x-axis)

R_x = I_{xef}/I_{xg}, where I_{xef} is to be calculated for a section in pure flexure

κ = reduction coefficient with respect to buckling, see Eq. 11b.W6

$\overline{\lambda}_x$ = slenderness parameter defined in Eq. 11b.W8

(B) Flexure in the plane of symmetry. Lateral-torsional buckling prevented. Moment varying along the member.

For a beam acted upon by a varying moment M_{eq} is substituted for M_x in Eq. 11d.W8.

$$M_{eq} = \beta\, M_{max} \qquad (11d.W9)$$

where β is the moment coefficient for taking into account the moment distribution along the member and may be taken from Table 11d.W1

(C) Flexure in the plane of symmetry. Lateral-torsional buckling not prevented. It shall be verified that

$$\frac{N}{N_{yd}} + \left(\frac{M_x}{M_{xd}}\right)^{\beta_{Mx}} \leq 1 \qquad (11d.W10)$$

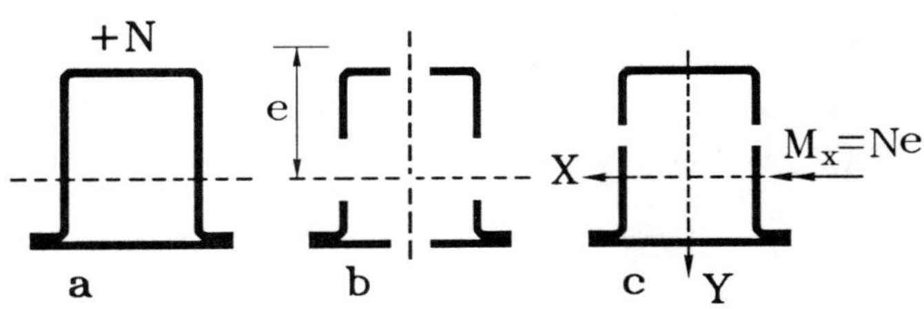

Fig. 11d.W1 For a Column of box section, gross cross section (a) and effective cross sections when subjected to uniform compression (b) and to flexure (c)

where

M_x = calculated moment about the x-axis without additional moment due to axial force

M_{xd} = design value of moment about the x-axis in pure flexure according to Eq. 11c.W2

N = calculated axial compressive force due to design load

N_{yd} = design value of concentric axial compressive force according to Eq. 11b.W5 (buckling about the y-axis)

β_{Mx} = coefficient taking account of moment distribution M_x. β_{Mx}= 1 may be used safely. See also Table 11d.W2

Table 11d.W1 Moment coefficients

Moment distribution	β	
M 	1.10	
M 	0.66	$\beta \geq 1 - \bar{\lambda}^2 \kappa N/N_d$
M  M	0.44	
M  ψM, $-1 < \psi < 1$	$0.66 + 0.44\psi \geq 0.44$	
For members with transverse load	1.00	

Table 11d.W2 Coefficients β_M

Moment distribution	β_M
M_{max} 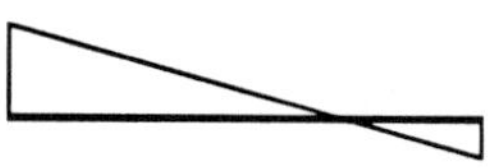 ψM	$1.8 - 0.7\psi$
M_{max} 	1.3
M_{max} 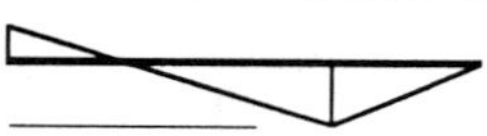	1.4

(D) Biaxial flexure (x–axis = strong axis)

It shall be verified that

$$\frac{M_x}{M'_{xd}} + \frac{M_y}{M'_{yd}} \leq 1 \tag{11d.W11}$$

where

$$M'_{xd} = \min\begin{cases} M_{xd}\,(1 - \frac{N}{N_{xd}})\left(1-(\kappa\overline{\lambda}_x)^2\,\frac{N}{N_{xd}}\,\frac{1}{Q\,R_x}\right) & \text{(11d.W12)} \\ M_{xd}\,(1 - \frac{N}{N_{yd}})^{1/\beta_{Mx}} & \text{(11d.W13)} \end{cases}$$

$$M'_{yd} = M_{yd}\,(1 - \frac{N}{N_{yd}})\left(1-(\kappa\overline{\lambda}_y)^2\,\frac{N}{N_{yd}}\,\frac{1}{Q\,R_y}\right) \tag{11d.W14}$$

with M_{xd} and M_{yd} defined in Section 11c(W).

INTERNATIONAL COMPARISONS & COMMENT

WHAT ARE THE DIFFERENCES? and *WHY ARE THERE DIFFERENCES?*

For all the referenced specifications, the beam-column behavior is expressed as an interaction formula, which contains the parameters determined for beams and columns. As a consequence the same "What" and "Why" as for beams and columns are relevant.

For a detailed comparison see Sections 11b and 11c.

11e. Other Topics Within Cold-Formed Members

SPECIFICATION PROVISIONS & EXPLANATION

AUSTRALIA

CHAPTER EDITOR
Greg J. Hancock, The University of Sydney

AS 1538 88

Bracing Requirements for Wall Studs

It shall be permissible to calculate the safe load-carrying capacity of a wall stud on the basis that either

(A) Intermediate brace (Noggins) at points along the stud as well as the top and bottom members (wall plates) provide lateral and torsional restraint in the plane of the wall, if the restraint members and connections comply with the following requirements:

(1) Intermediate brace shall be connected to the studs so as to resist lateral and torsional deformations of the stud at the point of connection.

(2) Noggins and their connections shall be designed to resist lateral forces in accordance with AS 1250 81.

(3) The stud shall be connected to the top and bottom members to resist lateral and torsional deformations of the stud at its ends; or

(B) Wall material or sheeting provides lateral support to the stud in the plane of the wall, if the wall material and its attachments to the stud comply with the following requirements.

(1) The wall material or sheeting shall be connected to the top and bottom members of the wall assembly to enhance the restraint provided to the stud and to stabilize the overall assembly; and either

(2) For studs having identical material attached to both flanges, the axial stress shall not exceed the lowest value computed by the following criteria.

- The maximum permissible stress for column buckling between the fasteners in the plane of the wall shall not be exceeded.
- The maximum permissible stress for flexural and/or torsional buckling of the whole stud computed using the design shear rigidity of the wall material or sheeting determined from tests shall not be exceeded.
- The maximum permissible stress shall not exceed half of the value at which the maximum permissible shear strain in the wall material or sheeting determined from tests, is computed to be exceeded; or

(3) For studs with wall material or sheeting on one side only, a rational flexural-torsional buckling analysis of the stud shall be performed accounting for the shear and torsional rigidity of the wall material, and the wall material shall be designed to resist the forces resulting from flexural and torsional deformation.

NOTE: Equations for computing the stresses in (B)(2) above are given in the AISI 86

CHINA

CHAPTER EDITORS
Shi-Ji Wang, Hunan University
Zhong-Quan Zhang, Zhongnan Building Design Institute

GBJ 18 87

Beneficial Effect of Cold Forming

The increase in yield stress due to cold forming is utilized for checking the tensile, compressive and flexural strength of the member provided that its cross section is fully effective. The formula for calculating the modified design strength f′ is

$$f' = \left(1 + \frac{\eta(12\gamma - 10)\,t}{L} \sum_{i=1}^{n} \frac{\theta_i}{2\pi}\right) \cdot f \qquad (11e.C1)$$

where

L = total mid-line length of the cold-formed section
n = number of corners
t = steel core thickness
γ = f_u/f_y; f_u, f_y, f are ultimate tensile strength, specified minimum yield stress and design strength of the virgin material, respectively
η = a coefficient depending on the cold-forming process
$\eta = 1.7$ for cold-rolled square and rectangular tubes
$\eta = 1.0$ for high frequency welded circular pipes and open sections
θ_i = central angle of the ith corner (in radian)

Following items have also been specified in GBJ 18 87:

- Channel and Z-purlins
- Lattice joists
- Proof trusses
- Rigid frames
- Profiled steel sheets

EAST EUROPE

CHAPTER EDITORS
Victor Gioncu, Polytechnic Institute of Timisoara
Jindrich J. Melcher, Technical University of Brno

TGL 13503/02 82

The web crippling strength

For simply supported flexural members with the unreinforced single solid webs connecting top and bottom flanges (Fig. 11e.E1a) the allowable concentrated load or reaction per web is

$$Q = \sigma_{all} \cdot t^2(5.2 + 0.11\ c/t)\ (1.33 - 0.33\frac{R_y}{R_y^*})\ \sin\alpha \qquad (11e.E1)$$

where

$c \leq h$ = the length of the concentrated load distribution
R_y = yield stress
R_y^* = 240 MPa
α = the angle of inclination of the web

For beams with two or more adjacent webs (Fig. 11e.E1b), Q_1 shall be computed for each individual web and the results added to obtain the allowable load or reaction for multiple web.

For load effected at the end of the beam

$$Q_1 = \sigma_{all} \cdot t^2(7.4 + 0.9\sqrt{c/t}) \qquad (11e.E2)$$

and for the load located at least 1.5 h away from the end of the end of the beam is

$$Q = \sigma_{all} \cdot t^2(11.1 + 2.4\sqrt{c/t}) \qquad (11e.E3)$$

In Eqs. 11e.E1 to 11e.E3, σ_{all} is the allowable stress of the steel grade used.

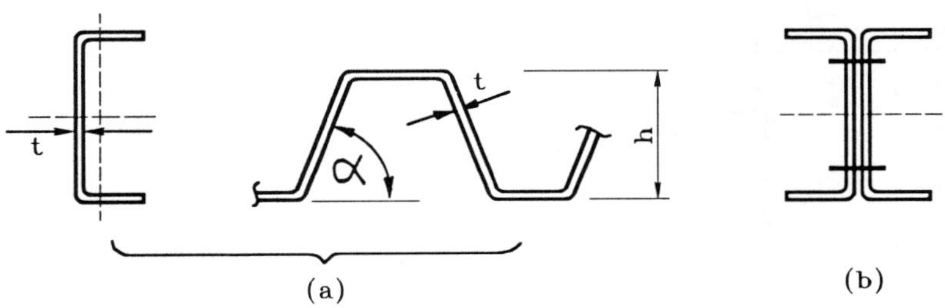

Fig. 11e.E1 Cross sections with single (a) and double (b) webs with regard to web crippling strength

In SNiP II-23-81 82 the critical stress concept for the webs of flexural members working in the elastic range is used. For compression webs the effective width approach is applied.

In TGL 13503/01 82 the maximum flat-width-to-thickness ratios B according to the Table 11e.E1 are required from the point of view that deflection limitation of compressed members shall be observed.

For aluminum thin-walled structures detailed provisions in SNiP II-24-74 75 are given. Special rules for compressed plate elements and especially the design provisions and criteria for stability of trapezoidal (ribbed) and corrugated (wave-formed) thin walled panels are presented.

The appropriate formulas have been described in detail in Chapter 10 – Cold-Formed Steel (part EE) of the 1st Edition of the World View. (SSRC, 1982)

COMMENTARY

The unified approach to the design of steel stiffened and unstiffened elements using the effective web concept is based on the Cornell University research.

In the specifications, the general design criteria for tension members, flexural members, lateral buckling, compression single and built-up members and for the connections and joints are given. Of course, appropriate attention is given to the problem of torsional and flexural-torsional buckling of thin walled members.

Table 11e.E1 Types of cross section with corresponding β $(=\frac{b}{t})$ as a criteria for deflection limitations of compressed members

Type of cross section	β
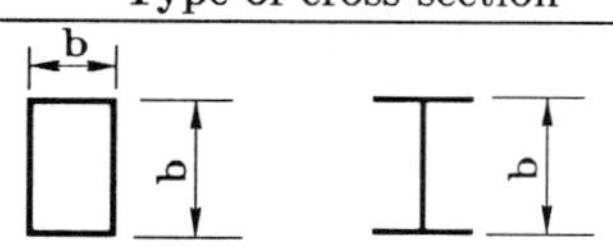	250
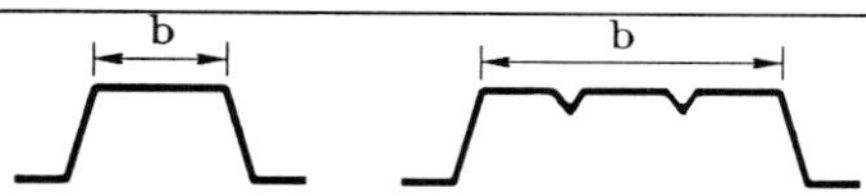	500
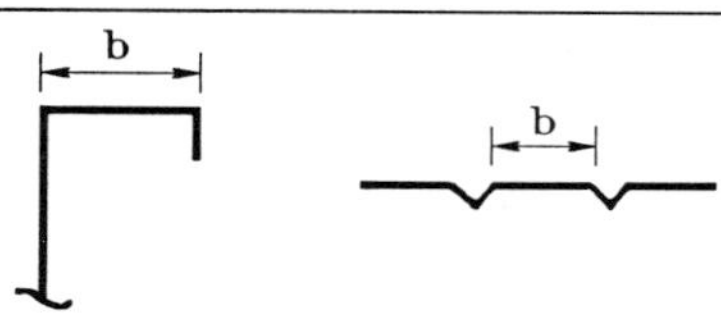	60
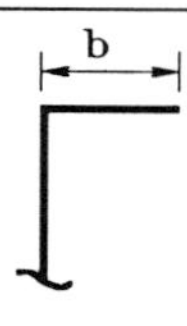	30

NORTH AMERICA

CHAPTER EDITOR
Teoman Pekoz, Cornell University

AISI 86

Wall Studs and Other Braced Members
Rather detailed provisions are given for the behavior of lipped channel and Z-section wall studs braced by wall board. Provisions also exist for channels braced by discrete braces.

Rack Manufacturers Institute Specifications
This specification contains provisions for types of members and structural systems found in industrial storage racks. These include perforated columns, columns and beam-columns partially braced against deflection and twisting and space frames with semi-rigid connections.

Effects of Cold Forming
Provisions are given for considering the beneficial effects of cold forming.

WEST EUROPE

CHAPTER EDITORS
Joachim Lindner, Technische Universitat Berlin
Ton Toma, TNO Institute for Building Materials and Structures

ECCS 87

The following items have been treated in the European recommendations:

- Utilization of the effect of cold working for members in tension
- Simplified treatment of structural systems such as
    - Z purlins (including torsional restraint of sheeting)
    - Z sheeting rails
    - Wall studs
    - Lattice joists

Design rules and their application limits are described.

INTERNATIONAL COMPARISONS & COMMENT

WHAT ARE THE DIFFERENCES? **and** ***WHY ARE THERE DIFFERENCES?***

The information given in this paragraph by the chapter editors, is of such a different nature that it is difficult to make comparisons. This paragraph has to be regarded as additional information for the readers of the Second Edition of the Stability of Metal Structures: A World View.

Table 11e.1 Comparison of the availability of rules in the different regions for a number of topics within cold-formed members

Region	Effects of Cold-forming	Wall Studs	Purlins	Web crippling Formulas	Perforated Elements	Open web joists	Rack Specs.
A	Y	Y	N	Y	N	N	Y
C	Y	N	Y	Y	N	Y	N
E	Y	Y	Y	Y	N	N	N
J	Y	N	Y	Y	Y	Y	N
N	Y	Y	Y	Y	Y	Y	Y
W	Y	Y	Y	Y	Y	Y	Y

Y = Rules are provided in some countries in the region
N = Rules are not provided for the region

Concerning the status of the given information in this chapter the following comments are important.

The specifications discussed are current specifications considered to be representative for each region at the time of preparation of this chapter. There are several additional national specifications not covered here.

For West Europe, the ECCS Recommendations were discussed in detail since they provide the general basis for the Annex A to the Eurocode 3 on Cold-formed Steel Sheeting and Members.

The following specifications are being prepared in various regions for approval in 1990.

Australia – Industrial Storage Rack Specifications
China – Specification for the Design of Steel Storage Rack Structures
North America – An addendum to AISI 86
– An LRFD version of AISI 86
– New edition of CSA S136 M89 84c
– Industrial storage rack specifications

In recent years, differences between different national and international specifications have been narrowed due to extensive exchange of technical information. However, differences still exist primarily due to the development of different approaches in the research conducted in different countries.

Symbol List for Chapter 11 -- Cold-Formed Members

A = area of gross cross section (11a.C), (11b.C), (11b.E)&(11b.N)
A = sum of sectional areas of components (11b.J)
A_e = area of effective cross section at yield stress level (11b.1)
A_e = area of the effective cross section (11a.C), (11b.C)&(11b.W)
A_e^* = effective area computed at failure stress F_n (11b.N) & (11b.1)
A_{ef} = area of the effective cross section of compression members (11d.E)
A_{ef} = area of effective cross section at uniform compression (11b.E)
A_{ef} = area of effective section (11d.C)
A_{ef}^B = area of the effective cross section due to bending taken into account different imperfection directions (11b.W)
A_{ef}^D = area of the effective cross section assuming uniform compression at yield stress (11b.W)
A_{eff} = net section based on effective width described in Eq. 11a.J5 (11b.J)
A_{eu} = effective area computed at the ultimate load (11b.N)
A_g = area of gross cross section (11b.W) & (11b.1)
A_o = coefficient used to determine effective width given by Eq. 11a.C5 (11a.C)
B = bimoment at the section corresponding to M_{max} (11c.C)
B = bimoment (11d.E)
B_o = coefficient used to determine effective width given by Eq. 11a.C5 (11a.C)
b = the width of the straight part of the web (10a.E)
b = width of stiffened plate (cm), that is, the distance between fillet tips for bent stiffener, and the distance between rows of welds for riveted or welded stiffeners (11b.J)
b = width of flat compression elements (11a.A) & (11a.N)
b_e = effective design width of flat compression elements (11a.A)
b_f = width of flange for I-shaped section, width of flange divided by 2 for box or V-shaped section, and width of flange plus height of lip for section with lip (see Fig. 11a.J2) (11a.E)
C_b = correction to reflect the moment gradient (11c.E)
C_{mx} = correction to reflect the moment gradient about x-axis in the member (11d.N)
C_{my} = correction to reflect the moment gradient about y-axis in the member (11d.N)
C_o = coefficient used to determine effective width given by Eq. 11a.C5 (11a.C)
c = length of the concentrated load distribution (11e.E)
c = coefficient used to determine strength reduction factor, given in Eq. 11b.E2 (11b.E)
d = distance between shear center and point of application of the transverse load, being positive/negative when load is applied below/above the shear center (11c.C)
d = height of web (11a.J)
E = modulus of elasticity (11b.C) & (11b.J)
e_o = abscissa of the shear center (11b.C)
e^* = equivalent eccentricity of initial imperfections (11d.C)
F = yield stress (11b.J)

F_a = maximum permissible average axial stress (11b.A)
F_b = maximum permissible stress in bending (11c.A)
F_{cr} = plate buckling stress (11a.N)
F_e = elastic flexural buckling stress of column (11b.N)
F_n = failure stress of column (11b.N)
F_o = elastic critical stress for flexural-torsional buckling (11c.A) & (11d.A)
F_{oc} = elastic buckling stress (11b.A)
F_{oy} = elastic flexural and torsional buckling stress about y-axis (11d.A)
F_{oz} = elastic flexural and torsional buckling stress (11d.A)
F_y = yield stress (11b.A) & (11c.A)
f = design strength (11b.C) & (11b.N)
f′ = the modified design strength due to cold forming (11e.C)
f_b = allowable bending stress (t/cm^2) (11c.E)
f_c = allowable compression stress (t/cm^2) (11b.J)
f_c = elastic critical stress (11c.1)
f_p = proportional limit (11a.A)
f_t = allowable tensile stress (11c.J)
f_{ty} = design value of yield stress (11c.W)
f_y = yield strength (11a.A), (11b.C) & (11b.E)
f_y = design yield stress (11a.W), (11b.W), (11b.1)&(11c.1)
g = distance from y—axis to center of gravity (11b.J)
h = height of the cross-section (11c.C)
I_{ef} = moment of inertia of effective cross section (11c.W)
I_{min} = required minimum moment of inertia about gravity axis for a lip considered as an edge stiffener of a compressive plate element or for any other stiffener (11a.J)
I_{st} = minimum moment of inertia of the stiffener about its own centroidal axis parallel to the element to be stiffened (11a.E)
I_t = St. Venant torsion constant (11b.C)
I_y = moment of inertia about the minor axis (11c.A) & (11c.C)
I_ω = warping constant (11b.C)
i = radius of gyration in the plane of web of tee section (11c.E)
i_{ef} = radius of gyration of effective cross section (11b.W)
i_g = radius of gyration of gross cross section (11b.W) & (11b.1)
i_x = radius of gyration about the symmetric principal axis (11b.C)
i_y = radius of gyration about the non-symmetric principal axis (11b.C)
i_1 = radius of gyration of component about 1—1 axis (cm) (11b.J)
J = St. Venant torsion constant (11c.A)
K_1 = coefficient determined by experiments (11a.C)
K_2 = coefficients determined by experiments (11a.C)
k = buckling coefficient (11a.A), (11a.E), (11a.W) & (11c.W)
L = equivalent span (11a.E)
L = total mid-line length of the cold-formed section (11e.C)
L_o = lateral unsupported length of the beam (11c.C)
L_ω = largest center to center spacing of batten plates for battened members (11b.C)
$L\omega$ = effective length of warping torsion, evaluated in such a manner: for members without intermediate batten plate (11b.C)
$L\omega$ = effective length of flexural buckling about the symmetric axis; for members, without intermediate batten plate (11b.C)

l_1 = spacing of separators (11b.J)
l = effective length of member (11b.W) & (11b.1)
l_b = unsupported length of compression flange (11c.E)
M = maximum bending moment caused by the design load (11c.E)
M_a = allowable moment for flexural member (11c.N)
M_c = lateral buckling moment (11c.N)
M_d = maximum bending moment caused by the design load (11c.1)
M_e = elastic lateral buckling moment (11c.N)
M_{eq} = equivalent moment for moment varying along the member (11d.W)
M_{ki} = elastic critical moment for lateral torsional buckling of the gross section (11c.W)
$M_{ki,ef}$ = elastic critical moment for lateral torsional buckling of the effective cross section (11c.W)
$M_{ki,PB}$ = elastic critical plate buckling moment (11c.W)
M_{max} = maximum bending moment (11c.C)
M_n = nominal lateral buckling moment (11c.N)
M_u = bending moment corresponding to the ultimate state (11a.A)
M_u = yield moment of effective cross section (11d.W)
$M_{u,B}$ = bending moment corresponding to the ultimate state (11c.W)
$M_{u,b}$ = bending moment corresponding to the ultimate state (11d.W)
M_x = calculated moment about the x-axis (11d.C), (11d.N)&(11d.W)
M_x^I = calculated moment about the x-axis according to the first order theory (11d.W)
M_x^{II} = calculated moment about x-axis, taking into account second order effects, especially initial imperfections and the shift of the centroid between the gross and the effective cross section (11d.W)
M_{xd} = design value of moment about the x-axis in pure flexure according to 11c.W (11d.W)
M'_{xd} = ultimate moment defined by Eq. 11d.W12 and 11d.W13 (11d.W)
M_{xo} = failure moment for bending about the x-axis in the absence of an axial load or bending about the y-axis (11d.N)
M_y = design value of bending moment about y-axis (11d.C) & (11d.N)
M_y = yield moment (11c.N)
M_{yd} = design value of moment about y-axis in pure flexure according to 11c.W (11d.W)
M'_{yd} = ultimate moment defined by Eq. 11d.W14 (11d.W)
M_{yo} = failure moment for bending about the y-axis in the absence of an axial load or bending about the x-axis (11d.N)
M_z = the minor axis bending moment (11d.E)
m_o = imperfection parameter depending on the type of cross section (11b.E)
N = design value of axial load (11b.C), (11b.W)&(11d.W)
N_d = axial load on member caused by design load (11b.1)
N_{Ey} = Euler buckling load about x-axis (11d.C)
N_{Ex} = Euler buckling load about y-axis (11d.C)
N_u = axial force corresponding to the ultimate state (11a.A), (11b.W)&(11d.W)
N_u^B = buckling load based on an effective area belonging to bending taking into account different imperfection parameters (11b.W)
N_u^D = buckling load based on an effective area belonging to uniform compression at yield stress (11b.W)

N_{xd}= design value of concentric axial compressive force (buckling about x-axis) (11b.W)
N_{yd}= design value of concentric axial compressive force (buckling about y-axis) (11d.W)
n = number of corners (11e.C)
P = axial force due to the applied loading (11d.N)
P_o = failure load in the absence of any moment (11d.N)
P_u = ultimate load of column (11b.N)
P_x = flexural buckling loads about x-axis (11d.N)
P_y = flexural buckling loads about y-axis (11d.N)
Q = ratio of effective design area to full area of the cross section (11a.A), (11b.A), (11b.W) & (11b.1)
Q = modification factor for yield stress, given in Eq. 11b.J3 (11b.J)
Q = allowable concentrated load for simply supported flexural members with unreinforced single solid webs connecting top and bottom flanges (11e.E)
R_d = design strength of material (11a.E), (11b.E) & (11d.E)
R_s = allowable shear force of welds, spot welds, rivets, or bolts (t) (11b.J)
R_y = yield stress (11e.E)
r_d = distance of the compression flange to the centroid of the gross (11b.W)
$r_{d,ef}$ = distance of the compression flange to the centroid of effective cross section (11b.W)
r_{o1} = polar radius of gyration about the shear center (11d.A)
$S_{x,ef}$ = effective section modulus with the outer fiber at actual stress or yield stress (11c.W)
s_f = section modulus of the full unreduced section for the compressive fiber (11c.N)
t = steel core thickness of plate element (11a.A),(11a.E),(11a.J),(11b.J)&(11e.C)
W = section modulus of gross area for bending (11a.C)
W_e = effective section modulus for bending (11a.C)
W_{ef} = effective section modulus for bending (11c.C), (11c.E), (11c.W)&(11c.1)
W_{efx} = effective section modulus about x-axis (11d.C)
W_{efy} = effective section modulus about y-axis (11d.C)
$W_{ef,y}$= section modulus of the effective cross section for flexural members (see Part 11c.E) (11d.E)
$W_{ef,z}$ = the minor axis section modulus (11d.E)
W_g = section modulus of the gross cross section (11c.W) & (11c.1)
W_x = section modulus about the major axis (11c.C)
W_ω = section modulus for warping torsion (11c.C) & (11d.E)
y_c = distance from the principle axis to the most extremely compressed fiber (11c.W)
Z_{xmin} = smallest value of the section modulus computed about the x-axis (11c.A)
Z'_{xmin} = smaller value of the section modulus computed about the x′-axis perpendicular to the web (11c.A)
α = coefficient depending on the type of cross-section (11b.1)
α = coefficients reflecting supporting conditions at the ends of the member, whose values are given in Table 11b.C1 (11b.C)
α = reduction factor of the European buckling curves, see Table 1a.W1 (11b.W)
α = the angle of inclination of the web (11e.E)

α = the ratio of stress range between the maximum stress and the minimum stress to the maximum stress (11a.A)

β = coefficients reflecting supporting conditions at the ends of the member, whose values are given in Table 11b.C1 (11b.C)

β = moment coefficient taking account for the moment distribution along the member (11d.W)

β_{Mx} = coefficient taking account of moment distribution M_x. $\beta_{Mx} = 1$ may be used safely. See also Table 11d.W2 (11d.W)

β_{mx} = equivalent uniform moment factors about x-axis (11d.C)

β_{my} = equivalent uniform moment factors about y-axis (11d.C)

β_x = monosymmetry section constant (11c.A) & (11d.A)

γ = ratio of ultimate tensile strength to specified minimum yield stress (11e.C)

γ = load factor (11a.2)

ΔW_o = distance between the gross and the effective section centroids for uniform compression at yield stress (11b.W)

δ = coefficient used to determine elastic critical stress, given in Eq. 11c.A2 (11c.A)

ϵ_o = a coefficient that is a function of λ and depends on the type of steel (11b.1)

ϵ_o = equivalent initial eccentricity ratio considering the effect of imperfection (11b.W)

η = a coefficient depending on the cold-forming process (11e.C)

η = imperfection parameter (11b.A)

η_A = coefficient used to determine design strength (11b.1)

η_w = coefficient used to determine design strength (11b.1)

θ_i = central angle of the ith corner (in radian) (11e.C)

κ = reduction coefficient with respect to buckling, see Eq. 11b.W6 (11d.W)

κ^1 = reduction factor as defined by Eq. 11b.W4 (11b.W)

κ_m = reduction factor (11c.W)

Λ = critical slenderness ratio (11b.J)

λ = slenderness ratio for the full unreduced section (11b.E)

λ = slenderness ratio of compression member (11b.J)

λ = strength reduction factor for compression members (see Section 11b(E)) (11d.E)

λ = effective slenderness ratio of the member (11b.C)

$\overline{\lambda}$ = limiting of slenderness ratio (11b.W)

$\overline{\lambda}_{ef}$ = slenderness parameter, given in Eq. 11b.E2 (11b.E)

λ_{lat} = lateral buckling reduction factor for the critical beam slenderness ratio determined for the full unreduced section (see Chapter 3 - Beams, East Europe) (11c.E)

λ_{lat} = lateral buckling reduction factor for flexural members (see Section 11c(E)) (11d.E)

$\overline{\lambda}'_m$ = slenderness parameter used to determine reduction factor, given in Eq. 11c.W1 (11c.W)

λ_o = slenderness parameter defined in Eq. 11b.C5 (11b.C)

$\overline{\lambda}_p$ = slenderness parameter defined in Eq. 11a.W1 (11a.W)

λ_{pl} = ?

λ_w = equivalent slenderness ratio (11b.C)

λ_x = slenderness ratio for flexural buckling about the x-axis (symmetrical axis) (11b.C)

$\overline{\lambda}_x$ = slenderness parameter defined in Eq. 11b.W8 (11d.W)
λ_y = slenderness ratio for flexural torsional buckling about the y-axis (11c.E)
ξ = coefficient depending on the supporting conditions of the unloaded edges, the location of σ_{max}, and the value of α (11a.C)
ξ_1 = coefficients regarding to the loading and supporting conditions of the beam, whose values are tabulated in the appendix of the specification (11c.C)
ξ_2 = coefficients regarding to the loading and supporting conditions of the beam, whose values are tabulated in the appendix of the specification (11c.C)
ξ_3 = coefficients regarding to the loading and supporting conditions of the beam, whose values are tabulated in the appendix of the specification (11c.C)
ρ = reduction factor (11a.E), (11a.N)&(11a.W)
ρ_1 = rivet or weld pitch (11b.J)
σ = compression stress caused by the unfactored load (11a.A)
σ_{all} = allowable stress of the steel grade (11e.E)
σ_{cr} = elastic critical stress of the plate element (11a.C)
σ_d = governing normal stress in design (11a.C)
σ_d = actual compressive stress caused by design load (11a.W)
σ_d = stress in the element considered calculated on the basis of the effective section (11a.N)
σ_E = Euler stress (11b.W) & (11b.1)
σ_{crlb} = elastic critical stress for local buckling of the compression flange (11c.W)
σ_{crM} = theoretical critical stress for elastic lateral buckling based on the gross cross-section (11c.W)
σ_e = elastic critical stress of plates (11c.W)
$\sigma_{ki,ef,m}$ = elastic critical moment for lateral torsional buckling of the effective cross section (11c.W)
σ_{max} = maximum compressive stress at one of the unloaded edge (11a.C)
σ_{min} = the stress at another edge corresponding to the maximum stress (11a.C)
σ_u = nominal ultimate stress obtained from the conversion of ultimate load (11a.C)
$c\sigma_b$ = compression stress caused by bending (11d.J)
$t\sigma_b$ = tension stress caused by bending (11d.J)
τ = plastic reduction factor (11a.C)
ϕ = a strength reduction factor depending upon the slenderness ratio and steel grade (11b.C)
ϕ_b = strength reduction factor for beams determined by the following equation (11c.C)
ϕ'_b = modified strength reduction factor (11c.C)
ϕ_{bx} = strength reduction factors for beams about x-axis (11d.C)
ϕ_{by} = strength reduction factors for beams about y-axis (11d.C)
ϕ_x = strength reduction factors for columns about x-axis (11d.C)
ϕ_y = strength reduction factors for columns about y-axis (11d.C)
φ = strength reduction factor depending upon the slenderness ratio and steel grade (11b.E)
χ = coefficient used to determine effective width given by Eq. 11a.C5 (11a.C)
ψ = coefficient used to determine effective width given by Eq. 11a.C5 (11a.C)
Ω = load factor (11b.A)

Chapter 12

Composite Members

TOPICAL EDITORS
Patrick J. Dowling, Imperial College
Amr S. Elnashai, Imperial College

CHAPTER EDITORS
Peter Ansourian, The University of Sydney
Reinhard Bergmann, Bochum University
Russell Q. Bridge, The University of Sydney
Saif Daresalemwala, Howard Needles Tammen & Bergendoff
Patrick J. Dowling, Imperial College
Amr S. Elnashai, Imperial College
Richard W. Furlong, University of Texas
Jozef Lapos, Slovak Technical University
Karlheinz Roik, Bochum University
Nikolaj Streleckij, CNIPSK
Minoru Wakabayashi, General Building Research Corporation
Shan-tong Zhong, Harbin Architectural & Civil Engineering Institute

12a. Axial Strength

SPECIFICATION PROVISIONS & EXPLANATION

AUSTRALIA

CHAPTER EDITORS
Peter Ansourian, The University of Sydney
Russell Bridge, The University of Sydney

There is no Australian Standard for the design of composite columns. However, several papers have been published including draft proposals for design, as presented below.

The design is based on the ultimate limit state with load and resistance factors, as in North American practice. Two methods are proposed, one based on reinforced concrete column design procedures (moment magnifier) while the other is closer to steel design practice, termed strength reduction method.

The moment magnifier method is given by

$$\delta_b = \frac{C_m}{(1 - N/N_c)} \geq 1.0 \tag{12a.A1}$$

where

$C_m = (0.6 - 0.5\beta) \geq 0.4$
L_e = effective length
N = applied axial load
$N_c = \pi^2 EI/L_e^2$
β = ratio of end moments

The strength reduction method is the opposite of the moment magnifier approach, where the capacity is reduced for increased slenderness. The reduction factor is given by

$$m = 1-1.1\left[\frac{\lambda-\lambda_s}{\lambda_x-\lambda_x}\right]^{1.3} \leq 1.0 \tag{12a.A2}$$

where

$N_E = \pi^2 \left[E_c I_c + E_s I_s\right]/L_e^2$

$\lambda = \sqrt{N_u/N_E}$

λ_c = the slenderness of a column that would collapse under the design axial load with no bending. If λ is greater than λ_c, then a new section should be chosen

$\lambda_s = 0.37\ (1 + \beta)\lambda_c$

CHINA

CHAPTER EDITOR

Shan-tong Zhong, Harbin Architectural & Civil Engineering

SDGJ 64 89

For axially compressed members

$$N \leq \varphi f_{sc} A_{sc} \quad (12a.C1)$$

where

A_{sc} = total area of member, equals to $A_s+A_c=\pi r^2$
f_{sc} = design compression strength of composite material
r = radius of steel tube (outside)
φ = strength reduction factor

Table 12a.C1 Design Strength of Composite Material f_{sc} (MPa)

Steel	Concrete	α=0.04	0.06	0.08	0.10	0.12	0.14	0.16	0.18	0.20
A3	C30	26.0	30.5	35.0	39.0	43.0	46.5	50.0	53.0	56.5
	C40	32.0	36.0	40.5	44.5	48.0	52.0	55.0	58.0	61.0
	C50	36.5	40.5	45.0	49.0	52.5	56.0	59.0	62.0	65.0
16Mn	C30	31.5	38.0	44.5	50.0	55.0	61.0	65.5	69.5	73.5
	C40	37.0	43.5	49.5	55.5	60.5	65.5	69.5	73.5	77.0
	C50	41.5	48.0	54.0	59.5	64.5	69.0	73.0	76.5	79.5
15MnV	C30	33.5	41.0	48.0	54.5	60.5	66.0	71.0	75.5	79.5
	C40	39.0	46.5	53.0	59.5	65.5	70.5	75.0	79.0	82.5
	C50	43.5	50.5	57.5	63.5	69.0	73.5	78.0	81.5	84.5

Table 12a.C2 The Strength Reduction Factor φ

Steel	λ=10	20	30	40	50	60	70
A3	1.0000	0.9904	0.9777	0.9439	0.9080	0.8715	0.8352
16Mn	1.0000	0.9894	0.9755	0.9400	0.9020	0.8630	0.8236
15MnV	1.0000	0.9897	0.9761	0.9410	0.9037	0.8654	0.8269
λ=	80	90	100	110	120	130	140
A3	0.7989	0.7622	0.7237	0.6821	0.6141	0.5233	0.4512
16Mn	0.7838	0.7429	0.6587	0.5524	0.4721	0.4023	0.3409
15MnV	0.7881	0.7448	0.6098	0.5114	0.4371	0.3724	0.3211
λ=	150	160	170	180	190	200	
A3	0.3931	0.3455	0.3060	0.2730	0.2450	0.2211	
16Mn	0.3022	0.2656	0.2352	0.2098	0.1883	0.1700	
15MnV	0.2797	0.2459	0.2178	0.1943	0.1744	0.1574	

EAST EUROPE

CHAPTER EDITORS
Jozef Lapos, Slovak Technical University
Nikolaj Streleckij, CNIPSK

ČSN 73 1206 86

For concentrically loaded composite columns with tubular cross section (Fig. 12a.E1-a) the proposed formula is

$$N_d \leq (\gamma_s A_s R_{sd} + \gamma_c A_c R_{cd})\varphi \tag{12a.E1}$$

where

d = diameter of concrete core

$$\gamma_s = 0.25\,(16 - 3\alpha^2 - \alpha) \tag{12a.E2}$$

$\varphi = \dfrac{1}{1.1 + \frac{1}{6}\left(\frac{L}{10d}\right)^2}$, reduction factor to account for buckling depending on slenderness

The value of α depends on the slenderness ratio.

for $75 < \lambda \leq 180$

$$\gamma_s = \gamma_c = 1$$

for $\lambda \leq 75$

$$\alpha = \frac{2}{1 + \left[\frac{\lambda}{75}\right]^2} - 1 \tag{12a.E3}$$

where

D = diameter of tube

t = thickness of tube

$\beta = 3.725 + 0.055\,R_{cd}$

$$\gamma_c = 1 + \left(\frac{\alpha\beta}{\chi_d}\right)\left(\frac{t}{D}\right) \tag{12a.E4}$$

$$\lambda = L\sqrt{\frac{A_s + A_c\chi_d}{I_s + I_c\chi_d}} \tag{12a.E5}$$

$$\chi_d = \frac{R_{cd}}{R_{sd}}$$

Then the column curve is expressed as

$$\varphi = \eta - \sqrt{\eta^2 - 0.875\varphi_c\,\frac{\sigma_e}{R_{sd}}} \tag{12a.E6}$$

where

A_c = concrete area

A_s = steel area

E_s = modulus of elasticity of steel

R_{cd} = design strength of concrete

R_{sd} = design strength of steel

$\eta = 0.5[\varphi_c + 0.875(1 + \chi)\frac{\sigma_e}{R_{sd}}]$

$\sigma_e = \dfrac{\pi^2 E_s}{\lambda^2}$, Euler critical stress

φ_c = creep factor

$\chi = 0.17 \left[\frac{\lambda}{93}\right]^2 \left(\frac{R_{sd}}{210}\right)$

For concentrically loaded columns with cross section as shown in Fig. 12a.E1b γ_s and γ_c are equal to unity.

MSZ 15022/1 71

For concentrically loaded tubular composite columns (Fig. 12a.E1-a) the Hungarian specification gives the formula

$$N_d \leq (\gamma_c A_c R_{cd} + A_s R_{sd}).\varphi \tag{12a.E7}$$

where

A_c = concrete area
A_s = steel area
d = diameter of concrete core
N_d = axial design load
R_{cd} = design strength of concrete
R_{sd} = design strength of steel
t = thickness of tube
$\gamma_c = 1 + (25 + \frac{R_{sd}}{R_{cd}}) \frac{t}{d}$, but γ_c is not greater than 2

The column curve is given by the formula

$$\varphi = \frac{1}{1.1 + \frac{1}{6}\left[\frac{L}{10d}\right]^2} \tag{12a.E8}$$

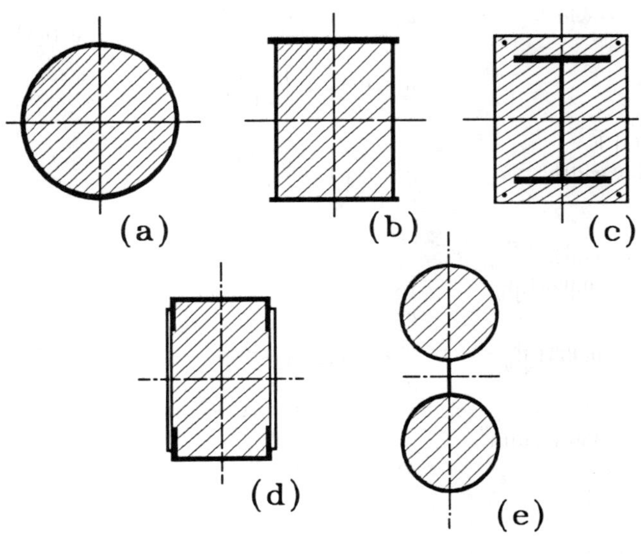

Fig. 12a.E1 Cross Sections of Composite Columns

USSR*

The column strength is given by

$$N_d \leq \varphi\,(A_c R_{cd} + A_s R_{sd}) \tag{12a.E9}$$

where

A_c = concrete area
A_s = steel area
N_d = axial design load
R_{cd} = design strength of concrete
R_{sd} = design strength of steel
φ = buckling curve factor

The buckling curve φ depends on the slenderness ratio λ and the load eccentricity ratio ρ. The slenderness ratio is given by

$$\lambda = \frac{L}{d}\,\frac{4}{\sqrt{2 - \dfrac{1}{1 + \dfrac{\mu\nu}{\chi_y}}}} \tag{12a.E10}$$

where

d = diameter of concrete core
L = length of column
R_{cy} = yield stress of concrete
R_{sy} = yield stress of steel
ϵ_{cy} = yield strain of concrete
ϵ_{sy} = yield strain of steel

$$\mu = \frac{A_s}{A_c}$$

$$\nu = \frac{\epsilon_{cy}}{\epsilon_{sy}}$$

$$\chi_y = \frac{R_{cy}}{R_{sy}}$$

The load eccentricity ratio ρ for eccentrically loaded columns is

$$\rho = \frac{2e}{d} \tag{12a.E11}$$

where e is the eccentric distance of compressive load

For centrically loaded columns the eccentricity ratio is given by the formula

$$\rho = \frac{\lambda}{1000} + 0.05 \tag{12a.E12}$$

---

* The specifications are in the preparation phase

JAPAN

CHAPTER EDITOR
Minoru Wakabayashi, General Building Research Corporation

AIJ 86

For Sections with Rolled Shape Encased in Concrete (ENC)

Allowable strength is given by

$$N = {}_rN_c + {}_sN = {}_rN_c + {}_sA \cdot {}_sf_c \qquad (12a.J1)$$

where ${}_rN_c$ is $\mathrm{Min}(A_e \cdot f'_c,\ A_e \cdot {}_mf_c/n)$

Ultimate strength

$$N_u = {}_rN_{cu} + {}_sN_u = {}_cN_u + {}_mN_u + {}_sN_u = {}_cr_u \cdot F_c \cdot bD + 2\,{}_mA_t \cdot {}_m\sigma_y + {}_sA \cdot {}_s\sigma_y \qquad (12a.J2)$$

For Tubular Sections Filled with Concrete (FIL)

Allowable strength is given by

$$N = {}_cN_c + {}_sN = {}_cAf_c + {}_sA_sf_c \qquad (12a.J3)$$

For encased hollow sections: same as for ENC sections.
For encased-filled hollow sections: (omission)
For short columns where ℓ_x/h and $\ell_y/h < 12$, the strengths are calculated by the formulas for the sections given above, but allow minimum eccentricity.

where

		Prefix Symbols	Suffix Symbols
N	= axial force	c = concrete	c = compressive strength
A	= area of cross section	m = main reinforcement	e = effective
f	= allowable stress		
B, b	= width of cross section	r = reinforced concrete	t = tensile
D, d	= depth of cross section		
σ	= axial stress	s = steel	u = ultimate
r	= reduction factor		Y, y = yield
F	= strength		
n	= modular ratio		

for ENC sections

$$f'_c = f_c(1-15\,{}_sP_c)$$

for FIL sections

$$f'_c = f_c$$

where

f_c = concrete cylinder strength
${}_sP_c$ = area ratio of the compression flange of the steel shape to the gross section
${}_cr_u$ = $0.85-2.5\,{}_sP_c$

NORTH AMERICA

CHAPTER EDITORS

Saif Daresalemwala, Howard Needles Tammen & Bergendoff
Richard W. Furlong, University of Texas

ACI 318 83

The axial capacity of a composite section is given by

$$N_o = (A_s f_y + A_r f_{yr} + 0.85\, f'_c A_c)\, \varphi \qquad (12a.N1)$$

Upper limit 0.85 N_o

$$\varphi = 0.75\ \text{(FIL)} \qquad (12a.N2)$$
$$\varphi = 0.70\ \text{(ENC)}$$

where

A_c = area of concrete
A_r = area of longitudinal reinforcing bars
A_s = area of steel shape
f_c = characteristic cylinder strength of concrete
f_y = yield strength of steel shape
f_{yr} = yield strength of longitudinal reinforcing bars
N_o = axial design capacity
ϕ = strength reduction factor

AISC LRFD 86

Axial strength is given by

$$N_o = A_s f_{ym} \qquad (12a.N3)$$

$$F_{cr} = (0.887/\lambda_c^2) F_{ym},\ \lambda_c > 1.5 \qquad (12a.N4)$$

$$= (0.658^{\lambda_c^2}) F_{ym},\ \lambda_c \le 1.5$$

$$\lambda_c = \frac{K\ell}{r_m \pi} \sqrt{\frac{F_{ym}}{E_m}} \qquad (12a.N5)$$

where

A_c = area of concrete
A_r = area of longitudinal reinforcing bars
A_s = area of steel shape
C_1, C_2, C_3 = numerical coefficients
For concrete-filled pipe and tubing:
C_1=1.0, C_2=0.85 and C_3=0.4
For concrete-encased shapes
C_1=0.7, C_2=0.6 and C_3=0.2
E = modulus of elasticity of steel shape
E_c = modulus of elasticity of concrete
E_m = $E + C_3 E_c (A_c/A_s)$
F_{ym} = $f_y + C_1 f_{yr} (A_r/A_s) + C_2 f'_c (A_c/A_s)$
f_y = yield stress of steel shape

f_{yr} = yield stress of longitudinal reinforcing bars
f'_c = cylinder strength of concrete
r_m = modified radius of gyration for composite section

CAN 3 A23.3 M84

Axial strength is given by

$$P_o=[0.85\phi_c f'_c(A_g-A_{st}-A_t)+\phi_s f_y A_{st}+\phi_a f_y A_t] \qquad (12a.N6)$$

$P_r = 0.85\ P_o$ FIL or spiral reinforcement in ENC
$P_r = 0.80\ P_o$ ENC with tie reinforcement

where
A_g = gross area of section
A_{st} = total area of longitudinal reinforcing bars
A_t = area of steel shape
f_y = yield strength of steel
P_o = factored axial load resistance at zero eccentricity
P_r = design axial load

WEST EUROPE

CHAPTER EDITORS
Reinhard Bergmann, Bochum University
Patrick J. Dowling, Imperial College
Amr S. Elnashai, Imperial College
Karlheinz Roik, Bochum University

BS 5400 82

Axial strength is given by

for ENC sections

$$N_u = 0.91\ A_s f_y + 0.87\ A_r f_{ry} + 0.45\ A_c f_{cu} \qquad (12a.W1)$$

for FIL sections

$$N_u = 0.91\ A_s f'_y + 0.45\ A_c f_{cc} \qquad (12a.W2)$$

for short columns where ℓ_x/h or $\ell_y/h < 12$

$$N \le N_{ay} = 0.85\ k_{1y}\ N_u \qquad (12a.W3)$$

$$e \ge 0.03b \qquad (12a.W4)$$

where

- A_c = area of concrete
- A_r = area of longitudinal reinforcing bars
- A_s = area of steel shape
- b = least lateral dimension of cross-section
- e = eccentricity of load
- f_{cu} = 28-day cube strength of concrete
- f_{ry} = yield strength of longitudinal reinforcing bars
- f_y = yield strength of steel shape
- h = greatest lateral dimension of cross-section
- ℓ_x = effective length about the major axis
- ℓ_y = effective length about the minor axis
- k_{1y} = coefficient determined from Appendix C in BS 5400 depending on the type of steel section used
- N_{ay} = axial load at failure

DIN 18806 P.1

Axial strength is given by

$$N_{pl} = A_a \cdot \beta_{S,a} + A_b \cdot \beta_r + A_s \cdot \beta_{s,s} \text{ - Encased} \qquad (12a.W5)$$

For circular hollow section (CHS) the influence of confinement may be taken into account for $e < \frac{d}{8}$

$$N_{pl} = A_b\, \beta_{RL} + A_a\, \beta_{SL} \tag{12a.W6}$$

where

$$e = \frac{M}{N}$$

$$\beta_{RL} = \beta_R \left(1 + \eta_1\, \frac{S}{d}\, \frac{\beta_{S,a}}{\beta_R}\right)$$

$$\beta_{SL} = \eta_2\, \beta_{S,a}$$

Axial strength:

$$N_{Kr} = \mathfrak{K}\, N_{pl} \tag{12a.W7}$$

$\mathfrak{K}$ = reduction coefficient from the relevant buckling curve: a, b or c
a = FIL
b = ENC bending about major axis of the profile
c = ENC bending about minor axis of the profile

where

A_a = area of steel profile
A_b = area of concrete
d = greatest diameter of tube
e = eccentricity of load
N_{kr} = axial design strength
N_{pl} = squash load of the column
S = tube thickness
β_{wn} = 28-day cube strength of concrete
β_R = design value of concrete strength
= 0.6 B_{wn} for ENC
= 0.7 B_{wn} for FIL
$\beta_{S,a}$ = yield strength of steel profile
$\beta_{s,s}$ = yield strength of longitudinal reinforcing bars

Table 12a.W1 Values for η_1 and η_2 depending on slenderness ratio $\bar{\lambda}$

$\bar{\lambda}$	η_1	η_2
0.1	3.20	0.80
0.2	1.91	0.85
0.3	0.90	0.90
0.4	0.24	0.95
0.5	0.00	1.00

EC 4 85

The squash load N_{pl} of a composite cross-section may be calculated by the squash loads of its components

$$N_{pl} = A_a f_y/\gamma_a + A_c (0.85 f_{ck}/\gamma_c) + A_s f_{sk}/\gamma_s \qquad (12a.W8)$$

For (CHS) and, if verified by tests, for rectangular hollow sections (RHS), the contribution to N_{pl} from the concrete may be increased to $A_c f_{ck}/\gamma_c$.

Alternatively for FIL (CHS) when $\bar{\lambda} < 0.5$ and $M_{max} < N_d\, d/8$ then

$$N_{pl} = A_a n_2 f_y/\gamma_a + A_c(f_{ck}/\gamma_c)\,[1 + n_1\,(t/d)\,(f_y/f_{ck})] + A_s f_{sk}/\gamma_s \qquad (12a.W9)$$

where e is equal to M_{max}/N_d

for $e = 0$

$$\eta_{10} = 4.9 - 18.5\,\bar{\lambda} + 17\bar{\lambda}^2 \qquad (\text{but} \geq 0)$$
$$\eta_{20} = 0.25\,(3 + 2\bar{\lambda}) \qquad (\text{but} \leq 1.0) \qquad (12a.W10)$$

for $0 < e \leq d/8$

$$\eta_1 = \eta_{10}\,(1 - 8e/d)$$
$$\eta_2 = \eta_{20} + (1 - \eta_{20})\,(8e/d) \qquad (12a.W11)$$

For Design by Method 1

$$N_u = \chi N_{pl} \quad N_d \leq N_u \qquad (12a.W12)$$

χ is a reduction coefficient obtained from buckling curves in Clause 5.3.2 of EC3.

For Design of Short Braced Columns by Method 2

(A) Assuming no significant bending N_u should not exceed the lesser of

$$N_u = [1.03 - 0.03\lambda_{max}]N_{pl} \quad \text{and} \quad N_u = 0.85\,N_{pl} \qquad (12a.W13)$$

where λ_{max} is the greater of ℓ_{ez}/b_c and $\ell_{ez}/h_c \leq 12$

(B) Short columns with small eccentricity of loading may be designed ignoring such eccentricity when
(1) the beams at both ends are approximately symmetrical about the center line of the column,
(2) the beams are designed for uniformly-distributed imposed load, and
(3) the span of the beams on opposite sides of the column and at the same level do not differ by more than 15%.

N_u should not exceed the lesser of

$$N_u = [0.80 - 0.025\,\lambda_{max}]\,N_{pl} \text{ or } N_u = 0.68\,N_{pl} \qquad (12a.W14)$$

where

A_a = area of steel
A_c = area of concrete
f_{ck} = characteristic cylinder strength of concrete
f_{sk} = yield strength of longitudinal reinforcement bars
f_y = yield strength of steel
N_{pl} = squash load
γ_a = material factor of safety for steel
γ_c = material factor of safety for concrete
γ_s = material factor of safety for longitudinal reinforcement bars

INTERNATIONAL COMPARISONS & COMMENT

WHAT ARE THE DIFFERENCES?

The codes included in this quantitative assessment are ACI 318 83, AISC LRFD 86, EC 4 85, BS 5400 P.5 79 and DIN. Load factors were not included in this study. Both factored and unfactored capacities were considered.

The cross section considered is an Encased Universal Column 356 x 368 x 129 with h/hc and b/bc ratios of 0.74 and 0.75, respectively. The steel is of 340 N/mm^2 yield strength and the casing is of Grade 35 cylinder strength. For the bridge code BS 5400, the resistance factor for steel and the minimum eccentricity were modified to 1.0 and 0.04, respectively. These modifications would be required to use the code for building design. The Grade 35 cylinder strength was assumed equivalent to Grade 40 cube strength in lieu of ISO 3893 recommendations. For the methods of ACI 318 83 and EC 4 85 the maximum live load was assumed equal to the maximum dead load in the calculation of the effective stiffness. The value of modulus of elasticity for concrete was estimated in accordance with the requirements given in each code. For the methods of ACI and AISC the moment magnification required for slenderness considerations was equivalently included as a reduction in the moment capacity.

The unfactored axial capacity vs. slenderness (expressed as the ratio between pin-ended column length to the cross-sectional dimension in the direction of minor axis bending) is shown in Fig. 12a.1. Values are normalized by the EC 4 capacity.

At very low slenderness ratios, BS 5400 gives the lowest capacity, about 19% less than EC 4, and 4% less than ACI and AISC. The ACI and AISC give 15% and 16% below EC 4, respectively. For high slenderness ratios, the difference between

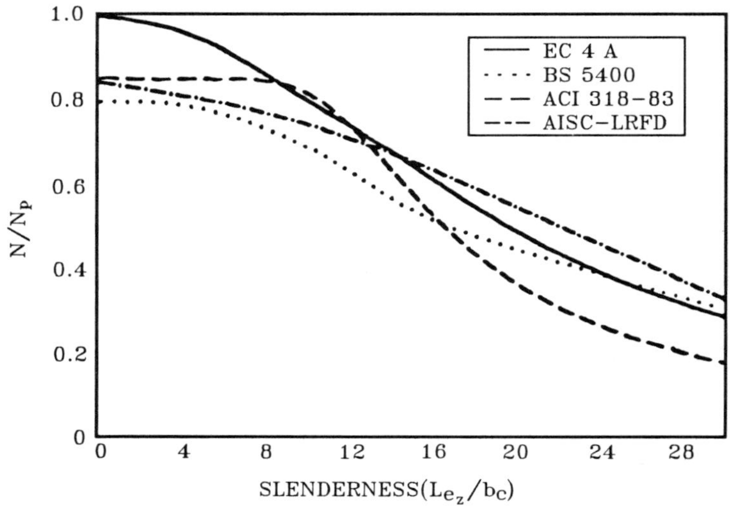

Fig. 12a.1 Unfactored Axial Capacity

EC 4 and BS 5400 drops. The capacity given by ACI is in reasonable agreement with other codes in the range of practical slenderness, but drops very rapidly for high slenderness. In the extreme case, ACI is 48% and 41% below AISC and EC 4, respectively.

The factored axial capacity as a function of slenderness is plotted in Fig. 12a.2. This includes capacity reduction factors in ACI and AISC, and the resistance factors in EC 4 and BS 5400. The values are normalized by the factored axial capacity of EC 4. The observations given above regarding unfactored capacities apply here, with the exception if ACI. The latter code gives a very substantial reduction in strength prediction for high slenderness, up to 65% less than all other codes.

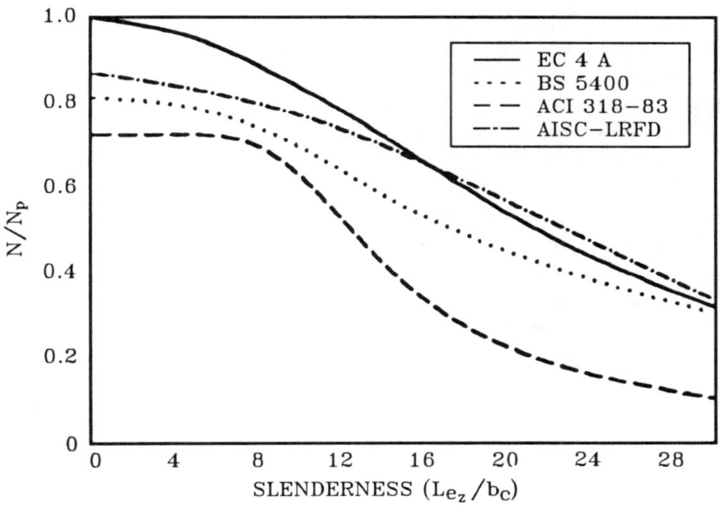

Fig. 12a.2 Factored Axial Capacity

WHY ARE THERE DIFFERENCES?

For the case of unfactored axial capacity, the low strength of BS 5400 (for low slenderness) compared to EC 4 is due to

i) reduced concrete strength of $0.6F_{cu}$ (0.45 factored), to take account of creep,

ii) the minimum eccentricity requirement for short columns calculated from the limiting axial force. The difference is reduced for higher slenderness since equivalent stiffness, with creep, comes into play.

ACI gives values lower than EC 4 due to the minimum eccentricity requirements, leading to the 0.85 factor on the axial plastic capacity. At intermediate values of slenderness, ACI gives values in good agreement with others. However, at high slenderness, ACI drops rapidly due to

i) the use of a cracked section for concrete,

ii) minimum eccentricity requirement accompanied with moment magnification.

The difference between AISC and EC 4 for low slenderness is due to the reduced concrete strength recommended by AISC. In general, good agreement is observed between BS 5400, AISC and EC 4, since they all use steel bucking curves. ACI is different, because of the considerations given above, arising from the fact that it is a concrete design code.

The same conclusions are drawn for the factored axial strength comparison. This is a consequence of the equivalence of the resistance factors used by EC 4 and the strength reduction factor used in AISC. The large drop in strength calculated from ACI is due to the capacity reduction factor of 0.7 used. It is noteworthy that ACI uses a factored Euler load in the moment magnification expression.

12b. Strength Under Combined Axial Force and Bending Moments

SPECIFICATION PROVISIONS & EXPLANATION

CHINA

CHAPTER EDITOR
Shan-tong Zhong, Harbin Architectural & Civil Engineering

SDGJ 64 89

Strength of a member is checked by

for eccentrically compressed members

$$\frac{N}{\varphi A_{sc}} + \frac{\beta_{mx} N_e}{1.5(W_{sc}(1-0.4N/N_{ex}))} \le f_{sc} \qquad (12b.C1)$$

for latticed members

$$\frac{N}{\varphi A_{sc}} + \frac{\beta_{mx} N_e}{W_{sc}(1-\varphi N/N_{ex})} \le f_{sc} \qquad (12b.C2)$$

where

A_{sc} = area of cross-section
e = eccentric distance of load
f_{sc} = design compressive strength
N = nominal axial load
N_c = critical axial load
N_{ex} = $\pi^2 E_{sc} A_{sc}/\lambda_x^2$
W_{sc} = section modulus of member: $W_{sc}=\frac{\pi}{4} r^3$

β_{mx} = equivalent moment factor according to GBJ 17 88
ϕ = strength reduction factor

For latticed members, must check the stability of member, which carried the maximum central compressive load.

Table 12b.C1 Modulus elasticity of composite material E_{sc} (MPa)

Steel	Concrete	α=0.04	0.066	0.08	0.10	0.12	0.14	0.16	0.18	0.20
A3	C30	30685	35135	39445	43615	47635	51520	55260	58855	62310
	C40	37770	42185	46440	50540	54480	58260	61885	65350	68660
	C50	44870	49260	53485	57540	61425	65150	68705	72095	75315
16Mn	C30	27660	32900	37905	42670	47205	51495	55555	59375	62957
	C40	33095	38275	43190	47840	52235	56360	60225	63830	67170
	C50	38555	43695	48555	53140	57445	61475	65225	68700	71895
15MnV	C30	27260	32835	38135	43150	47885	52340	56515	60410	64025
	C40	32285	37790	42980	47860	52430	56690	60635	64275	67600
	C50	37336	42795	47920	52720	57190	61330	65145	68625	71780

JAPAN

CHAPTER EDITOR
Minoru Wakabayashi, General Building Research Corporation

AIJ 86

Uniaxial Bending

(A) Allowable Strength for ENC Sections

(1) Simple superposed strength method

(a) Bending about major axis

for ${}_rN_t \leq N \leq {}_rN_c$ or $M \geq {}_sM_o$

$$N = {}_rN \qquad M \leq {}_sM_o + {}_rM \tag{12b.J1}$$

for $N > {}_rN_c$ or $M < {}_sM_o$

$$N \leq {}_rN_c + {}_sN \qquad M = {}_sM \tag{12b.J2}$$

(b) Bending about minor axis

for ${}_sN_t \leq N \leq {}_sN_c$ or $M \geq {}_rM_o$

$$N = {}_sN \qquad M \leq {}_rM_o + {}_sM \tag{12b.J3}$$

for $N > {}_sN_c$ or $M < {}_rM_o$

$$N \leq {}_sN_c + {}_rN \qquad M = {}_rM \tag{12b.J4}$$

(2) Generalized superposed strength method

$$N = {}_sN + {}_rN \qquad M \leq {}_sM + {}_rM \tag{12b.J5}$$

The quantities ${}_sN_c$, ${}_sM_o$, ${}_sN$ and ${}_sM$ are calculated by the formulas below

$${}_sN_c = {}_sA \cdot {}_sf_c$$

$${}_sM_o = {}_sZ \cdot {}_sf_t$$

$$\frac{{}_sN}{{}_sA} + \frac{{}_sM}{{}_sZ} = {}_sf_e \tag{12b.J6}$$

The expression ${}_rN_c$ is calculated by the formulas below

$${}_rN_c = \min({}_rN_{c1},\ {}_rN_{c2})$$

$${}_rN_{c_1} = A_e f'_c$$

$${}_rN_{c2} = \frac{A_e\ {}_mf_c}{n} \tag{12b.J7}$$

Calculation of ${}_rN$ and ${}_rM$ is based on AIJ Standard for Reinforced Concrete Structures. If the section is rectangular with symmetrical arrangement of reinforcing bars, Table 12b.J1 is used.

The concrete strength is given by

$$f'_c = f_c(1-15\,{}_sp_c)$$

where ${}_sp_c$ is the area ratio of the compression flange of the steel shape to the gross section.

(B) Allowable Strength for FIL Sections

(1) Simple superposed strength method

for $0 \le N \le {}_cN_c$ or $M \ge {}_sM_o$

$$N = {}_cN$$
$$M \le {}_sM_o + {}_cM \qquad (12b.J8)$$

for $N > {}_cN_c$ or $M < {}_sM_o$

$$N \le {}_cN_c + {}_sN$$
$$M = {}_sM \qquad (12b.J9)$$

Table 12b.J1 ${}_rN$ and ${}_rM$ for Rectangular Section with Symmetrical Re-bars

	x_{n1}	${}_rN$ ${}_rM$
concrete	$0 < x_{n1} \le 1$	${}_rn = \dfrac{{}_rN}{bD \cdot f'_c} = \dfrac{x_{n1}}{2} + n \cdot {}_mp_t\left(2 - \dfrac{1}{x_{n1}}\right)$ ${}_rm = \dfrac{{}_rM}{bD^2 \cdot f'_c} = \dfrac{x_{n1}}{12}(3-2x_{n1}) + \dfrac{n \cdot {}_mp_t}{2x_{n1}}(1-2\,{}_rd_{t1})^2$
	$x_{n1} > 1$	${}_rn = \dfrac{{}_rN}{bD \cdot f'_c} = (1+2\,n \cdot {}_mp_t)\left(1 - \dfrac{1}{2\,x_{n1}}\right)$ ${}_rm = \dfrac{{}_rM}{bD^2 \cdot f'_c} = \dfrac{1}{2\,x_{n1}}\left[\dfrac{1}{6} + n \cdot {}_mp_t(1-2\,{}_rd_{t1})^2\right]$
compression bars		$\dfrac{{}_rN}{bD \cdot {}_mf_c} = \dfrac{x_{n1}}{n(x_{n1} - {}_rd_{c1})}\,{}_rn$ $\dfrac{{}_rM}{bD^2 \cdot {}_mf_c} = \dfrac{x_{n1}}{n(x_{n1} - {}_rd_{c1})}\,{}_rm$
tension bars	$0 \le x_{n1} \le 1$	$\dfrac{{}_rN}{bD \cdot {}_mf_t} = \dfrac{x_{n1}}{n(1 - {}_rd_{t1} - x_{n1})}\,{}_rn$ $\dfrac{{}_rM}{bD^2 \cdot {}_mf_t} = \dfrac{x_{n1}}{n(1 - {}_rd_{t1} - x_{n1})}\,{}_rm$
	$x_{n1} < 0$	$\dfrac{{}_rN}{bD^2 \cdot {}_mf_t} = \dfrac{{}_mp_t(2\,x_{n1} - 1)}{1 - {}_rd_{t1} - x_{n1}}$ $\dfrac{{}_rM}{bD^2 \cdot {}_mf_t} = \dfrac{{}_mp_t(1-2\,{}_rd_{t1})^2}{1 - {}_rd_{t1} - x_{n1}}$

Note: x_n is the location of neutral axis

(2) Generalized superposed strength method

$$N = {}_sN + {}_cN$$
$$M \leq {}_sM + {}_cM \qquad (12b.J10)$$

For encased hollow sections, the formulas for ENC sections are used, and encased-filled hollow sections are proportioned by a method similar for ENC sections.

where

$f'_c = f_c$
${}_cN_c = {}_cA f'_c$

Calculation of ${}_cN$ and ${}_cM$ is based on AIJ Standard for Reinforced Concrete Structures. Table 12b.J2 is used for ${}_cN$ and ${}_cM$ of filled concrete.

(C) Ultimate Strength of ENC Sections

(1) Simple superposed strength method

for ${}_rN_{tu} \leq N_u \leq {}_rN_{cu}$

$$N_u = {}_rN_u$$
$$M_u = {}_sM_{uo} + {}_rM_u \qquad (12b.J11)$$

for $N_u > {}_rN_{cu}$

$$N_u = {}_rN_{cu} + {}_sN_u$$
$$M_u = {}_sM_u \qquad (12b.J12)$$

For a rectangular cross section with a full-web type of steel and symmetrical reinforcing bars, Table 12b.J3 is used for the calculation of M_u in addition to the formulas above.

(2) Ultimate strength of reinforced concrete part

for $0 \leq {}_rN_u \leq {}_cN_{cu}$

$$ {}_rN_u = {}_cN_u$$
$$ {}_rM_u = {}_cM_u + {}_mM_{uo} \qquad (12b.J13)$$

for ${}_rN_u > {}_cN_{cu}$

$$ {}_rN_u = {}_cN_{cu} + {}_mN_u$$
$$ {}_rM_u = {}_mM_u \qquad (12b.J14)$$

For a rectangular cross section with symmetrical reinforcing bars, Table 12b.J4 is used for the calculation of ${}_rM_u$ in addition to the formulas above.

${}_cM_u$ is calculated by Table 12b.J5 with ${}_cr_u$ below

$$ {}_cr_u = 0.85 - 2.5\,{}_sp_c \qquad (12b.J15)$$

where

${}_mM_u$ is calculated by Table 12b.J6
${}_sM_u$ is calculated by Table 12b.J7

(3) Generalized superposed strength method

$$N_u = {}_cN_u + {}_mN_u + {}_sN_u$$
$$M_u = {}_cM_u + {}_mM_u + {}_sM_u$$
$$N_u = {}_rN_u + {}_sN_u$$
$$M_u = {}_rM_u + {}_sM_u \qquad (12b.J16)$$

The ultimate strength of a girder is calculated by one of the formulas shown below

$$M_u = {}_sZ_p \cdot {}_s\sigma_y + {}_m\alpha_t \cdot {}_m\sigma_y \cdot {}_m\mathrm{d}$$
$$M_u = {}_sZ_p \cdot {}_s\sigma_y + \frac{{}_s\alpha_w}{8}\,{}_s\sigma_y \cdot {}_m\mathrm{d} + (\tfrac{3}{4}\,{}_m\alpha_t + \tfrac{1}{4}\,{}_m\alpha_c)\,{}_m\sigma_y \cdot {}_m\mathrm{d} \quad (12b.J17)$$

(D) Ultimate Strength for FIL Sections

Ultimate strength for FIL sections is calculated by the simple or the generalized superposed strength method for ENC sections with ${}_cM_u$ and ${}_sM_u$ shown in Tables 12b.J5 and 12b.J7 respectively, where ${}_cr_u = 0.85$.

Biaxial Bending

(A) Allowable Strength for ENC Sections

$$N = {}_sN + {}_rN$$
$$M_x \leq {}_sM_x + {}_rM_x$$
$$M_y \leq {}_sM_y + {}_rM_y \quad (12b.J18)$$

(B) Allowable Strength for FIL Sections

$$N = {}_sN + {}_rN$$
$$M_x \leq {}_sM_x + {}_cM_x$$
$$M_y \leq {}_sM_y + {}_cM_y \quad (12b.J19)$$

Superposed Strength Formulas For Slender Beam-Columns

(A) Allowable Strength for ENC Beam-Columns

(1) Bending about major axis

for $N \leq {}_rN_c$ or $M \geq {}_sM_o(1 - {}_r\nu \cdot {}_rN_c/N_k)$

$$N = {}_rN$$
$$M \leq {}_sM_o(1 - \frac{{}_r\nu \cdot {}_rN}{N_k}) + {}_rM \quad (12b.J20)$$

for $N > {}_rN_c$ or $M < {}_sM_o(1 - {}_r\nu \cdot {}_rN_c/N_k)$

$$N \leq {}_rN_c + {}_sN$$
$$M = {}_sM\,(1 - \frac{{}_r\nu \cdot {}_rN_c}{N_k}) \quad (12b.J21)$$

(2) Bending about minor axis

for $N \leq {}_sN_c$ or $M \geq {}_rM_o\,(1 - {}_s\nu \cdot {}_sN_c/N_k)$

$$N = {}_sN$$
$$M \leq {}_rM_o(1 - \frac{{}_s\nu \cdot {}_sN}{N_k}) + {}_sM \quad (12b.J22)$$

for $N > {}_sN_c$ or $M < {}_rM_o(1 - {}_s\nu \cdot {}_sN_c/N_k)$

$$N \leq {}_sN_c + {}_rN$$
$$M = {}_rM(1 - \frac{{}_s\nu \cdot {}_sN_c}{N_k}) \quad (12b.J23)$$

where

${}_r\nu = 3.0$ and ${}_s\nu = 1.5$ for long-term stress, and

${}_r\nu = 1.5$ and ${}_s\nu = 1.0$ for short-term stress

$$N_k = \frac{\pi^2}{\ell_k^2}\left[\frac{{}_cE \cdot {}_cI}{5} + {}_sE \cdot {}_sI\right] \quad (12b.J24)$$

${}_sN$ and ${}_sN_c$ are the strengths of the steel part treated as a slender beam-column, and calculations are based on AIJ Standard for steel structures.

${}_rN$ and ${}_rN_c$ are the strengths of the reinforced concrete part treated as a slender beam-column, and equal to the strengths of the cross section subjected to ${}_r\delta_r M$ and ${}_rN$, with a minimum axial load eccentricity more than 5% of the depth of the section. The quantity ${}_r\delta$ is given below.

$$ {}_r\delta = \frac{1}{1 - \dfrac{{}_r\nu \cdot {}_rN}{{}_rN_k}} \qquad {}_rN_k = \frac{2\pi^2 \cdot {}_cE \cdot {}_cI}{5\ell_k^2} \tag{12b.J25} $$

(E) Allowable Strength for FIL Beam-Columns

for $N \le {}_cN_c$ or $M \ge {}_sM_o(1 - \dfrac{{}_c\nu \cdot {}_cN_c}{N_k})$

$$ \begin{aligned} N &= {}_cN \\ M &\le {}_sM_o(1 - \frac{{}_c\nu \cdot {}_cN}{N_k}) + {}_cM \end{aligned} \tag{12b.J26} $$

for $N > {}_cN_c$ or $M < {}_sM_o\,(1 - \dfrac{{}_c\nu \cdot {}_cN_c}{N_k})$

$$ \begin{aligned} N &\le {}_cN_c + {}_sN \\ M &= {}_sM\,(1 - \frac{{}_c\nu \cdot {}_cN_c}{N_k}) \end{aligned} \tag{12b.J27} $$

where

${}_sN, {}_sN_c, {}_cN, {}_cN_c$ = calculated according to the first two paragraphs on the top of this page for ENC beam-columns shown above, in which prefix r is replaced by c and ${}_cN_k$ is taken equal to half the value of ${}_rN_k$.

${}_c\nu$ = 3.0 for long-term stress
= 1.5 for short-term stress

Table 12b.J2 Allowable Flexural Strength of Filled Concrete

	N.A.	${}_eN/({}_eD^2 \cdot f'_e)$, ${}_eM/({}_eD^3 \cdot f'_e)$	
CHS	inside section	${}_eN/({}_eD^2 \cdot f'_e)$	$\{\sin\theta_n(2+\cos^2\theta_n)/3-\theta_n\cos\theta_n\}/(8\,x_{n1})^*$
		${}_eM/({}_eD^3 \cdot f'_e)$	$\{\theta_n+\sin 2\theta_n\,(\cos^2\theta_n-5/2)/3\}/(64\,x_{n1})^*$
	outside section	${}_eN/({}_eD^2 \cdot f'_e)$	$\pi\{1-1/(2\,x_{n1})\}/4$
	$(e \le {}_eD/8)$	${}_eM/({}_eD^3 \cdot f'_e)$	$\pi/(64\,x_{n1})$
SHS	inside section	${}_eN/({}_eD^2 \cdot f'_e)$	$x_{n1}/2$
		${}_eM/({}_eD^3 \cdot f'_e)$	$x_{n1}(3-2\,x_{n1})/12$
	outside section	${}_eN/({}_eD^2 \cdot f'_e)$	$1-1/(2\,x_{n1})$
	$(e \le {}_eD/6)$	${}_eM/({}_eD^3 \cdot f'_e)$	$1/(12\,x_{n1})$

$^*\theta_n=\cos^{-1}(1-2\,x_{n1})$, $x_{n1}=x_n/{}_cD$, $f'_e=f_e$

Table 12b.J3 Ultimate Flexural Strength for ENC Section with Full-web H-shape

N_u	M_u
$\frac{{}_s a_w}{2}\,{}_s\sigma_y + {}_e r_u \cdot F_e \cdot bD \le N_u \le {}_e r_u \cdot F_e \cdot bD + {}_s A \cdot {}_s\sigma_y + 2\,{}_m a_t \cdot {}_m\sigma_y$	$\frac{{}_s Z_p \cdot {}_s\sigma_y + {}_m a_t \cdot {}_m\sigma_y \cdot {}_m d}{\left(-{}_s A + \frac{{}_s a_w}{2}\right){}_s\sigma_y - 2\,{}_m a_t \cdot {}_m\sigma_y} \times (N_u - {}_s A \cdot {}_s\sigma_y + 2\,{}_m a_t \cdot {}_m\sigma_y - {}_e r_u \cdot F_e \cdot bD)$
$\frac{{}_s a_w}{2}\,{}_s\sigma_y + \frac{{}_e r_u}{2} F_e \cdot bD \le N_u \le \frac{{}_s a_w}{2}\,{}_s\sigma_y + {}_e r_u F_e \cdot bD$	$\frac{D}{2}\left(N_u - \frac{{}_s a_w}{2}\,{}_s\sigma_y\right)\left(1 - \frac{N_u - \frac{{}_s a_w}{2}\,{}_s\sigma_y}{{}_e r_u \cdot F_c \cdot bD}\right) + {}_s Z_p \cdot {}_s\sigma_y + {}_m a_t \cdot {}_m\sigma_y \cdot {}_m d$
$-\frac{{}_s a_w}{2}\,{}_s\sigma_y + \frac{{}_e r_u}{2} F_e \cdot bD \le N_u \le \frac{{}_s a_w}{2}\,{}_s\sigma_y + \frac{{}_e r_u}{2} F_e \cdot bD$	$\frac{{}_e r_u}{8}\, b D^2 \cdot F_e + {}_s Z_p \cdot {}_s\sigma_y + {}_m a_t \cdot {}_m\sigma_y \cdot {}_m d$
$-\frac{{}_s a_w}{2}\,{}_s\sigma_y \le N_u \le -\frac{{}_s a_w}{2}\,{}_s\sigma_y + \frac{{}_e r_u}{2} F_e \cdot bD$	$\frac{D}{2}\left(N_u + \frac{{}_s a_w}{2}\,{}_s\sigma_y\right)\left(1 - \frac{N_u + \frac{{}_s a_w}{2}\,{}_s\sigma_y}{{}_e r_u\, F_e bd}\right) + {}_s Z_p \cdot {}_s\sigma_y + {}_m a_t \cdot {}_m\sigma_y \cdot {}_m d$
$-{}_s A \cdot {}_s\sigma_y - 2\,{}_m a_t \cdot {}_m\sigma_y \le N_u \le -\frac{{}_s a_w}{2}\,{}_s\sigma_y$	$\frac{{}_s Z_p \cdot {}_s\sigma_y + {}_m a_t \cdot {}_m\sigma_y \cdot {}_m d}{\left({}_s A - \frac{{}_s a_w}{2}\right){}_s\sigma_y - 2\,{}_m a_t \cdot {}_m\sigma_y} \times (N_u + {}_s A \cdot {}_s\sigma_y + 2\,{}_m a_t \cdot {}_m\sigma_y)$

Table 12b.J4 Ultimate Flexural Strength of Reinforced Concrete Part

${}_r N_u$	${}_r M_u$
${}_e r_u \cdot F_e \cdot bD \le {}_r N_u \le {}_e r_u \cdot F_e \cdot bD + 2\,{}_m a_t \cdot {}_s\sigma_y$	${}_m a_t \cdot {}_m\sigma_y \cdot {}_m d - \frac{{}_m d}{2}({}_r N_u - {}_e r_u \cdot F_e \cdot bD)$
$0 \le {}_r N_u \le {}_e r_u \cdot F_e \cdot bD$	${}_m a_t \cdot {}_m\sigma_y \cdot {}_m d + \frac{{}_r N_u \cdot D}{2}\left(1 - \frac{{}_r N_u}{{}_e r_u \cdot F_e \cdot bD}\right)$
$-2\,{}_m a_t \cdot {}_m\sigma_y \le {}_r N_u \le 0$	${}_m a_t \cdot {}_m\sigma_y \cdot {}_m d + \frac{{}_m d}{2}\,{}_r N_u$

Table 12b.J5 Ultimate Flexural Strength of Concrete Part

RHS	${}_eM_u/(bD^2 \cdot F_e)$	
	$\frac{1}{2}\frac{{}_eN_u}{bD \cdot F_e}\left(1-\frac{1}{{}_er_u}\frac{{}_eN_u}{bD \cdot F_e}\right)$	
CHS	${}_eN_u/({}_eD^2 \cdot F_e)$	${}_eM_u/({}_eD^3 \cdot F_e)$
	$\frac{{}_er_u}{4}(\theta_n - \sin\theta_n \cdot \cos\theta_n)$	$\frac{{}_er_u}{12}\sin^3\theta_n$

Table 12b.J6 Ultimate Flexural Strength of Reinforcing Bars

	${}_mN_u$	${}_mM_u$
symmetrical	$0 \leq {}_mN_u \leq 2\,{}_ma_t \cdot {}_m\sigma_y$	${}_md({}_ma_t \cdot {}_m\sigma_y - {}_mN_u/2)$
bar-arrangement	$-2\,{}_ma_t \cdot {}_m\sigma_y \leq {}_mN_u \leq 0$	${}_md({}_ma_t \cdot {}_m\sigma_y + {}_mN_u/2)$
symmetrical bar-arrangement	${}_ma_m \cdot {}_m\sigma_y \leq {}_mN_u \leq (2\,{}_ma_t + {}_ma_m)\,{}_m\sigma_y$	${}_md\left({}_ma_t \cdot {}_m\sigma_y - \frac{1}{2}({}_mN_u - {}_ma_m \cdot {}_m\sigma_y)\right)$
with middle bars	$-{}_ma_m \cdot {}_m\sigma_y \leq {}_mN_u \leq {}_ma_m \cdot {}_m\sigma_y$	${}_md \cdot {}_ma_t \cdot {}_m\sigma_y$
	$-(2\,{}_ma_t + {}_ma_m)\,{}_m\sigma_y \leq {}_mN_u \leq -\,{}_ma_m \cdot {}_m\sigma_y$	${}_md\left({}_ma_t \cdot {}_m\sigma_y + \frac{1}{2}({}_mN_u + {}_ma_m \cdot {}_m\sigma_y)\right)$

Notations

E = modulus of elasticity
h = height
I = moment of inertia
j = internal lever arm
ℓ = effective length
Z = plastic modulus
M = bending moment
p = reinforcement ratio, steel ratio
t = plate thickness
x = distance
δ = magnification factor
ν = factor of safety

Postscript Symbols

k = critical
n = neutral axis
o = zero axial force

Table 12b.J7 Ultimate Flexural Strength of Steel Shape

Cross section	${}_sN_u$	${}_sM_u$
open-web H major-axis bending	$0 \le {}_sN_u \le {}_sA \cdot {}_s\sigma_y$	${}_sZ_p \cdot {}_s\sigma_y - {}_sN_u \cdot {}_sd/2$
	$-{}_sA \cdot {}_s\sigma_y \le {}_sN_u \le 0$	${}_sZ_p \cdot {}_s\sigma_y + {}_sN_u \cdot {}_sd/2$
open-web cruciform	${}_sa_m \cdot {}_s\sigma_y \le {}_sN_u \le {}_sA \cdot {}_s\sigma_y$	${}_sZ_p \cdot {}_s\sigma_y - ({}_sd/2)({}_sN_u - {}_sa_m \cdot {}_s\sigma_y)$
	$-{}_sa_m \cdot {}_s\sigma_y \le {}_sN_u \le {}_sa_m \cdot {}_s\sigma_y$	${}_sZ_p \cdot {}_s\sigma_y$
	$-{}_sA \cdot {}_s\sigma_y \le {}_sN_u \le {}_sa_m \cdot {}_s\sigma_y$	${}_sZ_p \cdot {}_s\sigma_y + ({}_sd/2)({}_sN_u + {}_sa_m \cdot {}_s\sigma_y)$
full-web H major-axis bending & RHS	$\frac{{}_sa_w}{2} \cdot {}_s\sigma_y \le {}_sN_u \le {}_sA \cdot {}_s\sigma_y$	${}_sZ_p \cdot {}_s\sigma_y - ({}_sd/2)({}_sN_u - \frac{1}{2}{}_sa_w \cdot {}_s\sigma_y)$
	$-\frac{{}_sa_w}{2} \cdot {}_s\sigma_y \le {}_sN_u \le \frac{{}_sa_w}{2} \cdot {}_s\sigma_y$	${}_sZ_p \cdot {}_s\sigma_y$
	$-{}_sA \cdot {}_s\sigma_y \le {}_sN_u \le -\frac{{}_sa_w}{2} \cdot {}_s\sigma_y$	${}_sZ_p \cdot {}_s\sigma_y + ({}_sd/2)({}_sN_u + \frac{1}{2}{}_sa_w \cdot {}_s\sigma_y)$
full-web H minor-axis bending	$({}_sa_f + {}_sa_w){}_s\sigma_y \le {}_sN_u \le {}_sA \cdot {}_s\sigma_y$	${}_sZ_p \cdot {}_s\sigma_y - ({}_sb/2)({}_sN_u - {}_sa_w \cdot {}_s\sigma_y)$
	$-({}_sa_f + {}_sa_w){}_s\sigma_y \le {}_sN_u$ $\le ({}_sa_f + {}_sa_w){}_s\sigma_y$	${}_sZ_p \cdot {}_s\sigma_y$
	$-{}_sA \cdot {}_s\sigma_y \le {}_sN_u$ $\le -({}_sa_f + {}_sa_w){}_s\sigma_y$	${}_sZ_p \cdot {}_s\sigma_y + (b_f/2)({}_sN_u + {}_sa_w \cdot {}_s\sigma_y)$
full-web cruciform	$(2{}_sa_f + \frac{3\,{}_sa_w}{2}){}_s\sigma_y \le {}_sN_u$ $\le {}_sA \cdot {}_s\sigma_y$	${}_sd \cdot \frac{{}_sA}{2} \cdot {}_s\sigma_y - \frac{{}_sd}{2} \cdot {}_sN_u$
	$({}_sa_f + \frac{3\,{}_sa_w}{2}){}_s\sigma_y \le {}_sN_u$ $\le (2{}_sa_f + \frac{3\,{}_sa_w}{2}){}_s\sigma_y$	${}_sZ_p \cdot {}_s\sigma_y - (b_f/2)({}_sN_u - \frac{3}{2}{}_sa_w \cdot {}_s\sigma_y)$
	$-({}_sa_f + \frac{3\,{}_sa_w}{2}){}_s\sigma_y \le {}_sN_u$ $\le ({}_sa_f + \frac{3\,{}_sa_w}{2}){}_s\sigma_y$	${}_sZ_p \cdot {}_s\sigma_y$
	$-(2{}_sa_f + \frac{3\,{}_sa_w}{2}){}_s\sigma_y \le {}_sN_u$ $\le -({}_sa_f + \frac{3\,{}_sa_w}{2}){}_s\sigma_y$	${}_sZ_p \cdot {}_s\sigma_y + (b_f/2)$ $\times ({}_sN_u - \frac{3}{2}{}_sa_w \cdot {}_s\sigma_y)$
	$-{}_sA \cdot {}_s\sigma_y \le {}_sN_u$ $\le -(2{}_sa_f + \frac{3\,{}_sa_w}{2}){}_s\sigma_y$	${}_sd \cdot \frac{{}_sA}{2} \cdot {}_s\sigma_y + \frac{{}_sd}{2} \cdot {}_sN_u$
CHS	$0.2\,{}_sA \cdot {}_s\sigma_y \le {}_sN_u \le {}_sA \cdot {}_s\sigma_y$	$1.25\left(1 - \frac{{}_sN_u}{{}_sA \cdot {}_s\sigma_y}\right){}_sZ_p \cdot {}_s\sigma_y$
	$-0.2\,{}_sA \cdot {}_s\sigma_y \le {}_sN_u$ $\le 0.2\,{}_sA \cdot {}_s\sigma_y$	${}_sZ_p \cdot {}_s\sigma_y$
	$-{}_sA \cdot {}_s\sigma_y \le {}_sN_u$ $\le -0.2\,{}_sA \cdot {}_s\sigma_y$	$1.25\left(1 + \frac{{}_sN_u}{{}_sA \cdot {}_s\sigma_y}\right){}_sZ_p \cdot {}_s\sigma_y$

NORTH AMERICA

CHAPTER EDITORS
Saif Daresalemwala, Howard Needles Tammen & Bergendoff
Richard Furlong, University of Texas

ACI 318 83

Similar to reinforced concrete design, with strain compatibility and a compression block of depth

$$0.65 \leq \frac{a}{y_c} = 0.85 - \frac{(f'_c - 4000)}{20000} \leq 0.85 \qquad (12b.N1)$$

Minimum eccentricity for axial loading, e = (0.6 + 0.03H) inches

where

- a = depth of rectangular stress block from compression face
- f'_c = concrete cylinder strength (psi)
- H = greater lateral dimension in inches
- y_c = depth of neutral axis from compression face

AISC LRFD 86

The general interaction formula for uniaxial or biaxial bending is given by

for $\frac{N_u}{\varphi_c N_n} \geq 0.2$

$$\frac{N_u}{\varphi_c N_n} + \frac{8}{9}\left[\frac{M_{ux}}{\varphi_b M_{nx}} + \frac{M_{uy}}{\varphi_b M_{ny}}\right] \leq 1.0 \qquad (12b.N2a)$$

for $\frac{N_u}{\varphi_c N_n} < 0.2$

$$\frac{N_u}{2\varphi_c N_n} + \left[\frac{M_{ux}}{\varphi_b M_{nx}} + \frac{M_{uy}}{\varphi_b M_{ny}}\right] \leq 1.0 \qquad (12b.N2b)$$

where

- M_{nx}, M_{ny} = nominal flexural strength using plastic stress distribution about major and minor axes, respectively
- M_{ux} = design moment about major axis
- M_{uy} = design moment about minor axis
- N_n = nominal compressive strength
- N_u = design compressive force
- φ_b = capacity (resistance) factor for bending
- φ_c = capacity (resistance) factor for compression

CAN 3 A23.3 M84

The resistance of the section is computed for the same limiting conditions applicable to ordinary reinforced concrete sections. As in ACI 318 83 code, interaction diagrams are constructed using a maximum concrete strain of 0.003 and rectangular stress blocks. The depth of the concrete stress block as in ACI 318 83.

Minimum eccentricity e = (15 + 0.03h) (mm), where h is the overall thickness of the member in millimeters.

WEST EUROPE

CHAPTER EDITORS
Reinhard Bergmann, Bochum University
Patrick J. Dowling, Imperial College
Amr S. Elnashai, Imperial College
Karlheinz Roik, Bochum University

BS 5400 82

Uniaxial Bending
Strength of a member is checked by

$$N < N_y = N_u\left[K_{1y} - (K_{1y} - K_{2y} - 4K_3)\frac{M_y}{M_{uy}} - 4K_3\left[\frac{M_y}{M_{uy}}\right]^2\right] \quad (12b.W1)$$

Uniaxial Bending about Major Axis Restrained from Failure about Minor Axis
Strength of a member is checked by

$$N < N_x = N_u\left[K_{1x} - (K_{1x} - K_{2x} - 4K_3)\frac{M_x}{M_{ux}} - 4K_3\left[\frac{M_x}{M_{ux}}\right]^2\right] \quad (12b.W2)$$

(A) For ENC: Plastic neutral axis outside the steel section (see Fig. 12b.W1a)

$$\rho b d_s \geq A_s$$

$$d_c = \frac{A_s}{b\rho} \quad (12b.W3)$$

and $$M_{ux} = 0.91\, f_y A_s \frac{(h-d_c)}{2} + 0.87 f_{ry} \frac{A_r}{2} d_r \quad (12b.W4)$$

$$M_{uy} = 0.91\, f_y A_s \frac{(b-d_c)}{2} + 0.87 f_{ry} \frac{A_r}{2} d_r \quad (12b.W5)$$

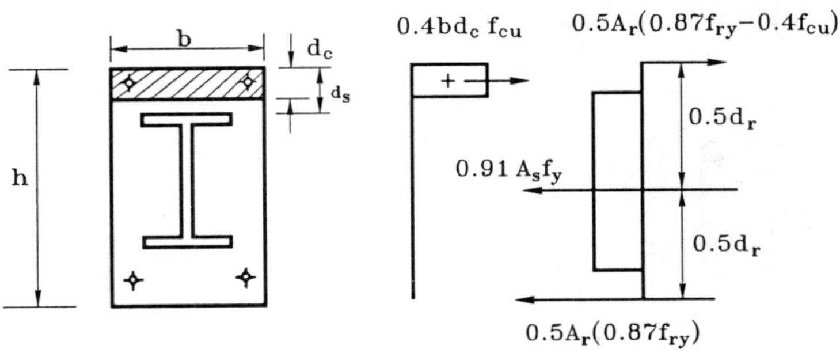

(a)

Fig. 12b.W1a Force Diagram for Calculating M_u (Plastic neutral axis outside the steel section)

(B) For ENC: Plastic neutral axis within top flange major axis bending (see Fig. 12b.W1b)

$$\rho p d_s < A_s \text{ and } 0.91 A_s f_y \leq 0.4 f_{cu} [b d_s + t_f (b - b_f) + 1.82\, A_f f_y]$$

$$d_c = (A_s + 2 b_f d_s)/(b\rho + 2 b_f) \quad (12b.W6)$$

$$M_u = 0.91\, f_y [A_s (h - d_c) 2 - b_f d_s (d_c - d_s)] + 0.87 f_{ry} \frac{A_r}{2} d_r \quad (12b.W7)$$

(C) For ENC: Plastic neutral axis in web/major axis bending (see Fig. 12b.W1c)

where

$$d_c = h t_w/(b\rho + 2 t_w) \quad (12b.W8)$$

$$M_u = 0.91 f_y [A_s (h - d_c)/2 - b_f t_f (d_s - d_w) - t_w d_w (d_c - d_w)] + 0.87 f_{ry} \frac{A_r}{2} d_r \quad (12b.W9)$$

(D) For ENC: Plastic neutral axis in flanges/minor axis bending (see Fig. 12b.W1d)

where

$$d_c = (A_s + 4 t_f d_s)/b\rho + 4 t_f \text{ since } \rho b d_d < A_s \quad (12b.W10)$$

$$M_u = 0.91\, f_y\, [A_s \frac{h - d_c}{2} - 2 t_f d_s (d_c - d_s)] + 0.87 f_{ry} \frac{A_r}{2} d_r \quad (12b.W11)$$

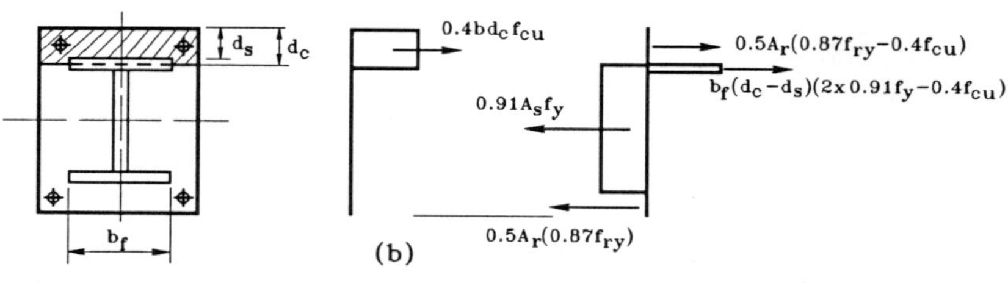

Fig. 12b.W1b Force Diagram for Calculating M_u (Plastic neutral axis within top flange/major axis bending)

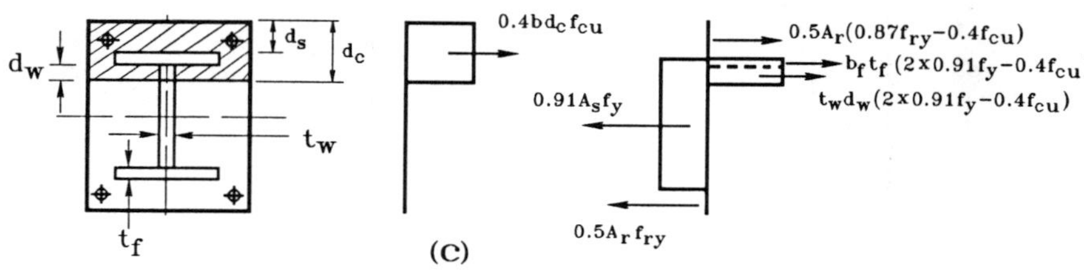

Fig. 12b.W1c Force Diagram for Calculating M_u (Plastic neutral axis in web/major axis bending)

(E) For FIL (RHS) (see Fig. 12b.W1e)

$$d_c = \frac{(A_s - 2b_f t_f)}{(b\rho + 4t_f)} \qquad (12b.W12)$$

$$M_u = 0.91 f_y \left[A_s \frac{(h-d_c)}{2} + b_t t_f (t_f + d_c)\right] \qquad (12b.W13)$$

(F) For FIL (CHS)

$$M_u = 0.91\ S f_y (1 + 0.01m) \quad (\text{m - Appendix C, BS 5400}) \qquad (12b.W14)$$
$$S = t^3 (D_e/t - 1)^2$$

Biaxial Bending

$$M_x < M_{ux} \qquad N \leq N_{xy} \qquad (12b.W15)$$
$$M_y < M_{uy} \qquad N_{ax} = K_{1x} N_u \qquad (12b.W16)$$
$$\frac{1}{N_{xy}} = \frac{1}{N_x} + \frac{1}{N_y} + \frac{1}{N_{ax}} \qquad (12b.W17)$$

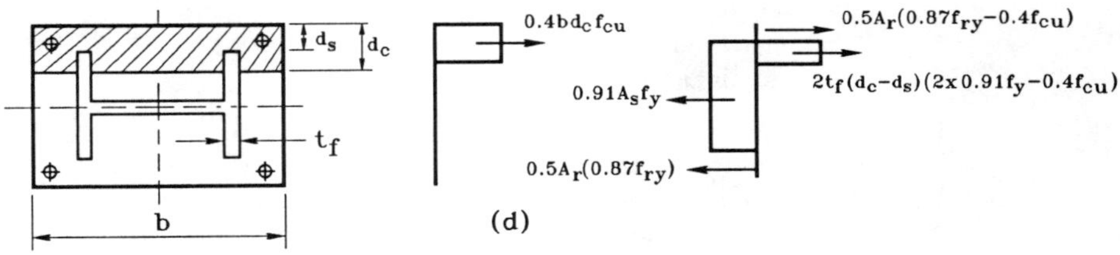

Fig. 12b.W1d Force Diagram for Calculating M_u (Plastic neutral axis in flanges/minor axis bending)

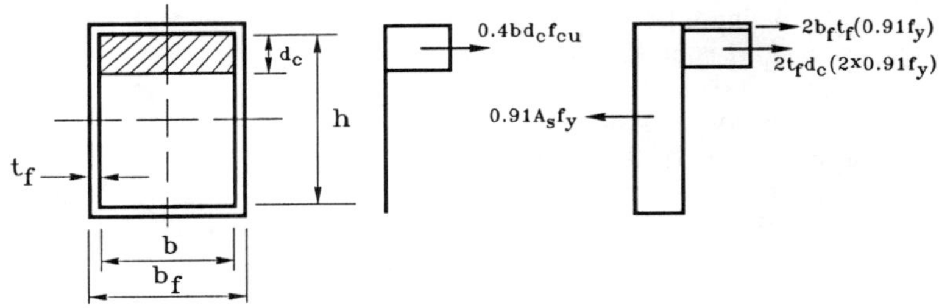

Fig. 12b.W1e Force Diagram for Calculating M_u in the Case of Filled Sections

where

A_c	= area of concrete in the cross section
A_f	= area of the top flange of the steel section
A_r	= cross-sectional area of reinforcement
A_s	= cross-sectional area of rolled or fabricated structural steel section
a_c	= concrete contribution factor
b	= least lateral dimension of concrete in cross-section
b_f	= breadth of steel flange of I-section or the external dimension of a rectangular hollow section
b_s	= external dimension of the wall of the RHS
D_e	= outside diameter of the CHS
d_c	= distance of neutral axis from the more compressed face of concrete
d_r	= distance between symmetrically placed reinforcing bars measured perpendicular to the axis of bending
d_s	= thickness of concrete cover to encased steel section
d_w	= depth of steel web in compression
E_s	= Young's modulus of elasticity of steel
e	= accidental eccentricity of axial load
f_{cc}	= characteristic strength of triaxially contained concrete under axial load
f_{cu}	= 28-day cube characteristic compressive strength
f_{ry}	= characteristic yield strength of reinforcement
f_y	= nominal yield strength of the steel
f'_y	= reduced nominal yield strength of steel casing
h	= greatest lateral dimension of concrete in cross-section
I_c, I_s, I_r	= second moment of area of the uncracked cross-section, the steel section, and the reinforcement, respectively, above the axis of the composite column
ℓ	= effective length
ℓ_E	= length of column for which Euler load equals the squash load
ℓ_e	= effective length of the actual column in the plane of bending considered
M_{ux}	= design ultimate moment of resistance in x axis with N=0
M_{uy}	= design ultimate moment of resistance in y axis with N=0
N_{ay}	= short column squash load
N_u	= squash load
t	= thickness of steel casing
t_f	= average thickness of the flange of a steel section
t_w	= thickness of web at steel section
ρ	= ratio of the average compressive stress in the concrete at failure to the design yield strength of the steel taken as $0.4\ f_{cu}/0.91F_y$

DIN 18806 P.1

Interaction relations between axial force and moment are illustrated in Figs. 12b.W2 and 12.W3.

Uniaxial Bending

$$M = 0.9 \cdot s \cdot M_{pl} \qquad (12b.W18)$$

$$\mathfrak{S}_n = \mathfrak{S}\left(\frac{1-\Psi}{4}\right) \qquad (12b.W19)$$

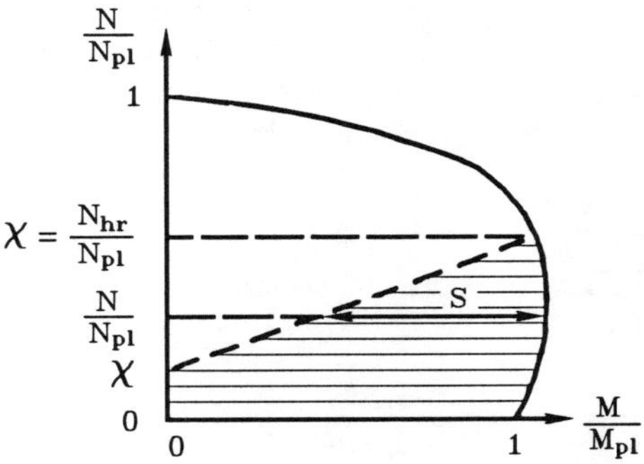

Fig 12b.W2 Interaction Curve for Compression and Uniaxial Bending

Biaxial Bending

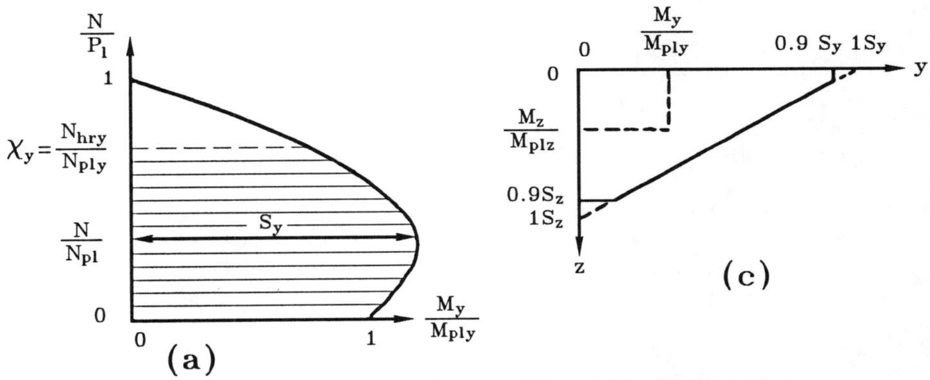

Fig. 12b.W3a Design for Compression and Biaxial Bending

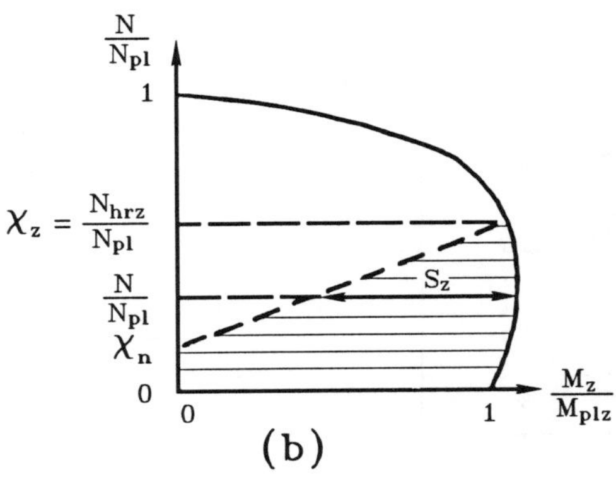

Fig. 12b.W3b Design for Compression and Biaxial Bending

where

M_{pl} = plastic moment of the composite cross-section
S = as defined in figure above
χ = as defined in figure above = N
Ψ = moment distribution factor

$$\frac{M_y}{S_y\, M_{ply}} + \frac{M_z}{S_z\, M_{plz}} \leq 1 \qquad (12b.W20)$$

Imperfection has to be taken into account only for the axis more endangered to fail.

EC 4 85

The ultimate resistance of a cross-section subjected to compression combined with bending above a principal axis may be calculated from first principles, neglecting slip and using stress-strain curves for materials from EC 2 and EC 3.

If the rectangular stress blocks are used as shown below the ultimate resistance in bending is over-estimated, and allowance should be made elsewhere.

For Design by Method 1

For columns with $\bar{\lambda} \leq 0.2$, the bending moment may be calculated from the values at the end of the column length and the transverse loading without considering second-order effects.

For columns with $0.2 < \bar{\lambda} \leq 2.0$, second-order effects may be neglected if there is no transverse loading within the column length. Otherwise, bending moments should be calculated by second-order elastic theory, using flexural stiffnesses in accordance with next section.

(A) Uniaxial Bending

The interaction design curve is defined by Fig. 12b.W5. The design procedure is given in step-by-step form, with reference to the curve. It is assumed that $N_{p\ell}$, M_{pi} and the interaction curve for cross-sections are known.

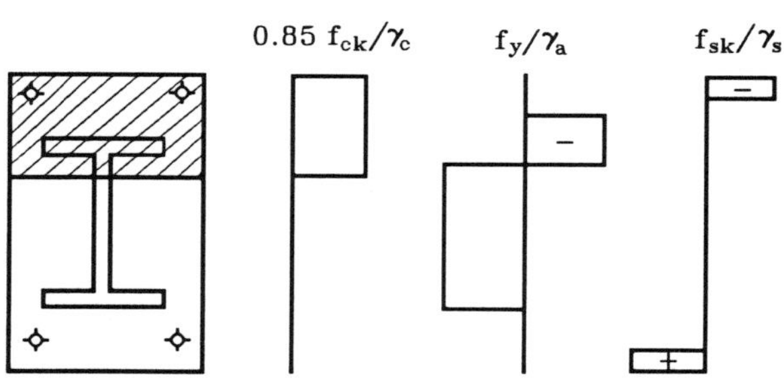

Fig. 12b.W4 Full Plastic Stress Distribution for Cross Section

(1) Let $K = N_u/N_{p\ell}$ and obtain μ_k

(2) Let $K_d = N_d/N_{p\ell}$ and obtain μ_d

(3) When the variation of bending moment along the column length is approximately linear, $k_n = k(1 - \psi)/4$ but $k_n \leq k_d$.

where ψ is the ratio of the lesser to the greater end moment. (Fig. 12b.W6)

(4) The length μ is calculated from $\mu = \mu_d - \mu_k (k_d - k_n)/(k - k_n)$ $\mu \leq 1.0$ unless the bending moment M_d is due solely to the action of the force N_d

(5) The member has sufficient resistance if $M_d \leq 0.9\ \mu M_{pc}$

(B) Biaxial Bending (Fig. 12b.W7)

Design for compression and biaxial bending.

The following design method may be used for a design axial force N_d combined with design bending moments M_y and M_z.

(1) The values of μ_y and μ_z are calculated independently as for μ in the uniaxial case by using Fig. 12b.W7a for minor axis bending

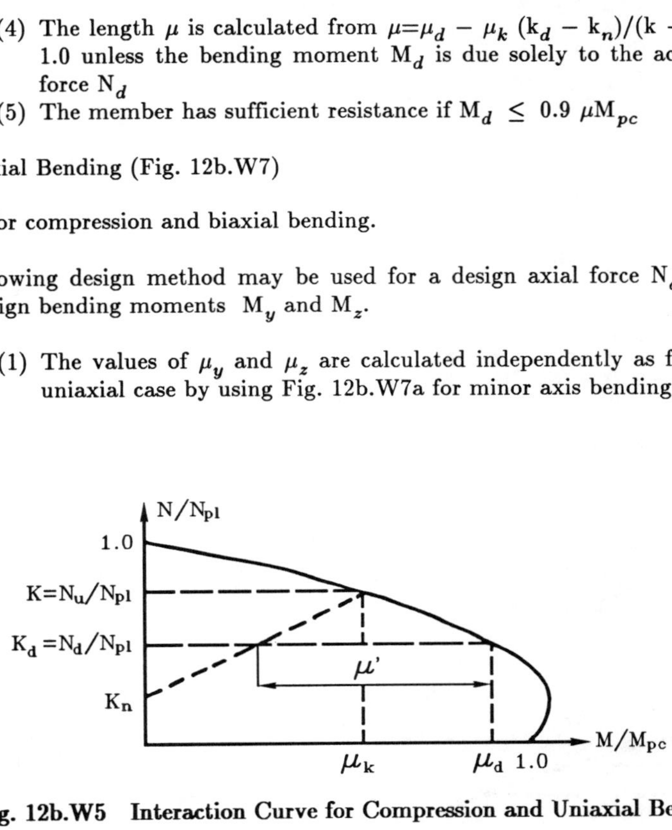

Fig. 12b.W5 Interaction Curve for Compression and Uniaxial Bending

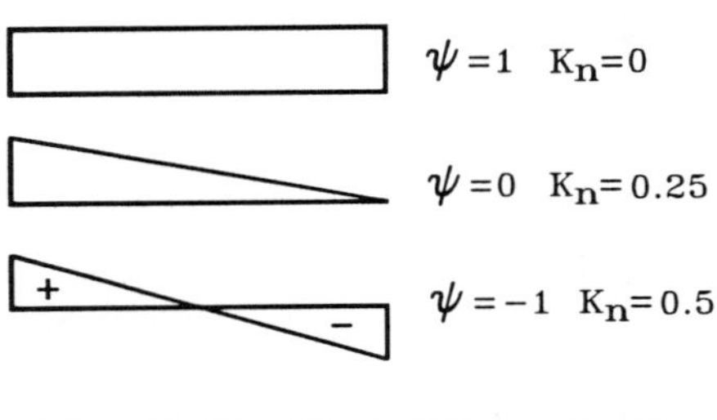

Fig. 12b.W6 Typical Values for K_n

(2) The member is strong enough if

$M_y \leq 0.9\ \mu_z M_{pcy}$

$M_z \leq 0.9\ \mu_z M_{pcz}$ and $M_y/\mu_y M_{pcy} + M_z/\mu_z M_{pcz} \leq 1$

as shown in Fig. 12b.W7b above.

For Design by Method 2

Short braced columns can be designed as follows if

$e_y \leq 1.5h_c,\ e_z \leq 1.0b_c$

and $h/h_c \leq 0.5,\ b/b_c \leq 0.5$

The design eccentricity $e > 0.04h$

(A) Major axis bending

M_d should not exceed

$$N_{uy} = N_{p\ell}k_1\ (1 + k_2) \qquad (12b.W21)$$

where k_2 is $\dfrac{2.3e_y/h_c}{\delta h/h_c + (1-\delta)/3}$

k_2 of the lower of k_{1y} and k_{1z}

(B) Minor axis bending

N_d should not exceed

$$N_{uz}=N_{p\ell}\ k_1/(1+k_3) \qquad (12b.W22)$$

where k_3 is $\dfrac{4e_z/b_c}{\delta b/b_c+(1-\delta)/2}$

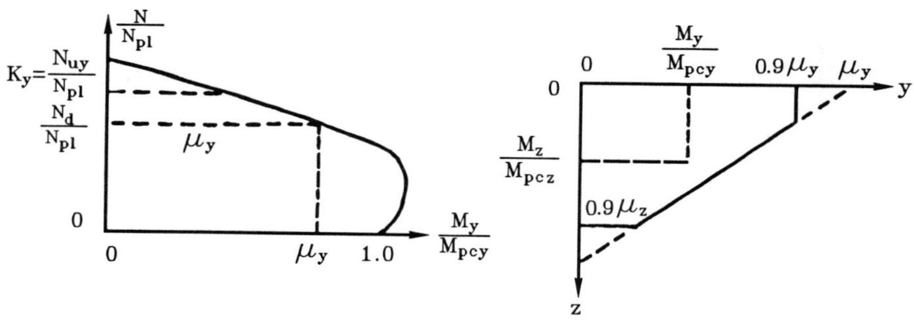

(a) Value of μ_y for failure by bending

(b) Interaction diagram for resistance in bending minor-axis

Fig. 12b.W7 Design for Compression and Biaxial Bending

(C) Biaxial Bending

N_d should not exceed

$$N_{uyz} = N_{p\ell}K_1/(1 + k_2 + k_3) \tag{12b.W23}$$

N_{uy}, N_{uz} and N_{uyz} may be increased by

5% when $\ell_{ey}/h_c \leq 8$ and $\ell_{ez}/b_c \leq 8$
10% when $\ell_{ey}/h_c \leq 8$ and $\ell_{ez}/b_c \leq 12$

(D) ENC slender columns
It may be designed as for short columns by substituting k_2 and k_3 by k_4 and k_5, respectively

$$k_4 = k_2\,[1 - (\lambda_{max} - 12)/37] \tag{12b.W24}$$
$$k_5 = k_3\,[1 - (\lambda_{max} - 12)/37] \tag{12b.W25}$$

for $k_{1z} \leq k_{1y}$ λ_{max} is the greater of ℓ_{ez}/b_c and 12

for $k_{1z} > k_{1y}$ λ_{max} is the greater of $0.7\ell_{ey}/h_c$ and 12

INTERNATIONAL COMPARISONS & COMMENT

WHAT ARE THE DIFFERENCES?

The same composite column section and assumptions given in Section 12a (What are the differences?) apply to the code differences for strength under combined axial load and flexure.

At a slenderness of zero, major axis bending, EC 4 and BS 5400 are with 5% difference, while ACI and AISC are 20% and 32% below EC 4 for intermediate axial forces. At slenderness of 12 and 20, ACI and BS 5400 show better agreement but are both about 25% less than EC 4. Except at extreme values of axial force, AISC gives about 36% less than EC 4. At a slenderness of 30, the resistance given by EC 4 and BS 5400 are close, within 10%. On the other hand, ACI and AISC are about 33% less than EC 4, except at extreme values of axial load. Observations regarding minor axis bending curves are, in general, along the same lines given above.

Figs. 12b.1 and 12b.2 give plots of unfactored interaction curves for major and minor axis bending with no slenderness effects. Capacities are normalized with respect to the EC 4 value. It is shown that the capacities from BS 5400 and AISC are the same for zero axial load, while BS5400 is about 14% and 11% less than EC 4 for major and minor axis bending. The corresponding drop for AISC is 32% and 36%.

The factored capacities under combined loading for major and minor axis bending are given in Figs. 12b.3 and 12b.4, with no consideration for slenderness, where the moment magnification.

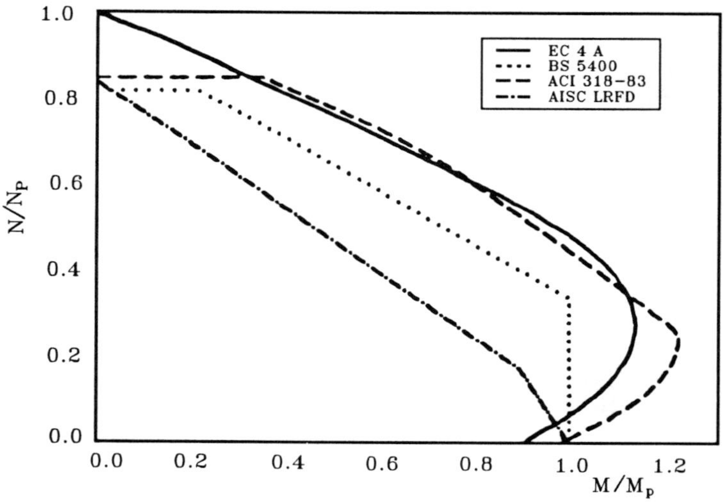

Fig. 12b.1 Unfactored Interaction Curve for Zero Slenderness (Major Axis)

For compactness, only major axis bending is given thereafter, for values of slenderness of 12, 20, and 30. This is given in Figs. 12b.5, 12b.6, and 12b.7.

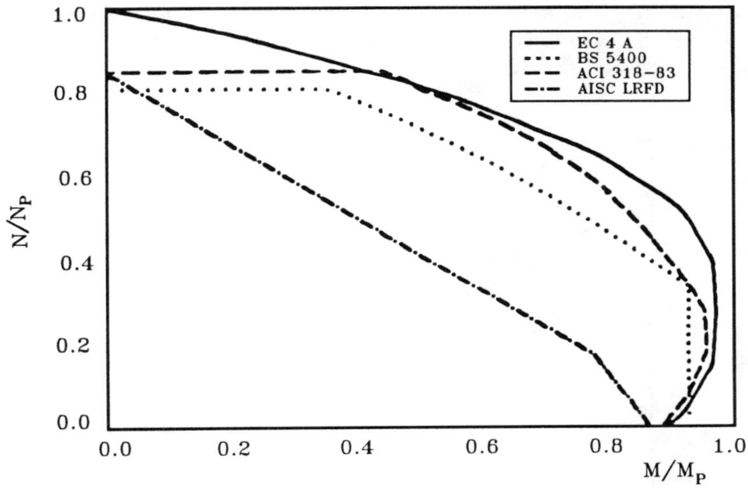

Fig. 12b.2 Unfactored Interaction Curve for Zero Slenderness (Minor Axis)

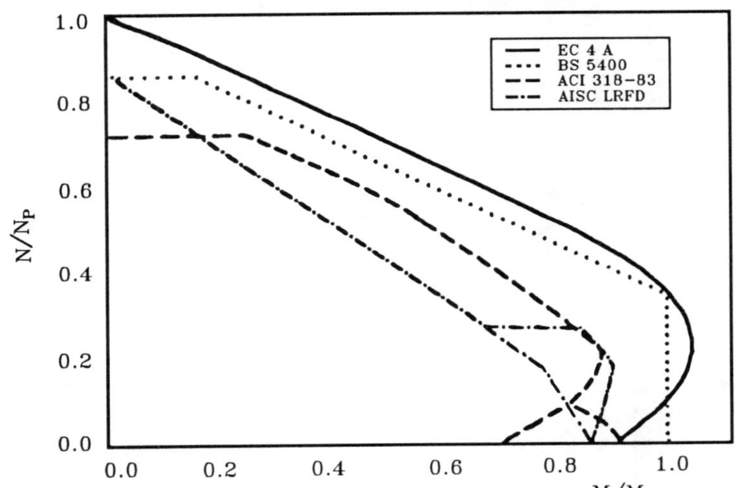

Fig. 12b.3 Factored Interaction Curve for Zero Slenderness (Major Axis)

WHY ARE THERE DIFFERENCES?

Under combined loading (not including load factors), ACI considers a full interaction curve calculated from conditions of equilibrium and compatibility of strains with a strain limit in concrete of 0.3%. On the other hand, in EC 4 the resistance may be calculated from first principles using appropriate stress-strain curves. Alternatively, the Eurocode considers a full interaction curve but with full

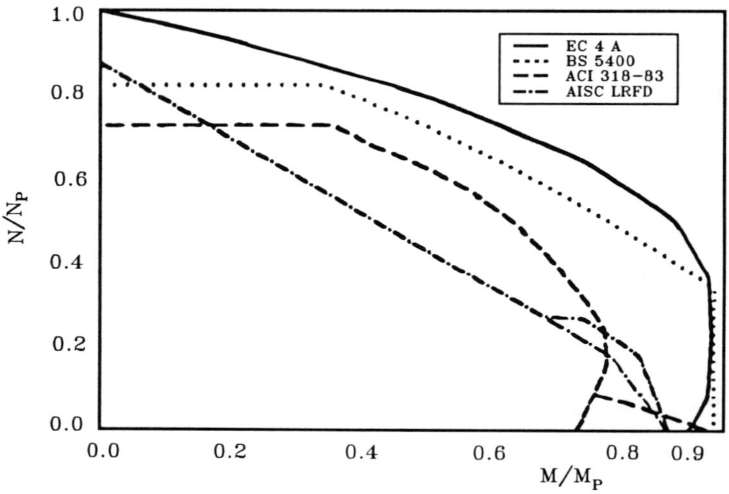

Fig. 12b.4 Factored Interaction Curve for Zero Slenderness (Minor Axis)

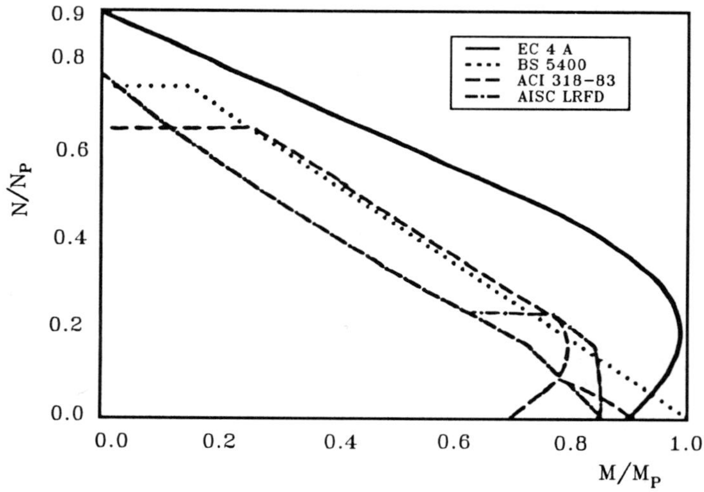

Fig. 12b.5 Factored Interaction Curve for Slenderness 12 (Major Axis)

plastic distribution using rectangular stress blocks that leads to an overestimate of the ultimate resistance. However, the moment capacity in EC 4 is accompanied by a reduction factor of 0.9 on the moment capacity to account for two assumptions: (i) use of uncracked stiffness for concrete and (ii) use of rectangular stress blocks without consideration to concrete strain restrictions.

For zero axial loading the capacities given by BS 5400 and AISC are similar to that given by ACI because in the major axis direction at this position of the

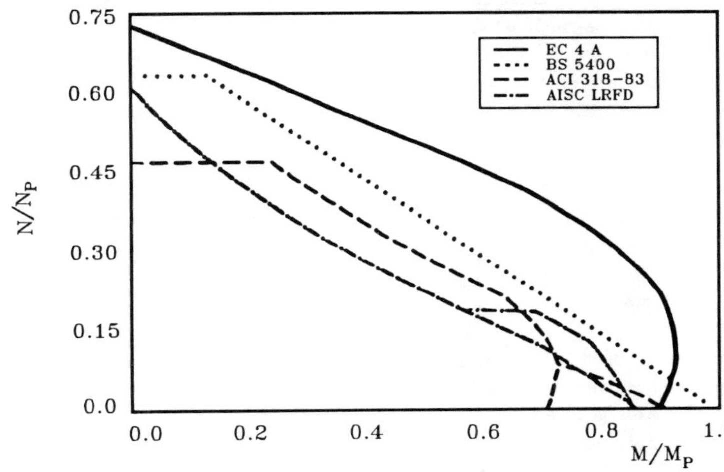

Fig. 12b.6 Factored Interaction Curve for Slenderness 20 (Major Axis)

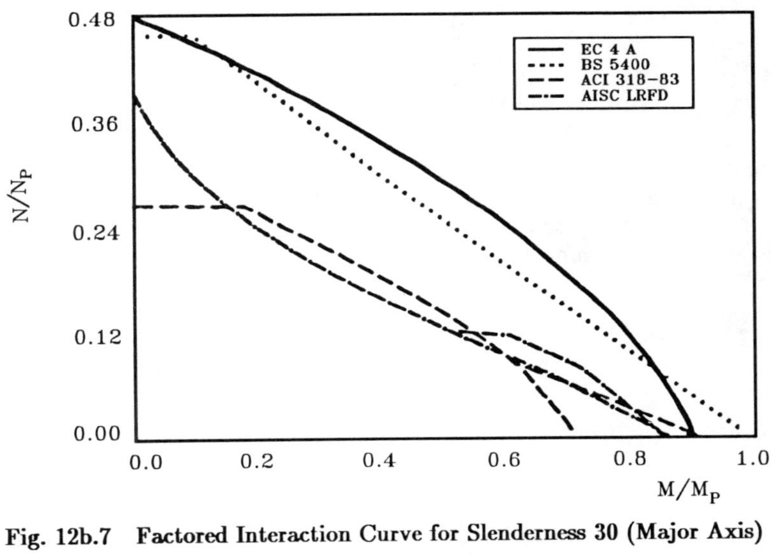

Fig. 12b.7 Factored Interaction Curve for Slenderness 30 (Major Axis)

neutral axis, most of the steel would have yielded, and therefore, the plastic distribution does not significantly over-estimate the strength. Also, the concrete contribution is not significant under pure bending; therefore variability in concrete strength does not affect the calculated capacities significantly. This is not as applicable to the case of minor axis bending, and therefore slightly higher differences are observed.

With regard to the shape of the interaction curve, EC 4 is comparable to ACI by virtue of the 90% reduction, which compensates for overestimating the strength. BS 5400 and AISC use approximate interaction curves, which are very simple to construct but with some sacrifice of strength. In BS 5400 no account of the increasing moment capacity at low axial loads is taken because its availability depends on the loading history of the column. BS 5400 gives lower capacity, compared to EC 4 and ACI, due to the use of simplified interaction curves. The simplification in the AISC design procedure is achieved at some sacrifice in strength.

Comparison between unfactored and factored resistances for major and minor axes (at zero slenderness) shows that the capacity reduction factors are, in general, more conservative. This is particularly true for ACI, which used a factor of 0.7 derived from reinforced concrete design practice.

For higher values of slenderness, the agreement between AISC and BS 5400 is still observed, and the discrepancies with EC 4 decreases. The very low resistance observed from ACI is due to the use of a very low value of stiffness, less than half that used by other codes.

In summary, the differences in capacities under combined loading, shown in Figs. 12b.1 through 12b.7, are due to discrepancies in the following factors

Young's modulus of concrete
Equivalent stiffness
Concrete compressive strength
Minimum eccentricities
Method of section analysis

The above list is not exhaustive, and other factors, such as concrete contribution factors and dimensional limitations, may affect the results. It is, however, to be noted that the above assessment does not include factors on loads and load combination. Inclusion of these may affect the above observations.

12c. Stiffness and Slenderness

SPECIFICATION PROVISIONS & EXPLANATION

AUSTRALIA

CHAPTER EDITORS
Peter Ansourian, University of Sydney
Russell Q. Bridge, University of Sydney

For section stiffness the Austalian procedures use the secant stiffness EI corresponding to the maximum moment capacity M_{ub} of the cross-section ("balanced" moment for reinforced concrete sections). For encased I-sections, the maximum moment occurs when the neutral axis is at mid-depth D/2 and it can be shown that

$$EI = 167\ M_{ub}D \tag{12c.A1}$$

where

D = depth of the cross-section in the plane of bending
M_{ub} = moment capacity

CHINA

CHAPTER EDITOR
Shan-tong Zhong, Harbin Architectural & Civil Engineering Institute

SDGJ 64 89

Stiffness:

$$E_{sc}\,A_{sc} \text{ and } E_{sc}\,I_{sc}$$

where

A_{sc} = area of total cross-section=πr^2
B = $7.483 \times 10^{-4} f_y + 0.974$
C = $-3.63 \times 10^{-3} f_{cu} + 0.0328$
E_s = elastic modulus of steel=2.06×10^3 MPa
E_{sc} = modulus of composite material = f^p_{sc}/ϵ^p_{sc}
f_c = axial compressive strength of concrete
f_{cu} = cubic strength of concrete
f_y = yield point of steel
f^p_{sc} = $(8.17 \times 10^{-4}\, f_y + 0.488) f^y_{sc}$
f^y_{sc} = $(1.143 + B\,\xi + C\,\xi^2) f_c$
I_{sc} = $\frac{\pi}{4} r^4$, r is outside radius of steel tube
t = thickness of steel tube
α = $A_s/A_c \approx 2t/r$
ϵ^p_{sc} = $0.67 f_y/E_s$
ξ = $\alpha f_y/(0.7 f_{cu})$

The Equivalent Slenderness Ratio of Latticed Members

(A) Two limbs with batten plates
The slenderness ratio is given by

$$\lambda^2_{oy} = \lambda^2_y + 17\lambda^2_1 \qquad (12c.C1)$$

where

λ_y = slenderness ratio about y-y axis of the member
λ_1 = slenderness ratio of main components in one panel

(B) Two limbs with lacing bars
The slenderness ratio is given by

$$\lambda^2_{oy} = \lambda^2_y + 13.5\,\Sigma E_{sc}\,A_{sc}/(E_s\,A_1) \qquad (12c.C2)$$

where A_1 is the area of a hollow steel lacing bar

(C) Three limbs with lacing bars
The slenderness ratio is given by

$$\lambda_{oy}^2 = \lambda_y^2 + 27\,\Sigma \mathrm{E}_{sc}\,\mathrm{A}_{sc}/(\mathrm{E}_s\,\mathrm{A}_1) \qquad (12c.C3)$$

(D) Four limbs with lacing bars
The slenderness ratio is given by

$$\lambda_{oy}^2 = \lambda_y^2 + 6.75\,\Sigma \mathrm{E}_{sc}\,\mathrm{A}_{sc}/(\mathrm{E}_s\,\mathrm{A}_1) \qquad (12c.C4)$$

$$\lambda_{oy}^2 = \lambda_x^2 + 6.75\,\Sigma \mathrm{E}_{sc}\,\mathrm{A}_{sc}/(\mathrm{E}_s\,\mathrm{A}_1) \qquad (12c.C5)$$

where λ_x is the slenderness ratio about x-x axis of member

JAPAN

CHAPTER EDITOR

Minoru Wakabayashi, General Building Research Corporation

AIJ 86

The stiffness follows the superposition principle, used also in design

Slenderness limitations (ENC, FIL)

for columns

$$\ell/b < 30$$

for compression members

$$\ell/b < 50$$

where

b = least lateral dimension of cross-section
ℓ = effective length

NORTH AMERICA

CHAPTER EDITORS

Saif Daresalemwala, Howard Needles Tammen & Bergendoff
Richard W. Furlong, University of Texas

ACI 318 83

Stiffness is defined by

$$\left.\begin{aligned} EI &= E_s(I_s + I_r) + \frac{0.2E_cI_g}{(1 + 1.4M_o/M_u)} \\ r &= \sqrt{\frac{0.2E_cI_g + E_s(I_s + I_r)}{0.2E_cA_g + E_s(A_s + A_r)}} \end{aligned}\right\} \text{(FIL)} \qquad (12c.N1)$$

For ENC, eliminate the terms E_sI_r and E_sA_r from the above two expressions.

where

A_g = gross area of section
A_r = area of longitudinal reinforcing bars
A_s = area of steel shape
E_c = modulus of elasticity of concrete
E_g = modulus of elasticity of gross section
E_s = modulus of elasticity of steel
EI = equivalent stiffness of composite section
I_r = moment of inertia of longitudinal reinforcing bars
I_s = moment of inertia of steel shape
r = radius of gyration of cross-section
M_o = smaller factored end moment
M_u = larger factored end moment

AISC LRFD 86

Stiffness is defined by

$$E_m = E_s + 0.4E_cA_c/A_s \quad \text{(FIL)} \qquad (12c.N2)$$
$$E_m = E_s + 0.2\, E_cA_c/A_s \quad \text{(ENC)} \qquad (12c.N3)$$

where

A_c = area of concrete
A_s = area of steel
E_c = modulus of elasticity of concrete
E_s = modulus of elasticity of steel
EI = $E_m\, I_s$, equivalent stiffness of composite section

CAN 3 A23.3 M84

The stiffness is determined from

$$EI = \frac{0.2\,E_c I_g}{1+\beta_d} + E_s I_t \tag{12c.N4}$$

using the moment magnification concept

$$r = \sqrt{\frac{(E_c I_g/5) + E_s I_t}{(E_c A_g/5) + E_s A_t}} \tag{12c.N5}$$

where

- A_g = gross area of section
- E_c = modulus of elasticity of concrete
- E_s = modulus of elasticity of reinforcement
- EI = flexural stiffness of compression member
- I_g = moment of inertia of gross concrete section neglecting reinforcement
- I_t = moment of inertia of steel shape
- r = radius of gyration of the cross section
- β_d = absolute value of ratio of maximum factored dead load moment to the maximum factored total load

WEST EUROPE

CHAPTER EDITORS
Reinhard Bergmann, Bochum University
Patrick J. Dowling, Imperial College
Amr S. Elnashai, Imperial College
Karlheinz Roik, Bochum University

BS 5400 P.5 79

Slenderness Limits

for ENC $\ell_e/b < 30$

for FIL (CHS) $\ell_e/b < 55$

for FIL (RHS) $\ell_e/b < 65$

$$\lambda = \frac{\ell_e}{\ell_E} \tag{12c.W1}$$

$$\ell_E = \pi \sqrt{\frac{E_c I_c + E_s I_s + E_r I_r}{N_u}} \tag{12c.W2}$$

where

- b = least lateral dimension
- E_c = modulus of elasticity of concrete
- E_s = modulus of elasticity of steel shape
- E_r = modulus of elasticity of longitudinal reinforcing bars
- I_c = moment of inertia of concrete
- I_s = moment of inertia of steel shape
- I_r = moment of inertia of longitudinal reinforcing bars
- ℓ_E = effective length
- N_u = squash load
- λ = slenderness factor

DIN 18806 P.1

Effective stiffness

$$EI_W = E_a I_a + E_{bi} I_b + E_s I_s \tag{12c.W3}$$

where E_{bi} is equal to $500\,\beta_{WN}$

Slenderness

$$\overline{\lambda} = \sqrt{\frac{N_{pl}}{N_{ki}}} \tag{12c.W4}$$

where

B_{wn} = 28-day cube strength of concrete
E_a = modulus of elasticity of steel shape
E_{bi} = modulus of elasticity of concrete
E_s = modulus of elasticity of longitudinal reinforcing bars
I_a = moment of inertia of steel shape
I_b = moment of inertia of concrete
I_s = moment of inertia of longitudinal reinforcing bars

N_{ki} $= \frac{\pi^2}{S_k^2}(EI)_W$

N_{pl} = squash load

S_k = effective length

$\overline{\lambda}$ ≤ 2.0

EC 4 85

If redistribution of moments arising from plasticity is assumed in the analysis of the structure, the flexural stiffness of any restraining member, which is loaded to more than 90% of its plastic moment capacity at any cross-section should be taken as zero.

Short-term elastic flexural stiffness

$$(EI)_e = E_a\, I_a + E_{ce}\, I_c + E_s\, I_s \tag{12c.W5}$$

$$E_{ce} = 600\ f_{ck} \tag{12c.W6}$$

If $\overline{\lambda}$ exceeds the values of Table 12c.W1 or $e/d < 2$, long-term stiffness $E_{c\infty}$ should be used instead of E_{ce}.

where $E_{c\infty}$ is equal to $E_{ce}\,(1 - N_{Gd}/2\,N_d)$

Relative slenderness $\overline{\lambda}$

$$N_{cr} = \pi^2 (EI)_e / \ell_e^2 \tag{12c.W7}$$

$$\overline{\lambda} = \sqrt{(N_{pl}/N_{cr}} \tag{12c.W8}$$

Table 12c.W1 Limiting Values of $\overline{\lambda}$

	Braced frames	Sway frames
Concrete-encased sections	$\overline{\lambda} = 0.8$	$\overline{\lambda} = 0.5$
Concrete-filled tubes	$\overline{\lambda} = 0.8/(1-\delta)$	$\overline{\lambda} = 0.5/(1-\delta)$

For Design by Method 1

$\bar{\lambda} < 2$

For Design by Method 2

Short column is defined when $\ell_{ey}/h_c < 12$ and $\ell_{ez}/b_c < 12$

where

b_c = lateral dimension about minor axis
E_a = modulus of elasticity of steel shape
E_{ce} = modulus of elasticity of concrete
E_s = modulus of elasticity of longitudinal reinforcing bars
e = eccentricity of load
h_c = lateral dimension about major axis
I_a = moment of inertia of steel shape
I_c = moment of inertia of concrete
I_s = moment of inertia of longitudinal reinforcing bars
ℓ_{ey} = effective length about major axis
ℓ_{ez} = effective length about minor axis
N_{Gd} = permanent load
N_d = total load
N_{pl} = squash load
δ = steel contribution to the fully plastic normal face

INTERNATIONAL COMPARISONS & COMMENT

WHAT ARE THE DIFFERENCES?

Table 12c.1 below gives recommended values for concrete modulus of elasticity.

For section stiffness, the values are used in the following equation for equivalent stiffness

$$E_{eq} = a_1 E_c I_c + E_s I_s + a_2 E_r I_r \tag{12c.1}$$

where a_1 and a_2 are factors given in Table 12c.2 below.

For the example used in Sections 12a and 12b, the values of E_c and $(EI)_{eq}$ given by various codes are given in Table 12c.2.

Comparison of the effect of slenderness consideration from various codes is implicitly included in the comparison presented in Sections 12a and 12b.

Table 12c.1 Recommened Values for Concrete Modulus of Elasticity

Code	E_c	a_1	a_2
BS 5400	$450\ F_{cu}$	1	1
EC 4	$600\ F_{cu}$	1	1
DIN	$500\ F_{cu}{}^{*}$	1	1
ACI (ENC)	$0.043\sqrt{(W_c^3\ F_{cyl})}$	0.2	0
ACI (FIL)		0.4	0

* reduced by $\left(\dfrac{1}{1 - 0.5N_s/N_u}\right)$

Table 12c.2 Recommended Values for Equivalent Stiffness

Code	E_c (in MPa)	$(EI)_{eq}$
EC4	21000	102×10^6 $(126 \times 10^6)^*$
BS5400	18000	113×10^6
ACI	28200	cracked: 48×10^6 uncracked: 88×10^6 $(163 \times 10^6)^*$
AISC	25200	99×10^6

* value for $(EI)_{eq}$ if no account is taken of creep factor

WHY ARE THERE DIFFERENCES?

The variations in Young's modulus for concrete and section equivalent stiffness (EI) are due to one or both of the following

Inclusion of a creep factor for E_c.
Use of a cracked section, with or without limitation on radius of gyration.

The BS 5400 value is deliberately low to account for creep and the use of an uncracked concrete section. The value recommended by EC 4 is slightly higher, since creep is considered separately. The ACI and AISC codes give significantly higher values, since a cracked section is used in the analysis. Also, the ACI equivalently stiffness expression ignores the reinforcement steel contribution, which is included in the other codes.

Symbol List for Chapter 12 -- Composite Members

A = area of cross section (12a.J)
A_a = area of steel profile (12a.W)
A_b = area of concrete in the cross section (12a.W)
A_c = area of concrete in the cross section (12a.E),(12a.N),(12a.W),(12b.N)&(12c.N)
A_e = effective area of cross section (12a.J)
A_f = area of the top flange of the steel section (12b.N)
A_g = gross area of section (12a.N) & (12c.N)
A_r = area of longitudinal reinforcing bars (12a.N), (12a.W) (12b.N) & (12c.N)
A_s = area of steel shape (12a.E), (12a.N), (12a.W) (12b.N) & (12c.N)
A_{sc} = total area of member (12a.C) & (12c.C)
A_{st} = total area of longitudinal reinforcing bars (12a.N)
A_t = area of steel shape (12a.N)
A_1 = area of a hollow steel lacing bar (12c.C)
${}_cA$ = area of cross section of concrete (12a.J) & (12b.J)
${}_mA_t$ = area of cross section of main tensile reinforcement (12a.J)
${}_sA$ = area of cross section of steel (12a.J) & (12b.J)
a = depth of rectangular stress block from compression face (12b.N)
a_c = concrete contribution factor (12b.N)
a_1 = coefficient used to determine equivalent stiffness (12c.1)
a_2 = coefficient used to determine equivalent stiffness (12c.1)
${}_ma_t$ = area of main reinforcement (12b.J)
${}_sa_f$ = area of steel profile flange (12b.J)
${}_sa_w$ = area of steel profile web (12b.J)
B = width of cross section (12a.J)
B_{wn} = 28 day cube strength of concrete (12c.W)
b = width of cross section (12a.J)
b = least lateral dimension of cross-section (12a.W), (12b.N), (12c.J)&(12c.W)
b_c = lateral dimension about minor axis (12c.W)
b_f = breadth of steel flange of I-section or the external dimension of a rectangular hollow section (12b.N)
b_s = external dimension of the wall of the RHS (12b.N)
C_m = coefficient used to consider the effect of varying moment along the length of member (12a.A)
C_1 = numerical coefficients (12a.N)
C_2 = numerical coefficients (12a.N)
C_3 = numerical coefficients (12a.N)
c = compressive strength (12a.J)
D = diameter of tube (12a.E)
D = depth of the cross-section in the plane of bending (12a.J) & (12c.A)
D_e = outside diameter of the CHS (12b.N)
${}_cD$ = depth of concrete cross section (12b.J)
${}_eD$ = effective depth of cross section (12b.J)
d = diameter of concrete core (12a.E)
d = depth of cross section (12a.J)
d = greatest diameter of tube (12a.W)
d_c = distance of neutral axis from the more compressed face of concrete (12b.N)

d_r = distance between symmetrically placed reinforcing bars measured perpendicular to the axis of bending (12b.N)
d_s = thickness of concrete cover to encased steel section (12b.N)
d_w = depth of steel web in compression (12b.N)
${}_m d$ = depth of main reinforcement (12b.J)
${}_r d_{c1}$ = depth of the most extremely compressive reinforced concrete (12b.J)
${}_r d_{t1}$ = depth of the most extremely tensile reinforced concrete (12b.J)
E = modulus of elasticity (12a.N) & (12b.J)
E_a = modulus of elasticity of steel shape (12c.W)
E_{bi} = modulus of elasticity of concrete (12c.W)
E_c = modulus of elasticity of concrete (12a.N), (12c.N), & (12c.W)
E_{ce} = modulus of elasticity of concrete (12c.W)
E_g = modulus of elasticity of gross section (12c.N)
E_m = equivalent steel modulus of composite section (12a.N) & (12c.N)
E_r = modulus of elasticity of longitudinal reinforcing bars (12c.W)
E_s = modulus of elasticity of steel (12a.E), (12b.N), (12c.C), (12c.N)&(12c.W)
E_{sc} = modulus of composite material (12c.C)
${}_c E$ = elastic modulus of concrete (12b.J)
e = eccentric distance of compressive load (12a.E), (12a.W), (12b.C), (12b.N)&(12c.W)
e_y = eccentric distance of load about y-axis (12b.W)
e_z = eccentric distance of load about z-axis (12b.W)
F_c = concrete strength (12a.J)
F_{cr} = buckling load of equivalent section (12a.N)
F_{cu} = 28-day cube characteristic compressive strength (12c.1)
F_{cyl} = concrete cylinder strength (12c.1)
F_e = effective strength (12b.J)
F_{ym} = modified yield stress for the composite section (12a.N)
f_c = concrete cylinder strength (12a.J) & (12a.N)
f_c = axial compressive strength of concrete (12c.C)

f'_c = modified allowable concrete stress for tubes (12a.J)
f_{cc} = strength of triaxially contained concrete (12a.W) & (12b.N)
f_{ck} = characteristic cylinder strength of concrete (12a.W)
f_{ck} = effective compressive allowable stress of concrete (12c.W)
f_{cu} = 28 day cube characteristic compressive strength (12a.W), (12b.N)&(12c.C)

f'_e = effective allowable stress (12b.J)
f_{ry} = yield strength of longitudinal reinforcing bars (12a.W) & (12b.N)
f_{sc} = design compression strength of composite material (12a.C)
f_{sk} = yield strength of longitudinal reinforcement bars (12a.W)
f_y = yield strength of steel (12a.N), (12a.W), (12b.N) & (12c.C)

f'_y = reduced nominal yield strength of steel casing (12b.N)
f_{yr} = yield stress of longitudinal reinforcing bars (12a.N)

f'_c = cylinder strength of concrete (12a.N) & (12b.N)
${}_m f_c$ = allowable compressive stress of main reinforcement (12a.J) & (12b.J)
${}_m f_t$ = allowable tensile stress of main reinforcement (12b.J)
${}_s f_c$ = allowable compression stress of steel (12a.J) & (12b.J)
${}_s f_e$ = allowable effective stress of steel (12b.J)
${}_s f_t$ = allowable tensile stress of steel (12b.J)

H = greater lateral dimension in inches (12b.N)
h = greatest lateral dimension of cross-section (12a.W) & (12b.N)
h = height (12b.J)
h_c = lateral dimension about major axis (12c.W)
I = moment of inertia (12b.J)
I_a = moment of inertia of steel shape (12c.W)
I_b = moment of inertia of concrete (12c.W)
I_c = second moment of area of the uncracked cross-section, the steel section, and the reinforcement, respectively above the axis of the composition column (12b.N)
I_g = moment of inertia of gross concrete section neglecting reinforcment (12c.N)
I_r = second moment of area of the uncracked cross-section, the steel section, and the reinforcement, respectively above the axis of the composition column (12b.N)
I_r = moment of inertia of longitudinal reinforcing bars (12c.N) & (12c.W)
I_s = second moment of area of the uncracked cross-section, the steel section, and the reinforcement, respectively above the axis of the composition column (12b.N)
I_s = moment of inertia of steel shape (12c.N) & (12c.W)
I_t = moment of inertia of steel shape (12c.N)
${}_cI$ = moment of inertia of concrete (12b.J)
j = internal lever arm (12b.J)
K = ratio of design compressive force to squash load (12b.W)
K_d = ratio of total load to squash load (12b.W)
K_{1x} = coefficient of bending to account for slenderness about x-axis (12b.W)
K_{1y} = coefficient to account for slenderness about y-axis of bending (12b.W)
K_{2x} = coefficient to account for slenderness about y-axis of bending (12b.W)
K_{2y} = coefficient to account for slenderness about y-axis of bending (12b.W)
K_3 = coefficient to account for slenderness (12b.W)
k_n = coefficient to account for moment gradient (12b.W)
k_1 = coefficient used to determine the axial strength (12b.W)
k_{1y} = coefficient determined from Appendix C of BS 5400 depending on the type of steel section used (12a.W)
k_{1z} = coefficient determined from Appendix C of BS 5400 depending on the type of steel section used (12b.W)
k_2 = coefficient used to determine the axial strength of major axis bending (12b.W)
k_3 = coefficient used to determine the axial strength of minor axis bending (12b.W)
k_4 = coefficient used to determine axial strength of composite members (12b.W)
k_5 = coefficient used to determine axial strength of composite members (12b.W)
L = length of column (12a.E)
L_e = effective length of column (12a.A)
ℓ = effective length (12b.J), (12b.N) & (12c.J)
ℓ_E = length of column for which Euler load equals the squash load (12b.N)&(12c.W)
ℓ_e = effective length of the actual column in the plane of bending considered (12b.N) & (12c.W)
ℓ_{ey} = effective length about major axis (12c.W)
ℓ_{ez} = effective length about minor axis (12c.W)

ℓ_k = effective critical length (12b.J)
ℓ_x = effective length about the major axis (12a.W)
ℓ_y = effective length about the minor axis (12a.W)
M = bending moment (12b.J)
M_{nx} = nominal flexural strength using plastic stress distribution about major axis (12b.N)
M_{ny} = nominal flexural strength using plastic stress distribution about minor axis (12b.N)
M_o = smaller factored end moment (12c.N)
M_{pc} = plastic moment of the composite cross section with axial force (12b.W)
M_{pcy} = plastic moment of the composite cross section with axial force about y-axis (12b.W)
M_{pcz} = plastic moment of the composite cross section with axial force about z-axis (12b.W)
M_{pl} = plastic moment of the composite cross-section (12b.N)
M_{ply} = plastic moment of the composite cross section about y-axis (12b.W)
M_{plz} = plastic moment of the composite cross section about z-axis (12b.W)
M_u = larger factored end moment (12c.N)
M_u = ultimate bending strength (12b.J)
M_{ub} = moment capacity (12c.A)
M_{ux} = design moment about major axis (12b.N)
M_{uy} = design moment about minor axis (12b.N)
M_x = bending moment about x-axis (12b.J)
M_y = bending moment about y-axis (12b.J) & (12b.W)
M_z = applied moment about z-axis (12b.W)
${}_cM$ = bending strength of concrete (12b.J)
${}_cM_u$ = ultimate bending strength of concrete (12b.J)
${}_cM_x$ = bending strength of concrete about x-axis (12b.J)
${}_cM_y$ = bending strength of concrete about y-axis (12b.J)
${}_eM$ = effective bending moment (12b.J)
${}_mM_u$ = ultimate bending strength of main reinforcement (12b.J)
${}_mM_{uo}$ = ultimate bending strength of main reinforcement at zero axial force (12b.J)
${}_rM$ = bending strength of reinforced concrete (12b.J)
${}_rM_o$ = bending strength of reinforced concrete at zero axial force (12b.J)
${}_rM_u$ = ultimate bending strength of reinforced concrete (12b.J)
${}_rM_x$ = bending strength of reinforced concrete about x-axis (12b.J)
${}_rM_y$ = bending strength of reinforced concrete about y-axis (12b.J)
${}_sM$ = bending strength of steel (12b.J)
${}_sM_o$ = bending strength of steel at zero axial force (12b.J)
${}_sM_u$ = ultimate bending strength of steel (12b.J)
${}_sM_{uo}$ = ultimate bending strength of steel at zero axial force (12b.J)
${}_sM_x$ = bending strength of steel about x-axis (12b.J)
${}_sM_y$ = bending strength of steel about y-axis (12b.J)
N = applied axial load (12a.A) & (12a.J)
N_{ax} = short column squash load about x-axis (12b.W)
N_{ay} = short column squash load (12a.W) & (12b.N)
N_c = Euler buckling load (12a.A)
N_{cr} = Euler load of effective section (12c.W)
N_d = axial design load (12a.E)

N_d = total load (12c.W)
N_E = Euler load of effective section (12a.A)
N_{ex} = Euler load of effective section (12b.C)
N_{Gd} = permanent load (12c.W)
N_k = citical axial load of column (12b.J)
N_{ki} = Euler load of effective section (12c.W)
N_{kr} = axial design strength (12a.W)
N_n = nominal compressive strength (12b.N)
N_o = axial design capacity (12a.N)
N_{pl} = squash load of the column (12a.W) & (12c.W)
N_u = squash load (12b.N) & (12c.W)
N_u = ultimate axial strength (12a.J) & (12a.W)
N_{uy} = axial strength of major axis bending composite members (12b.W)
N_{uyz} = axial strength of biaxial bending composite members (12b.W)
N_{uz} = axial strength of minor axis bending composite members (12b.W)
N_{xy} = ultimate strength of column in biaxial bending (12b.W)
N_y = axial yield load of column (12b.W)
${}_cN_c$ = axial compressive strength of concrete (12a.J) & (12b.J)
${}_cN_{cu}$ = ultimate axial compressive strength of concrete (12b.J)
${}_cN_k$ = critical axial load of concrete (12b.J)
${}_cN_u$ = ultimate axial compressive strength of concrete (12a.J)
${}_eN$ = effective axial force (12b.J)
${}_mN_u$ = ultimate axial compressive strength of main reinforcement (12a.J)&(12b.J)
${}_rN$ = axial strength of reinforced concrete (12b.J)
${}_rN_c$ = axial compression strength of reinforced concrete (12a.J)
${}_rN_{cu}$ = ultimate axial compressive strength of reinforced concrete (12a.J)&(12b.J)
${}_rN_{c1}$ = axial compressive strength of concrete (12b.J)
${}_rN_{c2}$ = axial compressive strength of main reinforcement (12b.J)
${}_rN_k$ = critical axial load of reinforced concrete (12b.J)
${}_rN_t$ = axial tensile strength of reinforced concrete (12b.J)
${}_rN_{tu}$ = ultimate axial tensile strength of reinforced concrete (12b.J)
${}_rN_u$ = ultimate axial strength of reinforced concrete (12b.J)
${}_sN$ = axial strength of steel (12a.J) & (12b.J)
${}_sN_c$ = axial compressive strength of steel (12b.J)
${}_sN_t$ = axial tensile strength of steel (12b.J)
${}_sN_u$ = ultimate axial strength of steel (12a.J) & (12b.J)
n = modular ratio (12a.J)
P_o = factored axial load resistance at zero eccentricity (12a.N)
P_r = design axial load (12a.N)
${}_mP_t$ = main tensile reinforcement ratio (12b.J)
${}_sP_c$ = area ratio of the compression flange of the steel shape to the gross section (12a.J) & (12b.J)
p = reinforcement ratio, steel ratio (12b.J)
${}_sp_c$ = area ratio of the compression flange of the steel shape to the gross section (12b.J)
R_{cd} = design strength of concrete (12a.E)
R_{cy} = yield stress of concrete (12a.E)
R_{sd} = design strength of steel (12a.E)
R_{sy} = yield stress of steel (12a.E)
r = radius of steel tube (outside) (12a.C)

r = reduction factor (12a.J)
r = radius of gyration of cross-section (12c.N)
r_m = modified radius of gyration for composite section (12a.N)
${}_c r_u$ = ultimate strength reduction factor for concrete (12a.J) & (12b.J)
${}_e r_u$ = reduction factor of effective strength (12b.J)
S = tube thickness (12a.W)
S_y = section modulus about y-axis (12b.W)
S_z = section modulus about z-axis (12b.W)
Sk = effective length (12c.W)
t = thickness of tube (12a.E) & (12c.C)
t = thickness of steel casing (12b.N)
t = plate thickness (12b.J)
t_f = average thickness of the flange of a steel section (12b.N)
t_w = thickness of web at steel section (12b.N)
W_{sc} = section modulus of member (12b.C)
$\mathfrak{X}$ = reduction coefficient from the relevant buckling curve (12a.W)
x_n = location of neutral axis (12b.J)
x_n = modified reduction coefficient, given by Eq. 12b.W19 (12b.W)
x_{n1} = location of neutral axis (12b.J)
y_c = depth of neutral axis from compression face (12b.N)
Z = plastic modulus (12b.J)
Z_s = pastic section modulus of steel (12b.J)
${}_s Z_p$ = plastic section modulus of steel (12b.J)
α = ratio of steel area to concrete area (12c.C)
α = coefficient used to determine the material factor for steel (12a.E)
${}_m \alpha_c$ = area of main reinforcement in compression (12b.J)
${}_m \alpha_t$ = area of main reinforcement in tension (12b.J)
${}_s \alpha_w$ = area of steel web (12b.J)
β = ratio of end moments (12a.A)
β_d = absolute value of ratio of maximum factored dead (12c.N)
β_{mx} = equivalent moment factor (12b.C)
β_R = design value of concrete strength (12a.W)
β_{RL} = coefficient used to determine the squash load of column (12a.W)
$\beta_{S,a}$ = yield strength of steel profile (12a.W)
β_{SL} = coefficient used to determine the squash load of column (12a.W)
$\beta_{s,s}$ = yield strength of longitudinal reinforcing bars (12a.W)
β_{wn} = 28 day cube strength of concrete (12a.W)
γ_a = material factor of safety for steel (12a.W)
γ_c = material factor of safety for concrete (12a.W)
γ_s = material factor of safety for steel (12a.E) & (12a.W)
δ = magnification factor (12b.J)
δ = steel contribution to the fully plastic normal face (12c.W)
δ_b = moment magnification factor (12a.A)
${}_r \delta$ = magnification factor of reinforced concrete (12b.J)
ϵ_{cy} = yield strain of concrete (12a.E)
ϵ_{sy} = yield strain of steel (12a.E)
η_1 = coefficient used to determine squash load of columns (12a.W)
η_2 = coefficient used to determine squash load of columns (12a.W)
η_{10} = coefficient used to determine the coefficient η_1 (12a.W)
η_{20} = coefficient used to determine the coefficient η_2 (12a.W)

θ_n = coefficient used to determine allowable flexural strength of filled concrete (12b.J)
λ = slenderness factor (12c.W)
λ = slenderness ratio (12a.E)
$\overline{\lambda}$ = slenderness ratio (12a.W)
$\overline{\lambda}$ = relative slenderness, defined by Eq. 12c.W8 (12c.W)
λ_c = the slenderness of a column that would collapse under the design axial load with no bending (12a.A)
λ_c = slenderness ratio (12a.N)
λ_{max} = maximum slenderness (12a.W) & (12b.W)
λ_{oy} = equivalent slenderness ratio of latticed members (12c.C)
λ_x = the slenderness ratio about x-x axis of member (12c.C)
λ_y = slenderness ratio about y-y axis of the member (12c.C)
λ_1 = slenderness ratio of main components in one panel (12c.C)
μ = ratio of steel area to conrete area (12a.E)
ν = ratio of yield strain of concrete to yield strain of steel (12a.E)
ν = factor of safety (12b.J)
${}_c\nu$ = factor of safety of concrete (12b.J)
${}_r\nu$ = factor of safety of reinforcement (12b.J)
${}_s\nu$ = factor of safety of steel (12b.J)
ρ = ratio of the average compressive stress in the concrete at failure to the design yield strength of the steel taken as $0.4\ f_{cu}/0.91F_y$ (12b.N)
ρ = load eccentricity ratio (12a.E)
σ = axial stress (12a.J)
σ_e = Euler critical stress (12a.E)
${}_m\sigma_y$ = yield stress of main reinforcement (12a.J) & (12b.J)
${}_s\sigma_y$ = yield stress of steel (12a.J)
ϕ = strength reduction factor (12a.N)
φ = strength reduction factor (12a.C)
φ = column curve factor (12a.E)
φ_b = capacity (resistance) factor for bending (12b.N)
φ_c = creep factor (12a.E)
φ_c = capacity (resistance) factor for compression (12b.N)
χ = reduction coefficient obtained from buckling curves (12a.W)
χ_d = ratio of design strength of concrete to design strength of steel (12a.E)
χ_y = ratio of yield stress of concrete to yield stress of steel (12a.E)
Ψ = moment distribution factor (12b.N)
ψ = ratio of the lesser to the greater end moment (12b.W)

Stability under Seismic Loading

TOPICAL EDITOR
Franklin Y. Cheng, University of Missouri-Rolla

CHAPTER EDITORS
J. W. Butterworth, The University of Auckland
Shao-Fan Chen, Xian Institute of Metallurgy & Construction Engineering
Franklin Y. Cheng, University of Missouri-Rolla
Emilia Juhásová, Slovak Academy of Sciences
Federico M. Massolani, Universita Di Napoli
Sheng-Taur Mau, University of Houston
Koichi Takanashi, The University of Tokyo
Nicholas S. Trahair, The University of Sydney

SPECIFICATION PROVISIONS & EXPLANATION

AUSTRALIA & NEW ZEALAND

CHAPTER EDITORS
Nicholas S. Trahair, The University of Sydney
J. W. Butterworth, The University of Auckland

AS 1250 81

A structure and its elements are designed for a total horizontal earthquake force on the structure and a horizontal earthquake force on individual elements. The magnitude of these horizontal forces depends on the earthquake zone (Zones A, 1 and 2 in order of increasing earthquake severity) and the specific structural system used to resist the forces (expressed in terms of a horizontal force factor K). Some typical values are listed below.

Structural System	K-value
Ductile moment resisting	0.67
Moment resisting	1.00
Dual bracing system	0.80
Braced	1.00
Incomplete braced	1.33
Free standing tank	2.50
Free standing pendulum type	3.50

In Seismic Zone A (much of Australia is in this Zone), ductile structures need not be designed to resist earthquake forces. The ductility must be provided by continuity, bracing or shear walls, yield stress not exceeding 350 MPa, attention to welding requirements, and connection details, and excess capacity in diagonal braces.

In Seismic Zone 1 the same ductility requirements apply but the structure must be designed to resist the earthquake forces generated in this Zone.

In Seismic Zone 2 there are many additional requirements to Seismic Zone 1 concerning plastic hinge capacity, full moment connections, ultimate-to-yield strength ratio, non-destructive examination of connections, the slenderness of members in ductile moment-resisting frames, and the capacity of framing elements to accommodate additional moments resulting from amplified distortions of the frame.

Story drift calculated from the application of the unfactored seismic forces is not to exceed 0.005 times the story height unless it can be shown that greater drift can be tolerated.

Where sections of a structure are separated, the distance must be sufficient to avoid contact under deflection caused by seismic forces. The deflection can be calculated by either a dynamic analysis or a static analysis with magnified static deflections.

NZS 3404 89, NZS 4203 84, DZ 4203 89

General Comments

Design is governed principally by a materials code NZS 3404 89 and a loadings code NZS 4203 84 (currently under revision to limit states format DZ 4203 89) using a strength method design philosophy. Analysis is usually elastic with a requirement for consideration of nonlinear effects where necessary. Inclusion of inelastic effects in the analysis may be introduced as a result of the revision referred to above. Plastic analysis is permitted only in restricted circumstances. The revised loadings code will introduce serviceability limit state and in addition to the normal strength limit state, a severe seismic limit state. Members will be proportioned to remain nominally elastic under the serviceability limit state loadings with a limitation on inter-story deflection of 0.0035 intended to provide protection against damage to non-structural components.

The severe seismic limit state will require dynamic analysis unless the structure satisfies specified horizontal regularity and torsional stability requirements. The magnitudes of the response of the structure to the seismic loading will depend on the ductility of the structure (see below) and the members will be proportioned to have reliable strengths adequate to resist the resulting actions. In addition, capacity design is required for all category I and most category II structures (in capacity design, energy dissipating elements of mechanisms are chosen and designed and detailed while all other structural elements are provided with sufficient reserve strength to ensure that the chosen energy dissipating mechanisms are maintained throughout the deformations that may occur in a severe seismic event). Capacity design is not required if the structure is designed as a system elastically responding to high intensities of seismic loading.

Steel structures are classified into one of three categories depending on ductility:

I Fully Ductile (capable of sustaining global displacement ductility demands sufficient to strain plastic hinges in the primary seismic resisting elements into the strain hardening range).

II Limited Ductile (capable of sustaining global displacement ductility demands sufficient to form plastic hinges in the primary seismic resisting elements).

III Nominally or fully elastic.

Members are similarly classified into three categories according to their ability to sustain the ductility demands specified above.

The main stability requirements associated with seismic design are related to the maintenance of stable energy dissipating mechanisms under conditions of large inelastic strains. The additional requirements relating to cross section stability, member stability and overall structural stability (P-Δ effects) are outlined in the following paragraphs.

To control local buckling of sections under severe seismic deformation conditions, limitations on section geometry as shown in the following table are specified in NZS 3404 89. The background to these limits is discussed by Walpole and Butcher, 1985 and in the commentary to NZS 3404 89.

Lateral Bracing Requirements

Overall stability of flexural and compressive members which may be required to sustain the fully plastic state with inelastic rotations into the strain hardening

Table SL.A1 Section Geometry and Local Buckling Requirements

	Category I	Category II	Category IIIA	Category IIIB
Flanges & plates in compression with one unstiffened edge	$\frac{b_1\sqrt{F_y}}{T} \leq 136$	$\frac{b_1\sqrt{F_y}}{T} \leq 136$	$\frac{b_1\sqrt{F_y}}{T} \leq 156$	$\frac{b_1\sqrt{F_y}}{T} \leq 256$
Flanges of welded box sections in compression	$\frac{b_2\sqrt{F_y}}{T} \leq 500$	$\frac{b_2\sqrt{F_y}}{T} \leq 512$	$\frac{b_2\sqrt{F_y}}{T} \leq 560$	$\frac{b_2\sqrt{F_y}}{T} \leq 560$
Flanges of rect-angular hollow sections in comp.	$\frac{b_2\sqrt{F_y}}{T} \leq 350$	$\frac{b_2\sqrt{F_y}}{T} \leq 420$	$\frac{b_2\sqrt{F_y}}{T} \leq 635$	$\frac{b_2\sqrt{F_y}}{T} \leq 635$
Webs under flexural compression	$\frac{d_1\sqrt{F_y}}{t} \leq 1120$	$\frac{d_1\sqrt{F_y}}{t} \leq 1120$	$\frac{d_1\sqrt{F_y}}{t} \leq 1600$	$\frac{d_1\sqrt{F_y}}{t} \leq 2550$
Webs under uniform axial compression	$\frac{d_1\sqrt{F_y}}{t} \leq 512$	$\frac{d_1\sqrt{F_y}}{t} \leq 512$	$\frac{d_1\sqrt{F_y}}{t} \leq 635$	$\frac{d_1\sqrt{F_y}}{t} \leq 635$

where

d, t, T = dimensions of the investigated element

F_y = yield stress

regime is assured by the lateral bracing requirements set out in the following table. The background of these values is discussed by Walpole and Butcher (1985) and by Butterworth and Spring (1985).

Axial Load Limitations on Columns and Braces

To prevent the formation of plastic hinges at unrestrained interior points of columns and to permit adequate plastic hinge rotation, the following limitations on axial load and slenderness are specified:

for category I columns $$\frac{P_u}{P_y} \leq 0.5 \text{ and } \leq \frac{1+\beta-\lambda}{1+\beta+\lambda}$$

for category II columns $$\frac{P_u}{P_y} \leq 0.7 \text{ and } \leq \frac{1+\beta-\lambda}{1+\beta+\lambda}$$

for category III columns $$\frac{P_u}{P_y} \leq 1.0$$

where

- L = actual length
- P_u = design axial force
- P_y = yielding axial force
- r = radius of gyration about the same axis as the applied moment
- β = ratio of end moments
- $\lambda = \frac{L}{\pi r}\sqrt{\frac{F_y}{E}}$

P-Δ Effects and Lateral Deflection Limits

P-Δ effects and lateral deflection limits are being further revised in DZ 4203 89 and final recommendations are not yet available (1990). Standard NZS 3404 89

Table SL.A2 Lateral Bracing Requirements

	Yielding regions of category I members		Yielding regions of category II members	
Flange length where $M_u > C_1 M_r$	$> \frac{480 r_y}{\sqrt{F_y}}$	$\leq \frac{480 r_y}{\sqrt{F_y}}$	$> \frac{640 r_y}{\sqrt{F_y}}$	$\leq \frac{640 r_y}{\sqrt{F_y}}$
Spacing of braces within length where $M_u > C_1 M_r$	$\leq \frac{480 r_y}{\sqrt{F_y}}$	one brace required	$< \frac{640 r_y}{\sqrt{F_y}}$	one brace required
Spacing to brace or eff. length adj. to length where $M_u > C_1 M_r$	$\leq \frac{720 r_y}{\sqrt{F_y}}$	$\leq \frac{720 r_y}{\sqrt{F_y}}$	$\leq \frac{960 r_y}{\sqrt{F_y}}$	$\leq \frac{960 r_y}{\sqrt{F_y}}$

where

- $C_1 = 0.85$ for $\frac{P_u}{P_y} \leq 0.15$
- $C_1 = 0.75$ for $\frac{P_u}{P_y} > 0.15$
- M_u = design moment
- r_y = radius of gyration about the y-axis

allows P-Δ moments to be ignored if interstory deflections, calculated by scaling equivalent static of modal analysis values by the structure ductility factor, are less than 0.01 of the story height. For structures located outside the most severe seismic zone the interstory deflection limits are either 0.0083 or 0.0067 (depending on zone) but these limits may be exceeded up to the 0.01 level provided the corresponding P-Δ moments in the columns do not exceed 5%, 8%, or 11% (for categories I, II, and III) of the flexural or shear strength of the beam or beams stabilizing the column. Panel zone distortion is included in the lateral deflection calculation as is the contribution of column axial deformation where this exceeds 10%.

The objectives of the lateral deflection limits under severe seismic loading are two-fold:

(a) To limit potential energy loss and the member forces generated by P-Δ effects to acceptable limits.
(b) To avoid unnecessary secondary damage under both severe and moderate seismic loading.

Shortcomings in the present methods of calculating lateral deflection, such as scaling deflections derived from equivalent static or modal analyses are recognized. The deflected shape in the presence of inelastic deformations is unlikely to be well approximated by a scaled elastic or first mode shape, and there is also good reason to believe that P-Δ effects are more significant in structures of high ductility than in those proportioned to respond elastically or in a limited ductile manner.

The use of inelastic time-history dynamic analysis with a suitable range of three or more earthquake records is permitted.

Concentrically Braced Frames

Due to the particular problems associated with this structural form under seismic loading, limitations are placed on the maximum height or number of stories. These range from 10 stories or 45m for an X-braced frame with brace slenderness < 40 down to 2 stories for an inverted V-braced (chevron) frame with brace slenderness > 80. Braced frames employing braces which are effective only in tension may not exceed two stories without special study due to the high probability of unstable soft story mechanisms.

Brace slenderness is defined as $\frac{k\,L}{r}\sqrt{\frac{F_y}{250}}$

where

F_y = yield stress (MPa)
k = buckling length factor
L = member length
r = radius of gyration

Eccentrically Braced Frames

Detailed provisions for the seismic design of eccentrically braced frames are contained in part 2 of standard NZS 3404 89. Stability considerations relate mainly to the stabilizing of the web in the energy dissipating active link mechanisms.

CHINA

CHAPTER EDITOR
Shao-Fan Chen, Xian Institute of Metallurgy & Construction Engineering

GBJ 11 89

Provisions for Single-Story Steel Mill Buildings

The slenderness ratio of columns shall not exceed the following specified values:

Table SL.C1 Maximum Slenderness Ratio of Columns

Steel Grade	$\rho < 0.15$	$\rho \geq 0.15$
$f_y = 235$ MPa	150	$120(1-\rho)$
$f_y = 345$ MPa		$100(1-\rho)$

Note: ρ is the ratio of column compression force N to yield load $N_y = A f_y$

In regions of earthquake intensity 9, the width-thickness ratio of plate elements of columns shall not exceed the following specified values:

Table SL.C2 Maximum Width-Thickness Ratio of Plate Elements

Steel Grade	Edge Restraining Conditions	
	one edge free	two edges restrained
$f_y = 235$ MPa	9.0	43
$f_y = 345$ MPa	7.5	37

Provisions for Steel Bracings of Reinforced Concrete Columns of Single-Story Mill Buildings

The slenderness of bracing members should preferably not exceed the following specified values:

Table SL.C3 Maximum Slenderness Ratio of X-Bracing

Position	Earthquake Intensity			
	6	7	8	9
upper portion of columns	250	250	200	150
lower portion of columns	200	200	150	150

The seismic effect on column x-bracings may be checked according to the following:
(1) The horizontal deflection of column bracing system with slenderness ratio of diagonals not exceeding 200, when subjected to a unit lateral force, may be determined from

$$u = \Sigma_i \frac{1}{1-\varphi_i} \cdot u_{ti} \qquad \text{(SL.C1)}$$

where

- u = deflection at the point of application of the unit force
- u_{ti} = relative deflection of the ith panel subject to a unit lateral force assuming that only the tensile bracing is active
- φ_i = strength reduction factor of a compression member for the ith panel diagonal in accordance with the Specification for the Design of Steel Structures

(2) The cross-sectional area of diagonals whose slenderness ratio does not exceed 200 may be checked as tension members with consideration of the effects of the compression member. The tensile force may then be determined from

$$N_i = \frac{l_i}{(1-\psi_c \varphi_i) S_c} \cdot V_{bi} \qquad \text{(SL.C2)}$$

where

- l_i = full length of the ith panel diagonal
- N_i = design tensile force of the ith panel diagonal
- S_c = clear spacing of columns where the ith panel bracing is located
- V_{bi} = design value of earthquake shear force of the ith panel bracing
- ψ_c = deterioration factor of compression members, taken as 0.7, 0.6 and 0.5 for bracing slenderness ratio 60, 100, and 200, respectively

COMMENTARY

GBJ 11 89 does not cover multi-story steel buildings. As single-story steel mill buildings can withstand heavy earthquake shaking without severe damage, provisions related to stability of steel members in this code are limited.

EAST EUROPE

CHAPTER EDITOR
Emilia Juhásová, Slovak Academy of Sciences

ČSN 73 0036 75 and ČSN 73 0036 89

The primary elements should have appropriate ductility to avoid brittle failure. The columns and girders should be designed to develop plastic hinges in girders first. Tall frame systems should be stiffened either by braces or by stiffening shear walls.

Ductility of steel elements can be secured by the suitable design of section shape in the places of expected plastic hinges and by using steel with higher ductility.

NiP 565/86 87

The structural carrying systems should be designed so that plastic hinges will develop first in horizontal elements and not in columns. In the design of buildings taller than 16 stories, the deformation and overturning analysis should consider the deformations from bending and shear combined with the effect of axial forces in columns with the effect of deformations in the subsoil. The stability control is necessary to be sure that the carrying capacity of individual elements is sufficient.

Indicativ P.100-81 82

The position of stiffening shear walls or braces in steel frame structure should be chosen proportionally in both horizontal directions.

To avoid the concentration of stresses the skew reduction of sections near places of connections is recommended.

In seismic regions with the seismicity $> 8^\circ$, the slenderness of steel elements that influence the stability of a structural system must be limited. The beams in the plane of roof are limited to slenderness $\lambda_{max}=180$; the girders above the openings are limited to slenderness $\lambda_{max}=100$.

Frame structures, namely those with larger distances between columns, should be stiffened either by braces or by stiffening shear walls. When designing the connections the development of plastic deformations must be allowed avoiding high concentration in fillet welds.

Steel must have suitable ductility with ratio of strength to yield stress not less than 1.3 and with ultimate strains of 18%.

SNiP II-7-81 82

In the design of tall structures, it is necessary to account in the column analysis for bending, shear, and axial forces. Also the deformations in subsoil must be considered to make sure that the danger of overturning of the whole structure does not exist.

In the seismic standards the use of direct seismic response analysis in time domain with account of stability effects is at the very beginning. The stability control is considered in basic standards for steel structures.

COMMENTARY

Seismic loading in general acts on a structure in all directions, but the prevailing components are the horizontal with the vertical components. The vertical component is usually about half the horizontal. If the structural system lacks enough stiffening elements in the horizontal direction, there exists the danger of stability loss in vertical slender columns and consequently the whole structure may collapse. During seismic disturbances the columns are loaded by vertical compression force with slight modification from vertical seismic loading and by varying bending and shear from horizontal seismic accelerations. The column failure is usually due to its capacity reaching ultimate strength or by the loss of stability. The stability limit can be either caused by local web buckling or by total sidesway of columns. Local web buckling is usually followed by total sidesway of the whole thin wall steel columns.

It should be mentioned that steel columns have similar behavior for both static and dynamic loading. For seismic loading there is a large dynamic bending which causes some differences between static and dynamic response behavior. In the case of steel thin wall columns, the symmetrical or asymmetrical wave mechanism may develop in local buckling of webs. The asymmetrical wave mechanism appears in slender columns with the changes in stiffness and damping; the natural frequencies are decreasing and the damping is increasing. The result is a total collapse in spite of the damping increases during the vibration in the elasto-plastic region.

The symmetrical wave mechanism can appear in columns with more slender webs where the critical stresses are reached alternately on both edges of a column. When the starting yield strains in webs are reached, they are followed by sudden changes of web shapes with the nearly symmetrical local buckling and the initiation of sidesway. But sidesway is limited at a certain value of story drift. The reason is that the symmetrical buckling waves protect columns against continuing in sidesway as far as the sidesway is connected with the asymmetrical buckling mode. Therefore the successive chain collapse of the whole structure is prevented. From the point of view of structural safety, it is more convenient to use such structural elements in the design and construction because they have larger resistance against the loss of total stability.

The horizontal elements under seismic loading are not influenced by buckling as frequently as the vertical elements. In usual ceiling structure their compression webs or flanges are stiffened and protected against buckling by filling materials such as concrete or other stiffening elements. In open frames the buckling of horizontal elements is more probable from the resulting effect of bending, torsion and compression.

In Central and East Europe, the tectonic seismic effects are prescribed to be considered in the Standards of Bulgaria, Czechoslovakia, Romania, Yugoslavia and USSR.

The aforementioned Standards have a few similar features. First, the response spectra approach is recommended for the seismic design. In such a case the linear behavior is assumed which is modified through a set of coefficients for quasi static load. Next, the computation as well as the stability analysis are realized by application of adequate structural design standards for steel, concrete, or masonry structural systems.

To secure the seismic resistance the Standards also include design rules concerning structural details, story drift limits, eccentricity, slenderness, etc.

JAPAN

CHAPTER EDITOR
Koichi Takanashi, The University of Tokyo

MOC 80

General

1.1 Purpose

The purpose of this aseismic design method is that the buildings shall withstand moderate earthquake motions with almost no damage, which would occur several times during the use of the buildings, and shall not collapse nor harm human lives by the severe earthquake motions, which would occur less than once during the use of the buildings.

1.2 Scope

Buildings shall satisfy one or more of the design procedures specified in the following section, according to the structural type, floor area, height, etc. (See Table SL.J1)

Buildings exceeding 60 meters in height shall have special permission by the Minister of Construction through the detailed review of the dynamic behavior of the structure by the board of technical members.

Design Procedure

2.1 Structural Requirements

Buildings shall meet the relevant structural requirements specified by the Building Standard Law Enforcement Order and the Specifications of Architectural Institute of Japan, etc.

Table SL.J1 Design Procedures Required

<table>
<tr><th colspan="3">Buildings</th><th>Design Procedures Required*</th></tr>
<tr><td>A</td><td colspan="2">(1) One or two story wooden buildings not exceeding 500 m^2 in total floor area
(2) Single story buildings other than wooden, not exceeding 200 m^2 in floor area
(3) Special buildings used for school, hospital, etc., not exceeding 100 m^2 in floor area</td><td>2.1</td></tr>
<tr><td rowspan="3">B</td><td rowspan="3">Buildings not higher than 31 meters</td><td>(1) Building listed in Table SL.J2</td><td>2.1 and 2.2</td></tr>
<tr><td rowspan="2">(2) Others</td><td>i) 2.1, 2.2, 2.3 and 2.4</td></tr>
<tr><td>ii) 2.1, 2.2, 2.3 and 2.5</td></tr>
<tr><td>C</td><td colspan="2">Buildings higher than 31 meters</td><td>2.1, 2.2, 2.3 and 2.5</td></tr>
<tr><td>D</td><td colspan="2">Buildings higher than 60 meters</td><td>Special permission by the Minister of Construction</td></tr>
</table>

* Refers to the sections in the design specifications.

2.2 Stresses

The stresses caused by the lateral seismic shear for moderate earthquake motions prescribed in sections 3.1, 3.2 and 3.3 of the design specifications shall not exceed the allowable stresses for temporary loads.

2.3 Story drift

Each story drift of the building caused by the lateral seismic shear for moderate earthquake motions prescribed in 3.1 shall not exceed 1/200 of the story height. This value can be increased to 1/120 in case the nonstructural members shall have no severe damage at the increased story drift limitation.

2.4 Eccentricity, Stiffness, etc.

(i) The following eccentricity of stiffness R_e of each story shall be less than 0.15.

$$R_e = \frac{e}{r_e} \quad \text{(SL.J1)}$$

where

- e = the eccentricity of the center of the stiffness from the center of gravity
- r_e = the elastic radius (that can be defined as the square root of the torsional stiffness R_s of each story) shall be greater than 0.6

(ii) The following variation of lateral stiffness R_s of each story shall be greater than 0.6.

$$R_s = \frac{r}{\bar{r}} \quad \text{(SL.J2)}$$

where

- r = lateral stiffness that shall be defined as the value of the story height divided by the story drift caused by the lateral seismic shear for moderate earthquake motions prescribed in 3.1
- $\bar{r}$ = mean lateral stiffness that shall be defined as the arithmetic mean of r values above ground level

Table SL.J2 Buildings That May Be Designed by Procedure 2.1 and 2.2

1	One, two or three story buildings of conventional wooden construction, reinforced concrete block construction or masonry construction that shall meet the structural requirements stipulated in the relevant regulations.
2	Buildings not exceeding 20 meters in height of reinforced concrete construction or steel encased reinforced concrete construction where each story shall meet the following formula in the longitudinal and transverse directions. $25\,A_w + 7\,A_c > Z\,W_i$ (SL.J3) where A_c = the sum of horizontal cross-sectional area in square centimeters of columns A_w = the sum of horizontal cross-sectional area in square centimeters of reinforced concrete shear walls in the direction concerned W_i = the weight in kilograms of the building above the story concerned Z = the seismic hazard zoning coefficient as shown in Fig. SL.J1
3	Buildings specially permitted by the Minister of Construction.

(iii)-a Ultimate shear strength of each reinforced concrete member shall be greater than ultimate flexural strength of the member.
(iii)-b The lateral seismic shear for moderate earthquake motions of steel structures shall be increased according to the formula

$$Q_b = (1 + 0.7\beta)\, Q \tag{SL.J4}$$

where

- Q = lateral seismic shear for moderate earthquake motions prescribed in 3.1
- Q_b = increased lateral seismic shear
- β = ratio of the lateral shear of braces to the total lateral seismic shear of the story and this ratio shall be assigned by the structural designer

Each bracing member shall meet the requirement

$$_{J}P_u \geq 1.2\ _{M}P_y \tag{SL.J5}$$

where

- $_{J}P_u$ = the ultimate strength of the joint of the brace
- $_{M}P_y$ = the yield strength of the brace

2.5 Ultimate Lateral Shear Strength
The ultimate lateral shear strength of each story shall not be less than the required lateral shear Q_r determined by the formula

$$Q_r = D_s \cdot F_{es} \cdot Q \tag{SL.J6}$$

where

- D_s = structural coefficient given in Table SL.J3
- Q = lateral seismic shear for severe earthquake motions prescribed in 3.1

F_{es} = shape factor that shall be determined by the following

$$F_{es} = F_e \cdot F_s \tag{SL.J7}$$

where F_e is given by Table SL.J4 as a function of eccentricity of stiffness R_e defined in 2.4 (i).

Table SL.J3 Structural Coefficient, D_s

The value of D_s	Types of Structures
Not less than 0.3	Moment frame structures with excellent ductility
Not less than 0.4	Structures with ductile shear walls or braces
Not less than 0.5	Structures with shear walls or braces with poor ductility

Table SL.J4 Shape Factor F_e by Eccentricity of Stiffness R_e

R_e	F_e
less than 0.15	1.0
$0.15 \leq R_e \leq 0.3$	linear interpolation
more than 0.3	1.5

The value F_s can be obtained from Table SL.J5 as a function of variation of lateral stiffness R_s defined in 2.4 (ii).

Lateral Seismic Shear

3.1 Lateral Seismic Shear above the Ground Level

The lateral seismic shear Q of each story above the ground level shall be determined by the formula

$$Q = C_i \cdot W_i \qquad \text{(SL.J8)}$$

where

C_i = lateral seismic shear coefficient of the i-th story as determined from Eq. SL.J9

W_i = weight of the building above the i-th story

The weight of the building shall be the sum of dead load and the applicable portion of live load. In heavy snow districts, the effect of snow load shall be considered.

The lateral seismic shear coefficient of the i-th story, C_i, shall be determined from

$$C_i = Z \cdot R_T \cdot A_i \cdot C_o \qquad \text{(SL.J9)}$$

where

A_i = lateral shear distribution factor, which shall be determined by the fundamental natural period and the weight distribution of the buildings, as shown in Fig. SL.J3

C_o = standard shear coefficient, which shall be not less than 0.2 and 1.0 for moderate earthquake motions and for severe earthquake motions, respectively

R_T = design spectral coefficient, which shall be determined by the type of soil profile and the fundamental natural period of the buildings, as illustrated in Fig. SL.J2

Z = seismic hazard zoning coefficient as shown in Fig. SL.J1

The fundamental natural period of the building T to determine the design spectral coefficient and the lateral shear distribution factor, shall be determined by one of the following

for steel frame buildings

$$T = 0.028\ H \qquad \text{(SL.J10)}$$

for other buildings

$$T = 0.020\ H \qquad \text{(SL.J11)}$$

where

H = height of the building in meters

T = fundamental natural period of the building in seconds

Table SL.J5 Shape Factor F_s by Variation of Lateral Stiffness R_s

R_s	F_s
less than 0.6	1.0
$0.3 \leq R_s \leq 0.6$	linear interpolation
more than 0.3	1.5

3.2 Lateral Seismic Shear of Appendages

The lateral seismic shear q of the penthouses, chimneys, towers, cisterns, parapets and other appendages on buildings shall be determined from

$$q = k \cdot w \qquad \text{(SL.J12)}$$

where

q = lateral seismic shear of the appendage

k = seismic design coefficient of appendages and shall be 1.0, but the value can be minimized down to 0.5 in case no harm to human lives will occur

w = weight of the appendage

3.3 Lateral Seismic Shear of the Basement

The lateral seismic shear of basement, Q_b, shall be determined from

$$Q_b = Q_p + K \cdot W_b \qquad \text{(SL.J13)}$$

where

K = seismic design coefficient of the basement as determined by Eq. SL.J14

Q_p = portion of the lateral seismic shear of the first story that will act to the basement

W_b = weight of the basement

The seismic design coefficient of basement, K, shall be determined from

$$K = 0.5\left(1 - \frac{H}{40}\right) Z \cdot C_o \qquad \text{(SL.J14)}$$

where

H = depth of the basement in meters and the value of H is fixed as 20 meters in case the depth exceeds 20 meters

Z and C_o are the same as defined in 3.1

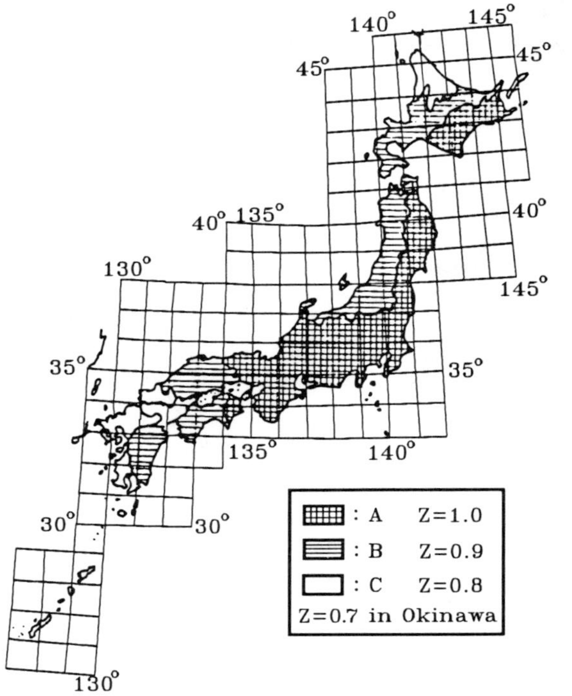

Fig. SL.J1 Seismic Hazard Zoning Coefficient, Z

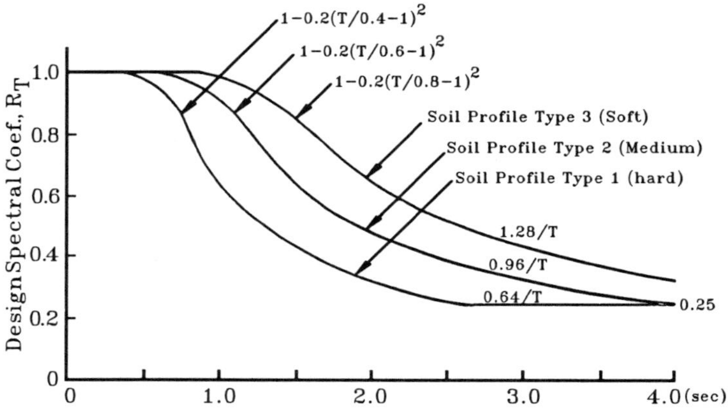

Fundamental Natural Period, T

Fig. SL.J2 Design Spectral Coefficient, R_T

Through the precise analysis of the structure, the foundation, the soil, etc., the value of R_T can be reduced to 0.75 of the value given by this figure. The value shall not be less than 0.25.

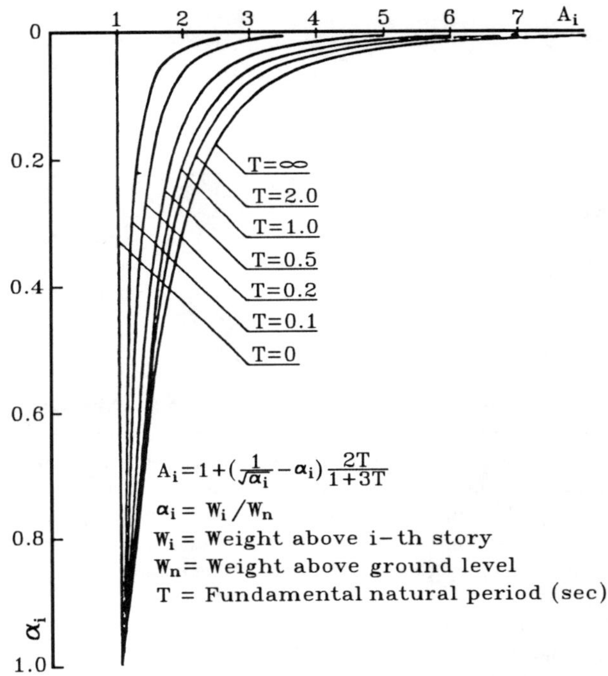

Fig. SL.J3 Lateral Shear Distribution Factor, A_i

NORTH AMERICA

CHAPTER EDITORS
Franklin Y. Cheng, University of Missouri-Rolla
Sheng-Taur Mau, University of Houston

UBC 88

Stability under seismic loading is considered at both the structural level and the member level in design. At the structural level stability is considered through the P-delta effect. At the member design level it is considered through loading and strength requirements and local buckling prevention. In the following text, Seismic Zone 3 and Zone 4 refer to the high risk seismic zones in the U.S. The provisions of UBC are for allowable stress design, although the LRFD is acceptable as an alternative.

P-delta Effects

The resulting member forces and moments and the story drifts induced by P-delta need not be considered when the ratio of secondary moment to primary moment does not exceed 0.10; the ratio may be evaluated for any story as the product of the total dead and live load above the story times the seismic drift in that story divided by the product of the seismic shear in that story times the height of that story. In Seismic Zones 3 and 4, designs conforming to the following drift limitations may be deemed to satisfy this requirement.

Calculated story drift shall not exceed $0.04/R_w$ nor 0.005 times the story height for buildings less than 65 feet in height. For buildings greater in height, the calculated story drift shall not exceed $0.03/R_w$ nor 0.004 times the story height. Calculated drift shall include translational and torsional deflections. The coefficient R_w is a structural system factor that roughly reflects the structure's ductility capacity. For a steel or concrete special moment-resisting space frame (SMRSF), the value of R_w is 12.

Column Requirements

(A) Column strength. In Seismic Zones 3 and 4, columns in frames shall have the strength to resist the axial compression resulting from the following load combination

$$1.0\ P_{DL} + 0.7\ P_{LL} + 3(R_w/8)\ P_E \qquad \text{(SL.N1)}$$

where

P_{DL} = the axial dead load
P_E = earthquake load
P_{LL} = the live load
R_w = structural system factor reflecting ductility (see above)

(B) Slenderness evaluation. In determining the permitted axial and bending stresses of columns of moment frames resisting earthquake forces in Seismic

Zones 3 and 4, the factor K in the plane of the earthquake forces may be taken as unity when all the following conditions are met:

1. The column is either continuous or is fixed at each joint.
2. The maximum axial compressive stress does not exceed 0.4 F_y under design load, where F_y is the specified minimum yield stress.
3. The calculated story drift ratios are less than the story drift limits described in the above.

Special Moment-resisting Space Frame (SMRSF) Requirements
(For Seismic Zones 3 and 4 only)

(A) Panel zone thickness. The thickness of the panel zone of the joint t_z shall conform to

$$t_z \geq (d_z + w_z)/90 \qquad \text{(SL.N2)}$$

where

d_z = the panel zone depth between continuity plates
w_z = the panel zone width between column flanges

In computing the thickness, any double plate thickness shall not be included unless the double plate is connected to the column web with plug welds adequate to prevent local buckling of the plate.

(B) Flange width-thickness ratio. The flange width -thickness ratio, $b_f/2t_f$, shall not exceed $52/\sqrt{F_y}$.

(C) Trusses in SMRSF. Trusses may be used as horizontal members in SMRSF if the sum of the truss seismic force flexural strength exceeds the sum of the column seismic force flexural strength immediately above and below the truss by a factor of at least 1.25. In buildings of more than one story, the column axial stress shall not exceed 0.4 F_y and the ratio of the unbraced column height to the least radius of gyration shall not exceed 60.

(D) Unrestrained girder-column joint. Columns without lateral support transverse to a joint shall not have a L/r exceeding 60. At truss frames the column shall be braced at each truss chord for a lateral force equal to one percent of the compression yield strength of the chord.

(E) Beam bracing. Both flanges of beams shall be braced directly or indirectly. The beam bracing between column center lines shall not exceed 96 times the lesser of the radius of gyration. In addition, braces shall be placed at concentrated loads where a hinge may form.

Requirements for Braced Frames
(A) Slenderness. In Seismic Zones 3 and 4 the L/r ratio for bracing members shall not exceed $720/\sqrt{F_y}$, except for bracing members in buildings not over two stories in height and in roof structures if they are designed for forces of $3(R_w/8)$ times the code equivalent static forces at code allowable seismic stresses.

(B) Stress reduction. The allowable stress F_{as} for bracing members resisting seismic forces in compression shall be determined from the following formula

$$F_{as} = \beta\, F_a \qquad \text{(SL.N3)}$$

where

F_a = the regular allowable axial compression stress
β = the stress reduction factor determined from the following formula

$$\beta = 1/[1+([kL/r]/2C_c)] \qquad \text{(SL.N4)}$$

where

C_c = the column slenderness ratio separating elastic and inelastic buckling

$$C_c = \sqrt{\frac{2\pi^2 E}{F_y}} \qquad \text{(SL.N5)}$$

(C) Built-up members. The L/r of individual parts of built-up bracing members between stitches, when computed about a line perpendicular to the axis through the parts, shall not be greater than 75 percent of the L/r of the member as a whole.

(D) Bracing configuration. Bracing members in a Chevron bracing shall be designed for 1.5 times the otherwise prescribed forces. K bracing is prohibited in Seismic Zones 3 and 4 except in buildings not over two stories in height and in roof structures if they are designed for forces of $3(R_w/8)$ times the code equivalent static forces at code allowable seismic stresses.

Eccentrically Braced Frame (EBF) Requirements

(A) Link beam. There shall be a link beam provided at least at one end of each brace. Beams in EBF should have a flange width-thickness ratio not exceeding $52/\sqrt{F_y}$. Link beams shall have full depth web stiffeners on both sides of the beam web at the brace end of the link beam. In addition, for link beams with clear lengths between $1.6M_s/V_s$ and $2.6M_s/V_s$, full-depth stiffeners shall be placed at a distance b_f from each end of the link. The stiffeners shall have a combined width not less than $b_f - 2t_w$ and a thickness not less than 0.75 t_w nor 3/8 inch. In the above, the M_s and V_s are the flexural and shear strength of the member, respectively, the b_f is the flange width and t_w is the thickness of the web.

(B) Intermediate stiffeners. Intermediate full-depth web stiffeners shall be provided in either of the following conditions:

(1) Where the link beam strength is controlled by V_s.
(2) Where the link beam strength is controlled by flexure and the shear determined by applying the reduced flexural strength, M_{rs}, exceeds 0.45 $F_y$dt. In the above the reduced flexural strength M_{rs} is defined as $Z(F_y - F_a)$, with Z being the plastic section modulus, d is the depth of the beam and t is the thickness of the web of the beam.

The spacing of the stiffener shall conform to the following requirements:

(1) For link beams with a rotation angle of 0.06 radians the spacing shall not exceed $38t_w - d/5$.

(2) For link beams with a rotation angle of 0.03 radians or less the spacing shall not exceed $56t_w - d/5$. Interpolation may be used for rotation angles between 0.03 and 0.06 radians.

For beams 24 inches in depth and greater, intermediate full-depth web stiffeners are required on both sides of the web. Such web stiffeners are required only on one side of the beam web for beams less than 24 inches in depth. The stiffener thickness, t_w, of one-sided stiffeners shall be not less than 3/8 inch and the width shall be not less than $(b_f/2)-t_w$.

(C) Brace and column strengths. Each brace shall have a compression strength at least 1.5 times the axial force corresponding to the controlling link beam strength. The controlling link beam strength is either the shear strength, V_s, or the reduced flexural strength, M_{rs}, whichever results in the lesser force in the brace. Columns shall be designed to remain elastic at 1.25 times the strength of the EBF bay, as defined by the controlling link beam strength.

(D) Beam flanges. Top and bottom flanges of EBF frame beams shall be laterally braced at the ends of link beams and at intervals not exceeding $76/\sqrt{f_y}$ times the beam flange width. End bracing shall be designed to resist 1.5 percent of the beam flange strength, defined as $F_y b_f t_f$. Intermediate bracing shall be designed to resist 1.0 percent of the beam flange force at the bracing point using the controlling link beam strength.

(E) Story drift limitation. Calculated drift shall include translational and torsional deflections and shall not exceed $0.04/R_W$ nor 0.005 times the story height for buildings less than 65 feet in height. For buildings greater in height, the calculated drift shall not exceed $0.03/R_W$ nor 0.004 times the story height.

(F) P-Δ effects. The resulting member forces and moments and the story drift included by P-Δ effects shall be considered in the evaluation of overall structural frame stability. P-Δ need not be considered when the ratio of secondary moment to primary moment does not exceed 0.10. In zones 3 and 4, designs conforming to the story drift limitation may be deemed to satisfy this requirement.

WEST EUROPE

CHAPTER EDITOR
Federico M. Massolani, Universita Di Napoli

PREFACE

From the examination of the main national codes on seismic design in West Europe, it appears that the problem of stability under seismic actions is not always specifically covered. In fact, many codes (France, Italy, Portugal, Spain, Switzerland) mainly deal with general calculation rules and, therefore, for specific design problems of stability they refer the reader to the common provisions for steel structures. Only a few seismic codes provide special rules for local buckling and column buckling (Germany, Turkey). On the contrary, the new seismic codes, which are based on the concept of reducing the response spectrum by means of appropriate ductility parameters (q-factors), extensively deal with stability problems for dissipative structures. The ECCS Recommendations have been substantially adopted in the Part 3 "Steel Structures" of EC 8 and in the new proposed Italian seismic code (CNR-GNDT). In these codes the assessment of local ductility leads to a definition of ductility classes based on appropriate limitations of b/t ratios, which are correlated to the ones specified in the standards for design of steel structures in non seismic areas (that is, EC 3).

ECCS 88

General Statement
The structure should be stable when it is subjected to seismic forces and jointly acting loads.

Assessment of Local Ductility
Sufficient local ductility of the structural members is assured by restricting the width-thickness ratio b/t in compression parts of the cross-sections. In non-dissipative zones (where no ductility requirements have to be satisfied and therefore assume $q=1$), the width-thickness ratios should conform to the limit specified in standards governing steel structures in non seismic areas (EC 3). In dissipative zones of the structure (where ductility requirements have to be satisfied and therefore assume $q > 1$), the maximum width-thickness ratios (b/t) are given in Table SL.W2, in relation to the three ductility classes given in Table SL.W1.

Table SL.W1 Maximum Width-Thickness Ratio

Behavior factor q	cross sectional class
$4 < q \leq 6$	class A
$2 < q \leq 4$	class B
$q \leq 2$	class C

For q > 6, the b/t ratios of class A can be used, provided

$$\frac{N}{N_y} + 0.8\,\frac{\lambda}{\lambda_o} \leq 1 \qquad \text{(SL.W1)}$$

where

$$\lambda_o = \pi\sqrt{\frac{E}{f_y}} \qquad \text{(SL.W2)}$$

Class A - values are necessary where sufficient rotation capacity of the plastic zones is required.
Class B - values must be reached where plastic resistance shall be obtained.
Class C - values are necessary where the resistance is limited by yielding of extreme fibers.

The values of b/t ratios given for class A and class B corresponds to the ones of class 1 (plastic sections) and class 2 (compact sections) of EC 3, respectively. The values of class C are intermediate between the ones of class 2 and class 3 (semicompact sections) of EC 3 (Fig. SL.W1). The values in Table SL.W1 come from an engineering evaluation; their more refined values should be deduced from the experimental evidence.

Frame Structures
(A) Beams
Beams in which dissipative zones can arise must be constrained in terms of lateral buckling. Whenever a beam is not transversally held in place by a continuous horizontal diaphragm, special retainers must be employed that impede movements of the dissipative part of the beam in a direction perpendicular to the bending plane.

The spacing should not be greater than 40 $i_y(235/f_y)^{1/2}$, i_y is the radius of inertia of the beam in a direction perpendicular to the bending plane, and should be placed at a distance from the column not greater than L/4 or h/4, where L is the beam span and h is its depth.

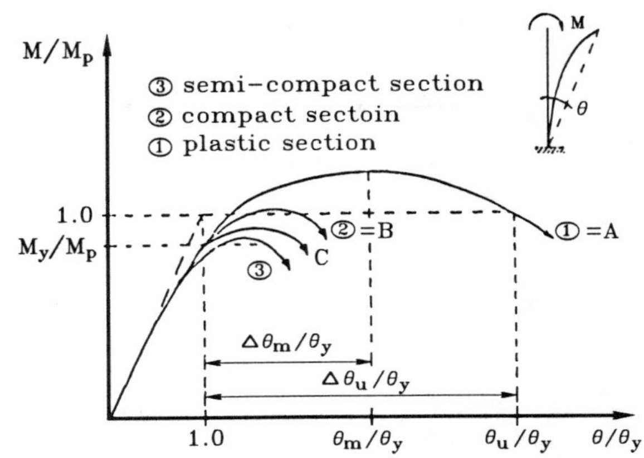

Fig. SL.W1 Effect on Compactness on Moment-Rotation Relationship

Table SL.W2 Maximum b/t ratio of cross-sections for compression members

Cross section and boundary condition	stress distribution (compression positive)	class A	class B	class C
Rectangular hollow section 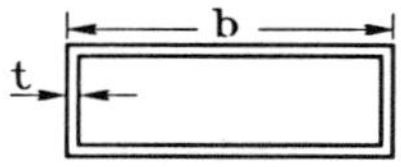	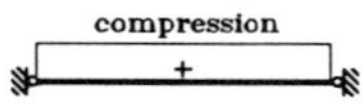	$33\ \epsilon$	$37\ \epsilon$	$41\ \epsilon$
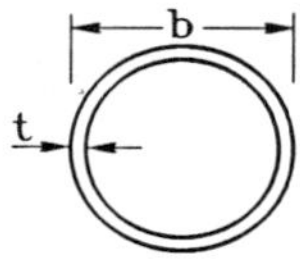	compression	$50\ \epsilon^2$	$70\ \epsilon^2$	$85\ \epsilon^2$
Webs of I-Profiles Webs of flanges of welded sections 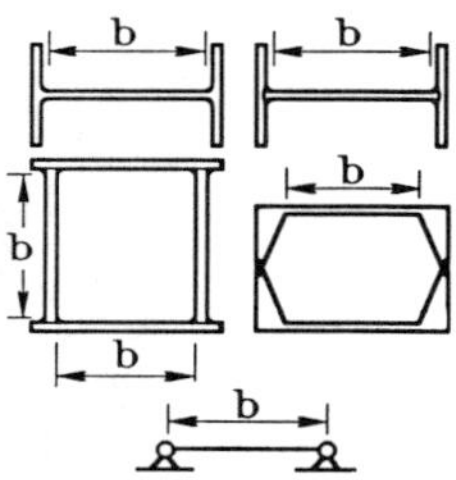	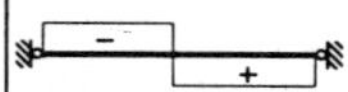  plastic distribution, elastic distribution	$66\ \epsilon$	$78\ \epsilon$	$90\ \epsilon$
	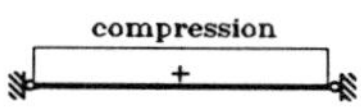	$33\ \epsilon$	$39\ \epsilon$	$41\ \epsilon$
	combined bending and compression: plastic distribution 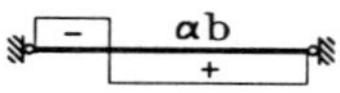 elastic distribution 	$\frac{33}{\alpha}\ \epsilon$	$\frac{39}{\alpha}\ \epsilon$	$\frac{41}{\alpha}\ \epsilon$
Outstanding flanges of welded box section or flanges of I-Profiles 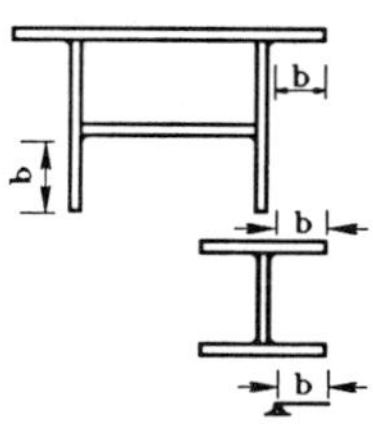	compression 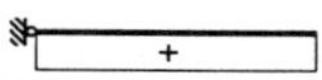	$9\ \epsilon$	$10\ \epsilon$	$12\ \epsilon$
	combined bending and compression 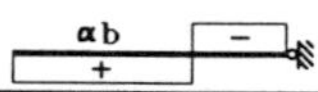	$\frac{9}{\alpha}\ \epsilon$	$\frac{10}{\alpha}\ \epsilon$	$\frac{12}{\alpha}\ \epsilon$
	combined bending and compression 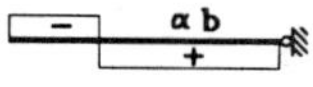 	$\frac{9}{\alpha\sqrt{\alpha}}\epsilon$	$\frac{10}{\alpha\sqrt{\alpha}}\epsilon$	$\frac{12}{\alpha\sqrt{\alpha}}\epsilon$
Flanges of I-Profiles 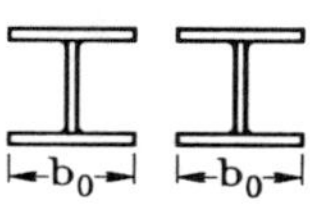	compression 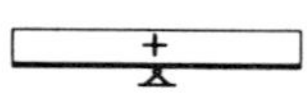	$20\ \epsilon$	$22\ \epsilon$	$26\ \epsilon$

where

$$\epsilon = \sqrt{235/f_y}$$

f_y	235	275	355
ϵ	1	0.92	0.81

(B) Columns

Columns should be designed such that hinges form in the beams before yielding in the columns occurs and that the shear in the columns does not hinder full plastic hinges to develop. This can be achieved by means of an amplificator factor α to increase bending moments in the column.

The combined compression and bending of a column must be checked by considering the worst case combination of axial force N and bending moments M_x and M_y.

Beam-column interaction formulae, as the one given in EC3, should be used for checking stability.

Concentric Truss Bracings

(A) Diagonal Members

The slenderness of diagonal members should be limited to

$$\frac{\lambda}{\lambda_o} \leq 1.5 \quad \text{(SL.W3)}$$

where

$$\lambda_o = \pi\sqrt{\frac{E}{f_y}} \quad \text{(SL.W4)}$$

The following equation must be verified

$$\frac{1}{\gamma_p} N_R \geq N(\gamma_n E, G, \Sigma\psi_i Q_i) \quad \text{(SL.W5)}$$

where

- E = seismic action
- G = characteristic value of dead loads
- N_R = the tensile strength of the cross-section computed, assuming as A the gross area of the profiles
- $N(.)$ = the axial tensile force produced by external forces
- Q_i = characteristic value of live loads
- γ_n = importance factor
- γ_p = 1.0
- ψ_i = combination factor

Once the diagonals have been verified, one can compute the amplification factor α, equal to the smaller of the ratios between strength $N_{R,j}$ of the j-th diagonals and the corresponding force $N_j(.)$ present in the same diagonal member.

$$\alpha = \min\,[N_{R,j}/N_j(.)] \quad \text{(SL.W6)}$$

The amplification factor is used in the verification of beams and columns and of diagonal member connections in concentric truss bracings. It must guarantee that the axial yielding of the tensile diagonals precedes the ultimate strength of their connections as well as the collapse of beams and columns.

(B) Beams and Columns

Both beams and columns must meet the requirement

$$\frac{1}{\gamma_p} N_R \geq \alpha\gamma_y N(\gamma_n E, G, \Sigma\psi_i Q_i) \quad \text{(SL.W7)}$$

where

N_R = ultimate strength of the element
N = axial force produced by external forces
α = amplification factor as defined in Eq. SL.W6
γ_p = 1.2
γ_y = f_{ymax}/f_y (f_{ymax}=maximum yield limit)

The ultimate strength of the element should be

$N_R = A f_y$, for tension
$N_R = \chi A f_y$, for compression, χ being the appropriate reduction coefficient for buckling

Eccentric Truss Bracings

(A) Beams

Besides the usual checks, it must be verified that the beam should not be subject to flexural-torsional buckling and local buckling phenomena.

(B) Columns and Diagonal Members

The column and diagonal members must be verified considering the worst case combination of the axial force and bending moments. The design values of axial force and bending moment are obtained by multiplying the computed values by the amplification factor α. In any case, α should be assumed not less than 1.2.

The amplification factor is used in the verification of column and diagonal members in the eccentric truss bracings. It must guarantee that the yielding in bending or in shearing of beams precedes the collapse of columns and diagonal members. The value of α to be used for checks should be

$$\alpha = \min_{i,j}\left(\frac{V_{R,i}}{V_i}, \frac{M_{R,j}}{M_j}\right) \qquad \text{(SL.W8)}$$

where

M, V= moment and shear due to external force
M_R = bending moment at the elastic limit
V_R = shear based on the design values of the material strength

Cantilever Structures

(A) Dimensional Restrictions

The columns must satisfy the following restrictions

period $T \leq$ 2.5 sec.
slenderness in the plane of seismic action $\lambda < 1.5\,\lambda_o$, where $\lambda_o = \pi\sqrt{\frac{E}{f_y}}$

(B) Column

The following equations must be satisfied for the column

$$0.20\, N_{cr} \geq N(\gamma_n E,\, G,\, \Sigma\psi_i Q_i) \qquad \text{(SL.W9)}$$

$$\frac{1}{\gamma_p} M_R \geq \frac{M_l}{1-N/N_R} \qquad \text{(SL.W10)}$$

where

M_l = $M(\gamma_n E,\, G,\, \Sigma\psi_i Q_i) + q\, \Sigma F_j\, v_j$, is the bending moment at the base of the column including the geometrical effects of the 2nd order

N_{cr} = $\pi^2 EA/\lambda^2$ is the critical Euler load of the column in the bending plane under consideration

N_R, M_R = ultimate values of the axial force and the bending moment, respectively

N(.), M(.) = axial force and the bending moment at the base of the column produced by external forces

γ_p = 1.0

$F_j v_j$ = product of the j-th force applied to the column multiplied by the value of the horizontal displacement of the corresponding application point, caused by external forces

EC 8 88

Structures in Seismic Regions

Chapter 3 "Steel Structures" of EC 8 has been prepared by the ECCS Committee on Seismic Design, which acted as a consulting body of the EC 8 drafting panel. Only some small formal differences exist between both texts. From a stability point of view, we can mention the following points are briefly discussed.

The evaluation of the q-factors for frame structures is given by the formula

$$q = \chi \frac{\alpha_u}{\alpha_1} \qquad \text{(SL.W11)}$$

where

α_u = multiplier of the horizontal seismic actions, by keeping constant the other design loads, which corresponds to the point, where the structure reaches the maximum bearing capacity due to the formation of plastic hinges in the assumed dissipative zones sufficient to transform the structure into a mechanism and/or due to the presence of some element that become unstable

α_1 = multiplier of the horizontal seismic actions, by keeping constant the other design loads, which corresponds to the point, where the structure reaches its elastic limit in one section

χ = 5 for regular buildings

χ = 4 for irregular buildings

The ratio α_u/α_1 must be taken not greater than 1.2, except the columns with

$$N/N_{pl} \leq 0.1$$

where

N = total weight acting over the considered story

N_{pl} = axial load carrying capacity

$\alpha_u/\alpha_1 \leq 1.6$

This limitation, which has been introduced to take indirectly into account the P-Δ effect, does not exist in the ECCS Recommendations. The slenderness should be limited to avoid secondary effects of instability under seismic forces.

In all the verifications the resistance to buckling of the element should be calculated as shown in EC 3.

CNR-GNDT 84

Technical Rules for Constructions in Seismic Areas

The assessment of buckling checks is very similar to ECCS Recommendations. The main differences are related to the limitation of the width-thickness ratios of the compressed parts of the section: in non dissipative zones the b/t ratios will be taken from the codes on steel structures in non seismic zones. In dissipative zones the maximum b/t ratios must be limited by the values given for plastic sections (class 1) in the EC 3. Profiles with section HEA in Fe360 and Fe430 can be used even if their b/t ratios exceed those in the table.

The aforementioned table is the same of the one given in EC 3 for class 1, which is coincident with class A of ECCS.

DIN 4149 81

Steel Construction

Stability analyses (buckling, lateral buckling, bulging) are to be submitted according to DIN 4114 Part 1 and Part 2. The safety factors required there for the load condition HZ may be reduced to 0.75 times. In connection with the application of the procedure, the allowable stresses may be correspondingly raised to 0.8 of the yield stress. The slenderness of the compressed members may not exceed λ=150. Load condition HZ corresponds to the permanent loads plus earthquake.

INTERNATIONAL COMPARISONS & COMMENT

WHAT ARE THE DIFFERENCES?

All of the codes reviewed in the previous section deal with the stability problem in a different manner. Some codes consider both the structural level and the member level, and some consider the member level only. At the structural level, stability is controlled through the P-delta and drift criteria; at the member level, stability is controlled through the design by including loading, strength, and local buckling requirements. The U.S. code may represent the former and the Chinese code represents the latter. The reasons are apparent that the Chinese provisions are for single-story steel mill buildings, stability at structural level may not be considered as significant as that for tall buildings. The code provisions on stability in Australia and East Europe are mainly related to basic steel design standards with a certain amount of modification on ductility, slenderness, and deflection requirements, which have various limitations for different countries. In West Europe, main national codes dealing with the stability problem also rely on the basic steel design standards; only a few seismic codes (Germany, Turkey) provide special rules for local buckling and column buckling. The new code, ECCS Recommendation, has extensively dealt with stability problems by reducing the response spectrum in terms of ductility parameters.

The U.S. code has a specific criterion (0.1) on the P-delta effects. The other countries' codes do not specify it in the same manner. It is interesting to note that the U.S. code specifies that, for zones 3 and 4, the designs conforming to the drift limits may be deemed to satisfy P-delta effects. For zones 1 and 2 of lower seismic force level, the code requires both drift limits and P-delta effects to be satisfied.

No specific comparisons between codes can be included here due to their dependence on many different factors. Meaningful comparisons may possibly be made by using several specific numerical analyses to detect the trends. The Table SL.1 is just a simplified attempt to review some of the main factors considered in mentioned standards.

Table SL.1 Topics considered in selected specifications

TOPICS \ SPECIFICATIONS			A	A (NZ)			C	E				J	N
			AS 1250 81	**NZS 3404 89**	**NZS 4203 84**	**DZ 4203 89**	**GBJ 11 89**	**NiP 565/86 87**	**ČSN 73 0036 75 89**	Indicativ	**SNiP II-7-81 82**	**MOC 80**	**UBC 88**
Reliable assessment method	ASD				X							X	X
	LSD				X			X	X		X		X
Scope	single story buildings				X		X	X	X				
	multi-story buildings				X			X	X			X	X
	not specified		X							X	X		
General arrangement of a structure	specified		X						X			X	X
	not specified						X	X		X	X		
Considered	struct. & member levels		X		X			X	X		X	X	X
	member level only						X			X			
Provisions	modif. basic steel design		X					X		X		X	X
	special rules						X		X		X		
Ductility parameter	specified & applied		X		X					X		X	X
	not covered						X	X			X		
Drift Limit	specified		X		X		X						X
	not covered							X		X	X		
P-Δ effect	special criterion				X							X	X
	not covered		X				X	X		X	X		
Special Slenderness Limitations	columns	specified			X		X		X		X		X
		not covered	X					X		X		X	
	girders & bracings	specified			X		X		X	X	X		X
		not covered	X					X				X	
Material limitations	specified		X						X				X
	not covered				X		X	X		X	X		
Number of cycles (accum. of damage)	considered				X								
	not considered		X				X	X	X	X	X		X
Level of tolerable damage due to earthquake	specified				X							X	
	not covered		X				X	X	X	X	X		

Table SL.1 (CONT.) Topics considered in selected specifications

TOPICS \ SPECIFICATIONS			W			
			ECCS 88	EC 8 88	CNR GNDT 84	DIN 4149 81
Reliable assessment method	ASD					X
	LSD		X	X		
Scope	single story buildings					
	multi-story buildings		X	X	X	X
	not specified					
General arrangement of a structure	specified		X			X
	not specified					
Considered	struct. & member levels		X	X	X	X
	member level only					
Provisions	modif. basic steel design					
	special rules		X	X	X	X
Ductility parameter	specified & applied		X	X	X	
	not covered					
Drift Limit	specified		X	X		
	not covered					
P-Δ effect	special criterion		X			
	not covered			X		
Special Slenderness Limitations	columns	specified	X	X	X	X
		not covered				
	girders & bracings	specified	X	X	X	X
		not covered				
Material limitations	specified					
	not covered					
Number of cycles (accum. of damage)	considered					
	not considered		X	X	X	X
Level of tolerable damage due to earthquake	specified					
	not covered		X			

WHY ARE THERE DIFFERENCES?

Differences in handling design and stability assessment of steel structures exposed to earthquake loading seem to be attributable to:

(a) Differences in reliability assessment (allowable stress method and limit states method) and their various interpretations in specifications,

(b) Different levels of probability of earthquakes in different countries,

(c) Different qualitative levels of design models used for transformation of the earthquake loading to the response of the structure to the loading,

(d) The experience, "know-how", and level of the application of computer technology in different countries.

Symbol List -- Stability under Seismic Loading

A_c = sum of horizontal cross-sectional area in square centimeters of columns (SL.J)
A_i = lateral shear distribution factor (SL.J)
A_w = sum of horizontal cross-sectional area in square centimeters of reinforced concrete shear walls in the direction concerned (SL.J)
b = one-half flange width of I-section (SL.W)
b_1 = flange or plate width in compression with one unstiffened edge (SL.A)
b_2 = flange width of welded box sections (SL.A)
C_c = column slenderness ratio separating elastic and inelastic buckling (SL.N)
C_i = lateral seismic shear coefficient of the i-th story as determined by Eq. SL.J9 (SL.J)
C_o = standard shear coefficient (SL.J)
C_1 = modification factor (SL.A)
D_s = structural coefficient given in Table SL.J3 (SL.J)
d_z = panel zone depth between continuity plates (SL.N)
d_1 = web width (SL.A)
E = seismic action (SL.W)
e = eccentricity of the center of the stiffness from the center of gravity (SL.J)
F_a = regular allowable axial compression stress (SL.N)
F_{as} = allowable stress for bracing members resisting seismic forces in compression (SL.N)
F_{es} = shape factor that shall be determined by Eq. SL.J7 (SL.J)
F_j = the j-th axial force applied to the column (SL.W)
F_y = yield stress (SL.A) & (SL.C)
f_y = yield stress (SL.W)
f_{ymax} = maximum yield stress (SL.W)
G = characteristic value of dead loads (SL.W)
H = height of the building in meters (SL.J)
H = depth of the basement in meters (SL.J)
h_g = distance between the inner faces of I-section flanges (SL.W)
K = seismic design coefficient of the basement as determined by Eq. SL.J14 (SL.J)
k = buckling length factor (SL.C)
k = seismic design coefficient of appendages (SL.J)
L = member length (SL.A) & (SL.C)
l_i = full length of the ith panel diagonal (SL.C)
M_j = moment of member j due to external force (SL.W)
M_l = bending moment at the base of the column including the geometrical effects of the 2nd order (SL.W)
M_R = bending moment at the elastic limit (SL.W)
$M_{R,j}$ = bending moment at the elastic limit of member j (SL.W)
M_u = design moment (SL.A)
M(.) = bending moment at the base of the column produced by external forces (SL.W)
N = total weight acting over the considered story (SL.W)
N_c = actual axial force due to earthquake actions (SL.W)

N_{cr} = $\pi^2 EA/\lambda^2$ is the critical Euler load of the column in the bending plane under consideration (SL.W)
N_i = design tensile force of the ith panel diagonal (SL.C)
N_p = yielding axial force (SL.W)
N_{pl} = axial load capacity (SL.W)
N_R = ultimate values of the axial force (SL.W)
$N_{R,j}$ = ultimate axial strength of the element of member j (SL.W)
N(.) = axial force produced by external forces (SL.W)
P_{DL} = axial dead load (SL.N)
P_E = earthquake load (SL.N)
P_{LL} = live load (SL.N)
P_u = design axial force (SL.A)
P_y = yielding axial force (SL.A)
${}_JP_u$ = ultimate strength of the joint of the brace (SL.J)
${}_MP_y$ = yield strength of the brace (SL.J)
Q = lateral seismic shear force (SL.J)
Q_b = increased lateral seismic shear (SL.J)
Q_i = characteristic value of live loads (SL.W)
Q_p = portion of the lateral seismic shear of the first story that will act to the basement (SL.J)
q = lateral seismic shear of the appendage (SL.J)
R_T = design spectral coefficient, as illustrated in Fig. SL.J2 (SL.J)
R_w = structural system factor reflecting ductility (SL.N)
r = lateral stiffness (SL.J)
r = radius of gyration (SL.A) & (SL.C)
$\bar{r}$ = mean lateral stiffness (SL.J)
r_e = elastic radius (SL.J)
r_y = radius of gyration about the y-axis (SL.A)
S_c = clear spacing of columns where the ith panel bracing is located (SL.C)
T = fundamental natural period of the building in seconds (SL.J)
T = plate thickness (SL.A)
t = web thickness (SL.A)
t_b = thickness of I-section flange (SL.W)
t_g = thickness of I-section web (SL.W)
t_z = thickness of the panel zone of the joint (SL.N)
u = deflection at the point of application of the unit force (SL.C)
u_{ti} = relative deflection of the ith panel subject to a unit lateral force assuming that only the tensile bracing is active (SL.C)
V_{bi} = design value of earthquake shear force of the ith panel bracing (SL.C)
V_R = shear based on the design values of the material strength (SL.W)
$V_{R,i}$ = shear at the elastic limit of member i (SL.W)
V_i = shear of member i due to external force (SL.W)
v_j = horizontal displacement of the corresponding application point of the j-th axial force applied to the column (SL.W)
W_b = weight of the basement (SL.J)
W_i = weight in kilograms of the building above the story concerned (SL.J)
w = weight of the appendage (SL.J)
w_z = panel zone width between column flanges (SL.N)
Z = seismic hazard zoning coefficient as shown in Fig. SL.J1 (SL.J)
α = amplification factor as defined in Eq. SL.W6 (SL.W)

α_u = multiplier of the horizontal seismic actions (SL.W)
α_1 = multiplier of the horizontal seismic actions (SL.W)
β = ratio of end moments (SL.C)
β = ratio of the lateral shear of braces to the total lateral seismic shear of the story (SL.J)
β = stress reduction factor determined from Eq. SL.N4 (SL.N)
γ_n = importance factor (SL.W)
γ_y = ratio of maximum yield limit to material yield strength (SL.W)
λ = slenderness ratio (SL.C)
ρ = ratio of column compression force to yield load (SL.C)
χ = modification factor for the q-factor (SL.W)
φ_i = strength reduction factor of a compression member for the ith panel diagonal (SL.C)
ψ_c = deterioration factor of compression members (SL.C)
ψ_i = combination factor (SL.W)

General Provisions and Design Requirements

TOPICAL EDITORS
John E. Harding, University of Surrey
Pavel J. Marek, San Jose State University
John Springfield, Carruthers & Wallace Limited

REGIONAL CONTRIBUTORS
Shao-Fan Chen, Xian Institute of Metallurgy and Construction Engineering
John E. Harding, University of Surrey
Jerome S.B. Iffland, Iffland Kavanagh Waterbury, P.C.
Ben Kato, Toyo University
Jindrich J. Melcher, Technical University of Brno
John Springfield, Carruthers & Wallace Limited
Koichi Takanashi, University of Tokyo
Nicholas S. Trahair, The University of Sydney

Introduction

The first edition of "Stability of Metal Structures - A World View," provided historical information for four world regions. These were Japan, North America, West Europe, and East Europe. Some of the following information was adapted from that publication. It is important to note that code and specification development is directly related to the particular socio-political nature of each country. Centralization of a government and a structured socio-political system is reflected in the development of individual specifications and in building, bridge, and other structural design codes.

The second edition of the World View covers a total of 120 specifications according to the following regions:

Australia	9
China	6
East Europe	34
Japan	14
North America	25
West Europe	32
Total	120

In addition, if a particular country had the motivation, leadership, and necessary consensus to establish a national standards organization responsible for specification development, this status obviously is reflected in their code and specification development. There are two extremes represented by China (one country) reporting on 6 specifications and North America (two countries) reporting on 25 specifications. The former reflects a minimum number of specifications and a consistent philosophical approach as well as format, whereas the latter covers many specifications, diverse philosophies of design, and reveals little consistency in format.

Following are some generalized comments about specification development in the several regions of the world covered by the World View.

Australia

Except for a specification covering structural design of steel bins for bulk solids, all of the specifications are developed by the Standards Association of Australia (SAA). This fact, and because only one country is being covered, results in only 7 specifications being treated in the World View. In addition, there is the obvious consistency in philosophy of design bases and format. Although Australia has a limit states specification for design of building structures, all of the other specifications are based on allowable stress design.

China

This region also covers only one country and has the further advantage of all specifications being developed by one organization, the China Technical Committee for Standards of Steel Structures. The result is that only 6 specifications are needed for full representation in the World View document. Four of these specifications follow the instruction of the Unified Code for the Design of Building Structures which has adopted the limit states design approach based on probability theory. The specification for design of highway and railway structural members utilizes the allowable stress approach.

East Europe

Because of the number of countries covered by the region designated as East Europe, the number of specifications is large, totaling 34. Although progress is slow, there is an effort underway aimed at unification of national codes and specifications. This was initiated in the 1970's under the framework of the Council of Mutual Economic Aid (CMEA). This organization has adopted and approved two unified specifications, one covering steel structures which is included in the list of 34 for the region. At present, those specifications covered for each country are those most commonly used. However, because of the existence of two approved regional standards (with more underway) there is a definite trend for consistency and unification. Most of the East European countries covered have national Boards of Standardization so that, at least in an individual country, there is consistency in philosophy of the design bases and format between the several national specifications. While there are some exceptions for specific types of members and structures, in general, most of the East European specifications adopt some form of limit states design approach based on a probabilistic theory. Use of divided (multiple) safety factors to express uncertainties due to loads and uncertainties due to resistance is common.

Japan

All of the Japanese specifications reflect the dominating load effect caused by earthquakes. Accordingly both ultimate states (permitting significant damage without collapse) and serviceability limit states (restricting damage) are addressed for extreme earthquakes of low probability and moderate earthquakes of reasonable probability, respectively. All of the Japanese building specifications are developed by the Architectural Institute of Japan (AIJ) whereas the railway bridge specifications are developed by the Japan Society of Civil Engineers (JSCE) and the highway bridge specifications are developed by the Japan Road Association (JRA). Involvement by this minimum number of specification bodies helps in formulating consistent specifications. Although limit states specifications are available for building design, the allowable stress approach is still most commonly used.

North America

In the U.S.A., highway bridge specifications are developed by the American Association of State Highway and Transportation Officials (AASHTO). Railway bridge specifications are developed by a users group, the American Railway Engineering Association (AREA). Other specifications in the U.S.A., except for loading, are generally industry initiated.

Highway bridge specifications in Canada are developed by both the Canadian Standards Association (CSA) and the Provincial governments. The technical aspects of other specifications are developed under the CSA. Loading and load combinations for buildings are formulated by the National Building Code with the support of the National Research Council.

The U.S.A. is characterized by the local autonomy that modifies specifications for building structures for adaptation into city building codes and modifies specifications for highway bridge structures for adaptation into state highway bridge specifications. Although these modifications are not necessarily substantial, every city and every state make enough modifications to express their local independence, and an engineer has to keep up with all the local variations in the areas in which he practices.

The situation in Canada is more uniform. Building and bridge regulations are under the jurisdiction of the Provinces, some of which impose uniform standards throughout the province. In reality, the technical aspects of CSA Standards are accepted coast to coast. The National Building Code is also widely adopted, but the legalities of its adoption tend to delay each new edition.

The situation in the United States is further complicated by the existence of private organizations that develop model building codes which are often adapted by smaller municipalities. The model building code groups, generally modify any industry-developed specifications and then incorporate them into their model codes. In establishing the 25 specifications reviewed in this document for North America, a list of 47 specifications was culled. Except for the highway bridge specifications for one Canadian Province, no local autonomous building codes or specifications were considered. However, one model code for earthquake design was included. Both building and bridge design specifications in Canada are limit state based whereas specifications in the U.S.A. are based on allowable stress design or alternatively limit states design for buildings and allowable stress design or alternatively load factor design for bridge structures.

West Europe

The region designated as West Europe has the largest number of specifications, 32, covered in the World View. However, because this group makes up the Common Market Community, the future of unification and reduction in the number of design specifications appears assured. Several of the Eurocode specifications are included in the total of 32 although most of the specifications represent national codes of the countries covered by this region. Whereas the prospects for unification look good, the date is still in the future and the specifications for each country will continue to be used for some time after the Eurocodes become official. The Eurocodes will be mandatory for common market activities, but each country may continue to use their own design specifications within their country however changes will not be made and eventually these specifications will become outdated. In general, all West European specifications adopt limit states design concepts based on probabilistic theory although there are still some allowable stress design specifications in use.

Early in the development of the outline for the 2nd edition of the World View, common sense suggested that comparisons of differences and the reasons for these differences between design provisions for various structural components and systems might be of limited value because of philosophical differences about loads, combinations of loads, use of materials and designated bases. Accordingly, a section of the document covering these differences, was included and designated as "General Provisions and Design Requirements".

Although the 1st edition of the World View also recognized the need for some general discussion by providing a section of "Philosophical Background and Safety Concepts," no quantitative data was given. In addition, there was not a consistent approach to the discussion between the various world regions. These shortcomings have been specifically addressed in the 2nd edition.

GPa. Limits of Applicability

The initial concept for this section on "Limits of Applicability" was to provide verbatim extracts from each specification to define the limits of application in the exact words of the specification writers. This would eliminate misinterpretation and possible unintentional expansion of areas not intended. This approach was attempted for the North American Region covering some 25 specifications. The result was an 18 page document. If this were expanded to cover all regions, this section alone would be completely out of proportion to the rest of the document. Since most of the philosophical statements on "Limits of Applicability" in the individual specifications could be covered quantitatively in other sections of the General Provisions, it was decided to limit this section to a simple matrix covering types of structural members and structural systems covered by each of the specifications.

SPECIFICATION PROVISIONS & EXPLANATION

AUSTRALIA

REGIONAL CONTRIBUTOR
Nicholas S. Trahair, The University of Sydney

Specification	Buildings	Bridges	Other Structures
AS 1250 81	X	X	X
AISC 83	–	–	X[1]
AS 1538 81	X	–	X
AS DR87164 87	X	X	X

[1] Steel Bins

CHINA

REGIONAL CONTRIBUTOR
Shao-Fan Chen, Xian Institute of Metallurgy and Construction Engineering

Specification	Buildings	Bridges	Other Structures
GBJ 17 88	X	–	X
GBJ 18 87	X	–	X[2]
HRS	–	–	X[1]
SDGJ 64 89	X	–	–
TBJ 2 85	–	X	–

[1] Towers, masts and chimneys
[2] Cold-formed steel structures

EAST EUROPE

REGIONAL CONTRIBUTOR
Jindrich J. Melcher, Technical University of Brno

Specification	Buildings	Bridges	Other Structures
CMEA 83	X	X	X
ČSN 73 1401 84	X	X	X
ČSN 73 6205 84	–	X	–
ČSN 73 1206 86	X[2]	X[2]	X[2]
ČSN 73 1402 88	X[1]	–	–
ME 04 180 83	X[1]	–	–
MSZ 15024/3 85	X	–	–
MSZ 15024/4 85	X	–	–
MSZ 15024/1 85	X	–	–
MSZ 5132 88	–	X	–
NiP 565/86 87	X	X	X
PN 80/B-03200 87	X	X	X
SNiP II-24-74 75	X[3]	–	–
SNiP II-23-81 82	X	X	X
SNiP 2.05.03 84	–	X	–
STAS 10108/0 78	X	X	X
TGL 13503/01 82	X	–	–
TGL 13503/02 82	X	–	–

[1] Cold-formed structures
[2] Composite structures
[3] Aluminum structures

The State of the Common Activities in Eastern Europe

From the point of view of the designer, the effort aimed at the unification of national codes and specifications has brought limited results.

Early in the 1970's the formulation of unified provisions for the design of steel structures within the framework of the Council of Mutual Economic Aid (CMEA) was started. The work was organized by the Rumanian specialists, but because of substantial differences in national approaches to individual regulations, a common consistent proposal was not accepted. It became evident that the process of code unification should begin considering general regulations first. In 1976 the standard of CMEA "ST SEV 384 76 Structures and Foundations - Basic design requirements" was approved during the 40th Session of the Permanent Commission for the Building Industries (PCBI) of CMEA.

During the 53rd Session of PCBI-CMEA in 1983 the proposal of unified standard "ST SEV 3972 83 Steel Structures - Basic design requirements" was approved.

Both of these codes contain only general provisions without any detailed formulas or numerical data. Their purpose is to present a universal basis for the formulation of national specifications. In the meantime, the semiprobabilistic method of limit states was introduced for structural design. Both the standards ST SEV 384 76 and ST SEV 3972 83 were under revision in 1990. It is expected that through various supplements to the revised specification of ST SEV 3972 83 the design provisions for different special problems will be presented.

Due to the current political and economical development in East Europe, some countries are considering the adoption of the Eurocode No. 3 specification.

National Approaches to Problems of Structural Stability

As stated in the previous paragraph, the common unified design rules for structural stability are not yet approved in the East European region. Thus, the essential design provisions will be presented considering selected characteristic examples of different national specifications.

Several national codes for the design of metal structures include special stability provisions. The following basic documents are cited:

Bulgaria	NiP 565/86 87
Czechoslovakia	ČSN 73 1401 84 ČSN 73 6205 84
German Democratic Republic[1]	TGL 13503/01 82 TGL 13503/02 82 Both of these specifications are based on allowable stress design concept. At present, the modification of these codes to the limit state design concept has been presented (proposal of 1987).
Hungary	MSZ 15024/1 85 MSZ 15024/3 85 MSZ 15024/4 85
Poland	PN 80/B-03200 87
Romania	STAS 10108/0 78
U.S.S.R.	SNiP II-24-74 75 SNiP II-23-81 82

The above mentioned technical documents represent the essential specifications that include the basic stability provisions. Nevertheless, many additional special codes give particular design procedures for bridges, cranes, tubular members, etc. In this connection, the design provisions for the light-gauge structural members

[1]For information on development until 1990 only.

should be emphasized. The corresponding information includes some characteristic East European specifications. For example:

ČSN 73 1402 88
ME 04 180 83
STAS 10108/2 83
SNiP II-23-81 82

In addition to these typical examples, specifications for bridge design can be noted. For example

ČSN 73 6205 84
MSZ 5132 88
SNiP 2.05.03 84

COMMENTARY

At present, different stability design provisions are used in the national specifications of the East European countries. The previous effort aimed at code unification turned out to be less productive than had been hoped. Most of the national codes are presently under revision. Many East European specialists are of the opinion that the specification EUROCODE No. 3 "Design of Steel Structures" (Draft 1989) is a convenient document for code integration, especially when this specification has common design philosophy and a research basis with the specification "ISO/TC 167/SC 1 Steel Structures - Materials and Design" (Draft of 1988 revision) of the International Organization for Standardization. It is expected that the cooperation of the East European design specification committee with the authors of EUROCODE standards can contribute to the arrangement of an acceptable international design code with detailed stability provisions.

JAPAN

REGIONAL CONTRIBUTORS
Ben Kato, Toyo University
Koichi Takanashi, University of Tokyo

Specifications	Buildings	Bridges	Other Structures
AIJ 70	X		X
AIJ 80	X		X
AIJ 85			X
AIJ 86	X		
AIJ 89a	X		X
AIJ 89b	X		
AIJ 90	X		
JRA 80		X[1]	
JRA 90		X	
JSCE 83		X	
JSCE 87a		X	X
JSCE 87b		X	X

[1] Bridges with span of less than 200m

NORTH AMERICA

REGIONAL CONTRIBUTORS
Jerome S.B. Iffland, Iffland Kavanagh Waterbury, P.C.
John Springfield, Carruthers & Wallace Limited

Specification	Buildings	Bridges	Other Structures
AA 86	X	X	X
AASHTO 89	–	X[3]	–
ACI 318 83	X	–	–
AISC ASD 89	X	–	–
AISC LRFD 86	X	–	–
AISI 76	X[5]	–	–
AISI 86	X[6]	X	X[6]
API RP2A 89	–	–	X[2]
AREA 89	–	X[4]	–
ASME N284 80	–	–	X[1]
BSSC NEHRP 85	X	–	–
CSA A23.3 M84 77	X	–	–
CSA S6 M89 84a	–	X[3]	–
CSA S16.1 M89 84b	X	–	–
CSA S136 M89 84c	X	–	–
FHWA TS 80	–	X	–
OHBDC 83	–	X	–

[1] Shells
[2] Circular cylindrical shells
[3] Highway bridges
[4] Railway bridges
[5] Tubular sections
[6] Cold-formed sections

WEST EUROPE

REGIONAL CONTRIBUTOR

John E. Harding, University of Surrey

Specifications	Buildings	Bridges	Other Structures
BS 5400 P.3 82	–	X	–
BS 5950 P.5 85	X	–	–
BS 5500 85	–	–	X[1]
CNR 10111 85	X	X	X
DIN 18800 P.2 86	X	X	X
EC 3 88	X[5]	–	–
EC 3 89 (Part 1)	X	–	–
EC 4 85	X[2,5]	–	–
NBN B51002 88	X	X	X
NEN 3851 72	X	–	–
NEN 6770	X	X	–
NEN 3854	X	–	–
NS 3472 84	X	X	X
ÖNORM B4600 78	X	X	X
RKP 88	X	–	–
SIA 161 79	X	–	–
StBK N1 80	X		
ECCS 84	–	–	X[3]
ECCS 87	X[4]	–	X[4]
CM 6	X		

[1] Pressure vessels
[2] Composite structures
[3] Shell structures
[4] Cold-formed structures
[5] Current draft – wider coverage prospected

INTERNATIONAL COMPARISONS & COMMENT

WHAT ARE THE DIFFERENCES?

A review of the Limits of Applicability shows that each region covers all applications in general (buildings, bridges, and other structures). Although a few specifications cover both buildings and bridges, the more usual practice to have different specifications for these two major applications. Within each application there are individual variations as noted in the prior tables.

WHY ARE THERE DIFFERENCES?

The differences in the limits of applicability reflect the development of fabrication and construction of bridges, buildings, etc., in individual countries. Standards are in essence a summary of the current "know-how," expressed by incomplete set of instructions. The level, scope, and applicability of standards depend on the need and on requirements related to efficient application of technology and theoretical achievements.

GPb. Material*

SPECIFICATION PROVISIONS & EXPLANATION

AUSTRALIA

REGIONAL CONTRIBUTOR
Nicholas S. Trahair, The University of Sydney

AS 1250 81, AISC 83

Same as for AS DR87164 87.

AS DR87164 87

Specification and Grade	Product	Thickness (mm)	Yield stress** (MPa)	Tensile strength (MPa)
AS 1163-350	Hollow sections	All	350	450
AS 1204-350	Hot-rolled plates	≤ 12	360	480
		13 - 20	350	480
		21 - 80	340	480
		81 - 180	330	480
AS 1204-350	Hot-rolled sections	≤ 12	360	480
		13 - 40	340	480
		> 40	330	480
AS 1205-350	Weather resistant plates and sections	All	340	480
AS 1163-250	Hollow sections	All	250	350
AS 1204-250	Hot-rolled plates	≤ 8	280	410
		9 - 12	260	410
		13 - 20	250	410
		21 - 80	240	410
		81 - 180	230	410
AS 1204-250	Hot-rolled sections	≤ 12	260	410
		13 - 40	250	410
		> 40	230	410
AS 1163-200	Hollow sections	All	200	320
AS 1204-200	Plates	≤ 12	200	300

*Following is for the material of steel. Material for composite members is listed in the section on International Comparison and Comments

**General Comment: The individual magnitudes of the yield stress contained in specifications cannot be compared in a fully precise manner without considering the corresponding testing procedure applied and the method used for evaluation of the corresponding test results.

AS 1538 81

Specification	Product	Grade	Yield Stress (MPa)	Tensile Strength (MPa)
AS 1163	Hollow sections	C200 & H200	200	320
		C250 & H250	250	350
		C350	350	430
		H350	350	450
AS 1397	Coated sheet and strip	G550	550	550
		G500	500	520
		G450	450	500
		G350	350	420
		G300	300	370
		G250	250	320
AS 1594	Hot-rolled plate sheet and strip	HX340	340	420
		HX280	280	380
		HX240	240	340
		HX200	200	300
		HX1	(Note 2)	(Note 2)
		HX2	(Note 2)	(Note 2)
		HX3	(Note 2)	(Note 2)
		HX4	(Note 2)	(Note 2)
AS 1595 (Note 3)	Cold-rolled sheet and strip	CX85T	(Note 4)	550
		CX70T	(Note 4)	380
		CX60T	(Note 4)	310
		CX50T	(Note 4)	300
		CX1	(Note 4)	280
		CX2	(Note 4)	280
		CX3	(Note 4)	280
		CX4	(Note 4)	280

Notes:

1. Values in parenthesis are for information only.
2. For design purposes, yield stress and tensile strength will approximate the structural grade HX200
3. Steel is made to a nominal chemical composition.
4. Yield stress is not specified, but for design purposes these values may be obtained by the following:
   (a) from the manufacturer
   (b) by testing in accordance with AS 1391.
   (c) by using 170 MPa

CHINA

REGIONAL CONTRIBUTOR

Shao-Fan Chen, Xian Institute of Metallurgy and Construction Engineering

Materials Commonly Used

Specification	Steel grades				
	A3F, A3	16Mn	15MnV	16q	16Mnq
GBJ 17 88	X	X	X	X	X
GBJ 18 87	X	X	–	–	–
HRS	X	X	X	–	X
SDGJ 64 89	X	X	X	–	–
TBJ 2 85	–	–	–	X	X

Mechanical Properties of the Materials

Grade	Thickness (mm)	Yield Stress (MPa)	Tensile Strength (MPa)
A3F	Group 1	235	
	Group 2	215	370-460
	Group 3	205	
A3	Group 1	235	
	Group 2	225	370-460
	Group 3	215	
16Mn	$\leq$ 16	345	$\geq$ 510
	17-25	325	$\geq$ 490
	26-36	315	$\geq$ 470
	38-50	295	$\geq$ 470
15MnV	5-16	390	$\geq$ 530
	17-25	370	$\geq$ 510
	26-36	355	$\geq$ 490
16q	Not specified	225	$\geq$ 370
16Mnq	$\leq$ 16	345	$\geq$ 510
	17-25	325	$\geq$ 490
	26-36	305	$\geq$ 470
	38-50	285	$\geq$ 470

Grouping steel grades A3F and A3

Thickness (mm)			
Group	Bars	Shapes*	Plates
1	$\leq$ 40	$\leq$ 15	$\leq$ 20
2	40-100	15-20	20-40
3		over 20	40-60

* web thickness

EAST EUROPE

REGIONAL CONTRIBUTOR

Jindrich J. Melcher, Technical University of Brno

Different steel materials are used in the East European countries. The steel grades and designation are not yet unified and their system was developed mostly on a national basis. Thus, because of the large number and types of steel materials, only characteristic examples will be presented.

Comment on the Definition of Material Resistance (see SNiP 2.01.07-85)
The specified material resistance (design stress) R_d is defined as

$$R_d = \frac{R_y}{\gamma_m}$$

where R_y is the specified yield stress (with the statistical guarantee no less than 0.95) and γ_m is the material reliability factor considered in the range of 1.025 to 1.15 according to the type and kind of the steel used. The material reliability factor covers both the influence of the dispersion of the actual yield stress and the dispersion of the cross-sectional areas.

ČSN 73 1401 84

Steel Grade	Designation	Product	Thickness (mm)	Yield Stress (MPa)	Tensile Strength (MPa)
	11 373	Sheet & Plate	≤ 25	235	363 to 441
	11 373	Sheet & Plate	26 - 40	226	363 to 441
37	11 373	Shapes	≤ 13	235	363 to 441
	11 373	Shapes	> 13	226	363 to 441
	11 375	Sheet & Plate	≤ 25	235	363 to 441
	11 375	Plate	26 - 60	226	363 to 441
	11 375	Plate	61 - 80	216	343 to 441
37	11 375	Plate	81 - 120	206	343 to 441
	11 375	Sheet	26 - 50	226	363 to 441
	11 375	Shapes	≤ 13	235	363 to 441
	11 375	Shapes	> 13	235	363 to 441
	11 378	Sheet & Plate	≤ 25	235	363 to 441
	11 378	Plate	26 - 60	226	363 to 441
	11 378	Plate	61 - 100	216	343 to 441
37	11 378	Sheet	26 - 50	225	363 to 441
	11 378	Shapes	≤ 13	235	365 to 460
	11 378	Shapes	> 13	225	365 to 460
	11 523	Sheet & Plate	≤ 16	353	510 to 628
	11 523	Sheet & Plate	17 - 25	343	510 to 628
	11 523	Sheet & Plate	26 - 60	333	510 to 638
52	11 523	Plate	61 - 80	314	490 to 628
	11 523	Plate	81 - 100	304	490 to 628
	11 523	Shapes	≤ 13	353	510 to 628
	11 523	Shapes	> 13	343	510 to 628

For material selection the structures are divided into four groups according to their service conditions and type of connections.

The proper choice of the kind of steel depends also on the factor of guaranteed weldability (WG). Thus, for plates made of steel grade 37 the WG applies up to thickness of 16 mm for steel designation 11 373, up to 25 mm for material 11 375 and up to 60 mm for material 11 378. For plates made of steel 11 523 the WG applies up to a thickness of 25 mm.

Among other kinds of steel materials that can be mentioned are steel grade designated 11 353 for round tubing and the materials of steel grade 52 with the increased resistance under atmospheric corrosion 15 217 (ATMOFIX A) and 15 127 (ATMOFIX B).

MSZ 15024/1 85

Steel Grade	Designation	Product	Thickness (mm)	Yield Stress (MPa)	Tensile Strength (MPa)
37	A O	Sheet & Plate Shapes	≤ 40	220	305
37	A 34, A 34 X A 34 B	Sheet & Plate Shapes	≤ 40	220	330
37	A 38, A 38 X A38 B 37B, 37 C LK 37 C A 35 , A 35 K	Sheet & Plate Shapes	≤ 50	240	350
45	A 44, A 44 B	Sheet & Plate Shapes	≤ 40	280	400
45	45 B, 45 C, 45 D LK 45 C LK 45 D A 45, A 45 K	Sheet & Plate Shapes	≤ 40	300	420
52	52 C, 52 D LK 52 C LK 52 D A 52 K	Sheet & Plate Shapes	≤ 50	360	490
58	E 420 E E 420 D	Sheet & Plate Shapes	≤ 50	420	520
64	E 460 C E 460 D	Sheet & Plate Shapes	≤ 50	460	550

Detailed data are contained in Specifications, MSZ 500, MSZ 6280, MSZ 6259, MSZ 29/2 and MSZ 29/3.

PN 80/B-03200 84

Designation	Product	Thickness (mm)	Yield Stress (MPa)	Tensile Strength (MPa)
St OS		≤ 16	195	min. 315
St OS		17-40	185	min. 315
St3SX, St3SY, St3S	Sheet	≤ 16	235	375 to 460
St3SX, St3SY, St3S	Sheet	17 - 40	225	375 to 460
St3SX, St3SY, St3S	Sheet	41 - 100	215	375 to 460
St4VX, St4VY, St4V, St4W	Plate	≤ 16	255	410 to 510
St4VX, St4VY, St4V, St4W		≤ 40	245	410 to 510
18 G2, 18 G2A	Shapes	≤ 16	355	490 to 630
18 G2, 18 G2A		17 to 30	345	490 to 630
18 G2, 18 G2A		31 to 50	335	490 to 630
18 G2AV		≤ 16	440	560 to 720
18 G2AV		17 to 30	430	560 to 720
18 G2AV		31 to 50	420	560 to 720
10 HAV		≤ 20	390	min. 510

Also the materials for round tubing are given, namely R 35, R 45, and 10 BX with the yield stress 235, 255, and 205 MPa, respectively, and with the minimum tensile strength 345, 440, and 335, respectively.

SNiP II-23-81 82

Over 130 different kinds of materials according to the designation, type of product, and corresponding thickness are given in the Specification. Some of typical examples are presented here.

Designation	Product	Thickness (mm)	Yield Stress (MPa)	Tensile Strength (MPa)
18 kp	Plate	4 - 20	225	365
18 kp	Plate	21 - 40	215	365
18 ps	Plate	4 - 16	235	370
18 sp	Plate	4 - 20	235	370
18 Gps	Plate	4 - 20	235	370
18 Gps	Plate	21 - 30	225	370
18 Gps	Plate	31 - 40	235	390
18 kp	Shape	4 - 20	235	365

SNiP II-23-81 82 (CONTINUED)

Designation	Product	Thickness (mm)	Yield Stress (MPa)	Tensile Strength (MPa)
18 ps, 18 sp, 18G ps	Shape	4 - 20	245	370
18 Gps	Shape	21 - 30	225	370
18 Gps	Shape	31 - 40	235	390
VSt3 kp 2-1	Plate	4 - 10	225	355
VSt3 kp 2-1	Plate	11 - 20	215	345
VSt3 ps 6-1	Plate	4 - 10	235	365
VSt3 ps 6-1	Plate	11 - 20	235	355
VSt3 ps 6-2	Plate	4 - 10	275	370
VSt3 ps 6-2	Plate	11 - 20	265	365
VSt3 kp 2-1	Shape	4 - 10	235	365
VSt3 kp 2-1	Shape	11 - 20	225	355
VSt3 kp 2-1	Shape	21 - 30	215	345
VSt3 ps 6-1	Shape	4 - 10	245	370
VSt3 ps 6-1	Shape	11 - 20	245	365
VSt3 ps 6-1	Shape	21 - 30	225	355
VSt3 ps 6-2	Shape	4 - 10	275	380
VSt3 ps 6-2	Shape	11 - 20	275	370
VSt3 kp 2	Plate	41 - 100	205	365
VSt3 kp 2	Plate	$>$ 100	185	365
VSt3 kp 2	Shape	41 - 100	205	365
VSt3 kp 2	Shape	$>$ 100	185	365
10 G2 S1	Plate	5 - 9	345	490
10 G2 S1	Shape	5 - 9	345	490
10 G2 S1	Plate	10 - 20	335	480
10 G2 S1	Shape	10 - 20	335	480
10 G2 S1	Plate	21 - 32	325	470
10 G2 S1	Shape	21 - 32	325	470
10 G2 S1	Plate	33 - 60	325	450
10 G2 S1	Shape	33 - 60	325	450
10 G2 S1	Plate	61 - 100	295	430
10 G2 S1	Shape	61 - 100	295	430
15 Ch SND	Plate	4 - 32	345	490
15 Ch SND	Shape	4 - 9	345	490
15 Ch SND	Shape	10 - 32	325	470
10 Ch SND	Shape	4 - 15	390	530
10 Ch SND	Plate	4 - 32	390	530

SNiP II-23-81 82 (CONTINUED)

Designation	Product	Thickness (mm)	Yield Stress (MPa)	Tensile Strength (MPa)
10 Ch SND	Plate	33 - 40	390	510
15 G2 SMF	Plate	10 - 32	590	685
12 GN2 MFAJu	Plate	16 - 40	590	685

TGL 13503/01 82

This specification deals with the problems of stability without description of special kinds of material used.

According to the Specification TGL 13503/01 82 the four groups of steel strength are allowed for use in civil engineering structures, namely

Grades (Designation)	Yield Stress (MPa)	Tensile Strength (MPa)
S 38/24	240	380
S 45/30	300	450
S 52/36	360	520
S 60/45	450	600

Mechanical and other properties of mentioned steel grades are contained in the specifications TGL 7960, TGL 22 426 and TGL 28 192.

REGIONAL COMPARISON

The examples of materials presented cover the range of steel parameters used in related countries of the East European region.

The range of the yield stress of the steel materials commonly used for structural members is 185 to 590 MPa, and the range of the corresponding tensile strength is 315 to 685 MPa.

The comparison of the steel grades listed would require additional information, as for example information on chemical composition and on weldability of the steel.

So for instance, the following steel grades can be considered as equivalent:

11 373	–	Czechoslovakia
VSt 3kp	–	U.S.S.R.
St 3 S	–	Poland
37 B	–	Hungary
OL 38	–	Romania
St 38 u-2	–	G.D.R.

JAPAN

REGIONAL CONTRIBUTORS
Ben Kato, Toyo University
Koichi Takanashi, University of Tokyo

AIJ 70

Designation	Product	Thickness (mm)	Yield Stress (MPa)	Tensile Strength (MPa)
SS41, SM41		≤ 16	245	402
		17-40	235	402
		> 40	216	402
SS41	Plates & Shapes			
SM41A, SM41B	Plates & Shapes	≤ 100		
SM41C	Plates	≤ 50		
SMA41A, SMA41B	Plates & Shapes	6-50		
SMA41C	Plates	6-50		
SS50	Plates & Shapes	≤ 16	284	490
		17-40	275	490
		> 40	255	490
SM50		≤ 16	324	490
		17-40	314	490
		> 40	294	490
SM50A, SM50B	Plates & Shapes	≤ 100		
SM50C	Plates	≤ 50		
SM50Y		≤ 16	363	490
		17-40	353	490
		> 40	333	490
SM50YA, SM50YB	Plates & Shapes	≤ 50		
SMA50A, SMA50B	Plates & Shapes	6-50		
SMA50C	Plates	6-50		
SM53		≤ 16	363	520
		17-40	353	520
		> 40	333	520
SM53B	Plates & Shapes	≤ 50		
SM53C	Plates	≤ 50		
SS55	Plates & Shapes	≤ 16	402	539
		17-40	390	539
SM58	Plates	≤ 16	461	569
		17-40	451	569
		> 40	431	569

Remarks:

SS : Rolled Steel for General Applications
SM : Rolled Steel for Welded Structures
SMA : Hot-Rolled Atmospheric Corrosion Resisting Steel for Welded Structures

Designation	Product	Thickness (mm)	Yield Stress (MPa)	Tensile Strength (MPa)
STK41	Tubes	2-22	235	402
STK50	Tubes	2-22	314	490
STKR41	Square hollow	1.6-12	245	402
STKR50	Square hollow	1.6-12	324	490
SSC41	Cold form	$\leq$ 3.2	245	402

AIJ 90

Designation	Product	Thickness (mm)	Yield Stress (MPa)	Tensile Strength (MPa)
SS41, SM41, SMA41		$\leq$ 16	245	402
		17-40	235	402
		$>$ 40	216	402
SS41	Plates & Shapes			
SM41A, SM41B	Plates & Shapes	$\leq$ 100		
SM41C	Plates	$\leq$ 50		
SMA41A, SMA41B	Plates & Shapes	6-25		
SMA41C	Plates	6-50		
SM50		$\leq$ 16	324	490
		17-40	314	490
		$>$ 40	294	490
SM50A, SM50B	Plates & Shapes	$\leq$ 100		
SM50C	Plates	$\leq$ 50		
SM50Y, SMA50		$\leq$ 16	363	490
		17-40	353	490
		$>$ 40	333	490
SM50YA, SM50YB	Plates & Shapes	$\leq$ 50		
SMA50A, SMA50B	Plates & Shapes	6-50		
SMA50C	Plates	6-50		
SM53		$\leq$ 16	363	520
		17-40	353	520
		$>$ 40	333	520
SM53B	Plates & Shapes	$\leq$ 50		
SM53C	Plates	$\leq$ 50		
STK41	Tubes	2-22	235	402
STK50	Tubes	2-22	314	490
STKR41	Square hollow	1.6-12	245	402
STKR50	Square hollow	1.6-12	324	490
SSC41	Cold form	$\leq$ 3.2	245	402

Remarks:

SS : Rolled Steel for General Applications
SM : Rolled Steel for Welded Structures
SMA : Hot-Rolled Atmospheric Corrosion Resisting Steel for Welded Structures

JSCE 83

Designation	Product	Thickness (mm)	Yield Stress (MPa)	Tensile Strength (MPa)
SM41, SMA41		≤ 16	245	402
		17-40	235	402
		> 40	216	402
SM41A, SMA41A	Plates & Shapes	9-16		
SM41B, SMA41B	Plates & Shapes	17-25		
SM41C, SMA41C	Plates & Shapes	26-50		
SM50		≤ 16	324	490
		17-40	314	490
		> 40	294	490
SM50B	Plates & Shapes	9-25		
SM50C	Plates	26-50		
SM50Y, SMA50		≤ 16	363	490
		17-40	353	490
		> 40	333	490
SM50YB, SMA50B	Plates & Shapes	9-25		
SM53		≤ 16	363	520
		17-40	353	520
		> 40	333	520
SM53C	Plates	25-50		
SM58, SMA58		≤ 16	461	569
		17-40	451	569
		> 40	431	569
SM58Q, SMA58Q	Plates	9-50		

Remarks:

SM : Rolled Steel for Welded Structures

SMA : Hot-Rolled Atmospheric Corrosion Resisting Steel for Welded Structures

JRA 90

Designation	Product	Thickness (mm)	Yield Stress (MPa)	Tensile Strength (MPa)
SS41, SM41, SMA41		≤ 16	245	402
		17-40	235	402
		> 40	216	402
SS41	Plates & Shapes	8-22		
SM41A	Plates & Shapes	8-32		
SM41B	Plates & Shapes	8-38		
SM41C	Plates	8-50		
SMA41AW	Plates & Shapes	8-25		
SMA41BW	Plates & Shapes	8-38		
SMA41CW	Plates	8-50		

JRA 90 (CONTINUED)

Designation	Product	Thickness (mm)	Yield Stress (MPa)	Tensile Strength (MPa)
SM50		≤ 16	324	490
		17-40	314	490
		> 40	294	490
SM50A	Plates & Shapes	8-25		
SM50B	Plates & Shapes	8-38		
SM50C	Plates	8-50		
SM50Y, SMA50		≤ 16	363	490
		17-40	353	490
		> 40	333	490
SM50YA	Plates & Shapes	8-16		
SM50YB	Plates & Shapes	8-32		
SMA50AW	Plates & Shapes	8-16		
SMA50BW	Plates & Shapes	8-32		
SMA50CW	Plates	8-50		
SM53		≤ 16	363	520
		17-40	353	520
		> 40	333	520
SM53B	Plates & Shapes	8-38		
SM53C	Plates	8-50		
SM58, SMA58W		≤ 16	461	569
		17-40	451	569
		> 40	431	569
SM58	Plates	8-50		
SMA58W	Plates	8-50		

Remarks:

SS : Rolled Steel for General Applications
SM : Rolled Steel for Welded Structures
SMA : Hot-Rolled Atmospheric Corrosion Resisting Steel for Welded Structures

NORTH AMERICA

REGIONAL CONTRIBUTORS

Jerome S.B. Iffland, Iffland Kavanagh Waterbury, P.C.
John Springfield, Carruthers & Wallace Limited

AA 86

Products have been limited to those commonly used for structural members.

Designation	Product	Thickness (mm)	Yield Stress (MPa)	Tensile Strength (MPa)
B209-5086-H116, H32	Sheet & Plate	$\leq$ 50	195	275
B209-5086-H34	Sheet & Plate	$\leq$ 25	235	300
B210-5086-H34	Drawn Tube	$\leq$ 11.5	235	300
B209-5154-H38	Sheet	0.15 - 3.20	240	310
B209-5456-H16	Sheet & Plate	4 - 30	230	315
B209-5456-H321	Sheet & Plate	4 - 12.5	230	315
B209-5456-H321	Sheet & Plate	$\leq$ 40	215	305
B221-6005-T5	Extruded Shapes	$\leq$ 25	240	260
B209-6061-T6, T651	Sheet & Plate	$\leq$ 100	240	290
B221-6061-T6, T6510, T6511	Extruded Shapes	$\leq$ 25	240	260
B211-6061-T6, T651	Rolled Rod & Bar	$\leq$ 200	240	290
B210-6061-T6	Drawn Tube	0.6 - 12.5	240	290
B429-6061-T6	Extruded Pipe	$\leq$ 6.0	241	262
B221-6063-T5	Extruded Shapes	$\leq$ 12.5	110	150
B221-6063-T5	Extruded Shapes	$\leq$ 25	105	145
B221-6063-T6	Extruded Shapes	$\leq$ 25	170	205
B429-6063-T6	Extruded Pipe	$\leq$ 25	172	207
B221-6351-T5	Extruded Shapes	$\leq$ 25	240	260

AASHTO 89

Designation	Product	Thickness (mm)	Yield Stress (MPa)	Tensile Strength (MPa)
A36	Shapes	Groups 1,2,3,4 & 5	250	400
A36	Plates & Bars	$\leq$ 200	250	400
A572-Grade 50	Shapes	Groups 1,2,3,4 & 5	345	450
A572-Grade 50	Plates & Bars	$\leq$ 100	345	450
A514	Plates & Bars	$\leq$ 65	690	760
A514	Plates & Bars	$\leq$ 100	620	690
A517	Plates & Bars	$\leq$ 65	690	795
A517	Plates & Bars	66 - 100	620	725

AASHTO 89 (CONTINUED)

Designation	Product	Thickness (mm)	Yield Stress (MPa)	Tensile Strength (MPa)
A588	Shapes	Groups 1,2,3,4 & 5	345	485
A588	Plates & Bars	$\leq$ 100	345	485
A852	Plates & Bars	$\leq$ 100	485	590

AISC ASD 89

Materials for this Specification are identical to those of AISC LRFD 86.

AISC LRFD 86

Designation	Product	Thickness (mm)	Yield Stress (MPa)	Tensile Strength (MPa)
A36	Shapes	Groups 1,2,3,4 & 5	250	400
A36	Plates & Bars	$\leq$ 200	250	400
A36	Plates & Bars	$>$ 200	220	400
A53-Grade B	Pipe	$\leq$ 63.5	240	
A242	Shapes	Groups 1 & 2	345	485
A242	Shapes	Group 3	315	460
A242	Shapes	Groups 4 & 5	290	435
A242	Plates & Bars	$\leq$ 20	345	480
A242	Plates & Bars	21 - 40	315	460
A242	Plates & Bars	41 - 100	290	435
A441	Shapes	Groups 1 & 2	345	485
A441	Shapes	Group 3	315	460
A441	Shapes	Groups 4 & 5	290	435
A441	Plates & Bars	$\leq$ 20	345	485
A441	Plates & Bars	$\leq$ 40	315	460
A441	Plates & Bars	$\leq$ 100	290	435
A441	Plates & Bars	$\leq$ 200	275	415
A500-Grade A	Round Tubing	$\leq$ 15.88	228	310
A500-Grade B	Round Tubing	$\leq$ 15.88	290	400
A500-Grade C	Round Tubing	$\leq$ 15.88	317	427
A500-Grade A	Shaped Tubing	$\leq$ 15.88	269	310
A500-Grade B	Shaped Tubing	$\leq$ 15.88	317	400
A500-Grade C	Shaped Tubing	$\leq$ 15.88	345	427
A501	Tubing	$\leq$ 25.40	250	400
A514	Plates	$\leq$ 65	690	760
A514	Plates	$\leq$ 100	620	690
A529	Shapes	Group 1	290	415
A529	Plates & Bars	$\leq$ 13	290	415
A570-Grade 40	Sheet & Strip	$\leq$ 6.0	275	380
A570-Grade 45	Sheet & Strip	$\leq$ 6.0	310	415
A570-Grade 50	Sheet & Strip	$\leq$ 6.0	345	450

AISC LRFD 86 (CONTINUED)

Designation	Product	Thickness (mm)	Yield Stress (MPa)	Tensile Strength (MPa)
A572-Grade 42	Shapes	Groups 1,2,3	290	415
A572-Grade 50	Shapes	Groups 1,2,3,4 & 5	345	450
A572-Grade 60	Shapes	Groups 1 & 2	415	520
A572-Grade 65	Shapes	Group 1	450	550
A572-Grade 42	Plates & Bars	$\leq$ 150	290	415
A572-Grade 50	Plates & Bars	$\leq$ 100	345	450
A572-Grade 60	Plates & Bars	$\leq$ 32	415	520
A572-Grade 65	Plates & Bars	$\leq$ 32	450	550
A588	Shapes	Groups 1,2,3,4 & 5	345	485
A588	Plates & Bars	$\leq$ 100	345	485
A588	Plates & Bars	$\leq$ 125	315	460
A588	Plates & Bars	$\leq$ 200	290	435
A606-Hot Rolled Cut Lengths	Sheet & Strip	1.11 - 5.84	340	480
A606-Hot Rolled Coils & Annealed Metal	Sheet & Strip	1.11 - 5.84	310	450
A606- Cold Formed	Sheet & Strip	$\leq$ 6.0	310	450
A607-Grade 45	Sheet & Strip	$\leq$ 6.0	310	410
A607-Grade 50	Sheet & Strip	$\leq$ 6.0	340	450
A607-Grade 55	Sheet & Strip	$\leq$ 6.0	380	480
A607-Grade 60	Sheet & Strip	$\leq$ 6.0	410	520
A607-Grade 65	Sheet & Strip	$\leq$ 6.0	450	550
A607-Grade 70	Sheet & Strip	$\leq$ 6.0	480	590
A618-Grades Ia, Ib and II	Tubing	$\leq$ 19.05	345	485
A618-Grades Ia, Ib and II	Tubing	$\leq$ 38.1	315	460
A618-Grade III	Tubing	$\leq$ 19.05	345	450
A852	Plate	$\leq$ 100	485	590

AISI 76

Designation	Product	Thickness (mm)	Yield Stress (MPa)	Tensile Strength (MPa)
A53 Type F	Pipe	1.7 - 59.4	172	310
A53 Type E-Grade A	Pipe	1.7 - 59.4	207	330
A53 Type E-Grade B	Pipe	1.7 - 59.4	241	413
A53 Type S-Grade B	Pipe	1.7 - 59.4	241	413
A500-Grade A	Round Tubing	$\leq$ 15.88	228	310

AISI 76 (CONTINUED)

Designation	Product	Thickness (mm)	Yield Stress (MPa)	Tensile Strength (MPa)
A500-Grade B	Round Tubing	$\leq$ 15.88	290	400
A500-Grade C	Round Tubing	$\leq$ 15.88	317	427
A500-Grade A	Shaped Tubing	$\leq$ 15.88	269	310
A500-Grade B	Shaped Tubing	$\leq$ 15.88	317	400
A500-Grade C	Shaped Tubing	$\leq$ 15.88	345	427
A501	Tubing	$\leq$ 25.40	250	400
A618-Grades Ia, Ib & II	Tubing	$\leq$ 19.05	345	485
A618-Grades Ia, Ib & II	Tubing	$\leq$ 38.1	315	460
A618-Grade III	Tubing	$\leq$ 19.05	345	450
Type W-300	Hollow Sections	$\leq$ 16	300	410
Type W-350	Hollow Sections	$\leq$ 16	350	450
Type W-380	Hollow Sections	$\leq$ 16	380	480
Type WT-350	Hollow Sections	$\leq$ 16	350	480
Type WT-380	Hollow Sections	$\leq$ 16	380	480
Type A-350	Hollow Sections	$\leq$ 16	350	480
Type AT-350	Hollow Sections	$\leq$ 16	350	480

AISI 86

Designation	Product	Thickness (mm)	Yield Stress (MPa)	Tensile Strength (MPa)
A446-Grade A	Coated Sheet	$\leq$ 6.0	230	310
A446-Grade B	Coated Sheet	$\leq$ 6.0	255	360
A446-Grade C	Coated Sheet	$\leq$ 6.0	275	380
A446-Grade D	Coated Sheet	$\leq$ 6.0	345	450
A446-Grade E	Coated Sheet	$\leq$ 6.0	550	570
A446-Grade F	Coated Sheet	$\leq$ 6.0	345	480
A500-Grade A	Round Tubing	$\leq$ 15.88	228	310
A500-Grade B	Round Tubing	$\leq$ 15.88	290	400
A500-Grade C	Round Tubing	$\leq$ 15.88	317	427
A500-Grade A	Shaped Tubing	$\leq$ 15.88	269	310
A500-Grade B	Shaped Tubing	$\leq$ 15.88	317	400
A500-Grade C	Shaped Tubing	$\leq$ 15.88	345	427
A529	Plates & Bars	$\leq$ 13	290	415
A570-Grade 30	Sheet & Strip	$\leq$ 6.0	205	340
A570-Grade 33	Sheet & Strip	$\leq$ 6.0	230	360
A570-Grade 36	Sheet & Strip	$\leq$ 6.0	250	365
A570-Grade 40	Sheet & Strip	$\leq$ 6.0	275	380
A570-Grade 45	Sheet & Strip	$\leq$ 6.0	310	415
A570-Grade 50	Sheet & Strip	$\leq$ 6.0	345	450
A606- Hot Rolled Cut Lengths	Sheet & Strip	6.0	340	480

AISI 86 (CONTINUED)

Designation	Product	Thickness (mm)	Yield Stress (MPa)	Tensile Strength (MPa)
A606-Hot Rolled Coils & Annealed Met.	Sheet & Strip	6.0	310	450
A606-Cold Formed	Sheet & Strip	$\leq$ 6.0	310	450
A607-Grade 45	Sheet & Strip	$\leq$ 6.0	310	410
A607-Grade 50	Sheet & Strip	$\leq$ 6.0	340	450
A607-Grade 55	Sheet & Strip	$\leq$ 6.0	380	480
A607-Grade 60	Sheet & Strip	$\leq$ 6.0	410	520
A607-Grade 65	Sheet & Strip	$\leq$ 6.0	450	550
A607-Grade 70	Sheet & Strip	$\leq$ 6.0	480	590
A611-Grade A	Sheet	$\leq$ 6.0	170	290
A611-Grade B	Sheet	$\leq$ 6.0	205	310
A611-Grade C	Sheet	$\leq$ 6.0	230	330
A611-Grade D	Sheet	$\leq$ 6.0	275	360
A611-Grade E	Sheet	$\leq$ 6.0	550	565
A715-Grade 50	Sheet & Strip	$\leq$ 6.0	345	415
A715-Grade 60	Sheet & Strip	$\leq$ 6.0	415	485
A715-Grade 70	Sheet & Strip	$\leq$ 6.0	485	550
A715-Grade 80	Sheet & Strip	$\leq$ 6.0	550	620
A715-Grade 50	Sheet	$\leq$ 6.0	345	415
A715-Grade 60	Sheet	$\leq$ 6.0	415	485
A715-Grade 70	Sheet	$\leq$ 6.0	485	550
A715-Grade 80	Sheet	$\leq$ 6.0	550	620
A792-Grade 33	Coated Sheet	$\leq$ 5.0	230	310
A792-Grade 37	Coated Sheet	$\leq$ 5.0	255	360
A792-Grade 40	Coated Sheet	$\leq$ 5.0	275	380
A792-Grade 50	Coated Sheet	$\leq$ 5.0	345	450
A792-Grade 80	Coated Sheet	$\leq$ 5.0	550	570

Note: AISI 86 also permits use of cold formed members of materials with ASTM Designations A36, A242, A441, A572 and A588. For these Designations see Specification AISC LRFD 86.

API RP2A 89

Designation	Product	Thickness (mm)	Yield Stress (MPa)	Tensile Strength (MPa)
A36	Shapes	Groups 1,2, & 3	250	400
A36	Plates & Bars	$\leq$ 50.5	250	400
A53-Grade B	Pipe	1.7 - 59.5	240	415
A106-Grade B	Pipe	$\leq$ 63.5	240	415
A131-Grade A	Shapes	Group 1	235	400
A131-Grade A	Shapes	Groups 2,3,4 & 5	220	400
A131-Grades B,D C S & E	Plates & Bars	$\leq$ 200	235	400

API RP2A 89 (CONTINUED)

Designation	Product	Thickness (mm)	Yield Stress (MPa)	Tensile Strength (MPa)
A131-Grade AH32	Shapes	Groups 1,2,3,4 & 5	315	470
A131-Grade AH36	Shapes	Groups 1,2,3,4 & 5	350	490
A131-Grades DH32 & EH32	Plates & Bars	≤ 200	315	470
A131-Grades DH36 & EH36	Plates & Bars	≤ 200	350	490
A135-Grade B	Pipe	≤ 12.7	240	415
A139-Grade B	Pipe	≤ 25.4	240	415
A285-Grade C	Plates	≤ 19	205	380
A333-Grade 6	Pipe	≤ 8	240	415
A333-Grade 6	Tubing	≤ 8	240	415
A500-Grade A	Round Tubing	≤ 15.88	230	310
A500-Grade A	Shaped Tubing	≤ 15.88	270	310
A500-Grade B	Round Tubing	≤ 15.88	290	400
A500-Grade B	Shaped Tubing	≤ 15.88	320	400
A501	Tubing	≤ 25.4	250	400
A516-Grade 65	Plates	≤ 205	240	450
A524-Grade I	Pipe	≤ 9.52	240	415
A524-Grade II	Pipe	≤ 63.50	205	380
A537-Class I	Plates	≤ 65	345	485
A537-Class II	Plates	≤ 65	415	550
A572-Grade 42	Plates & Bars	≤ 50.4	290	415
A572-Grade 50	Plates & Bars	≤ 12.7	345	450
A572-Grade 42	Shapes	Groups 1,2, & 3	290	415
A573-Grade 65	Plates	≤ 40	240	450
A618-Grades Ia, Ib & II	Tubing	≤ 38.1	345	485
A633-Grade A	Shapes	Groups 1,2,3,4 & 5	290	430
A633-Grade C & D	Shapes	Groups 1,2 & 3	345	485
A633-Grade C & D	Shapes	Groups 4 & 5	315	450
A633-Grade A	Plates & Bars	≤ 100	290	430
A633-Grades C & D	Plates & Bars	≤ 65	345	485
A633-Grades C & D	Plates & Bars	66 - 100	315	450
A678-Grade A	Plates	≤ 40	345	485
A678-Grade B	Plates	≤ 65	415	550
A709-Grade 36T2	Shapes	Groups 1,2,3,4 & 5	250	400
A709-Grades 50T2 & 50T3	Shapes	Groups 1,2,3,4 & 5	345	450
A709-Grades 36T2	Plates & Bars	≤ 102	250	400
A709-Grades, 50T2 & 50T3	Plates & Bars	≤ 102	345	450
A710-Grade A, Class 3	Shapes	Groups 1, 2, & 3	515	585
A710-Grade A, Class 3	Shapes	Groups 4 & 5	450	515

API RP2A 89 (CONTINUED)

Designation	Product	Thickness (mm)	Yield Stress (MPa)	Tensile Strength (MPa)
A710-Grade A, Class 3	Plates & Bars	≤ 50	515	585
A710-Grade A, Class 3	Plates & Bars	51 - 100	450	515
A710-Grade A, Class 3	Plates & Bars	101 - 200	415	485

Designation (API)				
52-Grade X 42.2% max	Pipe	≤ 25.4	290	415
52-Grade X 52.2% max	Pipe	≤ 25.4	360	455
52-Grade X 52	Pipe	≤ 25.4	360	455
52-Grade B	Pipe	≤ 25.4	240	415
Spec. 2H-Grade 42	Plates	≤ 25.4	290	430
Spec. 2H-Grade 50	Plates	≤ 63.5	345	483
Spec. 2H-Grade 50	Plates	63.6 - 100	325	483
Spec. 2W-Grade 42	Plates	≤ 25.4	290 - 462	427
Spec. 2W-Grade 42	Plates	25.5 - 100	290 - 427	427
Spec. 2W-Grade 50	Plates	≤ 25.4	345 - 517	448
Spec. 2W-Grade 50	Plates	25.5 - 100	345 - 483	448
Spec. 2W-Grade 50T	Plates	≤ 25.4	345 - 522	483
Spec. 2W-Grade 50T	Plates	25.5 - 100	345 - 517	483
Spec. 2W-Grade 60	Plates	≤ 25.4	414 - 621	517
Spec. 2W-Grade 60	Plates	25.5 - 100	414 - 586	517
Spec. 2Y-Grade 42	Plates	≤ 25.4	290 - 462	427
Spec. 2Y-Grade 42	Plates	25.5 - 100	290 - 427	427
Spec. 2Y-Grade 50	Plates	≤ 25.4	345 - 517	448
Spec. 2Y-Grade 50	Plates	25.5 - 100	345 - 483	448
Spec. 2Y-Grade 50T	Plates	≤ 25.4	345 - 572	483
Spec. 2Y-Grade 50T	Plates	25.5 - 100	345 - 517	483
Spec. 2Y-Grade 60	Plates	≤ 25.4	414 - 621	517
Spec. 2Y-Grade 60	Plates	25.5 - 100	414 - 586	517

AREA 89

Designation	Product	Thickness (mm)	Yield Stress (MPa)	Tensile Strength (MPa)
A36	Shapes	Groups 1,2,3,4 & 5	250	400
A36	Plates & Bars	≤ 200	250	400
A36	Plates & Bars	> 200	220	400
A441	Shapes	Groups 1 & 2	345	485
A441	Shapes	Group 3	315	460
A441	Shapes	Groups 4 & 5	290	435
A441	Plates & Bars	≤ 20	345	485

AREA 89 (CONTINUED)

Designation	Product	Thickness (mm)	Yield Stress (MPa)	Tensile Strength (MPa)
A441	Plates & Bars	$\leq$ 40	315	460
A441	Plates & Bars	$\leq$ 100	290	435
A441	Plates & Bars	$\leq$ 200	275	415
A572-Grade 42	Shapes	Groups 1,2,3,4 & 5	290	415
A572-Grade 50	Shapes	Groups 1,2,3,4 & 5	345	450
A572-Grade 60	Shapes	Groups 1 & 2	415	520
A572-Grade 42	Plates & Bars	$\leq$ 150	290	415
A572-Grade 50	Plates & Bars	$\leq$ 100	345	450
A572-Grade 60	Plates & Bars	$\leq$ 32	415	520
A588	Shapes	Groups 1,2,3,4 & 5	345	485
A588	Plates & Bars	$\leq$ 100	345	485
A588	Plates & Bars	$\leq$ 125	315	460
A588	Plates & Bars	$\leq$ 200	290	435
A709-Grade 36	Shapes	Groups 1,2,3,4 & 5	250	400
A709-Grade 50	Shapes	Groups 1,2,3,4 & 5	345	450
A709-Grade 36	Plates & Bars	$\leq$ 102	250	400
A709-Grade 50	Plates & Bars	$\leq$ 102	345	450

ASME N 284 80

This Specification is not applicable to materials.

BSSC NEHRP 85

This specification covers all applications and materials of AISC ASD 89 and AISI 86.

CSA S6 M89 84a

Materials for this specification are identical to those for OHBDC 83.

CSA S16.1 M89 84b

Materials for this specification are identical to those for OHBDC 83 and CSA S136 M89 84c.

CSA S136 M89 84c

Materials in this specification are identical to those of AISI 86 plus the following additions:

Designation	Product	Thickness (mm)	Yield Stress (MPa)	Tensile Strength (MPa)
A816-Grade 50	Coated Sheet	$\leq$ 6.0	340	
A816-Grade 60	Coated Sheet	$\leq$ 6.0	410	
A816-Grade 70	Coated Sheet	$\leq$ 6.0	480	
A816-Grade 80	Coated Sheet	$\leq$ 6.0	550	

Note: CSA S136 M89 84c also permits use of cold formed members of materials with ASTM designations A36, A242, A441, A572 and A588 and also all CSA designation G40.21 materials listed under specification OHBDC 83.

OHBDC 83

Designation	Product	Thickness (mm)	Yield Stress (MPa)	Tensile Strength (MPa)
Type G-230	Shapes	Groups 1,2,3,4 & 5	230	380
Type G-350	Shapes	Groups 1,2 & 3	350	480
Type G-400	Shapes	Groups 1,2 & 3	400	550
Type G-230	Plates & Bars	$\leq$ 300	230	380
Type G-350	Bars	$\leq$ 30	350	480
Type G-400	Bars	$\leq$ 30	400	550
Type W-260	Shapes	Groups 1,2,3 & 4	260	410
Type W-300	Shapes	Groups 1 & 2	300	450
Type W-300	Shapes	Groups 3 & 4	290	450
Type W-350	Shapes	Groups 1 & 2	350	450
Type W-400	Shapes	Groups 1 & 2	400	520
Type W-480	Shapes	Groups 1 & 2	480	590
Type W-260	Plates & Bars	$\leq$ 65	260	410
Type W-260	Plates & Bars	$\leq$ 100	250	410
Type W-300	Plates & Bars	$\leq$ 40	300	450
Type W-300	Plates & Bars	$\leq$ 65	290	450
Type W-300	Plates & Bars	$\leq$ 100	280	450
Type W-350	Plates & Bars	$\leq$ 40	350	450
Type W-350	Plates & Bars	$\leq$ 65	330	450
Type W-350	Plates & Bars	$\leq$ 100	320	450
Type W-400	Plates & Bars	$\leq$ 20	400	520
Type W-480	Plates & Bars	$\leq$ 20	480	590
Type WT-260	Shapes	Groups 1,2,3 & 4	260	410
Type WT-260	Shapes	Group 5	250	410
Type WT-300	Shapes	Groups 1 & 2	300	450
Type WT-300	Shapes	Groups 3 & 4	290	450
Type WT-300	Shapes	Group 5	280	450
Type WT-350	Shapes	Groups 1 & 2	350	480
Type WT-350	Shapes	Groups 3 & 4	330	480
Type WT-400	Shapes	Groups 1 & 2	400	520
Type WT-480	Shapes	Groups 1 & 2	480	590
Type WT-260	Plates & Bars	$\leq$ 65	260	410

OHBDC 83 (CONTINUED)

Designation	Product	Thickness (mm)	Yield Stress (MPa)	Tensile Strength (MPa)
Type WT-260	Plates & Bars	≤ 100	250	410
Type WT-300	Plates & Bars	≤ 40	300	450
Type WT-300	Plates & Bars	≤ 65	290	450
Type WT-300	Plates & Bars	≤ 100	280	450
Type WT-350	Plates & Bars	≤ 40	350	480
Type WT-350	Plates & Bars	≤ 60	330	480
Type WT-400	Plates & Bars	≤ 20	400	590
Type WT-480	Plates & Bars	≤ 20	480	590
Type R-350	Shapes	Groups 1 & 2	350	480
Type R-350	Plates & Bars	≤ 14	350	480
Type A-350	Shapes	Groups 1,2,3 & 4	350	480
Type A-350	Shapes	Group 5	320	480
Type A-400	Shapes	Groups 1 & 2	400	520
Type A-350	Plates & Bars	≤ 100	350	480
Type A-400	Plates & Bars	≤ 40	400	500
Type A-480	Plates & Bars	≤ 20	480	590
Type Q-700	Plates & Bars	≤ 50	700	800
Type QT	Plates & Bars	≤ 50	700	800

REGIONAL COMPARISONS

Mechanical Properties of Material listed in Specifications

Specification	M1	M2	M3	M4	M5	M6	M7	M8
AA 86	X	X	X	–	–	–	–	–
AASHTO 89	X	X	X	–	–	–	–	–
AISC ASD 89	X	X	–	–	–	–	–	–
AISC LRFD 86	X	X	–	–	–	–	–	–
AISI 76	X	X	–	–	–	–	–	–
AISI 86	X	X	–	–	–	–	–	–
API RP2A 89	X	X	–	–	–	–	–	–
AREA 89	X	X	X	–	–	–	–	–
ASME N284 80				Not Applicable				
BSSC NEHRP 85	–	–	–	–	–	–	–	–
CSA S6 M89 84a	X	X	X	–	–	–	–	–
CSA S16.1 M89 84b	X	X	–	–	–	–	–	–
CSA S136 M89 84c	X	X	–	–	–	–	–	–
FHWA TS 80	–	–	–	–	–	–	–	X
OHBDC 83	X	X	X	–	–	–	–	–

Key:
M1 = yield stress
M2 = minimum tensile strength
M3 = fracture toughness
M4 = crack propagation rate
M5 = properties corresponding to elevated temperature
M6 = properties corresponding to low temperature
M7 = properties corresponding to high response rate
M8 = by reference to other standards

WEST EUROPE

REGIONAL CONTRIBUTOR
John E. Harding, University of Surrey

EC 3 88

Structural steels in accordance with EN 10025

Designation	Product	Application	Thickness (mm)	Yield Stress (MPa)	Tensile Strength (MPa)
Fe E 235	Hot-rolled shapes	All structures	≤ 40	235	360
Fe E 235	Hot-rolled shapes	All structures	41 - 100	215	340
Fe E 275	Hot-rolled shapes	All structures	≤ 40	275	430
Fe E 275	Hot-rolled shapes	All structures	41 - 100	255	410
Fe E 355	Hot-rolled shapes	All structures	≤ 40	255	510
Fe E 355	Hot-rolled shapes	All structures	41 - 100	335	490

Note: Use data in EN 10025 for under range of thickness for plates analysis, $F_u/F_y \geq 1.2$

BS 5950 P.1 85

Structural steels in accordance with BS 4360. Maximum thickness requirements for plates in tension.

Designation	Product	Application	Thickness (mm)	Tensile Strength (MPa)
43A,B & C	Steel section, plates, hollow section	Buildings	≤ 16	275
43A,B & C		Buildings	17 - 40	265
43A,B & C		Buildings	41 - 100	245
50 B & C	Steel section, plates, hollow section	Buildings	≤ 16	355
50 B & C		Buildings	17 - 63	340
50 B & C		Buildings	64 - 100	325
55 C	Steel section, plates, hollow section	Buildings	≤ 16	450
55 C		Buildings	17 - 25	430
55 C		Buildings	26 - 40	415

INTERNATIONAL COMPARISONS & COMMENT

WHAT ARE THE DIFFERENCES?

Steel & Concrete Material Limitations

The second section of the General Provisions is designated "Material". Since it was agreed that this would be for comparison purposes, it was decided to provide quantitative data on the materials covered by each specification. For each specification, for each steel material covered, the thickness limitations, the minimum yield stress, and the minimum tensile strength are tabulated. The comparison, in matrix form, provides information on the specification requirements for material properties. A similar coverage is next provided for steel-concrete composite materials for all regions.

Material Properties

Specification	Concrete Strength (MPa)		Min. Concrete Density (kg/m3)**	Steel Strength (MPa)
	Minimum	Maximum		Maximum
AIJ 86	Fcyl=14.7 EC	–	2,500	343
	Fcyl=17.6 EC	–	2,500	392
	Fcyl=14.7 FC	–	2,500	343
	Fcyl=17.6 FC	–	2,500	392
AISC LRFD 86	Fcyl=20	Fcyl=50	–	365
ACI 318 83	Fcyl=17	–	–	330
BS 5400 P.5 82	Fcu=20 EC	–	–	–
	Fcu=25 FC	–	–	–
CSA A23.3 M84 77	Fcyl=20 EC	–	–	400 EC
DIN 18800 P.2 86	Fcu=25	–	–	–
EC 4 85	Fcyl=20	–	–	450 (methods A&B)
	Fcyl=20	–	–	355 (method C)

**(Including bars and shapes)

where

EC = encased columns
FC = filled columns
Fcu = cube strength
Fcyl = cylinder strength = approx. 0.80 Fcu

Composite Members – Concrete and Structural Reinforcing Steel Limitations

Specification	Steel Capacity / Total Capacity	Concrete Capacity / Total Capacity	Max. Steel Area / Gross Area	Min. Steel Area / Gross Area
AIJ 86	–	–	–	0.008
AISC LRFD 86	–	–	–	0.04
ACI 318 83	–	–	0.08(FC) 0.08(EC)	– 0.01 (EC)
BS 5400 P.5 82	–	0.15 Min., 8 Max. (EC) 0.10 Min., 8 Max. (FC)	–	–
CSA A23.3 M84 77	–	–	0.20(EC)	–
DIN 18800 P.2 86	0.2 Min., 0.9Max.	–	–	0.03
EC 4 85	0.2 Min., 0.9Max.	–	–	–

where

EC = encased columns
FC = filled columns

WHY ARE THERE DIFFERENCES?

The differences related to material depend on:

(a) different levels of "know-how" in the steel industry, and in technology in individual countries,

(b) the individual magnitudes of the yield stress and other properties of material contained in specifications cannot be compared without considering the corresponding testing procedure, definitions of individual properties, and the methods used for evaluation of the test results.

GPc. Design Basis

SPECIFICATION PROVISIONS & EXPLANATION

AUSTRALIA

REGIONAL CONTRIBUTOR
Nicholas S. Trahair, The University of Sydney

AS 1250 81

Allowable Stress Design[2]

Member factors of safety $= 1/0.6 \approx 1.67$ generally

Load combination

$$X_D + X_L \qquad \text{(GPc.A1)}$$
$$0.75\,(X_D + X_L + X_W)$$
$$X_D{}^1 + X_W{}^1$$

where

X_D = dead load
X_L = live load
X_W = wind load
$X_D{}^1$ = minimum dead load
$X_W{}^1$ = wind load opposing dead load

Plastic Design

Load factor $= 1/0.6 \approx 1.67$

AISC 83

Same as for AS 1250 81, allowable stress design, using appropriate allowable loads.

AS 1538 81

Same as for AS 1250 81, Allowable Stress Design.

[2]General Comment: The individual magnitudes of Safety Factors, Loads and Load Factors, Resistance Factors, etc. cannot be compared without considering the corresponding reliability format applied in the standard.

AS DR87164 87

Method of Design

Limit states for strength and serviceability

$$S \leq \phi R \qquad \text{(GPc.A2)}$$

where

R = nominal design capacity
S = design action effect
ϕ = capacity factor

Design Action Effects

(A) Ultimate (strength) limit states

S is determined from the most severe of the following possible load combinations

$1.25\ G + 1.5\ Q$
$1.25\ G + W_u + \psi_c\ Q$
$0.8\ G + 1.5\ Q$
$0.8\ G + W_u$

where

G = nominal dead load
Q = nominal live load
W_u = nominal ultimate wind load
ψ_c = 0, 0.4, 0.6 depending on type of load

(B) Serviceability limit states

S is determined from the appropriate one of the load combinations

$G,\ W_s,\ \psi_s Q,\ \psi_l Q$
$G + W_s,\ G + \psi_s Q,\ G + \psi_l Q$
$G + \psi_l Q + W_s$

where

W_s = nominal serviceability wind load
ψ_s = short term live load factor (0.7, 1.0)
ψ_l = long term live load factor (0, 0.4, 0.6)

Strength Capacity Factors

Members, piles in bearing, butt welds	$\phi =$ 0.90
Individual bolts, fillet, plug and shot welds, weld groups	0.80
Pins, bolt groups, other connection components	0.70

CHINA

REGIONAL CONTRIBUTOR

Shao-Fan Chen, Xian Institute of Metallurgy and Construction Engineering

GBJ 17 88, GBJ 18 87, HRS 87, and SDGJ 64 89

All specifications follow the instructions of the UNIFIED CODE FOR THE DESIGN OF BUILDING STRUCTURES (GBJ 68 84) which has adopted the limit states design approach based on probabilistic theory. The reliability index β for members ranges from 2.7 to 3.7 with 3.2 for regular building structures. The general expression for member capacity assessment in case of basic load combination is

$$\gamma_o[\,1.2\ S_{GK} + 1.4(S_{Q1K} + \sum \Psi_{ci}\ S_{QiK})] \le f_y/\gamma_R \qquad \text{(GPc.C1)}$$

where

- f_y = specified minimum yield stress of steel
- S_{GK} = dead load effect
- S_{QiK} = other live load effects
- S_{Q1K} = first (dominant) live load effect
- γ_o = coefficient equal to 1.1, 1.0, 0.9 in accordance to the importance of the structure
- γ_R = resistance factor for member
  - = 1.087 for all members made of hot-rolled A3F, A3, 16Mn and 16Mnq steel
  - = 1.111 for members made of hot-rolled 15MnV steel
  - = 1.165 for cold-formed steel members
- Ψ_{ci} = combination factor for the ith variable load effect, taken as 0.6 when wind load is included in the combination and as 1.0 without wind load

The magnitude of nominal dead loads and live loads are given in the "SPECIFICATION OF BUILDING STRUCTURE LOADINGS" (GBJ 9 87) and the load factors specified in this document are 1.2 and 1.4 for permanent and variable loads, respectively, as shown in Eq. GPc.C1.

For extraordinary combinations which include accidental loads, reference is made to a special code for load magnitude, load combinations, and member capacity.

TBJ 2 85

This specification refers to the allowable stress method with a basic safety factor of 1.70. For beams and girders, the allowable stress is increased by 5% as compared to axially loaded members.

Loads acting on bridges and culverts are classified as:

1. Primary force, including DL, LL and impact
2. Secondary force, including wind, and transverse and longitudinal horizontal forces
3. Special forces, including seismic and construction loads

For different load combinations, different correction factors for the allowable stress shall be taken as follows

Primary loads only	1.0
Primary plus braking loads	1.25
Primary plus wind or other horizontal transverse loads	1.20
Primary loads plus secondary stress plus braking load (or wind or horizontal transverse load)	1.45
Primary loads (construction phase)	1.20
Primary loads and wind (construction phase)	1.30-1.40

EAST EUROPE

REGIONAL CONTRIBUTOR
Jindrich J. Melcher, Technical University of Brno

A semiprobabilistic Limit States Design method has been introduced in the specifications for structural steel design in all East European countries during the last three decades. The general principles of reliability assessment ("rules of the game") are contained in special documents (that is, ČSN 73 0031 Design of Building Structures and Foundations, ÚNM Czechoslovakia) common for all building materials and all types of load carrying structures.

The data on loading and load combinations are available in documents also common for structural systems in steel, concrete, composite, and timber. For building structures the loading, load factors, and load combinations are specified in SNiP 2.01.07-85 86 and ČSN 73 0035 86.

The differences among the national design methods are not substantial. Thus the basic principles will be explained using SNiP 2.01.07-85.

Classification of the Loads

In general, the loads have two characteristic values:

- specified load F_s
  (actual maximum magnitude of load under actual service conditions)
- design load $F_d = \gamma_f F_s$
  (F_d = factored specified load; γ_f = load factor)

The classification includes these types of the loads:

(A) Permanent (dead) loads F_p

- dead loads of metal structure	$\gamma_f = 1.05$ or 0.9
- concrete construction	$\gamma_f = 1.1$ or 0.9
- cladding, insulation, roofing and fill to form roof slopes	$\gamma_f = 1.3$ or 0.9
- earth fill	$\gamma_f = 1.15$ or 0.9

(B) Temporary (live) Loads

(1) Long-term temporary loads F_{tl}

- stationary machinery and appliances	$\gamma_f = 1.05$
- vibration and sound isolating materials	$\gamma_f = 1.2$
- live load (on mechanical floors, computer centers)	$\gamma_f = 1.3$(for the $F_s < 2$ kPa) $\gamma_f = 1.2$(for the $F_s \geq 2$ kPa)

(2) Short-term temporary loads F_{ts}

- live load (in residential buildings)	$\gamma_f = 1.3$ (for the $F_s < 2$ kPa) $\gamma_f = 1.2$ (for the $F_s \geq 2$ kPa)
- snow load	$\gamma_f = 1.4$
- wind load	$\gamma_f = 1.2$ to 1.4 (according to the shape, equipment, and size of the building)
- ice load	$\gamma_f = 1.3$
- crane loads	$\gamma_f = 1.1$ to 1.6 (according to the type of the crane and its lifting capacity)

(3) Extraordinary temporary load F_{te}

- earthquake loads
- explosion

} γ_f is given in special codes and technical documents

Load Combinations[3]

(A) The Basic Combination (BC)

The BC takes into account the permanent loads F_p and the temporary loads F_{tl}, F_{ts} according to the following scheme (the load factors γ_f correspond to the particular type of loading)

$$
\begin{aligned}
(S)_{BC} = {} & (\gamma_f\, F_{p1} + \gamma_f\, F_{p2} + \cdots + \gamma_f\, F_{pi}) + \\
& + \psi_1\, (\gamma_f F_{tl1} + \gamma_f\, F_{tl2} + \cdots + \gamma_f\, F_{tlj}) + \\
& + \psi_2\, (\gamma_f\, F_{ts1} + \gamma_f\, F_{ts2} + \cdots + \gamma_f\, F_{tsk})
\end{aligned}
$$

for $(j + k) \geq 2$ is $\psi_1 = 0.95$, $\psi_2 = 0.9$

for $(j + k) < 2$ is $\psi_1 = 1.0$, $\psi_2 = 1.0$

where ψ_1, ψ_2 are the coefficients of load combinations.

(B) Extraordinary Combination

The appropriate scheme is:

$$
\begin{aligned}
(S)_{EC} = {} & (\gamma_f\, F_{p1} + \gamma_f\, F_{p2} + \cdots + \gamma_f\, F_{pi}) \\
& + \psi_1\, (\gamma_f\, F_{tl1} + \gamma_f F_{tl2} + \cdots + \gamma_f\, F_{tlj}) \\
& + \psi_2\, (\gamma_f\, F_{ts1} + \gamma_f F_{ts2} + \cdots + \gamma_f\, F_{tsk}) + \gamma_f\, F_{te1}
\end{aligned}
$$

where

$\psi_1 = 0.95$

$\psi_2 = 0.8$

This combination takes into account just one decisive type of extraordinary temporary load.

[3]The Czechoslovak code ČSN 73 0035 88 also allows one to consider application of probabilistic methods (Marek and Guštar, 1989)

Other Reliability Factors

(A) Coefficient of the use of structure (Importance factor): γ_a

According to the significance of the building or construction the three categories of structures with different importance factor γ_a are considered, namely

Group I: $\gamma_a = 1.0$
Buildings and structures with very special social and economic importance

Group II: $\gamma_a = 0.95$
Structures of somewhat less importance

Group III: $\gamma_a = 0.9$
Structures of secondary importance

(B) Coefficient of the behavior conditions (Adjustment factor): γ_b

Special cases of behavior of elements and the deviations from the current analysis of structures are covered by the so-called adjustment factor, the purpose of which is to consider possible inaccuracies of the analysis.

The range of the value of the adjustment factor is

$$\gamma_b = 0.70 \text{ to } 1.1$$

according to the type of member and its connections (for lateral buckling of beams for example, $\gamma_b = 0.95$; for compression angle struts with one-leg connections $\gamma_b = 0.75$, for columns in buildings $\gamma_b = 0.95$ etc).

The coefficients γ_a and γ_b can be considered as the correction factors of the design stress according to the formula

$$R_{d,ef} = R_d \frac{\gamma_b}{\gamma_a} \qquad \text{(GPc.E1)}$$

The Reliability Conditions

(A) Ultimate Limit States

The maximum effect of factored loads (considering their combination) S_{max} must be less or equal to the defined minimum strength

$$S_{max} \leq R_d \frac{\gamma_b}{\gamma_a} \qquad \text{(GPc.E2)}$$

where

R_d = design yield stress

γ_b, γ_a = coefficients defined earlier

(B) Serviceability Limit States

The load factor $\gamma_f = 1.0$

JAPAN

REGIONAL CONTRIBUTORS
Ben Kato, Toyo University
Koichi Takanashi, University of Tokyo

AIJ 70

This specification provides for allowable stress design of building structures.

Factor of Safety

Tensile stress	
F.S. on yield strength	1.5
Shear stress	
F.S. on yield stress	1.5
Compressive stress	
F.S. on buckling strength	1.5 ~ 2.13
Compressive stress for bending members	
F.S. on buckling strength	1.5

50% increase in allowable stresses is provided for short duration loads, such as snow, wind, and seismic forces.

Load Combinations

(A) Long Duration loads

$$D + L$$
$$D + L + SW$$

(B) Short duration loads

$$D + L + SW$$
$$D + L + W$$
$$D + L + SW + W$$
$$D + L + EQ$$
$$D + L + SW + EQ$$

where

D = dead load
EQ = seismic load
L = live load
SW = snow load
W = wind pressure

AIJ 90

This specification provides for design of building structures, and adopts the limit states design concept. The limit states are classified as ultimate limit states and the serviceability limit states. The reliability index β ranges from 1.5 ~ 2.5 for the ultimate limit states and assigns 1.0 for the serviceability limit states.

Ultimate Limit State

$$\phi R_n \geq 1.1\,D + 1.6\,L$$
$$\phi R_n \geq 1.1\,D + 1.6\,SW + 0.6\,L$$
$$\phi R_n \geq 1.1\,D + 2.0\,EQ + 0.4\,L$$
$$\phi R_n \geq 1.1\,D + 1.6\,W + 0.6\,L$$
$$\phi R_n \geq 0.9\,D - 1.6\,W$$

For heavy snowfall regions

$$\phi R_n \geq 1.1\,D + 1.5\,SW + 0.5\,W + 0.4\,L$$
$$\phi R_n \geq 1.1\,D + 1.7\,EQ + 0.4\,SW + 0.4\,L$$

where

D = dead load
EQ = earthquake
L = live load
W = wind load on structure
SW = snow load
ϕ = 0.85, 0.9 for flexural and lateral buckling, and web crippling, respectively
= 0.9 for local buckling
= 0.9 for others

Serviceability Limit States

$$\phi R_n \geq 1.0\,D + 1.0\,L$$
$$\phi R_n \geq 1.0\,D + 0.9\,SW + 0.6\,L$$
$$\phi R_n \geq 1.0\,D + 0.4\,EQ + 0.4\,L$$
$$\phi R_n \geq 1.0\,D + 0.9\,W + 0.4\,L$$
$$\phi R_n \geq 1.0\,D - 0.9\,W$$

For heavy snowfall regions

$$\phi R_n \geq 1.0\,D + 0.9\,SW + 0.5\,W + 0.4\,L$$
$$\phi R_n \geq 1.0\,D + 0.4\,EQ + 0.4\,SW + 0.4\,L$$

where

ϕ_d = 1.0 for deflection and drift
ϕ_F = 1.0 for fatigue
ϕ_s = 0.9 for yielding, buckling, and slipping

JRA 90

This specification provides for the allowable stress design of highway bridges. Tensile, compressive, and building allowable stresses are directly assigned. Safety factors are not explicit.

Principal loads (P)

1. Dead load D
2. Live load L
3. Impact load I
4. Prestressed load PS
5. Effect of creep CR
6. Effect of shrinkage SH
7. Earth pressure E
8. Water pressure HP
9. Buoyancy U

Secondary loads (S)	10. Wind pressure W
	11. Effect of temperature T
	12. Earthquake EQ
Supplementary loads (PP)	13. Snow load SW
	14. Effect of earth movement GD
	15. Support movement SD
	16. Wave pressure WP
	17. Centrifugal force CF
Exceptional loads (PA)	18. Brake reaction BK
	19. Erection: temporary loads ER
	20. Vehicle collision loads CO
	21. Others

For load combinations, the sum of the loads can be reduced by the reduction factors below:

P + PP + T	Reduction factor	1.15	
P + PP + W	Reduction factor	1.25	
P + PP + T + W	Reduction factor	1.35	
P + PP + BK	Reduction factor	1.25	
P + PP + CO	Reduction factor	1.70	(for steel structure members)
P + PP + CO	Reduction factor	1.50	(for reinforced concrete structural members)
W	Reduction factor	1.20	
BK	Reduction factor	1.20	
P(except L+I) + EQ + T	Reduction factor	1.70	
ER	Reduction factor	1.25	

NORTH AMERICA

REGIONAL CONTRIBUTORS
Jerome S.B. Iffland, Iffland Kavanagh Waterbury, P.C.
John Springfield, Carruthers & Wallace Limited

AA 86

Allowable stress design method is used.
Factors of Safety

	Buildings and similar type structures	Bridges and similar type structures
Tension Members		
F.S. on tensile strength, n_u	1.95	2.20
F.S. on yield strength, n_y	1.65	1.85
Columns		
F.S. on buckling strength, n_y	1.95	2.20
F.S. on crippling strength of thin sections, n_u	1.95	2.20
F.S. on yield strength for short columns, n_y	1.65	1.85
Beams		
F.S. on tensile strength, n_u	1.95	2.20
F.S. on tensile yield strength, n_y	1.65	1.85
F.S. on compressive yield strength for short beams, n_y	1.65	1.85
F.S. on buckling strength, n_y	1.65	1.85
F.S. on crippling strength of thin sections, n_y	1.65	1.85
F.S. on shear buckling of webs, n_a	1.20	1.35

Load Combinations

33% Increase in allowable stresses for wind and seismic forces.

AASHTO 89

Allowable stress design method is used.
Factors of Safety Based on Yield Stress (F_y or ultimate stress F_u as noted)

Axial tension in members with no holes for high strength bolts or rivets

A36, A572, A588, A852	F_y	1.82
A514, A517	F_u	2.17

Axial tension in members with holes for H.S. bolts or rivets and tension in rolled beams and built-up members subject to bending.

A36, A572, A588, A852

Gross section	F_y	1.82
Net section	F_u	2.00

A514, A517		
Gross section	F_u	2.17
Net section	F_u	2.17
Axial compression		
Gross section, Stiffeners, Splices	F_y	1.82
Compression in extreme fibers of members subjected to bending		
Fully supported	F_y	1.82
Partially supported	F_y	1.82
Compression in columns based on buckling load	F_y	2.12
Shear in girder webs	F_y	3.00

Load Factor Design

Members shall be proportioned so that their combined minimum strengths shall be at least equal to the total effect of design loads multiplied by the respective correction factors given under load combinations.

Load Combinations (see Table of Coefficients γ and β)

$$\text{Group (N)} = \gamma[\beta_D D + \beta_L(L+I) + \beta_C CF + \beta_E E + \beta_B B + \beta_S SF + \beta_W W + \beta_{WL} WL + \beta_L LF + \beta_R(R+S+T) + \beta_{EQ} EQ + \beta_{ICE} ICE]$$

where

B = buoyancy
CF = centrifugal force
D = dead load
E = earth pressure
EQ = earthquake
I = impact
ICE = ice pressure
L = live load
LF = longitudinal force from live load
(N) = group number (see Table of Coefficients)
R = rib shortening
S = shrinkage
SF = stream flow pressure
T = temperature
W = wind load on structure
WL = wind load on live load – 100 pounds per linear foot
β = coefficient (see Table of Coefficients)
γ = load factor (see Table of Coefficients)

Table of Coefficients γ and β

Col. No.	1	2	3	3A	4	5	6	7	8	9	10	11	12	13	14
						β FACTORS									
GROUP	γ	D	$(L+I)_n$	$(L+I)_p$	CF	E	B	SF	W	WL	LF	R+S+T	EQ	ICE	%
(ALLOWABLE STRESS DESIGN)															
I	1.0	1	1	0	1	β_E	1	1	0	0	0	0	0	0	100
IA	1.0	1	2	0	0	0	0	0	0	0	0	0	0	0	150
IB	1.0	1	0	1	1	β_E	1	1	0	0	0	0	0	0	—
II	1.0	1	0	0	0	1	1	1	1	0	0	0	0	0	125
III	1.0	1	1	0	1	β_E	1	1	0.3	1	1	0	0	0	125
IV	1.0	1	1	0	1	β_E	1	1	0	0	0	1	0	0	125
V	1.0	1	0	0	0	1	1	1	1	0	0	1	0	0	140
VI	1.0	1	1	0	1	β_E	1	1	0.3	1	1	1	0	0	140
VII	1.0	1	0	0	0	1	1	1	0	0	0	0	1	0	133
VIII	1.0	1	1	0	1	1	1	1	0	0	0	0	0	1	140
IX	1.0	1	0	0	0	1	1	1	1	0	0	0	0	1	150
X	1.0	1	1	0	0	β_E	0	0	0	0	0	0	0	0	100
(LOAD FACTOR DESIGN)															
I	1.3	β_D	1.67	0	1.0	β_E	1	1	0	0	0	0	0	0	
IA	1.3	β_D	2.20	0	0	0	0	0	0	0	0	0	0	0	
IB	1.3	β_D	0	1	1.0	β_E	1	1	0	0	0	0	0	0	
II	1.3	β_D	0	0	0	β_E	1	1	1	0	0	0	0	0	
III	1.3	β_D	1	0	1	β_E	1	1	0.3	1	1	0	0	0	
IV	1.3	β_D	1	0	1	β_E	1	1	0	0	0	1	0	0	
V	1.25	β_D	0	0	0	β_E	1	1	1	0	0	1	0	0	
VI	1.25	β_D	1	0	1	β_E	1	1	0.3	1	1	1	0	0	
VII	1.3	β_D	0	0	0	β_E	1	1	0	0	0	0	1	0	
VIII	1.3	β_D	1	0	1	β_E	1	1	0	0	0	0	0	1	
IX	1.20	β_D	0	0	0	β_E	1	1	1	0	0	0	0	1	
X	1.30	1	1.67	0	0	β_E	0	0	0	0	0	0	0	0	

ACI 318 83

Load & resistance factor design method is used.

$$\phi\ U = 1.4\ D + 1.7\ L$$
$$\phi\ U = 0.75\ (1.4\ D + 1.7\ L + 1.7\ W)$$
$$\phi\ U = 0.9\ D + 1.3\ W$$
$$\phi\ U = 1.4\ D + 1.7\ L + 1.7\ H$$
$$\phi\ U = 0.75\ (1.4\ D + 1.4\ T + 1.7\ L)$$
$$\phi\ U = 1.4\ (D + T)$$

where

D = dead load
H = lateral earth pressure
L = live load
W = wind load

T = combined effect of temperature, creep, shrinkage and differential settlement
U = strength
ϕ = 0.90 for flexure with or without axial tension
ϕ = 0.90 for axial tension
ϕ = 0.75 for axial compression with or without flexure for members with spiral reinforcement
ϕ = 0.70 for other members for axial compression with or without flexure

AISC ASD 89

Allowable stress design method is used
Factors of Safety – based on yield stress and gross section

Tension and compression in extreme fibers of symmetrical members subjected to

Bending	1.52
Tension	1.67
Shear	2.50

Compression in columns varies from 1.67 to 1.92 (buckling strength).

Plastic Design

$$\begin{aligned}\text{Resistance} &\geq \text{(Load Factor) (Load)}\\ &\geq 1.7\,(L + D)\\ &\geq 1.3\,(L + D + W)\\ &\geq 1.3\,(L + D + E)\\ &\geq 1.3\,(L + D + W + E)\end{aligned}$$

where
D = dead Load
E = earthquake load
L = live load
W = wind load

AISC LRFD 86

Load & resistance factor design method is used.

$$\phi\, R_n \geq 1.4\, D \qquad (1)$$

$$\phi\, R_n \geq 1.2\, D + 1.6\, L + 0.5\, (L_r, \text{ or } S \text{ or } R) \qquad (2)$$

$$\phi\, R_n \geq 1.2\, D + 1.6\, (L_r, \text{ or } S \text{ or } R) + (0.5\, L \text{ or } 0.8\, W) \qquad (3)$$

$$\phi\, R_n \geq 1.2\, D + 1.3\, W + 0.5\, L + 0.5(L_r, \text{ or } S \text{ or } R) \qquad (4)$$

$$\phi\, R_n \geq 1.2\, D + 1.5\, E + (0.5\, L \text{ or } 0.2\, S) \qquad (5)$$

$$\phi\, R_n \geq 0.9\, D - (1.3\, W \text{ or } 1.5\, E) \qquad (6)$$

Exception: The live load factor in combinations (3), (4) and (5) shall equal 1.0 for garages, areas occupied as places of public assembly, and all areas where the live load is greater than 100 psf.

where

- D = dead load due to the weight of the structural elements and the permanent features on the structure
- E = earthquake load
- L = live load due to occupancy and movable equipment
- L_r = roof live load
- R = load due to initial rainwater or ice exclusive of the ponding contribution
- R_n = nominal resistance
- S = snow load
- W = wind load
- ϕ = 0.90 for tension members for yielding in gross section
- ϕ = 0.75 for tension members for fracture in net sections
- ϕ = 0.85 for compressions members with appropriate width-thickness ratios
- ϕ = 0.90 for flexural members
- ϕ = 0.85 for unsymmetric members under torsion flexure and axial force for the limit state of buckling
- ϕ = 0.90 for local flange bending
- ϕ = 1.00 for local web yielding
- ϕ = 0.75 for web crippling
- ϕ = 0.85 for sidesway web buckling
- ϕ = 0.90 for all other conditions

AISI 76

Allowable stress design method is used.

Factors of Safety are identical with those for AISC ASD 89 and AISI 86 (whichever of these specifications is applicable to the tubular member being considered).

AISI 86

Allowable stress design method is used.

Factors of Safety:

Tension member	1.67	against yielding
Flexural members		
Flexure only	1.67	against yielding and buckling
Shear	1.44	against shear yielding
	1.67	against inelastic shear buckling
	1.71	against elastic shear buckling
Web crippling	1.85	for single unreinforced webs
	2.0	for I-beams

Concentrically loaded compression members	1.92	except when Fe is determined for fully effective sections having wall thickness greater than 0.09 in. and $F_e > F_y/2$. In this case the Factor and Safety varies from 1.92 to 1.67.
Wall studs	1.92	against column buckling
	1.92	for computing the permissible shear strain in sheeting material.
Wind or earthquake load		A twenty five percent reduction of nominal safety factor is permissible provided that the section thus designed is not less than that required for combination of dead and live load.

API RP2A 89

Allowable stress design method is used.

Factors of Safety are identical to AISC-ASD-89.

AREA 89

Allowable stress design method is used.

Factors of Safety Based on Yield Stress (except for compression in columns)

Axial tension	1.82
Tension in extreme fibers of members subjected to bending	1.82
Axial compression	1.82
Compression in columns based on buckling load	1.67
Compression in extreme fibers of members subjected to bending. Based on buckling load with maximum based on yield strength	1.82
Diagonal tension in webs of bending members	1.82

Load Combinations

Members which are subject to stresses resulting from lateral loads other than centrifugal load and/or longitudinal load, may be proportioned for stresses 25% greater than the allowable stresses.

BSSC NEHRP 85

Allowable stress design method is used.

Factors of Safety are identical to AISC ASD 89 and AISI 86 whichever of these specifications is applicable to the member being considered.

CSA A23.3 M84 77

Limit states design method is used.

$$\phi R \geq 1.25\,D + 1.5\,L$$
$$\phi R \geq 1.25\,D + 1.5\,Q$$
$$\phi R \geq 1.25\,D + 1.25\,T$$
$$\phi R \geq 1.25\,D + 0.7\,(1.5\,L + 1.5\,Q)$$
$$\phi R \geq 1.25\,D + 0.7\,(1.5\,L + 1.25\,T)$$
$$\phi R \geq 1.25\,D + 0.7\,(1.5\,Q + 1.25\,T)$$
$$\phi R \geq 1.25\,D + 0.6\,(1.5\,L + 1.5\,Q + 1.25\,T)$$

where

R = member factored resistance
D = dead load
L = live load
T = influences of temperature, shrinkage and creep
Q = wind or earthquake
ϕ = 0.85 for reinforcing bars
ϕ = 0.90 for structural steel, and prestressing tendons
ϕ = 0.60 for concrete

CSA S6 M89 84a

Service Load Design

The following loading combinations shall be used for service load design at the percentage of the basic stress indicated.

		% of Basic Stress
Group I	= D + L+ I + CF + E + B + SF	100
Group II	= D +E+ B + SF + W	125
Group III	= Group I + 0.30W + WL	125
Group IV	= Group I + R + S + T + LF	140
Group V	= Group II + R + S + T	140
Group VI	= Group III + R + S + T + LF	140
Group VII	= D + E + B + SF + Q	130
Group VIII	= Group I + ICE	140
Group IX	= Group II + ICE	150
Group X	= W, or (0.3W + WL)	100

where

B = buoyancy
CF = centrifugal force
D = dead load
E = earth pressure
I = live load impact
ICE = ice pressure
L = live load
LF = longitudinal force from live load
Q = earthquake
R = rib shortening
S = shrinkage

SF = stream flow pressure
T = temperature
W = wind load on structure
WL = wind load on live load

Both 1.0E and 0.5E shall be checked for lateral earth loads on rigid frames to determine which loading governs.

Limit states design method is used.
The following group loading combinations shall be used for limit states design:

Group I = 1.3 [D + 1.67 (L + I) + CF+ E*+ B + SF]
Group II = 1.3 [D + E* + B + SF + W]
Group III = 1.3 [D + L + I + CF + B+ SF + 0.30W + WL]
Group IV = 1.3 [D + L + I + LF + CF +E* + B + SF + R + S +T]
Group V = 1.25 [D + E + B + SF +W + R + S + T]
Group VI = 1.25 [D + L + I + LF+ CF + E* + B + SF+ 0.3W + WL + R + S + T]
Group VII = 1.3 [D + E* + B + SF +Q]
Group VIII = 1.3 [D + L + CF + E* + B +SF + ICE]
Group IX = 1.2 [D- + E* + B +SF + W +ICE]
Group X = 1.67 [W], or 1.67 [0.3W+ WL]

where D, L, etc., shall be in accordance with above for service load design.

*For earth pressure, the load factor shall be 1.0 for vertical earth pressure and 1.3 for lateral earth pressure, except that 0.5E shall be used for lateral earth pressure when checking positive moments in rigid frames.

CSA S16.1 M89 84b

Limits states design method is used.

$$0.90\ R \geq 1.25\ D + 1.5\ L$$
$$0.90\ R \geq 1.25\ D + 1.5\ Q$$
$$0.90\ R \geq 1.25\ D + 1.25\ T$$
$$0.90\ R \geq 1.25\ D + 0.7\ (1.5\ L + 1.5\ Q)$$
$$0.90\ R \geq 1.25\ D + 0.7\ (1.5\ L + 1.25\ T)$$
$$0.90\ R \geq 1.25\ D + 0.7\ (1.5\ Q + 1.25T)$$
$$0.90\ R \geq 1.25\ D + 0.6\ (1.5\ L + 1.5\ Q + 1.25\ T)$$

where

R = member factored resistance
D = dead load
L = live load
T = influence of temperature, shrinkage, and creep
Q = wind or earthquake

The resistance factor ϕ is taken as 0.9 for members and elements, but is 0.67 for bolts and welds.

CSA S136 M89 84c

Limits states design method is used.

$$\phi R \geq 1.25\,D + 1.5\,L$$
$$\phi R \geq 1.25\,D + 1.5\,Q$$
$$\phi R \geq 1.25\,D + 1.25\,T$$
$$\phi R \geq 1.25\,D + 0.7\,(1.5\,L + 1.5\,Q)$$
$$\phi R \geq 1.25\,D + 0.7\,(1.5\,L + 1.25\,T)$$
$$\phi R \geq 1.25\,D + 0.7\,(1.5\,Q + 1.25\,T)$$
$$\phi R \geq 1.25\,D + 0.6\,(1.5\,L + 1.5\,Q + 1.25\,T)$$

where

R = member factored resistance
D = dead loads
L = live loads
T = influences of temperature, shrinkage, and creep
Q = wind or earthquake
ϕ = 0.90 for axial tension, bending and shear
ϕ = 0.90 for axial compression of doubly symmetric sections, double angles restrained against flexural-torsional buckling and wall studs
ϕ = 0.75 for bearing stiffeners and other members in axial compression
ϕ = 0.75 for limit states determined by the tensile strength of the material
ϕ = 0.80 for web crippling in beams with single unreinforced webs and deck sections
ϕ = 0.67 for web crippling of other webs
ϕ = 0.67 for connections

FHWA TS 80

Load factor design method is used. Load factors and load combinations are identical to AASHTO-1989 sections (2) and (3).

OHBDC 83

Limit states design method is used.

$$\phi \text{ (Factored Resistance)} \geq \Sigma \text{ (Load Factor)(Load)}$$

where

ϕ = 0.95 for shapes fabricated from plates.
ϕ = 0.90 all other shapes and sections and with Load Factor given in the following table:

	D	E	P	L	K	W	V	S	Q	F	A	H
Serviceability Limit States												
Type I	1.00	1.00	1.00	0.80	0	1.00	0	0	0	0	0	0
Type II	1.00	1.00	1.00	0.75	0.80	0.70	0	1.00	0	0	0	0

	D	E	P	L	K	W	V	S	Q	F	A	H
Ultimate Limit States												
ULS Combination 1	αD	αE	αp	1.40	0	0	0	0	0	0	0	0
ULS Combination 2	αD	αE	αp	1.25	1.15	0	0	0	0	0	0	0
ULS Combination 3	αD	αE	αp	1.15	1.00	0.40	0.40	0	0	0	0	0
ULS Combination 4	αD	αE	αp	0	1.25	1.30	0	0	0	0	0	0
ULS Combination 5	αD	αE	αp	0	0	0.70	0	0	1.30	1.30	1.30	1.40

where

A = ice accretion load

D = dead load

E = loads due to earth pressure and hydrostatic pressure other than dead load. Factors for surcharges shall be those of the loads causing them.

F = loads due to stream flow and ice pressure

H = collision load

K = limiting friction in sliding bearings, or the combined effect of all strains including those due to temperature change and differential, concrete shrinkage, differential shrinkage and creep, but excluding the effects of elastic distortion

L = live load. All applicable loads including dynamic load allowance when applicable

P = secondary prestress effects

Q = earthquake load

S = load due to differential foundation settlement

V = wind load on live load

W = wind load on structure

WEST EUROPE

REGIONAL CONTRIBUTOR
John E. Harding, University of Surrey

BS 5400 P.3 82

Limit States

$$\gamma_{fL} \leq \frac{1}{\gamma_{f3}\,\gamma_m}$$

where

$\gamma_m = \gamma_{m1}\,\gamma_{m2}$
$\gamma_{m1} =$ factor on characteristic yield stress
$\gamma_{m2} =$ factor for modeling uncertainties

Ultimate Limits

$\gamma_m = 1.05$ $\gamma_{f3} = 1.1$

except $\gamma_m = 1.2$ for
a) Bending resistance of beams (compression limits)
b) Combined bending and shear (beams with intermediate transverse stiffeners) deflection of moment resistance due to flanges
c) Compressive strength of unstiffened flanges
d) Strength of longitudinal flange stiffeners
e) Strength of longitudinal web stiffeners
f) Buckling of transverse web stiffeners
g) Buckling of bearing stiffeners
h) Buckling of diaphram stiffeners
i) Strength of diaphram at junction

Serviceability Limits

Load combination ($R_{fL}\,Q_K$)

combination 1. basic load combination for highway and foot/cycle bridge and roadway bridge
combination 2. live loads + wind and temporary erection loads (construction)
combination 3. above loads + temperature loads and temporary erection loads (construction)
combination 4. above loads + loading on highway bridge due to traffic
combination 5. above loads (construction) + loading due to bearing function

BS 5950 P.1 85

Limit States

$$\frac{f_k}{\gamma_m} \geq \text{effects of } \gamma_f Q_k \text{ (dead loads)} + \gamma_f Q_k \text{ (imposed loads)}$$

where

f_k = characteristic strength of materials
Q_k = nominal loading
γ_m = 1
γ_f = values for various load combinations

Load Factor γ_f for Different Loads

dead load	1.4
dead load restraining uplift and overturning	1.0
dead load acting with wind and imposed load combinations	1.2
imposed load	1.6
imposed load acting with wind load	1.2
wind load	1.4
wind load acting with imposed load or crane load	1.2
forces due to temperature effects	1.2
vertical crane load	1.6
vertical crane load acting with longitudinal loads	1.4
horizontal crane load	1.6
horizontal crane load acting with vertical loads	1.4
crane load acting with wind load	1.2

INTERNATIONAL COMPARISONS & COMMENT

WHAT ARE THE DIFFERENCES?

Originally a separate section of the General Provisions was planned under the designation "Loads and Load Combinations." Not only did this approach reveal itself to be overly ambitious, it was not even possible. Specifications for design of highway and railway structural members provided voluminous information on dead, live, wind, earthquake, temperature and other loads. Although those specifications that addressed building structures provided almost nothing, they covered the topic by reference to other standards which in turn were also voluminous. It was decided that the actual design loads themselves and the philosophy behind their derivations was a subject too vast to be covered in this document. The important issue was the load combinations.

Accordingly, it was decided to eliminate a separate section in the General Provisions on loads and to include the subject of load combinations together with the section on design basis with which they were intricately linked to factors of safety or member factored resistances. However, since design loading remains a subject of great importance, the directions being followed in recent developments of several types of loading have been summarized.

Thus, the third and last section of the General Provisions was designated "Design Basis". In this section, quantitative information is provided for each specification on factors of safety, load factors, load combinations, resistance factors, load combination factors, importance factors and other parameters that form the design basis for each specification. Some specifications provide more than one design basis and most have specific serviceability checks. Thus, by providing this quantitative information, the specification philosophical approach is presented.

Table GPc.1 Reliability Assessment*

Specification	D1	D2	D3	S1	S2	S3	S4	S5	P1	P2	P3	P4
	Australia											
AISC 83	X	–	–	–	–	–	–	–	–	–	–	–
AS 1210 77	X	–	–	–	–	–	–	–	–	–	–	–
AS 1664 79	X	–	–	–	–	–	–	–	–	–	–	–
AS 2327-P.1 80	X	X	–	–	–	–	–	–	–	–	–	–
AS 1250 81	X	X	–	–	–	–	–	–	–	–	–	–
AS DR87164 87	–	–	X	–	X	–	–	–	–	–	–	–
AS 1538 88	X	–	–	–	–	–	–	–	–	–	–	–
	China											
GBJ 18 87	–	–	–	–	X	–	–	–	–	–	–	–
GBJ 17 88	–	–	–	–	X	–	–	–	–	–	–	–
GBJ 11 89	–	–	–	–	X	–	–	–	–	–	–	–
HRS 87	–	–	–	–	X	–	–	–	–	–	–	–
SDGJ 64 89	–	–	–	–	X	–	–	–	–	–	–	–
TBJ 2 85	X	–	–	–	–	–	–	–	–	–	–	–

Table GPc.1 Reliability Assessment* (Cont.)

Specification	D1	D2	D3	S1	S2	S3	S4	S5	P1	P2	P3	P4
						East Europe						
CMEA 83	–	–	–	X	–	–	–	–	–	–	–	–
ČSN 73 1401 84	–	–	–	X	–	–	–	–	–	–	–	–
ČSN 73 6205 84	–	–	–	X	–	–	–	–	–	–	–	–
ĞSN 73 1206 86	–	–	–	X	–	–	–	–	–	–	–	–
CSN 73 1402 88	–	–	–	X	–	–	–	–	–	–	–	–
ME 04 180 83	–	–	–	X	–	–	–	–	–	–	–	–
MI 04.184 78	–	–	–	X	–	–	–	–	–	–	–	–
MI 15024/3 85	–	–	–	X	–	–	–	–	–	–	–	–
MI 15024/4 85	–	–	–	X	–	–	–	–	–	–	–	–
MSZ 15024/1 85	–	–	–	X	–	–	–	–	–	–	–	–
MSZ 5-132 88	–	–	–	X	–	–	–	–	–	–	–	–
NiP 565/86 87	–	–	–	X	–	–	–	–	–	–	–	–
PN 80/B-03200 87	–	–	–	X	–	–	–	–	–	–	–	–
SNiP II-24-74 75	–	–	–	X	–	–	–	–	–	–	–	–
SNiP II-23-81 82	–	–	–	X	–	–	–	–	–	–	–	–
SNiP 2.05.03 84	–	–	–	X	–	–	–	–	–	–	–	–
STAS 10108/0 78	–	–	–	X	–	–	–	–	–	–	–	–
TGL 13503/01 82	–	–	X	–	–	–	–	–	–	–	–	–
TGL 13503/02 82	–	–	X	–	–	–	–	–	–	–	–	–
						Japan						
AIJ 70	X	–	–	–	–	–	–	–	–	–	–	–
AIJ 80	X	–	–	–	–	–	–	–	–	–	–	–
AIJ 85	X	–	–	–	–	–	–	–	–	–	–	–
AIJ 86	X	–	–	–	–	–	–	–	–	–	–	–
AIJ 89	X	–	–	–	–	–	–	–	–	–	–	–
AIJ 90	–	X	–	–	X	–	–	–	–	–	–	–
JRA 90	X	–	–	–	–	–	–	–	–	–	–	–
JSCE 83	X	–	–	–	–	–	–	–	–	–	–	–
						North America						
AA 86	X	–	–	–	–	–	–	–	–	–	–	–
AASHTO 89	X	–	X	–	X	–	–	–	–	–	–	–
ACI 318 83	–	–	X	–	–	–	–	–	–	–	–	–
AISC ASD 89	X	X	–	–	–	–	–	–	–	–	–	–
AISC LRFD 86	–	X	–	?	X	–	–	–	–	–	–	–
AISI 76	X	–	–	–	–	–	–	–	–	–	–	–
AISI 86	X	–	–	–	–	–	–	–	–	–	–	–
API RP2A 89	X	–	–	–	–	–	–	–	–	–	–	–
AREA 89	X	–	–	–	–	–	–	–	–	–	–	–
BSSC NEHRP 85	X	X	–	–	–	–	–	–	–	–	–	–
CSA A23.3 M84 77	–	X	–	–	X	–	–	–	–	–	–	–
CSA S6 M89 84a	–	X	–	–	X	–	–	–	–	–	–	–
CSA S16.1 M89 84b	–	X	–	–	X	–	–	–	–	–	–	–
CSA S136 M89 84c	–	X	–	–	X	–	–	–	–	–	–	–
FHWA TS 80	X	–	X	–	X	–	–	–	–	–	–	–
OHBDC 83	–	X	–	–	X	–	–	–	–	–	–	–

Key:

Deterministic:	D1	Factor of Safety	: One value (for ASD)
	D2	Load Factor	: One value (for Plastic Design)
	D3	Other	: Two or more values
Semiprobabilistic:	S1	Partial Factors of Reliability	
	S2	Reliability Index	
	S3	Upper and Lower Fractiles	
	S4	Central Factor of Reliability	
	S5	Other	
Probabilistic:	P1	Reliability Index	
	P2	Upper and Lower Fractiles	
	P3	Central Reliability Factor	
	P4	Other	

* Deterministic: All variables are evaluated using deterministic approach. Semiprobabilistic: Some variables are evaluated using statistics and probability. Probabilistic: All variables are evaluated using statistics and probability.

Loads[4]

(A) Dead Load

Dead load and its variability are discussed fully in the Tall Building Monograph Volume CL (Council on Tall Buildings, Group CL, 1980) and warrants no further discussion.

(B) Live Load

Live load requirements for 10 countries are tabulated comprehensively in the Tall Building Monograph, Vol. CL. Although the logic by which these live loads were derived seems to be lost in antiquity, a striking feature is their uniformity, especially for assembly areas. Office buildings range from 2.0 to 2.9 kPa (41 to 61 psf) but for valid comparison two other factors need to be considered also, the minimum allowance for partitions and the live load reduction allowance. With large tributary areas, some tributary area formulas reduce live load to only one third whereas the lower limit for others is one half.

For designers to be able to make sound judgment on the adequacy of special loading situations, it is essential that code writers formulate proper guidelines. For various types of storage, such as libraries, filing areas, manufacturing and warehousing, it is frequently necessary to determine appropriate live loads for which to design new structures or to evaluate existing structures. Lacking guidelines, designers use their own rationale. Such an approach appears satisfactory until the loading for parking garages is considered. Whereas the actual loading imposed by parked automobiles is about 1 kPa (20 psf), the specified live load for the 10 countries listed in the Monograph ranges from 1.4 to

[4]In the preparation of the discussion on loading, the assistance of Dr. A.G. Davenport of the University of Western Ontario, Canada, on wind loading, Dr. D.A. Taylor and Dr. D.E. Allen of the Institute for Research in Construction, National Research Council of Canada, on Snow Loads, and Live Loads and Load Combinations, respectively, is gratefully acknowledged.

5.4 kPa (30 to 113 psf)! Clearly, the different countries have no consistent approach.

Aspects which, it has been suggested, should be considered in determining live loads are a practical minimum value, which is of special concern during construction, the degree of control which can be exercised over the occupancy, possible change of occupancy, the environment, and its influence on deterioration.

Two other loading factors affect overall structural stability, namely, live load reduction and the proportion of live load to be used in seismic analyses. The former affects column loads and second order (P-Δ) analyses. The latter affects the seismic inertia forces. As an example, automobile parking garages are again treated illogically by North American Codes in these matters. Although the required live load is extraordinarily high, only storage type live load reduction is permitted in the determination of column loads. Furthermore, some codes require the full live load to be included in lateral seismic inertia forces.

(C) Wind Load*

During the past 50 years the description of wind loading for structural design has moved from relatively simple, straightforward notions of static drag forces to much more sophisticated models involving all the manifold questions of climate, meteorology, aerodynamics, structural mechanics and dynamics and, more recently, reliability. It is worth reflecting on some of these changes and why they happened.

During the 1930's structures were, by present standards, relatively massive structural members themselves were heavier due to the relatively weak materials, and dead loads were higher due, for example, to the heavy masonry and stone facades on buildings and the use of heavy, reinforced concrete deck systems on bridges. These massive structures were frequently much stiffer than they were predicted to be due to the participation of the non-structural components, the contribution of which was difficult to estimate. The massiveness of these structures did little to emphasize the importance of wind forces - in questions of overturning for example - and effectively disguised the latent dynamic problems which were later to become important.

Wind loads were formulated along simple lines as the product of the velocity pressure $\rho v^2/2$ and an aerodynamic coefficient related either to the drag (C_D) or the local pressures (C_p). The wind speed v was commonly based on the highest recorded gust wind speed (or some other convenient measure) and allowance was made for the increase of wind speed with height, normally based on open country conditions. The drag and pressure coefficients were derived from uniform flow tests in aeronautical wind tunnels. The very simple rules for wind loading which resulted were sometimes streamlined further to obviate unnecessary calculations. An example was the New York City code in which wind loads below 200 ft were ignored and were only 20 lb/ft^2 above this height. The 33% increase in design stress under wind action, commonly allowed by codes, may have had similar origins.

* Contributed by Alan G. Davenport, The University of Western Ontario

Wind tunnel testing at this time was mostly carried out in aeronautical wind tunnels in smooth uniform flow. Although there were significant differences in the pressures between uniform and boundary layer flow, it appeared to be of academic interest only.

In tracing the further development of wind loading from the 1930's to the present there appear to have been several forces at work. First there have been some radical changes in structural properties - mass, stiffness, and damping. The average densities of tall buildings and long span bridges, for example, have fallen by roughly a factor of 2 - from around 20 lb/ft^3 to 10 lb/ft^3 in the case of buildings, and from 200 lb/ft^2 to 100 lb/ft^2 of roadway in the case of long span bridges. In parallel, the substantial increases in material strengths (by factors of 2 and 3 in the cases of structural steel and concrete) has led to reductions in member sizes and consequently stiffness; this has been exaggerated further by the omission of heavy masonry cladding and frame infill and the loss therefore of their contributions to stiffness. To this we can add significant improvements such as the enormous increase in analytical capability contributed by the digital computer and our better understanding of structural behavior. Significant in this latter aspect was the 1930's full scale monitoring of steel frameworks by J.F. Baker. Lastly there have been significant reductions in damping due to extensive use of welding, pre-stressing and the omission of heavy masonry elements. This led Southwell to remark, prophetically in 1949, that the endowment of structures with damping would become an urgent problem for the future. All of these changes in structural form, it turns out, directly increase susceptibility to wind and hence the need for better models of wind loading.

A second factor was the stimulus provided by some dramatic failures. The Tacoma Narrows Bridge was the notable example, but other large bridges, the Golden Gate included, survived some hair-raising episodes of aerodynamic instability. Twenty-five years later the collapse of the Ferrybridge cooling towers provided a similar spur to research efforts. Theodore von Karman, who was involved in the post mortems of the Tacoma Bridge drew attention to the dynamic problems and chided bridge engineers for their reluctance to recognize their effects.

A highly significant development during the 1950's was due to Jensen. He undertook a comparison of the mean pressures on small buildings in full scale and in wind tunnel model experiments, in a variety of boundary layers. From this Jensen developed his model law:

> 'The current model test for phenomena in the wind must be carried out in a turbulent boundary layer and the model law requires the boundary layer to be scaled as regards the velocity profile.'

The extension of Jensen's essential concept beyond the time-averaged effects to considerations of the fluctuations of velocity and pressures and of the turbulence scales was the next important stage.

(1) A Contemporary View of Wind Loading

It can be helpful to consider the wind loading process as a chain of interconnected components as shown in Fig. GPc.1. This notion can be adapted to the prediction of the variety of responses important in wind loading, ranging from serious structural damage either to the entire structure or cladding, to factors affecting serviceability, such as excessive deflections or vibration causing discomfort to occupants.

The first two links of the chain - the wind climate and the influence of the terrain - are primarily meteorological; they concern the strength of the wind and the manner in which the terrain shapes the structure of wind in the boundary layer as it approaches the structure. The third link concerns the translation of flow into forces and the fourth, the forces into mechanical responses such as deflections, accelerations, and stresses. The final link relates these measures of performance to the criteria and estimates of safety.

(2) Structural Responses to Wind
The integrated effect of the external wind pressures separate into mean and fluctuating components. The treatment of the time-independent mean forces presents no special problem and will not be discussed further.

The dynamic components, however, do present special problems for a variety of reasons. First, the pressure fluctuations are time- and space-dependent and direct integration cannot be used. Second, the structures themselves are usually flexible and pronounced resonance effects are often produced at the natural frequencies, since the damping in most structures is normally low, of the order of 1% of critical damping. Third, the dynamic response may be coupled with special design questions such as accelerations causing discomfort to occupants of tall buildings.

The origins of the dynamic forces can normally be traced back to one or several of the following in partnership:

a) Buffeting effects due to turbulence;
b) Forces (sometimes termed 'vortex shedding' forces), produced by the structure's own wake acting on the 'afterbody' of the structure behind the 'separation points'; and
c) Aeroelastic forces induced by the structure's motion through the wind.

(3) The Prediction of Risk
The engineering motivation for many studies of wind action on structures springs from the need to minimize the risks involved. These risks are, on occasion, vital questions (such as the survival of structures under wind load), in other instances related to significant nuisances (such as the discomfort of occupants and the annoyance of pedestrians). On this point, and also perhaps on the question of the complexity of the model which is appropriate, the scientific and engineering approaches may diverge.

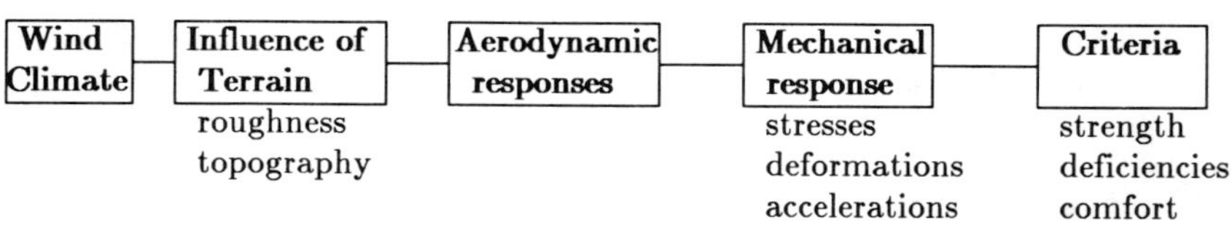

Fig. GPc.1 The Wind Loading Chain

Earlier the wind loading process was described as a chain. Symbolically this wind loading chain can be represented by the equation

$$W = q\, C_e\, C_p\, C_g \qquad \text{(GPc.1)}$$

where

q = reference wind pressure
$= \frac{\rho v^2}{2}$, essentially a climate component
C_e = exposure factor to adjust for the terrain conditions and the height
C_g = gust response factor
C_p = pressure coefficient

C_e, and to some extent C_g, represent wind effect coefficient on structures. Those codes which represent wind loads in this manner provide a particularly convenient frame for refinement.

If the wind velocity v is the 20 minute to hourly mean, the gust factor may be taken to be 2 or 2.5. However, a much more detailed dynamic approach to the derivation of the gust factor may be undertaken, including the effects of frequency, height and size of the building, as well as the approach terrain, all within the same basic format.

The engineering objective in most cases is to ensure survival. This means that if the structure is to last a certain time - say 50 years - then our purpose is to predict the largest load in that time span. If all the factors in the above expression could be nailed down precisely, this task would offer no difficulty. Unfortunately this is not the case. Each term contains uncertainty, arising from our imprecise knowledge of climate, of the exposure, the aerodynamic properties, and the mechanical characteristics. Often our state of knowledge is compromised further by simplifications introduced to streamline the design procedures and to circumvent intractable detail. Regardless of these simplifications, studies show that the greatest component of the uncertainty in estimating the wind loading lies in the estimation of the wind climate and the exposure at the site.

Specific structural aspects of wind loading which may be significant but which are not necessarily given consideration by codes include:

- concurrent but uncoupled downwind and across wind oscillation of buildings. This phenomenon has a significant effect on columns that participate in orthogonal wind resisting systems and is approximated in the Canadian Code by taking 75% of the sum of the two separate effects.

- the enhancement of lateral displacement or decrease in effective lateral stiffness caused by the P-Δ effect. In flexible structures this may amount to a 20% increase, or even more.

- the non-uniform character of turbulent wind which results in non-uniform wind loading. This may contribute to torsional response of buildings. It has a profound effect on structures such as guyed towers since it has been the practice in some countries to design these structures statically as continuous beams under uniform load, tailoring the structural capacity as precisely as possible to follow the theoretical (and obviously improbable) bending moment diagram.

- recognition of the dependence of structural loading on structural response. This characteristic of wind loading on flexible structures may require a reappraisal of researchers' practice of applying factored gravity load to a structure followed by incremental application of static wind load through the inelastic response phase to failure.

In looking to the future, it appears certain that the continuing requirements for lighter, more flexible structures will continue to make wind loading important and dynamic effects a major component of the problem. Improvements in our understanding of, and the response of structures to turbulence, the better description of wind climate by improved instrumentation and further developments in wind tunnel technology are probable trends for the future. However, the inclusion of reliability and safety principles into the mainstream of the wind engineering development will introduce an important perspective on the utility of future research and provide a framework for consistent safety.

(D) Snow Loads

The almost universal basis for design snow loads on roofs is ground snow loads. In contrast with the relatively few direct measurements of roof snow available, a vast amount of data on ground snow has been recorded by meteorological and hydrological stations. Consequently, the data, in some instances going back 20 to 30 years and even longer, are used to establish statistical distributions to predict ground snow loads with an annual probability of exceedance of between 1/25 and 1/50 for many locations.

With increasing collection of data, it is becoming possible to define ground snow loads to reflect more local variations due to microclimates. Indeed, significant changes in snow load can occur in relatively short distances. The suggestion in the Tall Buildings Criteria and Loading Monograph, Volume CL, that too many variations in loading may be burdensome for designers, has proved not to be the case. Experience shows that designers are not loath to take advantage of this information when it is available in a convenient form.

Climate has a marked effect on ground and roof snow loads. In coastal areas subject to freeze-thaw conditions, the critical uniformly distributed loads on roofs may result from single storms, whereas in very cold dry areas snow may accumulate throughout the entire winter. However, where drifting is prevalent there have been many observations of large drifts forming on roofs during single storms.

While drifting is caused primarily by wind, other major factors are the characteristics of the snow, temperature, and rainfall. Because of the primary influence of wind, it has long been recognized that roof shape and building form are crucial factors in snow loading. Due to lack of data, it was not possible to give explicit information to designers until about 20 to 25 years ago. Even now many design standards give little aid in determining drift loads on fairly common building shapes. As a result some designers continue the alarming practice of ignoring such loads on domes and arches.

It has become well recognized that dramatic drifts can occur in the lee of higher roofs in multi-level roofed buildings due to the transport of snow by wind from higher exposed areas to zones of aerodynamic shade where lower velocities cannot support transport. It may not be as well known that the wind has great impact on the uniform snow on flat roofs also. Generally wind removes a certain amount of snow from roofs, a phenomenon recognized by the reduction factors in design standards. More perceptive engineers are beginning to question whether the reductions should be as great for very large flat roofs or indeed, whether any reduction is warranted. Although snow may appear to be uniformly distributed, measurements on very large flat roofs suggest that this may not be the case. Of concern is the almost imperceptible build up of snow into large areas of unexpectedly high loading. Complete bays of roof framing may be loaded in this manner. Very large multi-level roofs create their own special problems. A large fetch area at high level can supply snow for enormous drifts when it is wind swept onto a lower roof.

Formation of ice on flat roofs due to freeze/thaw cycles can prevent drainage, resulting in higher loads and possibly leakage. Ice and snow sliding from upper level roofs can load lower levels heavily; such impact load may be expressed in terms of kinetic energy.

Instability due to snow loading arises in arches, domes, and continuous beams. Unanticipated or exaggerated non-uniform load may produce reversal of bending moment which, while well understood in the context of floor beams or bridge girders, may not be foreseen in roof structures. An example of a less widely appreciated possibility is the dramatic reversal in bending moment in the rafter of a ridged roof portal frame which occurs if snow accumulates on one side only. Almost the entire length of the lower flange on the unloaded side, which under uniform load would be in tension and deemed not to require lateral bracing, will be in compression. Since in some countries so many of these frames are designed plastically under uniform loading, this is a situation which may be overlooked. Limit states design is a particularly valuable approach in such situations since the potential for moment reversal and the corresponding need for additional lateral bracing is more forcibly identified. Indeed, the need for a cautious approach to bracing to control instability is perhaps more important than the precise load intensity for which the structure is designed.

Recently highly sophisticated procedures have been used to predict snow loads on important long span or unusually shaped roofs. These procedures involve wind tunnel model testing and Monte Carlo simulations utilizing extensive meteorological snow and wind data from the last 25 years or so.

Load Factors and Load Combinations

(A) Load Factors

Since there are little data on the extreme values of loads other than those naturally occurring in the environment, most codes have not strayed too far from the load factors implied in allowable stress design (ASD). With factors for dead load of 1.2 to 1.3, it is understandable that live load factors, in combination with resistance factors, should be selected to achieve similar strength to that obtained with an allowable stress of 0.67 times the yield strength. The factors in the American LRFD, identical to the American National Standard, ANSI A58.1, are typical of a large group of countries. Values of 1.2 and 1.6 are used for dead and live loads, respectively. Australia, Canada, and Britain use similar factors. The U.S.S.R. (SNiP) and Japan use significantly lower dead load factors, only 1.1. For live load, the U.S.S.R. uses only 1.2 or 1.3, China 1.4, but Japan adopts 1.6.

Clearly, there is a wide divergence which stems of course from the difficulty in assessing the magnitude of probable overload and the inclusion of other factors that code writers consider should be taken into account. It is said that since dead load is treated as a constant, the live load should be sufficiently generous to cover variability of dead load (Tall Building Monograph Vol. CL). Reliability of the structural analysis may be a component of some code load factors. Quality of workmanship is rarely treated explicitly and almost never discussed. As an example of the need to quantify these other factors, in determining the appropriate design load for a storage area, an engineer may well question why a load factor greater than unity should be required if the loading is based on the occupant's product being stacked to the roof?

Several countries obviously view snow load with concern since they adopt especially high load factors for snow, presumably a result of a high incidence of roof failures. Canadian experience does not support this. If a rational approach to snow loads is taken, that is, long return period loads, treatment of snow as a non-uniformly distributed load and proper recognition of the effects of roof shape on snow redeposition, then load factors need not be extraordinarily high.

In addition to lack of knowledge and concern, convenience sometimes influences codes. For example, Canadian values were selected so that the ratio of live to dead load factors was about the same as the ACI load factors of 1.7 and 1.4. The LSD steel design code, CSA S16.1, was to co-exist with the ACI derived CSA 23.3 for 10 years before a new independent limit states design standard for reinforced concrete was finalized.

(B) Load Combinations

Load combinations are much more difficult to compare since the result will vary with different contributions of load from dead, live, wind, and earthquake. Since dead load is nearly constant throughout the life of a structure, a combination of dead load with any other constitutes a fundamental combination in which the basic safety or load factors are applicable. When dead load is combined with two or more other loads, the simultaneous occurrence of the full design values of each of the load effects is less likely to occur than the basic combination. This may be taken into consideration by allowing reductions in the total effect due to combinations of the dead load effect with two or more other load effects.

The rule for determining load combinations can be determined by means of the following principle, suggested by Turkstra (1970). Each load can be separated into two components, one a sustained or frequently occurring component (such as, dead load, storage), the other a transient component which acts rarely and for a short time only (such as, impact, wind, earthquake, rare accumulation of people or equipment). Since transient loading components are unlikely to occur simultaneously, the critical load combination for any structural effect can be estimated by combining the load having the largest transient component with the sustained or frequent components of all other loads. AISC LRFD 86 follows this principle and it is typical of the majority of countries in its load combination procedure. Each variable load, such as live load or wind, is assumed to act at its full factored value with factored dead load but combined with a much reduced second variable load, usually about one half of its nominal value.

The British and Canadian approaches are unique among the countries listed. When two variable loads are considered, they are both reduced by a common factor reflecting a reduced probability of both occurring at their maximum values simultaneously. The British codes use reduced values for dead load also whereas the Canadian retains this at a constant value except when acting to stabilize structures. The British and Canadian approaches have much the same result as the old allowable stress approach of permitting stress increase under wind or earthquake loading.

From the viewpoint of design practice the current procedures for load effect combination analysis contained in standards do not utilize to the full extent the possibilities offered by available computer technology. However, some standards already allow the application of the probability approach in the analysis considering statistical data, an adequate probability level of exceedance for response combinations, and the application of personal computers (see ČSN 73 0035 - Czechoslovakia). A procedure has been developed based on simplified representations of the time-dependent responses of a structure to individual loadings by "marginal curves" and on the extreme magnitudes of responses. To create a resulting histogram corresponding to several simultaneously appearing responses of a structure to time-dependent loads and to obtain the extreme (maximum and minimum) magnitudes, the Monte Carlo method has been applied and the computer program RESCOM is utilized (Marek and Guštar, 1989). The resulting histogram depends on both extreme magnitudes of individual responses and on marginal curves expressing the time-dependency of each response. The proposed procedure allows for the consideration of different design situations (erection, temporary structures, strengthening of existing structures, fire) by adjusting the shapes and the extreme magnitudes of the marginal curves, and/or by selecting the level of probability of exceedance. The so-called service or nominal magnitudes of loads are not at all introduced in the proposed procedure. Those magnitudes may be considered as a heritage from the era of Allowable Stress Design and their significance and application in the design process may not be justified by considering the proposed method of response combination analysis. However, while the time dependent relationship of climatological loading such as wind and snow and in some cases, manufacturing loading as cranes and storage vessels may be established, time dependent characteristics for many other loadings appear to be ill defined at this time.

Table GPc.2 Review of Differences in Loading and Load Combinations

Specification	L1	L2	L3	L4	L5	T1	T2	T3	T4	T5	C1	C2	C3	C4	C5
	Australia														
AISC 83	X	–	–	–	–	X	–	–	–	–	–	–	–	–	X
AS 1210 77	X	–	–	–	–	X	–	–	–	–	X	–	–	–	–
AS 1664 79	X	–	–	–	–	X	–	–	–	–	X	–	–	–	–
AS 2327-P.1 80	X	–	–	–	–	X	–	–	–	–	X	–	–	–	–
AS 1250 81	X	–	–	–	–	X	–	–	–	–	X	–	–	–	–
AS DR87164 87	–	X	–	–	–	X	–	–	–	–	–	X	–	–	–
AS 1538 88	X	–	–	–	–	X	–	–	–	–	X	–	–	–	–
	China														
GBJ 18 87	–	–	–	X	–	X	–	–	–	–	–	–	X	–	–
GBJ 17 88	–	–	–	X	–	X	–	–	–	–	–	–	X	–	–
GBJ 11 89	–	X	–	–	–	X	–	–	–	–	–	–	X	–	–
HRS 87	–	–	–	X	–	X	–	–	–	–	–	–	X	–	–
SDGJ 64 89	–	–	–	X	–	X	–	–	–	–	–	–	X	–	–
TBJ 2 85	X	–	–	–	–	X	–	–	–	–	–	X	–	–	–
	East Europe														
CMEA 83	–	X	–	–	–	X	–	–	–	–	–	X	–	–	–
ČSN 73 1401 84	–	X	–	–	–	X	–	–	–	–	–	X	–	–	–
ČSN 73 6205 84	–	X	–	–	–	X	–	–	–	–	–	X	–	–	–
ČSN 73 1206 86	–	X	–	–	–	X	–	–	–	–	–	X	–	–	–
ČSN 73 1402 88	–	X	–	–	–	X	–	–	–	–	–	X	–	–	–
ME 04 180 83	–	X	–	–	–	X	–	–	–	–	–	X	–	–	–
MI 04.184 78	–	X	–	–	–	X	–	–	–	–	–	X	–	–	–
MI 15024/3 85	–	X	–	–	–	X	–	–	–	–	–	X	–	–	–
MI 15024/4 85	–	X	–	–	–	X	–	–	–	–	–	X	–	–	–
MSZ 15024/1 85	–	X	–	–	–	X	–	–	–	–	–	X	–	–	–
MSZ 5-132 88	–	X	–	–	–	X	–	–	–	–	–	X	–	–	–
NiP 565/86 87	–	X	–	–	–	X	–	–	–	–	–	X	–	–	–
PN 80/B-03200 87	–	X	–	–	–	X	–	–	–	–	–	X	–	–	–
SNiP II-24-74 75	–	X	–	–	–	X	–	–	–	–	–	X	–	–	–
SNiP II-23-81 82	–	X	–	–	–	X	–	–	–	–	–	X	–	–	–
SNiP 2.05.03 84	–	X	–	–	–	X	–	–	–	–	–	X	–	–	–
STAS 10108/0 78	–	–	–	–	–	X	–	–	–	–	–	X	–	–	–
TGL 13503/01 82	X	–	–	–	–	X	–	–	–	–	–	X	–	–	–
TGL 13503/02 82	X	–	–	–	–	X	–	–	–	–	–	X	–	–	–
	Japan														
AIJ 70	X	–	–	–	–	X	–	–	–	–	–	X	–	–	–
AIJ 80	–	–	–	–	X	–	–	–	–	X	–	–	–	–	X
AIJ 85	–	–	–	–	X	–	–	–	–	X	–	–	–	–	X
AIJ 86	–	–	–	X	–	–	–	–	–	X	–	–	–	–	–
AIJ 89	–	–	–	–	X	–	–	–	–	X	–	–	–	–	X
AIJ 90	–	X	–	–	–	X	–	–	–	–	–	–	X	–	–
JRA 90	X	–	–	–	–	X	–	–	–	–	–	X	–	–	–
JSCE 83	X	–	–	–	–	X	–	–	–	–	–	X	–	–	–

Table GPc.2 Review of Differences in Loading and Load Combinations (Cont.)

Specification	L1	L2	L3	L4	L5	T1	T2	T3	T4	T5	C1	C2	C3	C4	C5
	North America														
AA 86	X	–	–	–	–	X	–	–	–	–	X	–	–	–	–
AASHTO 89	–	X	–	–	–	X	–	–	–	–	–	X	–	–	–
ACI 318 83	–	X	–	–	–	X	–	–	–	–	–	X	–	–	–
AISC ASD 89	X	–	–	–	–	X	–	–	–	–	X	–	–	–	–
AISC LRFD 86	–	X	–	–	–	X	–	–	–	–	–	X	–	–	–
AISI 76	–	–	–	–	X	X	–	–	–	X	–	–	–	–	X
AISI 86	–	–	–	–	X	X	–	–	–	X	–	–	–	–	X
API RP2A 89	–	–	–	–	X	X	–	–	–	X	–	–	–	–	X
AREA 89	–	X	–	–	–	X	–	–	–	–	–	X	–	–	–
BSSC NEHRP 85	–	–	–	X	–	X	–	–	–	–	–	X	–	–	–
CSA A23.3 M89 77	–	X	–	–	–	X	–	–	–	–	–	X	–	–	–
CSA S6 M89 84a	–	X	–	–	–	X	–	–	–	–	–	X	–	–	–
CSA S16.1 M89 84b	–	X	–	–	–	X	–	–	–	–	–	X	–	–	–
CSA S136 M89 84c	–	X	–	–	–	X	–	–	–	–	–	X	–	–	–
FHWA TS 80	–	X	–	–	–	X	–	–	–	–	–	X	–	–	–
OHBDC 83	–	X	–	–	–	X	–	–	–	–	–	X	–	–	–

Key:

Loads are specified:	L1	By nominal magnitudes only
	L2	By nominal magnitudes and by load factors
	L3	By extreme magnitudes only
	L4	By reference to other standards
	L5	Not at all
Time dependency of loadings:	T1	Is not expressed in specification
	T2	Is expressed in simplified forms
	T3	Is expressed in forms of loading histories
	T4	By reference to other standards
	T5	Not at all
Load combinations:	C1	Only very simplified instructions are included in specification
	C2	Deterministic evaluation of loading without detail description of load histories
	C3	Evaluation is based on statistics and probability of simultaneous occurrence of loads
	C4	Specification consider combinations of responses of the structure to the loading
	C5	Not at all

Table GPc.3 Serviceability

Specification	G1	G2	G3	C1	C2	C3	C4
			Australia				
AISC 83	–	–	X	–	–	X	–
AS 1210 77	–	X	–	–	X	–	–
AS 1664 79	–	X	–	–	X	–	–
AS 2327-P.1 80	–	X	–	–	X	–	–
AS 1250 81	–	X	–	–	X	–	–
AS DR87164 87	–	X	–	–	X	–	–
AS 1538 88	–	X	–	–	X	–	–
			China				
GBJ 18 87	X	–	–	X	–	–	–
GBJ 17 88	X	–	–	X	–	–	–
GBJ 11 89	X	–	–	X	–	–	–
HRS 87	X	–	–	X	–	–	–
SDGJ 64 89	X	–	–	X	–	–	–
TBJ 2 85	X	–	–	X	–	–	–
			East Europe				
CMEA 83	–	X	–	–	–	X	–
ČSN 73 1401 84	X	–	–	X	–	–	–
ČSN 73 6205 84	X	–	–	X	–	–	–
ČSN 73 1206 86	X	–	–	–	–	X	–
ČSN 73 1402 88	–	X	–	–	–	–	X
ME 04 180 83	–	X	–	–	–	–	X
MI 04.184 78	–	X	–	–	–	–	X
MI 15024/3 85	–	X	–	–	–	–	X
MI 15024/4 85	–	X	–	–	–	–	X
MSZ 15024/1 85	–	X	–	–	–	–	X
MSZ 5-132 88	X	–	–	–	X	–	–
NiP 565/86 87	X	–	–	X	–	–	–
PN 80/B-03200 87	X	–	–	X	–	–	–
SNiP II-24-74 75	X	–	–	X	–	–	–
SNiP II-23-81 82	X	–	–	X	–	–	–
SNiP 2.05.03 84	X	–	–	X	–	–	–
STAS 10108/0 78	X	–	–	X	–	–	–
TGL 13503/01 82	–	–	X	–	–	–	X
TGL 13503/02 82	–	–	X	–	–	–	X
			Japan				
AIJ 70	X	–	–	X	–	–	–
AIJ 80	–	–	X	–	–	X	–
AIJ 85	X	–	–	–	X	–	–
AIJ 86	–	–	X	–	–	X	–
AIJ 89	–	–	X	–	–	X	–
AIJ 90	X	–	–	X	–	–	–
JRA 90	X	–	–	X	–	–	–
JSCE 83	X	–	–	X	–	–	–

Table GPc.3 Serviceability (Cont.)

Specification	G1	G2	G3	C1	C2	C3	C4
	North America						
AA 86	X	–	–	X	–	–	–
AASHTO 89	X	–	–	X	–	–	–
ACI 318 83	X	–	–	X	–	–	–
AISC ASD 89	X	–	–	X	–	–	–
AISC LRFD 86	X	–	–	X	–	–	–
AISI 76	X	–	–	–	–	X	–
AISI 86	X	–	–	–	–	X	–
API RP2A 89	–	–	X	–	–	X	–
AREA 89	X	–	–	X	–	–	–
BSSC NEHRP 85	X	–	–	X	–	–	–
CSA A23.3 M89 77	X	–	–	X	–	–	–
CSA S6 M89 84a	X	–	–	X	–	–	–
CSA S16.1 M89 84b	X	–	–	X	–	–	–
CSA S136 M89 84c	X	–	–	X	–	–	–
FHWA TS 80	X	–	–	X	–	–	–
OHBDC 83	X	–	–	X	–	–	–

Key:

General Rules:	G1	Specifications contain general rules for assessment of serviceability
	G2	Very limited instructions on assessment are available
	G3	Not at all
Criteria:	C1	Sufficient information is available for specific cases
	C2	Partial information is included in the specifications
	C3	Not at all
	C4	Included in other specifications

WHY ARE THERE DIFFERENCES?

The answer to why the difference has to be tht, with imperfect knowledge, engineering judgment will vary. One group of engineers in one country will not arrive at the same decision as a group in another country when faced with similar imponderables.

Because of the nature of the approach required for the assessment of reliability of steel structures and structural elements, it is not possible to compare the different provisions in a direct quantitative fashion. The number of variables involved in the reliability assessment, many of which interact with one another, means that a parametric study would not be possible and only by selecting specific examples could direct comparison be made.

Symbol List -- General Provisions and Design Requirements

A = ice accretion load (GPc.N)
B = buoyancy (GPc.N)
C_D = the drag (GPc.W)
C_e = exposure factor to adjust for the terrain conditions and the height (GPc.W)
C_g = gust response factor (GPc.W)
C_p = pressure coefficient (GPc.W)
CF = centrifugal force (GPc.N)
D = dead load (GPc.J) & (GPc.N)
E = earth pressure (GPc.N)
E = earthquake load (GPc.N)
EQ = earthquake force (GPc.N)
F = loads due to stream flow and ice pressure (GPc.N)
F_d = design load (GPc.E)
F_e = Euler's elastic buckling stress (GPc.N)
F_p = permanent (dead) loads (GPc.E)
F_s = factored load (GPc.E)
F_{te} = extraordinary temporary load (GPc.E)
F_{tl} = long-term temporary loads (GPc.E)
F_{ts} = short-term temporary loads (GPc.E)
F_u = ultimate stress (GPb.W)
F_y = yield stress (GPb.W) & (GPc.N)
Fcu = cube strength of concrete (GPb)
Fcyl = cylinder strength of concrete (GPb)
f_k = characteristic strength of materials (GPc.W)
f_y = specified minimum yield stress of steel (GPc.A)
G = nominal dead load (GPc.A)
H = lateral earth pressure (GPc.N)
H = collision load (GPc.N)
I = impact (GPc.N)
ICE = ice pressure (GPc.N)
K = limiting friction in sliding bearings, or the combined effect of all strains (GPc.N)
L = live load (GPc.J) & (GPc.N)
L_r = live roof load (GPc.N)
LF = longitudinal force from live load (GPc.N)
n_a = shear buckling of webs (GPc.N)
n_u = tensile strength (GPc.N)
n_u = crippling strength (GPc.N)
n_y = yield strength (GPc.N)
n_y = buckling strength (GPc.N)
P = secondary prestress effects (GPc.N)
Q = nominal live load (GPc.A)
Q = seismic load (GPc.J)
Q = wind or earthquake force (GPc.N)
Q_k = nominal loading (GPc.W)
q = reference wind pressure (GPc.W)

R = nominal design capacity (GPc.A)
R = rib shortening (GPc.N)
R = load due to initial rainwater or ice exclusive of the ponding contribution (GPc.N)
R = factored member resistance (GPc.N)
R_d = design yield stress (GPb.E) & (GPc.E)
$R_{d,ef}$ = modified design yield stress (GPc.E)
R_{fL} = load factor (GPc.W)
R_n = nominal resistance (GPc.N)
R_y = specified yield stress (GPb.E)
S = design action effect (GPc.A)
S = snow load, or load due to differential foundation settlement, or shrinkage (GPc.N)
S_{GK} = dead load effect (GPc.A)
S_{max} = maximum effect of factored loads (GPc.E)
S_{QiK} = other live load effect (GPc.A)
S_{Q1K} = first (dominant) live load effect (GPc.A)
SF = stream flow pressure (GPc.N)
SW = snow load (GPc.J)
T = combined effect of temperature, creep, shrinkage and differential settlement (GPc.N)
T = temperature (GPc.N)
T = influences of temperature, shrinkage and creep (GPc.N)
U = strength (GPc.N)
V = wind load on live load (GPc.N)
v = wind speed (GPc.W)
W = wind pressure (GPc.J) & (GPc.N)
W_s = nominal serviceability wind load (GPc.A)
W_u = nominal ultimate wind load (GPc.A)
WL = wind load on live load (GPc.N)
X_D = dead load (GPc.A)
X_D^1 = minimum dead load (GPc.A)
X_L = live load (GPc.A)
X_W = wind load (GPc.A)
X_W^1 = wind load opposing dead load (GPc.A)
β = reliability index for members (GPc.A)
β = coefficient of load combination (GPc.N)
β_B = coefficient of load combination for buoyancy (GPc.N)
β_C = coefficient of load combination for centrifugal forces (GPc.N)
β_D = coefficient of load combination for dead loads (GPc.N)
β_E = coefficient of load combination for earth pressure (GPc.N)
β_{EQ} = coefficient of load combination for earthquake (GPc.N)
β_{ICE} = coefficient of load combination for ice pressure (GPc.N)
β_L = coefficient of load combination for live loads (GPc.N)
β_R = coefficient of load combination for rib shortening, shrinkage and temperature (GPc.N)
β_S = coefficient of load combination for stream flow pressure (GPc.N)
β_W = coefficient of load combination for wind load pressure on structures (GPc.N)

β_{WL} = coefficient of load combination for wind load on live load (GPc.N)
γ = load factor (GPc.N)
γ_a = coefficient of the use of structure (Importance factor) (GPc.E)
γ_b = coefficient of the behavior conditions (Adjustment factor) (GPc.E)
γ_f = load factor (GPc.E) & (GPc.W)
γ_{fL} = resistance factor (GPc.W)
γ_{f3} = resistance factor (GPc.W)
γ_m = material reliability factor (GPb.E)
γ_m = resistance factor (GPc.W)
γ_{m1} = resistance factor on characteristic yield stress (GPc.W)
γ_{m2} = resistance factor for modeling uncertainties (GPc.W)
γ_o = coefficient in accordance to the importance of the structure (GPc.A)
γ_R = resistance factor for member (GPc.A)
ρ = air density (GPc.W)
ϕ = capacity factor (GPc.A)
ϕ = resistance factor (GPc.J) & (GPc.N)
ϕ_d = resistance factor for deflection and drift (GPc.J)
ϕ_F = resistance factor for fatigue (GPc.J)
ϕ_s = resistance factor for yielding, buckling, and slipping (GPc.J)
Ψ_{ci} = combination factor for the ith variable load effect (GPc.A)
ψ_c = coefficient depending on type of load (GPc.A)
ψ_l = long term live load factor (GPc.A)
ψ_s = short term live load factor (GPc.A)
ψ_1 = coefficients of load combinations (GPc.E)
ψ_2 = coefficients of load combinations (GPc.E)

NOMENCLATURE

Symbols

A list of symbols is included at the end of each chapter. The actual symbols used in specifications are retained in the original form as much as possible since this will help when one wants to refer to the original document. However, in regional comparisons or in the international comparison sections, symbols are unified to facilitate comparison between the different specifications.

Glossary

Alignment chart for columns. A nomograph for determining the effective length factor K of a column from the joint bending stiffness ratios at each end of the column.

Allowable load. The load that induces the computed maximum allowable or permitted stress at the critical section.

Allowable stress. As used in allowable stress design, it is the maximum stress permitted under service loads. (Values may be obtained from codes and specifications).

Allowable Stress Design. A method of proportioning structures which is based on service loads such that computed stresses do not exceed prescribed values.

Amplification. An increase in value of moment or deflection in the unbraced length of an axially loaded member to reflect the secondary values generated from the eccentricity of the applied load.

Amplification factor. A multiplier of the value of moment or deflection in the unbraced length of an axially loaded member to reflect the secondary values generated from the eccentricity of the applied load.

Arch. A structure curved in the vertical plane that carries load primarily by compression as a result of horizontal end constraints.

Aspect ratio. In any rectangular configuration, the ratio of the lengths to the sides.

Batten plate. A plate element used to join two chord components of a built-up column, girder, or strut rigidly connected to the chord components and designed to transmit shear between them.

Bauschinger effect. A softening of the stress-strain curve of a metal after prior yielding in the opposite direction.

Bay instability. Buckling of the stringers together with the attached shell plate between rings (or ends of the cylinders if there are no ring stiffeners).

Beam. A structural member whose primary function is to carry loads transverse to its longitudinal axis.

Beam-column. A structural member whose primary function is to carry loads both transverse and parallel to its longitudinal axis.

Biaxial bending. Simultaneous bending of a member about two perpendicular axes.

Bifurcation. The phenomenon whereby a system or a member may assume different deflected shapes under the same load.

Box girder. A hollow beam, usually rectangular in section, fabricated from steel plates and shapes.

Braced frame. A frame in which the resistance to lateral load or frame instability is primarily provided by a diagonal, a K-brace, or other system of bracing.

Bracing. A member or a system of members provided in a structure to primarily resist lateral loads and to provide member or frame stability.

Bracing element. A strut or tie acting to increase the buckling strength of a structure.

Bracing system. A portion of a structure designed to resist lateral loads and to provide frame stability and to control the drift.

Brittle fracture. Abrupt cleavage with little or no prior ductile deformation.

Buckling. The state of collapse of a compressed element, member, or frame.

Buckling load. The load at which a perfectly straight compressed element, member, or frame assumes a deflected position.

Buckling curve. *see* Column curve.

Built-up column. A column made up of chord components connected together by batten plates or latticed members.

Built-up member. A member made up of structural metal elements that are welded, bolted, or riveted together.

Catenary arch. *see* Funicular arch.

Central safety factor. The ratio of the mean resistance to the mean.

Checkerboard loading. An assumed pattern of live loading in multistory frames wherein the live loading in alternate bays has different values of load factors. This pattern may be assumed for design purposes in order to cause the most severe bending on columns and in beams.

Circular arch. An arch whose longitudinal axis is circular.

Cladding. The exterior or interior covering of the structural components of a building.

Coefficient of variation. The ratio of the standard deviation of a variable to its mean value.

Cold-formed and stress-relieved tube. A tube manufactured by cold-forming and stress-relieved after the final sizing.

Cold-formed members. Structural members formed from steel by rolling or press breaking under room temperature.

Cold-formed tube. A tube manufactured by the process in which a plate is cold-formed by rolling into a circular shape followed by welding of the edges. It is finished to the exact circular, rectangular, or square section by sizing.

Collapse load. *see* Plastic limit load.

Column. A structural member whose primary function is to carry loads parallel to its longitudinal axis.

Column curve. A curve expressing the relationship between axial column strength and slenderness ratio.

Column deflection curve. A curve defining the deformed equilibrium shape of a column under loading.

Compact section. A section capable of developing a fully plastic stress distribution and which can develop a rotation capacity of approximately 3 before the onset of local buckling. *see also* Plastic design section, Noncompact section, Slender section.

Composite beam. A steel beam structurally connected to a concrete slab so that the beam and slab respond to loads as a unit. *see also* Concrete-encased beam.

Composite column. A steel column fabricated from rolled or built-up steel shapes and encased in structural concrete or fabricated from steel pipe or tubing and filled with structural concrete in which both steel and concrete components participate significantly in supporting the load.

Concrete confined strength. The compressive strength of concrete under the effect of a triaxial state of stress.

Concrete cover. The least distance between the surface of the reinforcement and the outer surface of the concrete.

Concrete creep. The continuing time-dependent part of strain resulting from member forces.

Concrete-encased beam. A beam totally encased in concrete cast integrally with the slab.

Concrete limiting strain. The compressive strain corresponding to the peak compressive stress in a uniaxial compression test.

Concrete shrinkage. The volume decrease in concrete caused by drying and chemical changes.

Connection. A structural element or combination of elements used to join structural members utilizing fasteners.

Coupled buckling. One buckling mode occurs with another buckling mode at the same time.

Cracked section. In reinforced concrete, a section which is either designed or analyzed on the assumption that the concrete has no resistance to tensile stress.

Critical load. The load at which bifurcation occurs as determined by a theoretical stability analysis.

Crown. The highest point of an arch.

Cube strength. Compressive strength determined by testing a standard concrete cube using standardized procedures.

Curvature. The rotation per unit length due to bending.

Cyclic load. *see* Repeated load, Reversed load.

Cylinder strength. Compressive strength determined using standardized procedures by testing a standard concrete cylinder.

Damping. The resistance which reduces vibration of a structure. The resistance may be provided by internal frictional resistance of the structure, friction in the connections, the drag effects of the surrounding medium, or contact with or connection to an adjacent structure.

Dead load. The weight of a structure itself, including the weight of fixtures or equipment permanently attached to it.

Deck type arch. An arch bridge in which the road surface is at or above the arch.

Degrees of freedom. The minimum number of possible independent parameters needed to define the deflected shape of a member or structure.

Design load. The total nominal load on a structural system for the worst combination of loads and forces which it is designed to sustain. (In Allowable Stress Design it is the service load. In Load and Resistance Factor Design it is the nominal load times the load factor).

Design moment. The property, determined under the design load, by which a force tends to cause a body, to which it is applied, to rotate about a point or line; equal in magnitude to the product of the force and the perpendicular distance of the point, or line, from the line of action of the force.

Design strength. Resistance (force, moment, stress, as appropriate) provided by an element or a connection; the product of nominal strength and resistance factor.

Diagonal bracing. Inclined, straight structural members employed to transform a structural frame into a truss to resist shear forces.

Diaphragm. A two-dimensional structural element possessing a large in-plane shear stiffness.

Distribution factor. The probability density function of a random variable.

Double curvature of columns. A bending condition in which end moments acting on a column cause it to attain an S-shape.

Doubly symmetric member. A member in which the cross-sectional area is symmetrical about two mutual perpendicular axes through the centroid of the area.

Drift. Lateral deflection of a building.

Drift index. The ratio of lateral deflection to the height of the building.

Ductility factor. The ratio of the total deformation at maximum load to the elastic-limit deformation.

Dynamic instability of columns. The buckling of an axially loaded column or frame subjected to time-dependent applied axial loads characterized by unrestricted amplitudes of lateral vibration.

Dynamic response. The time-dependent load-deflection or moment-strain relationship of an element, member, or frame under load.

Eccentricity. The perpendicular distance between the applied load and a plane passing through the centroidal axis of a member.

Eccentricity ratio. The ratio of the eccentricity of the applied load to the corresponding section dimension in the plane of eccentricity.

Effective length. The equivalent length KL used in compression formulas and determined by a bifurcation analysis.

Effective length factor K. The ratio between the effective length and the unbraced length of the member measured between the centers of gravity of the bracing members.

Effective moment of inertia. The moment of inertia of the cross section of a member that remains elastic when partial plastification of the cross section takes place, usually under the combination of residual stress and applied stress. Also, the moment of inertia based on effective widths of elements that buckle locally. Also, the moment of inertia used in the design of partially composite members.

Effective stiffness. The stiffness of a member computed by using the effective moment of inertia of its cross section.

Effective width. The reduced width of a plate or slab which, with an assumed uniform stress distribution, produces the same effect on the behavior of a structural member as the actual plate width with its nonuniform stress distribution.

Elastic analysis. Determination of load effects (force, moment, or stress as appropriate) on members and connections based on the assumption that material deformation disappears on removal of the force that produced it.

Elastic-perfectly plastic. A material which has an idealized stress-strain curve that varies linearly from the point of zero stress up to the yield stress of the material, and then increases in strain at the value of the yield stress without any further increases in stress.

Encased section. A steel member completely encased in cast-in-place concrete.

End-moment ratio. The ratio between the applied moments at the opposite ends of a member.

Equivalent column. A single column used to replace all other columns in a multistory building frame, used to simplify the analysis of the building, with the cross-sectional properties of the equivalent column subject to variation dependent upon the varying assumptions.

Equivalent column method. A method of stability analysis of a multistory building in which all columns of the building are replaced by an equivalent column amenable to a simplified bifurcation analysis.

Equivalent frame. A single-bay substitute frame used to replace one or more stories of a multi-bay, multistory frame for the purpose of simplifying the analysis.

Equivalent stiffness. The stiffness of an encased section computed in terms of the stiffness of the steel member, with the properties of the steel member, steel reinforcement, and concrete factored appropriately.

Euler formula. The mathematical relationship expressing the value of the Euler load in terms of the modulus of elasticity, the moment of inertia of the cross section, and the length of a column.

Euler load. The critical load of a perfectly straight, centrally loaded, pin-ended column.

Factored capacity. The capacity of a member including a load factor.

Factored load. The product of the nominal load and a load factor.

Factor of safety. The ratio of a limit of structural usefulness (be it load, stress , or deformation) to the working or service condition.

Fastener. A generic term for welds, bolts, rivets, or other connecting devices.

Fatigue. A fracture phenomenon resulting from a fluctuating stress cycle.

Filled section. A circular, square, or rectangular steel section filled with concrete.

First-order analysis. Analysis based on first-order deformations in which equilibrium conditions are formulated on the undeformed structure and in which linear properties of material are assumed.

First-order (second moment) theory. A statistics-based methodology in which only the mean and the standard deviation of a random function are used.

Fixed arch. An arch fixed against rotation at the supports.

Flame-cut plate. A plate in which the longitudinal edges have been prepared by oxygen cutting from a wider plate.

Flexible connection. A connection permitting a portion, but not all, of the simple beam rotation of a member end.

Flexural torsional buckling. A mode of buckling involving bending and twisting simultaneously under axial load.

Frame. A system of interconnected members used to transmit loads from the points of their application to points of frame support.

Frame buckling. A condition under which a symmetrical frame subjected to symmetric axial loading at the joints may suddenly deflect in an unsymmetrical mode and be unable to carry any increase in load. Frame buckling is the frame equivalent to the Euler load on a column.

Frame instability. A condition in which a symmetrical frame subjected to unsymmetrical gravity and/or horizontal loads or an unsymmetrical frame subjected to gravity or horizontal loads deforms with increasing lateral deflection accompanying increase of loads until it reaches a critical value of the load called the "stability limit load" where the frame will continue to sway without further increase in load.

Frame stability. The capacity of a frame to remain in position and support load, even if forced slightly out of line or position by an added lateral force.

Funicular arch. An arch in which the shape follows the dead load equilibrium polygon.

Gaussian distribution. Normal distribution.

General instability. A buckling mode for shells in which the shell wall and both the stringers and the rings participate.

Half-through arch. An arch whose road surface lies between the crown and the supports.

Hanger. Tension members in arches or trusses connecting the arch or the top chord of the truss to the roadway girder of floor beams. Any tension member supporting gravity load.

Hinged arch. An arch with one or more hinges.

Hot-finished tube. A tube manufactured by cold-forming and heating, before the final sizing process.

Hot-formed tube. A tube manufactured using a heating process by piercing a bar or forming a heated continuous strip.

Hybrid beam. A fabricated steel beam composed of flanges with yield stress different from that of the web plate.

Hysteresis loop. A plot of force versus deflection of a structure or member subjected to reversed, repeated load into the inelastic range in which the path followed during release and removal of load falls below the path for the addition of load over the same range of deflection.

Impulsive loading. Rapidly applied loading causing a dynamic response to the member or system.

Inelastic behavior. Deformation of a material that does not completely disappear when the force that produced it is removed.

Initial imperfection. A deviation from perfect geometry; for example, initial crookedness of a member, initial out-of-plumb of a story, initial out-of-flatness of a plate, or initial denting or bulging of a shell, (which is within the accepted practical tolerance of the particular fabrication technology).

Instability. A condition reached in the loading of a compressed element or structure in which continued deformation results in a decrease of load-resisting capacity.

Interaction equation. A mathematical expression defining the relationship between two or more variables and usually associated with the design equations used for design of a beam-column.

Joint. Area where two or more ends, surfaces, or edges are attached. Categorized by type of fastener or weld used and method of force transfer.

K-bracing. A system of struts used in a braced frame in which the pattern of the struts resembles the letter K, either normal or on its side.

Knock-down factor. A factor accounting for the detrimental effect of shape imperfections, residual stresses and edges disturbances on the load carrying capacity of a structural element, member or system, a shell for example. In practice, the knock-down factor is taken to be the lower limit of the ratio of the experimental buckling loads to some theoretical critical load of the same structural element, member or system, assumed to be geometrically and physically perfect.

Lacing. A triangular system of bars connecting main elements.

Lamellar tearing. Separation in highly restrained base metal caused by through-thickness strains induced by shrinkage of adjacent weld metal.

Lateral bracing member. A member utilized individually or as a component of a lateral bracing system to prevent buckling of members or elements and/or to resist lateral loads.

Lateral (or lateral-torsional) buckling. Buckling of a member involving lateral deflection and twist.

Latticed member. A built-up member connected by lacing bars or batten plates.

Limit states. Limits of structural usefulness according to ultimate carrying capacity, such as brittle fracture, fatigue, instability and plastic collapse or according to serviceability, such as deformation and durability.

Linear. A mathematical relationship between two or more variables in which all terms are first order. For relationships between two or three variables, the mathematical relationship can be represented as a straight line.

Live load. The moving or movable external load in a structure including the weight of furnishings of a building, people, and of equipment usually identified with nominal load.

Load and Resistance Factor Design (LRFD). A method of proportioning structural components (members, connectors, connecting elements and assemblages) such that no applicable limit state is exceeded when the structure is subjected to all appropriate load combinations.

Load effect. An internal response of a member, a connection, or a structure to loads.

Load factor. A factor that accounts for unavoidable deviations of the actual load from the nominal value and for uncertainties in the analysis.

Loads. Forces or other actions that arise on structural systems from the weight of all permanent construction, occupants and their possessions, environmental effects, differential settlement and restrained dimensional changes. Permanent loads are those loads in which variations in time are rare or of small magnitude. All other loads are variable. *see* Nominal loads.

Local buckling. The partial buckling of a compression element.

Local buckling imperfection factor. A modification coefficient of ideal local buckling stress to consider the effect of imperfections.

Local shell buckling. Buckling of the shell plate between stiffeners.

Local shell instability. Buckling of a shell panel between stiffeners, not involving buckling of stiffeners.

Local stringer instability. Torsional buckling of a stringer stiffening a shell, or buckling of a stiffener flange, or buckling of a stiffener web.

Low-cycle fatigue. Fracture resulting from a relatively high stress range resulting in a relatively small number of cycles to fatigue.

Lower bound load. A load computed on the basis of an assumed equilibrium moment diagram in which the moments are not greater than M_p. It is less than or at best equal to the true ultimate load.

Mechanism. An articulated system able to deform without an increase in load, used in the special sense that the linkage may include real hinges or plastic hinges, or both.

Mechanism method. A method of plastic analysis in which equilibrium between external forces and internal plastic hinges is calculated on the basis of an assumed mechanism. The failure load so determined is an upper bound.

Moving loads. A load for which the position of the load changes with time.

Multiple column curve. A series of column curves, each dependent upon material, manufacture, fabrication and cross-sectional properties of the members for which they are applicable.

Nominal load. The magnitude of the load specified by the applicable code.

Nominal strength. The capacity of a structure or component to resist the effects of loads, as determined by computations using specified material strengths and dimensions and formulas derived from accepted principles of structural mechanics or by field tests or laboratory tests of scaled models, allowing for modeling effects and differences between laboratory and field conditions.

Noncompact section. A section which can develop yield stress in compression elements before local buckling occurs, but will not resist inelastic local buckling at strain levels required for a fully plastic stress distribution.

Non-linear. A mathematical relationship between two or more variables. The mathematical relationship usually cannot be represented as a straight line.

Optimum design. A design which makes the most efficient use of the available resources in performing a desired task. The criterion is often minimum cost.

Parabolic arch. An arch whose longitudinal axis is parabolic.

Parametric resonance. The buckling of an axially loaded column or frame subjected to harmonically applied loading characterized by unrestricted level of vibration.

Panel zone. The zone in a beam-to-column connection that transmits moments through a shear panel.

P-Delta effect. Secondary effect of column axial loads and lateral deflection on the moments in members.

Piece-wise linear. Behavior which can be described graphically by a series of connected straight line segments.

Plane frame. A structural system assumed to be two-dimensional for the purpose of analysis and design.

Plastic analysis. Determination of load effects (force, moment, stress, as appropriate) on members and connections based on the assumption of rigid-plastic behavior, that is, that equilibrium is satisfied throughout the structure and yield is not exceeded anywhere. Second order effects may need to be considered.

Plastic design. A design method for continuous steel beams and frames which defines the limit of structural usefulness as the maximum load. (The term plastic comes from the fact that the maximum load is computed from a knowledge of the strength of steel in the plastic range.)

Plastic design section. The cross section of a member which can maintain a full plastic moment through large rotations so that a mechanism can develop; a section suitable for plastic design.

Plastic hinge. A yielded zone which forms in a structural member when full plasticity is attained. The member rotates as if hinged at the plastic moment M_p.

Plastic limit. The limiting condition at which full section plasticity is attained.

Plastic-limit load. The load that is attained when a sufficient number of yield zones have formed to permit the structure to deform plastically without further increase in load. It is the largest load a structure will support, when perfect plasticity is assumed and when such factors as instability, second-order effects, strain hardening, and fracture are neglected.

Plastic mechanism. *see* Mechanism.

Plastic modulus. The section modulus of resistance to bending of a completely yielded cross-section.

Plastic moment. The resisting moment of a fully yielded cross section.

Plastic strain. The difference between total strain and elastic strain.

Plastic zone. The yielded region of a member.

Plastification. The process of successive yielding of fibers in the cross section of a member as bending moment is increased.

Plate. A flat structural element, the depth (thickness) of which is substantially smaller than its length and width.

Plate girder. A built-up structural beam.

Portal frame. A stiff cross frame at the end of a through type of bridge connecting parallel trusses or arch ribs.

Post. Compression members in arches or trusses connecting the arch or the top chord of the truss to the roadway girder or floor beam.

Post-buckling (Post-critical) strength. The load which can be carried by a plate element or structural member after bifurcation.

Proportional limit. The load or stress beyond which there is a significant amount of deviation from a prior linear load-deformation or stress-strain relationship. The term is usually used in connection with a tensile or compressive test, and the sensitivity of the strain or deformation measuring device is a determining factor in the evaluation.

Proportional loading. A system of loading in which all loads increase in a constant ratio, one to the other.

Random variable. A numerical variable associated with random events, and as such its specific value cannot be determined before an experiment.

Redistribution of moment. A process which results in the successive formation of plastic hinges so that less-highly-stressed portions of a structure may carry increased moments.

Reliability analysis. Structural analysis using probability theory as one of its components.

Required strength. Load effect (force, moment, stress, as appropriate) acting on an element or connection determined by structural analysis from the factored loads (using most appropriate critical load combinations).

Residual stress. The initial stresses that remain in an unloaded member after it has been formed and/or fabricated into a finished product. (Examples of such stresses include, but are not limited to, those induced by cold bending, cooling after rolling or welding, or quenching during heat treatment).

Resistance. The capacity of a structure or component to resist the effects of loads. It is determined by computations using specified material strengths, dimensions, and formulas derived from accepted principles of structural mechanics, or by field tests of laboratory tests of scaled models, allowing for modeling effects and differences between laboratory and field conditions. In Limit States Design, resistance is a generic term that includes both strength and serviceability limit states.

Resistance factor. A factor that accounts for unavoidable deviations of the actual strength (or other limit state) from the nominal value and the manner and consequences of failure.

Resonance. Vibrations of large amplitude that occur when the frequency of the vibration is close to the natural frequency of the member or structure.

Response spectrum. A set of curves or diagrams which gives the maximum value of the response which can be expected for a given applied variable load or disturbance for a member or structure with prescribed values of mass, spring constant, and damping characteristics.

Reticulated structure. A structural frame consisting of pin-connected axial elements arranged in a triangular pattern.

Reversed loading. A cyclic loading condition in which the load is first applied in one direction, then completely removed and reapplied in the opposite direction.

Reversed repeated loading. The cyclic application of a series of loads of the same magnitude but reversed in direction.

Rib. The curved component of an arch.

Rigid frame. A structure made up of beam and column members joined together in such a manner that the full rotation at the end of a member can be transmitted from one member to the other.

Rigid-plastic analysis. An idealized analysis which assumes that a structure has no deflection under increasing load until the plastic limit is reached.

Ring-stiffener. A stiffener attached to a shell to prevent local buckling of the shell.

Rise. The vertical distance in an arch between the chord and the crown.

Rotation capacity. The angular rotation that a given shape can accept at the plastic moment value without local failure.

Second-order analysis. Analysis based on second-order deformation, in which equilibrium conditions are formulated on the deformed structure.

Section modulus. The moment of inertia of the area of the cross section of a structural member divided by the distance from the center of gravity to the farthest point of the section; a measure of the flexural strength of the beam.

Section slenderness. *see* Slenderness ratio.

Service load. Load expected to be supported by the structure under normal usage.

Serviceability limit states. Limiting condition affecting the ability of a structure to preserve its appearance, maintainability, durability, the comfort of its occupants, or the function of machinery under normal usage.

Shape factor. The ratio of the plastic moment to the yield moment, or the ratio of the plastic modulus to the section modulus for a cross section.

Shear center. Of any cross section of a beam, that point in the plane of the cross section through which a transverse load must be applied in order that there will be only bending of the section and no twisting.

Shear modulus. In an elastic material which has been subjected to stress, the ratio of the shearing stress to the shearing strain.

Shear panel. A panel zone which mainly transmits shear forces.

Shear wall. A wall which, in its own plane, carries shear forces and bending moments from applied wind, earthquake, or other transverse loads.

Shell. A three-dimensional structure defined by a curved surface with a thickness that is much less than the radii of curvatures.

Sidesway. The lateral movement of a structure under the action of lateral loads, vertical loads, or due to unsymmetrical properties of the structure.

Sidesway buckling. The buckling mode of a frame precipitated by the relative lateral displacement of joints, leading to failure by lateral deformation of the frame.

Single curvature (of a column). A deformed shape which describes one smooth continuous arc as opposed to double curvature in which reversal is experienced.

Slender section. The cross section of a member which will experience local buckling in the elastic range.

Slenderness ratio. The ratio of the effective length of a member to the radius of gyration of the member section, both with respect to the same axis of bending.

Snap-through buckling. A buckling phenomenon wherein the deflected shape of the member or system under load changes suddenly from a stable shape of single curvature to another stable shape of double curvature under the application of an incremental increase in load.

Solid rib. An arch rib made up of a continuous plate web.

Space frame. A three-dimensional structural framework (as contrasted to a plane frame).

Span. The length of a structural member between two consecutive points of support.

Squash load. An idealized load on a compression member in which all fibers would reach the yield stress without buckling or other instability.

St. Venant torsion. That portion of the torsion in a member that induces only shear stresses in the member.

Stability. The capacity of a compression member or element to remain in position and support load, even if forced slightly out of line or position by an added lateral force. In the elastic range, removal of the added lateral force would result in a return to the prior loaded position, unless the disturbance yielding to commence.

Stability functions. Expressions used to compute stiffness factors, carry-over factors, fixed-end moment coefficient factors, or similar corresponding terms in the slope-deflection or three-moment equations, that are dependent upon the presence of axial load in the member and evaluated by use of a parameter which is the ratio of the actual column load to the Euler buckling load.

Stability-limit load. Maximum (theoretical) load a structure can support when second-order instability effects are included.

Stabilizing force. The force required at each end of a column to prevent a sidesway buckling mode under axial load.

Standard deviation. A statistical quantity used to measure the dispersion of a set of values equal to the root-mean-square deviation of the values from their mean value.

Standard nominal strength of material. The capacity of material to resist the effect of loads, as determined by tests.

Static yield stress. The stress within the yield plateau in a unaxial compression or tension test, measured at a zero strain rate.

Stationary loads. *see* Dead load.

Stepped column. A column with changes from one cross section to another occurring at abrupt points within the length of the column.

Stiffened panel buckling or bay instability. A buckling mode in which the wall and the stringers of cylindrical shells participate in buckling but without the participation of ring stiffeners (if any).

Stiffener. A member, usually an angle, a tee, or plate, attached to a plate or web of a beam or girder to distribute load, to transfer shear or to prevent local buckling of the member to which it is attached.

Stiffness. The parameter expressing the resistance to deformation of a member or structure measured by the ratio of the applied force to the corresponding deformation.

Story drift. The difference in horizontal deflection at the top and bottom of a story.

Strain hardening. A phenomenon wherein ductile steel, after undergoing considerable yielding deformation, exhibits the capacity to resist substantially higher loading than that which caused initial yielding.

Strain-hardening strain. For structural steels that have a flat (plastic) region in the stress-strain relationship, the value of the strain at the onset of strain hardening.

Stress. Force per unit area.

Stress concentration. Localized stress considerably higher than average due to abrupt changes in geometry or localized loading.

Strong axis. The major principle axis of a cross section.

Structural system. An assemblage of load-carrying components which are joined together to provide regular interaction or interdependence.

Stub column. A short compression-test specimen, long enough for use in measuring the stress-strain relationship for the complete cross section, but short enough to avoid buckling as a column in the elastic and plastic ranges.

Subassemblage. A truncated portion of a structural frame.

Supported frame. A frame which depends upon adjacent braced or unbraced frames for resistance to lateral load or frame instability. (This transfer of load is frequently provided by the floor or roof system through diaphragm action or by horizontal cross bracing in the roof).

Sway stiffness. The translational stiffness of a frame free to sidesway.

Symmetrical member. A member in which the cross-sectional area is symmetrical about at least one perpendicular axis through the centroid of the area.

Tangent modulus. The slope of the stress-strain curve of material in the inelastic range, at any given stress level, as determined by the compression test of a small specimen under controlled conditions. The "effective tangent modulus" (as determined by a stub-column test) is modified by nonhomogeneity of material properties and by residual stresses.

Tangent-modulus load. The critical column load obtained by substituting E_t, the tangent modulus, in place of E in the Euler formula.

Tensile strength. The maximum stress that a material can sustain under tension.

Tension-field action. A description of the postbuckling behavior of a plate girder panel under shear force, during which diagonal compressive stresses cause the web to form diagonal waves. The tensile stresses, parallel to the wave troughs, induce compressive stresses in the transverse stiffeners.

Three-dimensional analysis. Structural analysis where a three-dimensional model is used to approximate the actual structure.

Through arch. An arch whose road surface is at the level of the chord.

Thrust. Axial compression in a member. Usually a term associated with arches.

Tie. A tension member joining points of support.

Tied arch. An arch whose supports are connected by a tie.

Triangular structure. *see* Reticulated structure.

Truss. A frame built up from members designed primarily to resist tension or compression.

Trussed rib. An arch rib made up of a trussed web.

Two-dimensional analysis. Method of analysis wherein a three-dimensional structure is approximated as a series of planar structures.

Ultimate strength. The maximum strength of a cross section, member, or structure.

Unbraced frame. A frame in which the resistance to lateral load is provided by the bending resistance of frame members and their connections.

Unbraced length. The distance between braced points of a member.

Uncracked section. In reinforced concrete, a section which is either designed or analyzed on the assumption that the concrete has resistance to tensile stress.

Unfactored load. *see* Nominal load.

Universal mill plate. A plate in which the longitudinal edges have been formed by a rolling process during manufacturing.

Upper bound load. The load on a structure computed on the basis of an assumed mechanism. The load will always be greater than or at least equal to the plastic limit load.

Vertical bracing system. A system of shear walls, braced frames or both, extending throughout one or more floors of a building.

Virgin material. The steel strip which is used for the cold-forming of hollow sections.

Von Mises yield criterion. A theory which states that inelastic action at any point in a body under any combination of stresses begins only when the strain energy of distortion per unit volume absorbed at the point is equal to the strain energy of distortion absorbed per unit volume at any point in a simple tensile bar stressed to the elastic limit under a state of uniaxial stress. It is often called the maximum strain-energy-of-distortion theory. Accordingly, shear yield occurs at 0.58 times yield strength.

Warping resistance. The resistance against warping of the cross section as distinguished from pure torsional resistance (St. Venant torsion), the sum of the two values equaling the total resistance.

Warping torsion. That portion of the total resistance to torsion that is provided by resistance to warping of the cross section.

Weak axis. The minor principal axis of a cross section.

Web buckling. The buckling of a web plate.

Web crippling. The local failure of a web plate in the immediate vicinity of a concentrated load or reaction.

Working load. *see* Service load.

Yield moment. In a member subjected to bending, the moment at which an outer fiber first attains the yield stress.

Yield plateau. The portion of the stress-strain curve for uniaxial tension or compression in which the stress remains essentially constant during a period of substantially increased strain.

Yield point. The first stress in a material at which an increase in strain occurs without an increase in stress.

Yield strength. The stress at which a material exhibits a specified limiting deviation from the proportionality of stress to strain. The deviation is expressed in terms of strain.

Yield stress. Yield point, yield strength, or yield-stress level as defined.

Yield-stress level. The average stress during yielding in the plastic range, the stress determined in a tension test when the strain reaches 0.005.

Abbreviations

A	Australia
AA	Aluminum Association
AASHTO	American Association of State Highway and Transportation Officials
ACI	American Concrete Institute
AIJ	Architectural Institute of Japan
AIJ LSD	Architectural Institute of Japan Limit State Design
AISC	American Institute of Steel Construction or Australian Institute of Steel Construction
AISI	American Iron and Steel Institute
API	American Petroleum Institute
AREA	American Railway Engineering Association
ASD	Allowable Stress Design
ASME	American Society of Mechanical Engineers
ASTM	American Standards for Testing and Materials
AWWA	American Waterwork Association
BS	British Standards
BSSC	Building Seismic Safety Committee
C	China
CEC	Committee of European Communities
CF	Cold formed
CHS	Circular hollow section
CIDECT	Comite International pour le Development et Etude de la Construction Tubulaire
CMEA	Council for Mutual Economic Aid
CRC	Column Research Council
CSA	Canadian Standards Association
ČSN	Ceskoslovenska Statni Norma (Czechoslovakian State Standard)
DNV	Det Norske Veritas (Norwegian Classification Notes)
E	East Europe
EC	Eurocode
ECCS	European Convention for Constructional Steelwork
EEC	European Economic Community
ENC	Section with rolled shape encased in concrete
ESDU	Engineering Sciences Data Unit (London)
FHWA	Federal Highway Administration
FIL	Tubular section filled with concrete
FLB	Flange Local Buckling
FRG	Federal Republic of Germany (for historic reference only)
GDR	German Democratic Republic (for historic reference only)
GPDR	General provision and design requirements
HE	Wide flange column type section
HR	Hot-rolled or hot finished
HW	Heavily welded
HW	Highway
INCERC	Institute de Cercetari in Constructii si Economia Constructiilor
IPE	Narrow flange I-shaped beam section
ISO	International Organization for Standardization
J	Japan

JEC	Japanese Electrical Committee
JRA	Japan Road Association
JSCE	Japan Society of Civil Engineers
LFD	Load Factor Design
LFDM	Load Factor Design Method
LRFD	Load and Resistance Factor Design
LSD	Limit States Design
LSP	Limiting slenderness for full plastic moment
LSPA	Limiting slenderness for plastic analysis
LSY	Limiting slenderness for yield moment
LW	Light welded
ME	Müszaki Elöirás, Hungarian Technical Recommendations
MI	Müszaki Irányelvck, Hungarian Technical Instructions
MOC	Ministry of Construction
MSZ	Magyar Szebvány, Hungarian State Standard
N	North America
NASA	National Aeronautics and Space Administration
NEHRP	National Earthquake Hazards Reduction Program
NDSS	Numerical Data-Base for Steel Structures
NiP	Normü i Pravila, Bulgarian Standard
NZS	New Zealand Standard
OHBDC	Ontario Highway Bridge Design Code
PCBI	Permanent Commission for the Building Industries
PD	Plastic design
PRC	People's Republic of China
R	Rolled beam
RC	Reinforced concrete
RHS	Rectangular hollow section
RMI	Rack Manufactures Institute
RR	Rail road
SAA	Standards Association of Australia
SEC	State Electricity Commission
SHS	Square hollow section
SL	Seismic loading
SLDM	Service Load Design Method
SNiP	Stroitelnüe Normü i Pravila, USSR State Standard
SR	Stress relieved
SSRC	Structural Stability Research Council
ST SEV	Recommendation for Steel Structures, Council for Mutual Economic Assistance
STAS	Institul Roman de Standardizare, Romanian State Standard
TGL	Technische Norme, Gütervorschriften und Lieferbedingungen, GDR State Standard
UB	Universal beam
UBC	Uniform Building Code
UC	Universal column
USA	United States of America
W	Welded beam
W	West Europe
WLB	Web Local Buckling

Units Conversion Table

In the table below are given conversion factors for commonly used units. The numerical values have been rounded off to the values shown. The British (Imperial) System of units is the same as the American System except where noted. Le Système International d'Unités (abbreviated "SI") is the name formally given in 1960 to the system of units partly derived from, and replacing, the old metric system.

	SI	American	Old Metric
Length	1 mm	0.03937 in.	1 mm
	1 m	3.28083 ft	1 m
		1.093613 yd	
	1 km	0.62137 mile	1 km
Area	1 mm^2	0.00155 in.^{2}	1 mm^2
	1 m^2	10.76392 ft^2	1 m^2
		1.19599 yd^2	
	1 km^2	247.1043 acres	1 km^2
	1 hectare	2.471 acres [(1)]	1 hectare
Volume	1 cm^3	0.061023 in^3	1 cc
			1 ml
	1 m^3	35.3147 ft^3	1 m^3
		1.30795 yd^3	
		264.172 gal[(2)] liquid	
Velocity	1 m/sec	3.28084 ft/sec	1 m/sec
	1 km/hr	0.62137 miles/hr	1 km/hr
Acceleration	1 m/sec^2	3.28084 ft/sec^2	1 m/sec^2
Mass	1 g	0.035274 oz	1 g
	1 kg	2.2046216 lb[(3)]	1 kg
Density	1 kg/m^3	0.062428 lb/ft^3	1 kg/m^3
Force, Weight	1 N	0.224809 lbf	0.101972 kgf
	1 kN	0.1124045 tons[(4)]	
	1 MN	224.809 kips	
	1 kN/m	0.06853 kips/ft	
	1 kN/m^2	20.9 lbf/ft^2	
Torque, Bending Moment	1 N-m	0.73756 lbf-ft	0.101972 kgf-m
	1 kN-m	0.73756 kip-ft	101.972 kgf-m
Pressure, Stress	1 N/m^2 = 1 Pa	0.000145038 psi	0.101972 kgf/m^2
	1 kN/m^2 = 1 kPa	20.8855 psf	
	1 MN/m^2 = 1 MPa	0.145038 ksi	
Viscosity (Dynamic)	1 N-sec/m^2	0.0208854 lbf-sec/ft^2	0.101972 kgf-sec/m^2
Viscosity (Kinematic)	1 m^2/sec	10.7639 ft^2/sec	1 m^2/sec
Energy, Work	1 J = 1 N-m	0.737562 lbf-ft	0.00027778 w-hr
	1 MJ	0.37251 hp-hr	0.27778 kw-hr

	SI	American	Old Metric
Power	1 W = 1 J/sec 1 kW	0.737562 lbf ft/sec 1.34102 hp	1 w 1 kw
Temperature K = 273.15 + °C K = 273.15 + 5/9(°F−32) K = 273.15 + 5/9(°R−491.69)		°F = (°C × 1.8) + 32	°C = (°F − 32)/1.8

(1) Hectare as an alternative for km^2 is restricted to land and water areas.
(2) 1 m^3 = 219.9693 Imperial gallons
(3) 1 kg = 0.068522 slugs
(4) 1 American ton = 2000 lb. 1 kN = 0.1003612 Imperial ton 1 Imperial ton = 2240 lb.

Abbreviations for Units

°C	degree Celsius (centigrade)
cc	cubic centimeters
cm	centimeter
°F	degree Fahrenheit
ft	foot
g	gram
gal	gallon
hp	horsepower
hr	hour
Imp	British Imperial
in.	inch
J	joule
K	kelvin
kg	kilogram
kgf	kilogram-force
kip	1000 pound force
km	kilometer
kN	kilonewton
kPa	kilopascal
ksi	kips per square inch
kW	kilowatt
lb	pound
lbf	pound force
MJ	megajoule
MPa	megapascal
m	meter
ml	milliliter
mm	millimeter
MN	meganewton
N	newton
oz	ounce
Pa	pascal
psf	pounds per square foot
psi	pounds per square inch
°R	degree Rankine
sec	second
slug	14.594 kg
W	watt
yd	yard

CODES, SPECIFICATIONS, AND STANDARDS

Australia

AISC 83*
Structural Design of Steel Bins for Bulk Solids (Trahair, N.S., Abel, A., Ansourian, P., Irvine, H.M., and Rotter, J.M.), Australian Institute of Steel Construction

AS 1210 77
SAA Unfired Pressure Vessels Code, Standards Association of Australia

AS 1664 79
SAA Aluminum Structures Code, Standards Association of Australia

AS 2327 P.1 80
Composite Construction in Structural Steel and Concrete, Simply-Supported Beams (AS 2327, Part 1-1980), Standards Association of Australia

AS 1163 81
SAA Structural Steel Hollow Sections, Standards Association of Australia

AS 1250 81
SAA Steel Structures Code, Standards Association of Australia

AS DR87164 87
Draft Australian Standard for Steel Structures, Standards Association of Australia

AS 1538 88
SAA Cold-Formed Steel Structures Code, Standards Association of Australia

DZ 4203 89
General Structural Design and Design Loadings for Buildings, Second Draft of Revised Limit States Format, Standards Association of New Zealand

NZS 4203 84
Practice for General Structural Design and Design Loadings for Buildings, Standards Association of New Zealand

NZS 3404 P.1 89
Steel Structures Code, Part 1, New Zealand Amendments to AS 1250 81 (with Commentary), Standards Association of New Zealand

NZS 3404 P.2 89
Steel Structures Code, Part 2, Sections 12, 13, 14, Means of Compliance (with Commentary), Standards Association of New Zealand

---

* A brief citation for codes, specifications and standards under letter designation of specification further identification for the documents (if applicable) and the year indicated by the last two digits.

SEC 81
Code of Practice for Design of Transmission towers, State Electricity Commission, Victoria

China

GBJ 18 87
Specification for the Design of Cold-formed Thin-walled Steel Structures, PRC National Standard

GBJ 17 88
Specification for the Design of Steel Structures, PRC National Standard

GBJ 11 89
Specification for the Design of Earthquake Resistant Buildings, PRC National Standard

HRS 87
Specification for the Design of High-Rise Structures, Draft for a Forthcoming Standard, Ministry of Construction

SDGJ 64 89
Technical Regulation for the Design of Building Structures of Thermal Power Plants, Ministry of Energy

TBJ 2 85
Specification for the Design of Railroad Bridges and Culverts, Ministerial Standard, Ministry of Railways

East Europe

CMEA 72
Design of Building Structures and Foundations. General Principles, RS-242/72, Council for Mutual Economic Aid, Moscow

CMEA 76
Design of Building Structures and Foundations. General Principles, ST SEV 384/76, Council for Mutual Economic Aid, Moscow

CMEA 83
Steel Structures - Basic Requirements for Design, ST SEV 3972-83, Council for Mutual Economic Aid, Moscow

ČSN 73 0031 83
Úřad pro normalizaci a měření, (Design of Building Structures and Foundations), Praha

ČSN 73 0035 88
Zatížení stavebníck konstrukcí, (Loading of Building Structures), Úřad pro normalizaci a měření, Praha

ČSN 73 0036 89
Seizmická zatížení a odezva stavebnich objektu. TZÚS (Seismic loading of structures), Praha

ČSN 73 0036 75
Seizmická zatížení staveb, (Seismic loading of structures), Úřad pro normalizaci a měření, Praha

ČSN 73 1401 84
Navrhování ocelových konstrukcí (Design of Steel Structures), Úřad pro normalizaci a měření, Praha

ČSN 73 6205 84
Navrhování ocelových mostních konstrukcí (Design of steel bridge structures), Úřad pro normalizaci a měření, Praha

ČSN 73 1206 86
Spřažené ocelobetonové konstrukce. Základní ustanovení pro navrhování (Composite steel-concrete structures. Basic design requirements) Úřad pro normalizaci a měření, Praha

ČSN 73 1402 88
Navrhování tenkostěnných profilů v ocelových konstrukcích (Design of light-gauge cold-formed profiles in steel structures), Úřad pro normalizaci a měření, Praha

Indictiv P. 100-81 82
Normativ pentru proiectarea antiseismica a Constructilor de Locuinte, Social-Culturale, Agrozootehnice Si Industriale, (Seismic Design Standard for Structures), Romania

ME 04 180 83
Vékonyfalú acélszerkezetek méretezése, csomóponti kialakitása és ellenörző vizsgálata (Design of Thin-Walled Structures), Ministry of Building Industry, Budapest

MI 04.184 78
Acélszerkexzetü termenytároló silók méretezése (Design of Steel Bins for the Storage of Bulk Solids), Information Center of Building Industry, Budapest

MI 15024/3 85
Épitmények acélszerkezeteinek erötani tervezése, Méretezési eljárások (Design of Steel Structures for Buildings, Design Methods), Hungarian Instruction, Budapest

MI 15024/4 85
Épitmények acélszerkezeteinek erőtani tervezése, Méretezési irányelvek különleges esetekre (Design of Steel Structures for Buildings, Directives for special cases), Hungarian Instruction, Budapest

MSZ 07 3702 87
Acélhidak tervezése (Design of Steel Bridges), Hungarian State Standard, Budapest

MSZ 15024/1 85
Épitmények acélszerkezeteinek erőtani tervezése, Altalanos eljárások (Design of Steel Construction for Buildings, Design requirements) Hungarian State Standard, Budapest

MSZ 5132 88
Design of Steel Bridge Structures, Hungarian State Standard, Budapest

NiP 565/86 87
Normi za proektirane na stomaneni konstrukcii (Specification for the Design of Steel Structures), Bulgarian Standard, Sofia

PN 80/B-03200 84
Konstrukcje Stalowe, Obliczenia Statyczne i Projektowanie (Steel Structures, Design Rules), Polska norma, Warsaw

PN 80/B-03200 87
Konstrukcje Stalowe, Obliczenia Statyczne i Projektowanie (Steel Structures, Design Rules), Polska norma, Warsaw

SNiP II-24-74 75
Alyuminiyevye konstrukcii (Aluminum Structures), U.S.S.R. Standard, Stroyizdat, Moscow

SNiP II-7-81 82
Strojitelstvo v sejsmičeskich rajonach, (Construction in Seismic Regions), Gosstroj SSSR, Moskva Strojizdat, Moscow

SNiP II-23-81 82
Stalnye konstrukcii (Steel Structures), U.S.S.R Standard, Stroyizdat, Moscow

SNiP 2.05.03 84
Mosty i truby (Bridges and Pipe Culverts), U.S.S.R. Standard, Gosstroy, Moscow

ST SEV 3972 83
Steel Structures–Basic Requirements for the Design, ST SEV 3972-83, Council of Mutual Economic Aid, Moscow

STAS 763/1 71
Constructu Civile, industiale si agricole Prescripti Pentru Proiectarea Constructulor Metalice (Non-industrial, Industrial and Farm Buildings, Design Rules for Metal Structures) Standardizarea Romania, Bucuresti

STAS 1911 75
Design of Steel Railway Bridges, Romanian State Standard, Bucuresti

STAS 10108/0 78
Civil, Industrial and Agricultural Buildings–Design of Steel Structural Members, STAS 10108/0–1978, Romanian State Standard, Bucuresti

STAS 10108/2 83
Steel Buildings - Design of the Cold-formed Steel Structural Members. Romanian State Standard, Bucuresti

TGL 13503/01 82
Stahlbau, Stabilität von Stahltragwerken, Grundlagen (Steel Structures, Stability of Steel Supporting Structures, Fundamentals) DDR-Standard, Amt für Standardisierung, Meβwesen und Warenprüfung, Berlin

TGL 13503/02 82
Stahlbau, Stabilität von Stahltragwerken, Erläuterungen und Berechnungsmöglichkeiten (Steel Structures, Stability of Steel Supporting Structures, Explanations and Methods of Calculation) DDR-Standard, Amt für Standardisierung, Meβwesen und Warenprüfung, Berlin

Japan

AIJ 70
Design Standard for Steel Structures, Architectural Institute of Japan

AIJ 75
Guide for Plastic Design of Steel Structures, Architectural Institute of Japan

AIJ 80
Guide for Stability Design of Steel Structures, Architectural Institute of Japan

AIJ 85
Recommendation for the Design and Fabrication of Light Weight Steel Structures, Architectural Institute of Japan

AIJ 86
Standard for Structural Calculation of Steel Reinforced Concrete Structures, Architectural Institute of Japan

AIJ 89a
Specifications for Steel Tubular Structures, Architectural Institute of Japan

AIJ 89b
Standard for Limit State Design of Steel Structures (Draft), Architectural Institute of Japan

JEC 79
Standards of the Japanese Electrical Committee-127, Institute of Electrical Engineers of Japan

JRA 80
Specification for Highway Bridges, Part II, Steel Bridges, Japan Road Association

JRA 89
Specification for Highway Bridges, Part II, Steel Bridges, Japan Road Association

JSCE 83
Standard Regulation for Design of Structures–Steel Railway Bridges, Japan Society of Civil Engineers

JSCE 87a
Guidelines for Stability Design of Steel Structures, SS Series 2, Japan Society of Civil Engineers

JSCE 87b
Design Code for Steel Structures, SS Series 3A and 3B, Japan Society of Civil Engineers

MOC 80
Building Standard Law and Enforcement Order, Ministry of Construction

North America

AA 86
Aluminum Construction Manual - Specifications for Aluminum Structures, 4th Ed., Aluminum Association, Washington, D.C.

AASHTO 83
Standard Specifications for Highway Bridges, The American Association of State Highway and Transportation Officials, Washington, D.C.

AASHTO 87
Standard Specifications for Highway Bridges, The American Association of State Highway and Transportation Officials, Washington, D.C.

AASHTO 89
Standard Specifications for Highway Bridges, The American Association of State Highway and Transportation Officials, Washington, D.C.

ACI 318 83
Building Code Requirements for Reinforced Concrete, ACI 318-83, American Concrete Institute, Detroit

AISC ASD 78
Manual of Steel Construction, Allowable Stress Design and Plastic Design, American Institute of Steel Construction Inc., Chicago

AISC ASD 89
Manual of Steel Construction, Allowable Stress Design and Plastic Design, American Institute of Steel Construction Inc., Chicago

AISC LRFD 86
Manual of Steel Construction, Load and Resistance Factor Design, American Institute of Steel Construction Inc., Chicago

AISI 76
Tentative Criteria for Structural Applications of Steel Tubing and Pipe, SP 604-876-7-5 M-MP, American Iron and Steel Institute, Washington D.C.

AISI 86
Specification for the Design of Cold-Formed Steel Structural Members, American Iron and Steel Institute, Washington, D.C.

API 620 82
Recommended Rules for Design and Construction of Large Welded, Lower Pressure Storage Tanks, API Standards 620, 7th Edition, September, American Petroleum Institute

API RP2A 89
Recommended Practice for Planning, Designing and Constructing Fixed Offshore Platforms, API-RP2A. 18th Edition, American Petroleum Institute

AREA 85
Specifications for Steel Railway Bridges, American Railway Engineering Association, Washington, D.C.

AREA 87
Specifications for Steel Railway Bridges, American Railway Engineering Association, Washington, D.C.

AREA 89
Specifications for Steel Railway Bridges, American Railway Engineering Association, Washington, D.C.

ASME N284 80
Boiler and Pressure Vessel Code, American Society of Mechanical Engineers

AWWA D100 73
Standard for Welded Steel Elevated Tanks, Standpipes, and Reservoirs for Water Storage, American Waterworks Association.

BSSC NEHRP 85
NEHRP Recommended Provisions for the Development of Seismic Regulations for New Buildings, Building Seismic Safety Committee, Washington, D.C.

CSA A23.3 M84 77
Design of Concrete Structures for Buildings, CAN3-A23.3-M89, Canadian Standards Association, Ontario

CSA S6 M89 84a
Design of Highway Bridges, CAN3-S6-M89, Canadian Standards Association, Ontario

CSA S16.1 M89 84b
Steel Structures for Buildings, (Limit States Design) CAN3-S16.1-M89, Canadian Standards Association, Ontario

CSA S136 M89 84c
Cold Formed Steel Structural Members, CAN3-S136-M89, Canadian Standards Association, Ontario

FHWA TS 80
Proposed Design Specifications for Steel Box Girder Bridges, FHWA-TS-80-205, Federal Highway Administration, Washington, D.C.

OHBDC 83
Ontario Highway Bridge Design Code, Highway Engineering Division, Ministry of Transportation and Communications, Ontario

UBC 88
Uniform Building Code, International Conference of Building Officials, Whittier, California

West Europe

BS 5400 P.3 82
Steel, Concrete and Composite Bridges, British Standards Institute, Part 3, Code of Standards for Steel Bridges, Great Britain

BS 5400 P.5 82
Steel, Concrete and Composite Bridges, British Standards Institute, Part 5, Code of Standards for Steel Bridges, Great Britain

BS 5500 85
BS 5500; 1985 -UK

BS 5950 P.1 85
Structural Use of Steelwork in Building, British Standard Institution, BS 5950, Part 1

BS 5950 P.5 85
Structural Use of Steelwork in Building, British Standard Institution, BS 5950, Part 1

BSK 87
Code for Steel Structures, Bestammelser for Stalkonstruktioner, Statens Planverk, Swedish Institute of Steel Construction, National Board of Physical Planning and Building, Stockholm, Sweden

CNR 10111 85
Steel Structures, Italy

CNR GNDT 84
Consiglio Nazionale Delle Ricerche: Gruppo Nazionale Per La Difesa Dai Terremoti, "Norme tecniche per le costruzioni in zone sismiche", Italy

CTICM 80
Règales de calcul de constructions en acier-modification of: Recommendations pour le calcul en plastic des constructions en acier (1975), ADDITIF France

DASt 013 80
DASt–Richtlinie 013: Beulsicherheitsnachweise für Schalen (Recommendations for Steel Structures), Deutscher Ausschuβ für Stahlbau, Köln, Germany

DASt 015 80
DASt–Richtlinie 015: Beulsicherheitsnachweise für Schalen (Recommendations for Steel Structures), Deutscher Ausschuβ für Stahlbau, Köln, Germany

DASt 016 80
DASt–Richtlinie 016: Beulsicherheitsnachweise für Schalen (Recommendations for Steel Structures), Deutscher Ausschuβ für Stahlbau, Köln, Germany

DIN 4149 81
Deutsche Institut Für Normung, "Bauten in deutschen Erdbebengebieten"

DIN 18800 P.2 86
Steel Structures Stability Cases, Buckling of Members and Frames, Part 2, West Germany

DIN 18806 84
Teil 1, Verbundkonstruktionen, Verbundstutzen, Ausgabe Marz, Beuth – Verlag, Berlin

EC 3 84
Eurocode n.3: Common Unified Rules for Steel Structures, EUR 8849 EN

EC 3 88
Eurocode n.3: Common Unified Rules for Steel Structures, EUR 8849 EN

EC 3 89
Eurocode n.3: Design of Steel Structures, Part 1 - General Rules and Rules for Buildings, Vol. 1 and 2, CEC Industrial Processes, Building and Civil Engineering

EC 4 85
Eurocode n.4: Common Unified Rules for Composite Steel and Concrete Structures, EUR 9886 EN

EC 8 88
Eurocode n.8: Structures in Seismic Regions, Commission of European Communities

ECCS 87
European Recommendations for the Design of Light Gauge Steel Members, European Convention for Constructional Steelwork, Brussels

ECCS CECM EKS 88
European Recommendations for Steel Structures in Seismic Zones, Technical Working Group 1.3: Seismic Design, N.54

ECCS No. 29 84
Buckling of Shells, European Convention for Constructional Steelwork, 3rd Edition, Brussels

ECCS No. 49 88
European Convention for Constructional Steelwork, publication No. 49.

ECCS No. 56 80
European Convention for Constructional Steelwork, publication No. 56.

NBN B51002 88
Steel Structures, Belgium Standards Institute, Belgium

NEN 2018 74
Cranes, Loads and Combinations of Loads, Netherlands

NEN 3851 70
Regulations for the Calculation of Buildings Structures, General Considerations and Loading, Netherlands

NEN 3851 72
Technical Foundations for the Design of Steel Structures, Netherlands

NEN 3851 74
Regulations for the Calculation of Buildings Structures, Design of Steel Structures, Netherlands

NEN 3854 83
Regulations for the Calculation of Buildings Structures, Design of Aluminum Structures, Netherlands

NEN 6702
Structural Principles for Building Structures, Part: Loads, Forces, and Other Actions and Deformations, In preparation (will replace NEN 3850), Netherlands

NEN 6770
Structural Principles for Building Structures, Part: Steel – Basic Requirements and Simplified Calculation Rules, In preparation (will replace NEN 3851), Netherlands

NS 3472 84
NS 3472: Projektering av Stalkonstruksjoner, beregning og dimensjonering (Steel Structures, Design Rules), Norges Standardise rings forbund, Oslo, Norge 1984 Norway

ÖNORM B4600 78
Stahlbau, Stabilitätsnachweise, Grundfälle, part 4 (Steel Structures, Stability, Basic problems), Österreichisches Normeninstitut, Wien, Austria, 1978

RÈGALES CM 66
Règales de Calcul des Constructions en Acier, Additif 80, Paris, France

RKP 88
Recommendations for the Calculation and Testing of Linear Cold-Formed Sections (1988, in preparation), Netherlands

SIA 161 79
Steel Structures, Swiss Institute of Engineers and Architects, Switzerland

StBK N1 70
Stalbayggnadsnorm 70 (Code for Steel Structures), Statens Stalbyggnadskommitte, Stockholm, Sweden

StBK N1 80
Stalbayggnadsnorm 70 (Code for Steel Structures), Statens Stalbyggnadskommitte, Stockholm, Sweden

REFERENCES

Adams, P. F., 1974
THE DESIGN OF STEEL BEAM-COLUMNS, Canadian Steel Industries, Construction Council, Toronto.

AISC, 1986
LOAD AND RESISTANCE FACTOR DESIGN SPECIFICATION FOR STRUCTURAL STEEL BUILDINGS, (Part 6 of AISC LRFD Manual, 1st Edition), AISC, Chicago.

Alekseev, S. A., 1956
POST BUCKLING OF THIN PLATES AND SHELLS, Ph. D. Thesis , Moscow.

Anslijn, R. and Massonnet, C. E., 1978
NEW TESTS ON STEEL I BEAM-COLUMNS IN MILD STEEL, SUBJECTED TO THRUST AND BIAXIAL BENDING, Festschrift Otto Jungbluth-60 Jahre, Lehrstuhl für Stahlbau, Technische Hochschule Darmstadt, pp 103-123.

Arbocz, J. and Abramovich, H., 1979
THE INITIAL IMPERFECTION DATA BANK AT THE DELFT UNIVERSITY OF TECHNOLOGY-PART 1, Rep. LR-290, Delft University of Technology, The Netherlands

Arda, S., 1985
DESIGN OF STRUCTURES UNDER SEISMIC CONDITIONS IN TURKEY, International Colloquium on Stability of Steel Structures, Tihany, Hungary.

ASCE - WRC Joint Committee, 1969
PLASTIC DESIGN IN STEEL (A GUIDE AND COMMENTARY), ASCE Manual, No. 41, 2nd Edition, ASCE, New York.

Baker, E. H., Kovalevsky, L., and Rish, F. L., 1972
STRUCTURAL ANALYSIS OF SHELLS, McGraw-Hill, New York.

Beaulieu, D. and Adams, P. F., 1977
A STATISTICAL APPROACH TO THE PROBLEMS OF STABILITY RELATED TO STRUCTURAL OUT-OF-PLUMBS, 2nd. International Colloquium on Stability of Steel Structures, ECCS-IABSE, Liege, Belgium.

Beer, M. and Schulz, G. W., 1970
BASES THEORETIQUES DES COURBES EUROPEENNES DE FLAMBEMENT, Construction Metallique, No. 3, pp 37-55.

Belskij, G. E., 1982
O RASCHETACH STALNYCH KONSTRUKCIJ V SVJAZI S VYCHODOM NOVOJ GLAVY SNiP, (About the Design of Steel Structures in Connection with the New Edition of the SNiP Specification), Strojitelnaja, Mechanika i Raschet Soorushenij, No. 2.

Bijlaard, P. P., 1957
PLASTIC BUCKLING OF SIMPLY SUPPORTED PLATES SUBJECTED TO COMBINED SHEAR AND BENDING OR ECCENTRICAL COMPRESSION IN THEIR PLANE, Journal of Applied Mechanics, Vol. 23, pp 291-303.

Bild, S. and Kulak, G. L., 1986
LOCAL BUCKLING RULES FOR STRUCTURAL STEEL MEMBERS, Report No. 137, Department of Civil Engineering, University of Alberta, Canada, May.

Bjorhovde, R., 1972
DETERMINISTIC AND PROBABILISTIC APPROACHES TO THE STRENGTH OF STEEL COLUMNS, PhD Dissertation, Lehigh University, University Microfilms, Ann Arbor.

Bjorhovde, R. and Tall, L., 1971
MAXIMUM COLUMN STRENGTH AND THE MULTIPLE COLUMN CURVE CONCEPT, Fritz Laboratory, Report 337.29, Lehigh University, October.

Bjorhovde, R. and Tall, L., 1975
DEVELOPMENT OF MULTIPLE COLUMN CURVES, Proceedings, 1972, International Colloquium on Column Strength, Paris, IABSE, Vol. 23.

Bjorhovde, R. and Tall, L., 1975
THE PROBABILISTIC CHARACTERISTICS OF MAXIMUM COLUMN STRENGTH, Proceedings, 1972 International Colloquium on Column Strength, Paris, IABSE, Vol. 23.

Bridge, R. Q., 1986
ELASTIC CRITICAL BUCKLING LOADS - THEIR ASSESSMENT AND USE, Tenth Australian Conference on the Mechanics of Structures and Materials, Adelaide, pp 27-32.

Bridge, R. Q. and Trahair, N. S., 1987
LIMIT STATE DESIGN RULES FOR STEEL BEAM-COLUMNS, Steel Construction, Australian Institute of Steel Construction, Vol. 21, No. 2, pp 2-11.

Broude, B. M., 1963
INFLUENCE OF INITIAL IMPERFECTIONS ON THE BUCKLING OF CYLINDRICAL SHELLS, Stroitelnaya Mechanika i Raschet Sooruzeni, pp 32-40.

Bushnell, D., 1985
COMPUTERIZED BUCKLING ANALYSIS OF SHELLS, Martines NijHoff Publishers.

Butterworth, J.W. and Spring, K.C.F., 1985
COLUMN DESIGN, SECTION D OF THE DELIBERATIONS OF THE STUDY GROUP FOR THE SEISMIC DESIGN OF STEEL STRUCTURES, Bulletin N.Z. National Society for Earthquake Engineering, Vol. 18, No. 4.

Chen, Ji., 1982
THE INTERACTION FORMULA FOR CALCULATING STABILITY CAPACITY OF STEEL BEAM-COLUMNS IN THE PLANE OF MOMENT BY PARTIALLY UTILIZING PLASTICITY, Selections from Research Papers and Reports on Steel Structures, Vol. 1.

Chen, S. F., 1982
THE INTERACTION FORMULAS FOR LATERAL-TORSIONAL BUCKLING OF BEAM-COLUMNS, Selections from Research Papers and Reports on Steel Structures, Vol. 1.

Chen, W. F. and Atsuta, T., 1976
THEORY OF BEAM-COLUMNS, Vol. 1, In-Plane Behavior and Design, McGraw-Hill, New York.

Chen, W. F. and Atsuta, T., 1977
THEORY OF BEAM COLUMNS, Vol. 2, Space Behavior and Design, McGraw-Hill, New York.

Chen, W. F. and Lui, E. M., 1985
STRUCTURAL STABILITY: THEORY AND IMPLEMENTATION, Elsevier, New York.

Chuvikin, G. M., 1962
ON THE INELASTIC STABILITY OF BEAM-COLUMNS OF THIN-WALLED OPEN SECTION, "Study on Steel Structures," Gosstroiizdat (in Russian).

Constrado, 1985
STEELWORK DESIGN GUIDE TO BS 5950, P.1, SECTION PROPERTIES MEMBER CAPACITIES, Vol. 1, 1st Edition, The Steel Construction Institute, Sept.

Council on Tall Buildings, Group CL, 1980
TALL BUILDING CRITERIA AND LOADING, VOL. CL OF MONOGRAPH ON PLANNING AND DESIGN OF TALL BUILDINGS, ASCE, New York.

Cuk, P. E., Bradford, M. A. and Trahair, N. S., 1986
INELASTIC LATERAL BUCKLING OF STEEL BEAM-COLUMNS, Canadian Journal of Civil Engineering, Vol. 13., No. 6, Dec., pp 693-699.

Davids, A. J. and Hancock, G. J., 1985
THE STRENGTH OF LONG LENGTH I-SECTION COLUMNS FABRICATED FROM SLENDER PLATES, Civil Engineering Transactions, Institution of Engineers, Australia, Vol. CE27, No. 4, pp 347-352.

Davies, J. M. and Thomasson, P. O., 1983
LOCAL AND OVERALL BUCKLING OF LIGHT GAUGE MEMBERS, Contribution to the Michael R. Horne Conference, The University of Manchester, Sept.

Djubek, J., Kodnar, R. and Škaloud, M., 1983
LIMIT STATE OF THE PLATE ELEMENTS OF STEEL STRUCTURES, Birkhauser Verlag, Basel Boston - Stuttgart.

Donnell, L. H., 1934
NEW THEORY FOR THE BUCKLING OF THIN CYLINDERS UNDER AXIAL COMPRESSION AND BENDING, Trans. ASME Vol. 56, No. 11.

Duan, L. and Chen, W. F., 1988
DESIGN RULES OF BUILT-UP MEMBERS IN LOAD AND RESISTANCE FACTOR DESIGN, Journal of Structural Engineering, ASCE, Vol. 114, No. 11, pp 2544-2554.

Dubas, P., 1986
RESISTANCE A LA RUINE DES TREILLIS PLANS, Construction Metallique, No. 3.

Dunkerley, S., 1984
ON THE WHIRLING AND VIBRATION OF SHAFTS, Philosophical Transactions of the Royal Society of London, Ser. A 185, pp 279-360.

Dwight, J., 1975
ADOPTION OF PERRY FORMULA TO REPRESENT THE NEW EUROPEAN STEEL COLUMN-CURVES, Steel Construction, Vol. 9, No.1, pp 10-23.

ECCS
DOCUMENT NO. 33, ULTIMATE LIMIT STATE CALCULATION OF SWAY FRAMES WITH RIGID JOINTS, Brussels.

ECCS, 1978
EUROPEAN RECOMMENDATIONS FOR STEEL CONSTRUCTION.

Ellinas, C. P. and Croll, J. G. A., 1983
EXPERIMENTAL AND THEORETICAL CORRELATIONS FOR ELASTIC AXIALLY-COMPRESSED STRINGER STIFFENED CYLINDERS, J-Strain Analysis, 18., pp 41-67.

Ellinas, C. P., Supple, W. J., and Walker, A. C., 1983
BUCKLING DESIGN OF STRINGER AND ORTHOGONALLY STIFFENED CYLINDERS, Final Report, Third International Colloquium, "Stability of Metal Structures", Paris, pp 145-156.

Ellinas, C. P. and Williams, K. A. I., 1987
ASSESSMENT OF EUROPEAN AND OTHER SHELL BUCKLING RECOMMENDATIONS, ECCS Colloquium on Stability of Plate and Shell Structures, Ghent University, 6-8 April, pp 637-644.

ESDU, 1974
BUCKLING STRESS COEFFICIENTS FOR THIN-WALLED UNSTIFFENED CIRCULAR CYLINDERS UNDER COMBINED AXIAL COMPRESSION AND INTERNAL PRESSURE, Eng. Sciences Data Unit, Sheet 04.01.01A, London.

Esslinger, M. Geier, B., and Wood, J. G. M., 1977
SOME COMPLEMENTS TO THE ECCS DESIGN CODE CONCERNING ISOTROPIC CYLINDERS, Preliminary Report, Second International Colloquium on Stability of Steel Structures, Liege.

Feng, D. G., 1973
EFFECTIVE LENGTH OF COMPRESSION MEMBERS FOR TRUSSES, Research Report, Xian Institute of Metallurgy and Construction Engineering. (in Chinese)

Fukumoto, Y., Editor, 1987
GUIDELINES FOR STABILITY DESIGN OF STEEL STRUCTURES, Committee on Steel Structures, Japan Society of Civil Engineers, Tokyo. (in Japanese)

Fukumoto, Y. and Itoh, Y., 1983
EVALUATION OF MULTIPLE COLUMN CURVES FROM THE EXPERIMENTAL DATA BASE APPROACH, Journal of Constructional Steel Research, Vol. 3, No. 3, pp 2-19.

Fukumoto, Y. and Itoh, Y., 1984
LATERAL-TORSIONAL BUCKLING STRENGTH OF STEEL BEAMS FROM TEST DATA, Proceedings of JSCE, No. 341, pp 137-146.

Fukumoto, Y. and Itoh, Y., 1984
BASIC COMPRESSIVE STRENGTH OF STEEL PLATES FROM TEST DATA, Proceedings of JSCE, No. 341, pp 129-140.

Fukumoto, Y. and Kubo, M., 1977
AN EXPERIMENTAL REVIEW OF LATERAL BUCKLING OF BEAMS AND GIRDERS, Proceedings, International Colloquium on Stability of Structures under Static and Dynamic Loads, SSRC-ASCE, Washington, pp 541-562.

Fung, Y. C. and Sechler, E. E., 1957
BUCKLING OF THIN-WALLED CIRCULAR CYLINDERS UNDER AXIAL COMPRESSION AND INTERNAL PRESSURE, J. Aero. Sci., No. 5, pp 351-356.

Galambos, T. V., 1968
STRUCTURAL MEMBERS AND FRAMES, Prentice-Hall Inc., Englewood Cliffs, New Jersey.

Galambos, T. V., 1989
IMPACT OF CODE DIFFERENCES ON THE STABILITY DESIGN OF BEAMS, Proceedings of the SSRC Annual Technical Session, New York.

Galambos, T. V., ed., 1988
GUIDE TO STABILITY DESIGN CRITERIA FOR METAL STRUCTURES, 4th Edition, Structural Stability Research Council, John Wiley & Sons, New York.

Geitzelt, R. and Nethercot, D. A., 1983
ZUM BIEGEDRILLKNICKEN VON WABENTRÄGERN, Stahlbau, 52, pg 346.

Hancock, G. J., Davids, A. J., Key, P. W., and Rasmussen, K. J., 1987
STRENGTH TEST OF THIN-WALLED HIGH TENSILE STEEL COLUMNS, Proceedings, International Conference on Steel and Aluminum Structures, Cardiff, July.

Harrison, H. B., 1973
COMPUTER METHODS IN STRUCTURAL ANALYSIS, Prentice-Hall Inc., New Jersey.

He, B. K., 1985
EXPERIMENTAL STUDY OF LOCAL BUCKLING OF WEBS AND FLANGES IN I-SECTION COLUMNS, Proceedings of 1985 Annual Technical Session, Structural Stability Research Council, pp 1-12.

Hoff, N. J. and Soong, T. C., 1967
BUCKLING OF AXIALLY COMPRESSED CIRCULAR CYLINDRICAL SHELLS WITH NON-UNIFORM BOUNDARY CONDITIONS, Symposium on Thin-Walled Steel Structures, University College, Swansea, pp 61-80.

Horne, M. R. and Morris, L. J., 1981
PLASTIC DESIGN OF LOW-RISE FRAMES, Collins, London.

Horne, M. R. and Morris, L. J., 1985
PLASTIC DESIGN OF LOW-RISE FRAMES, Constrado Monograph, Collins, London.

Hunyadi, F. and Ivanyi, M., 1977
SOME REMARKS ON STABILITY PROBLEMS OF ORTHOTROPIC CYLINDRICAL SHELLS, Regional Colloquium on "Stability of Steel Structures", Hungary, Proceedings, pp 329-335.

Hutchinson, J., 1965
AXIAL BUCKLING OF PRESSURIZED IMPERFECT CYLINDRICAL SHELLS, AIAA Journal 3, pp 1461-1466.

Jetteur, Ph., Maquoi, R., Massonnet, Ch. et Škaloud, M., 1983
CALCUL DES ÂMES ET SEMELLES RAIDIES DES PONTS EN ACIER, Construction Metallique, No. 4.

Jin, C. C., Hao, P. Z. and Liu, S. P., 1983
YIELD STRENGTH OF COLD-FORMED STEEL SHAPES, Selections from Research Papers and Reports on Steel Structures, Vol. 2. (in Chinese)

Johnston, B. G., Editor, 1976
GUIDE TO STABILITY DESIGN CRITERIA FOR METAL STRUCTURES, 3rd Edition, Structural Stability Research Council, John Wiley & Sons, New York.

Kanemitshu, S. and Nojima, H., 1939
AXIAL COMPRESSION TESTS OF THIN CIRCULAR CYLINDERS, Thesis, California Institute of Technology.

Kennedy, D. J. L., 1984
LIMIT STATES DESIGN OF STEEL STRUCTURES IN CANADA, Journal, Structural Engineering, ASCE, Vol. 110, February.

Key, P. W. and Hancock, G. J., 1985
AN EXPERIMENTAL INVESTIGATION OF THE COLUMN BEHAVIOR OF COLD-FORMED SQUARE HOLLOW SECTIONS, Research Report No. P493, School of Civil and Mining Engineering, University of Sydney, June.

Key, P. W. and Hancock, G. J., 1985
STRENGTH TESTS OF COLD-FORMED SQUARE HOLLOW SECTION COLUMNS, Metal Structures Conference, Institution of Engineers, Australia, Melbourne, pp 32-36.

Key, P. W. and Hancock, G. J., 1985
STRENGTH TESTS OF COLD-FORMED SQUARE HOLLOW SECTION COLUMNS, Civil Engineering Transactions, Institution of Engineers, Australia, Vol. CE27, pp 34E1-346.

Key, P. W., Hasan, S. W. and Hancock, G. J., 1986
THE COLUMN BEHAVIOR OF COLD-FORMED HOLLOW SECTIONS, Proceedings, Eighth International Specialty Conference on Cold-Formed Steel Structures, St. Louis, Missouri, Nov.

Kikin, A. I., Sanzharovsky, R. S., and Trull, V.A., 1974
CONCRETE FILLED TUBULAR STEEL STRUCTURES (in Russian), Stroyisdat, Moscow.

Kinra, R. K. and Miller, C. D., 1987
DEVELOPMENT OF API BULLETIN 2U FOR STABILITY DESIGN OF CYLINDRICAL SHELLS, Structural Stability Research Council, Annual Technical Session, Proceedings, pp 329-341.

Klöppel K. and Winkelmann, E., 1962
EXPERIMENTELLE UND THEORETISCHE URTERSUCHUNGEN ÜBER DIE TRAGLAST VON ZWEIACHSIG AUSSERMITIG GEDRÜCTEN STAHLSTÄBEN (EXPERIMENTAL AND THEORETICAL INVESTIGATION ON THE STRENGTH OF STEEL COLUMNS WITH BIAXIALLY ECCENTRIC LOAD), Der Stahlbau, V. 31. pp 33-47, 78-87, 109-119.

Koiter, W. T., 1945
ON THE STABILITY OF ELASTIC EQUILIBRIUM (in Dutch), Ph. D. Thesis, Delft University.

Kromm, A., 1940
JAHRBUCH DER DEUTSCHEN LUFTFAHRT-FORSCHUNG, p. 1832.

Lai, S. M. A., Macgregor, J. G. and Hellesland, J., 1983
GEOMETRIC NON-LINEARITIES IN NON-SWAY FRAMES, Journal of the Structural Division, ASCE, Vol. 109, No. St12, pp 2770-2785.

Lai, S. M. A. and Macgregor, J. G., 1983
GEOMETRIC NON-LINEARITIES IN UNBRACED MULTISTORY FRAMES, Journal of the Structural Division, ASCE, Vol. 109, No. ST11, pp 2528-2545.

LeMessurier, W. J., 1977
A PRACTICAL METHOD OF SECOND ORDER ANALYSIS PART 2. RIGID FRAMES, AISC Engineering Journal of Steel Construction, Vol. 4, No. 2.

Li, C. Q., 1984
BUCKLING OF THE WEB OF BEAM-COLUMNS, Journal of Xian Institute of Metallurgy and Construction Engineering, No.1, pp 1-17. (in Chinese)

Li, K. X. and Xiao, Y. H., 1982
DETERMINATION OF CRITICAL LOAD OF FLEXURAL INSTABILITY OF COLUMNS BY REVERSAL CALCULATION ELEMENT LENGTH METHOD, Journal of Chongqing Institute of Architecture and Engineering, No. 4. (in Chinese)

Li, K. X. and Xiao, Y. H., 1982
THE CALCULATION OF THE STRENGTH OF STEEL COLUMNS WITH UNIAXIAL BENDING BY THE CALCULATION ELEMENT METHOD, Journal of Chongqing Institute of Architecture and Engineering, No.4. (in Chinese)

Lindner, J., 1989
SPECIFICATION AND CODES IN WESTERN EUROPE, Proceedings, SSRC 4th International Colloquium on Structural Stability, New York.

Lindner, J. and Gietzelt, R., 1980
BIEGEDRILLKNICKEN-ERLAUTERUNGEN, VERSUCHE, BEISPIELE, DASt H 10, K01n, Stahlbau-Verlag.

Lindner, J. and Gietzelt, R., 1983
DISCUSSION OF INTERACTION EQUATIONS FOR MEMBERS IN COMPRESSION AND BENDING, Third International Colloquium on Stability of Metal Structures, Final Report, Paris, pp 45-54.

Lindner, J. and Gietzelt, R., 1985
ZUR TRAGFAHIGKEIT AUSGEKLINKTER TRAGER, Stahlbau 54, pp 39-45.

Lindner, J. and Gietzelt, R., 1985
ZWEIACHSIGE BIEGUNG UND LANGSKRAFT - EIN ERGANZTER BEMESSUNGSVORSCHLAG, Stahlbau, 54, pp 265-271.

Lindner, J. and Gregull, T., 1986
TORSIONAL RESTRAINT COEFFICIENT OF PROFILED SHEETING, Proceedings of the IABSE Colloquium Stockholm, "Thin-Walled Metal Structures in Buildings", Vol. 49, Zurich, pp 161-168.

Lu, L. W., Ozer, E., Daniels, J. H., Okten, O. S. and Morino, S., 1975
FRAME STABILITY AND DESIGN OF COLUMNS IN UNBRACED MULTISTORY STEEL FRAMES, Fritz Engineering Lab. Report No. 375.2, Lehigh University, Bethlehem.

Marek, P. and Guštar, M., 1989
ASSESSMENT OF THE COMBINATION OF RESPONSES OF A STRUCTURE TO LOADING, Pozemní stavby No. 2, 1989, pp 78-82, Praha.

Matsui, C., etal, 1986
EXPERIMENTAL STUDY ON LATERAL-TORSIONAL BUCKLING OF TRUSS BEAM-COLUMNS WITH CIRCULAR TUBE SECTIONS, Journal of Structural and Construction Engineering, Trans. AIJ, No. 363, pp 12-21, May.

McGuire, W., 1968
STEEL STRUCTURES, Prentice Hall, New Jersey, p 1112.

Melcher, J., 1982
THE BEHAVIOR AND THE TRANSVERSE DEFLECTIONS PROBLEM OF COMPRESSION MEMBERS, Proceedings of the 3rd International Colloquium on Stability, Timisoara.

Melcher, J., 1987
TORSION AND BUCKLING BEHAVIOR OF DOUBLE-ANGLE STRUTS: STEEL STRUCTURES - ADVANCES, DESIGN AND CONSTRUCTION, Ed. R. Narayanan, Elsevier Applied Science, London-New York.

Miller, C. D., 1983
CURRENT PRACTICE AND RECENT ACTIVITIES IN THE UNITED STATES RELATED TO DESIGN RULES FOR SHELL STRUCTURES, Proceedings of Third Colloquium on Stability of Metal Structures, SSRC, Toronto, Canada, pp 509-520.

Moriwaki, Y. and Fujino, M., 1977
EXPERIMENTAL STUDY ON PURE BENDING STRENGTH OF PLATE GIRDERS WITH INITIAL IMPERFECTIONS, Proceedings of JSCE, No. 264, pp 1-15.

Mou, Z. M. and Others, 1983
SAFETY ANALYSIS OF COLD-FORMED THIN-WALLED STEEL SECTIONS, Selections from Research Papers and Reports on Steel Structures, Vol. 2. (in Chinese)

Mrázik, A. and Sadovský, Z., 1986
NOTES TO THE DESIGN OF STEEL CONSTRUCTIONS AT ELEVATED TEMPERATURE, (in Slovak). Pozemní Stavby 34, pp 352-356, Praha.

Mushtari, H. M. and Savtshenkov, A. V., 1954
STABILITY OF CYLINDRICAL AND CONICAL SHELLS SUBJECTED TO AXIAL AND RADIAL PRESSURE, Prikladnaja Matematika i Mechanika, T.18., No.6.

Narayanan, R., 1982
AXIALLY COMPRESSED STRUCTURES, Chapter 5, Cold-Formed Welded Steel Tubular Members, Elsevier Applied Science, London.

O'hEachteirn, P. and Nethercot, D. A., 1985
LATERAL-TORSIONAL BUCKLING OF PLATE-GIRDERS CONTAINING WELDING INDUCED RESIDUAL STRESSES, Conference on "Effects of fabrication related stresses on product manufacture and performance", Welding Institute, Sept., pp 24-1 – 24-14.

O'hEachteirn, P. and Nethercot, D. A., 1988
LATERAL BUCKLING TESTS ON MONOSYMMETRIC PLATE GIRDERS, Journal of Constructional Steel Research, Vol. 11, No. 4, pp 241-260.

O'hEachteirn, P. and Nethercot, D. A., 1988
LATERAL BUCKLING ON MONOSYMMETRIC PLATE GIRDERS, Journal of Constructional Steel Research, Vol. 11, No. 4, pp 261-282.

Odland, J., 1978
BUCKLING RESISTANCE OF UNSTIFFENED AND STIFFENED CIRCULAR CYLINDRICAL SHELL STRUCTURES, Norwegian Maritime Research, 6, No. 3, pp 2-22.

Pillai, S. U., 1981
AN ASSESSMENT OF CSA STANDARD EQUATIONS FOR BEAM COLUMN DESIGN, Canadian Journal of Civil Engineering, 8, pp 130-136.

Rondal, J. and Maquoi, R., 1979
SINGLE EQUATION FOR SSRC COLUMN STRENGTH CURVES, Journal of the Structural Division, ASCE, Vol. 105, No. ST1, pp 247-250.

Rotter, J. M., 1982
A SIMPLE APPROACH TO MULTIPLE COLUMN CURVES, Civil Engineering Transactions, Institution of Engineers, Australia, Vol.CE24, No. 1, pp 19-23.

Rotter, J. M., 1982
MULTIPLE COLUMN CURVES BY MODIFYING FACTORS, Journal of the Structural Division, ASCE, Vol. 108, No. ST7, pp 1665-1669.

Saal, H., Kahmer, H., and Reif, A., 1979
BEULLASTEN AXIAL GEDRUCKTER KREISZLINDERSCHALEN MIT INNENDRUCK - NEUE VERSUCHE UND VORSHIFTEN, Der Stahlbau, 48, No. 9, pp 262-269.

Sadovský, Z., 1983
SIMILARITY IN BEHAVIOR OF THIN PLATE GIRDER PANELS OF SINGLY SYMMETRIC I CROSS-SECTION SUBJECTED TO COMPRESSION AND BENDING MOMENT, ACTA Technica ČSAV, 1, pp 4-15.

Samuelson, L. A., Vandepitte, D. and Paridaens, R., 1987
THE BACKGROUND TO THE ECCS RECOMMENDATIONS FOR BUCKLING OF STRINGER STIFFENED CYLINDERS, ECCS Colloquium on Stability of Plate and Shell Structures, Ghent University, 6-8 Apr., pp 513-522.

Sanzharovsky, R. S., 1984
CONCRETE INFILLED PIPE STRUCTURES IN BUILDINGS (in Russian), promy hennoye, stroyitelstvo 57, Nr 5/79, pp 22-23.

Schulz, V., 1981
DER STABILITATSNACHWEIS BEI SCHALEN, Berichte, 4, Folge-Hefte 2. Univ. Karlsruhe.

Seide, P., 1965
AXISYMMETRICAL BUCKLING OF CIRCULAR CONES UNDER AXIAL COMPRESSION, Journal Appl. Mech. 23, pp 625-628.

Sherman, D. R., 1986
INELASTIC FLEXURAL BUCKLING OF CYLINDERS, Steel Structures: Recent Research Advances and Their Application to Design, (International Conference, Budva) edited by M.N. Pavlovic, Elsevier Applied Science Publishers, London.

Shen, Z. Y. and Chen, Y. Y., 1986
THE MODIFIED IN-PLANE INTERACTION FORMULA OF BEAM-COLUMNS, The Structural Engineers, No.2. (in Chinese)

Shen, Z. Y.
STABILITY DESIGN FORMULA FOR COLUMNS WITH WIDE FLANGE SECTIONS, Research Report No. 82271, Tongji University. (in Chinese)

Singer, J., Abramovich, H., and Yaffe, R., 1978
INITIAL IMPERFECTIONS MEASUREMENT OF INTEGRITY STRINGER-STIFFENED SHELLS, TAE Rep. 330, Technion - Israel Inst. of Techn., Dept. of Aeron. Eng., Haifa, Israel.

Škaloud, M., 1970
POST-BUCKLING BEHAVIOR OF STIFFENED WEBS, Transactions of the Czechoslovakian Academy of Science, Technological Sciences, No. 1, pp 1-154.

Škaloud, M., Zörnerová, M., 1972
POST-BUCKLING BEHAVIOR OF WEBS IN SHEAR, Transactions of the Czechoslovakian Academy of Science, Technological Sciences, No. 3, pp 1-96.

Škaloud, M., Novák, P., 1975
POST-BUCKLING BEHAVIOR OF WEBS UNDER PARTIAL EDGE LOADING, Transactions of the Czechoslovakian Academy of Science, Technological Sciences, No. 3, pp 1-94.

Škaloud, M., Náprstek, J., 1977
LIMIT STATE OF THIN-WALLED STEEL COLUMNS, Transactions of the Czechoslovakian Academy of Science, Technological Sciences, No. 2, pp 1-146.

Škaloud, M., Křístek, V., 1981
STABILITY PROBLEMS OF STEEL BOX GIRDER BRIDGES, Transactions of the Czechoslovakian Academy of Science, Technological Sciences, No. 1.

Škaloud, M., Karníková, J., 1985
EXPERIMENTAL RESEARCH OF THE LIMIT STATES OF THE PLATE ELEMENTS OF STEEL BRIDGES, Transactions of the Czechoslovakian Academy of Science, Technological Sciences, No. 1.

Škaloud, M. and Drdácký, M., 1975
ULTIMATE LOAD DESIGN OF WEBS OF STEEL PLATE GIRDERS, Part 3. Webs under Concentrated Loads. Stavebnicky Casopis 23, 1975, 3, pp 140-160.

Smith, R. G., 1984
SIMPLIFIED ANALYSIS OF THE P-Δ EFFECT IN SLENDER BUILDINGS, Top-Tier Design Methods in the Draft Unified Concrete Code, Lecture 6, University of Sydney, pp 6.1.-6.49.

Springfield, J. and P. F. Adams, 1972
ASPECTS OF COLUMN DESIGN IN TALL STEEL BUILDINGS, Journal of Structural Division, Proc. ASCE, Volume 98, No. ST5, May.

SSRC, 1982
STABILITY OF METAL STRUCTURES, A WORLD VIEW, 1st Edition. American Institute of Steel Construction, Chicago.

Steinhardt, O. and Schulz, V., 1971
ZUM BEULVERHALTEN VON KREISZULINDERSCHALEN, Schweiz Bauztg., Jan.

Stevens, L. K., 1967
ELASTIC STABILITY OF PRACTICAL MULTI-STORY FRAMES, Proceedings of the Institute of Civil Engineering, V. 36, Jan.

Sun, Z. L. and He, H. X. , 1982
INVESTIGATION OF EXPERIMENT AND CALCULATION FOR BUCKLING STRENGTH OF THE RECTANGULAR PLATE WITH THREE EDGES SIMPLY SUPPORTED AND ONE EDGE FREE UNDER NONUNIFORM COMPRESSION, Selections from Research Papers and Reports on Steel Structures, Vol. 1 (in Chinese)

Suzuki, T., etal, 1986
EXPERIMENTAL STUDY ON ULTIMATE STRENGTH AND LOAD-BEARING CAPACITY OF STEEL TRUSS BEAMS, Journal of Structural and Construction Engineering, Trans. AIJ, No. 360, pp 75-83, Feb.

Suzuki, T., etal, 1986
LOAD-DEFORMATION RELATIONSHIP AND EARTHQUAKE RESISTANCE DESIGN OF STEEL TRUSS FRAMES, Journal of Structural and Construction Engineering, Trans. AIJ, No. 365, pp 28-39, July.

Takanashi, K., 1970
PLASTIC BUCKLING AND POST-BUCKLING BEHAVIORS OF STEEL PLATES, Report of IIS., University of Tokyo, Vol. 20, No. 4.

Tall, L., 1964
RECENT DEVELOPMENTS IN THE STUDY OF COLUMN BEHAVIOR, Journal, Institution of Engineers, Australia, Vol. 36.

Tall, L., Editor, 1974
STRUCTURAL STEEL DESIGN, 2nd Edition, Wiley/Ronald, New York.

Tebedge, N. and Chen, W. F., 1974
DESIGN CRITERIA FOR H-COLUMNS UNDER BIAXIAL LOADING, Journal of the Structural Division, ASCE, Vol. 100, No. ST3, pp 579-599.

Trahair, N., 1986
DESIGN STRENGTH FOR STEEL BEAM-COLUMNS, Canadian Journal of Civil Engineering, Vol. 13., No. 6, Dec., pp 639-646.

Trahair, N. S., Abel, A., Ansourian, P., Irvine, H. M., Rotter, J. M., 1983
STRUCTURAL DESIGN OF STEEL BINS FOR BULK SOLIDS, Australia Institute of Steel Construction.

Trahair, N. S. and Bradford M. A., 1988
THE BEHAVIOR AND DESIGN OF STEEL STRUCTURES, Chapman and Hall, London.

Turkstra, C.J., 1970
THEORY OF STRUCTURAL SAFETY, SM Study No.2, Solid Mechanics Division, University of Waterloo, Waterloo, Ontario.

Vinnakota, S., 1977
THEORY OF BEAM-COLUMNS, Chapter 10, pp 451-503. McGraw-Hill, New York.

Vogel, U., 1985
CALIBRATING FRAMES, Stahlbau, October 1985, Wilhelm Ernst & Sohn, Berlin.

Volmir, A. S., 1967
STRUCTURAL STABILITY OF SYSTEMS (in Russian), Nauka, Moscow.

von Kármán, T., Sechler, E. E., and Donnell, L. H., 1932
STRENGTH OF THIN PLATES IN COMPRESSION, Transactions, American Society of Mechanical Engineers, Vol. 54, No. APM-54-5, p 53.

Walker, A. C. and Sridharan, S., 1980
ANALYSIS OF THE BEHAVIOR OF AXIALLY COMPRESSED STRINGER-STIFFENED CYLINDRICAL SHELLS, Proc. ICE, Part 2, 69., pp 447-472.

Walpole, W.R. and Butcher, G.W., 1985
BEAM DESIGN, SECTION C OF THE DELIBERATIONS OF THE STUDY GROUP FOR THE SEISMIC DESIGN OF STEEL STRUCTURES, Bulletin N.Z. National Society for Earthquake Engineering, Vol. 18, No. 4.

Wang, S. J., 1985
TORSIONAL-FLEXURAL BUCKLING OF OPEN THIN-WALLED COLUMNS WITH BATTENS, Thin-Walled Structures, Vol. 3, No. 4, pp 323-344.

Wei, M. Z., Li, K. X. and Xia, Z. Z., 1985
ANALYSIS APPLICATION ABOUT STABILITY OF STEEL COLUMNS, Proceeding of 1985 Annual Technical Session, Structural Stability Research Council, pp 71-81.

Weingarten, V. I. & al., 1968 (NASA)
BUCKLING OF THIN-WALLED CIRCULAR CYLINDERS, NASA SP-8007.

Weingarten, V. I., Morgan, E. J., and Seide, P., 1965
ELASTIC STABILITY OF THIN-WALLED CYLINDRICAL AND CONICAL SHELLS UNDER COMBINED INTERNAL PRESSURE AND AXIAL COMPRESSION, AIAA Journal 3., No. 6, pp 1118-1125.

Weingarten, V. I., Morgan, E. J., and Seide, P., 1965
ELASTIC STABILITY OF THIN-WALLED CYLINDRICAL AND CONICAL SHELLS UNDER AXIAL COMPRESSION, AIAA Journal 3., No. 3, pp 500-505.

Wilson, W. M. and Newmark, N. M., 1933
THE STRENGTH OF THIN CYLINDRICAL SHELLS AS COLUMNS, University of Illinois Engineering Exp. Sta. Bull., No. 255.

Wilson, W. M., 1937
TESTS OF STEEL COLUMNS, University of Illinois Engineering Exp. Sta. Bull., No. 292.

Winter, G., 1947
STRENGTH OF THIN STEEL COMPRESSION FLANGES, Transactions, American Society of Civil Engineers, Vol. 112, pp 527.

Wu, J. and Hao, J. S., 1981
EXPERIMENTAL STUDY OF CROSS BRACINGS FOR TRANSMISSION TOWERS, Research Institute of Power Engineering Construction. (in Chinese)

Xia, Z. Z. and Chen, G. X., 1984
THE SUGGESTION OF COLUMN CURVES AND THE PROBABILITY ANALYSIS, Journal of Chongqing Institute of Architecture and Engineering, No. 2. (in Chinese)

Yabuki, T. and Vinnakota, S., 1984
STABILITY OF STEEL ARCH BRIDGES: A STATE-OF-ART REPORT, Solid Mechanics Archives, Vol. 9, No. 2, pp 115-158.

Zahn, C. J. and Haaijer, G., 1987
EFFECT OF CONNECTOR SPACING ON DOUBLE ANGLE COMPRESSIVE STRENGTH, Proceedings from AISC National Engineering Conference and Conference of Operating Personnel, New Orleans, Louisiana, Apr. 30 - May 2.

Zandonini, R., 1985
STABILITY OF COMPACT BUILT-UP STRUTS: EXPERIMENTAL INVESTIGATION AND NUMERICAL SIMULATION, Construzione Metalliche, No. 4.

Zhang, Z. Q., 1982
STABILITY OF AXIALLY LOADED COMPRESSION MEMBERS OF COLD-FORMED STEEL SECTIONS, Selections from Research Papers and Reports on Steel Structures, Vol. 1. (in Chinese)

Zhang, Z. Q., 1986
COLD-FORMED STEEL CONSTRUCTION IN THE P. R. CHINA, Proceedings of the Eighth International Specialty Conference on Cold-Formed Steel Structures.

Zhao, H. Q., 1964
DETERMINATION OF EFFECTIVE LENGTH OF WEB MEMBERS WITH PRINCIPAL AXES OUTSIDE THE TRUSS PLANE, Research Report No. 6418, Xian Institute of Metallurgy and Construction Engineering. (in Chinese)

Zhao, W. W., 1981
THE APPLICATION OF EQUIVALENT PRINCIPLE IN CALCULATION FOR STABILITY OF STEEL STRUCTURES, Metallurgy and Construction, No. 1. (in Chinese)

Zhong, S. T., 1989
INTRODUCTION ON STABILITY CALCULATION OF MEMBERS IN DESIGN REGULATION OF CONCRETE FILLED STEEL TUBULAR STRUCTURES IN CHINA, Proceeding, 4th International Colloquium-Code Differences Around the World, New York Session, Structural Stability Research Council, pp 419-428.

NAME INDEX

SUBJECT INDEX